CAMBRIDGE LIBRARY COLLECTION

Books of enduring scholarly value

Botany and Horticulture

Until the nineteenth century, the investigation of natural phenomena, plants and animals was considered either the preserve of elite scholars or a pastime for the leisured upper classes. As increasing academic rigour and systematisation was brought to the study of 'natural history', its subdisciplines were adopted into university curricula, and learned societies (such as the Royal Horticultural Society, founded in 1804) were established to support research in these areas. A related development was strong enthusiasm for exotic garden plants, which resulted in plant collecting expeditions to every corner of the globe, some-times with tragic consequences. This series includes accounts of some of those expeditions, detailed reference works on the flora of different regions, and practical advice for amateur and professional gardeners.

Flora Capensis

This seminal publication began life as a collaborative effort between the Irish botanist William Henry Harvey (1811–66) and his German counterpart Otto Wilhelm Sonder (1812–81). Relying on many contributors of specimens and descriptions from colonial South Africa – and building on the foundations laid by Carl Peter Thunberg, whose *Flora Capensis* (1823) is also reissued in this series – they published the first three volumes between 1860 and 1865. These were reprinted unchanged in 1894, and from 1896 the project was supervised by William Thiselton-Dyer (1843–1928), director of the Royal Botanic Gardens at Kew. A final supplement appeared in 1933. Reissued now in ten parts, this significant reference work catalogues more than 11,500 species of plant found in South Africa. Volume 6 comprises sections that were published individually between 1896 and 1897, covering Haemodoraceae to Liliaceae.

Cambridge University Press has long been a pioneer in the reissuing of out-of-print titles from its own backlist, producing digital reprints of books that are still sought after by scholars and students but could not be reprinted economically using traditional technology. The Cambridge Library Collection extends this activity to a wider range of books which are still of importance to researchers and professionals, either for the source material they contain, or as landmarks in the history of their academic discipline.

Drawing from the world-renowned collections in the Cambridge University Library and other partner libraries, and guided by the advice of experts in each subject area, Cambridge University Press is using state-of-the-art scanning machines in its own Printing House to capture the content of each book selected for inclusion. The files are processed to give a consistently clear, crisp image, and the books finished to the high quality standard for which the Press is recognised around the world. The latest print-on-demand technology ensures that the books will remain available indefinitely, and that orders for single or multiple copies can quickly be supplied.

The Cambridge Library Collection brings back to life books of enduring scholarly value (including out-of-copyright works originally issued by other publishers) across a wide range of disciplines in the humanities and social sciences and in science and technology.

Flora Capensis

*Being a Systematic Description
of the Plants of the Cape Colony,
Caffraria & Port Natal,
and Neighbouring Territories*

VOLUME 6:
HAEMODORACEAE TO LILIACEAE

WILLIAM H. HARVEY *ET AL.*

CAMBRIDGE
UNIVERSITY PRESS

CAMBRIDGE
UNIVERSITY PRESS

University Printing House, Cambridge, CB2 8BS, United Kingdom

Cambridge University Press is part of the University of Cambridge.
It furthers the University's mission by disseminating knowledge in the pursuit of
education, learning and research at the highest international levels of excellence.

www.cambridge.org
Information on this title: www.cambridge.org/9781108068147

© in this compilation Cambridge University Press 2014

This edition first published 1896–7
This digitally printed version 2014

ISBN 978-1-108-06814-7 Paperback

FLORA CAPENSIS.

DATES OF PUBLICATION OF THE SEVERAL PARTS OF
THIS VOLUME.

PART I., pp. 1–192, was published *April*, 1896.

PART II., pp. 193–384, was published *September*, 1896.

PART III., pp. 385–563, was published *June*, 1897.

FLORA CAPENSIS:

BEING A

Systematic Description of the Plants

OF THE

CAPE COLONY, CAFFRARIA, & PORT NATAL

(AND NEIGHBOURING TERRITORIES)

BY

VARIOUS BOTANISTS.

EDITED BY

W. T. THISELTON-DYER, C.M.G., C.I.E., LL.D., F.R.S.

DIRECTOR, ROYAL GARDENS, KEW.

*Published under the authority of the Governments of the
Cape of Good Hope and Natal.*

VOLUME VI.

HÆMODORACEÆ TO LILIACEÆ.

LONDON

L. REEVE & CO., 6, HENRIETTA STREET, COVENT GARDEN.

Publishers to the Home, Colonial, & Indian Governments.

1896-1897.

PREFACE.

THE third volume of the Flora Capensis was published in 1865. The following year Professor Harvey, who had been its principal author and guiding spirit, died. Although in the preface the fourth volume is referred to as "shortly to be in preparation for the press," practically nothing available relating to it was found amongst Professor Harvey's papers. Nor did his coadjutor, Dr. Sonder, who died in 1881, undertake any further part in the work.

Its continuation was urged upon Kew by Sir Henry Barkly, G.C.M.G., K.C.B., F.R.S., who was Governor of the Cape of Good Hope from 1870 to 1877. During a long official career in different parts of the Empire, this enlightened administrator, himself an ardent naturalist, never failed to foster the scientific interests of the colonies committed to his charge. Sir Joseph Hooker, at that time Director of the Royal Gardens, entrusted the task of continuing the work of Harvey and Sonder to me. But the pressure of official duties in which I almost immediately found myself immersed, left me little time for the task. It became evident that it could only be accomplished by the co-operation of numerous workers. Another difficulty was the rapid expansion of British South Africa. This led to a continuous influx to Kew of new material, which had to be determined and made available for future working up in the Flora. It was soon obvious that it would be necessary to largely extend the area comprised by the published volumes, and it was ultimately determined to do this still further so as to include, as far as possible, all known flowering plants occurring in the area between the Tropic of Capricorn and the Ocean. To the north, the present and future volumes will therefore be supplemented by the "Flora of Tropical Africa."

During the last twenty years the time of one member of the Kew staff has been almost exclusively occupied with the determination of South African plants. Upwards of 10,000 specimens have been named and catalogued for South African botanists and collectors, and a considerable number have been figured and described. These labours were a necessary preparation for the continuation of the Flora on its extended scale.

This extension necessitated breaking up the whole area into smaller regions, the physical characters of which will probably be found tolerably well marked. They have been adopted in great part from the important paper, " Sketch of the Flora of South Africa," by Harry Bolus, F.L.S., published in the Cape of Good Hope ' Official Handbook ' for the Colonial and Indian Exhibition, 1886 (pp. 286-317), and also printed with a separate pagination.

These regions may be briefly defined as follows :—

i. **Coast Region.**—Includes the narrow belt lying between the South-western and Southern coasts and the Zwarte Bergen range, from the Oliphants to the Kei rivers.

ii. **Central Region.**—Can only be roughly defined as lying between the Coast and the Kalahari regions.

iii. **Western Region.**—Extends from the Tropic to the Oliphants river, and includes Great and Little Namaqualand.

iv. **Kalahari Region.**—Includes the Kalahari, Bechuanaland, Griqualand West, Transvaal, Orange Free State, and Basutoland.

v. **Eastern Region.**—Includes the belt lying between the Eastern coast and the Drakens Berg range, from the Kei river to the Tropic. It therefore comprises Natal, Zululand, Griqualand East, &c.

The plants of the older collectors, which are often destitute of precise localities, have been simply referred to under the general head of **South Africa.**

For reasons of convenience it has been found advisable to publish the present volume in anticipation of the fourth and fifth, which are also in preparation, and to which it is hoped that Mr. Bolus, the well-known South African botanist, who has paid several visits to Kew for the purpose, will largely contribute. The present instalment will be found probably of more than ordinary interest to horticulturists as well as to botanists, as it includes the whole of the plants known familiarly as "Cape Bulbs." The cultivation of these was popular on their introduction to Kew by Masson at the end of the last, and by Bowie at the beginning of the present century, and the taste for them has of late years revived.

The volume has been in preparation for several years, but its publication has been from time to time delayed by the desire to include in it the novelties which have been continually received and published as new territories have been explored.

Even while passing through the press sufficient have accumulated to render an appendix necessary. The whole has been elaborated by Mr. John Gilbert Baker, F.R.S., the Keeper of the Herbarium and Library of the Royal Gardens, who has long been the accepted authority on the Petaloid Monocotyledons. I must add my obligations to Mr. C. H. Wright, Assistant in the Herbarium, who has greatly helped me in reading the proofs.

The distribution of the localities under the different regions has been a laborious and intricate task. It will afford a basis for at any rate a partial analysis of the Flora of South Africa, which will no doubt bring into prominence important facts as to its geographical distribution. It has been accomplished with much care and patience by Mr. N. E. Brown, A.L.S., Assistant in the Herbarium of the Royal Gardens. And finally it has been subjected to the invaluable revision of Mr. H. Bolus.

The orthography adopted for the local names has met with some criticism from South African botanists. It has been thought advisable, however, to adhere to the standard, no doubt in great measure conventional, of authoritative maps. Those which have been relied upon principally are:—

Cape of Good Hope. By J. Arrowsmith, 1834. (Useful for old names of localities.)

A Map of the Colony of the Cape of Good Hope and neighbouring Territories. By A. de Smidt, 1876.

Map of the Transvaal and the surrounding Territories. By F. Jeppe, 1880.

Spezial-Karte von Afrika. Gotha : Justus Perthes, 1885.

Reviewing the contents of the present volume, two points may seem to invite some criticism. A considerable number of species appear never to have been collected but once. Many are still only known from descriptions and figures published in the last century, and are unrepresented in herbaria. It is difficult, however, to believe that they are really extinct. The fact is more probably accounted for by the extremely local limitation of species in South Africa, which is hardly paralleled in this respect by any other Flora in the world.

In the case of succulent genera, such as *Aloe* and *Haworthia*, herbarium specimens are lamentably deficient. But Mr. Baker has had the advantage of having had under observation for years the collection of succulent plants at Kew, which in extent is undoubtedly unique. Many of these have been, in all probability, under cultivation in the Royal Gardens since their introduction in the last century. The advantage of consulting living specimens is of peculiar importance in describing the Petaloid Monocotyledons. But in the case of the succulent genera, it may be safely said that, without it, the task would not be possible at all. Unfortunately, when the majority of these plants were introduced, little importance was attached to their exact localization ; and this, therefore, for the present, must remain for the most part unrecorded.

It only remains to follow the example of my predecessors, and give some account of those among a great body of contributors who have supplied Kew with the most important recent collections.

Two names will be for ever memorable in the history of South African Botany.

More than thirty years have rolled away since Professor Harvey bore eloquent testimony to the indefatigable services of Peter MacOwan, Esq., B.A., F.L.S., then Principal of Shaw's College, Grahamstown, now Government Botanist. Time has not staled his enthusiasm for the beautiful Flora amidst which he has spent the best years of his life, nor his energy in investigating it. Without his self-sacrificing aid the present undertaking would have been miserably incomplete. By a correspondence which has never intermitted, he has done all in his power to keep Kew abreast of the progress of botanical discovery in South Africa. And he possesses the happy art of communicating some touch of his enthusiasm to others, and has thus secured the investigation of many parts of the area of the Flora which might otherwise have remained all but unknown.

To HARRY BOLUS, Esq., F.L.S., the gratitude of Kew is no less due for aid and encouragement of the most varied kind. His admirable researches into the difficult problem of the geographical distribution of South African plants, and his patience and accurate investigation of the *Orchideæ* and other groups, will, it may be hoped, always ensure his close personal association with the present work. Mr. Bolus has further contributed to Kew many hundreds of specimens—a large proportion of which were new to science, and many of great interest and rarity.

At the risk of seeming to make an invidious choice amongst a formidable list of Kew contributors, I cannot but further single out the following for particular acknowledgment :—

Sir HENRY BARKLY, G.C.M.G., K.C.B., F.R.S., was indefatigable while Governor of Cape Colony in procuring for Kew many of the rarer and more remarkable of South African plants. He paid especial attention to those of a succulent habit. Amongst many interesting introductions to European cultivation, the remarkable Tree-Aloe (*Aloe dichotoma*) deserves especial mention. And it was due to his support that the approval and aid of the Legislatures of Cape Colony and Natal was secured for the continuation of this work.

JOHN MEDLEY WOOD, Esq. A.L.S., the Curator of the beautiful Botanic Gardens at Berea, Durban, in the Colony of Natal, and the only institution of the kind in South Africa, has investigated the Flora of Natal with conspicuous energy, and has done more than any other botanist to reveal its riches.   Kew is indebted to him for large and invaluable collections.

The Rev. LEOPOLD RICHARD BAUR has sent to Kew a large and interesting collection of Tembuland plants, chiefly from the neighbourhood of Bazeia.

MAURICE S. EVANS, Esq., of Durban, has furnished collections which, though not numerically large, have proved very rich in new species.

H. G. FLANAGAN, Esq., has especially studied the rich local Flora of the Kei River Basin.   Beautifully preserved specimens from him have reached Kew chiefly through Mr. Harry Bolus.

ERNEST E. GALPIN, Esq., of Queenstown, has sent collections rich in undescribed species from the Transvaal, Swaziland, and the Queenstown district.

Dr. EMIL HOLUB contributed the entire collection made by him during his travels in South Africa between the years 1872 and 1879.

WILLIAM NELSON, Esq., of Johannesburg, has sent an extensive collection of plants from the Transvaal and adjoining territory.

The Rev. WILLIAM MOYLE ROGERS, of Bournemouth, has contributed a parcel of plants from various parts of Cape Colony, containing several novelties.

Mrs. KATHARINE SAUNDERS has communicated from time to time interesting plants from Natal, Zululand, and the Lobombo Mountains.

WILLIAM TYSON, Esq., of Kokstad, Griqualand East, has sent a large and interesting collection of plants from the Eastern districts of Cape Colony, Griqualand East, and Pondoland, containing numerous new species.   He is commemorated in the Boragineous genus, *Tysonia*.

It only remains to add that the expense of preparation and publication of the present volume has been aided by grants from the Governments of Cape Colony and Natal.

W. T. T. D.

Kew, May, 1897.

SEQUENCE OF ORDERS CONTAINED IN VOL. VI.
WITH BRIEF CHARACTERS.

Continuation of Series II. EPIGYNÆ. Ord. CXXXIII—CXXXVII.

CXXXIII. HÆMODORACEÆ (page 1). *Flowers* hermaphrodite, regular or slightly irregular. *Perianth* corolline, in the Cape genera inferior, half-inferior or superior, glabrous or hairy outside. *Stamens* 6 or 3. *Ovary* superior, half-inferior or inferior. *Fruit* usually a 3-celled capsule with loculicidal dehiscence. *Seed* with embryo placed in a marginal hollow of the fleshy albumen.

CXXXIV. IRIDEÆ (page 7). *Flowers* hermaphrodite, regular or irregular. *Perianth* corolline, superior, glabrous. *Stamens* 3, placed opposite the outer lobes of the perianth. *Ovary* inferior. *Fruit* a 3-celled capsule, with loculicidal dehiscence. *Seed* with embryo surrounded by horny albumen.

CXXXV. AMARYLLIDEÆ (page 171). *Flowers* hermaphrodite, regular, or nearly so. *Perianth* corolline, superior, glabrous outside, except in *Hypoxis* and *Vellosia*. *Stamens* usually 6. *Ovary* inferior. *Fruit* a 3-celled capsule with loculicidal dehiscence or indehiscent. *Seed* with embryo enclosed in the usually fleshy albumen.

CXXXVI. DIOSCOREACEÆ (page 246). *Flowers* regular, diœcious. *Perianth* small, green, glabrous, superior. *Stamens* 6 or 3. *Fruit* in the Cape genera an acutely-angled triquetrous capsule, with loculicidal dehiscence. *Seed* with embryo enclosed in the fleshy albumen.

CXXXVII. LILIACEÆ (page 253). *Flowers* regular, hermaphrodite in the Cape genera, except in *Smilax*. *Perianth* inferior, glabrous. *Stamens* 6. *Ovary* superior. *Fruit* a capsule, with loculicidal or septicidal dehiscence or a berry. *Seed* with embryo enclosed in fleshy or horny albumen.

FLORA CAPENSIS.

Order CXXXIII. HÆMODORACEÆ.

(By J. G. Baker).

Flowers hermaphrodite, regular or slightly irregular. *Perianth* corolline, with or without a tube, the segments usually persistent and biseriate. *Stamens* 6, all perfect, or 3, opposite the inner segments of the perianth; filaments usually free; anthers 2-celled, versatile or erect, dehiscing longitudinally, rarely by terminal pores. *Ovary* wholly or partially free, 3-celled; ovules axile, one or many in a cell; style usually filiform, with a capitate stigma. *Fruit* usually a 3-celled capsule with loculicidal dehiscence, rarely 1-celled and 1-seeded by abortion. *Seeds* 1, few or many in a cell, globose or compressed; testa thin or coriaceous; embryo placed in a marginal hollow of the fleshy albumen.

Perennial herbs, often densely pilose, rootstock short, tuberous, never bulbous. Leaves often distichous and firm in texture. Inflorescence various, usually racemose or panicled.

DISTRIB. A widely-spread small Order, concentrated in Australia, forming a connecting link between *Irideæ*, *Amaryllideæ* and *Liliaceæ*.

Tribe 1. EUHÆMODOREÆ. *Perianth* persistent, pilose.
I. **Wachendorfia.**—*Stamens* 3. *Fruit* free, 3-celled.
II. **Barberetta.**—*Stamens* 3. *Fruit* free, 1-celled.
III. **Dilatris.**—*Stamens* 3. *Fruit* inferior, 3-celled.
IV. **Lanaria.**—*Stamens* 6. *Fruit* inferior, 1-celled.

Tribe 2. OPHIOPOGONEÆ. *Perianth* glabrous. *Pericarp* bursting at an early stage. *Seeds* fleshy. *Pedicels* articulated.
V. **Sansevieria.**—*Perianth* with a long tube. *Fruit* free.

Tribe 3. CONANTHEREÆ. *Perianth* glabrous, deciduous. *Fruit* a 3-celled loculicidal capsule.
VI. **Cyanella.** *Stamens* all perfect, but unequal. *Ovary* ½-inferior.

I. WACHENDORFIA, Linn.

Perianth rather oblique, funnel-shaped; segments oblong, subequal, 3 outer firmer, hairy outside, 3 upper shortly joined and sometimes obscurely spurred at the base. *Stamens* 3, opposite the inner segments; filaments included, filiform, declinate; anthers small, ovate-sagittate. *Ovary* free, 3-celled; ovules solitary in the cells; style filiform, declinate; stigma capitate. *Capsule* acutely 3-lobed, dehiscing loculicidally. *Seeds* solitary in each cell, laterally affixed.

Rootstock tuberous. Leaves lanceolate, plicate. Inflorescence copiously panicled. Flowers usually bright yellow. Stains paper red.
DISTRIB. Endemic.

Panicle dense, cylindrical; seed smooth ... ... (1) **thyrsiflora.**
Panicle lax, deltoid; seed pilose ... ... (2) **paniculata.**

1. W. thyrsiflora (Linn. Sp. Plant. 59); rootstock a fleshy tuber; leaves lorate, firm, plicate, glabrous, the lower 2–3 ft. long including the channelled petiole, 2–3 in. broad at the middle; stem stout, erect, 1–2 ft. long, furnished with several reduced leaves; panicle dense, cylindrical, a foot or more long; axis and branches densely pilose; bracts lanceolate, persistent, scariose, $\frac{1}{2}$–2 in. long; perianth bright yellow, $\frac{5}{8}$–$\frac{3}{4}$ in. long; segments oblong; outer posticous hairy on the back all over and closely ribbed; stamens included; capsule acutely 3-lobed, $\frac{1}{2}$ in. long and broad; seeds smooth. *Burm. Monog.* 2, t. 2; *Thunb. Prodr.* 12; *Fl. Cap.* i. 306; *Red. Lil.* t. 93; *Ker in Bot. Mag.* t. 1060; *Ait. Hort. Kew.* edit. 1, i. 75; *Roem. et Schultes Syst. Veg.* i. 485.

Coast Region: Dutoit's Kloof and Genadendal, *Drège!* near George, *Burchell*, 5984! near Melville, *Burchell*, 5449! Kromme River Heights, *Burchell*, 4808! Uitenhage Div., Galgebosch, *Mac Owan*, 2078!

2. W. paniculata (Linn. Sp. Plant. 59); rootstock a globose tunicated corm; basal leaves 3–4, lanceolate, nearly glabrous, 3-nerved, $\frac{1}{2}$–1 ft. long, $\frac{3}{4}$–1 in. broad, narrowed to the dilated clasping base; peduncle a foot or more long, green, with several reduced leaves; inflorescence a lax deltoid panicle $\frac{1}{2}$–1 ft. long, with finely hairy branches; upper bracts ovate; perianth $\frac{1}{2}$–1 in. long, bright yellow, turning reddish-black when dried; 3 upper segments spotted inside at the base; 2 upper inner not contiguous; outer upper hairy and ribbed, all over; stamens included; capsule hairy, acutely 3-lobed at the sides and emarginate at the top; seeds covered all over with brown bristly hairs. *Burm. Monog.* 4; *Thunb. Prodr.* 12; *Fl. Cap.* i. 307; *Ker in Bot. Mag.* t. 616; *Smith Ic. Pict.* i. t. 5; *Roem. et Schultes Syst. Veg.* i. 485.

Var. β, hirsuta (Thunb. Prodr. 12, sp.); leaves more hairy; stem more slender, reddish-brown; branches of the panicle more hairy and more divaricated; perianth bright yellow, the upper inner segments contiguous. *Fl. Cap.* i. 308; *Ker in Bot. Mag.* t. 614; *Roem. et Schultes Syst. Veg.* i. 485. *W. villosa, Andr. Bot. Rep.* t. 398.

Var. γ, brevifolia (*Ker in Bot. Mag.* t. 1166, sp.); stem shorter, panicle closer and leaves shorter and broader than in the type; flowers dull greenish purple, with the two upper inner segments contiguous. *Roem. et Schultes Syst. Veg.* i. 486; *Ait. Hort. Kew.* edit. 2. i. 107.

Var. δ, tenella (Thunb. Prodr. 12, sp.); a small slender variety, with narrow 3-nerved leaves, $\frac{1}{6}$–$\frac{1}{3}$ in. broad. *Fl. Cap.* i. 308. *W. graminea, Thunb. Prodr.* 12?; *Fl. Cap.* i. 309? *W. graminifolia, Linn. fil. Suppl.* 101?

South Africa: without locality, var. β, *Harvey*, 88! *Wright! Burchell!* var. γ, *Thom*, 770!

Coast Region: Malmesbury Div., *Thunberg!* Dutoit's Kloof and Drakenstein Mts., *Drège*, 8573! Riversdale Div., *Burchell*, 6597! 6865! Var. δ, Simons Bay, *Wright!*

II. BARBERETTA, Harv.

Perianth funnel-shaped, petaloid; tube 0; segments oblanceolate, equal, 3–5-nerved. *Stamens* 3, attached to the base of the outer segments; filaments filiform, longer than the perianth; anthers small, ovate, dorsifixed. *Ovary* free, glabrous, oblique; 1-cell perfect, containing a single ovule; 1–2 small, empty; style filiform; stigma capitate. *Fruit* unknown.

DISTRIB. Endemic.

1. **B. aurea** (Harv. Gen. S. Afr. Pl. edit. 2, 377); rootstock tuberous; leaves about 3, superposed, sessile, lanceolate, glabrous, membranous, $\frac{1}{2}$–1 ft. long, $\frac{1}{2}$–1 in. broad at the middle, narrowed very gradually to both ends, furnished with five very distinct vertical ribs, with several finer ones between each pair; stem weak, $\frac{1}{2}$–1 ft. long, slightly hairy upwards; raceme dense, simple, 2–3 in. long; pedicels ascending, lower $\frac{1}{2}$–$\frac{3}{4}$ in. long; bract lanceolate, persistent, wrapped round the pedicel; perianth bright orange, glabrous, $\frac{1}{6}$ in. long; stains paper red.

EASTERN REGION: Kaffraria; Isomo Valley, *Bowker* and *Mrs. Barber*, 880! Natal; York Bush, in damp shady places, *McKen!*

III. DILATRIS, Berg.

Perianth slit down to the ovary; segments subequal, linear or oblong. *Stamens* 3, attached to the base of the inner segments; filaments filiform; anthers small, versatile. *Ovary* globose, inferior, 3-celled; ovules solitary in the cells; style filiform; stigma minute, terminal. *Capsule* globose, hairy, rigid, indehiscent or finally dehiscing loculicidally. *Seeds* solitary, discoid.

Rootstock short, woody. Leaves several in a distichous basal rosette, rigid. Peduncle hairy, elongated, with a few reduced leaves. Flowers numerous, forming a short congested panicle.

DISTRIB. Endemic.

Perianth-segments oblong; hairs not viscous ... ... (1) corymbosa.
Perianth-segments linear; hairs gland-tipped... ... ... (2) viscosa.

1. **D. corymbosa** (Berg. Cap. 9 t. 3, fig. 5); basal leaves numerous, linear, rigid, subglabrous, 6–15 in. long, $\frac{1}{8}$–$\frac{1}{6}$ in. broad, acuminate; ribs immersed; peduncle 1–1$\frac{1}{2}$ ft. long, densely hairy upwards, with a few small erect linear leaves; flowers very numerous, arranged in a very dense panicle 2–3 in. diam.; main branches umbellate; ovary globose, clothed like the branches with short, dense, soft, spreading, whitish hairs; perianth-segments oblong, much imbricated, purplish, $\frac{1}{4}$–$\frac{1}{3}$ in. long; stamens shorter than the perianth-segments; capsule globose, very hairy, the size of a small pea, crowned by the accrescent scariose segments; seed orbicular, hairy, discoid. *Thunb. Prodr.* 10; *Smith, Exot. Bot.* i. 29 t. 16. *D. umbellata, Linn. fil. Suppl.* 101; *Roem. et Schultes Syst. Veg.* i. 483. *D. ixioides, Lam. Ill. Gen.* 127. *Ixia hirsuta, Linn. Mant.* 27.

COAST REGION: Camps Bay, *Pappe! Burchell*, 323! Simons Bay, *MacGillivray*, 472! *Milne*, 176! Cape Div. and Dutoit's Kloof, *Drège!* Baviaan's Kloof, near Gonadendal, *Burchell*, 7805! Worcester Div., *Cooper*, 1658!

2. **D. viscosa** (Linn. fil. Suppl. 101); rootstock very thick, short, vertical; basal leaves ensiform, glabrous, coriaceous, 6–9 in. long, $\frac{1}{4}$–$\frac{1}{3}$ in. broad; stem very hairy, $\frac{1}{2}$–1 ft. long, with a few

clasping, erect, lanceolate leaves ; flowers very numerous, arranged in a dense, short, corymbose panicle ; ovary densely clothed, as are the segments and branches, with soft, spreading, brown, gland-tipped hairs ; perianth-segments linear, brownish, $\frac{1}{3}$–$\frac{1}{2}$ in. long ; stamens as long as the perianth-segments ; capsule ovoid, densely hairy, $\frac{1}{4}$ in. long ; seed oblong, discoid, dull black, closely punctate.   *Thunb. Prodr.* 10 ; *Fl. Cap.* i. 259 ; *Lam. Ill. Gen.* t. 34 ; *Roem. et Schultes Syst. Veg.* i. 483.

COAST REGION : Table Mountain and vicinity, *Burchell*, 530! *Milne*, 183! *Pappe ! Villette !* Riversdale Div., Langebergen, *Burchell*, 7071!

IV. LANARIA, Ait.

Perianth regular, with a tube above the ovary ; segments 6, sub-equal.   *Stamens* 6, inserted at the throat of the perianth-tube ; filaments filiform, shorter than the segments : anthers small, ovate-sagittate.   *Ovary* inferior, except at the apex, 3-celled ; ovules two in a cell, collateral ; style filiform ; stigma capitate. *Capsule* subglobose, crowned by the persistent perianth, 1-celled, 1-seeded.   *Seed* globose ; testa shining, black, crustaceous.

DISTRIB. Endemic.

1. **L. plumosa** (Ait. Hort. Kew, edit. 1, i. 462) ; rootstock woody, with a dense tuft of fleshy cylindrical root-fibres ; leaves several, in a basal rosette, surrounded by the fibrous relics of the old ones, linear, glabrous, rigid, closely ribbed, 1–1$\frac{1}{2}$ ft. long; stems 1–2 ft. long, with 2–3 reduced leaves; flowers in a dense panicle, with scorpioid cymose branches, both branches and perianth densely and persistently coated with spreading, soft, plumose hairs ; perianth, including ovary, $\frac{1}{3}$–$\frac{1}{2}$ in. long.   *Roem. et Schultes Syst.* ` Veg.* vii. 294 ; *Schnitzl. Iconogr.* i. t. 62, fig. 5.   *Argolasia, Juss. Gen.* 60. *Augea, Retz. Obs. v.* 3.

COAST REGION: Port Elizabeth Div., Leadmine River, *Burchell*, 4606! Riversdale Div., Langebergen, *Burchell*, 6871! Uitenhage Div., Vanstadens-hoogte, *MacOwan*, 2088!

V. SANSEVIERIA, Thunb.

Perianth glabrous ; tube long, cylindrical ; segments 6, subequal, spreading, linear.   *Stamens* 6, inserted at the throat of the perianth-tube ; filaments filiform ; anthers oblong, versatile.   *Ovary* free, 3-celled ; ovules solitary in the cells ; style long filiform ; stigma capi-tate.   *Pericarp* bursting before the seeds ripen.   *Seeds* 1–3, globose ; testa fleshy.

Rhizome thick, wide-creeping. Leaves rosulate from the nodes, rigid, flat, subterete, or terete, containing an abundant supply of strong fibre. Peduncle elongated, with only a few scariose bract-leaves. Flowers in a dense cylindrical raceme or spike, whitish ; pedicels articulated, usually fascicled.

DISTRIB. Species about 12, spread through the warmer regions of the Old World.

Leaves nearly flat :
  Inflorescence racemose ... ... (1) **thyrsiflora.**
  Inflorescence subspicate ... ... (2) **subspicata.**
Leaves subterete ... ... (3) **zeylanica.**

1. **S. thyrsiflora** (Thunb. Prodr. 65) ; leaves 6–12 in a cluster, nearly flat, oblanceolate, rigid, 6–18 in. long, $1\frac{1}{2}$–2 in. broad above the middle, narrowed gradually to a concave base, obscurely fasciated with cross-bands of white, margined with a distinct red-brown line ; peduncle a foot or more long; raceme cylindrical, 9–12 in. long ; pedicels 2–6-nate, $\frac{1}{8}$–$\frac{1}{4}$ in. long, articulated at the middle ; bracts lanceolate, scariose ; perianth $1\frac{1}{4}$–$1\frac{1}{2}$ in. long; segments rather shorter than the tube; anthers oblong, pale yellow; style exserted beyond the segments. *Fl. Cap.* edit. *Schult.* 329 ; *Baker in Journ. Linn. Soc.* xiv. 547. *S. spicata, Haw. Syn. Succ.* 66 ; *Kunth Enum.* v. 20. *S. fulvo-cincta, Haw. Suppl. Succ.* 30. *S. guineensis, Kunth Enum.* v. 16, *ex parte. S. angustiflora, Lindb. in Act. Soc. Fenn.* x. 130, t. 5. *S. rufo-cincta, Hort. Salmia spicata, Cav. Ic.* iii. 24, t. 246.

COAST REGION : Uitenhage Div., *Thunberg!* Zwartkops River, *Zeyher*, 612 ! Albany Div., *Cooper*, 3269 ! Enon., *Baur*, 1099 ! Stockenstrom Div., *Scully*, 163 !

2. **S. subspicata** (Baker in Gard. Chron. 1889, ii. 436) ; leaves up to ten in a cluster, rigid, plain green, pale, not red on the edge, oblanceolate, nearly flat, 8–9 in. long, $1\frac{1}{2}$–2 in. broad at the middle, narrowed gradually to the concave base ; peduncle terete, stiffly erect, shorter than the leaves ; inflorescence dense, subspicate ; flowers all solitary ; bracts minute, lanceolate, scariose ; perianth 2 in. long ; tube cylindrical, more than twice as long as the segments ; stamens a little longer than the perianth-segments ; anthers small, oblong ; style much overtopping the stamens.

EASTERN REGION : Delagoa Bay, *Mrs. Monteiro !*
Described from a plant that flowered at Kew, Oct., 1889.

3. **S. zeylanica** (Willd. Sp. Plant. ii. 159); leaves 8–15 in a cluster, semiterete, $\frac{1}{2}$–1 ft. long, an inch broad at the base, $\frac{1}{3}$–$\frac{1}{2}$ in. thick at the middle, rounded on the back, deeply channelled down the face, margined with a distinct red line ; peduncle a foot or more long ; raceme dense, cylindrical, $\frac{1}{2}$–1 ft. long; pedicels 3–6-nate, $\frac{1}{12}$–$\frac{1}{6}$ in. long, articulated above the middle ; perianth $1\frac{1}{4}$ in. long; segments about as long as the tube; stamens shorter than the segments; anthers oblong, $\frac{1}{8}$ in. long ; style exserted beyond the tip of the segments. *Red. Lil.* t. 290 ; *Bot. Reg.* t. 160 ; *Baker in Journ. Linn. Soc.* xiv. 548. *Aletris hyacinthoides,* var. *zeylanica, Linn. Sp. Plant.* 456. *Aloe zeylanica, Jacq. Hort. Vind.* 310. *Aletris zeylanica, Mill. Dict.* edit. 8, No. 4. *S. æthiopica, Thunb. Prodr.* 65 ; *Kunth Enum.* v. 19.

COAST REGION : Uitenhage, *Burchell*, 4420 ! Zuurbergen, *Cooper*, 3267 ! Albany Div., *Cooper*, 3268 !

CENTRAL REGION: Near Graaff Reinet, *Bolus,* 720!
KALAHARI REGION: Griqualand West, near Griqua Town, *Burchell,* 1824!
Also in Tropical Africa and Tropical Asia.

8. **paniculata** (Schinz Conspect. Fl. Afric. 141), from Port Alfred, *Schoenland,* 290! proves to be *Dracæna Hookeriana,* K. Koch.

VI. CYANELLA, Linn.

Perianth cut down to the ovary; segments subequal, acute, or obtuse, petaloid, glabrous, laxly nerved. *Stamens* 6, inserted at the base of the segments, 3–5 arcuate, the others declinate; filaments short; anthers dehiscing by two large terminal pores. *Ovary* adnate towards the base, globose, 3-celled; ovules numerous and superposed; style filiform; stigma capitate. *Capsule* loculicidal. *Seeds* ovoid, turgid.

Corm with matted fibrous tunics. Leaves mostly in a basal rosette, terete, or linear, persistent. Flowers usually racemose; bracts persistent.

DISTRIB. Endemic.
Leaves terete ...   . (1) **alba.**
Leaves flat :
    Stamens 3 arcuate, perianth-segments obtuse   ... (2) **orchidiformis.**
    Stamens 5 arcuate, perianth-segments acute :
        Perianth $\frac{1}{4}$–$\frac{1}{2}$ in. long   ...   . (3) **capensis.**
        Perianth $\frac{1}{2}$–$\frac{3}{4}$ in. long   ...   . (4) **lutea.**

1. **C. alba** (Linn. fil. Suppl. 201); corm globose, with a neck 3–4 in. long; basal leaves numerous, erect, slender, terete, 4–6 in. long; peduncles 3–6 in a cluster, simply 1-flowered, 6–9 in. long, leafless, or with a single small leaf; perianth spreading, whitish, $\frac{1}{2}$–$\frac{3}{4}$ in. long; segments oblong, acute, 3 outer with green cusps; stamens half as long as the perianth, 5 arcuate, 1 declinate, all with short filaments and anthers $\frac{1}{4}$ in. long; capsule globose, adnate only at the very base. *Thunb. in Act. Holm.* 1794, t. 7, fig. 2; *Prodr.* 65; *Kunth Enum.* iv. 640; *Baker in Journ. Linn. Soc.* xvii. 497. *Pharetrella alba,* *Salisb. Gen.* 47.

COAST REGION: Clanwilliam Div., *Masson! Mader,* 135!

2. **C. orchidiformis** (Jacq. Collect. iv. 211; Ic. t. 447); corm globose, $\frac{1}{2}$ in. diam., with a short neck; basal leaves 3–4, oblong-lanceolate, chartaceous, glabrous, 3–6 in. long, $\frac{3}{4}$–$1\frac{1}{2}$ in. broad; peduncle branched, a foot long, including the lax raceme; racemes 2–4 in. long; pedicels cernuous, $\frac{1}{2}$–1 in. long, sometimes bracteolate; bracts lanceolate; perianth bright red, $\frac{1}{2}$ in. long; segments obovate-cuneate, with a yellow claw; stamens 3, arcuate, with longer filaments and small anthers, 3 declinate, with shorter filaments and anthers $\frac{1}{8}$ in. long; capsule adnate at the very base only. *Kunth Enum.* iv. 637; *Baker in Journ. Linn. Soc.* xvii. 497. *Trigella orchidiformis, Salisb. Gen.* 46.

SOUTH AFRICA: without locality, *Pappe!* WESTERN REGION: Namaqualand, *Drège!*

3. C. capensis (Linn. Sp. Plant. 443); corm globose, with a neck 3–6 in. long; basal leaves 6–8, linear, moderately firm, ½–1 ft. long, finely ribbed; peduncle 1–1½ ft. long, including the inflorescence, often with several arcuate branches; racemes very lax; pedicels arcuate, lower 1–1¼ in. long, with a linear bracteole at the middle; bracts lanceolate; perianth white or pale red, ¼–⅓ in. long; segments oblong, outer 5-nerved, inner 1-nerved; stamens 5, arcuate, with shorter anthers, 1 declinate with an anther ⅛ in. long; capsule globose, adnate in the lower third. *Thunb. Prodr.* 65; *Fl. Cap.* edit. *Schult.* 330; *Bot. Mag.* t. 568; *Andr. Bot. Rep.* t. 141; *Jacq. Hort. Vind* iii. t. 35; *Red Lil.* t. 373; *Kunth Enum.* iv. 636; *Baker in Journ. Linn. Soc.* xvii. 498. *C. cœrulea, Eckl. Top. Verz.* 4.

COAST REGION: Clanwilliam Div., Ebenezer, *Drège*, 8606! Lion Mt., *Burchell*, 129! 135! Table Mt., *Macgillivray*, 478! Mountains above Simons Bay, *Milne*, 142! Groenekloof, *Zeyher!* 1718! Riversdale Div., between Valsche R. and Zoetmelks R., *Burchell*, 6532! 6534!

4. C. lutea (Linn. fil. Suppl. 201); corm ovoid, ½ in. diam., with a neck 1–6 in. long; basal leaves linear, 4–6 in. long, about ½ in. broad; peduncles 6–18 in. long, including the inflorescence, often branched; racemes many-flowered, lax or dense, 2–4 in. long; lower pedicels 1–1½ in. long, with a bracteole at the middle, and a large lanceolate bract at the base; perianth milk-white, ½–¾ in. long; segments oblong-lanceolate; stamens 5 arcuate, with anthers ⅙ in. long; 1 declinate, with an anther ¼ in. long; capsule globose, adnate in the lower third. *Thunb. Prodr.* 65; *Act. Holm.* 1794, 175, t. 7, fig. 1; *Fl. Cap.* edit. *Schult.* 330; *Ker in Bot. Mag.* t. 1252; *Kunth Enum.* iv. 639; *Baker in Journ. Linn. Soc.* xvii. 498. *C. lineata, Burchell Travels* ii. 589. *C. odoratissima, Lindl. in Bot. Reg.* t. 1111.

VAR. β, rosea (Baker in Saund. Ref. Bot. t. 259). Flowers pale red. *C. rosea, Eckl. MSS.*

COAST REGION: Malmesbury Div., *Drège!* Swellendam Div., *Zeyher*, 4256! Mossel Bay Div., *Burchell*, 6174! George Div., *Burchell*, 6100! Bathurst Div., *Burchell*, 4148! (C. lineata, Queenstown, *Cooper*, 270!) Var. β, Zoetmelks R., *Gill!* Near Grahamstown, *Galpin*, 369!
CENTRAL REGION: Somerset Div., *Drège*, 8604!
WESTERN REGION: Var. β, Namaqualand, *Scully*, 177!
KALAHARI REGION: Near the Orange R., *Burchell*, 1630! Bechuanaland, *Burchell*, 2346! (C. lineata, Bechuanaland, near Moshowa R., *Burchell*, 2256·2! Basutoland, *Cooper*, 2273!)

ORDER CXXXIV. IRIDEÆ.

(By J. G. BAKER.)

Flowers hermaphrodite, regular or irregular. *Perianth* superior, corolline, with or without a tube above the ovary; lobes biseriate, equal or unequal. *Stamens* 3, placed opposite the outer lobes of the perianth, divaricated equally from the style, or more or less distinctly unilateral; filaments filiform and free, or sometimes united in a tube;

anthers 2-celled, dehiscing longitudinally extrorsely, or down the margin. *Ovary* inferior, almost invariably 3-celled; ovules axile, anatropal, usually many in a cell, and superposed; style filiform; branches 3, variously stigmatose, filiform, cuneate, or petaloid, sometimes bifid. *Fruit* a 3-celled capsule with loculicidal dehiscence. *Seeds* globose, angled or discoid; testa thin, membranous; albumen horny, with the minute embryo enclosed in it near the hilum.

Perennial herbs, very rarely annual herbs or undershrubs. Leaves narrow, sessile, dry, usually firm in texture and persistent, often equitant and distichous. Flowers very various in colour, either comparatively persistent, each sessile in a 2-valved spathe and arranged in simple or panicled spikes, or fugitive, stalked and clustered in many-valved spathes from which they successively emerge.

DISTRIB. About half the known species are concentrated at the Cape; the others are spread widely through both hemispheres, mainly in temperate regions.

Sub-Order 1. MORÆEÆ. *Inflorescence* corymbose; flowers comparatively fugitive, generally more than one to a spathe. *Stamens* opposite the style-branches and adpressed to them.

I. **Moræa.**—*Style-branches* large and petaloid, transversely stigmatose at the base of the large crests.

II. **Homeria.**—*Style-branches* petaloid, with two small divergent crests, papillose round their edges.

III. **Ferrari**.—*Style-branches* small, petaloid, bifid, densely ciliated on the edge.

IV. **Hexaglottis.**—*Style-branches* not petaloid, deeply forked.

Sub-Order 2. SISYRINCHIEÆ. *Inflorescence* corymbose. *Stamens* alternate with the style-branches.

Tribe 1. *GALAXIEÆ.* *Spathes* 1-flowered.

V. **Galaxia.**—*Spathes* sessile in the centre of the rosette of leaves. *Stamens* monadelphous. *Stigma* peltate.

VI. **Syringodea.**—*Spathes* sessile in the centre of the rosette of leaves. *Stamens* free. *Style* with three clavate branches.

VII. **Romulea.**—*Spathes* peduncled.

Tribe 2. *ARISTEÆ.* *Spathes* usually more than 1-flowered.

VIII. **Bobartia.**—*Style-branches* long, subulate.

IX. **Witsenia.**—*Style-branches* short. *Perianth* lobes equal, shorter than the tube.

X. **Cleanthe.**—*Style-branches* short. *Perianth* lobes unequal.

XI. **Aristea.**—*Style-branches* short. *Perianth* lobes subequal, longer than the tube, not unguiculate.

XII. **Klattia.**—*Style-branches* short. *Perianth* lobes subequal, longer than the tube, unguiculate.

Sub-Order 3. IXIEÆ. *Inflorescence* spicate; flowers not fugitive, solitary, each subtended by a pair of spathe-valves.

* *Style-branches simple. Flowers regular. Stamens equilateral.*

XIII. **Schizostylis.**—Like *Hesperantha,* but rootstock not thickened into a corm.

XIV. **Hesperantha.**—*Style* short; branches long, subulate. *Spathe-valves* green.

XV. **Geissorhiza.**—*Style* longer than in the last; branches short, subulate. *Spathe-valves* all green or membranous at the tip.

XVI. **Ixia.**—*Style* long; branches short, subulate. Outer *spathe-valve* short, brown, emarginate.

XVII. **Streptanthera.**—*Style* long; branches short, clavate. *Spathe-valves* both membranous, lacerated. *Leaves* short.

XVIII. **Dierama.**—*Style* long; branches short, clavate. *Spathe-valves* both membranous, entire. *Leaves* long, rigid.

** *Style-branches bifid. Stamens unilateral.*

XIX. **Lapeyrousia.**—*Perianth-tube* slender, with the stamens inserted at its throat. *Ovules* many, superposed.

XX. **Micranthus.**—*Perianth-tube* cylindrical, with the stamens inserted at its throat. *Ovules* 2, erect, collateral.

XXI. **Freesia.**—*Perianth-tube* broad, with the stamens inserted below its throat. *Spathe-valves* small, green.

XXII. **Watsonia.**—*Perianth-tube* widened at the middle where the stamens are inserted. *Spathe-valves* moderately large, rigid.

*** *Style-branches simple. Stamens unilateral, arcuate.*

XXIII. **Dallaua.**—Differs from the following genera by its plicate hairy leaves. *Perianth* irregular or subregular.

XXIV. **Melasphœrula.**—*Perianth* without any tube ; segments acuminate.

XXV. **Sparaxis.**—*Perianth* regular, with a short funnel-shaped tube. *Spathe-valves* membranous, deeply lacerated.

XXVI. **Tritonia.**—*Perianth* subregular, with a short cylindrical tube. *Spathe-valves* small, oblong, brown, emarginate.

XXVII. **Crocosma.**—*Perianth* subregular, with a cylindrical tube. *Spathe-valves* short, oblong. *Capsule* inflated, deeply 3-lobed.

XXVIII. **Acidanthera.**—*Perianth* subregular, with a long subcylindrical tube. *Spathe-valves* long, green.

XXIX. **Synnotia.**—*Perianth* irregular. *Spathe-valves* membranous, deeply lacerated.

XXX. **Gladiolus.**—*Perianth* irregular, with a funnel-shaped tube. *Spathe-valves* large, green, lanceolate.

XXXI. **Antholyza.**—*Perianth* irregular ; tube dilated at the middle. *Spathe-valves* oblong-lanceolate.

I. MORÆA, Linn.

Perianth funnel-shaped, without any proper tube ; segments unequal, more or less distinctly unguiculate, 3 outer obovate-oblong, 3 inner always smaller, usually oblanceolate. *Stamens* opposite the style-branches ; filaments nearly always more or less connate ; anthers linear. *Ovary* 3 celled, obtuse or rostrate ; ovules crowded, superposed ; style-branches large and petaloid, as in *Iris,* with the stigma at the base of the petaloid crests. *Capsule* ovoid or ellipsoid, dehiscing loculicidally. *Seeds* ovoid or subglobose, often angled by pressure.

Rootstock, usually a tunicated corm, rarely a rhizome. Produced leaves usually few, linear sometimes forming a distichous rosette ; upper rudimentary sheathing. Spathes cylindrical ; outer valves green, more or less rigid. Flowers smaller and more fugitive than in *Iris,* usually lilac or yellow.

DISTRIB. Species about 60, many Tropical African ; the rest inhabiting Madagascar and Australia. The line of limit between *Iris* and *Morœa* has been differently drawn by different authors, but the two genera are extremely close.

Subgenus I. EUMORÆA. *Rootstock* a corm. Inner *segments* of the perianth oblanceolate-unguiculate. *Ovary* not produced into a beak.
ACAULES. *Spathes* sessile in the centre of the rosette of leaves.
Basal leaves 2–4 :
    Spathe 1½–2 in. long ...   . (1) **ciliata.**
    Spathe 2–4 in. long ...   . (2) **macrochlamys.**

Basal leaves many :
   Leaves linear-convolute :
      Flowers lilac ... ... (3) **galaxioides.**
      Flowers yellow ... ... (4) **fasciculata.**
   Leaves lanceolate ... ... (5) **falcifolia.**

BREVICAULES. Stems very short.
   Leaves hairy all over ... ... (6) **papilionacea.**
   Leaves hairy on the edges only ... ... (7) **fimbriata.**

MONOCEPHALÆ. Stems long, usually 1-headed.
   Leaves terete :
      Lowest leaf only produced, very long ... ... (8) **angusta.**
      Leaves 3–4 produced, all short ... ... (9) **Baurii.**
   Leaves linear :
      Stem slender. Flowers red-brown ... ... (10) **lurida.**
      Stem robust. Flowers yellow ... ... (11) **spathacea.**

CORYMBOSÆ. Stems elongated. *Inflorescence* laxly
   corymbose.
Stems viscous below the nodes ... ... (12) **viscaria.**
Stems glabrous :
   Comparatively dwarf, with small flowers :
      Spathes ½–¾ in. long :
         Leaves linear-complicate ... ... (13) **juncea.**
         Leaves linear crisped ... ... (14) **crispa.**
      Spathes ¾–1 in. long :
         Leaves glabrous ... ... (15) **arenaria.**
         Leaves strongly ciliated ... ... (16) **serpentina.**
      Spathes 1–1½ in. long :
         Leaves subulate ... ... (17) **Cooperi.**
         Leaves linear-subulate ... ... (18) **polyantha.**
         Leaves narrow linear :
            Basal leaf only produced ... ... (19) **iriopetala.**
            1–2 leaves produced ... ... (20) **mira.**
            2–3 leaves produced ... ... (21) **tristis.**
   More robust, with larger flowers :
      Flowers lilac ... ... (22) **polyanthos.**
      Flowers yellow :
         Spathes an inch long ... ... (23) **ramosa.**
         Spathes 1½ in. long ... ... (24) **gigantea.**

SUBRACEMOSÆ. Clusters of flowers few or many, sessile or shortly peduncled.
Stems with 1–2 produced leaves remote from the inflor-
   escence :
   Leaves subulate, very slender ... ... (25) **setacea.**
   Leaves linear :
      Flower lilac ... ... (26) **undulata.**
      Flower yellow ... ... (27) **Bolusii.**
Stems with only one leaf produced from the base of the
   inflorescence :
   Perianth ¾–1 in. long ... ... (28) **natalensis.**
   Perianth 1–1½ in. long ... ... (29) **edulis.**

Subgenus II. HELIXYRA. *Rootstock* a corm. Inner *segments* of the perianth
oblanceolate-unguiculate. *Ovary* produced into a long beak at the top, like a
perianth-tube.
   Leaf terete :
      Spathes ½–⅝ in. long ... ... (30) **spiralis.**
      Spathes 1–1¼ in. long :
         Stem scarcely any ... ... (31) **Rogersii.**
         Stem shortly produced ... ... (32) **Burchellii.**

Spathes 1½–2 in. long:
 Stem very short ... ... (33) longiflora.
 Stem elongated ... ... (34) simulans.
Leaf linear ... ... (35) cladostachya.

Subgenus III. VIEUSSEUXIA. *Rootstock* a corm. Inner *segments* of the perianth much smaller than the outer, usually tricuspidate.
Inner segments of perianth entire:
 Inner segments linear, minute ... ... (36) tripetala.
 Inner segments oblanceolate obtuse ... (37) Elliotii.
Inner segments of the perianth with a large central
  cusp and two lateral lobes:
 Outer segments with a short distinct claw ... ... (38) glaucopis.
 Outer segments without any distinct claw:
  Flower coloured ... ... (39) pavonia.
  Flower white ... ... (40) candida.
Inner segments of the perianth with three large cusps:
 Claw of the outer segments about as long as the
  blade:
  Leaf linear-subulate ... ... (41) unguiculata.
  Leaf linear ... ... (42) tricuspis.
 Claw of the outer segments half as long as the
  blade ... ... (43) tenuis.
Subgenus IV. DIETES. *Rootstock* a short creeping rhizome.
 Flowers white ... ... (44) iridoides.
 Flowers yellow ... ... (45) bicolor.

1. **M. ciliata** (Ker in Konig and Sims' Ann. i. 241); corm small, globose, coarsely honeycombed; basal leaves 3–4, lanceolate, acuminate, finally ½ ft. long, firm, sometimes crisped, glabrous on the faces, distinctly ciliated on the edge; spathes solitary, rather inflated sessile in the centre of the tufts of leaves, 1–3-flowered; valves lanceolate, 1½–2 in. long; perianth lilac or yellow, very fugitive, 1¼ in. long; outer segments oblong, inner oblanceolate, both distinctly unguiculate, with a spreading limb; filaments united in a column in the lower part; ovary clavate, ¼ in. long; crests of stigma lanceolate; capsule clavate, an inch long. *Bot. Mag.* t. 1061; *Ait. Hort. Kew.* edit. 2, i. 114; *Ker, Gen. Irid.* 42; *Klatt in Linnæa* xxxiv. 558; *Ergänz.* 31; *Baker in Journ. Linn. Soc.* xvi. 130; *Handb. Irid.* 48. *Iris ciliata, Linn. fil. Suppl.* 98; *Thunb. Diss.* No. 1; *Prodr.* 11; *Fl. Cap.* i. 284.

VAR. β, **M. barbigera** (Salisb. in Trans. Hort. Soc. i., 306). Flowers red, outer segments bearded on the claw. *Klatt in Linnæa* xxxiv. 557; *Ergänz.* 31; *Baker in Journ. Linn. Soc.* xvi. 130. *M. ciliata, Ker in Bot. Mag.* t. 1012. *M. pilosa, Wendl. Obs. Bot.* 42? *M. hantamensis, Klatt, Ergänz.* 31.

VAR. γ, **M. tricolor** (Andr. Bot. Rep. t. 83); leaves broader, flowers red; outer segments with a yellow spot at the base of the blade. *Salisb. in Trans. Hort. Soc.* i. 306; *Baker in Journ. Linn. Soc.* xvi. 130; *Klatt, Ergänz.* 31.

VAR. δ, **M. minuta** (Ker in Konig and Sims' Ann. i. 241). A dwarf form, with a yellow flower smaller than in the type; leaves not more than 2–3 in. long, obscurely ciliated on the edge. *Ker, Gen. Irid.* 43; *Baker in Journ. Linn. Soc.* xvi. 130. *Iris minuta, Linn. fil. Suppl.* 98; *Thunb. Diss.* No. 2; *Fl. Cap.* i. 285.

SOUTH AFRICA: without locality, *Masson! Pappe! Villette!* Var. γ, known only from Andrews' figure.

2. M. macrochlamys (Baker, Handb. Irid. 49) ; corm not seen ;
basal leaves 3–4, lanceolate-acuminate, nearly an inch broad at the
dilated base ; ½–1 ft. long, moderately firm, glabrous on both surfaces,
minutely ciliated on the edge ; spathes 1–2, nearly or quite sessile
in the centre of the rosette of leaves, inflated, 2–4 in. long ; outer
valves lanceolate, foliaceous ; perianth fugitive (colour uncertain),
an inch long ; outer segments oblong, inner oblanceolate, both distinctly unguiculate ; filaments connate at the base only ; ovary
clavate, ¼–⅓ in. long ; style-crests lanceolate.

CENTRAL REGION : On the Sneeuwberg range, 4000–6000 ft. alt., *Drège,*
2186 !

3. M. galaxioides (Baker in Journ. Linn. Soc. xvi. 130) ; corm
small, ovoid ; outer tunics with wiry parallel strands connected by
short transverse fibres ; basal leaves a dozen or more, narrow linear,
convolute, 1–1½ in. long, falcate, glabrous ; spathe under an inch
long, 1-flowered, sessile in the centre of the rosette of leaves ;
perianth lilac, fugacious, ½–¾ in. long ; outer segments oblong, inner
oblanceolate, both distinctly unguiculate ; filaments connate in the
lower half ; ovary clavate, ½ in. long ; style-crests small, lanceolate.
Handb. Irid. 49.

KALAHARI REGION : Griqualand West, in rocky places, Dutoit's pan, *Tuck,*
13 ! Klip Drift, *Mrs. Barber,* 14 !

4. M. fasciculata (Klatt, Ergänz. 32) ; corm small, ovoid ; outer
tunics of thick fibres ; leaves 6–10 in a basal rosette, narrow linear
acuminate, convolute, glabrous, falcate, 2–3 in. long ; spathes
1-flowered, short, sessile, hidden in the centre of the rosette of
leaves ; perianth yellow, fugacious, ½ in. long ; outer segments oblong, inner oblanceolate, both distinctly unguiculate ; ovary clavate,
⅙ in. long ; style-column ⅛ in. long ; crests of stigmas lanceolate.
Baker, Handb. Irid. 49. *M. polyphylla, Baker in Journ. Linn. Soc.*
xvi. 130 (name only). *M. longiflora, Klatt in Linnæa* xxxiv. 725,
non Ker.

SOUTH AFRICA : without locality, *Drège,* 2301a !
COAST REGION : Clanwilliam Div., at Langevalei, *Drège,* 2600 !

5. M. falcifolia (Klatt, Ergänz. 32) ; corm small, ovoid ; tunics
rigidly fibrous ; leaves many, lanceolate, falcate, glabrous, ⅙–⅕ in.
long ; peduncle very short ; spathes 3–4 sessile, an inch long ; valves
unequal, the outer acute, herbaceous ; perianth reddish, fugacious,
under ½ in. long ; outer segments oblong, inner oblanceolate, both
unguiculate ; style-crests lanceolate. *Baker, Handb. Irid.* 49.

CENTRAL REGION : Calvinia Div., Hantam Hills, *Meyer.*

6. M. papilionacea (Ker in Bot. Mag. t. 750) ; corm globose, ½ in.
diam. ; outer coats of brown imbricated lanceolate processes ; basal

leaves 2–4, linear, rigid, 3–6 in. long, strongly ribbed, densely hairy ; stems 1–2 in. long, simple, or with 2–4 short arcuate branches ; spathes cylindrical, pubescent, 1–2 in. long; valves green, outer short; perianth very fugitive, lilac or red, $\frac{3}{4}$ in. long; outer segments oblong, inner oblanceolate, both with a distinct claw and spreading blade ; filaments connate in the lower half ; ovary clavate, glabrous, $\frac{1}{4}$ in. long ; crests of stigma linear, $\frac{1}{4}$–$\frac{1}{2}$ in. long; capsule small, clavate. *Ker, Gen. Irid.* 43 ; *Ait. Hort. Kew,* edit. 2, i. 114 ; *Klatt in Linnæa* xxxiv. 560 ; *Ergänz.* 31 ; *Baker in Journ. Linn. Soc.* xvi. 129 ; *Handb. Irid.* 49. *Iris. papilionacea, Linn. fil. Suppl.* 98 ; *Thunb. Diss.* No. 37, tab. 2, fig. 1 ; *Prodr.* 12 ; *Fl. Cap.* i. 298 : *Jacq. Collect. Suppl.* 159, tab. 3, fig. 2. *Moræa ciliata, Herb. Drège, non Ker. M. hirsuta. Ker, Gen. Irid.* 43. *Iris hirsuta, Licht. in Roem. et Schultes Syst. Veg.* i. 478. *Vieusseuxia ciliata* and *nervosa, Eckl. Top. Verz.* 11-12.

COAST REGION : Hills and Flats near Cape Town, *Thunberg! Bolus,* 2832! 3693! *Drège! Zeyher,* 1640! *Ecklon,* 818! *Wright!* Klein River Berg, *Zeyher,* 4071! Paarl Berg, *Drège!*

7. **M. fimbriata** (Klatt in Linnæa xxxiv. 561); corm globose, $\frac{1}{2}$ in. diam.; outer tunics thick, with close, wiry, parallel strands ; basal leaves 4–6, linear, 2–3 in. long, firm, much crisped, glabrous on the faces, minutely ciliated on the edge ; stems slender, 1–3 in. long, single or fascicled ; spathes cylindrical, 1$\frac{1}{2}$ in. long, 1–3-flowered ; valves lanceolate, scariose at the tip, outer short ; perianth lilac, very fugacious, $\frac{3}{4}$ in. long ; outer segments oblong, inner oblanceolate, both distinctly unguiculate ; ovary clavate, $\frac{1}{2}$ in. long ; style-column $\frac{1}{6}$ in. long ; crests of stigma small, lanceolate ; capsule clavate, $\frac{1}{2}$ in. long. *Klatt, Ergänz.* 33 ; *Baker in Journ. Linn. Soc.* xvi. 130 ; *Handb. Irid.* 50. *Vieusseuxia crispa, Eckl. Top. Verz.* 12.

COAST REGION : Klein River Berg, 1000–3000 ft. alt., *Zeyher,* 4091! A near ally of *M. papilionacea.*

8. **M. angusta** (Ker in Bot. Mag., t. 1276); corm globose, $\frac{1}{2}$–1 in. diam., with parallel, brown, wiry strands, and a crown of long bristles ; stem simple, wiry, flexuose, 1–2 ft. long, with one long, wiry terete leaf low down, and two or three others higher up, short and rudimentary; spathe cylindrical, 2–2$\frac{1}{2}$ in. long, 3–4-flowered ; valves rigid, pale green, obtuse ; outer short ; perianth bright yellow, fugacious, 1$\frac{1}{2}$ in. long; outer segments oblong-unguiculate, $\frac{1}{2}$ in. broad ; inner nearly as long, oblanceolate ; filaments united towards the base ; ovary clavate, $\frac{1}{2}$ in. long ; style-crests lanceolate, $\frac{1}{2}$ in. long; capsule oblong, an inch long. *Ker, Gen. Irid.* 37 ; *Klatt in Linnæa* xxxiv. 559 ; *Ergänz.* 32 ; *Baker in Journ. Linn. Soc.* xvi. 130 ; *Handb. Irid.* 50. *Iris angusta, Thunb. Diss.,* No. 28 ; *Prodr.* 12 ; *Fl. Cap.* i. 294 ; *Willd. Sp. Plant.* i. 235 ; *Roem. et Schultes Syst. Veg.* i. 473. *M. teretifolia, Soland. MSS.*

COAST REGION: vicinity of Cape Town, *Thunberg! Masson! Wright,* 240! *Bolus,* 3802! *Elliot,* 1095! 1198! Swellendam Div., *Zeyher,* 4076!

9. M. Baurii (Baker, Handb. Irid. 50); corm not seen; stem a foot long, simple, slender, 1-headed, bearing about four superposed sheathing leaves with short, free, erect, enrolled linear tips; spathes cylindrical, 1–2-flowered, 2 in. long; valves rigid, lanceolate-acuminate, nearly equal; perianth yellow; outer segments 1½ in. long, with a reflexing obovate-connate blade as long as the claw; inner shorter, oblong-unguiculate, erect; filaments connate only at the base; ovary clavate, ½ in. long; style-crests ⅓ in. long; capsule clavate, an inch long.

EASTERN REGION: Kaffraria, Bazeia Mts., 2000-3000 ft. alt., *Baur,* 247!

10. M. lurida (Ker in Bot. Reg., t. 312); corm globose; ½ in. diam.; stem a span long, very slender, glabrous, bearing one long, narrow linear, glabrous, produced leaf, ⅙ in. broad, and two rudimentary ones higher up; spathe single, terminal, cylindrical, ½ in. long; perianth fugacious, bright reddish-brown, an inch long; outer segments obovate-unguiculate, ⅓ in. broad; inner oblanceolate; filaments united at the base; style-crests linear. *Klatt in Linnæa* xxxiv. 559; *Ergänz.* 32; *Baker in Journ. Linn. Soc.* xvi. 130; *Handb. Irid.* 50.

SOUTH AFRICA: locality unknown.
Known only from the figure cited. Habit of a *Vieusseuxia.*

11. M. spathacea (Ker in Bot. Mag. sub t. 1103); corm large, ovoid, with a dense coat of brown-black, wiry fibres; stem short, 2–4 ft. long, simple or rarely branched, with a single leaf produced from the base, which is thick, rigid, linear, acuminate, flat, strongly ribbed, 1½–2 ft. long, ½–¾ in. broad, low down, and several rudimentary stem-leaves; spathes 3–5 in. long, 2–5-flowered; valves with a long cusp; outer much shorter than the inner; perianth bright yellow, 1½–2 in. long; outer segments oblong-unguiculate, ¾–1 in. broad; inner oblanceolate; filaments connate in the lower half; ovary clavate, ½ in. long; style ¼–⅓ in. long; capsule ellipsoid, obtusely angled, 1½ in. long. *Ker, Gen. Irid.* 39; *Baker in Journ. Linn. Soc.* xvi. 131; *Handb. Irid.* 51. *Iris spathulata, Linn. fil. Suppl.* 99. *Iris spathacea, Thunb. Diss.,* No. 23; *Prodr.* 12; *Fl. Cap.* i. 292; *Eckl. Top. Verz.* 11. *Moræa longispatha, Klatt in Linnæa* xxxv. 308. *Dietes Huttoni, Hook. fil. in Bot. Mag.* t. 6174.

VAR. β, natalensis (Baker); flowers large; segments more distinctly unguiculate; style-crests larger.

VAR. γ, Galpini (Baker); leaves long, very narrow, convolute; flowering stem not above ⅓ ft. long; flowers in August, the type in February.

COAST REGION: Paal Berg, *Drège!* Knysna Div., *Burchell,* 5552! *Drège,* 8301! Near George, *Burchell,* 60 5! Near Grahamstown, *MacOwan!* Stockenstrom Div., *Scully,* 12 ! 65! Queenstown Div., *Cooper,* 316!
EASTERN REGION: Kaffraria; Bazeia along streams 2000 ft. alt., *Baur,* 514! Var. β, Griqualand East, *Haygarth* in *Hb. Wood,* 4183! Natal, ascending to 5000 or 6000 ft., *Sanderson,* 355! *Wood,* 487! 4526! *Mrs. K. Saunders!*
KALAHARI REGION: Var. β, Transvaal; Houtbosch Berg, *Nelson,* 502! Var. γ, Transvaal, Saddleback Range, near Barberton, *Galpin,* 459!

12. **M. viscaria** (Ker in Konig and Sims' Ann. i. 240); corm ovoid, 1 in. diam.; outer tunics thick, with parallel wiry strands and close cross-bristles; produced leaves 1–2, narrow, linear, rigid, channelled, strongly-ribbed, glabrous, 1–2 ft. long; inflorescence a lax panicle, with 5–20 clusters of flowers; branches short, arcuate, viscose, subtended by rudimentary leaves; spathes cylindrical, 1–1½ in. long, 2–3-flowered; valves rigid, lanceolate, scariose at the tip; perianth very fugitive, brownish, ¾ in. long; outer segments oblong-unguiculate; innor, oblanceolate; filaments free nearly or quite to the base; ovary clavate, ⅓ in. long; style-crests small, lanceolate; capsule ellipsoid, obtusely angled, ¼–⅓ in. long. *Ker in Bot. Mag.* sub t. 696; *Ait. Hort. Kew.* edit. 2, i. 113; *Klatt in Linnæa* xxxiv. 567; *Ergänz.* 33; *Baker in Journ. Linn. Soc.* xvi. 131; *Handb. Irid.* 55. *Iris viscaria, Linn. fil Suppl.* 98; *Thunb. Diss.*, No. 41; *Prodr.* 12; *Fl. Cap.* i. 304. *Vieusseuxia viscaria, Eckl. Top. Verz.* 12.

VAR. β, **M. bituminosa** (Ker in Konig and Sims' Ann. i. 240); taller, with broader leaves, bright yellow flowers, and green branchlets. *Ker in Bot. Mag.* t. 1045. *Iris bituminosa, Linn. fil. Suppl.* 98; *Thunb. Diss.*, No. 42, tab. 2, fig. 2; *Protr.* 12; *Fl. Cap.* i. 305. *Vieusseuxia bituminosa, Eckl. Top. Verz.* 14.

COAST REGION: Saldanha Bay, *Thunberg!* Paarl Div., *Drège!* Dutoit's Kloof, *Drège,* 8319! Near Cape Town, *Burchell,* 23! *Ecklon,* 819! *Bolus,* 3803! Baviaan's Kloof, near Genadendal, *Drège,* 8320! Swellendam Div., *Zeyher,* 1648ᵇ! Riversdale Div., *Burchell,* 6598! 6639! Mossel Bay, *Burchell,* 6296.

KALAHARI REGION: Orange Free State, between Rhenoster R. and Vaal R., *Zeyher,* 1648!

13. **M. juncea** (Linn. Sp. Plant, edit. ii. 59); corm globose; ½ in. diam.; outer tunics with thick, parallel, wiry strands; stem slender, 6–8 in. long, with two falcate, firm, linear-conduplicate leaves produced near its base, which are 2–3 in. long, ⅛ in. broad, not at all crisped; clusters of flowers 2–3, peduncled, corymbose, the leaves from their base rudimentary; spathes cylindrical, 2–3-flowered, about ½ in. long; valves lanceolate, green, the outer much the shortest; perianth lilac, fugacious, ½ in. long; outer segments oblong-unguiculate, inner oblanceolate; anthers oblong, 1/12 in. long, shorter than the filaments; ovary oblong, ⅛ in. long. *Baker in Journ. Linn. Soc.* xvi. 130; *Handb. Irid.* 52.

SOUTH AFRICA: without locality, seen only in the Linnean herbarium.

14. **M. crispa** (Ker in Bot. Mag. t. 1284); corm small; outer tunics very thick, with strong, parallel fibres, and short transverse strands; stems slender, ½–1½ ft. long, with about two produced linear leaves near the base, which are ½–1 ft. long, ⅓–½ in. broad, glabrous, in the type much crisped; clusters of flowers few or many, laxly corymbose, with long, ascending peduncles; spathes ½–¾ in. long; glabrous, moderately firm; valves lanceolate, outer short; perianth ½ in. long, fugacious, lilac or yellow, rarely whitish; outer segments oblong-unguiculate, inner oblanceolate; ovary oblong, ⅛ in. long; style-crests linear, ¼ in. long; filaments connate in the lower

half; capsule ellipsoid, $\frac{1}{3}$ in. long. *Gen. Irid.* 41; *Ait. Hort. Kew*, edit. 2, i. 114; *Klatt in Linnæa* xxxiv. 566; *Baker in Journ. Linn. Soc.* xvi. 131; *Handb. Irid.* 53. *M. decussata, Klatt, Ergänz.* 33. *Iris crispa, Linn. fil. Suppl.* 98; *Thunb. Diss.*, No. 36 tab. 1, fig. 1.

VAR. β, **rectifolia** (Baker); leaves narrower, more rigid, not crisped. Baker, Handb. Irid. p. 53. *M. crispa, Ker in Bot. Mag.* t. 759.

COAST REGION: near Cape Town, *Thunberg!* Between Paarl and Pont, *Drège*, 8318! Var. β, Between Paarl and Pont, *Drège*, 8328! Simons Bay, *Wright*, 257! Table Mt, 1200 ft. alt., *Bolus*, 4718! Swellendam Div., *Zeyher*, 4088!

15. M. arenaria (Baker, Handb. Irid. 52); corm not seen; stem, 3–6 in. long, very slender; glabrous; with 2–3 basal, terete, glabrous, spirally-curled leaves 2–3 in. long; clusters of flowers 2–4, laxly corymbose, with short erect peduncles; spathes cylindrical, $\frac{3}{4}$–1 in. long; valves lanceolate, strongly ribbed, outer small; perianth fugacious (colour uncertain), with very long claws to the segments; ovary clavate, glabrous, $\frac{1}{6}$ in. long; capsule ellipsoid, obtuse, $\frac{1}{4}$ in. long.

COAST REGION: Clanwilliam Div., Ebenezer Sandhills, under 500 ft. alt. *Drège*, 8324! Intermediate between *M. iriopetala* and *M. setacea.*

16. M. serpentina (Baker, Handb. Irid. 52); corm ovoid, $\frac{1}{4}$–$\frac{1}{3}$ in. diam., with thick outer tunics strongly cancellate towards the top; stem slender, 3–4 in. long, with two subulate, spreading, very thick leaves, curling up spirally and strongly ciliated on the margin; clusters of flowers 2–4, corymbose, on short erect peduncles; spathes cylindrical, $\frac{3}{4}$–1 in. long, 2–3-flowered; valves lanceolate, strongly ribbed, outer small; perianth fugacious (colour uncertain), $\frac{1}{2}$ in. long, with long claws to all the segments; capsule ellipsoid, $\frac{1}{4}$–$\frac{1}{3}$ in. long.

WESTERN REGION: Little Namaqualand, Roodeberg and Ezelskop, 4000-5000 ft. alt., *Drège*, 2599! Near Ookeep, *Bolus*, 6571! 6572!

17. M. Cooperi (Baker, Handb. Irid. 54); corm not seen; stem slender, terete, 1–1$\frac{1}{2}$ ft. long, with two long subulate, terete, wiry leaves, one from near its base, and the other from halfway up to the inflorescence; inflorescence of 6–20 clusters in a subcorymbose panicle, the ascending peduncles not over $\frac{1}{4}$–$\frac{1}{2}$ in. long; spathes cylindrical, 1–1$\frac{1}{4}$ in. long, 2–3-flowered, smooth, glossy, glabrous; valves acute, rigid to the tip; perianth fugitive, lilac, $\frac{3}{4}$–1 in. long; outer segments oblong, spathulate; inner oblanceolate; filaments connate in a cylindrical column $\frac{1}{8}$ in. long; ovary clavate; style-crests lanceolate; capsule not seen.

COAST REGION: Tulbagh, on Mount Winterhoek, 1500 ft., *Bolus*, 5248!
CENTRAL REGION: Worcester Div., *Cooper*, 1661!

18. M. polyanthos (Thunb. Diss., No. 14); corm ovoid, $\frac{1}{4}$–$\frac{1}{2}$ in. diam.; outer tunics thick, dark brown, cancellate; produced leaves about three, one from near the base of the stem, and the others from

the lower forks, linear-subulate, ½–1 ft. long; clusters of flowers
5–20, laxly corymbose; spathes cylindrical, glabrous, 2–3-flowered,
1–1½ in. long; valves firm, scariose at the tip, strongly ribbed, outer
small; perianth lilac, fugacious, ¾–1 in. long, outer segments obovate-
unguiculate, inner oblanceolate; filaments connate towards the base;
ovary clavate, ⅛ in. long; style-crests large, lanceolate; capsule
ellipsoid, obtuse, under ½ in. long.   *Thunb. Prodr.* 11; *Fl. Cap.* i.
276; *Linn. fil. Suppl.* 99; *Ker, Gen. Irid.* 33; *Klatt in Linnæa,*
xxxiv. 563; *Ergänz.* 33; *Baker in Journ. Linn. Soc.* xvi. 130.
Vieusseuxia graminifolia, Eckl. Top. Verz. 11.

SOUTH AFRICA: without locality, *Harvey,* 908! *Masson! Bowie!*
COAST REGION: Swellendam Div., *Zeyher,* 4080! Mossel Bay Div., *Burchell,*
6386! Albany Div., *Cooper,* 1529!
CENTRAL REGION: Graaff Reinet, *Bolus,* 522!
EASTERN REGION: Kaffraria, *Cooper,* 3216!

Burchell's 6149, collected on the west side of Great Brak R., Mossel Bay Div.,
differs by its clavate trigonous capsule, ½–¾ in. long. Plants from the Transvaal,
McLea in Hb. Bolus, 5789! and Apies R., *Burke!* and from Clanwilliam Div.,
Mader in Hb. MacOwan, 2168! are forms or near allies.

19. **M. iriopetala** (Linn. fil. Suppl. 100); corm ovoid, ¼–½ in.
diam., with thick, wiry coats; stems slender, erect, glabrous, ½ ft.
long, with one narrow, linear, long leaf of firm texture from the base,
and another shorter from the lowest fork; clusters of flowers 1–4,
laxly corymbose, with short, slender, erect peduncles; spathes cylindrical, 1½ in. long, 2–4-flowered; valves lanceolate, strongly ribbed,
green to the tip, outer short; perianth lilac, fugitive, ½ in. long,
outer segments oblong-unguiculate, inner oblanceolate; ovary clavate,
⅛ in. long; style-crests small, lanceolate; capsule ellipsoid, ¼ in.
long.   *M. vegeta, Linn. herb! non Mill. nec Jacq.   M. plumaria, Ker
in Konig and Sims' Ann.* i. 240; *Gen. Irid.* 40; *Baker in Journ. Linn.
Soc.* xvi. 130; *Handb. Irid.* 53.   *Iris plumaria, Thunb. Diss.* No.
16; *Prodr.* 11.   *Vieusseuxia plumaria, Eckl. Top. Verz.* 14.

SOUTH AFRICA: without locality, *Harvey,* 94! *Drège,* 2327a!
COAST REGION: Wynberg, *Bolus,* 2833! Devils Mountain, *Thunberg!*

20. **M. mira** (Klatt in Trans. Cape Phil. Soc. iii. pl. 2, 202);
corm small, globose; tunics scaly, reticulated; stem slender, flexuose,
glabrous, 1½–2 in. long, with a single, produced, narrow, linear,
strongly ribbed, glabrous leaf of firm texture, 3–4 in. long from its
base, and a single shorter one from a little above it; clusters 1–2;
peduncles erect, ½–¾ in. long; spathes cylindrical 1–1½ in. long,
3-flowered; outer valves green, firm in texture, with a narrow, membranous edge and tip; outermost shorter than the next, conspicuously
cuspidate; perianth bright violet, fugitive, ½ in. long, segments
oblong unguiculate; pollen bright scarlet; ovary turbinate, glabrous,
⅛ in. long; style-crests deeply fimbriated.   *Baker, Handb. Irid.*
53.

COAST REGION: Caledon Div.; in damp places on the Zwartberg, near the hot
springs, *Templeman in Hb. MacOwan,* 2612!

21. M. tristis (Ker in Konig and Sims' Ann. Bot. i. 241); corm globose, $\frac{1}{2}$ in. diam., outer tunics of matted parallel fibres; stems slender, $\frac{1}{2}$–1 ft. long, with 2–3 produced, linear, grass-like leaves from near the base, which are glabrous, 1–2 ft. long, $1\frac{1}{3}$ in. broad; clusters of flowers 4–6, laxly corymbose, 3–4-flowered, with peduncles 2–3 in. long, the lower subtended by short leaves; spathes cylindrical, 1–1$\frac{1}{4}$ in. long; valves green, firm, strongly ribbed, scariose at the tip; perianth fugacious, $\frac{3}{4}$ in. long, dull lilac, ochraceous or salmon-coloured, outer segments oblong-unguiculate, with a yellow eye, inner oblanceolate; filaments connate, high up; ovary clavate, $\frac{1}{6}$ in. long, glabrous or pilose; style-crests small, lanceolate; capsule ellipsoid, obtuse, $\frac{1}{3}$ in. long. *Ker in Bot. Mag.* sub t. 1103; *Gen. Irid.* 41; *Klatt in Linnæa*, xxxiv. 564; *Ergänz.* 33; *Baker in Journ. Linn. Soc.* xvi. 131; *Handb. Irid.* 53. *Iris tristis, Linn. fil. Suppl.* 97; *Thunb. Diss.* No. 39; *Prodr.* 12; *Fl. Cap.* i. 302; *Ker in Bot. Mag.* t. 577. *M. sordescens, Jacq. Ic.* t. 225; *Collect. Suppl.* 29; *DC. in Red. Lil.* t. 71. *Miller Ic.* t. 238, figs. 1–2.

COAST REGION: vicinity of Cape Town, *Thunberg! Rogers! Elliot,* 1097! 1148! *Bolus,* 3745!

22. M. polystachya (Ker in Konig and Sims' Ann. i. 240); corm ovoid, $\frac{3}{4}$–1 in. diam.; tunics thick, with parallel wiry strands, connected by short cross-bristles; stems stout, erect, 2–3 ft. long; produced leaves about 4, linear, glabrous, strongly ribbed, 1–2 ft. long, $\frac{1}{2}$ in. broad; panicle a lax corymb of 5–20 clusters; spathes cylindrical, 1$\frac{1}{2}$–2 in. long, 3–6-flowered; valves with very long, membranous cusps; perianth fugacious, bright lilac, 1–1$\frac{1}{2}$ in. long, outer segments oblong-unguiculate, $\frac{1}{2}$ in. broad, with a large, bright yellow spot at the base of the blade, inner oblanceolate; ovary clavate, $\frac{1}{4}$ in. long; crests of styles large, lanceolate; filaments connate towards the base; capsule clavate-oblong, obtuse, $\frac{1}{2}$ in. long. *Ker, Gen. Irid.* 39; *Baker in Journ. Linn. Soc.* xvi. 131. *Iris polystachya, Thunb. Diss.* No. 40; *Prodr.* 12; *Fl. Cap.* i. 303. *Iris lacera, Lam. Encyc.* iii. 304; *Illustr.* i. 125. *Vieusseuxia polystachya, Eckl. Exsic. M. multiflora, Solander MSS.*

COAST REGION: near Grahamstown, *MacOwan,* 10!
CENTRAL REGION: Uitenhage Div., *Zeyher!* Somerset Div., *Bowker!* Sneeuw Berg, *Burchell,* 2831! Near Graaff Reinet, *Burchell,* 2941! Between Sunday R. and Fish R., *Thunberg!*

Totally different from *M. catenulata,* with which Dr. Klatt has united it, *Ergänz.* 34.

23. M. ramosa (Ker in Bot. Mag. t. 771); corm small, globose, surrounded by dense, short, spreading, branched, spine-like fibres, bearing, as do the lower nodes of the stem, a copious supply of small ovoid bulbillæ; stem stout, erect, 2–3 ft. long; produced leaves about six, from the base of the stem, linear, firm in texture, strongly ribbed, glabrous, 1–1$\frac{1}{2}$ ft. long, $\frac{1}{2}$–$\frac{3}{4}$ in. broad; inflorescence a lax corymb of often 20–30 clusters, the lower branches copiously com-

pound; spathe cylindrical, 1 in. long, 2–3-flowered; outer valve
very small; perianth bright yellow, 1–1¼ in. long, outer segments
obovate, maculate at the base of the limb, inner oblong, both unguiculate; filaments free nearly or quite down to the base; ovary
small, globose; crests of the style large, lanceolate; capsule globose or
turbinate. *Ker, Gen Irid.* 40; *Ait. Hort. Kew.* ed. 2, i. 113; *Klatt,
Ergänz.* 33; *Baker, Handb. Irid.* 54. *M. bulbifera, Jacq. Hort.
Schoenbr.* t. 197; *Klatt in Linnæa,* xxxiv. 565; *Hook, fil. in Bot.
Mag.* t. 5785. *Iris ramosa, Thunb. Diss.* No. 24; *Prodr.* 12; *Fl.
Cap.* i. 293. *I. ramosissima, Linn. fil. Suppl.* 99; *Roem. et Schult.
Syst. Veg.* i. 470. *Moræa racemosa, Herb. Drège. Freuchenia
bulbifera, Eckl. Top. Verz.* 14.

SOUTH AFRICA: without locality, *Drège!* distributed as *M. racemosa.*
COAST REGION: Malmesbury Div., Zwartland, *Thunberg!* Mostertsberg,
Bolus, 5249! Swellendam Div., *Zeyher,* 1645! Riversdale Div., *Burchell,* 7143!
Near George, *Burchell,* 6098! At the foot of Wittklip, *MacOwan,* 399!

24. M. gigantea (Klatt in Linnæa, xxxv. 381); corm not seen;
stem 2 ft. or more long, stout, erect, terete; produced leaves about 3,
linear, glabrous, firm in texture, strongly ribbed, lower 1½–2 ft. long,
½–¾ in. broad at the base; inflorescence a dense corymb of 20–30
clusters, with short, crowded, erect peduncles; spathes cylindrical,
2–3-flowered, 1⅛ in. long; valves with a very long scariose cusp;
perianth very fugitive, yellow, above 1 in. long; ovary clavate, ⅓ in.
long; capsule subcylindrical, obtusely-angled, ½ in. long. *Baker in
Journ. Linn. Soc.* xvi. 131; *Handb. Irid.* 54; *Klatt, Ergänz.* 33.

CENTRAL REGION: Fraserburg Div.; between Great and Little Reed Rivers,
Burchell, 1388! near Little Quaggas Fontein, *Burchell,* 1431!

25. M. setacea (Ker in Konig and Sims' Ann. i. 240); corm ovoid,
⅓–½ in. diam., coats thick, fibrous; stems slender, ½–1 ft. long, with
1–2 very slender, subulate, glabrous leaves, ½–1 ft. long, springing
from it a distance from both base and inflorescence; clusters 1–3, erect,
nearly sessile, springing from the axils of lanceolate, rudimentary
leaves; spathes cylindrical, 2–3-flowered, about an inch long; valves
subequal, cuspidate, strongly ribbed, scariose only at the tip; perianth
fugitive, lilac, ½ in. broad, inner segments oblanceolate; crests of style
lanceolate; filaments connate at the base; ovary clavate; capsule
ellipsoid, ½ in. long. *Ker, Gen. Irid.* 39; *Klatt in Linnæa,* xxxiv.
562; *Ergänz.* 33; *Baker in Journ. Linn. Soc.* xvi. 130; *Handb.
Irid.* 56. *Iris setifolia, Linn. fil. Suppl.* 99. *Iris setacea, Thunb.
Diss.* No. 29, t. 1, fig. 2; *Prodr.* 12; *Fl. Cap.* i. 295. *Vieusseuxia
setacea, Eckl. Top. Verz.* 13.

SOUTH AFRICA: without locality, *Thunberg! Masson!*

26. M. undulata (Ker, Gen. Irid. 43); corm small, ovoid, with
thick coats of matted fibres; stem ½ foot long, with a single, long,
glabrous, linear leaf, spreading from near its base, ½–1 ft. long, plain,
or frilled along the edge; clusters of flowers 2–3, shortly peduncled,
erect; spathes 2–3-flowered, 1–1½ in. long; valves cuspidate, strongly

ribbed, glabrous, scariose at the tip ; perianth lilac, fugacious, $\frac{3}{4}$–1 in. long ; segments with long claws, outer oblong, inner oblanceolate ; filaments united at the base only ; ovary oblong, $\frac{1}{8}$ in. long ; style-crests very small. *Baker, Handb. Irid.* 56. *M. crispa, Thunb. Diss* No. 17 ; *Prodr.* 11 ; *Fl. Cap.* i. 279 ; *Willd. Sp. Plant.* i. 244 ; *Roem. et Schult. Syst. Veg.* i. 453. *Klatt, Ergänz.* 33, non *Ker.*

SOUTH AFRICA : without locality, *MacOwan,* 2102 !
CENTRAL REGION : Fraserburg Div. ; Roggeveld, *Thunberg !*

27. M. Bolusii (Baker, Handb. Irid. 57) ; corm not seen ; produced leaf solitary, spreading, basal, linear, coriaceous, much crisped, 6–8 in. long, $\frac{1}{4}$ in. broad ; flowering stem 4–6 in. long, bearing several erect, sessile clusters ; spathe 1$\frac{1}{4}$–1$\frac{1}{2}$ in. long ; outer valve short ; perianth fugacious, yellow, an inch long, outer segments obovate, inner smaller ; filaments connate ; style-crests lanceolate.

WESTERN REGION : Little Namaqualand, Ookiep, 3000 ft., *Bolus,* 6574 !

28. M. natalensis (Baker, Handb. Irid. 56) ; corm not seen ; stem slender, wiry, 1–1$\frac{1}{2}$ ft. long, without any produced leaf below the inflorescence ; leaf from the base of the inflorescence like a continuation of the stem, terete, rigid, glabrous, 3–6 in. long ; inflorescence of 2–5 ascending, subracemose clusters ; spathes cylindrical, 2–3-flowered, 1$\frac{1}{4}$–1$\frac{1}{2}$ in. long ; valves rigid, strongly ribbed, scariose only at the cusp ; perianth fugitive, lilac, $\frac{3}{4}$–1 in. long, outer segments obovate-unguiculate, with a yellow eye decurrent down the claw, inner shorter, oblanceolate ; ovary clavate, $\frac{1}{6}$ in. long ; crests of style lanceolate, $\frac{1}{8}$ in. long ; capsule ellipsoid, $\frac{1}{3}$–$\frac{1}{2}$ in. long, truncate.

EASTERN REGION : Natal, 2000–6000 ft., *Sanderson,* 253 ! Pietermaritzberg, *Sutherland !* Drakensberg, near the sources of the Tugela R., *Allison ! Wood,* 3442 ! Liddesdale, *Wood,* 4255 ! near Curry's Post, *Wood,* 3436 !

29. M. edulis (Ker in Konig and Sims' Ann. i. 241) ; corm globose, $\frac{1}{2}$–$\frac{3}{4}$ in. diam. ; coats with wiry strands connected by short, parallel fibres ; stems wiry, terete, 1–2 ft. long, without any produced leaves below the inflorescence ; leaf one from the base of the inflorescence, wiry, rigid, narrow linear, convolute, 1–2 ft. long ; clusters 2–20, on a short or elongated axis ; peduncles short, stiff, ascending ; spathes cylindrical, 1$\frac{1}{2}$–2 in. long, 2–4-flowered ; valves rigid, strongly ribbed, scariose at the tip ; perianth fugitive, lilac, 1–1$\frac{1}{2}$ in. long, fragrant, outer segments oblong-spathulate, with a bright, yellow eye, inner shorter, oblanceolate ; ovary cylindrical, $\frac{1}{2}$ in. long, narrowed to the top ; crests of style lanceolate, $\frac{1}{8}$ in. long ; filaments connate, in a cylindrical tube ; capsule cylindrical, an inch long, narrowed to the apex. *Ker in Bot. Mag.* t. 613 ; *Gen. Irid.* 38 ; *Klatt in Linnæa,* xxxiv. 564 ; *Ergänz.* 33 ; *Baker in Journ Linn. Soc.* xvi. 130 ; *Handb. Irid.* 56. *Iris edulis, Linn. fil. Suppl.* 98 ; *Thunb. Diss.* No. 38 ; *Prodr.* 12 ; *Fl. Cap.* i. 300. *Iris capensis, Burm. Prodr.* 2. *Vieusseuxia edulis, Eckl. Top. Verz.* 14. *M. vegeta, Jacq. Ic.* t. 224, non *Linn.* nec *Mill. M. fugax, Jacq.* Hort. Vind. iii. 14, t. 20.

VAR. β, **M. odora** (Salisb. Parad. t. 10). Flowers white.

VAR. γ, **M. longifolia** (Sweet, Hort. Brit. edit. 2, 497). Flowers yellow. *Iris longifolia,* Andr. Bot. Rep. t. 45. *M. edulis, var.,* Ker in Bot. Mag. t. 1238. *Vieusseuxia longifolia,* Eckl. Top. Verz. 14. *V. aristata, Delaroche,* Diss. 33.

VAR. δ, **M. umbe lata** (Thunb. Diss. No. 16). Main axis of inflorescence short; clusters numerous, erect. Prodr. 11; *Fl. Cap.* i. 279. *M. fasciculata, Soland. MSS. Bobartia umbellata, Ker, Gen. Irid.* 31. *Aristea umbellata, Spreng. Syst. Veg.* i. 158.

VAR. ε, **gracilis** (Baker). More slender than in the type, with acute spathes scarcely over an inch long, smaller flowers, and much smaller capsules (½ in. long) narrowed into a distinct beak.

SOUTH AFRICA: without locality; Var. γ, *Zeyher,* 4079! Var. δ, *Thunberg! Masson!*

COAST REGION: Malmesbury Div.; Groenekloof, *Thunberg! Zeyher,* 1646! Table Mountain and vicinity of Cape Town, *Ecklon,* 823! *Thunberg! Bolus,* 4676; Swellendam Div., *Zeyher,* 4087! Mossel Bay, *Burchell,* 6295! Var. γ, near Cape Town, *Bolus,* 2830! Var. ε, near Tulbagh, *Zeyher,* 1647!

30. M. spiralis (Baker, Handb. Irid. 57); corm not seen; stems very slender, 3–4 in. long, with a single, produced, subulate leaf, 3–6 in. long, springing from the stem above the base, spirally twisted, and sometimes very wavy; clusters 2–4, erecto-patent, close, nearly sessile; spathes cylindrical, 2–3-flowered, ½–⅝ in. long, glossy, glabrous, not ribbed, not cuspidate at the tip; perianth fugitive, lilac, under ½ in. long, outer segments oblong-spathulate, inner oblanceolate; ovary clavate, narrowed into a neck above ¼ in. long.

WESTERN REGION: Little Namaqualand; near Ookiep, *Morris in Hb. Bolus,* 5788! Kook Fontein Mountains, 3000–4000 ft., *Drège,* 2604!

31. M. Rogersii (Baker, Handb. Irid. 57); corm not seen; stem very short, with a single terminal cluster of flowers, and from the base of it a single, falcate, linear-subulate, glabrous leaf of firm texture, 3–4 in. long; spathe cylindrical, 1–1¼ in. long; valves pale and scariose, long-pointed, with a few strong ribs; perianth fugacious, ½ in. long, lilac; all the segments with long claws, outer obovate, with a yellow spot at the base of the blade; ovary cylindrical, with a short pedicel, narrowed into a filiform beak, ½ in. long.

COAST REGION: the Point, Mossel Bay, *Rogers!*

Habit of the dwarf forms of *Iris Sisyrinchium,* L.

32. M. Burchellii (Baker in Journ. Linn. Soc. xvi. 132); corm not seen; stems 3–6 in. long, with 1–3 clusters of flowers, the side ones sessile, and a single, slender, terete, glabrous leaf, ½–1 ft. long, produced from its base; spathes cylindrical, 1–1¼ in. long; valves long-pointed, pale, and scariose at the flowering season, strongly ribbed; perianth fugacious, pale lilac, ½ in. long, outer segments oblong-unguiculate, with a claw as long as the lamina, inner oblanceolate; filaments free nearly to the base; ovary cylindrical, with a short peduncle, narrowed into a filiform beak, an inch long; style-crests lanceolate. *Baker, Handb. Irid.* 57.

KALAHARI REGION: Bechuanaland; Jabiru Fountain, near Old Litaku, *Burchell,* 2250!

33. M. longiflora (Ker in Bot. Mag. t. 712); corm small, ovoid; produced leaves 1–2, subulate, glabrous, ½ ft. long; stem very slender, ending in two sessile clusters; spathes cylindrical, 1½–2 in. long; valves lanceolate, acute; perianth bright yellow, fugacious, about an inch long, outer segments obovate-unguiculate, ⅓–½ in. broad, inner shorter, oblanceolate; ovary with a long filiform beak, which is protruded from the spathe; style-crests small, lanceolate. *Ait. Hort. Kew,* edit. 2, i. 112; *Roem. et Schult. Syst. Veg.* i. 455; *Ker, Gen. Irid.* 39; *Baker in Journ. Linn. Soc.* xvi. 132; *Handb. Irid.* 57. *Helixyra flava, Salisb. in Trans. Hort. Soc.* i. 305.

SOUTH AFRICA: without locality.

Known only from the figure cited.

34. M. simulans (Baker, Handb. Irid. 58); corm small, ovoid; stems 6–15 in. long, apparently decumbent amongst sand; leaves 1–2, produced from the base of the inflorescence, often with large bulbillæ in their axils, linear-convolute, ½ ft. long, very thick in texture, glabrous, strongly ribbed; clusters 2–4 to a stem, the side ones sessile, or shortly peduncled; spathes cylindrical, 1½–2 in. long; valves pale and scariose at the flowering-time, long-pointed, laxly, strongly ribbed; perianth-limb lilac, fugacious, under an inch long, all the segments with long, distinct claws, blade of outer obovate, inner oblanceolate; filaments free nearly to base; ovary cylindrical, narrowed into a long beak; style-crests small.

SOUTH AFRICA: without locality, *Elliot,* 1250!
KALAHARI REGION: Transvaal; Bloemhof, *Nelson,* 203!

Habit of the well-known *M. Sisyrinchium,* Ker (*Bot. Mag.* t. 1407), of the Mediterranean region.

35. M. cladostachya (Baker, Handb. Irid. 58); corm not seen; stem above a foot long, with a single, produced leaf, with a large bulbil in its axil, springing from it a space below the inflorescence, which is narrow linear, flat, glabrous, strongly ribbed, 2–3 ft. long, ¼ in. broad; clusters 8–10, arranged in a panicle, with three ascending, spicate branches; spathes cylindrical, 1½–2 in. long; valves pale and scariose at the flowering time, with long points and few distant ribs; perianth fugacious, small, lilac; ovary cylindrical, with a short pedicel, narrowed into a slender beak, nearly an inch long.

SOUTH AFRICA: without locality, *Burchell!*
WESTERN REGION: Little Namaqualand; near Verleptpram, *Drège,* 2610 !
CENTRAL REGION: Eastern Frontier of Cape Colony, *Mrs. Barber!*

36. M. tripetala (Ker in Bot. Mag. t. 702); corm globose, ½–1 in. diam., coats thick, fibrous or cancellate; produced leaf one, sub-basal, long, narrow linear, firm, glabrous; stem very slender, 1–2 ft. long, simple, or with 1–3 short, erect branches; spathes cylindrical, 2–2½ in. long, 1–3 flowered; valves cuspidate, scariose only at the very tip; perianth-limb 1–1¼ in. long, fugitive, lilac, rarely blue or reddish, outer segments with a claw nearly as long as the spreading, oblong lamina, inner linear $\frac{1}{4}-\frac{1}{3}$ in. long; column very short; ovary clavate, $\frac{1}{4}-\frac{1}{3}$ in. long; style-crests linear, $\frac{1}{4}-\frac{1}{2}$ in. long; capsule clavate, obtusely angled, $\frac{1}{2}$ in. long. *Ker, Gen. Irid.* 35; *Baker, Handb. Irid.* 60. *Iris tripetala, Linn. fil. Suppl.* 97; *Thunb. Diss.* No. 14; *Prodr.* 11; *Fl. Cap.* i. 286; *Jacq. Collect.* iii. 272; *Ic. tab.* 221. *Vieusseuxia tripetaloides, DC. in Ann. Mus.* ii. 138; *Roem. et Schult. Syst. Veg.* i. 489; *Klatt in Linnæa,* xxxiv. 620; *Baker in Journ. Linn. Soc.* xvi. 133.

VAR. β, mutila (Baker). Leaf pilose. *Vieusseuxia mutila, Klatt in Linnæa,* xxxiv. 621. *Iris mutila, Licht. in Roem. et Schult. Syst. Veg.* i. 477.

COAST REGION: Hills and plains in the neighbourhood of Cape Town and Simons Bay, *Thunberg! Wright! Bolus,* 3393! 3732! *MacOwan* and *Bolus, Herb. Normale,* 251! Paarl Div., *Drège!* Caledon Div.; Klein River Berg, *Zeyher,* 1650! Var. β, Lion Mountain, *Bergius.*

37. M. Elliotii (Baker in Journ. Bot. 1891, 70); corm not seen; basal leaf single, rudimentary; stem slender, 1–1$\frac{1}{2}$ ft. long, bearing a single, long, linear leaf from the middle; clusters of flowers 1–2; spathe cylindrical, 2-flowered, 1$\frac{1}{4}$ in. long; valves lanceolate, rigid, the outer short; perianth bright lilac, outer segments oblong-unguiculate, $\frac{5}{8}$ in. long, with a pale spot at the base of the blade, inner about half as long, entire, oblanceolate-unguiculate, obtuse; style-crests lanceolate; capsule small, oblong-clavate. *Baker, Handb. Irid.* 58. *Vieusseuxia Elliotii, Klatt in Durand and Schinz Conspect. Fl. Afric.* 154.

KALAHARI REGION: Transvaal; marshes near Lake Chrissie, *Elliot,* 1592!

Habit of *M. tenuis,* but in the flower connects *Vieusseuxia* with *Eumorea.*

38. M. glaucopis (Drap. Enc. iv. (1836) cum ic.); corm globose, $\frac{1}{2}$ in. diam.; coats cancellate; produced leaf subbasal, long, firm, narrow linear, glabrous; stem 1$\frac{1}{2}$–2 ft. long, terete, with 2–3 sheaths, simple, or with 1–3 short, erect branches; spathes cylindrical, 2–2$\frac{1}{2}$ in. long, 2–3-flowered; valves cuspidate, scariose only at the very tip; outer much shorter than inner; perianth $\frac{3}{4}$–1$\frac{1}{4}$ in. long, fugitive, outer segments obovate or sub-orbicular, with a short claw, white, with a blue-black circular spot at the base of the blade, inner with a filiform claw, a large central cusp, and two small side ones; ovary clavate, $\frac{1}{3}$–$\frac{1}{2}$ in. long; style column short; crests of stigmas large, lanceolate; capsule clavate, obtusely-angled, larger than in *tricuspis* and *tripetala. Baker, Handb. Irid.* 59. *Iris Pavonia, Curt. in Bot. Mag.* t. 168, *non aliorum. I. tricuspis, Jacq. Collect.* iv. 99, t. 9, fig. 1. *Vieusseuxia glaucopis, DC. in Ann. Mus.* ii. 141, t. 42; *Red. Lil.* t. 42; *Flore des Serres,* t. 423; *Baker in Journ. Linn. Soc.* xvi. 133; *Klatt, Ergänz.* 34.

COAST REGION: Devils Mountain, *Thunberg! Bolus,* 4004! Table Mountain, *Rogers!*

39. M. Pavonia (Ker in Bot. Mag. t. 1247); corm small, globose; coats thick, cancellate; produced leaf one, subbasal, long, linear, pilose; stem long, slender, terete, with several sheaths, simple, or with 1–3 short, erect branches; spathes cylindrical, 2–2½ in. long, 2–3-flowered; valves lanceolate, cuspidate, scariose only at the very tip; outer much shorter than inner; perianth-limb fugitive, 1–1¼ in. long, outer segments orbicular-cuneate, without any distinct claw, bright red, with a large blue-black or greenish-black spot at the glabrous base, inner ⅓ in. long, with a linear, central cusp, and two small side ones; ovary clavate, ½ in. long; style-column cylindrical, ⅙ in. long; crests of stigma lanceolate; capsule clavate, ½ in. long. *Ker in Ann. Bot.* i. 240; *Gen. Irid.* 34; *Baker, Handb. Irid.* 59. *Iris Pavonia, Linn. fil. Suppl.* 98; *Thunb. Diss.* No. 35, t i. fig. 3; *Prodr.* 12; *Fl. Cap.* i. 296; *Jacq. Hort. Schoenbr.* i. 6, t. 10; *Andr. Bot. Rep.* t. 364. *Vieusseuxia Pavonia, DC. in Ann. Mus.* ii. 139; *Spreng. Syst.* i. 166; *Klatt in Linnæa,* xxxiv. 622; *Ergänz.* 34; *Baker in Journ. Linn. Soc.* xvi. 133.

VAR. β, villosa (Baker). Leaf pilose; outer segments of perianth bright purple, with a hairy claw decorated with a blue-black spot. *Iris villosa, Ker in Bot. Mag.* t. 571. *Vieusseuxia villosa, Spreng. Syst. Veg.* i. 166; *Klatt in Linnæa,* xxxiv. 623, excl. syn.; *Ergänz.* 34; *Baker in Journ. Bot.* xvi. 133. *Moræa villosa, Ker in Ann. Bot.* i. 240; *Gen. Irid.* 36.

VAR. γ, lutea (Baker). Leaf glabrous; flower yellow; outer segments without a distinct spot on the claw. *Vieusseuxia spiralis, Delaroche, Descr.* 31, t. 5; *DC. Ann. Mus.* ii. 140; *Roem. et Schult. Syst. Veg.* i. 490; *Baker in Linn. Journ.* xvi. 133; *Klatt, Ergänz.* 34. *Moræa tricuspis, var. lutea, Ker in Bot. Mag.* t. 772; *Gen. Irid.* 36. *V. Bellendeni, Sweet, Hort. Brit.* edit. ii. 498; *Klatt in Linnæa,* xxxiv. 624.

SOUTH AFRICA: without locality, *Masson!* *Harvey*, 93! Var. *lutea, Drège*, 2329! *Thom!*

COAST REGION: Malmesbury Div.; Zwartland, *Thunberg!* Var. *lutea,* near Wynberg, *Bolus,* 2831! Swellendam Div., *Zeyher,* 4081! 4082!

40. M. candida (Baker, Handb. Irid. 59); produced leaf one, grass-like, twice as long as the stem; stem simple, 1½–2 ft. long, with several sheaths; flowers white, twice the size of those of *M. Pavonia lutea*; outer segments with a purple, densely-bearded claw, crowned by a spot of two circles, the inner one yellow, the outer one purplish and dotted, inner segments with a large, straight, central cusp. *Vieusseuxia fugax, Delaroche, Descr.* 33; *DC. Ann. Mus.* ii. 139.

SOUTH AFRICA: without locality.

Known only from the description of Delaroche.

M. fugax, Jacq., to which it is referred by De Candolle, is totally different.

41. M. unguiculata (Ker in Bot. Mag. t. 593); corm globose, ½ in. diam.; coats coarsely cancellate; produced leaves 1–2, long, linear-subulate, glabrous; stem very slender, terete, 1–1½ ft. long, simple, or with 1–4 short, erect branches; spathes cylindrical, 1½ in. long, 2–3-flowered; valves cuspidate, scariose only at the very tip; perianth 1–1¼ in. long, very fugitive, outer segments with a claw about as long as the oblong lamina, which is white with reddish-purple spots, inner segments with three linear cusps, and a long filiform claw; ovary clavate, ¼ in. long; column cylindrical; crests of stigma linear; capsule small, clavate. *Ker, Gen. Irid.* 37; *Ait. Hort.*

Kew. ii. 112 ; *Klatt in Linnæa,* xxxiv. 562 ; *Baker, Handb. Irid.* 59. *Vieusseuxia unguicularis, Roem. et Schult. Syst. Veg.* i. 491 ; *Baker in Journ. Linn. Soc.* xvi. 133 ; *Klatt, Ergänz.* 34.

COAST REGION : Mossel Bay Div., *Burchell,* 6377! George Div., *Burchell,* 6102 ! Queenstown Div., *Cooper,* 317 !
WESTERN REGION : Little Namaqualand, *Bolus,* 6573 !

42. **M. tricuspis** (Ker in Bot. Mag. t. 696) ; corm small, globose; coats thick, fibrous ; produced leaf one, basal, long, firm, linear, glabrous ; stem 1 2 ft. long, with 2–3 short, sheathing leaves, simple, or with 2–4 short, erect branches ; spathes cylindrical, 2–2½ in. long, 2–3-flowered ; valves cuspidate, scariose only at the very tip ; perianth fugitive, ¾–1 in. long, whitish or lilac, outer segments obovate, obtuse, with a spreading lamina about as long as the claw, inner segments with a long claw, and three filiform cusps ; ovary clavate, ¼ in. long ; column cylindrical ; crests of stigmas linear ; capsule clavate, ½ in. long. *Ker, Gen. Irid.* 36, in part ; *Baker, Handb. Irid.* 58. *Iris tricuspidata, Linn. fil. Suppl.* 98. *Iris tricuspis, Jacq. Collect.* iv. 99 ; *Ic.* t. 222, *in part ; Thunb. Diss.* No. 15 ; *Prodr.* 11 ; *Fl. Cap.* i. 289 ; *Ait. Hort. Kew.* iii. 482. *Vieusseuxia tricuspis, Spreng. Syst. Veg.* i. 165 ; *Baker in Journ. Linn. Soc.* xvi. 133 ; *Klatt, Ergänz.* 34. *V. aristata, Houtt. Handl.* xii. 92, t. 80, f. 1, non *Delaroche.*

COAST REGION : Devils Mountain, and Malmesbury Div., *Thunberg !* Simons Bay, *Wright !* Dutoits Kloof, *Drège !*
EASTERN REGION : Faku's Territory, *Sutherland !* Natal ; Inanda, *Wood,* 414! 1116!
KALAHARI REGION : Orange Free State, *Cooper,* 884! 1028 !

43. **M. tenuis** (Ker in Bot. Mag. t. 1047) ; corm small, globose ; coats thick, fibrous or cancellate ; produced leaf one, basal, long, narrow linear, glabrous ; stem long, very slender, terete, simple, or with 1–2 short, erect branches ; spathes cylindrical, 1½ in. long, 2–3-flowered ; valves cuspidate, scarious only at the very tip ; outer much shorter than inner ; perianth fugitive, under an inch long, yellowish-brown, outer segments with a filiform claw about half as long as the spreading, oblong, cuspidate lamina, inner with a long claw, and three large linear cusps ; ovary clavate, ¼ in. long ; style-column cylindrical, as long as the claw of the perianth-segments ; crests of stigma large, lanceolate ; capsule small, clavate-oblong, obtusely-angled. *Ker, Gen. Irid.* 37 ; *Ait. Hort. Kew.* edit. 2, i. 112 ; *Baker, Handb. Irid* 58. *Iris tricuspis,* var. *Jacq. Ic.,* tab. 222. *Vieusseuxia tenuis, Roem. et Schult. Syst. Veg.* i. 491 ; *Spreng. Syst. Veg.* i. 165 ; *Klatt in Linnæa,* xxxiv. 622 ; *Ergänz.* 34 ; *Baker in Journ. Linn. Soc.* xvi. 133. *Iris gracilis, Licht. in Roem. et Schult. Syst. Veg.* i. 477 ?

COAST REGION : Swellendam Div., *Zeyher,* 4083! Mossel Bay Div., *Burchell,* 6202!
EASTERN REGION : Natal, *Buchanan !* near Nottingham Road Station, 4800 ft., *Wood,* 4396! Liddesdale, 5000 ft., *Wood,* 4252!

44. **M. iridioides** (Linn. Mant. 28) ; rhizome short-creeping, as thick as a man's finger ; leaves in crowded, fan-shaped, basal rosettes,

equitant, dark green, rigid in texture, 1–2 ft. long, $\frac{1}{2}$–$\frac{3}{4}$ in. broad; stems 1–2 ft. long, with many short, sheathing, lanceolate, rudimentary leaves, sometimes procumbent, elongated, and viviparous; clusters few, laxly corymbose; spathes cylindrical, $1\frac{1}{2}$–2 in. long, 3–4-flowered; valves firm in texture, not pointed; outer small; perianth white, about $1\frac{1}{2}$ in. long; all the segments spreading outer obovate-unguiculate, with a beard down the claw, and a yellow keel at the base of the blade inner narrower, concolorous; ovary cylindrical, $\frac{1}{2}$ in. long; style and filaments joined in a tube at the base; style-crests lanceolate, $\frac{1}{3}$–$\frac{1}{2}$ in. long; capsule ellipsoid, obtusely-angled, 1–$1\frac{1}{2}$ in. long. *Thunb. Diss.* No. 18; *Jacq. Hort. Schoenbr.* t. 196 (excl. syn.); *Ker in Bot. Mag.* t. 693. *Iris compressa, Linn. fil. Suppl.* 98; *Thunb. Diss.* No. 12; *Baker in Journ. Linn. Soc.* xvi. 147; *Handb. Irid.* 60. *Dietes iridifolia, Salisb. in Trans. Hort. Soc.* i. 307. *D. compressa, Klatt in Linnæa,* xxxiv. 584. *D. iridioides, Sweet, Brit. Flow. Gard.* edit. ii. 497. *Moræa vegeta, Mill. Dict.* edit. viii. No. 1; *Icones,* tab. 239, fig. 1, *non Linn. nec Jacq. Iris crassifolia, Lodd. Bot. Cab.* t. 1861. *I. moræoides, Ker in Bot. Mag.* t. 1407; *Gen. Irid.* 64. *Dietes crassifolia, Klatt, Ergänz.* 40.

VAR. β, **M. prolongata** (Hort. Leichtlin). Flowers pure white, with segments an inch long.

COAST REGION: George Div., *Drège,* 4555a! *Burchell,* 6047! Humansdorp Div., near Zeekoe R., *Thunberg;* Uitenhage, *Zeyher,* 797! Bathurst Div., *Burchell,* 4011! Albany Div., *Cooper,* 1528!

CENTRAL REGION: Somerset Div., Boschberg, 3000 ft., *MacOwan,* 1845!

KALAHARI REGION: Transvaal; Barberton, 3500–4000 ft., *Galpin,* 1206!

EASTERN REGION: Natal; Groenberg, *Wood,* 1099!

Also hills of Zambesi country, *Kirk!*

45. M. bicolor (Spae in Flore des Serres, tab. 744); habit of *M. iridioides;* rhizome short-creeping; leaves in a distichous, basal rosette, $1\frac{1}{2}$–2 ft. long, $\frac{1}{2}$ in. broad; stems 1–2 ft. long; clusters few, laxly corymbose; spathes cylindrical, 2–3-flowered, $1\frac{1}{4}$–$1\frac{1}{2}$ in. long; valves rigid in texture; outer short; expanded flower 2 in. diam., lemon-yellow, outer segments obovate-unguiculate, with no beard down the claw, and a blackish spot at the base of the blade, inner segments oblong-unguiculate, concolorous, $\frac{1}{3}$–$\frac{1}{2}$ in. broad; style-crests small, lanceolate-deltoid. *Baker, Handb. Irid.* 60. *Dietes bicolor, Klatt in Linnæa,* xxxiv. 584; *Ergänz.* 40. *Iris bicolor, Lindl. in Bot. Reg.* t. 1404; *Lodd. Bot. Cab.* t. 1886; *Paxt. Mag.* ix. 29.

COAST REGION: Bathurst Div.; among shrubs on the banks of the Kap River, *MacOwan,* 2986! and *Herb. Aust. Afr.,* 1538! King Williamstown Div.; near Komgha, *Flanagan,* 232!

It was originally described and figured from the garden of the Comte de Vandes at Bayswater in 1831.

II. HOMERIA, Vent.

Perianth infundibuliform, cut down very nearly to the ovary; segments oblanceolate-oblong, not distinctly unguiculate, subequal, or the three inner narrow. *Stamens* with filaments connate in a cylindrical column, and large linear anthers opposite the style-branches. *Ovary* clavate, with the ovules crowded and superposed; style filiform, with three small, petaloid, emarginate branches, papillose round the outer margin. *Capsule* clavate, operculate. *Seeds* small, globose, angled by pressure.

Rootstock a tunicated corm ; produced leaf usually one only, linear, overtopping the stem ; clusters one or few, the side ones shortly peduncled ; spathes cylindrical, with few successive fugacious flowers, yellow or fulvous.

DISTRIB. An endemic Cape genus.

Perianth-segments with a blotch in the middle.
　　Segments all obtuse, with a minute cusp ... ... (1) **elegans.**
　　Inner segments lanceolate, acute ... ... (2) **maculata.**
Perianth-segments not blotched in the middle.
　　Flowers large ; spathes 2½–3 in. long :
　　　　Leaf not vittate ... ... (3) **collina.**
　　　　Leaf with a white central band ... ... (4) **lineata.**
　　Flowers small ; spathes 1½–2 in. long :
　　　　Flower pale yellow ... ... (5) **pallida.**
　　　　Flower fulvous red ... ... (6) **miniata.**

1. H. elegans (Sweet, Hort. Brit. ed. 2, 498) ; corm globose, ½ in. diam., with thick brown tunics ; produced leaf one, subbasal, linear, firm in texture, strongly ribbed, a foot or more long, ¼–⅓ in. broad, overtopping the stem ; stem a foot or more long ; clusters 1–4, on short, erect peduncles ; perianth-limb broadly infundibuliform, 1–1½ in. long, the outer segments fulvous, and the inner yellow, or all the six yellow, oblanceolate-oblong, with a cusp ⅓–½ in. broad, the three outer with a large green or yellow-brown blotch in the middle ; stamens ½ in. long, the linear anthers equalling the cylindrical column ; ovary cylindrical-trigonous, ½ in. long ; style-branches with two small quadrate crests ; capsule clavate, ¾ in. long, with three rigid valves. *Baker, Handb. Irid.* 74. *Morœa elegans, Jacq. Hort. Schoenbr.* i. 6, t. 12 ; *Ker in Bot. Mag.* sub t. 1103 ; *Pers. Syn.* i. 49 ; *Ker, Gen. Irid.* 33. *Sisyrinchium elegans, Willd. Sp. Plant.* iii. 577 ; *Roem. et Schult. Syst. Veg.* i. 491, excl. syn. *H. spicata, Klatt in Linnœa,* xxxiv. 626, in part ; *Ergänz.* 52. *Morœa spicata, Ker in Bot. Mag.* sub t. 1103 ; *Bot. Mag.* t. 1283 ; *Gen. Irid.* 33.

SOUTH AFRICA : without locality, *Thunberg ! Stanger ! Mundt !*
COAST REGION : Caledon Div., Kleinriver Kloof, 1000–3000 ft., *Zeyher*, 4073 ! on the Zwartberg near Caledon, 400 ft., *MacOwan* and *Bolus, Herb. Norm.*, 799 !

2. H. maculata (Klatt in Linnæa, xxxiv. 627) ; corm globose ; tunics thick and cancellate ; produced leaf narrow linear, overtopping the stem, 1 in. broad ; stem ½–1 ft. long ; spathes cylindrical, peduncled, 2–2½ in. long ; perianth an inch long, the segments yellow, with a greenish claw, and a semilunar, greenish blotch above the base, the outer ½ in. broad, obtuse, the inner oblanceolate, acute ; forks of stigma linear, acute ; capsule, cylindrical, ½ in. long. *Ergänz.* 52 ; *Baker, Handb. Irid.* 74.

COAST REGION : Caledon Div., *Ecklon* and *Zeyher*, 38, 259, *fide Klatt.*

3. H. collina (Vent. Decad. 5); corm globose, $\frac{3}{4}$–1 in. diam., with thick, dark brown, cancellate tunics; produced leaf one, linear, rigid in texture, strongly ribbed, glabrous, $1\frac{1}{2}$–2 ft. long, $\frac{1}{4}$–$\frac{1}{2}$ in. broad; stem 1–$1\frac{1}{2}$ ft. long, with 1–4 clusters, the side ones on short, erect peduncles, and several lanceolate, sheathing, rudimentary leaves of firm texture; spathes cylindrical, 2–3-flowered, $2\frac{1}{2}$–3 in. long; valves cuspidate, firm in texture, except the tip, closely ribbed; perianth-limb bright red, $1\frac{1}{4}$–$1\frac{1}{2}$ in. long, the segments oblanceolate-oblong, obtuse, $\frac{3}{8}$–$\frac{1}{2}$ in. broad, with a yellow throat inside; stamens $\frac{1}{2}$–$\frac{5}{8}$ in. long, the anthers as long as the tube of filaments; ovary clavate, $\frac{3}{4}$–1 in. long; style-crests minute, subquadrate; capsule clavate, an inch long. *Salisb. in Trans. Hort. Soc.* i. 307; *Sweet, Hort. Brit.* ed. 2, 498; *Baker, Handb. Irid.* 74. *Moræa collina, Thunb. Diss.* No. 13; *Prodr.* 11; *Fl. Cap.* i. 273; *Jacq. Ic.* t. 226; *Ker in Bot. Mag.* t. 1033. *Sisyrinchium collinum, Cav. Diss.* vi. 346; *Red. Lil.* t. 250. *S. elegans, Red. Lil.* t. 171, non *Willd.*

VAR. β, **H. aurantiaca** (Sweet, Hort. Brit. ed. 2, 498); habit more slender; leaf narrower; segments of perianth light scarlet, with a yellow claw, narrower and more acute. *M. collina*, var., *Ker in Bot. Mag.* t. 1612.

VAR. γ, **H. ochroleuca** (Salisb. in Trans. Hort. Soc. i. 308); habit of the type, but flower pale yellow. *Sweet, Hort. Brit.* ed. 2, 498. *M. collina*, var., *Ker in Bot. Mag.* t. 1103. *M. grandiflora, Eckl. Top. Verz.* 14.

VAR. δ, **bicolor** (Baker, Handb. Irid. 75); habit of the type; flowers pale yellow; the segments flushed with lavender colour in the middle.

COAST REGION: vicinity of Cape Town, *Thunberg! Drège! Bolus*, 3790! Var. β. Cape district, *Cooper*, 3218! Caledon Div. *Zeyher*, 4074! Var. γ, Table Mountain, *Ecklon*, 534! 535! *Bolus*, 2811! Caledon Div., *Zeyher!* Var. δ, Stellenbosch, *Sanderson!*

4. H. lineata (Sweet, Brit. Flow. Gard. t. 178); corm large, globose, bristly round the neck; produced leaf rigid in texture, $\frac{1}{2}$ in. broad, strongly ribbed, and furnished with a distinct white band down the midrib on the face; stem $1\frac{1}{2}$–2 ft. long, bearing 4–5 clusters; spathes cylindrical, $2\frac{1}{2}$ in. long; flower copper-red, the segments an inch long, oblong, acute, $\frac{3}{8}$–$\frac{1}{2}$ in. broad, with a small, yellow-dotted blotch inside at the base; ovary clavate, pubescent, $\frac{1}{2}$ in. long; stamens $\frac{1}{3}$ in. long, with anthers shorter than the column. *Baker in Journ. Linn. Soc.* xvi. 105; *Handb. Irid.* 75.

SOUTH AFRICA: without locality.

Known only from the figure cited, which was drawn in Mr. Colvill's nursery in 1837, from a plant introduced by Mr. Synnot. *H. porrifolia*, Sweet, *Hort. Brit.* ed. 2, 498, is said to be like this in leaf and habit, but to have bright scarlet flowers.

5. H. pallida (Baker, Handb. Irid. 75); corm globose, $\frac{1}{2}$ in. diam., with hard, black, cancellate tunics; produced leaf one, subbasal, narrow linear, twice as long as the stem, firm in texture, strongly ribbed; stem a foot long, with 2–3 clusters, the side ones on short, erect peduncles; spathes cylindrical, $1\frac{1}{2}$–2 in. long, the valves firm in texture, closely ribbed, with long, scariose cusps; perianth-limb fugacious, pale yellow, $\frac{3}{4}$–1 in. long, the segments unspotted, subequal, oblanceolate-oblong, obtuse; stamens half as long as the perianth, the linear anthers as long as the tube of filaments; ovary clavate, under $\frac{1}{2}$ in. long; capsule much smaller than in *H. collina*.

KALAHARI REGION: Bechuanaland; at the Moshowa R., near Old Litaku, *Burchell*, 2252/1! Transvaal, Matebe Valley, *Holub*, 1543! 1544!

6. H. miniata (Sweet, Brit. Flow. Gard. t. 152); corm globose, with thick, rigid, black, cancellate tunics; produced leaves two, linear, rigid in texture, overtopping the stem, without a white central band, sometimes with bulbillæ in their axils; stems $\frac{1}{2}$–$1\frac{1}{2}$ ft. long; clusters several, on short, erect peduncles; spathes cylindrical, $1\frac{1}{2}$–2 in. long, the valves green and firm, except the long scariose tip; perianth fulvous, $\frac{3}{4}$–1 in. long, the segments subequal, oblanceolate, $\frac{1}{4}$ in. broad, obtuse; stamens half as long as the perianth; anthers much shorter than the column of filaments; ovary clavate-trigonous, $\frac{1}{2}$ in. long. *Baker, Handb. Irid. 75. Morœa miniata, Andr. Bot. Rep. t. 404.*

COAST REGION: Tulbagh, *Thom!* Caledon Div.; Zonder Einde R., *Gill!* Swellendam Div., *Zeyher*, 4075!
CENTRAL REGION: Worcester Div.; Klein Roggeveld, *Burchell*, 1301!

III. FERRARIA, Linn.

Perianth-segments joined in a tube at the very base, distinctly unguiculate above it, with a spreading limb much crisped at the edge, the blade of the outer a little broader than the inner. *Stamens* with filaments connate or connivent, in a cylindrical tube; anthers small, ovate, placed opposite the style-branches. *Ovary* cylindrical-trigonous; ovules crowded, superposed; style filiform inside the column of stamens, with three small, spreading, emarginate, petaloid branches strongly fimbriated on the margin. *Capsule* ellipsoid, membranous, acute, or rostrate. *Seeds* globose, angled by pressure.

Rootstock a large tuber-like corm; produced leaves few, linear, distichous, passing gradually into ovate-amplexicaul membranous bracts; spathes large, cylindrical; flowers large, fugitive, lasting only through the morning, lurid in colour, with an unpleasant scent like those of *Stapelia*.

DISTRIB. A single species in Angola.

Anther-cells distinctly divaricated:
    Expanded perianth $1\frac{1}{4}$–$1\frac{1}{2}$ in. diam., greenish-purple ... (1) **antherosa.**
    Expanded perianth 2 in. diam., brownish-purple ... (2) **divaricata.**
Anther-cells parallel:
    Perianth greenish, with narrow, very acuminate seg-
    ments ... ... (3) **uncinata.**
    Perianth dull, brownish-purple:
      Leaves long, narrowed to a point ... ... (4) **undulata.**
      Leaves short, obtuse ... ... (5) **obtusifolia.**
    Perianth dark purple ... ... (6) **atrata.**

1. F. antherosa (Ker in Bot. Mag. t. 751); stem stout, erect, $\frac{1}{2}$ ft. long, with two linear produced leaves of firm texture from its base, glabrous, strongly ribbed, glaucous, $\frac{1}{4}$ in. broad; ovate-

amplexicaul, upper bract-like leaves, firm in texture, acute, strongly
ribbed, 2 in. long; spathes inflated, 2–3½ in. long; valves closely ribbed,
pale green, firm in texture, flowers smaller than in *F. undulata*, dull,
greenish-purple, 1–1¼ in. diam. when expanded, the claw of the
segments about as long as the blade; anthers sagittate, with divaricating cells; capsule narrowed into a distinct beak.   *Baker in Journ.
Linn. Soc.* xvi. 106; *Handb. Irid.* 72. *F. viridiflora, Andr. Bot. Rep.*
t. 285. *F. Ferrariola, Willd. Sp. Plant.* iii. 581; *Dryand in Ait. Hort.
Kew.* ed. 2, iv. 136; *Ker, Gen. Irid* 28. *F. minor, Pers. Syn.* i. 50.
Morœa Ferrariola, Jacq. Collect. iv. 141; *Hort. Schoenbr.* t. 450.
F. viridis, Hort.

COAST REGION: Clanwilliam Div., *Zeyher!* Malmesbury Div.; Groenekloof,
Ecklon, 305!

F. angustifolia, Sweet, *Hort. Brit.* 499, is said to have a flower like *F. antherosa,* and linear-subulate leaves.

2. **F. divaricata** (Sweet, Brit. Flow. Gard. t. 192); stem flexuose,
attaining a height of 1½ ft.; produced leaves firm in texture, ensiform, glaucous, strongly ribbed; upper bract-like leaves, firm in
texture, ovate-amplexicaul, acute, 2 in. long; spathes cylindrical,
2 in. long, 3–4-flowered; perianth dull brownish purple, 2 in. diam.
when expanded, the segments 1½ in. long, with a claw nearly as
long as the blade, the three outer rhomboid, ½ in. broad, the three
inner narrow; ovary rostrate; filaments as long as the claw of the
segments, united in a tube nearly to the top; anther cells divaricated.   *Loudon, Ornam. Bulbs,* tab. v., fig. 3; *Baker in Journ. Linn.
Soc.* xvi. 106; *Handb. Irid.* 73.

SOUTH AFRICA: without locality.

Known only from the figure cited, which was drawn from a plant that flowered
in Mr. Colvill's nursery in 1838, sent from the Cape by Mr. Synnot.

3. **F. uncinata** (Sweet, Brit. Flow. Gard. t. 161); stem short,
little branched; produced leaves 2–3, linear, ¼ in. broad,
acute, overtopping the stem, pale green; upper rudimentary ovate-
lanceolate, acute, 1½ in. long; spathes ventricose, 1½ in. long,
2-flowered; perianth greenish, the segments an inch long, the claw
not more than half as long as the very acuminate blade, the outer
⅓ in. broad, the inner lanceolate, ¼ in. broad; ovary cylindrical, rostrate; filaments under ½ in. long, united in a tube nearly to the tip;
anthers small, with nearly parallel cells.   *Baker in Journ. Linn.
Soc.* xvi. 106; *Handb. Irid.* 73.

SOUTH AFRICA: without locality.

Known only from the figure cited, which was drawn in Mr. Colvill's nursery
in 1838, from a plant introduced from the Cape by Mr. Synnot.

4. **F. undulata** (Linn. Sp. Plant. 1353); stem stout, erect, sometimes
above a foot long; basal leaves ensiform, above a foot long, thinner
in texture, more finely veined than in *F. antherosa*, flat, ¼–½ in.
broad above the dilated clasping base; upper passing gradually into
the ovate-amplexicaul bracts, which are scariose in texture, 1½–2 in.

long, with a white margin; spathes ventricose, $1\frac{1}{2}$–2 in. long, 2–3-flowered; perianth dull purple, segments with a claw not more than half as long as the lamina, which in the outer ones is deltoid, $\frac{1}{2}$ 'in. broad, in the inner $\frac{1}{3}$ in. broad; stamens and styles $\frac{1}{4}$ in. long; anthers oblong, $\frac{1}{6}$ in. long; cells not diverging; ovary cylindrical, $1\frac{1}{2}$ in. long, narrowed to the apex. *Curt. Bot. Mag. t.* 144; *Jacq. Hort. Vind.* t. 63; *Cavan. Diss.* vi. 343; *Ic.* t. 190, fig. 1; *Red. Lil.* t. 28; *Ker, Gen. Irid.* 28; *Baker, Handb. Irid.* 73. *Moræa undulata, Thunb. Diss.* No. 21. *F. punctata, Pers. Syn.* i. 50. *Flos indicus, Ferr. Cult.* 170, t. 173. *Gladiolus indicus, Moris. Hist.* ii. 344, s. 4, t. 4, fig 7.—*Barrel. Ic.* 1216; *Rudb. Elys.* ii. 49. fig. 9; *Miller Ic.* 187, t. 280.

COAST REGION: Clanwilliam Div., Between Lange Kloof and Heeren Logement Berg, *Drège!* Piquetberg, *Drège!* vicinity of Cape Town, *Thunberg!* Mossel Bay Div., *Rogers!* *Burchell,* 6196!

Introduced into the Dutch gardens about 1640. *Lapeyrousia macrochlamys,* Baker in Journ. Bot. 1876, 338, proves to be this plant.

5. F. obtusifolia (Sweet, Brit. Flow. Gard. t. 148); stem stout, erect, densely branched, attaining a height of $1\frac{1}{2}$ ft.; produced leaves short, ensiform, obtuse; upper rudimentary, ovate-amplexicaul, obtuse, with a minute cusp; spathes ventricose, the valves green to the tip; perianth brownish-purple, $1\frac{1}{2}$ in. diam. when expanded, the claw of the segments not more than half as long as the blade, the outer segments deltoid, $\frac{1}{3}$ in. broad, the inner lanceolate, $\frac{1}{4}$ in. broad; ovary cylindrical, not rostrate; filaments $\frac{1}{3}$ in. long, united in a column nearly to the top; anthers small, with parallel cells. *Baker in Journ. Linn. Soc.* xvi. 106; *Handb. Irid.* 73.

SOUTH AFRICA: without locality.

Known only from the figure cited, which was drawn from a plant that flowered in Mr. Colvill's nursery at London, in 1838; introduced from the Cape by Mr. Synnot.

6. F. atrata (Lodd. Bot. Cab. t. 1356); stem $\frac{1}{2}$ ft. long, densely fastigiate; produced leaves about 4, ensiform, firm in texture, strongly ribbed, falcate, overtopping the stem, a foot long, $\frac{1}{4}$ in. broad; spathes ventricose, 3–4-flowered, $1\frac{1}{2}$–2 in. long; the valves acute, firm in texture, green to the apex, the flowers protruding from the side; perianth bright dark purple, $1\frac{1}{2}$–2 in. diam. when fully expanded, segments $1\frac{1}{2}$ in. long, the outer three deltoid, acuminate, $\frac{1}{2}$ in. broad, with a claw not more than half as long as the blade, three inner rather smaller; stamens $\frac{1}{2}$ in. long; filaments free in the upper half; anthers oblong, with parallel cells. *Baker in Journ. Linn. Soc.* xvi. 106; *Handb. Irid.* 73.

KALAHARI REGION: Victoria West; Buffels Bout, *Burchell,* 1599!

IV. HEXAGLOTTIS, Vent.

Perianth cut down to the ovary, rotate, twisting up spirally in fading; segments subequal, oblanceolate-oblong, obtuse. *Filaments*

very short, flattened, connivent in a tube; anthers large, linear.
Ovary cylindrical, 3-celled; ovules crowded, superposed; style
short, the three branches patent with two linear forks, which are
stigmatose at the tip. *Capsule* firm, cylindrical, 3-valved. *Seeds*
minute, oblong, turgid.

Rootstock a small tunicated corm. Produced leaves usually two, long, linear,
or terete. Inflorescence a panicle with spicate branches; flowers fugacious,
yellow, 3-4 to a spathe.

DISTRIB. An endemic Cape genus.

Leaf narrow linear ... ... (1) **longifolia.**
Leaf terete ... ... (2) **virgata.**

1. **H. longifolia** (Vent. Decad. 6); corm ovoid, ½–1 in. diam.;
tunics thick, of hard, black, parallel ribs; produced leaves 1–2, overtopping the stem, narrow linear, glabrous, strongly ribbed; stem
slender, erect, 1–1½ ft. long; inflorescence a panicle with long
spicate branches, the central one sometimes a foot long, bearing 6–8
clusters; spathes cylindrical, 1–1½ in. long, 3–4-flowered; valves
firm in texture, closely ribbed, with long scariose cusps; perianth
yellow, ½ in. long, segments ¼ in. broad, with a many-ribbed keel;
anthers linear, ⅛ in. long; style-branches ¼ in. long, forked halfway down; capsule clavate ½–¾ in. long. *Salisb. in Trans. Hort.
Soc.* i. 313; *Baker, Handb. Irid.* 76. *Morœa flexuosa, Linn. fil.
Suppl.* 100; *Thunb. Diss.* No. 12; *Prodr.* 11; *Fl. Cap.* i. 272;
Ker in Bot. Mag. t. 695; *Gen. Irid.* 33. *Homeria spicata, Klatt
in Linnœa.* xxxiv. 626, in part. *Sisyrinchium flexuosum, Spreng. Syst.*
i. 167. *Ixia longifolia, Jacq. Hort. Vind.* iii. 47, t. 90; *Morœa
longifolia, Pers. Syn.* i. 49. *Hexaglottis flexuosa, Sweet, Hort. Brit.*
edit. 2, 498. *Plantia flava, Herbert in Bot. Reg.* xxx. *Misc.* 89.

SOUTH AFRICA: without locality, *Zeyher,* 1643! 1644!
COAST REGION: Clanwilliam and Piquetberg Div., *Thunberg!* Vicinity of
Cape Town and Rondebosch, *Ecklon,* 9b! 536! *Burchell,* 34! 202! 252/1!
MacGillivray, 479! George Div., *Bolus,* 2484! *Burchell,* 6107! 6151! Uiten-
hage Div., Van Staden's Hoogte, *MacOwan,* 2055!

2. **H. virgata** (Sweet, Hort. Brit. edit. 2, 498); very near *H.
longifolia,* of which it is likely only a variety, and from which it
only differs by its more slender habit and terete leaves. *Baker,
Handb. Irid.* 76. *Morœa virgata, Jacq. Ic.* t. 228; *Collect.* iii. 194;
Ker in Bot. Mag., sub t. 1103; *Gen. Irid.* 33. *Ixia virgata, Willd.
Sp. Plant.* i. 202; *Vahl. Enum.* ii. 59. *Homeria spicata, Klatt in
Linnœa,* xxxiv. 626, in part.

COAST REGION: near Paarl, *Drège!* Table Mountain, *Ecklon,* 536! Near
Cape Town, *Bolus,* 3801!

V. GALAXIA, Thunb.

Perianth with a cylindrical tube, and a campanulate limb with
subequal obovate segments. *Stamens* inserted at the throat of the
perianth-tube; filaments connate in a cylindrical column; anthers

lanceolate, sagittate at the base. *Ovary* cylindrical, 3-celled ; ovules crowded in the cells ; style filiform, with a petaloid, peltate, 3-lobed stigma papillose on the margin. *Capsule* cylindrical, membranous. *Seeds* minute, subglobose.

Rootstock a small corm, tunicated with parallel lamellæ; leaves several, multifarious, linear or broad; spathes 1-flowered, cylindrical, membranous, sessile in the centre of the rosette of leaves; flowers small, fugacious, yellow or lilac.

DISTRIB. An endemic Cape genus.

Outer leaves broad...	...	...	...	...	...	(1) ovata
All the leaves linear	...	...	...	...	...	(2) graminea.

1. **G. ovata** (Thunb. Diss. Nov. Gen. ii. 51, cum icone) ; corm globose, ½ in. diam., thickly tunicated, furnished with a neck some-times 1–2 in. long below the rosette of leaves ; leaves several, in a dense rosette, firm in texture, with several distinct ribs, and a thickened, pale, cartilaginous border, outer ovate or oblong-lanceolate, ½–1½ in. long, inner narrower ; spathe cylindrical, ½–1 in. long ; perianth-tube filiform, exserted from the spathe ; perianth-limb bright yellow, about an inch in diameter, with imbricated obovate segments ; stigma exserted about ½ in. from the perianth-tube, overtopping the anthers. *Cav. Diss.* vi. 341, tab. 189, fig. 2 ; *Jacq. Ic.* tab. 291, upper figure ; *Andr. Bot. Rep.* t. 94 ; *Baker, Handb. Irid.* 97. *G. obtusa, Salisb. in Trans. Hort. Soc.* i. 315. *Ixia Galaxia, Linn. fil. Suppl.* 93. *Ovieda fasciculata, a* and *b, Herb. Drège.*

VAR. β, **G. grandiflora** (Salisb. in Trans. Hort. Soc. i. 314) ; flower larger, yellow, 1½ in. diam. when expanded, with imbricated obovate segments. *Andr. Bot. Rep.* t. 164. *G. ovata, a, Ker in Bot. Mag.* t. 1208.

VAR. γ, **G. mucronularis** (Salisb. loc. cit. 315) ; flower lilac, with a yellow throat, segments narrower than in the type, obtuse, with a minute cusp. *G. ovata, Jacq. Ic.* t. 291, lower left hand figure.

VAR. δ, **G. versicolor** (Salisb. loc. cit. 315) ; flower smaller than in the type, dark lilac, with a yellow throat; segments obtuse, without a cusp. *G. ovata, Jacq. Ic.* t. 291, lower right hand figure.

VAR. ε, **Eckloni** (Baker, Handb. Irid. 97) ; flowers lilac, about an inch in diam., with imbricated broad segments; leaves more numerous than in the type, the inner linear. *G. violacea, Eckl. Top. Verz.* 17.

COAST REGION: Clanwilliam Div., *Drège,* 2614a ! Cape Div., near Tyger Berg, *Burchell,* 971 ! Lion Mountain, *MacOwan,* 2552! and *Herb. Norm. Aust. Afr.,* 253! Near Cape Town, *Thunberg !* Var. ε, Devils Mountain, *Ecklon,* 308 !

2. **G. graminea** (Thunb. Diss. Nov. Gen. ii. 51, cum icone) ; corm ovoid, ¼–⅓ in. diam., with thick, much-raised ribs, and often furnished with a cylindrical neck, 1–2 in. long, below the rosette of leaves ; leaves numerous, in a rosette, all linear, convolute, 1–2 in. long, dilated and membranous at the base, and often furnished with bulbillæ in their axils ; spathe about ½ in. long ; perianth with a filiform tube 1–2 in. long, and a lilac or yellow limb 1–1½ in. diam. when expanded, with obovate obtuse segments ; style reaching half-way up the limb, overtopping the anthers. *Thunb. Prodr* 10 ; *Cav. Diss.* vi. 341, t. 189, fig. 3 ; *Ker in Bot. Mag.* t. 1292 ; *Jacq. Coll.* ii. 366, t. 18, fig. 2 ; *Ker, Gen. Irid.* 71 ; *Baker, Handb. Irid.* 97. *Ixia fugacissima, Linn. fil. Suppl.* 94.

COAST REGION: vicinity of Cape Town, *Thunberg!* *Burchell*, 8396! *Ecklon*, 306! *Bolus*, 4806! *MacOwan, Herb. Aust. Afric.* 1539! Caledon Div.; Klein River Berg, *Zeyher*, 4065!

VI. SYRINGODEA, Hook. fil.

Perianth hypocrateriform or infundibuliform, with a long cylindrical tube and subequal, entire or emarginate segments. *Stamens* inserted at the throat of the perianth-tube; filaments short, free; anthers large, linear, sagittate at the base. *Ovary* oblong-cylindrical; cells 3, many-ovuled; style long, filiform; style-branches slender, clavate. *Capsule* membranous, oblong-cylindrical. *Seeds* minute, turgid.

Rootstock a corm with many membranous tunics. Leaves usually narrow, multifarious. Spathes cylindrical, membranous, 1-flowered, subsessile in the centre of the rosette of leaves. Flowers small, usually purple.

DISTRIB. An endemic Cape genus, representing *Crocus* of the northern hemisphere.

Subgenus SYRINGODEA PROPER. Perianth hypocrateriform, with spreading emarginate segments.

The only species ... ... (1) **pulchella.**

Subgenus CROCOPSIS. Perianth infundibuliform, with ascending entire segments.

Leaves terete :
  Leaves several :
    Spathe ¼ in. long ... ... (2) **montana.**
    Spathe ⅓ in. long ... ... (3) **filifolia.**
    Spathe ½-¾ in. long ... ... (4) **Flanagani.**
    Spathe 1½-2 in. long ... ... (5) **bicolor.**
  Leaves single ... ... (6) **rosea.**
  Leaves linear ... ... (7) **minuta.**
  Leaves lanceolate ... ... (8) **latifolia.**

1. S. pulchella (Hook. fil. in Bot. Mag. t. 6072); corm globose, ½ in. diam., with thick, dull brown, outer tunics, 1-2-flowered; leaves 4-6, falcate, setaceous, glabrous, firm in texture, much overtopping the flowers, 3-4 in. long, ⅓ lin. diam.; spathe membranous, cylindrical, an inch long; perianth with a cylindrical tube 1½-2 in. long, much exserted from the spathe, and a rotate purple limb ⅓-½ in. long, with all the segments cuneate-unguiculate, and distinctly emarginate; anthers linear, ¼ in. long, much exceeding their filaments; style-branches cuneate at the tip, just overtopping the stamens. *Baker in Journ. Bot.* 1876, 67; *Handb. Irid.* 95.

CENTRAL REGION : Graaff Reinet Div., Sneeuw Berg, 4500 ft., *Bolus*, 1852!

2. S. montana (Klatt, Ergänz. 69); corm oblong, ½ in. diam., 1-3-flowered; tunics shining, castaneous; leaves setaceous, curved or spiral, hairy, ½ in. long, ¼ lin. diam.; valves of spathe equal, white, ovate, acute, ¼ in. long; perianth with a cylindrical tube ¾ in. long, a yellow throat, and obovate violet segments ⅓ in. long, ⅙ in. broad. *Baker, Handb. Irid.* 95.

CENTRAL REGION : Hantam Mountains, near Calvinia, *Meyer*.

3. S. filifolia (Baker in Journ. Bot. 1876, 67) ; corm oblong, ¼ in. diam., with dark brown shining tunics, and a neck an inch long ; leaves 6–8, setaceous, glabrous, 1–1½ in. long, very slender ; spathe membranous, cylindrical, ½ in. long ; perianth with a filiform tube an inch long, and a dark lilac limb ½ in. long, with a yellow throat, and obovate segments ; stamens half as long as the limb. *Baker, Handb. Irid. 96.*

SOUTH AFRICA : without locality, *Bowie!*

Described from a specimen at the British Museum, gathered by Bowie.

4. S. Flanagani (Baker in Kew Bullet. 1893, 158) ; corm oblong, ¼ in. diam., with brown, membranous tunics produced above its neck ; leaves 6–9, filiform, glabrous, recurved, 1–2 in. long, spathe ½–¾ in. long ; perianth-tube yellow, twice as long as the spathe, segments of the limb dark purple, ⅓ in. long, outer oblanceolate, inner oblanceolate-oblong ; stamens more than half as long as the perianth-segments.

COAST REGION : King Williamstown Div., summit of Gonubie Hill, near Komgha, 2400 feet, *Flanagan,* 720 !

5. S. bicolor (Baker in Journ. Bot. 1876, 67) ; corm globose, ¾ in. diam., 1–3-flowered, with brown, membranous tunics, and a neck 2–3 in. long ; leaves 6–8, setaceous, firm in texture, glabrous, much overtopping the flowers, 6–8 in. long, ½ lin. diam. ; spathe pale, membranous, cylindrical, 1–1½ in. long ; perianth with a cylindrical tube 1½–2 in. long, and a purple limb ¾ in. long, with a bright yellow throat, segments obovate-oblong, ¼–⅓ in. broad ; stamens half as long as the limb, the anthers exceeding the free filaments ; style-branches clavate, ⅛ in. long, overtopping the anthers. *Baker, Handb. Irid. 96. Trichonema longitubum, Klatt in Linnæa,* xxxiv. 665.

VAR. β, concolor (Baker, Handb. Irid. 96). Perianth-limb ½ in. long, dark purple, like the tube, without a yellow throat.

COAST REGION : near Grahamstown, 2000 ft., *MacOwan,* 827 !
CENTRAL REGION : Graaff Reinet, *Bowker,* 9 ! Somerset, *Bowker!* Coles-berg, *Shaw!* Var. β, Valley at Coetzier's Kraal, near Murraysburg, *Tyson,* 346 !
EASTERN REGION : Kaffraria, *Hutton!*

A plant gathered by Drège (No. 3498) on the Wittebergen, Aliwal North, at 7000-8000 ft., with much more slender leaves, and a concolorous lilac limb, ½ in. long, may be an alpine form of this species.

6. S. rosea (Klatt, Ergänz. 69, excl. syn.) ; corm 1–3-flowered ; leaf single, thick, terete, obtuse, curved, channelled down the face, 2–3 in. long, ½ lin. diam. ; spathes with subequal, membranous valves ½–1 in. long, 1–2 lines broad, with a one-nerved herbaceous midrib ; perianth with a cylindrical tube ¾ in. long, and a rose-red limb, with oblanceolate segments ½ in. long, ⅓ in. broad. *Baker, Handb. Irid. 96.*

CENTRAL REGION : Hantam Mountains, near Calvinia, *Meyer.*

7. S. minuta (Klatt, Ergänz. 69, excl. syn.) ; corm globose, 1–4 flowered ; leaves linear, flat, glabrous, erect ; spathe with two, small

linear-filiform valves; perianth with a white tube with purple
streaks, and a limb with concave white segments with two purple
streaks. *Baker, Handb. Irid.* 96.

SOUTH AFRICA: without locality or collector, but said to grow in sandy inundated places.

I only know this from Dr. Klatt's description. *Ixia minuta,* Linn. fil., which
he cites as a synonym, is *Pauridia minuta,* Harvey in *Hypoxideæ.* I have
examined the type specimens, both in the Linnean and Thunbergian herbaria.

8. S. **latifolia** (Klatt, Ergänz. 69); corm ovoid, $\frac{1}{6}$ in. diam.,
1-flowered, with brown, cuspidate tunics; leaves 4, lanceolate, flat,
patent, prominently 1-nerved, $\frac{1}{2}$–1 in. long, $\frac{1}{6}$ in. broad; spathes with
equal valves $\frac{1}{3}$ in. long, hairy on the keel; perianth with a
cylindrical tube nearly an inch long, purple and hairy at the throat,
and a limb $\frac{1}{2}$ in. long, with oblanceolate segments $\frac{1}{8}$ in. broad.
Baker, Handb. Irid. 96.

WESTERN REGION: Little Namaqualand; Kamiesbergen, 3000-4000 ft.,
Drège, 2633.

VII. ROMULEA, Maratti.

Perianth infundibuliform, with a short tube, and a regular limb
with subequal, imbricated, obovate segments. *Stamens* inserted at the
throat of the perianth-tube; filaments short, usually free; anthers
lanceolate, sagittate at the base. *Ovary* 3-celled; ovules crowded in
the cells; style filiform, with three bifid branches with clavate forks
papillose round the inner margins. *Capsule* globose or ellipsoid,
loculicidally 3-valved. *Seeds* globose or angled by pressure.

Rootstock a small globose corm, with tunicated, membranous, brown coats.
Leaves narrow, usually overtopping the flowers. Stems produced, simple or branched.
Spathe of two lanceolate valves. Flower lilac, purple, yellow or pale, always
solitary in the spathes.

DISTRIB. Species about 50; many Mediterranean, a few Central European, two
Tropical African.

Subgenus ROMULEA PROPER. Filaments free.

Luteæ. Flowers bright yellow.
 Perianth-tube very short :
  Leaves setaceous, long, straight ... ... (1) **sublutea.**
  Leaves setaceous, short, tortuose ... ... (2) **tortuosa.**
  Leaves flat ... ... (3) **bulbocodioides.**
 Perianth-tube $\frac{1}{3}$–$\frac{1}{4}$ in. long :
  Peduncle long ... ... (4) **filifolia.**
  Peduncle short ... ... (5) **citrina.**
 Perianth-tube $\frac{3}{4}$ in. long ... ... (6) **Macowani.**
Chloroleucæ. Flowers whitish, the outer segments
 green on the back.
  Leaves 1-1$\frac{1}{2}$ lin. broad ... ... (7) **latifolia.**
  Leaves setaceous, $\frac{1}{4}$–$\frac{1}{2}$ lin. broad :
   Flower large ... ... (8) **tridentifera.**
   Flower middle-sized :
    Leaves long, straight ... ... (9) **chloroleuca.**
    Leaves short, spirally twisted ... ... (10) **spiralis.**
   Flower small :
    Perianth-limb $\frac{1}{3}$ in. long ... ... (11) **similis.**
    Perianth-limb $\frac{1}{6}$ in. long ... ... (12) **minutiflora.**

Rubro-lilacinæ. Flowers bright rose-lilac.
  Peduncle short, sometimes only 1-flowered.
    Leaves very slender :
      Perianth-tube very short :
        Perianth bright orange in the lower half (13) **hirsuta.**
        Perianth yellow at the very base ... (14) **longifolia.**
        Perianth-tube half as long as the segments ... (15) **gracillima.**
    Leaves stouter and more rigid :
      All the perianth-segments rose-lilac ... ... (16) **rosea.**
      Inner segments white ; outer red, with a
        pale blotch in centre ... ... (17) **elegans.**
      Perianth yellow outside, with purple plumose
        stripes, fulvous and yellow inside ... ... (18) **cuprea.**
  Peduncle elongated, 2-4 flowered :
    Perianth-tube half as long as segments ... (19) **arenaria.**
Subgenus SPATHALANTHUS. Filaments connate in a
  tube to the top ... ... (20) **monadelpha.**

1. R. sublutea (Baker in Journ. Linn. Soc. xvi. 88) ; corm globose, $\frac{1}{4}-\frac{1}{2}$ in. diam., with rigid, brown tunics cut round at the base ; basal leaves 1-2, setaceous, 3-6 in. long, $\frac{1}{4}$ lin. broad, 1-ribbed, with convolute edges ; stem simple, 2-3 in. long, or forked low down, with 2-5 branches, with a small green lanceolate leaf from the base of each ; spathe of two green lanceolate valves about $\frac{1}{2}$ in. long ; perianth a uniform bright yellow, with a short, funnel-shaped tube, and a limb $\frac{3}{4}-1$ in. long, with oblong segments ; anthers bright yellow, $\frac{1}{6}$ in. long, about equalling the filaments ; styles reaching to the top of the anthers. *Klatt, Ergänz.* 65 ; *Baker, Handb. Irid.* 100. *Ixia crocea, Thunb., Fl. Cap.* edit. *Schult.* 55. *I. sublutea, Lam. Enc.* iii. 335. *Geissorhiza sublutea, Ker in Konig and Sims' Ann.* i. 224. *I. filifolia, Red. Lil.* tab. 251, fig. 2. *Trichonema filifolium, Ker, Gen. Irid.* 82. *R. aurea, Klatt, Ergänz.* 65.

SOUTH AFRICA : without locality, *Thunberg ! Grey !*
  COAST REGION : Simons Bay, *Wright,* 279 ! Zwartberg, near Caledon, *Zeyher,* 4045 !
  CENTRAL REGION : Hantam Mountains, near Calvinia, *Meyer.*

2. R. tortuosa (Baker in Journ. Linn. Soc. xvi. 88) ; corm ovoid, with brown, rigid tunics ; basal leaves 3-4, setaceous, tortuose, spreading, $1\frac{1}{2}-2$ in. long, firm in texture, 3-nerved, $\frac{1}{6}$ lin. diam. ; flowers 2-3, subsessile in the centre of the rosette of leaves ; spathe of two green lanceolate valves $\frac{1}{2}$ in. long ; perianth bright yellow, with a short tube, and a limb $\frac{1}{2}$ in. long, with oblong, lanceolate segments $\frac{1}{6}$ in. broad ; stamens reaching more than halfway up the limb, the filaments as long as the lanceolate-sagittate anthers. *Baker, Handb. Irid.* 100. *Ixia tortuosa, Licht. in Roem. et Schult. Syst. Veg.* i. 375. *Trichonema tortuosum, Ker, Gen. Irid.* 83 ; *Klatt in Linnæa,* xxxiv. 666.

CENTRAL REGION : Fraserburg Div., Middle Roggeveld, between Jakhals Fontein and Kuilenberg, near Sutherland, *Burchell,* 1343 ! *Lichtenstein.*

3. R. bulbocodioides (Baker in Journ. Linn. Soc. xvi. 88, non Eckl.) ; corm globose, $\frac{1}{3}-\frac{1}{2}$ in. diam., with pale brown, smooth, rigid tunics ; basal leaves 1–2, narrow linear, firm in texture, often a foot long, $\frac{1}{12}$ in. broad, with several distinct ribs ; peduncle usually short, bearing 3–5 flowers on long pedicels, with a large linear leaf from the base of each, $\frac{1}{8}$–$\frac{1}{6}$ in. broad ; outer spathe-valve green, lanceolate, closely ribbed, $\frac{3}{4}$–1 in. long, inner smaller, brown, membranous ; perianth bright yellow, with a funnel-shaped tube $\frac{1}{6}$–$\frac{1}{4}$ in. long, and obovate, obtuse segments $\frac{1}{4}$–$\frac{1}{3}$ in. broad, the outer tinged with green outside ; anthers yellow, $\frac{1}{6}$ in. long, equalling the filaments ; styles equalling or just overtopping the anthers. *Baker, Handb. Irid.* 101 ; *Klatt, Ergänz.* 68. *Ixia bulbocodioides, Delaroche, Descr.* 19. *I. recurva, Red. Lil.* tab. 251, fig. 1. *I. reflexa, Thunb., Fl. Cap.* edit. *Schult.,* 55. *Trichonema caulescens, Ker in Bot. Mag.* t. 1392 ; *Klatt in Linnæa,* xxxiv. 663. *T. hypoxidiflorum, Salisb. in Trans. Hort. Soc.* i. 316. *Ixia flava, Lam. Ill.* i. 109. *R. caulescens, Klatt, Ergänz.* 65. *R. chloroleuca, Eckl. Top. Verz.,* 20, excl. syn.

Var. β, **Elongata** (Baker). Peduncle a foot long.

SOUTH AFRICA : without locality, *Thunberg! Rogers! Villette! Bowie!* Var. β, *Thunberg!*

COAST REGION : near Tulbagh, *Burchell,* 986! Near Cape Town, *Bolus,* 2810! Caledon Div., Houwhoek Mts., 1000–3000 ft., *Zeyher,* 4046!

There are fine specimens in the Linnean herbarium, named *Ixia Bulbocodium.*

4. R. filifolia (Eckl. Top. Verz. 20, excl. syn. Red.) ; corm globose, $\frac{1}{2}$ in. diam., with rigid, bright brown tunics ; basal leaves 2, erect, setaceous, firm in texture, a foot long, $\frac{1}{4}$ lin. broad, 1-ribbed, with convolute edges ; peduncle 6–8 in. long, bearing 3–4 flowers on arcuate pedicels 1–1$\frac{1}{2}$ in. long, with a pair of leaves, one large and one rudimentary from the fork ; spathe-valves both green, lanceolate, firm in texture, $\frac{1}{2}$ in. long ; perianth bright orange-yellow towards the base, paler upwards, with a funnel-shaped tube, $\frac{1}{4}$ in. long, and, lanceolate segments $\frac{3}{4}$ in. long, $\frac{1}{6}$ in. broad ; anthers $\frac{1}{6}$ in. long, exceeding the filaments and overtopped by the stigmas. *Baker in Journ. Linn. Soc.* xvi. 88 ; *Handb. Irid.* 101. *Trichonema filifolium, Klatt in Linnæa,* xxxiv. 671.

COAST REGION : Uitenhage, *Zeyher!*

5. R. citrina (Baker, Handb. Irid. 100) ; corm small, globose ; stem short, bearing 2–3 flowers on erect or spreading peduncles 1–2 in. long ; leaves terete, much overtopping the flowers ; outer spathe-valve $\frac{1}{2}$ in. long ; perianth with a narrowly funnel-shaped tube $\frac{1}{3}$ in. long, and oblong, plain yellow segments $\frac{1}{2}$ in. long ; stamens less than half as long as the perianth-segments ; style-branches just overtopping the anthers.

WESTERN REGION : Little Namaqualand, near Modder Fontein, 3000 ft., *Bolus,* 6619!

6. R. Macowani (Baker in Journ. Bot. 1876, 236) ; corm globose, $\frac{1}{2}$ in. diam., with rigid, brown tunics ; basal leaves 4–6, falcate, setaceous, rigid in texture, 6–8 in. long, $\frac{1}{4}$–$\frac{1}{2}$ lin. broad, 3-nerved, with convolute edges ; stems very short, 1-flowered, or forked from the base, with 2–4 short, erect branches; spathe of two pale, almost membranous, lanceolate valves 1–$1\frac{1}{2}$ in. long; perianth with a tube $\frac{3}{4}$ in. long, cylindrical in the lower two-thirds, and a limb an inch long, bright golden yellow in the lower part, paler upwards, and sometimes tinged with red, segments oblong, $\frac{1}{4}$–$\frac{1}{3}$ in. broad; anthers linear, $\frac{1}{3}$ in. long, rather exceeding the filaments; styles considerably overtopping the anthers. *Baker in Journ. Linn. Soc.* xvi. 88; *Handb. Irid.* 101.

CENTRAL REGION: Somerset Div., *Bowker!* summit of Bosch Berg, 4500 ft., *MacOwan*, 1547!

7. **R. latifolia** (Baker in Journ. Bot. 1876, 237); corm globose, $\frac{1}{2}$ in. diam., with crustaceous, brown tunics; produced basal leaf 1, narrow linear, firm in texture, 1–$1\frac{1}{2}$ lin. broad, with several distinct ribs; peduncle 1-flowered, or branched from the base, with 2–3 pedicels 1–$1\frac{1}{2}$ in. long, with large flat leaves from their base; outer spathe-valve green, lanceolate, firm in texture, $\frac{3}{4}$ in. long, obtuse, inner smaller, scariose; perianth whitish, with a yellow throat, with a very short tube, and a limb $\frac{3}{4}$–1 in. long, with oblong-lanceolate segments $\frac{1}{6}$–$\frac{1}{4}$ in. broad, the outer tinged green on the back; stamens less than half as long as the perianth limb; stigmas overtopping the anthers. *Baker, Handb. Irid.* 101. *Trichonema latifolium, Herbert MSS. Ixia reflexa*, var., *Herb. Thunberg.*

COAST REGION: vicinity of Cape Town, *Pappe! Niven! Thunberg! Burchell*, 8533! *Bolus*, 3696! 3734, partly! *MacOwan* and *Bolus, Herb. Norm. Austr. Afr.*, 255!

This is, perhaps, merely a variety of *R. bulbocodioides*, with pale-coloured flowers.

8. **R. tridentifera** (Klatt, Ergänz. 64); corm conic, comose, flat at the base; produced leaf subsetaceous, overtopping the flower, flexuose, $\frac{1}{3}$ lin. broad, densely hairy on the midrib; peduncle 1-flowered, 2 in. long; outer valve of spathe $\frac{3}{4}$ in. long, striped with red on the back, inner herbaceous at the tip; perianth with a very short, funnel-shaped tube and a dull yellow, semi-lurid limb an inch long, with segments $\frac{1}{4}$ in. broad, the three outer with a 3-toothed violet blotch; styles as long as the stigmas. *Baker, Handb. Irid.* 101.

CENTRAL REGION: Hantam Mountains, near Calvinia, *Meyer.*

9. **R. chloroleuca** (Baker in Journ. Linn. Soc. xvi. 89); corm globose, $\frac{1}{2}$ in. diam., with dark brown, membranous tunics; produced basal leaves 2–4, setaceous, sometimes a foot long, erect, rigid in texture, $\frac{1}{3}$–$\frac{1}{2}$ lin. broad; flowers 1–2 to a corm, on erect peduncles 3–6 in. long; spathe of two lanceolate green valves of firm texture $\frac{3}{4}$ in. long; perianth whitish, with a yellow throat, with a very short, funnel-shaped tube, and oblong segments $\frac{1}{2}$–$\frac{3}{4}$ in. long; stamens half as long as the perianth limb; styles overtopping the stamens. *Baker, Handb. Irid.* 102. *Ixia chloroleuca, Jacq. Collect.* iv. 180; *Ic.* t. 272. *Trichonema chloroleuca, Ker in Bot. Mag.*, sub t. 575;

Gen. Irid. 82. *Ixia ochroleuca, Vahl. Enum.* ii. 50. • *Trichonema ochroleuca, Ker in Konig and Sims' Ann.* i. 223.

COAST REGION : near Villiersdorp, *Bolus,* 5245 ! Cape Flats, *Zeyher,* 1603 !

10. **R. spiralis** (Baker in Journ. Linn. Soc. xvi. 90); corm globose, ¼ in. diam., with shining, crustaceous, dark brown tunics, and a long neck ; leaves 4–5, linear-convolute, falcate, 1–1½ in. long, much twisted spirally; peduncle very short, 1-flowered ; spathe of two lanceolate valves, under ½ in. long, the outer scariose at the tip, and the inner throughout ; perianth whitish, tinged with red lilac, with a very short tube, and a limb under ½ in. long, with obovate-obtuse, imbricated segments ; stamens half as long as the perianth limb ; styles reaching to the top of the stamens. *Baker, Handb. Irid.* 102. *Trichonema spirale, Burchell, Travels* i. 260.

CENTRAL REGION : Fraserburg Div.; Middle Roggeveld, between Kuilenberg and Great Reed River, *Burchell,* 1356 !

11. **R. similis** (Eckl. Top. Verz. 19); corm globose, ¼ in. diam., with smooth, brown, crustaceous tunics ; produced leaf one, subsetaceous, firm in texture, 4–6 in. long, ½ lin. broad ; peduncle short ; pedicels 1–4, about an inch long, subtended by erect rudimentary leaves 1–2 in. long; outer spathe-valve green, lanceolate, firm in texture, ½ in. long, inner smaller, membranous ; perianth funnel-shaped, with a very short tube, and lanceolate segments ⅓ in. long, the inner white, the outer green ; stamens half as long as the perianth-segments ; styles as long as the stamens. *Baker in Journ. Linn. Soc.* xvi. 89 ; *Handb. Irid.* 102. *Ixia pumila, Herb. Banks. R. obscura, Klatt, Ergänz.* 65. *R. bulbocodioides, Eckl. Top. Verz.* 19? *Ixia Bulbocodium, Thunb. Diss. Ixia,* p. 6 and 22, and *Fl. Cap. edit. Schult.* p. 55, *non Linnæus.*

COAST REGION : Lion Mountain and ·Devils Mountain, *Thunberg!* Camps Bay, *Ecklon!* Near Cape Town, *Bolus,* 3734, partly ! This is the nearest representative at the Cape of the European R. *Columnæ.*

12. **R. minutiflora** (Klatt, Ergänz. 65) ; corm globose, ¼ in. diam., with crustaceous brown tunics ; leaves numerous, setaceous, terete, spreading, firm in texture, 3-nerved, ½ lin. broad ; flowers 2–4 to a corm, on pedicels under an inch long ; spathe of two lanceolate valves ¼ in. long ; perianth funnel-shaped, light lilac, ¼ in. long, with a short tube, and oblong-lanceolate segments 1/12 in. broad ; stamens more than half as long as the perianth-limb ; capsule ovoid, ⅙ in. long. *Baker, Handb. Irid.* 102.

COAST REGION : Hex River Mts., *Drège,* 538. Mossel Bay ; The Point, *Rogers!*

13. **R. hirsuta** (Ecklon, Top. Verz. 19); corm globose, ⅓ in. diam., with membranous, brown tunics cut off at the base; root-leaves very slender, setaceous, 1-nerved, 2–4 in. long, ¼ lin. diam., erect or spreading, obscurely pilose or glabrous; peduncles simple or branched at the base, with pedicels 1–2 in. long ; spathe of two

green, lanceolate valves of firm texture, $\frac{1}{3}-\frac{1}{2}$ in. long; perianth
$\frac{3}{4}-1\frac{1}{4}$ in. long, bright golden yellow in the lower half, bright red-
purple in the upper, the tube $\frac{1}{6}$ in. long, the segments $\frac{1}{4}$ in. broad,
the outer not distinctly striped on the back; stamens one-third as
long as the perianth-limb; styles reaching to the top of the anthers.
Baker in Journ. Linn. Soc. xvi. 89; *Handb. Irid.* 102; *Klatt,
Ergänz.* 64. *Trichonema hirsutum, Klatt in Linnæa,* xxxiv. 665.
Geissorhiza sublutea, Herb. Zeyher. R. uncinata, Klatt, Ergänz. 67.

COAST REGION : near Paarl, *Drège,* 8450a! vicinity of Cape Town, *Ecklon,*
708! *Bolus,* 3760! *Burchell! Zeyher,* 5007! *Tyson,* 2454! *MacOwan,*
2274! 2565! 2616! *MacOwan* and *Bolus, Herb. Norm.,* 254! 529!

14. **R. longifolia** (Baker in Journ. Linn. Soc. xvi. 89); corm
small, globose; leaves 2–4, setaceous, suberect, 6–10 in. long,
3-nerved; peduncle 1-flowered, 1–2 in. long; spathe of two lanceolate valves $\frac{1}{2}$ in. long, connate in the lower part; perianth with a
short tube, pale yellow inside, and a red-lilac limb $\frac{1}{2}$ in. long,
the outer segments with three purple lines down the back; stamens
half as long as the perianth-limb; anthers yellow, overtopping the
stigmas. *Baker, Handb. Irid.* 103. *Trichonema longifolium, Salisb.
in Trans. Hort. Soc.* i. 316. *T. cruciatum, Ker in Bot. Mag.* t. 575;
excl. syn. *Jacq.*

SOUTH AFRICA : without locality.

I have described this from the *Bot. Mag.* figure and an unpublished drawing of
Dean Herbert's. "There are several varieties in colour, varying from dark
reddish-purple to pale pink."

15. **R. gracillima** (Baker, Handb. Irid. 103); corm globose, $\frac{1}{4}$ in.
diam., with pale brown, membranous tunics; leaves 3–4, very slender,
setaceous, 1-nerved, 3–6 in. long, $\frac{1}{4}$ lin. broad; peduncle slender,
erect, 1-flowered, $1\frac{1}{2}$–2 in. long; spathe of two green, lanceolate
valves $\frac{1}{4}-\frac{3}{8}$ in. long; perianth infundibuliform, pale red, $\frac{1}{2}$ in. long,
with a tube as long as the three-nerved segments, which are $\frac{1}{8}$ in.
broad; anthers just exserted from the perianth-tube. *Trichonema
cruciatum, a, Herb. Drège.*

COAST REGION : Drakenstein Mountains, *Drège!*

16. **R. rosea** (Ecklon, Top. Verz. 19); corm globose, $\frac{1}{3}-\frac{1}{2}$ in.
diam., with smooth, brown, crustaceous tunics; root-leaves suberect,
linear-setaceous, reaching a foot or more in length, $\frac{1}{2}$ lin. broad, 3–5-
ribbed; peduncle 1-flowered, 2–4 in. long, or forked at the base,
with 2–3 pedicels 1–3 in. long; outer valve of the spathe lanceolate,
firm in texture, $\frac{3}{4}$ in. long, inner smaller, brown, scariose; perianth
with a short, funnel-shaped tube, and a red-lilac limb about an inch
long, with a yellow throat, and oblanceolate-oblong segments $\frac{1}{4}-\frac{1}{3}$ in.
broad, with three faint stripes on the back of the outer; anthers yellow,
$\frac{1}{4}$ in. long, exceeding the filaments; stigmas overtopping the anthers;
capsule subglobose, $\frac{1}{2}$ in. long. *Baker in Journ. Linn. Soc.* xvi. 88;
Handb. Irid. 103. *I. rosea, Linn. Syst.* edit. 12, ii. 75; *Ait. Hort.
Kew.* i. 56. *Trichonema roseum, Ker in Konig and Sims' Ann.* i. 223;

Bot. Mag. t. 1225; *Gen. Irid.* 81; *Klatt in Linnæa,* xxxiv. 663. *Crocus capensis, Burm. Prodr.* 2, in part. *R. vulgaris, Eckl. Top. Verz.* 18. *R. reflexa, Eckl. Top. Verz.* 18. *Ixia fugax, Hornem. Hort. Hafn.* i. 50. *R. rosea,* var. *Celsii, Planch. in Flore des Serres,* t. 799. *R. Celsii, Klatt, Ergänz.* 66. (*Bulbocodium pedunculis nudis uni-floris foliis subulatis linearibus, Mill. Ic.* 160, t. 140.)

VAR. β, R. pudica (Baker in Journ. Linn. Soc. xvi. 89; perianth with a red lilac limb an inch long, with a white throat; segments with a dark purple keel down the face at the bottom. *Ixia pudica, Solander in Herb. Banks!* *Tricho-nema pudicum, Ker in Bot. Mag.* t. 1244; *Gen. Irid.* 82; *Konig and Sims' Ann.* i. 223.

VAR. γ, R. speciosa (Baker in Journ. Linn. Soc. xvi. 89); leaves often 1 lin. diam.; flowers as large as in the type, with a bright yellow throat, the outer segments furnished with 3–5 very distinct purple stripes down the back, the outer ones beautifully feathered on the outer side. *Trichonema speciosum, Ker in Konig and Sims' Ann.* i. 223; *Bot. Mag.* t. 1476. *T. barbatum, Herb. MSS. Romulea barbata, Baker in Journ. Linn. Soc.* xvi. 89. *Ixia Bulbocodium,* var. *flore speciosissimo, Andr. Bot. Rep.* t. 170. *I. cruciata, Jacq. Ic.* t. 290? *Romulea tabularis* and *cruciata, Eckl. Top. Verz.* 18-19.

VAR. δ, R. parviflora (Baker, Handb. Irid. 104); flowers smaller; spathe-valves and perianth segments not more than ½ in. long, the latter ¼ in. broad. *R. parviflora, Eckl. Top. Verz.* 19? *Trichonema recurvum,* b, *Herb. Drège.*

VAR. ε, R. dichotoma (Baker in Journ. Linn. Soc. xvi. 89); peduncle elongated, bearing 3–4 flowers, with a large leaf from the fork. *Trichonema dichotomum, Klatt in Linnæa,* xxxiv. 666. *R. flexuosa, Klatt, Ergänz.* 66. *R. tubata, Klatt, Ergänz.* 67. *Trichonema fragrans, Herb. Zeyher. R. fragrans, Eckl. Top. Verz.* 19?

WESTERN REGION : Little Namaqualand, near Ookiep, *Bolus,* 6620!
COAST REGION : Near Port Elizabeth, *Bolus,* 2239! near Grahamstown, *Mac-Owan,* 246!
EASTERN REGION : Natal; top of Mount Erskine, *Evans,* 373!

Drège 2637a, from hills north of the Olifants River, alt. 2000–3000 feet, is not distinguishable in a dried state. *Ixia cruciata, Jacq. Ic.* t. 290, is not separable by the flowers, but the leaves are drawn as if cross-shaped in horizontal section. *Geissorhiza Zeyheri, Herb. Zeyher* (*Romulea Zeyheri, Eckl. Top. Verz.* 19 ?) differs from the type in having the perianth bright yellow in the lower half.

17. R. elegans (Klatt, Ergänz. 66) ; corm globose, ½ in. diam., with brown crustaceous tunics ; root-leaves setaceous, ½–1 ft. long, erect, 3-nerved, ¼ lin. diam. ; peduncle simple or forked at the base ; pedicels 1–3 in. long ; spathe of two lanceolate valves ½–⅝ in. long ; perianth with a funnel-shaped tube ¼ in. long, orange-yellow inside, and a limb an inch long with oblong acute segments ¼ in. broad, the inner plain cream-coloured, the outer red, with a yellow blotch in the centre ; stamens one-third as long as the perianth-limb ; anthers equalling the styles, exceeding the filaments. *Baker, Handb. Irid.* 103. *R. arenaria, Pappe MSS.,* non *Eckl.*

VAR. β, parviflora (Baker) ; leaves shorter, spreading ; spathe-valves smaller ; and perianth-limb not more than ½ in. long, with lanceolate segments.

COAST REGION : Cape Flats, *Zeyher,* 1602 ! Zwartberg, near Caledon, *Zeyher,* 4043 !

18. R. cuprea (Baker in Journ. Bot. 1876, 236) ; corm globose,

¼ in. diam; leaves 3, setaceous, suberect, 3–4 in. long, with a thick, square midrib; peduncle forked at the base; the pedicels 2–3 in. long, with a linear-setaceous rudimentary leaf overtopping the flowers; spathe of two lanceolate valves, the outer firm in texture, ½ in. long; perianth an inch long, with a very short tube, the divisions yellow on the back, with purple plumose stripes, fulvous on the face in the upper half, yellow in the lower half; stamens equalling the styles, half as long as the limb, the lanceolate anthers equalling the pilose filaments *Baker, Handb. Irid.* 104. *Trichonema cupreum, Herbert MSS.*

SOUTH AFRICA: without locality.

Described from an unpublished drawing of Dean Herbert's. Perhaps only a colour-variety of *R. rosea.*

19. R. arenaria (Eckl. Top. Verz. 18); corm globose, ½ in. diam., with brown crustaceous tunics; peduncle elongated, sometimes a foot long, with a single, long, narrow linear, erect leaf of firm texture below the inflorescence; flowers 2–4, on short, ascending pedicels, subtended by long or short leaves; outer spathe-valve lanceolate, green, firm in texture, ¾ in. long, inner smaller and membranous; perianth ¾–1 in. long, lilac, with a tube half as long as the segments, the three outer of which are distinctly striped with dark purple down the back; styles falling short of the stamens. *Baker in Journ. Linn. Soc.* xvi. 89; *Handb. Irid.* 104. *R. ramosa, Eckl. Top. Verz.* 19. *Trichonema arenarium, Klatt in Linnæa,* xxxiv. 667.

COAST REGION: Piquetberg Div.; Sandhills, Bergvallei, below 1000 ft., *Drège,* 8449! Cape Flats, *Zeyher!* Wynberg, *Ecklon.*

20. R. monadelpha (Baker, Handb. Irid. 104); corm ovoid, with crustaceous brown tunics; leaves about six, setaceous, 4–6 in. long; peduncle short, with about three branches 1½–2 in. long; outer spathe-valve firm in texture, green, lanceolate, an inch long; perianth 1–1¼ in. long, with a very short tube, and oblong segments ¼ in. broad, bright, coppery-red inside, with a yellow throat, the outer segments with 3–5 distinct stripes on a yellowish ground; filaments short, black, united to the top; anthers yellow, lanceolate, ½ in. long; style short, with three falcate branches. *Trichonema monadelphum, Sweet, Hort. Brit.* 399. *Spatalanthus speciosus, Sweet, Brit. Flow. Gard.* t. 300; *Baker in Journ. Linn. Soc.* xvi. 104.

Known only from a plant that flowered in Mr. Colvill's nursery, sent by Mr. Synnot.

Keitia natalensis (Regel. Descr. Pl. Nov. vi. 66), supposed to be a native of Natal, is probably identical with the American *Eleutherine plicata,* Herb. It was described from plants grown at Erfurt by Messrs. Haage and Schmidt.

VIII. BOBARTIA, Ker.

Perianth slit down to the ovary, with equal, spreading, obovate segments, not twisted up spirally as they fade. *Stamens* inserted at the base of the perianth-segments; filaments short, free, flattened; anthers lanceolate, sagittate at the base. *Ovary* turbinate-trigonous, 3-celled; ovules crowded in the cells; style short, triquetrous, with three spreading subulate branches. *Capsule* firm in texture, operculate, globose, or turbinate. *Seeds* small, angular.

Rhizome short-creeping. Stem produced at the base into a globose corm in *B. filiformis* only. Leaves rigid, terete, or ensiform. Clusters of flowers, one, few, or several in a head, usually overtopped by a rigid bract; flowers more than one to a spathe, fugitive, pale yellow.

DISTRIB An endemic Cape genus.

Clusters of flowers single :
   Spathe bracteated at the base ... ... (1) **filiformis.**
   Spathe not bracteated at the base ... ... (2) **macrospatha.**
Clusters few in a head :
   Stems terete :
     Spathe under an inch long ... ... (3) **Burchellii.**
     Spathe 1½–2 in. long ... ... (4) **aphylla.**
   Stems flat ... ... (5) **gladiata.**
Clusters many in a head :
   Stems terete :
     Capsule small, globose ... ... (6) **spathacea.**
     Capsule small, turbinate ... ... (7) **robusta.**
   Stem flat ... ... (8) **anceps.**

1. **B. filiformis** (Ker, Gen. Irid. 30); flowering-stem slender, erect, terete, 1½–2 ft. long, dilated at the base into a small, globose corm, and subtended by 3–4 slender, subterete, rigid, closely-ribbed leaves; clusters single, the main bract very small; spathe cylindrical, 1–1¾ in. long, subtended by 2–3 rigid, adpressed lanceolate bracts; perianth 1–1¼ in. long, the obovate-unguiculate segments nearly ½ in. broad; anthers and style-branches ⅙ in. long, exceeding the filaments; ovary turbinate-trigonous, ¼ in. long. *Klatt in Linnœa*, xxxiv. 555; *Baker in Journ. Linn. Soc.* xvi. 114; *Handb. Irid.* 119. *Morœa filiformis, Linn. fil. Suppl.* 100; *Thunb. Diss.* No. 10, tab. 1, fig. 2; *Prodr.* 11; *Fl. Cap.* i. 269. *Marica filiformis, Ker in Bot. Reg.* sub t. 229; *Eckl. Top. Verz.* 15. *Sisyrinchium filiforme, Spreng. Syst. Veg.* i. 166. *Hecaste filiformis, Soland. in Herb. Banks.*

SOUTH AFRICA : without locality, *Thunberg ! Masson ! Kitching !*
COAST REGION : Muizenberg, near Kalk Bay, *Bolus,* 3330 ! Caledon Div., Banks of Zonder Einde River, near Appelskraal, *Zeyher,* 4061 !

2. **B. macrospatha**(Baker, Handb. Irid. 119); stem terete, ending in a single cluster not subtended by a bract; leaves, none seen; spathe 1½ in. long; one oblong-navicular, obtuse, closely-ribbed, outer valve of firm texture, without any small ones adpressed to its base; perianth an inch long; anthers and style-arms ⅙ in. long, much exceeding the filaments.

COAST REGION : Riversdale Div., about the waterfall at Valley River's Poort, on the Lange Bergen, *Burchell,* 6987 !

3. **B. Burchellii** (Baker, Handb. Irid. 120); flowering stem slender, terete, 12–15 in. long, associated with 2–3 wiry, terete leaves; outer bract subulate, produced little or much beyond the

head of flowers; clusters not more than 2–4 to a head ; spathes cylindrical, under an inch long, subtended by small, rigid, adpressed, lanceolate bracts ; perianth ½ in. long ; stamens one-third as long as the perianth ; capsule smooth, globose, ⅙ in. diam.

COAST REGION: Riversdale Div., on or at the foot of the Lange Bergen, *Burchell*, 7145 !

4. **B. aphylla** (Ker, Gen. Irid. 30) ; flowering-stem wiry, tereto, several foot long, with a straight, rigid bract dilated at the base, equalling or overtopping the head of flowers ; leaves none ; clusters few to a head ; spathes cylindrical, 1½–2 in. long, subtended by rigid, adpressed, lanceolate bracts at the base ; perianth ¾–1 in. long ; anthers lanceolate-sagittate, ⅙ in. long, twice as long as the filaments ; ovary turbinate-trigonous, ¼ in. long ; style-branches subulate, ⅙ in. long. *Baker in Journ. Linn. Soc.* xvi. 114; *Handb. Irid.* 120. *Morœa aphylla*, *Linn. fil. Suppl.* 99 ; *Thunb. Diss.* No. 9, tab. 2, fig. 1 ; *Prodr.* 11 ; *Fl. Cap.* i. 269. *Marica aphylla*, *Ker in Bot. Reg.* sub t. 229 ; *Eckl. Top. Verz.* 15. *Sisyrinchium aphyllum*, *Spreng. Syst. Veg.* i. 166. *Marica gladiata*, c, *Herb. Drège*.

SOUTH AFRICA : without locality, *Thunberg!*
COAST REGION: Worcester Div., *Zeyher!* Drakenstein Mountains, *Drège!* Caledon Div., Donkerhoek, *Burchell*, 7968 ! Grahamstown, *Elliot!*

5. **B. gladiata** (Ker, Gen. Irid. 30) ; stems flattened, 1½–2 ft. long, associated with 3–4 distichous, closely-ribbed, rigid, ensiform leaves 2–3 ft. long ; outer bract rigid, long or short ; clusters 3–6 to a head ; spathes cylindrical, 1½–2 in. long, furnished with 2–3 small, rigid, adpressed, bracts at the base ; perianth 1–1¼ in. long, its oblong-unguiculate segments ½ in. broad ; anthers ⅓ in. long, much exceeding the filaments ; capsule turbinate, coriaceous, obtusely angled, ½ in. long. *Klatt in Linnæa*, xxxiv. 554 ; *Baker in Journ. Linn. Soc.* xvi. 114 ; *Handb. Irid.* 120. *Ixia gladiata*, *Linn. fil. Suppl.* 93. *Morœa gladiata*, *Thunb. Diss.*, No. 8 ; *Prodr.* 11 ; *Fl. Cap.* i. 268. *Marica gladiata*, *Ker in Bot. Reg.* t. 229 ; *Eckl. Top. Verz.* 15. *Sisyrinchium gladiatum*, *Spreng. Syst. Veg.* i. 166.

SOUTH AFRICA : without locality, *Zeyher*, 4062 !
COAST REGION : Table Mountain, *Thunberg!* Devils Mountain, *Bolus*, 3842 ! Wynberg, *Burchell*, 869 !

6. **B. spathacea** (Ker, Gen. Irid. 30) ; flowering-stem terete, 2–3 ft. long, associated with 2–3 rigid terete leaves ; main bract lateral, subulate, much longer than the head of flowers ; clusters 20 or more, in a dense, globose, terminal head ; spathes cylindrical, about an inch long, with 1–2 small ovate-lanceolate, rigid, adpressed bracts at the base ; perianth ½–¾ in. long ; anthers and style-branches ⅙ in. long, exceeding the filaments ; ovary turbinate-trigonous, ¼ in. long ; capsule subglobose, smooth, ⅙ in. diam. *Baker in Journ. Linn. Soc.* xvi. 114 ; *Handb. Irid.* 120. *Bobartia indica*, *Linn. Amoen. Acad.* i. 387, *ex parte*. *Morœa spathacea*, *Thunb. Diss.*, No. 11, tab. 1, fig. 1 ; *Prodr.* 11 ; *Fl. Cap.* i. 270 ; *Lam. Ill.* i. 114, tab. 31, fig. 2.

Marica spathacea, Ker in Bot. Reg. sub. t. 229 ; *Eckl. Top. Verz.*
15. *Aristea spathacea, Spreng. Syst. Veg.* i. 158. *Sisyrinchium*
spathaceum, Pers. Syn. i. 50. *Xyris altissima, Lodd. Bot. Cab.* t.
1900.

COAST REGION : near Cape Town, *Thunberg! Bolus,* 3805 ! Simons Bay,
Milne, 143 ! *MacGillivray,* 407 ! Near Rondebosch, *Burchell,* 224 ! Paarl
Mountains, *Drège!* Swellendam Div. ; between Sparbosch and Tradouw, *Drège,*
8305a! Riversdale Div., *Burchell,* 6601 ! Mossel Bay Div., *Burchell,* 6170 !
Humansdorp Div., *Burchell,* 4811 ! Albany Div., *Burchell,* 3489 ! *Cooper,*
1526 !

A form gathered by Villette has a rugose capsule. A plant sent to Herb.
Kew, by Wallich, with fewer clusters in a head, spathes 1½ in. long, and a larger
turbinate capsule, is likely a distinct species.

7. B. robusta (Baker, Handb. Irid. 120); habit of *B. aphylla.*
but stem much more robust ; main bract subulate, rigid, 4–6 in.
long, much dilated at the base ; clusters 10 or more, aggregated in a
dense head ; spathes 1–1½ in. long ; perianth an inch long ; stamens
one third as long as the perianth ; ovary turbinate, ⅙ in. long;
capsule turbinate, ½ in. long, the rigid valves very rugose on the back.

COAST REGION : George Div., between Malgat River and Great Brak River,
Burchell, 6124 !

8. B. anceps (Baker, Handb. Irid. 121); stems much flattened,
1½–2 ft. long, associated with 3–4 rigid, ensiform, closely-ribbed
leaves ; outer bract rigid, linear, a little overtopping the cluster of
flowers, clusters 6–10 to a head ; spathes an inch long, subtended by
2–3 rigid, adpressed, lanceolate bracts at the base ; perianth ½ in.
long. Style-branches ⅙ in. long; anthers ⅛ in. long, exceeding the
filaments.

COAST REGION : Riversdale Div., between Little Vet River and Kampsche
Berg, *Burchell,* 6913 !

IX. WITSENIA, Thunb.

Perianth with a long tube gradually dilated upwards and six
short ovate-lanceolate, connivent segments, the three outer pilose on
the back. *Stamens* inserted at the throat of the perianth-tube ;
filaments short, flat ; anthers small, lanceolate-sagittate. *Ovary*
oblong, 3-celled, crowned with an annular gland at the base of the
long filiform style, which is minutely tricuspidate at the stigmatose
tip. *Capsule* small, turbinate, glabrous, hard in texture, loculicidally
3-valved. *Seeds* few, angled.

DISTRIB. An endemic monotypic genus.

1. W. maura (Thunb. Diss. Nov. Gen. ii. 34, cum icone) ; stems
woody, 2–4 ft. long; branches woody, strongly ancipitous, densely
leafy up to the top ; leaves distichous, ensiform, acuminate,
amplexicaul, firm in texture, ascending, ½ ft. long, closely ribbed ;
flowers in one or several congested oblong terminal heads, formed of

closely imbricated, rigid, acute bracts; the outer smaller and sterile; the inner 1–1½ in. long, containing a shorter obtuse brown spathe in their axil, subtending a single flower; perianth-tube 2 in. long, brownish downwards, blue-black upwards, glabrous, ¼ in. diam. at the throat, segments ¼ in. long, the outer clothed on the back with yellow-brown tomentum, the inner pilose at the tip only; stamens shorter than the perianth-segments; style not exserted; capsule glossy, castaneous, ½–¾ in. long. *Thunb. Prodr.* 7; *Vahl. Enum.* ii. 47; *Red. Lil.* t. 245 and 463; *Lam. Ill.* i. 108, t. 30; *Ker in Konig and Sims' Ann.* i. 237; *Gen. Irid.* 8; *Ait. Hort. Kew.* edit. 2, i. 109; *Bot. Reg.* t. 5; *Maund. Bot.* t. 125; *Flore des Serres,* t. 72. *Baker, Handb. Irid.* 146; *Paxt. Mag.* viii. 221, *cum icone; Reich. Exot.* t. 23. *W. tomentosa, Salisb. in Trans. Hort. Soc.* i. 312. *Antholyza maura, Linn. Mant.* edit. 2, 175. *Ixia disticha, Lam. Encycl.* iii. 333.

COAST REGION : Cape Peninsula; Cape Point and Smitswinkel Bay, *Mac-Owan,* 2620! Noordhoek, *Thunberg!* Hottentots Holland Mountains, *Zeyher,* 3954! Swellendam Div., Tradouw Mountains, *Bowie!*

X. CLEANTHE, Salisb.

Perianth cut down very nearly to the ovary into six spreading, obovate-obtuse segments, the three inner larger than the three outer. *Stamens* inserted at the base of the perianth-segments; filaments short, free; anthers lanceolate. *Ovary* cylindrical-trigonous, 3-celled; ovules crowded in the cells; style cylindrical, with small, spreading, cuneate stigmas. *Capsule* cylindrical-trigonous. *Seeds* small, turgid.

DISTRIB. Endemic and monotypic.

1. C. bicolor (Salisb. in Trans. Hort. Soc. i. 312); rhizome short-creeping, cylindrical, sending out dense tufts of wiry root-fibres; leaves in a dense, basal, distichous rosette, surrounded by bristles, linear, rigid in texture, glabrous, closely-ribbed, 3–6 in. long; stems ½–1 ft. long, bearing only 1–2 rudimentary leaves; clusters terminal, single or few, arranged in a lax corymb, 2–3-flowered; outer spathe-valves lanceolate, acute, green, with a hyaline edge; pedicels ½–1 in. long; perianth fugitive, twisting up spirally, blue, inner segments entirely blue, 1–1¼ in. long, outer black, with a pale claw, not more than half as long as the inner; style ½ in. long, overtopping the stamens; capsule 1½–2 in. long. *Cleanthe melaleuca, Baker in Journ. Linn. Soc.* xvi. 112; *Baker, Handb. Irid.* 137. *Moræa melaleuca, Thunb. Diss.* No. 1, tab. 1, fig. 3; *Prodr.* 10; *Fl. Cap.* i. 261. *Aristea melaleuca, Ker in Konig and Sims' Ann.* i. 236; *Bot. Mag.* t. 1277; *Klatt in Linnæa,* xxxiv. 553. *Moræa lugens, Linn. fil. Suppl.* 90; *Ait. Hort. Kew,* i. 75.

COAST REGION : at the foot of Paarl Mountains, and on, and around Paarde Berg, *Thunberg!*

The plant distributed as *Sisyrinchium melaleucum* by Ecklon and Zeyher is not this, but a form of *Aristea spiralis.*

XI. ARISTEA, Soland.

Perianth rotate, with a short cylindrical tube and six subequal, patent, obovate segments, twisting up spirally after flowering. *Stamens* inserted at the throat of the perianth-tube ; filaments short, free ; anthers erect, linear-oblong. *Ovary* oblong or clavate, 3-celled; ovules crowded in the cells; style filiform, with three very short, spreading, flattened branches, stigmatose round the edges. *Capsule* rigid in texture, loculicidally 3-valved, oblong or cylindrical. *Seeds* small, globose or angled by pressure.

Rootstock never bulbous. Herbs with the leaves in a distichous basal rosette, rarely undershrubs, with leaves not condensed. Leaves firm in texture, closely-ribbed, linear or ensiform. Inflorescence very various ; flowers blue, clustered, appearing in succession ; clusters spicate, racemose or corymbose. Outer bracts wholly or partially firm in texture ; inner membranous, brown or white, often lacerated.

DISTRIB. Four outlying species in Tropical Africa, and several in Central Madagascar.

Subgenus ARISTEA PROPER. Herbs with leaves in a dense distichous basal rosette. Perianth-tube very short.
 Capsule oblong, obtusely angled :
  Clusters of flowers solitary terminal :
   Stem leafless ...   . (1) anceps.
   Stem with 2–3 bract leaves ...   . (2) montana.
  Clusters of flowers spicate or racemose :
   Clusters of flowers small   ...   . (3) torulosa.
   Clusters of flowers large   ...   . (4) schizolæna.
  Clusters many laxly panicled :
   Stems broadly winged :
    Bracts of lower branches small   ...   ... (5) compressa.
    Bracts of lower branches large   ...   ... (6) platycaulis.
   Stems only narrowly winged :
    Leaves very narrow   ...   . (7) majubensis.
    Leaves ¼–½ in. broad   ...   . (8) flexicaulis.
 Capsule oblong, acutely angled :
  Clusters of flowers racemose :
   Spathes ⅓ in. long   ...   . (9) racemosa.
   Spathes ½–¾ in. long ...   . (10) juncifolia.
  Clusters of flowers 1–2 :
   Leaves subterete   ...   . (11) Zeyheri.
   Leaves linear ...   . (12) oligocephala.
  Clusters of flowers corymbose :
   Spathe-valves not lacerated :
    Capsule distinctly peduncled   ...   . (13) dichotoma.
    Spathe-valves very much lacerated ...   . (14) cyanea.
  Clusters of flowers panicled :
   Many upper clusters sessile ...   . (15) capitata.
   All the clusters peduncled   ...   . (16) paniculata.
 Capsule cylindrical-trigonous :
  Clusters of flowers spicate :
   Flowers small ...   . (17) pusilla.
   Flowers large ...   . (18) spiralis.
  Clusters in a lax corymb ...   . (19) Eckloni.
Subgenus NIVENIA, *Vent.* Stems woody. Leaves not condensed into a basal rosette. Perianth-tube longer, cylindrical.
  Cluster of flowers solitary, on a short peduncle...   ... (20) fruticosa.
  Clusters of flowers many, in a dense corymb, with a
   long peduncle ...   . (21) corymbosa.

1. A. anceps (Eckl., ex Klatt in Linnæa, xxxiv. 550); tufts densely cæspitose; root-fibres slender, wiry; leaves all in a distichous basal rosette, linear, firm in texture, 4–8 in. long, $\frac{1}{12}$–$\frac{1}{8}$ in. broad; stems simple, $\frac{1}{2}$–1 ft. long, leafless, strongly ancipitous throughout; cluster single, few-flowered, usually overtopped by a rigid linear bract; inner spathe-valves entirely membranous, brown, $\frac{1}{3}$–$\frac{1}{2}$ in. long, slightly lacerated; perianth-segments oblong, dark blue, $\frac{1}{3}$ in. long; anthers oblong, equalling the filaments; style half as long as the limb; cap-sule oblong, obtusely angled, nearly sessile, torulose, $\frac{1}{4}$ in. long. *Baker in Journ. Linn. Soc.* xvi. 111; *Handb. Irid.* 140; *Klatt, Ergänz.* 47. *Marica bermudiana, Thunb., Fl. Cap.* edit. *Schult.* 69, *ex parte. A. Thunbergii, Pappe MSS.*

COAST REGION: Grahamstown, *Burchell*, 3562! Fort Beaufort Div., *Cooper*, 3213!

CENTRAL REGION: Albany Div.; Zwartwater Poort, *Burchell*, 3435! near Riebeek, *Burchell*, 3474!

KALAHARI REGION: Transvaal; Houtbosch, *Rehmann*, 5769!

EASTERN REGION: Griqualand East; Clydesdale, *Tyson*, 3070! Natal, *Sanderson*, 339! 386! 457! *Cooper*, 3214! *Wood*, 735!

This is, in Thunberg's Herbarium, confused with *A. spiralis* and American specimens of *Sisyrinchium anceps.*

2. A. montana (Baker); root-leaves linear, falcate, much shorter than the stem; stem flexuose, 1-headed, flattened, and distinctly winged, bearing 2–3 large bract leaves; head solitary, terminal; pedicels very short; inner spathe-valves brown, membranous, lacerated; outer firm, ovate; perianth blue, $\frac{1}{4}$ in. long; segments oblong; capsule oblong, obtusely angled.

EASTERN REGION: Natal; Amajuba Mountain, 6000–7000 ft., *Wood*, 4768!

3. A. torulosa (Klatt, Ergänz. 48); rhizome short, cylindrical; root-fibres very slender; leaves of the radical rosette linear, moderately firm in texture, 6–9 in. long, $\frac{1}{8}$–$\frac{1}{6}$ in. broad; stems erect, $1\frac{1}{2}$–2 ft. long, ancipitous in the lower half, with several small, erect, sheathing leaves, slender and subterete upwards, as is the axis of the inflor-escence; clusters several, small, arranged in a lax raceme, the upper ones sessile, the lower on short, alternate, erecto-patent peduncles; spathes $\frac{1}{3}$ in. long; outer valves with a firm centre, and broad, white, slightly-lacerated, membranous edge; inner entirely membranous; perianth-segments $\frac{1}{4}$ in. long; capsule small, oblong, nearly sessile, obtusely angled. *Baker, Handb. Irid.* 141.

VAR. β, **monostachya** (Baker). Leaves smaller, clusters of flowers smaller, arranged in a lax simple spike.

KALAHARI REGION: Transvaal; Houtbosch, *Rehmann*, 5770! near Barberton, 4500 feet, *Galpin*, 1209!

EASTERN REGION: Kaffraria, *Drège*, 4558! 4559! *Baur*, 447 partly! Natal, *Krauss!* *Wood*, 751! Var. β, Natal, *Buchanan!*

4. A. schizolæna (Harv. MSS.; Baker in Journ. Bot. 1876, 267); rhizome very short, as thick as a man's finger, erect; leaves of the radical rosette linear, moderately firm in texture, closely and finely

ribbed, a foot or more long, $\frac{1}{4}$–$\frac{1}{2}$ in. broad; stems stout, erect,
1$\frac{1}{2}$–2 ft. long, strongly ancipitous in the lower half, furnished with
several erect sheathing leaves, terete upwards, as is the axis of the
inflorescence; clusters several, large, globose, sessile, arranged in a lax
spike, each subtended by an ovate cuspidate bract of rigid texture;
spathes $\frac{1}{3}$ in. long; outer valves with a firm centre and brownish
margin, not lacerated; inner entirely membranous; perianth-limb
$\frac{1}{3}$ in. long; capsule oblong, subsessile, obtusely angled, $\frac{1}{4}$–$\frac{1}{3}$ in. long.
Baker in Journ. Linn. Soc. xvi. 111; *Handb. Irid.* 141; *Klatt,
Ergänz.* 47.

COAST REGION: Grahamstown, *MacOwan*, 195!
KALAHARI REGION: Transvaal; Saddleback Range, near Barberton, 4000–
5000 ft., *Galpin*, 1015!
EASTERN REGION: Griqualand East; near Clydesdale, *Tyson*, 2874! Natal,
Wood, 738!

5. **A. compressa** (Buching. ex Krauss in Flora 1845, 309, nomen);
rhizome cylindrical; fibres short, wiry; leaves of radical rosette
linear, moderately firm in texture, strongly ribbed, 6–12 in. long,
$\frac{1}{4}$–$\frac{1}{3}$ in. broad; stem 1–1$\frac{1}{2}$ ft. long, moderately stout, compressed and
ancipitous throughout, furnished with 2–3 sheathing short leaves;
clusters numerous, small, arranged in a lax deltoid panicle, with
alternate erecto-patent branches, all sessile, or one or two of the
lower peduncled; spathes $\frac{1}{4}$ in. long; outer valves with a green
centre and white-brown edge; inner entirely membranous, not
lacerated on the margin; perianth-limb $\frac{1}{4}$ in. long, segments oblong;
stamens half as long as the limb; anthers small, oblong; capsule
oblong, nearly sessile, black, glabrous, obtusely angled, $\frac{1}{4}$ in. long.
Baker in Journ. Linn. Soc. xvi. 111; *Handb. Irid.* 141.

EASTERN REGION: Griqualand East: near Clydesdale, *Tyson*, 2872! Natal,
Krauss, 358! *Sanderson*, 376! *Gerrard*, 114! 393!

6. **A. platycaulis** (Baker in Gard. Chron. 1887, i. 732); leaves firm,
ensiform, a foot long, an inch broad; peduncle as long as the leaves,
flat, and broadly winged; inflorescence a deltoid panicle 8–9 in.
long; main axis flattened, $\frac{1}{2}$ in. diam.; clusters dense, lateral,
sessile; spathe-valves small, lanceolate; perianth $\frac{1}{4}$ in. long; capsule
small, oblong, obtusely angled. *Handb. Irid.* 142.

EASTERN REGION: Coast of Pondoland.
Described from a plant flowered in April, 1887, by Mr. J. H. Tillett, at Sprowston,
near Norwich.

7. **A. majubensis** (Baker in Journ. Bot. 1891, 70); basal leaves
3–4, rigid, linear, 3–4 in. long; stem slender, obscurely ancipitous,
6–8 in. long, bearing 1–2 rudimentary leaves; panicle 1–2 in. long,
composed of 3–4 clusters on short, ascending peduncles; outer
spathe-valve oblong, $\frac{1}{6}$ in. long, with a brown centre and membranous
edge; perianth bright blue, $\frac{1}{6}$ in. long; capsule small, oblong, obtusely
angled, subsessile. *Handb. Irid.* 142. *A. Cooperi, Baker, Handb.
Irid.* 143.

KALAHARI REGION: Orange Free State, *Cooper*, 3212! Transvaal; near Barberton, 4500 ft., *Galpin*, 1210!

EASTERN REGION: Natal; Majuba, Inkwelo Mountain, *Elliot*, 1628! Tugela Valley, *Allison!*

8. **A. flexicaulis** (Baker, Handb. Irid. 143); basal leaves linear, moderately firm, a foot or more long, $\frac{1}{6}$–$\frac{1}{4}$ in. broad; peduncle as long as the leaves, subterete upwards; inflorescence a rhomboid panicle, 6–9 in. long, with a flexuose subterete axis and several ascending branches; lateral clusters sessile; outer spathe-valves $\frac{1}{4}$ in. long, ovate, green, with a membranous edge; inner not lacerated; perianth $\frac{1}{4}$ in. long; capsule subglobose, nearly sessile, $\frac{1}{6}$ in. long.

EASTERN REGION: Natal; near Howick, *Mudd*.

9. **A. racemosa** (Baker in Journ. Bot. 1876, 267); clusters densely tufted; radical leaves slender, linear-subulate, very rigid in texture, a foot or more long; stems about a foot long, slender, subterete, flexuose, with 3–4 small sheathing leaves; clusters several, arranged in a lax raceme, the upper ones sessile, the lower on short, ascending peduncles subtended by lanceolate, rigid bracts; spathes oblong, $\frac{1}{3}$ in. long, the outer valves with firm centres and narrow, brownish-white borders not lacerated, the inner entirely membranous; perianth-limb $\frac{1}{3}$ in. long; capsule oblong, acutely angled, $\frac{1}{2}$ in. long, on a short pedicel. *Baker in Journ. Linn. Soc.* xvi. 111; *Handb. Irid.* 143; *Klatt, Ergänz.* 47. *Witsenia spicata, E. Meyer in Herb. Drège.*

COAST REGION: Dutoits Kloof, *Drège!* Baviaans Kloof, near Genadendal, *Burchell*, 7883!

10. **A. juncifolia** (Baker in Journ. Bot. 1876, 267); tufts densely cæspitose; root-fibres long and wiry; leaves of radical rosette linear-subulate, 5–6 in. long, very rigid in texture, almost spinose at the tip, $\frac{1}{2}$–1 lin. broad; stems $\frac{1}{2}$–1 ft. long, subterete throughout, flexuose, furnished with 2–3 small sheathing leaves; clusters few, oblong, arranged in a short, simple raceme; bracts large, rigid, oblong or oblong-lanceolate, navicular, finely ribbed; spathes oblong, $\frac{1}{2}$–$\frac{3}{4}$ in. long; outer valves with a firm middle, and narrow, membranous, brown border, not lacerated; perianth-limb $\frac{1}{2}$ in. long; capsule oblong, acutely angled, $\frac{1}{2}$ in. long, with a peduncle as long as itself. *Baker in Journ. Linn. Soc.* xvi. 111; *Handb. Irid.* 143; *Klatt, Ergänz.* 47.

SOUTH AFRICA: without locality, *Thom.* 1005! *Grey!*
COAST REGION: Cape Peninsula; Muizenberg, 1500 ft., *Bolus*, 4626!

11. **A. Zeyheri** (Baker, Handb. Irid. 143); leaves slender, rigid, subterete, 4–6 in. long; peduncle slender, terete, 6–9 in. long, bearing 1–2 reduced leaves; clusters 1–2, the lateral one sessile, outer spathe-valves ovate, $\frac{1}{3}$–$\frac{1}{2}$ in. long, green, with a narrow membranous tip; inner lacerated; perianth $\frac{1}{2}$ in. long; ovary cylindrical, $\frac{1}{3}$ in. long, shortly pedicellate; mature capsule not seen.

COAST REGION: Hottentotts Holland Mountains, *Zeyher*, 4050!
May be identical with *A. juncea*, Eckl. MSS.

12. A. oligocephala (Baker, Handb. Irid. 144); leaves of the radical rosette linear, firm in texture, $\frac{1}{2}$ ft long, $\frac{1}{8}$ in. broad; stems erect, 6–9 in. long, slightly compressed; clusters of flowers 1–2, large, terminal; spathes $\frac{1}{2}$–$\frac{5}{8}$ in. long; outer valves oblong, firm in the centre, with a narrow brownish-white, obscurely lacerated, membranous border, inner entirely membranous; perianth-limb $\frac{1}{2}$ inch long; capsule oblong, acutely angled, $\frac{1}{3}$ in. long, with a pedicel of half its own length.

COAST REGION: Caledon Div.; Babylons Tower Mountains, *Zeyher*, 4049! *Pappe!*

13. A. dichotoma (Ker, Gen. Irid. 13); rootstock short-creeping, slender; root-fibres slender, wiry; leaves of the radical rosette narrow, linear, firm in texture, glaucous, 6–9 in. long, $\frac{1}{12}$ in. broad, closely ribbed; stems a foot or more long, slender, flexuose, forked low down, rarely simple, terete; clusters several, arranged in a lax corymb, the ascending long peduncles ancipitous at the top; outer bract small, rigid, lanceolate; spathes $\frac{1}{3}$ in. long; outer valves with a firm keel and broad, white, membranous border, not lacerated; perianth-limb $\frac{1}{2}$ in. long; capsule oblong, very acutely angled, $\frac{1}{4}$–$\frac{1}{3}$ in. long, on a short pedicel. *Baker in Journ. Linn. Soc.* xvi. 111; *Handb. Irid.* 144. *Morœa dichotoma, Thunb., Fl. Cap.* i. 266; edit. *Schult.* 69. *A. intermedia, Eckl. Top. Verz.* 16; *Klatt in Linnœa*, xxxiv. 549. *A. glauca, Klatt, Ergänz.* 47. *A bracteata, Zeyher in Linnœa*, xx. 231.

COAST REGION: Clanwilliam Div., *Drège*, 8337! *Mader in Herb. MacOwan*, 2179! vicinity of Cape Town, *Burchell*, 758! *Ecklon*, 67! *Bolus*, 2803! *Zeyher*, 1639! Muizenberg, *Bolus*, 4712! Caledon Div.; by the Zonder Einde River, *Burchell*, 7539!

A. bracteata, Zeyher, *Exsic.* 1639, *non* Pers., is a dwarf form with often only one cluster of flowers to a peduncle.

14. A. cyanea (Soland. in Ait. Hort. Kew. i. 67); clusters densely tufted; root-fibres long, slender, wiry; leaves of the radical rosette linear, moderately firm in texture, 3–6 in. long, $\frac{1}{12}$–$\frac{1}{8}$ in. broad, finely closely ribbed; stems 2–6 in. long, subterete; clusters of flowers 1–4, globose, laxly corymbose; spathes $\frac{1}{3}$ in. long, outer valves with a firm centre and brown membranous, conspicuously fimbriated margin; inner entirely membranous; perianth limb $\frac{1}{2}$ in. long; capsule oblong, acutely angled, $\frac{1}{3}$ in. long, with a very short pedicel. *Andr. Bot. Rep.* t. 10; *Red. Lil.* t. 462; *Curt. in Bot. Mag.* t. 458; *Klatt in Linnœa*, xxxiv. 548; *Baker, Handb. Irid.* 144. *Ixia africana, Linn. Sp. Plant.* 51; *Burm. Prodr. Cap.* 1. *Morœa africana, Murr. Syst. Veg.* xiv. 95; *Thunb. Diss. Morœa*, no. 3; *Prodr.* 10; *Fl. Cap.* i. 264. *Morœa Aristea, Lam. Ill.* 114; *Poir. Ency.* iv. 276. *A. eriophora, Pers. Syn.* i. 41; *Eckl. Top. Verz.* 17.

COAST REGION: near Paarl, *Drège!* vicinity of Cape Town, *Ecklon*, 58b! *Thunberg! Zeyher*, 4047! *Burchell*, 144! 727! *Bolus*, 2802! Swellendam Div., *Zeyher*, 4047b!

15. A. capitata (Ker in Bot. Mag. t. 605); rhizome short-creeping, as thick as a man's finger; root-fibres long, wiry; leaves of the radical rosette linear, very rigid in texture, closely ribbed, 2–4 ft. long, ⅓–½ in. broad; stems stout, terete, 3–4 ft. long including the inflorescence, with a few reduced, erect, sheathing leaves; clusters numerous, arranged in a long narrow panicle, sessile or the lower shortly peduncled; spathe-valves ½ in. long, outer with a firm centre and brownish-white membranous margin, not lacerated; inner entirely membranous, brownish white; perianth-limb ½ in. long; capsule oblong, rigid, very acutely angled, ¾–1 in. long, with a short pedicel. *Ker in Konig and Sims' Ann.* i. 236; *Gen. Irid.* 12; *Klatt in Linnæa*, xxxiv. 550; *Baker, Handb. Irid.* 144. *Gladiolus capitatus, Linn. Sp. Plant.* edit 2, 53; *Burm. Prodr. Cap.* 2. *A. major, Andr. Bot. Rep.* t. 160. *Moræa cærulea, Thunb. Diss.* No. 15, tab. 2, fig. 2; *Prodr.* 11; *Fl. Cap.* i. 277. *A. cærulea, Vahl. Enum.* ii. 124. *A. spicata, Pers. Syn* i. 41. *A. bracteata, Pers. Syn.* i. 41.

COAST REGION: Paarl Mountains, *Drège!* Table Mountain, *Bolus*, 4584! 4667! Wynberg, *Burchell*, 759! 865! mountains near Swellendam, *Burchell*, 7416! Riversdale Div., *Burchell*, 6879! George Div.; between Outeniqua and Langkloof, *Thunberg!* mountains north of George, *Burchell*, 6011!

An allied plant from the province of Clanwilliam, *Mader*, 195, with narrower leaves and a few large clusters in a simple spike, will likely prove a distinct species.

16. A. paniculata (Baker, Handb. Irid. 144); leaves of the radical rosette just like those of *A. capitata*; stems stout, terete, erect, 3–5 ft. long, including the inflorescence; clusters many, arranged in a lax deltoid panicle, with erecto-patent branches, all on short erecto-patent peduncles; spathes ⅓ in. long; outer valves with a firm centre and broad brownish-white membranous border, not lacerated; inner entirely membranous; perianth-limb ½ in. long; ovary clavate, shortly peduncled; capsule not seen.

COAST REGION: Uitenhage Div.; mountains near Vanstadens River, *MacOwan*, 2077!
EASTERN REGION: Natal; on the Drakensberg, 6000-7000 ft., *Evans*, 356!

17. A. pusilla (Ker in Konig and Sims' Ann. i. 236); rhizome oblique, cylindrical, with very slender, wiry, root-fibres; leaves of the basal rosette linear, firm in texture, ½ ft. long, ⅙–¼ in. broad; stem ancipitous throughout, ½–1 ft. long, flexuose, with 2–3 reduced leaves; clusters few, arranged in a lax simple spike, each subtended by a large lanceolate-navicular bract of firm texture; spathe-valves quite hidden by the bract; perianth-limb bright blue, ½ in. long; capsule cylindrical-trigonous, torulose, 1 in. long, shortly peduncled. *Ker in Bot. Mag.* t. 1231; *Gen. Irid.* 13; *Spreng. Syst. Veg.* i. 158; *Klatt in Linnæa*, xxxiv. 552; *Baker, Handb. Irid.* 145. *Moræa pusilla, Thunb. Diss.* No. 4; *Prodr.* 11; *Fl. Cap.* i. 265; *Vahl Enum.* ii. 154. *Sisyrinchium pusillum, Ecklon herb.*

SOUTH AFRICA: without locality, *Thunberg!*
COAST REGION: Swellendam Div., *Zeyher*, 4056! Uitenhage Div.; Vanstadens Berg, *Zeyher*, 343! 4055! Bathurst Div., *Burchell*, 3713! 3954!

18. A. spiralis (Ker in Konig and Sims' Ann. i. 236); rhizome oblique, cylindrical; root-fibres slender, wiry; leaves of the basal rosette linear, rigid in texture, acuminate, 6–9 in. long, $\frac{1}{4}$–$\frac{1}{3}$ in. broad; stem simple, ancipitous throughout, 2–3 ft. long, with several reduced sheathing leaves; clusters few, arranged in a very lax simple spike, each subtended by a large lanceolate acuminate bract; spathe-valves lanceolate, brown, membranous, $\frac{1}{2}$–$\frac{3}{4}$ in. long, hidden inside the large bract; perianth-limb 1–1$\frac{1}{4}$ in. long, segments oblong, $\frac{1}{2}$–$\frac{3}{4}$ in. broad, whitish, the outer furnished with a broad greenish-black keel; capsule cylindrical-trigonous, 2–2$\frac{1}{2}$ in. long, with a short pedicel. *Vahl Enum.* ii. 124; *Ait. Hort. Kew.* edit 2, i. 109; *Pers. Syn.* i. 41; *Klatt in Linnæa*, xxxiv. 552; *Baker, Handb. Irid.* 145. *Moræa spiralis, Linn. fil. Suppl.* 99; *Thunb. Diss.* No. 2; *Prodr.* 10; *Ker in Bot. Mag.* t. 520. *Sisyrinchium spirale* and *S. melaleucum, Eckl. Top. Verz.* 16.

SOUTH AFRICA: without locality, *Thunberg!*
COAST REGION: Table Mountain, *Bolus*, 4703! Wynberg, *Elliot*, 1085! Drakenstein Mountains, *Drège*, 8333! Robertson Div.; Reit Kuil, *Zeyher*, 4058! Riversdale Div., *Burchell*, 6612! 6710! Knysna Forest, *Bolus*, 2477!

19. A. Eckloni (Baker in Journ. Linn. Soc. xvi. 112); rootstock short, thick, oblique; root-fibres slender, wiry; leaves of the basal rosette linear, not rigid in texture, 6–18 in. long, $\frac{1}{3}$–$\frac{1}{2}$ in. broad; stems strongly ancipitous throughout, a foot or more long; clusters many, arranged in a lax corymbose panicle with ascending branches; spathe-valves small, lanceolate, brown, entirely membranous, not lacerated; perianth-limb bright blue, $\frac{1}{2}$ in. long; capsule cylindrical-trigonous, $\frac{1}{2}$–$\frac{3}{4}$ in. long, torulose, distinctly peduncled. *Baker, Handb. Irid.* 144. *A. dichotoma, Eckl. ex Klatt in Linnæa*, xxxiv. 551, *non Ker.*

COAST REGION: Uitenhage Div.; Zwartkop River and Vaustadens Berg, *Ecklon*, 235! near Uitenhage, *Burchell*, 4256! Near Grahamstown, *Galpin*, 331! *MacOwan*, 1207!
EASTERN REGION: Kaffraria, *Cooper*, 346! Natal; Inanda, *Wood*, 192!

20. A. fruticosa (Pers. Syn. i. 41); a very dwarf, much-branched undershrub; branches slender, ascending, rough downwards with the scars of fallen leaves, and above furnished with spaced out, distichous, rigid, erecto-patent, linear, amplexicaul leaves, much smaller than in *A. corymbosa*, 1$\frac{1}{2}$–2 in. long; flowers in a single, oblong, terminal cluster on a short peduncle; spathe-valves rigid in texture, lanceolate, acute, nearly an inch long; perianth with a cylindrical tube $\frac{1}{2}$ in. long, and a blue limb with oblong obtuse segments, shorter than the tube. *Baker, Handb. Irid.* 145. *Ixia fruticosa, Thunb. Diss.* No. 1; *Lam. Ill.* i. 108. tab. 31, fig. 4. *Witsenia fruticosa, Ker in Konig and Sims' Ann.* i. 237; *Gen. Irid.* 8. *W. ramosa, Vahl Enum.* ii. 47. *Thunb. Fl. Cap.* i. 256. *W. capitata, Klatt in Linnæa*, xxxiv. 546. *Nivenia fruticosa, Baker in Journ. Linn. Soc.* xvi. 109; *Klatt, Ergänz.* 48.

COAST REGION: Caledon Div.; Mountains of Nieuw Kloof, *Burchell*, 8096!

Swellendam Div., Tradouw Mountains, *Bowie!* Riversdale Div.; near the summit of Kampsche Berg, *Burchell*, 7100!

21. A. corymbosa (Benth. Gen. Plant. iii. 701); stems shrubby, erect, much-branched, ancipitous, branches rough with scars of fallen leaves; leaves spaced out along the stems, rigid, linear, distichous, amplexicaul, erecto-patent, 4–6 in. long, ⅙ in. broad, clusters very numerous, arranged in a dense corymb, with a long ancipitous erect peduncle; spathes ¼ in. long; outer bracts rigid, obtuse; inner membranous; perianth with a cylindrical tube exserted from the spathe, and bright blue limb ¼ in. long, with oblong segments; capsule rigid, oblong, ⅙ in. long. *Baker, Handb. Irid.* 145. *Witsenia corymbosa, Ker in Bot. Mag.* t. 895; *Gen. Irid.* 8 ; *Smith, Exot. Bot.* ii. 17, t. 68 ; *Red. Lil.* t. 453 ; *Lodd. Bot. Cab.* t. 254 ; *Reich. Exot.* t. 24 ; *Paxt. Mag.* iii. 269, *cum icone. Nivenia corymbosa, Baker in Journ. Linn. Soc.* xvi. 109. *N. stylosa, Salisb. in Trans. Hort. Soc.* i. 311. *N. binata, Klatt, Ergänz.* 48.

SOUTH AFRICA: without locality, *Cooper*, 3175!
COAST REGION: near Tulbagh, *Drège*, 1986 ! near Bains Kloof, *Bolus*, 4068! Caledon Div.; Appels Kraal, *Zeyher!* Oudtshoorn Div.; on the Great Zwarte Bergen, *Drège*, 2184!

Undescribed Species.

Wredowia pulchra, *Eckl. Top. Verz.* 16, from hills between Hemel and Aarde, in the division of Caledon, is said to have bright cinnabar-red flowers, and to be intermediate between *Sisyrinchium* (i.e. *A. spiralis*) and the typical *Aristeas.*

XII. KLATTIA, Baker.

Perianth blue, not twisted up spirally, with a short cylindrical tube and six very long subequal segments, with a small lanceolate blade and a long filiform claw. *Stamens* inserted at the throat of the perianth-tube; filaments long, free, filiform; anthers linear, sagittate at the base. *Ovary* turbinate, 3-celled; ovules few in a cell; style filiform, minutely tricuspidate at the stigmatose apex. *Capsule* turbinate, loculicidally 3-valved. *Seeds* one or few in a cell, angular or compressed.

DISTRIB. An endemic monotypic genus.

1. K. partita (Baker in Journ. Linn. Soc. xvi. 110); stems woody, much-branched, erect, 1–2 ft. long; branches ancipitous, leafy up to the top; leaves crowded, alternate, ensiform, amplexicaul, ascending, 6–9 in. long, firm in texture, acuminate, closely ribbed; flowers bright blue, 10–15 aggregated in a dense oblong terminal head, subtended by large lanceolate, rigid bracts; spathes 1–2-flowered, with lanceolate brown valves ¾–1 in. long. Perianth-tube cylindrical, ¼–⅓ in. long; segments 2–2½ in. long, with a lanceolate blade ½ in. long; stamens and style falling a little short of the perianth-segments; capsule ¼ in. long. *Baker, Handb. Irid.* 146. *Witsenia* *partita, Ker in Konig and Sims' Ann.* i. 237 ; *Gen. Irid.* 8 ; *Spreng. Syst. Veg.* i. 147, excl. syn. ; *Klatt in Linnæa,* xxxiv. 546.

COAST REGION : Swellendam Div. ; Tradouw Mountains, *Bowie!* on the Langeberg near Swellendam, *Burchell,* 7340! 7418! Riversdale Div. ; near Kampsche Berg, *Burchell,* 7050 ! 7159 !

XIII. SCHIZOSTYLIS, Backh. and Harv.

Perianth hypocrateriform, with a cylindrical tube and a campanulate limb with six equal, oblong, acute, petaloid segments. *Stamens* inserted at the throat of the perianth-tube ; filaments free, filiform ; anthers large, linear, basifixed. *Ovary* clavate, 3-celled ; ovules crowded in the cells ; style as long as the perianth-tube, with three long, spreading, subulate branches. *Capsule* obovoid-oblong, membranous, obtuse. *Seeds* small, angled.

Herbs with a few distichous grass-like leaves ; rootstock not bulbous ; flowers arranged in a lax simple equilateral spike ; spathe-valves large, lanceolate, green.

DISTRIB. An endemic Cape genus, differs only from *Hesperantha* by the rootstock not being bulbous.

Flowers deep crimson   ...   .:   ...   ...   (1) coccinea.
Flowers pale red ...   .   (2) pauciflora.

1. S. coccinea (Backh. and Harv. in Bot. Mag. t. 5422) ; root-fibres densely tufted, rather fleshy ; leaves of basal rosette 2–3, linear, glabrous, grass-like in texture, 1–1½ ft. long, ¼–½ in. broad, furnished with a distinct midrib ; stems slender, terete 1–2 ft. long, bearing 2–3 erect, sheathing, reduced leaves ; flowers 6–8, arranged in a lax distichous spike ; outer spathe-valve entirely herbaceous and green, lanceolate, 1–1¼ in. long ; inner rather smaller and more membranous ; perianth-tube straight, erect, 1–1¼ in. long ; limb deep crimson, as long as the tube, the oblong acute segments uniform in colour and texture ; anthers ⅛ in. long, equalling the filaments ; style-branches ¾ in. long ; capsule sessile, obtuse, ½ in. long. *Baker in Journ. Linn. Soc.* xvi. 108 ; *Handb. Irid.* 147 ; *Klatt, Ergänz.* 48.

COAST REGION : Stockenstrom Div., *Scully,* 14 ! *Elliot,* 400 !
KALAHARI REGION : Transvaal, *Atherstone!* Rhenoster Poort River, *Nelson,* 401 ! Lydenburg district, *Roe in Hb. Bolus,* 2653 !
EASTERN REGION : Kaffraria, hills up to 4000 ft.; Bazeia Mountains, *Baur,* 164 ! sources of Buffels River, *Murray in Herb. MacOwan,* 1892! Fakus territory, 5000 ft., *Sutherland!* Griqualand East, 5000 ft., *Tyson,* 1181 ! Natal, *Gerrard,* 1528 ! Swaziland, 3000-4500 ft., *Galpin,* 726 !

2. S. pauciflora (Klatt in Linnæa, xxxv. 380) ; root-fibres slender ; basal leaves 2–3, linear, moderately firm in texture, 1 foot long, ¼ in. broad ; stems erect, terete, 2–3 ft. long, bearing a few reduced, erect, sheathing leaves ; flowers 2–8, in a lax simple equilateral spike ; outer spathe-valve green, lanceolate, ¾–1 in. long ; inner rather smaller ; perianth-tube 1–1½ in. long ; limb purplish-pink, an inch long, with equal, membranous, oblong, acute segments ; stamens

and style branches as in the other species. *Klatt, Ergänz.* 48;
Baker in Journ. Linn. Soc. xvi. 108 ; *Handb. Irid.* 147. *S.
ixioides, MSS., Harv. ex Baker in Journ. Linn. Soc.* xvi. 108.

KALAHARI REGION : Orange Free State, *Cooper*, 1197! 3217! Transvaal,
Sanderson !
EASTERN REGION : Natal, *Sutherland !*

XIV. HESPERANTHA, Ker.

Perianth rotate, with a cylindrical tube and six subequal, oblong,
spreading segments. *Stamens* inserted at the throat of the perianth-
tube, divaricating ; filaments short, free ; anthers lanceolate. *Ovary*
3-celled ; ovules crowded in the cells ; style as long as the perianth
tube, with three long, subulate, falcate, entire branches. *Capsule*
small, turbinate, membranous, loculicidally 3-valved. *Seeds* small,
globose, or angled by pressure.

Rootstock a small tunicated corm, flat at the base ; leaves few, narrow,
distichous ; flowers small, arranged in lax spikes ; spathe-valves herbaceous
in texture, lanceolate, usually about as long as the perianth-tube. Only
differs from *Geissorhiza* by its longer style-branches and longer green spathe-
valves.

DISTRIB. Besides the Cape species there is one in Abyssinia and one in the
Cameroon Mountains.

Perianth-tube straight :
  Perianth-segments ¼–½ in. long :
    Inner segments white ; outer red outside :
      Leaves short, spreading :
        Leaves much crisped ... ...   (1) cinnamomea.
        Leaves not crisped :
          Stems short ... ...   (2) montana.
          Stems long ... ...   (3) falcata.
      Leaves long, erect :
        Leaves glabrous :
          Leaves linear-subulate ... ...   (4) flexuosa.
          Leaves linear ... ...   (5) graminifolia.
        Leaves pilose, linear ... ...   (6) pilosa.
      Inner perianth-segments yellow ; outer red
        outside ... ...   (7) lutea.
  Perianth-segments uniform, white :
    Perianth-tube shorter than the spathe ... ...   (8) namaquensis.
    Perianth-tube as long as the spathe ... ...   (9) leucantha.
  Perianth-segments uniform, lilac or reddish :
    Leaves short, lanceolate ... ...   (10) cucullata.
    Leaves long, linear :
      Perianth-tube half as long as the spathe ...   (11) gracilis.
      Perianth-tube as long as the spathe :
        Leaves three :
          Corm-tunics produced into long
            fibres ... ...   (12) fibrosa.
          Corm-tunics produced into short
            cusps ... ...   (13) erecta.
        Leaves two :
          Perianth-tube rather longer than
            the spathe ... ...   (14) modesta.

Perianth-segments $\frac{1}{2}$–1 in. long ... ... (15) **subexserta.**
Perianth-tube not longer than the spathe:
Acaulescent ... ... (16) **humilis.**
Stems elongated:
Flowers white:
Leaves rigid... ... ... (17) **lactea.**
Leaves grass-like ... ... (18) **candida.**
Flowers bright red:
Perianth-limb $\frac{1}{2}$ inch long... ... (19) **Baurii.**
Perianth-limb $\frac{3}{4}$–1 in. long:
Outer spathe-valve oblong-lanceolate ... ... (20) **pulchra.**
Outer spathe-valve lanceolate-acuminate ... ... (21) **Woodii.**
Perianth-tube longer than the spathe ... ... (22) **longituba.**
Perianth-tube curved:
Outer perianth-segments red ... ... (23) **radiata.**
Perianth-segments all white:
Leaves without bulbillæ in their axils:
Perianth-tube much shorter than the spathe (24) **angusta.**
Perianth-tube as long as the spathe... ... (25) **Tysoni.**
Leaves with bulbillæ in their axils ... ... (26) **bulbifera.**

1. H. cinnamomea (Ker in Konig and Sims' Ann. i. 225); corm conical, $\frac{1}{3}$–$\frac{1}{2}$ in. diam.; tunics dark brown, membranous; basal leaves 2, lanceolate, falcate, glabrous, firm in texture, obtuse, much crisped, 2–3 in. long, $\frac{1}{4}$–$\frac{1}{3}$ in. broad; stem slender, terete, simple, 3–9 in. long, with 1–2 sheathing, erect, strongly ribbed leaves at the base; flowers 3–12 in a short, often secund spike; outer spathe-valve oblong, obtuse, green, herbaceous, $\frac{1}{6}$ in. long; perianth-tube cylindrical, straight, or rather curved, equalling or a little exceeding the spathe segments oblong-lanceolate, patent, $\frac{1}{4}$ in. long, the three outer usually claret-red, and the three inner white; anthers $\frac{1}{8}$ in. long, 2–3 times as long as the filaments; style-branches shorter than the anthers. *Ker in Bot. Mag.* t. 1054; *Gen. Irid.* 91; *Ait. Hort. Kew.* edit. 2, i. 84; *Klatt in Linnæa,* xxxiv. 650; *Ergänz.* 60; *Baker in Journ. Linn. Soc.* xvi. 95: *Baker, Handb. Irid.* 148. *Ixia cinnamomea, Linn. fil. Suppl.* 92; *Thunb. Diss.* No. 9, *cum icone*; *Fl. Cap.* i. 227; *Vahl Enum.* ii. 56.

SOUTH AFRICA: without locality, *Pappe! Thom! Burchell!*

COAST REGION: Lions Rump, *Thunberg!* Rosebank, near Cape Town, *Bolus,* 3768!

2. H. montana (Klatt, Ergänz. 59); leaves lanceolate, obtuse, strongly ribbed, the lower spreading $1\frac{1}{2}$ in. long, $\frac{1}{4}$ in. broad, the upper narrower and longer, sheathing the stem; stems simple or forked, 3–4 in. long; flowers 1–3 on a flexuose rachis; outer spathe-valve ovate, acute, $\frac{1}{2}$–$\frac{3}{4}$ in. long; perianth with a straight tube $\frac{1}{3}$–$\frac{1}{2}$ in. long, and ovate, acute segments of the same length, the three inner white, the outer purplish. *Baker, Handb. Irid.* 148.

CENTRAL REGION: Calvinia Div.; Hantam Mountains, *Meyer.*

3. H. falcata (Ker in Konig and Sims' Ann. i. 225); corm conic, $\frac{1}{3}$–$\frac{1}{2}$ in. diam., with crustaceous black or brown tunics, with small cusps at the top; basal leaves 2–4, lanceolate, falcate, glabrous, moderately firm in texture, not crisped, 2–3 in. long, $\frac{1}{4}$–$\frac{1}{3}$ in. broad; stems slender, terete, $\frac{1}{2}$–1 ft. long, simple or forked, with 1–2 sheathing, erect, linear or lanceolate leaves below the middle; flowers 2–10 in a lax equilateral erect spike; outer spathe-valve oblong, obtuse, $\frac{1}{4}$–$\frac{1}{3}$ in. long, green, with a brown scariose edge; perianth with a cylindrical tube as long as or a little longer than the spathe, segments oblong, spreading, $\frac{1}{3}$–$\frac{1}{2}$ in. long, the three inner white, the outer claret-red; anthers lanceolate, $\frac{1}{4}$ in. long, three times as long as the filaments; style-branches $\frac{1}{8}$ in. long. *Ker in Bot. Mag.* sub t. 1254; *Gen. Irid.* 90; *Ait. Hort. Kew.* edit. 2, i. 84. *Klatt in Linnæa,* xxxiv. 647; *Ergänz.* 59; *Baker in Journ. Linn. Soc.* xvi. 96; *Handb. Irid.* 148. *Ixia falcata, Linn. fil. Suppl.* 92; *Thunb. Diss.* No. 23; *Prodr.* 10; *Fl. Cap.* i. 249; *Jacq. Ic.* t. 276; *Bot. Mag.* t. 566. *Ixia cinnamomea, Andr. Bot. Rep.* t. 44, *non Thunb.*

COAST REGION: vicinity of Cape Town, *Thunberg! Drège!* near Simons Town, *Bolus,* 4690! Paarl Div., *Drège!* Dutoits Kloof, *Drège!* Riversdale Div.; on the Langebergen, *Burchell,* 7045!

4. **H. flexuosa** (Klatt, Ergänz. 60); corm globose, squamose; scales castaneous, cuspidate at the apex; leaves narrow linear, superposed, $\frac{1}{4}$ lin. broad; stem flexuose, glabrous, forked, 9–10 in. long; flowers in a short flexuose spike; outer spathe-valve oblong, $\frac{1}{4}$ in. long, green, with a brown scariose margin; perianth-tube cylindrical, reddish, $\frac{1}{4}$ in. long, segments ovate, acute, the outer reddish, the inner white.

WESTERN REGION: Little Namaqualand; near Elbvagfontein, *Drège,* 2639; Nababeep, 3200 ft., *MacOwan* and *Bolus Herb. Norm. Austr. Afr.* 694!

5. **H. graminifolia** (D. Don in Sweet Hort. Brit. edit. 2, 503); corm small, globose, with crustaceous brown tunics; leaves 3–5, linear, glabrous, 4–6 in. long, $\frac{1}{12}$ in. broad, the upper sheathing the lower part of the stem; stems terete, slender, simple or rarely forked, $\frac{1}{2}$–1 ft. long; flowers 2–6 in a lax equilateral spike; outer spathe-valve green, oblong, $\frac{1}{4}$ in. long; perianth with a straight cylindrical tube a little larger than the spathe, segments oblong-lanceolate, spreading, $\frac{1}{4}$–$\frac{1}{3}$ in. long, the inner white, the outer reddish-brown or reddish green outside; anthers and style-branches as in *H. pilosa. Klatt in Linnæa,* xxxiv. 649; *Ergänz.* 59; *Baker in Journ. Linn. Soc.* xvi. 95; *Handb. Irid.* 148. *H. pilosa, var. nuda, Ker in Bot. Mag.* t. 1254; *Gen. Irid.* 90.

COAST REGION: at the foot of Table Mountain, *MacOwan,* 2386!

6. **H. pilosa** (Ker in Konig and Sims' Ann. i. 225); corm globose, $\frac{1}{4}$ in. diam.; tunics brown, crustaceous, with short cusps at the top; basal leaves 2, linear, erect, pilose, strongly ribbed, 3–6 in. long, $\frac{1}{12}$ in. broad; stems slender, erect, terete, $\frac{1}{2}$–1 ft. long, with a single sheathing erect leaf lower down; flowers 2–10 in a lax equilateral, distichous spike; outer spathe-valve lanceolate, acute, green, herbaceous, $\frac{1}{3}$ in. long; perianth with a cylindrical tube $\frac{1}{3}$–$\frac{1}{2}$ in. long, straight or rarely

curved in the lower flowers, segments of the limb oblong-lanceolate,
$\frac{1}{3}-\frac{1}{2}$ in. long, the inner white, the outer tinged outside with claret-
red or green; anthers $\frac{1}{6}$ in. long, much exceeding the filaments;
style-branches $\frac{1}{4}$ in. long. *Ker in Bot. Mag.* t. 1475; *Gen. Irid.* 90,
ex parte; *Klatt in Linnæa,* xxxiv. 648; *Ergänz.* 59; *Baker in Journ.
Linn Soc.* xvi. 95; *Handb. Irid.* 149. *Ixia pilosa, Linn. fil. Suppl.*
92; *Thunb. Diss.* No. 5; *Prodr.* 9; *Fl. Cap.* i. 222; *Vahl Enum.*
ii. 54.

SOUTH AFRICA : without locality, *Zeyher,* 1595! *Masson !*
COAST REGION: Malmesbury Div., near Groene Kloof, *Drège!* vicinity of
Cape Town, *Thunberg ! Ecklon,* 399! *Bolus,* 3767!

7. **H lutea** (Benth. Gen. Plant. iii. 703); corm not seen; leaves
3–4, linear, erect, acute, firm in texture, glabrous, 2–3 in. long,
$\frac{1}{8}-\frac{1}{6}$ in. broad, the upper sheathing the base of the stem; stems
slender, terete, 3–4 in. long, simple or forked low down; flowers
3–6 in a lax simple spike; outer spathe-valve oblong or oblong-
lanceolate, green, with a scariose edge, $\frac{1}{4}-\frac{1}{3}$ in. long; perianth with a
cylindrical tube as long as the spathe, and oblong ascending segments
$\frac{1}{3}$ in. long, the three outer claret-purple on the back; anthers lanceolate, $\frac{1}{6}$ in. long, much exceeding the filaments; style-branches overtopping the anthers. *Baker, Handb. Irid.* 149. *Geissorhiza lutea,
Eckl. Top. Verz.* 21; *Klatt in Linnæa,* xxxiv. 652; *Baker in Journ.
Linn. Soc.* xvi. 95.

COAST REGION : Swellendam Div.; Groote Vadersbosch, *Ecklon;* Caledon
Div., *Zeyher !*

8. **H. namaquensis** (Baker, Handb. Irid. 149); corm globose,
small; leaves 2–3, narrow linear, glabrous, 3–4 in. long; stem slender,
3–4 in. long, simple or branched low down; flowers 5–6 in a lax
distichous spike with a flexuose rachis; outer spathe-valve green,
lanceolate, $\frac{1}{3}-\frac{1}{2}$ in. long; perianth with a straight cylindrical tube
$\frac{1}{4}$ in. long, segments white, oblong, as long as the tube; stamens
reaching to the top of the perianth-segments.

COAST REGION : Little Namaqualand, *Scully,* 50!

9. **H. leucantha** (Baker, Handb. Irid. 150); corm not seen; leaves
3, linear, erect, glabrous, the largest 6–8 in. long; stem simple, very
slender, $\frac{1}{2}$ ft. long; flowers 3–4 in a short distichous spike; outer
spathe-valve oblong, green, $\frac{1}{4}-\frac{1}{3}$ in. long; perianth-tube straight, as
long as the spathe, segments oblong, white, $\frac{1}{3}$ in. long; anthers $\frac{1}{6}$ in.
long; style-branches nearly as long as the perianth-segments.

EASTERN REGION: Natal; Olivers Hoek Pass, *Wood,* 3437!

10. **H. cucullata** (Klatt, Ergänz. 59); corm oblong, squamose;
scales castaneous, equally slit at the top; leaves 3, the upper one
sheathing the stem, lanceolate acute, 2–3 in. long, $\frac{1}{4}-\frac{1}{3}$ in. broad;
stem erect, terete, 3–4 in. long; flowers 3, crowded on a flexuose
rachis; outer spathe-valve ovate, herbaceous, truncate at the apex,

½ in. long; perianth-tube cylindrical, straight, violet, brown at the
throat, $\frac{1}{4}-\frac{1}{3}$ in. long, segments lilac, oblong-lanceolate, under ½ in.
long. *Baker, Handb. Irid.* 149.

CENTRAL REGION : Calvinia Div.; Hantam Mountains, *Meyer.*

11. **H. gracilis** (Baker, Handb. Irid. 149); corm not seen; basal
leaves 3, narrow linear, glabrous, moderately firm in texture, 6–8 in.
long, $\frac{1}{12}$ in. broad, with only a single distinct rib; stem very slender,
terete, 1 ft. long, with a short sheathing leaf at the middle; flowers
2, distant; outer spathe-valve lanceolate, acute, $\frac{1}{2}-\frac{5}{8}$ in. long, firm
and green to the tip; perianth with a greenish, cylindrical, straight
tube ¼ in. long, and a bright red limb ½ in. long, with oblong uniform
segments; stamens half as long as the limb; anthers lanceolate, $\frac{1}{6}$ in.
long; style-branches ¼ in. long.

EASTERN REGION : Natal; in the bush at the base of perpendicular rocks at
Isangwaan, *Wood*, 923 !

Habit and perianth very like that of *Geissorhiza secunda.*

12. **H fibrosa** (Baker, Handb. Irid. 149); corm globose, ¼ in.
diam., with brown crustaceous tunics produced into long fibres at the
top; produced leaves 3, linear, superposed, glabrous, firm in texture,
4–6 in. long, ⅛ in. broad, obtuse, with revolute edges, the upper
sheathing the stem some distance from the base; stems slender,
terete, simple, 6–9 in. long; flowers 3–6 in a lax equilateral spike;
outer spathe-valve green, oblong, obtuse, $\frac{1}{3}-\frac{1}{2}$ in. long; perianth with
a cylindrical tube as long as the spathe, and six oblong concolorous
segments ¼ in. long; anthers $\frac{1}{6}$ in. long, with short filaments; style-
branches ⅛ in. long.

COAST REGION: Caledon Div.; Klein River Berg, *Zeyher,* 3960 !

13. **H. erecta** (Benth. Gen. Plant, iii. 703); corm globose, ¼ in.
diam.; tunics brown, crustaceous; leaves 3, linear, superposed,
glabrous, moderately firm in texture, erect, 3–4 in. long., $\frac{1}{12}-\frac{1}{8}$ in.
broad; stems slender, terete, 6–8 in. long, simple or forked; flowers
4–8, in a short, erect spike, with a very flexuose axis; spathe-valves
green, lanceolate, ⅛ in. long; perianth pale red, with a straight
tube ⅛ in. long, dilated into a funnel at the apex, and oblong,
ascending, concolorous segments not longer than the tube; anthers
$\frac{1}{6}$ in. long, exceeding the filaments; style-branches nearly as long as
the perianth-segments. *Geissorhiza erecta, Baker in Journ. Bot.*
1876, 238; *Journ. Linn. Soc.* xvi. 93.

WESTERN REGION : by the Olifants River, *Drège,* 8468 !

14. **H. modesta** (Baker, Handb. Irid. 150); corm not seen; basal
leaves 2, narrowly linear, erect, glabrous, 2–3 in. long; stem very
slender, ½ ft. long, with a small sheathing leaf at the middle;
flowers 2–4 in a lax distichous spike; outer spathe-valve oblong,
green, $\frac{1}{4}-\frac{1}{3}$ in. long; perianth-tube straight, as long as the spathe,

segments oblanceolate, pink, $\frac{1}{3}$–$\frac{1}{2}$ in. long ; anthers $\frac{1}{6}$ in. long ; style-branches $\frac{1}{4}$ in. long.

EASTERN REGION : Natal ; near Bevaan River, *Wood*, 3201!—not Beevari River, as given in the *Handbook of Irideæ.*

15. **H. subexserta** (Baker) ; corm not seen ; leaves 2, linear, erect, superposed, glabrous, 2–3 in. long ; stems flexuose, usually simple, 6–9 in. long, spike very lax, few-flowered ; outer spathe-valve green, oblong, acute, $\frac{1}{3}$ in. long ; perianth-tube $\frac{1}{2}$ in. long, segments oblong, pinkish, $\frac{1}{3}$ in. long ; anthers linear, large ; style-branches long.

EASTERN REGION : Natal ; in a valley near Bothas Hill, 2000 ft., *Wood*, 4543 !

16. **H. humilis** (Baker in Journ. Bot. 1876, 239) ; corm globose, 1 in. diam., with numerous very thick, rigid, dull brown tunics, lacerated from the base and ending in short cusps ; leaves 3–4, very falcate, linear-oblong, obtuse, firm in texture, glabrous, 1–2 in. long, $\frac{1}{8}$ in. broad, furnished only with a distinct midrib ; stem scarcely any ; flowers 2, remote ; outer spathe-valve ovate-navicular, very firm in texture from base to tip, brownish-green, not ribbed ; perianth with a straight cylindrical tube $\frac{1}{2}$ in. long, suddenly dilated into a distinct funnel at the apex, and ascending obovate segments of the same length ; colour uncertain ; style reaching to the top of the funnel ; branches $\frac{1}{4}$ in. long. *Baker in Journ. Linn. Soc.* xvi. 95 ; *Handb. Irid.* 150 ; *Klatt, Ergänz.* 59.

CENTRAL REGION : Fraserburg Div. ; in the Middle Roggeveld near Jakhals Fontein, *Burchell*, 1320 !

Habit and leaves of *Streptanthera*, but spathe very different.

17. **H. lactea** (Baker, Handb. Irid. 151) ; corm not seen ; leaves 3–4, superposed, linear, glabrous, firm in texture, erect, sheathing the stem, 4–8 in. long, $\frac{1}{12}$–$\frac{1}{8}$ in. broad ; stems slender, erect, terete, simple, 1–1$\frac{1}{2}$ ft. long ; flowers several, arranged in a lax erect equilateral spike with a flexuose rachis ; outer spathe ovate or oblong-lanceolate, green to the top, $\frac{1}{2}$–$\frac{3}{4}$ in. long ; perianth with a straight cylindrical tube $\frac{1}{4}$–$\frac{1}{3}$ in. long, and a yellowish-white limb $\frac{5}{8}$ in. long, with ascending, uniform, oblong-lanceolate segments ; anthers and style-branches half as long as the limb.

EASTERN REGION : Natal ; Durban Flat, *Sanderson*, 240 ! Fields Hill, *Wood*, 243 ! near Verulam, *Wood*, 1118!

May be a colour-variety of *H. Baurii*, with which it agrees in habit, leaves, and inflorescence.

18. **H. candida** (Baker, Handb. Irid. 151) ; corm globose, $\frac{1}{2}$ in. diam., with brown crustaceous tunics ; leaves 3–4, linear, grass-like, glabrous, the lower 6–9 in. long, $\frac{1}{8}$–$\frac{1}{3}$ in. broad, the upper small and stem-sheathing ; stems slender, terete, simple, 4–9 in. long ; flowers one or few, arranged in a lax erect spike ; outer spathe oblong-lanceolate, $\frac{3}{4}$–1 in. long, green, membranous towards the tip ; perianth with a straight cylindrical tube $\frac{1}{2}$ in. long, and a limb

$\frac{5}{8}$–$\frac{3}{4}$ in. long, with spreading, oblong, uniform segments; anthers and style-branches $\frac{1}{4}$ in. long.

VAR. β, **bicolor**, *Baker.* Outer segments of the perianth tinged and ribbed with claret-red on the outside.

KALAHARI REGION: Orange Free State, *Cooper*, 746!
CENTRAL REGION: Var. β, Somerset Div.; Boschberg, 4000 ft., *MacOwan*, 61 partly!

Habit of *H. longituba*, from which it differs by its short perianth-tube.

19. **H. Baurii** (Baker in Journ. Bot. 1876, 182); corm globose, $\frac{1}{2}$ in. diam.; tunics crustaceous, castaneous, ending in short cusps; leaves 3, linear, glabrous, superposed, firm in texture, $\frac{1}{2}$–1 ft. long, $\frac{1}{6}$–$\frac{1}{4}$ in. broad, strongly ribbed, sheathing the lower half of the stem; stems slender, terete, 1–2 ft. long; flowers several, arranged in a very lax equilateral erect spike, with a flexuose rachis; outer spathe-valve entirely green, firm in texture, obtuse, $\frac{1}{2}$–$\frac{3}{4}$ in. long; perianth with a cylindrical straight tube equalling the spathe, and a limb $\frac{1}{2}$ in. long, withuniform bright rose-red, ascending, oblong segments; anthers yellow, lanceolate, $\frac{1}{4}$–$\frac{1}{3}$ in. long, with very short filaments; style-branches filiform, $\frac{1}{4}$ in. long. *Baker in Journ. Linn. Soc.* xvi. 96; *Handb. Irid.* 151; *Klatt, Ergänz.* 60. *H. rubella, Baker in Journ. Bot.* 1876, 239; *Klatt, Ergänz.* 60. *H. disticha, Klatt, Ergänz.* 59.

KALAHARI REGION: Transvaal; summit of Saddleback Range near Barberton, 5000 ft., *Galpin*, 827! *Wood*, 4509! *Thorncroft*, 112! Orange Free State, *Cooper*, 1027!
EASTERN REGION: Kaffraria; between Umtata R. and Umzimvubu R., *Drège*, 4548! Bazeia Mts., *Baur*, 628! Natal, *Buchanan!* grassy hill, Eisdumbini, 1800 ft., *Wood*, 4346!

20. **H. pulchra** (Baker, Handb. Irid. 150); corm not seen; leaves 3–4, all superposed, narrow linear, glabrous, firm in texture, strongly ribbed, a foot or more long, $\frac{1}{12}$ in. broad, with very long sheaths; stems slender, terete, simple, erect, 1$\frac{1}{2}$–2 ft. long; flowers several, arranged in a lax erect equilateral spike; outer spathe-valve oblong-lanceolate, green to the tip, 1–1$\frac{1}{2}$ in. long; perianth with a straight cylindrical tube, $\frac{3}{4}$–1 in. long, and a bright pink limb $\frac{3}{4}$ in. long, with ascending oblong segments; anthers lanceolate, $\frac{1}{8}$ in. long, three times as long as the filaments; style-branches subulate, $\frac{1}{3}$ in. long.

EASTERN REGION: Kaffraria, Bazeia Mts., 2500–3000 ft., *Baur*, 159!

21. **H. Woodii** (Baker, Handb. Irid. 150); corm not seen; produced leaves 2, narrow linear, superposed, firm in texture, glabrous, 1 ft. long, $\frac{1}{12}$ in. broad; stem slender, terete, 1$\frac{1}{2}$ ft. long, with two long, sheathing leaves with short, free, linear points; flowers 3–6, bright mauve-purple, arranged in a very lax erect spike; outer spathe valve green, lanceolate, an inch long; perianth with a slender tube an inch long, and oblong segments nearly as long; anthers

lanceolate, bright yellow, $\frac{1}{3}$ in. long, twice as long as their filaments; style-branches as long as the anthers.

EASTERN REGION: Natal; stony places on the Peak of Byrne, 3500 ft., *Wood*, 1868!

Intermediate between *H. pulchra* and *H. longituba.*

22. H. longituba (Baker in Journ. Linn. Soc. xvi. 96); corm ovoid, $\frac{1}{2}$ in. diam., with black crustaceous tunics, ending in small cusps; leaves generally 3, linear, not rigid in texture, glabrous, 6–9 in. long, $\frac{1}{8}$–$\frac{1}{3}$ in. broad, the upper erect, sheathing the lower part of the stem; stems slender, terete, simple, $\frac{1}{2}$–2 ft. long; flowers few, arranged in a very lax equilateral spike; outer spathe-valve green to the tip, oblong-lanceolate, obtuse, $\frac{3}{4}$–1 in. long; perianth with a cylindrical tube 1–1$\frac{1}{4}$ in. long, and a spreading limb $\frac{5}{8}$–$\frac{3}{4}$ in. long, with oblong segments $\frac{1}{6}$–$\frac{1}{4}$ in. broad, the three inner usually white the outer tinged with claret-red; anthers lanceolate, $\frac{1}{3}$ in. long; style-branches $\frac{1}{3}$ in. long. *Handb. Irid.* 151; *Klatt, Ergänz.* 60. *Geissorhiza longituba, Klatt in Linnæa*, xxxv. 383.

CENTRAL REGION: near Fraserburg, *Burchell*, 1430! near Murraysburg, 4700 ft., *Tyson*, 303! Cave Mountain, near Graaff Reinet, *Bolus*, 686! Somerset Div.; *Bowker!* rocky places near Boschberg, *MacOwan*, 61 partly!
EASTERN REGION: Kaffraria, *Cooper*, 1810! Katberg, *Baur!* between Umzimvubu R. and Umtsikaba R., *Drège*, 4540!

H. acuta, Ker, Gen. Irid. 91 (*Ixia acuta*, Lichten. in Roem. et Schult. Syst. Veg. i. 383) from the Roggeveld, differs, according to the description, by its secund spike and acute perianth-segments.

23. H. radiata (Ker in Konig and Sims' Ann. i. 225); corm globose, $\frac{1}{2}$ in. diam., with thick, rigid, brown-black tunics, lacerated from the base; leaves 5–6, the lower, narrow linear, glabrous, 4–6 in. long, the upper with long sheaths and small free points; stem slender, simple, 1–1$\frac{1}{2}$ ft. long; flowers few or many in a long, secund spike; outer spathe-valve green, oblong-lanceolate, $\frac{1}{2}$–$\frac{3}{4}$ in. long; perianth with a much-curved tube $\frac{1}{2}$ in. long, and a limb $\frac{1}{3}$–$\frac{1}{2}$ in. long, with oblong-lanceolate segments, the inner usually white and the outer claret-red; anthers lanceolate, $\frac{1}{4}$ in. long, with very short filaments; style-branches $\frac{1}{4}$ in. long. *Ker in Bot. Mag.* sub 790 and 1254; *Gen. Irid.* 89; *Ait. Hort. Kew.* edit. 2, i. 84; *Klatt. in Linnæa*, xxxiv. 649; *Baker in Journ. Linn. Soc.* xvi. 96; *Handb. Irid.* 152. *Ixia radiata, Jacq. Ic.* t. 280; *Ker in Bot. Mag.* t. 573; *Red. Lil.* t. 441. *I. fistulosa, Andr. Bot. Rep.* t. 59. *Gladiolus recurvus, Thunb. Diss.* No. 9. *Ixia recurva, Vahl Enum.* ii. 58.

VAR. β, **caricina** (Ker in Bot. Mag. sub t. 573); lower leaves subterete; flowers smaller. *Ker in Bot. Mag.* t. 790. *H. caricina, Klatt, Ergänz.* 61. *H. setacea, Eckl. Top. Verz.* 23.

COAST REGION: Tulbagh, *Burchell*, 1037! near George, *Burchell*, 6086! Stellenbosch, *Sanderson*, 966! Fish River, *Gill*! VAR. β, Vicinity of Cape Town, *Drège! Bolus*. 3769! Swellendam Div., *Zeyher!* Grahamstown, *MacOwan! Galpin*, 283!

CENTRAL REGION: Victoria West; Karee Bergen, *Burchell*, 1548! near
Murraysburg, 4100 ft., *Tyson*, 305! Somerset Div., *Bowker!*
WESTERN REGION: Little Namaqualand, between Pedros Kloof and Lily
Fontein, 3000-4000 ft., *Drège*, 2638!
EASTERN REGION: Natal; Mooi River, 4000 ft., *Wood*, 4C56!

24. **H. angusta** (Ker in Konig and Sims' Ann. i. 225); corm
ovoid, $\frac{1}{4}-\frac{1}{2}$ in. diam; tunics brown, crustaceous, ending at the top in
short cusps; leaves 3-4, linear, glabrous, grass-like, lower 4-6 in.
long, $\frac{1}{12}-\frac{1}{4}$ in. broad, upper sheathing the stem half-way up; stems
slender, terete, simple or forked, $\frac{1}{2}-1$ ft. long; flowers few, arranged
in a lax spike; outer spathe-valve green, oblong, $\frac{1}{2}-\frac{3}{4}$ in. long;
perianth-tube curved, cylindrical, $\frac{1}{4}$ in. long; limb $\frac{1}{3}$ in. long,
with six oblong white segments, reflexed when fully expanded;
anthers lanceolate, $\frac{1}{4}$ in. long, three times as long as the filaments;
style-branches subulate, $\frac{1}{4}$ in. long. *Baker in Journ. Linn. Soc.* xvi.
96; *Handb. Irid.* 152; *Klatt, Ergänz.* 61. *H. radiata, var. angusta,
Ker, Gen. Irid.* 89. *Ixia radiata, var. angusta, Ker in Bot. Mag. sub
t.* 573. *I. angusta, Willd. Sp. Plant.* i. 202. *I. linearis, Jacq. Ic.
t.* 279; *Coll.* iv. 183, *non Thunb. H. virginea, Ker in Konig and
Sims' Ann.* i. 225; *Gen. Irid.* 91. *I. virginea, Soland. MSS. Diasia
iridifolia and D. reflexa, Harvey MSS.*

COAST REGION: Clanwilliam Div., *Zeyher!* Hex River Kloof, *Drège!*
CENTRAL REGION: near Somerset East, *MacOwan*, 345! 345b!
WESTERN REGION: Little Namaqualand, near Ookiep, *Morris in Hb. Bolus,*
5787!

25. **H. Tysoni** (Baker, Handb. Irid. 151); corm not seen; leaves
2-4, narrowly linear, erect, glabrous, $\frac{1}{2}$ ft. long; stem slender, $1\frac{1}{2}$ ft.
long; flowers 5-6 in a lax secund spike; spathe-valves oblong-lanceo-
late, $\frac{1}{2}$ in. long; perianth-tube cylindrical, curved, as long as the
spathe; segments oblong, whitish, $\frac{1}{2}$ in. long; stamens reaching
nearly to the tip of the segments.

EASTERN REGION: Griqualand East; banks of streams near Kokstad, 4700 ft..
Tyson, 1585!

26. **H. bulbifera** (Baker in Journ. Bot. 1876, 183); corm not
seen; leaves about 4, grass-like, linear, glabrous, above a foot long,
flaccid, $\frac{1}{4}-\frac{1}{3}$ in. broad, furnished with bulbillæ in their axils; stems
weak, simple, $1\frac{1}{2}$ ft. long; flowers few, arranged in a lax simple
spike; outer spathe-valve green, oblong-lanceolate, $\frac{3}{4}-1$ in. long;
perianth-tube curved, $\frac{1}{4}$ in. long; limb white, an inch long, with
uniform oblong segments; anthers nearly $\frac{1}{2}$ in. long, twice the length
of the filaments; style-branches $\frac{1}{3}$ in. long. *Baker in Journ. Linn.
Soc.* xvi. 96; *Handb. Irid.* 152; *Klatt, Ergänz.* 61.

CENTRAL REGION: Somerset Div.; Boschberg, in fissures of the cliff at the
cataract, *MacOwan*, 2215!
Perhaps only a form of *H. angusta*, grown in damp shade.

XV. GEISSORHIZA, Ker.

Perianth rotate, with a cylindrical tube and six spreading, subequal,
oblong segments. *Stamens* divaricating, inserted at the throat of the

perianth-tube; filaments short, free, filiform; anthers lanceolate, basifixed. *Ovary* oblong, 3-celled; ovules crowded in the cells; style longer than the perianth-tube, with three short, falcate-subulate, undivided branches. *Capsule* small, oblong, membranous, loculicidally 3-valved. *Seeds* globose or angled by pressure.

Rootstock a small tunicated corm; leaves few, narrow, distichous; flowers arranged in simple or forked lax spikes, very various in colour; outer spathe-valve oblong, obtuse, generally green and herbaceous, with a brown membranous edge.

Very near *Hesperantha*, from which it only differs by its longer style, with shorter branches, and by its more membranous spathe-valves.

DISTRIB. Besides the Cape species, there is only one, with a very different capsule from any of them, in Central Madagascar.

Perianth-tube shorter than the spathe :
  Flowers small; limb rarely above ⅓ in. long :
    Outer spathe-valve entirely green :
      Flower typically yellow; throat concolorous :
        Leaves and stem glabrous :
          Flowers permanently yellow ... (1) **humilis.**
          Flowers turning blue when dried (2) **ornithogaloides**
        Leaves and stem hairy ... .. (3) **flava.**
      Flower bright yellow, with a purple-black
        throat ... ... (4) **purpureo-lutea.**
      Flowers reddish, with a concolorous throat :
        Leaves filiform, long ... ... (5) **filifolia.**
        Leaves filiform, short ... ... (6) **Pappei.**
        Leaves linear ... ... (7) **Wrightii.**
    Outer spathe-valve brown and membranous in
    the upper half :
      Stem and leaves glabrous ... ... (8) **secunda.**
      Stem and leaves hairy :
        Leaves very short ... ... (9) **gracilis.**
        Leaves long ... ... (10) **graminifolia.**
  Flowers larger : limb ½-1 in. long :
    Flowers red or violet-purple :
      Cauline sheath slender ... ... (11) **furva.**
      Cauline sheath loose and ventricose :
        Leaves hairy ... ... (12) **hirta.**
        Leaves glabrous :
          Basal leaves narrow, few-ribbed :
            Perianth-segments with a
              nectary at the base ... (13) **rochensis.**
            Perianth-segments without
              any nectary ... ... (14) **Bellendeni.**
          Basal leaves broad, many-ribbed (15) **latifolia.**
    Flowers yellow, or outer segments flushed with
    red outside :
      Throat of perianth concolorous :
        Basal leaves subterete ... ... (16) **corrugata.**
        Basal leaves linear, with many close
        ribs ... ... (17) **imbricata.**
      Throat of perianth dark purple .. ... (18) **inflexa.**
    Doubtful species . ... (19) **quadrangula.**
Perianth-tube as long as or a little longer than the
spathe :
  Flowers one or few to a spike :
    Perianth-limb ⅓ in. long ... ... (20) **nana.**

Perianth-limb $\frac{1}{3}$ in. long :
  Dwarf, with narrow leaves... ... ... (21) **setacea.**
  Tall, with broader leaves ... ... (22) **bracteata.**
Flowers many to a spike :
  Flower small :
    Leaves concentrated at the base of the stem:
      Perianth-limb $\frac{1}{4}$–$\frac{1}{3}$ in. long ... ... (23) **Dregei.**
      Perianth-limb $\frac{1}{2}$ in. long ... ... (24) **foliosa.**
    Leaves distinctly superposed ... ... (25) **Bolusii.**
  Flower very large ... ... (26) **grandis.**
Perianth-tube much exserted from the spathe :
  Dwarf : perianth-limb shorter than the tube ... (27) **minima.**
  Tall : perianth-limb as long as tube :
    Leaf terete ... ... (28) **geminata.**
    Leaf lanceolate or oblong ... ... (29) **excisa.**

1. **G. humilis** (Ker in Konig and Sims' Ann. i. 224); corm globose, $\frac{1}{4}$–$\frac{1}{3}$ in. diam.; tunics brown, crustaceous, ending in short cusps; leaves 3, linear-subulate, glabrous, 4–6 in. long, $\frac{1}{2}$ lin. broad, thick in texture, with inflexed edges, upper much shorter, sheathing the lower part of the stem and dilated towards the base; stems slender, terete, flexuose, glabrous, 3–6 in. long; flowers few, in a lax spike with a flexuose axis; outer spathe-valve oblong, obtuse, green to the tip, closely ribbed, $\frac{1}{3}$–$\frac{1}{2}$ in. long; perianth-tube straight, cylindrical, $\frac{1}{6}$ in. long; limb bright lemon-yellow, $\frac{1}{3}$–$\frac{1}{2}$ in. long; segments oblong; anthers lanceolate, $\frac{1}{6}$ in. long, exceeding the filaments; style as long as the anthers, the falcate spreading branches $\frac{1}{8}$ in. long. *Ker, Gen. Irid.* 85; *Baker in Journ. Linn. Soc.* xvi. 93; *Handb. Irid.* 153. *Ixia humilis, Thunb. Diss., No.* 4; *Prodr.* 9; *Fl. Cap.* i. 221. *G. Brehmii, Eckl. Top. Verz.* 21; *Klatt in Linnæa* xxxiv. 653. *G. setacea, Ker in Bot. Mag. t.* 1255, *excl. syn. G. plicata, Pappe MSS.*

VAR. β, **grandiflora** (Baker, Handb. Irid. 153); more robust, with broader leaves and larger uniform yellow flowers.

VAR. γ, **bicolor** (Baker, loc. cit.); outer perianth-segments bright red down the centre of the back. *Ixia scillaris, β, Herb. Thunberg!*

VAR. δ, **juncea** (Baker, loc. cit.); stems longer, with two sheathing leaves, the two basal very long, slender, terete; flowers plain yellow. *G. juncea, Link. in Unio Itin. Exsic.* No. 314. *Ixia juncea, Link. Enum.* 50.

SOUTH AFRICA: without locality; Var. β, *Nivén!* Var. γ, *Thunberg!*

COAST REGION: Piquetberg, and near Cape Town, *Thunberg!* between Paarl and Pont, *Drège*, 8475! Cape Flats, *Zeyher*, 3964! near Malmesbury, *Bolus*, 4339! Simons Bay, *Wright*, 253! Var. β, Cape Flats, *Pappe!* near Cape Town, *Bolus*, 4602! Var. γ, Cape Flats, *Zeyher*, 1599! Simons Bay, *Wright*, 265! Muizenberg, *MacOwan*, 2474! *MacOwan and Bolus, Herb. Normale*, 257! Var. δ, by the Berg River near Paarl, *Drège*, 8472! Table Mountain, *Ecklon*, 314!

2. **G. ornithogaloides** (Klatt in Linnæa xxxiv. 656); corm globose, $\frac{1}{6}$–$\frac{1}{4}$ in. diam.; tunics brown, crustaceous, ending in a ring of short bristles; basal leaves 2–3, subsetaceous or narrowly linear, 1-nerved, weak, glabrous, falcate, 2–4 in. long, $\frac{1}{4}$–$\frac{1}{2}$ lin. broad; stems 1-flowered, simple or forked, very slender, with one short leaf with a dilated sheath; spathe-valves oblong, thin, green, $\frac{1}{4}$ in. long; flower blue when dry, but pale yellow when fresh; perianth-tube $\frac{1}{6}$ in. long; segments oblong, obtuse, $\frac{1}{4}$–$\frac{1}{3}$ in. long; anthers lanceolate, reaching

nearly to the tip of the perianth; style much exserted from the tube, with 3 short spreading branches. *Klatt, Ergänz.* 57; *Baker in Journ. Linn. Soc.* xvi. 94; *Handb. Irid.* 153. *G. romuleoides, Eckl. Top. Verz.* 21.

COAST REGION: Zwartberg, near Caledon, *Zeyher*, 3966!

3. **G. flava** (Klatt, Ergänz. 58); corm globose, flat at the base; tunics squamose, lacerated; leaves erect, narrow linear, 1-ribbed, those of the stem with a ventricose sheath; stem erect, hairy, terete, flexuose, branched, 4 in. long; branches 1–2-flowered; outer valve of spathe green, $\frac{1}{4}$ in. long; perianth greenish-yellow; tube cylindrical, $\frac{1}{8}$ in. long; segments obovate, obtuse, $\frac{1}{3}$ in. long; style overtopping the anthers. *Baker, Handb. Irid.* 153. *Waitzia flava, Reich. MSS.*

SOUTH AFRICA: without locality, *Breutel*. Described by Dr. Klatt from a specimen in the Lübeck herbarium, gathered by Breutel, which I have not seen.

4. **G. purpureo-lutea** (Baker in Journ. Bot. 1876, 238); corm ovoid, $\frac{1}{3}$ in. diam., with very thick, lacerated, rigid, dull brown tunics; leaves 3, linear, glabrous, strongly ribbed, thick in texture, the lower 2–3 in. long, $\frac{1}{12}$–$\frac{1}{8}$ in. broad, the upper one clasping the base of the stem, with a short free point; stems simple or forked from the base, slender, glabrous, flexuose, 2–4 in. long; flowers 1–2 to a stem; spathe-valves oblong, green, almost membranous, $\frac{1}{4}$–$\frac{1}{3}$ in. long; perianth-tube straight, purplish-black, $\frac{1}{8}$ in. long, suddenly dilated into a purplish-black throat, $\frac{1}{3}$–$\frac{1}{2}$ in. long; segments oblong, lemon-yellow; anthers yellow, lanceolate, $\frac{1}{6}$ in. long; filaments shorter than the anthers, yellow with a purplish-black base; style overtopping the anthers, with short falcate branches. *Baker in Journ. Linn. Soc.* xvi. 93; *Handb. Irid.* 154; *Klatt, Ergänz.* 56. *Ixia bicolor, a, Herb. Thunb.!*

SOUTH AFRICA: without locality, *Thunberg!*
COAST REGION: Malmesbury Div.; near Mamre, *MacOwan*, 2488! Paarl Div.; between Paarl and Pont, *Drège*, 8476!

5. **G. filifolia** (Baker in Journ. Bot. 1876, 238); corm the size of a pea; leaves 3, distant, filiform, erect, the lower 6–7 in. long, overtopping the flowers, the upper shorter, placed halfway up the stem; stem slender, straight, glabrous, $\frac{1}{2}$ ft. long; flowers 3, distant; outer spathe-valve green, lanceolate, $\frac{1}{4}$ in. long; perianth-tube cylindrical, $\frac{1}{8}$ in. long; limb with a purple throat $\frac{1}{3}$ in. long, and oblong, whitish segments; stamens one-third shorter than the perianth-limb. *Baker in Journ. Linn. Soc.* xvi. 94; *Handb. Irid.* 154; *Klatt, Ergänz.* 58.

SOUTH AFRICA: without locality, *Prior!* in the British Museum.

6. **G. Pappei** (Baker, Handb. Irid. 154); corm globose, $\frac{1}{6}$ in. diam., with brown membranous tunics and a bristly neck; stem very slender, filiform, glabrous, 2–3 in. long, with two superposed, erect, terete leaves 1–1$\frac{1}{2}$ in. long, dilated towards the base; flowers 3–4 in a short spike with a flexuose rachis; outer spathe-valve oblong, $\frac{1}{6}$ in. long, green with a hyaline edge; perianth reddish, concolorous; tube cylindrical, $\frac{1}{8}$ in. long; segments oblong, under $\frac{1}{4}$ in. long;

anthers slightly shorter than perianth-segments. *G. pusilla, Pappe MSS., no Klatt.*

COAST REGION : Zonder Einde Mountains, *Pappe! Zeyher,* 3965!

7. **G. Wrightii** (Baker in Journ. Bot. 1876, 238); corm globose, ⅓ in. diam.; tunics crustaceous, castaneous, ending in short cusps; basal leaves 2–3, narrow linear, glabrous, 6–9 in. long, ½–1½ lin. broad, closely and deeply ribbed; stems terete, ½ ft. long, simple or forked at the base, glabrous, sheathed at the base by a long erect leaf dilated downwards to a breadth of ¼–⅓ in. with 10–15 prominent close ribs; flowers 6–8 in a lax spike with a flexuose rachis; outer spathe-valve ovate-oblong, ¼–⅓ in. long, green to the tips; perianth-tube cylindrical, ⅛ in. long; limb dull red, ½ in. long, with uniform oblong segments; anthers lanceolate, ⅙ in. long, equalling the filaments; style overtopping the anthers, with 3 falcate branches, ⅛ in. long. *Baker in Journ. Linn. Soc.* xvi. 94; *Handb. Irid.* 154; *Klatt, Ergänz.* 58.

COAST REGION : Simons Bay, *Wright,* 243!

8. **G. secunda** (Ker in Konig and Sims' Ann. i. 224); corm globose, ¼–⅓ in. diam., with bright brown, crustaceous tunics; basal leaves 2, linear, glabrous, 4–6 in. long, ⅛–⅙ in. broad, not strongly or closely ribbed; stems slender, ½–1 ft. long, with 1–2 superposed leaves with strongly-ribbed ventricose sheaths; flowers 3–6 in a lax spike with a very slender flexuose rachis; outer spathe-valve oblong, ¼–⅓ in. long, green in the lower half, brown and scariose in the upper; perianth-tube cylindrical, ⅛ in. long; limb bright red, ½ in. long, with uniform oblong segments; anthers lanceolate, ⅙ in. long, exceeding the filaments; style reaching to the top of the anthers, the falcate branches ⅛ in. long. *Ker, Gen. Irid.* 85; *Ait. Hort. Kew, edit.* 2, i. 83; *Baker in Journ. Linn. Soc.* xvi. 93; *Handb. Irid.* 154; *Klatt in Linnæa* xxxiv. 658. *Ixia secunda, Delaroche Descr.* 17, *excl. syn.*; *Berg. Pl. Cap.* 6; *Thunb. Diss. No.* 7; *Prodr.* 9; *Fl. Cap.* i. 224; *Bot. Mag. t.* 597. *I. pusilla, Andr. Bot. Rep. t.* 245. *I. scillaris, a, Herb. Thunberg.*

VAR. β, **G. setifolia** (Eckl. Top. Verz. 22); leaves very slender, setaceous, flowers red. *Klatt, Ergänz.* 58. *G. secunda, var. setifolia, Baker in Journ. Linn. Soc.* xvi. 94; *Handb. Irid.* 155.

VAR. γ, **G. ramosa** (Klatt in Linnæa xxxiv. 657); taller than the type, with deeply forked stems with several flowers. *G. imbricata, E. Meyer in herb. Drège.*

VAR. δ, **G. pusilla** (Klatt, Ergänz. 58, ex parte); flowers pale rose or nearly white. *G. secunda, var. Ker in Bot. Mag. t.* 1105. *Ixia scillaris, Linn. herb.!*

VAR. ε, **bicolor** (Baker, Handb. Irid. 155); stems tall, simple; leaves very long, filiform; flowers smaller than in the type, white, the outer segments bright red down the keel.

SOUTH AFRICA : without locality. *Var. ε, Rogers!*
COAST REGION : Malmesbury Div.; Zwartland, *Thunberg!* near Groene Kloof, *Thunberg! Bolus,* 4338! Tulbagh, *Thom!* Paarl mountains, *Drège,* 8483a! between Paarl and Pont, *Drège,* 8483b and c! plains and mountains near Cape Town, *Ecklon,* 315! *Bolus,* 2804! 3321! 4804! near Humansdorp, *Tyson,* 306,! Var. β, near Groene Kloof, *MacOwan,* 2165! between Paarl and Pont, *Drège,* 8460! near Tulbagh Cataract, *Bolus,* 5388! Lions Rump, *Pappe!* near Ronde-

bosch, *Bolus*, 4609! Witkamp, *Zeyher*, 1598! Var. γ, Clanwilliam Div.; *Mader in Herb. MacOwan*, 2165! Witsenberg and Skurfdeberg, near Tulbagh, *Zeyher*, 1596! Dutoits Kloof, *Drège*, 8481a! mountains near Mitchell's Pass, *Bolus*, 5246! Genadendal, *Drège!* Riversdale Div.; on the Lange Bergen, near Kampsche Berg, *Burchell*, 7044!

9. **G. gracilis** (Baker, Handb. Irid. 155); corm not seen; stem ½–1 ft. long, simple, with a rudimentary leaf at the base and a tight hairy sheath at the middle, produced into a small, erect, rigid, linear, strongly-nerved hairy blade; flowers 2–3 in a lax spike with a flexuose rachis; outer spathe-valve oblong, ⅓ in. long, brown and membranous above the base; perianth pale lilac; tube very short; segments oblong, ½ in. long; anthers lanceolate, ⅙ in. long, as long as the filaments; style-branches short.

EASTERN REGION: Griqualand East; Zuur Berg Range, 5000 ft., *Tyson*, 1872!

10. **G. graminifolia** (Baker, Handb. Irid. 155); corm globose, ½ in. diam.; tunics entire, crustaceous, brown or squamose, cut up from below into slices; leaves 3, linear, the lower 6–9 in. long, ⅛–⅙ in. broad, firm in texture, pilose on the ribs, especially beneath, the upper much shorter, ventricose, sheathing the stem some distance above the base, with many close, prominent hairy ribs; stems slender, flexuose, simple, or forked, ½–1 ft. long; flowers few, arranged in a lax spike with a flexuose rachis; spathe-valves oblong, ¼–⅓ in. long, greenish towards the base, brown and membranous in the upper half; perianth deep or pale red, or nearly white; tube very short, cylindrical; segments oblong, ⅓–½ in. long; anthers lanceolate, ⅙ in. long, longer than the filaments; style reaching to the top of the anthers, with falcate branches ⅛ in. long. *G. hirta, Baker in Journ. Linn. Soc.* xvi. 94, *excl. syn. Hesperantha quinquangularis, Klatt, Ergänz.* 61, *excl. syn. H. rosea, Klatt, Ergänz.* 61?

VAR. β, **bicolor** (Baker, Handb. Irid. 155); flowers white, bright red on the back of the outer segments. *Ixia humilis*, δ, *Herb. Thunberg. G. pilosa, Pappe MSS.* Intermediate between *G. hirta* and *G. secunda.*

SOUTH AFRICA: without locality, *Thunberg.*

COAST REGION: near Swellendam, *Drège*, 3496! near George, *Burchell*, 6088! Var. β, near Cape Town, *Bolus*, 4805! *Pappe!* Hout Bay, *MacOwan*, 2649! and *MacOwan and Bolus, Herb. Norm.*, 261! Hottentots Holland, *Ecklon and Zeyher*, 214!

11. **G. furva** (Ker in Konig and Sims' Ann. i. 224); corm very small, globose; leaves filiform, 2 basal, 3–4 in. long, erect; cauline one, with a close sheath, and short, free point; stem slender, flexuose, 3–4 in. long, 1-flowered; outer spathe-valve oblong, ⅓ in. long, brown and membranous in the upper half; perianth bright red-purple; tube cylindrical, ⅛ in. long; limb ½–⅝ in. long, with oblong segments; anthers lanceolate, longer than the filaments; style reaching to the top of the filaments. *Ker, Gen. Irid.* 87; *Baker in Journ. Linn. Soc.* xvi. 94; *Handb. Irid.* 155. *Ixia furva, Solander MSS.*

SOUTH AFRICA: without locality, *Masson!*

COAST REGION: Between Paarl and Pont, *Drège*, 8478! Tulbagh, *Ecklon and Zeyher*, 217! Malmesbury Div.; near Groene Kloof, *Bolus*, 4341!

12. G. hirta (Ker in Konig and Sims' Ann. i. 224); corm globose, $\frac{1}{2}$ in. diam.; tunics dull brown, crustaceous, cut up into slices; leaves 3, linear, hairy, moderately firm in texture, the outer basal, 4–6 in. long, $\frac{1}{4}$ in. broad, the upper sheathing the lower part of the stem for 4–6 in., with a short, free point; sheath very ventricose, hairy and closely ribbed; stems simple or forked, slender, angular, $\frac{1}{2}$–1 ft. long; flowers 2–6, arranged in a lax spike with a flexuose rachis; outer spathe-valve oblong, $\frac{1}{2}$ in. long, brown and membranous almost to the base; perianth bright red, concolorous; tube very short, cylindrical, $\frac{1}{8}$ in. long; limb $\frac{3}{4}$–1 in. long, with obtuse uniform segments; stamens half as long as the perianth-limb; anthers lanceolate, $\frac{1}{4}$ in. long, equalling the filaments; style reaching to the top of the anthers; branches falcate, $\frac{1}{6}$ in. long. *Ker, Gen. Irid.* 86; *Baker, Handb. Irid.* 155. *Ixia hirta, Thunb. Diss. No.* 6; *Prodr.* 9; *Fl. Cap.* i. 223. *G. ciliaris, Salisb. in Trans. Hort. Soc.* i. 321! *Baker in Journ. Linn. Soc.* xvi. 94. *G. rosea, Eckl. Top. Verz.* 20. *Hesperantha kermesina, Klatt, Ergänz.* 61.

VAR. β, **G. quinquangularis** (Eckl. in Unio Itin No. 312); flowers a uniform pale rose-red. *Hesperantha quinquangularis, Eckl. Top. Verz.* 23. *H. ciliata, Klatt, Ergänz.* 60.

SOUTH AFRICA: without locality, *Thunberg! Auge! Rogers!* Var. β, *Thom!*

COAST REGION: near Tulbagh, *MacOwan*, 2678! Groene Kloof, *MacOwan and Bolus, Herb. Norm.*, 590! between Paarl and Pont, *Drège*, 8480! Var. β, Lion mountain, *Ecklon*, 312!

13. G. rochensis (Ker in Konig and Sims' Ann. i. 224); corm globose, $\frac{1}{4}$ in. diam., with brown-black, crustaceous, lacerated tunics; leaves 3, the two lower basal, subterete, 3–4 in. long, the upper sheathing the lower part of the stem, with a short, free point; sheath very ventricose, with 8–10 strong ribs; stems simple or forked low down, 3–6 in. long, 1-flowered; outer spathe-valve oblong, $\frac{1}{2}$ in. long, green in the lower half, brown and membranous in the upper; perianth dark bright violet-purple; tube cylindrical, $\frac{1}{4}$ in. long; limb $\frac{3}{4}$–1 in. long, with obovate segments $\frac{1}{3}$–$\frac{1}{2}$ in. broad, with a pale band across below the middle and a dark blotch at the base of each segment; anthers lanceolate-sagittate, $\frac{1}{4}$ in. long, with long filaments; style reaching to the top of the anthers; branches falcate, $\frac{1}{6}$ in. long. *Ker in Bot. Mag.* sub *t.* 1105; *Gen. Irid.* 84; *Ait. Hort. Kew. edit.* 2, i. 83; *Baker in Journ. Linn. Soc.* xvi. 94. *Ixia rochensis, Ker in Bot. Mag. t.* 598. *I. radians, Thunb. Fl. Cap.* i. 217. *G. rocheana, Sweet, Hort. Brit. edit.* 2, 503. *G. setacea, a, Ker in Bot. Mag. sub t.* 1255. *Rochea venusta, Salisb. in Trans. Hort. Soc.* i. 322. *G. tulipifera, Klatt, Ergänz.* 56.

VAR. β, **G. monantha** (Sweet, Hort. Brit. edit. 2, 503); one-flowered, flower small, darker purple, almost concolorous on the claw. *Ixia monanthos, Thunb. Fl. Cap.* i. 226. *G. monanthos, Eckl. Top. Verz.* 21.

VAR. γ, **spithamæa** (Ker in Bot. Mag. sub. t. 598); more robust; stem forked; sheath $\frac{1}{4}$-$\frac{1}{3}$ in. broad, with a long, free point; flowers sometimes 5-6 to a spike with a very flexuose rachis. *Ixia violacea, Soland. MSS. I. secunda, Houtt. Handl.* xii. *t.* 78, *fig.* 1, *non Delaroche.*

SOUTH AFRICA: without locality, *Thunberg!* Var. β, *Thunberg!* Var. γ, *Ma son! Thom!*

COAST REGION: near Paarl, *Drège,* 8486! Var. γ, Malmesbury Div.; Groene Kloof and vicinity, *Bolus,* 4340! *MacOwan,* 2281! *MacOwan and Bolus, Herb. Norm.,* 506! *Pappe!*

14. G. Bellendeni (MacOwan in Journ. Linn. Soc. xxv. 393); corm globose; outer tunics of imbricated scales; basal leaves 2, linear-subulate, 6–9 in. long; stem 4–8 in. long, simple or forked, bearing a broad, sheathing leaf at the middle; flowers 3–5 in a lax spike with a flexuose rachis; outer spathe-valve oblong, green, $\frac{1}{2}$ in. long; perianth-tube $\frac{1}{4}$–$\frac{1}{3}$ in. long; segments obovate, $\frac{3}{4}$ in. long, deep blue above, the rest subpellucid without any foveole in the claw; anthers lanceolate, $\frac{1}{3}$ in. long. *Baker, Handb. Irid.* 156.

COAST REGION: Malmesbury Div.; Groene Kloof, *MacOwan and Bolus, Herb. Norm.,* 810!

15. G. latifolia (Baker in Journ. Linn. Soc. xvi. 94); corm unknown; basal leaves ensiform, nervose, a little shorter than the stem; cauline with a long sheath; stem a foot long, simple, rather flattened; flowers in a simple spike; spathe-valves marcescent; perianth blue-purple; segments oblong. *Handb. Irid.* 156. *Ixia latifolia, Delaroche Descr.* 22.

SOUTH AFRICA: without locality.

A doubtful species, known only from Delaroche's description.

16. G. corrugata (Klatt, Ergänz. 57); corm ovoid; tunics shining, castaneous; leaves 4, linear-subulate, $\frac{1}{3}$ lin. broad, corrugated; stems simple, 1-flowered, $1\frac{1}{2}$–2 in. long; spathe-valves green, oblong, membranous, $\frac{1}{3}$ in. long; perianth-tube cylindrical, $\frac{1}{12}$ in. long, throat brown; limb bright yellow, $\frac{5}{8}$ in. long, with oblong segments; style overtopping the anthers. *Baker, Handb. Irid.* 156.

CENTRAL REGION: Calvinia Div.; Hantam Mountains, *Meyer.*

This may be a variety of *G. humilis.* I have not seen it.

17. G. imbricata (Ker in Konig and Sims' Ann. i. 224); corm globose, $\frac{1}{2}$ in. diam., with thick, dark brown, rigid, lacerated tunics; basal leaves 2, linear, glabrous, firm in texture, 3–8 in. long, $\frac{1}{8}$–$\frac{1}{4}$ in. broad, with very close, prominent ribs; cauline one, with a long, free point, and very broad, loose sheath, with close, prominent ribs; stems simple or forked, $\frac{1}{2}$–1 ft. long; flowers few or many, arranged in a spike with a very flexuose rachis; outer spathe-valve oblong, green nearly to the tip, $\frac{1}{3}$ in. long; perianth-tube straight, cylindrical, $\frac{1}{3}$ in. long; limb $\frac{1}{2}$–$\frac{3}{4}$ in. long, the segments oblong, the inner whitish, the outer bright red on the back; anthers lanceolate, $\frac{1}{6}$ in. long, exceeding the filaments; style reaching to the top of the filaments; branches falcate, $\frac{1}{8}$ in. long. *Ker, Gen. Irid.* 86; *Baker in Journ. Linn. Soc.* xvi. 94; *Handb. Irid.* 156. *Ixia imbricata, Delaroche Descr.* 17. *G. arenaria, Eckl. Top. Verz.* 21. *G. sabulosa, Klatt in Trans. S. Afric. Phil. Soc.* iii. *pt.* 2, 203.

VAR. β, **concolor** (Baker); flowers concolorous, pale sulphur-yellow or milk-white.

VAR. γ, **G. obtusata** (Ker in Bot. Mag. t. 672); perianth larger, with a limb ¾–1 in. long, the outer segments flushed red outside; habit more robust, and leaf broader and more obtuse. *Baker in Journ. Linn. Soc.* xvi. 94. *Ixia obtusata, Soland. MSS.*

SOUTH AFRICA: without locality, var. γ, *Roxburgh!*

COAST REGION: Cape Flats, *Ecklon*, 313! *Pappe! Zeyher*, 1597! Camps Bay, *MacOwan*, 2273! foot of Table Mountain up to 400 ft., *Bolus*, 4710! *MacOwan and Bolus, Herb. Norm.*, 258! Worcester Div., *Cooper*, 3205! Tulbagh, *Thom!* Algoa Bay? *Cooper*, 3203! Var. β, near Somerset West, *Bolus*, 5567! Hottentots Holland, *Ecklon and Zeyher*, 211!

18. G. inflexa (Ker in Konig and Sims' Ann. i. 224); corm globose, ½ in. diam., with thick, rigid, much lacerated tunics; basal leaves 2, linear, 4–6 in. long, ⅛–¼ in. broad, thick in texture, strongly and closely ribbed; cauline one, with a large, free point and very loose, strongly ribbed sheath; stems ½–1 ft. long, terete, simple or forked, low down; flowers several, arranged in a spike with a very flexuose axis; outer spathe-valve oblong, ⅓–½ in. long, brown and scariose towards the tip; perianth with a cylindrical tube ¼ in. long, throat dark purple; limb pale yellow, ¾–1 in. long; anthers yellow, lanceolate, ¼ in. long, exceeding the purple filaments; style reaching to the top of the anthers; branches falcate, ⅙ in. long. *Ker, Gen. Irid.* 87, *excl. syn.*; *Baker in Journ. Linn. Soc.* xvi. 94; *Handb. Irid.* 157. *Ixia inflexa, Delaroche Descr.* 15. *G. vaginata, Sweet, Brit. Flow. Gard. t.* 138.

SOUTH AFRICA: without locality, *Masson!*

COAST REGION: Tulbagh, *Thom!* near Malmesbury, *MacOwan, Herb. Aust. Afric.* 1568!

A near ally of *G. imbricata*, from the concolorous variety of which it differs mainly in the purple throat of its perianth.

19. G. quadrangula (Ker, Gen. Irid. 88); corms superposed, the outer tunics produced into bristles; leaves few, sheathing, rigid, subquadrangular, 1–2 lower overtopping the flowers; stem slender, simple; flowers few, arranged in a lax spike; spathe-valves persistent; perianth blue; tube much shorter than the spathe; segments ovate. *Baker, Handb. Irid.* 157. *Ixia quadrangula, Delaroche Descr.* 16; *Vahl., Enum.* ii. 52.

SOUTH AFRICA: without locality.

A doubtful plant, known only from the original description of Delaroche.

20. G. nana (Klatt, Ergänz. 57); corm very small, globose; leaves 3–4 at the base of the stem, linear, glabrous, moderately firm in texture, ¾–1 in. long, ½ lin. broad; stems 1–2 in. long, very slender, 1-flowered, simple or forked from the base; outer spathe-valve oblong, green, obtuse, ⅙ in. long; perianth-tube blue, reaching to the tip of the spathe; limb funnel-shaped, ⅙ in. long, with oblong segments, the outer bright blue, the inner white; anthers reaching halfway up the limb; style-branches overtopping the anthers. *Baker, Handb. Irid.* 157.

COAST REGION: Caledon Div.; banks of the Zonder Einde River, near Appels Kraal, *Zeyher*, 3967!

21. G. setacea (Baker in Journ. Linn. Soc. xvi. 95); corm ovoid,
¼ in. diam.; outer tunics ending in short bristles; basal leaves 3–4,
linear, glabrous, moderately firm in texture, 1–3 in. long, ½–1 lin.
broad; stems 2–3 in. long, very slender, simple or deeply forked,
1-flowered, with 1–2 reduced sheathing leaves below the middle;
spathe-valves oblong, ¼–⅓ in. long, firm in texture, green to the tip;
perianth-tube as long as, or rather longer than the spathe, funnel-
shaped at the top, and often blue; limb ½ in. long, with oblong
white segments, often tinted claret-red outside; stamens half·as long
as the limb; anthers lanceolate, ⅛ in. long; style reaching the top
of the anthers, with three falcate branches ⅛ in. long. *Handb. Irid.*
157; *Klatt, Ergänz.* 57, *non Ker. Ixia setacea, Herb. Linn.!*
Thunb. Diss. No. 13! *Fl. Cap.* i. 233. *G. recurvifolia, Klatt in*
Linnæa xxxiv. 655, *excl. syn.*

SOUTH AFRICA: without locality, *Villett!*
COAST REGION: near Cape Town, *Thunberg! Bolus,* 4803! mountains near
George, *Rogers!* along the rivulet at Grahamstown, *Burchell,* 3546!

This is the *Ixia setacea* of the Linnean, Thunbergian, and Smithian herbaria,
the plant figured by Ker being *G. humilis.*

22. G. bracteata (Klatt, Ergänz. 57); corm not seen; leaves 5–6,
concentrated at the base of the stem, linear, glabrous, thin in texture, 2–4 in. long, ¼–⅓ in. broad; stems slender, 6–8 in. long, with
1–2 distant branches, and 1–2 reduced, lanceolate, loosely-sheathing
leaves; flowers solitary, terminal; outer spathe-valve oblong, ⅓ in.
long, green to the tip; perianth-tube reaching to the tip of the
spathe, blue at the throat; limb white, ⅓ in. long, with oblong
segments; stamens and style half as long as the perianth-segments.
Baker, Handb. Irid. 158. *Weihea elatior, Eckl. Top. Verz.* 22,
teste Klatt. G. tabularis, var. elatior, Ecklon in Linnæa xx.
p. 222.

COAST REGION: Uitenhage, *Zeyher!* Duivelsbosch, near Swellendam, *Beil.*

23. G. Dregei (Baker, Handb. Irid. 158); corm ovoid, ¼ in.
diam.; tunics brown, membranous; leaves 5–6, concentrated at the
base of the stem, lanceolate, glabrous, moderately firm in texture,
¾–1 in. long, 1/12 in. broad; stems 2–3 in. long, very slender, simple,
flexuose, with one reduced, lanceolate, loosely-sheathing leaf below the
middle; flowers 4–6 in a short erect spike with a very flexuose
axis; outer spathe-valve oblong, obtuse, ¼ in. long, green to the
tip; perianth-tube straight, reaching to the tip of the spathe, bright
blue at the throat; limb pure white, ¼–⅓ in. long, with oblong
segments; stamens one-third as long as the perianth-limb; anthers
lanceolate; filaments very short; style-branches not overtopping
the anthers. *Ixia excisa, β, Herb. Thunberg! G. secunda, β, Herb.*
Drège.

SOUTH AFRICA: without locality, Thunberg!
COAST REGION: Paarl Mountains, *Drège!* Olifants Hoek, *Ecklon and Zeyher,*
221!

Intermediate between *G. secunda* and *G foliosa.*

24. G. foliosa (Klatt in Linnæa xxxiv. 658); corm globose, $\frac{1}{3}-\frac{1}{2}$ in. diam.; tunics brown, crustaceous, ending in short cusps; basal leaves 4–6, lanceolate, glabrous, firm in texture, 2–3 in. long, $\frac{1}{4}-\frac{1}{3}$ in. broad, with many prominent ribs; stems 4–6 in. long, simple or forked, with 1–2 reduced leaves with dilated sheaths; flowers 3–6 in a lax spike; outer spathe-valve oblong, green, $\frac{1}{3}$ in. long; perianth-tube erect, cylindrical, $\frac{1}{4}$ in. long; limb whitish, $\frac{1}{2}$ in. long, with oblong segments; anthers lanceolate, $\frac{1}{6}$ in. long, equalling the filaments; style much exserted from the tube; branches $\frac{1}{6}$ in. long. *Klatt, Ergänz.* 58; *Baker in Journ. Linn. Soc.* xvi. 95; *Handb. Irid.* 158.

COAST REGION: Swellendam Div.; near Riet Kuil, *Ecklon and Zeyher*, 218! *Zeyher*, 3961! Oudtshoorn Div.; near Olifants River, *Gill!*

Resembles *Hesperantha falcata* in habit and leaf.

25. G. Bolusii (Baker, Handb. Irid. 158); corm ovoid, $\frac{1}{4}$ in. diam.; tunics brown, membranous, ending in short bristles; leaves 4–6 to a stem, all superposed, thin in texture, linear, glabrous, $\frac{1}{8}-\frac{1}{6}$ in. broad, the lower 2–3 in. long, the upper becoming gradually shorter; stems simple, very slender, flexuose, terete, 4–8 in. long; flowers 6–8 to a spike; outer spathe-valve oblong, $\frac{1}{6}$ in. long, obtuse, green, with a brown scariose tip; perianth-tube cylindrical, as long as the spathe; limb white, $\frac{1}{4}-\frac{1}{3}$ in. long, with oblong segments; stamens half as long as the limb; anthers lanceolate, equalling the filaments; style not overtopping the anthers.

COAST REGION: On stony hills at Dutoits Kloof, alt. 2300 ft., *Bolus*, 5247!

26. G. grandis (Hook. fil. in Bot. Mag. t. 5877); corm globose, $\frac{3}{4}$ in. diam.; tunics of finely reticulated fibres; basal leaves ensiform, glabrous, 6–8 in. long, $\frac{1}{2}$ in. broad, strongly ribbed; stems about a foot long, with several reduced leaves; flowers 4–8 in a lax simple spike; outer spathe-valve oblong, green, acute, herbaceous, finely ribbed, $1\frac{1}{4}-1\frac{1}{2}$ in. long; perianth-tube cylindrical, 1 in. long; limb rather longer, whitish; segments ascending, oblong, obtuse, with a keel of claret-purple towards the base; anthers claret-purple, $\frac{1}{3}$ in. long, about as long as the filaments; style slightly exserted from the perianth-tube, the ascending falcate branches $\frac{1}{2}$ in. long. *Baker in Journ. Linn. Soc.* xvi. 95; *Handb. Irid.* 158.

COAST REGION: Port Elizabeth, *Wilson!*

Described from a plant that flowered at Kew in 1868. Differs from all the other species by its much longer spathes and flowers and longer style-branches, and approximates towards *Acidanthera*.

27. G. minima (Baker in Journ. Bot. 1876, 239); corm not seen; leaves 2–3, basal, terete, falcate, firm in texture, glabrous, $1-1\frac{1}{2}$ in. long, $\frac{1}{3}$ lin. diam.; scape very slender, shining, terete, 1 in. long, 1–2-flowered; spathe-valves oblong-lanceolate, acute, herbaceous, green with a reddish-brown tinge, $\frac{1}{6}$ in. long; perianth-tube straight, cylindrical, $\frac{1}{4}$ in. long; limb not more than half as long, with oblong white segments; anthers lanceolate $\frac{1}{12}$ in. long, exceeding the filaments. *Journ. Linn. Soc.* xvi. 95; *Handb. Irid.* 159.

WESTERN REGION : Little Namaqualand, mountains near Modderfontein 4000-5000 ft., *Drège*, 2632!

28. G. geminata (E. Meyer in herb. Drège); corm not seen; leaves 3–4, all distantly superposed and sheathing the stem, setaceous, glabrous, firm in texture, 3–6 in. long; stems slender, terete, 6–10 in. long, simple or deeply forked, the branches 1–2-flowered; spathe-valves green to the tip, oblong-lanceolate, acute, $\frac{1}{3}$ in. long; perianth-tube blue, funnel shaped, $\frac{1}{3}$–$\frac{1}{2}$ in. long; limb white, $\frac{1}{2}$ in. long, with oblong segments; anthers $\frac{1}{6}$ in. long, lanceolate, exceeding the filaments; style-branches not overtopping the anthers. *Baker in Journ. Linn. Soc.* xvi. 95; *Handb. Irid.* 159; *Klatt, Ergänz.* 58.

COAST REGION : between Slangenheuvel, Frenchhoek and Donkerhoek, *Drège!*

29. G. excisa (Ker in Konig and Sims' Ann. i. 223); corm globose, $\frac{1}{2}$ in. diam.; tunics brown, crustaceous, tipped with short cusps; basal leaves 2–3, oblong or lanceolate, erect, glabrous, firm in texture, obtuse or acute, $\frac{1}{2}$–1 in. long, not ribbed, furnished with copious minute black dots on both sides; stems flexuose, slender, glabrous, 2–6 in. long, rarely forked; flowers 2–5, distant, erect; outer spathe-valve oblong, green, obtuse, herbaceous, $\frac{1}{4}$–$\frac{1}{3}$ in. long; perianth-tube cylindrical, erect, $\frac{1}{2}$–$\frac{3}{4}$ in. long; segments oblong, spreading, $\frac{1}{3}$–$\frac{1}{2}$ in. long, the three inner white, the three outer claret-purple outside; anthers lanceolate-sagittate, $\frac{1}{6}$ in. long, exceeding the filaments; style overtopping the anthers, with short, spreading, falcate branches. *Ker in Bot. Mag. sub t.* 1105; *Ait. Hort. Kew. edit.* 2, i. 84; *Klatt in Linnæa* xxxiv. 651; *Baker in Journ. Linn. Soc.* xvi. 95; *Handb. Irid.* 159. *Ixia excisa, Linn. fil. Suppl.* 92; *Thunb. Diss. No.* 24, tab. 1, *fig.* 4; *Prodr.* 10; *Fl. Cap.* i. 250; *Ker in Bot. Mag. t.* 584. *Weihea excisa, Eckl. Top. Verz.* 22. *Ixia ovata, Burm. Prodr. Cap.* 1; *Houtt. Handl.* xii. 39.

COAST REGION : Between Paarl and Pont, *Drège!* Table Mountain up to 2000 ft., *Thunberg!* Ecklon, 831! Bolus, 4704! MacOwan, 2271! MacOwan and Bolus, Herb. Norm., 259! Lion Mountain, *Zeyher!* Simons Bay, *Wright!* Stellenbosch, *Sanderson!* Caledon Div.; Klein River Berg, *Zeyher*, 3959a! 3959b!

XVI. IXIA, Linn.

Perianth subrotate, with a straight, usually short, cylindrical tube sometimes dilated into a funnel at the top and six subequal, spreading, oblong segments. *Stamens* equally divaricating, inserted at the throat of the perianth-tube; filaments short, filiform, rarely connate; anthers lanceolate, sagittate at the base. *Ovary* oblong, 3-celled; ovules crowded, superposed; style filiform, exserted from the perianth-tube with three short, spreading, falcate branches. *Capsule* small, oblong, membranous, loculicidally 3-valved. *Seeds* globose or angled by pressure.

Rootstock a small globose corm, with fibrous tunics; leaves few, linear, distichous; flowers arranged in simple or panicled spikes, very various in colour; spathe of two short, oblong, membranous or chartaceous valves.

DISTRIB. An endemic Cape genus; the species very variable and indefinite in limitation.

Subgenus IXIA PROPER. Perianth-tube short, cylindrical; filaments free to the base, much shorter than the anthers.
Flowers small; limb ¼–½ in. long:
   Anthers lanceolate, ⅛ in. long:

Flower white, concolorous	(1) polystachya.
Flower brightly coloured	(2) flexuosa.
Flower pale with a black throat	(3) hybrida.
Anthers linear-oblong, ¹⁄₁₂ in. long	(4) micrandra.

Flowers larger, concolorous at the throat:
  Perianth-tube longer; limb white ... ... (5) aristata.
  Perianth-tube short:

Flower white	(6) leucantha.
Flower bright yellow	(7) lutea.
Flower bright lilac	(8) campanulata.
Flower pale red	(9) patens.
Flower dark crimson	(10) speciosa.

Flower larger, with a purple-black throat:

Flower white or yellow	(11) maculata.
Flower lilac	(12) columellaris.
Flower bright red	(13) ovata.
Flower green	(14) viridiflora.

Subgenus EURYOLICE. Perianth-tube short, cylindrical; filaments more or less connate.

The only species ... ... (15) monadelpha.

Subgenus MORPHIXIA. Perianth-tube short, dilated into a distinct funnel below the limb; filaments free, as long as the anthers.
Spathe-valves pale green with brown ribs:
  Leaves filiform ... ... (16) linearis.
  Leaves flat:
    Flowers lilac or reddish:

Leaves glabrous	(17) scariosa.
Leaves hairy	(18) brevifolia.
Flowers bright yellow	(19) odorata.

Spathe-valves brown and membranous down to the base (20) trichorhiza.

Subgenus HYALIS. Perianth-tube long, cylindrical, slightly and gradually dilated in the upper third.
Leaves terete ... ... (21) Cooperi.
Leaves linear:
  Spathe-valves pale green with brown ribs,chartaceous (22) paniculata.
  Spathe-valves brown and rigid ... ... (23) nervosa.

1. I. polystachya (Linn. Sp. Plant. edit. 2, i. 51); corm globose, ½–¾ in. diam.; tunics of fine parallel fibres; produced leaves about 4 at the base of the stem and 2 above, linear, glabrous, moderately firm in texture, ½–1 ft. long, ⅛–¼ in. broad, strongly ribbed, acuminate; stem slender, terete, 1–2 ft. long, simple or branched; flowers many, in erect, dense spikes with a flexuose rachis; outer spathe-valve oblong, tricuspidate, pale green, ¼ in. long; perianth-tube cylindrical, mostly exserted from the spathe; limb white, ⅓–½ in. long, with oblong segments; anthers lanceolate, ⅛ in. long; filaments short; style a little exserted with branches reaching to the top of the anthers.

Andr. Bot. Rep. t. 155 ; *Ait. Hort. Kew.* i. 58 ; *Baker, Handb. Irid.*
161. *I. erecta, Berg. Cap.* 5 ; *Thunb. Diss. No.* 18 ; *Prodr.* 10 ; *Fl. Cap.*
i. 239 ; *Bot. Mag. t.* 623 ; *Ker, Gen. Irid.* 100, *excl. syn. I. serotina,*
Salisb. Prodr. 34 (*Miller Ic.* 104, *tab.* 155, *fig.* 2).

VAR. β, ornata (Baker) ; segments tinged with red outside.
VAR. γ, I. bicolorata (Klatt, Ergänz. 62) ; flower pale yellow, tinged violet
outside.
VAR. δ, I. flavescens (Eckl. Top. Verz. 26) ; flowers pale yellow ; limb less
campanulate than in the type.

SOUTH AFRICA : without locality, *Sieber,* 130 ! VAR. β, Harvey, 912 !
COAST REGION : Cape Flats, *MacOwan,* 2555 ! near Rondebosch, *Burchell,*
184 ! Table Mountain, *MacGillivray,* 476 ! *MacOwan and Bolus, Herb.*
Norm., 264 ! Dutoits Kloof, *Drège,* 8370 ! Caledon Div. ; Valley of Palmiet
River, *Bolus,* 4214 ! Riversdale Div. ; near Zoetmelks River, *Burchell,* 6708 !
Mossel Bay Div. ; by the Great Brak River, *Burchell,* 6156 ! near George,
Burchell, 6092 ! Var. γ, near Cape Town, *Spielhaus.* Var. δ, Worcester,
Zeyher !

2. I. flexuosa (Linn. Sp. Plant. edit. 2. i. 51) ; differs only from *poly-*
stachya in the colour of the flowers, which is various shades of red or
lilac or sometimes red flushed lilac outside, without any distinct basal
blotch. *Ker in Bot. Mag. t.* 624 ; *Gen. Irid.* 97 ; *Ait. Hort. Kew.* i.
58 ; *Klatt in Linnæa* xxxiv. 642 ; *Baker, Handb. Irid.* 161. *I. erecta,*
Thunb. Diss. No. 18. *I. capitata, var. stellata, Andr. Bot. Rep. t.*
232. *I. thyrsiflora, Delaroche Descr.* 20. *I. tenella, Klatt, Ergänz.*
62 (*Miller Ic. tab.* 156, *fig.* 2). *I. polystachya, Red. Lil. t.* 126.

SOUTH AFRICA : without locality, *Thunberg !* *Drège,* 8365 !
COAST REGION : Malmesbury Div. ; Groene Kloof, *Ecklon,* 441 ! 1601 !
Worcester Div., *Cooper,* 1629 ! 1689 ! near Genadendal, *Drège,* 8377 ! Breede
River, near Swellendam, *Drège,* 8371 !

I. pallide-rosea, Eckl., is a form with pale red flowers with narrow segments
tinged with lavender outside. *I. tenella,* Klatt, a form with blueish flowers with
a pale yellow throat.

3. I. hybrida (Ker in Konig and Sims' Ann. i. 227) ; intermediate between *polystachya* and *maculata ;* flowers small, white with a
tinge of pink, with a distinct blotch of black at the throat. *Gen. Irid.*
98 ; *Baker Handb. Irid.* 161. *I. flexuosa, Ker in Bot. Mag. t.* 128,
non Linn.

COAST REGION : Camps Bay, *Pappe !* Paarl Mountains, *Drège,* 8362 !
Worcester Div., *Cooper,* 1611 ! 1882 !

4. I. micrandra (Baker in Journ. Bot. 1876, 237) ; corm globose,
½ in. diam. ; tunics of very fine fibres ; leaves very narrow ; stem
very slender, always simple ; flowers few, arranged in a short, erect
spike ; outer spathe-valve oblong, tinged brown, ⅙–¼ in. long ;
perianth-tube slender, about as long as the spathe ; limb pure white,
½ in. long, with a concolorous throat ; anthers linear-oblong, $\frac{1}{12}$ in.
long, with very short filaments ; style not exserted from the
perianth-tube. *Baker in Journ. Linn. Soc.* xvi. 90 ; *Handb. Irid.*
162 ; *Klatt, Ergänz.* 62. *I. filifolia, Pappe MSS. Agretta grandi-*
flora, Eckl. Top. Verz. 23 ?

SOUTH AFRICA: without locality, *Thunberg! Thom,* 357! *Stanger!*
COAST REGION: Caledon Div. 1000–3000 ft.; Zwartberg, *Pappe! Zeyher,*
4009! Klein River Berg, *Zeyher,* 4010! Houw Hoek Mts., *Zeyher,* 4011! near
Palmiet River and Houw Hoek, *Drège,* 8372!

Is in Herb. Thunberg as a doubtful variety of *aristata.*

5. I. aristata (Ker in Bot. Mag. t. 589, excl. syn. Thunb.);
corm globose, with strong fibrous tunics; basal leaves 3–4, linear,
firm in texture, strongly ribbed, $\frac{1}{4}$–$\frac{3}{8}$ in. broad; stems slender, 1–1$\frac{1}{2}$
ft. long, terete, simple or branched; flowers many, arranged in a lax
spike; spathe-valves green, $\frac{1}{6}$–$\frac{1}{4}$ in. long, herbaceous, with a membranous border; perianth-tube slender, $\frac{1}{2}$–$\frac{3}{4}$ in. long; limb whitish,
$\frac{1}{2}$–$\frac{3}{4}$ in. long, with oblong segments; anthers $\frac{1}{8}$ in. long, equalling
the free filaments; style reaching to the top of the filaments;
branches falcate, $\frac{1}{8}$ in. long. *Baker in Journ. Linn. Soc.* xvi. 91;
Handb. Irid. 162.

VAR. β, **elegans** (Baker, Handb. Irid. 162); leaves narrower and not so rigid
in texture; perianth-tube and limb each only $\frac{1}{2}$ in. long; anthers smaller.
Wurthia elegans, Regel *Gartenfl.* t. 46, *fig.* 2.

COAST REGION: Swellendam Div.; between Zonder Einde River and Breede
River, *Zeyher,* 4016! Var. β, Swellendam Div.; without locality, *Zeyher,* 4014!

I. aristata, Thunb., is a variety of *I. leucantha.* The plant so called by Aiton,
Willdenow and Vahl, is *Sparaxis grandiflora.*

6. I. leucantha (Jacq. Ic. t. 278); corm globose, $\frac{1}{2}$ in. diam.;
tunics of parallel fibres; leaves linear, glabrous, acuminate, firm in
texture, $\frac{1}{2}$–1 ft. long, $\frac{1}{8}$–$\frac{1}{4}$ in. broad; stem 1–2 ft. long, slender,
terete, simple or branched; flowers many, arranged in a dense, erect
spike; outer spathe-valve green, tricuspidate, $\frac{1}{4}$–$\frac{1}{3}$ in. long; perianth-
tube slender, cylindrical, as long as or a little longer than the spathe;
limb pure white, $\frac{1}{2}$–$\frac{3}{4}$ in. long, with spreading oblong segments; anthers
$\frac{1}{6}$–$\frac{1}{4}$ in. long, bright yellow; filaments shorter, free; style reaching
to the top of the filaments; branches $\frac{1}{8}$ in. long. *Jacq. Coll. Suppl.*
11; *Willd. Sp. Plant.* i. 204; *Klatt in Linnæa* xxxiv. 641; *Baker,*
Handb. Irid. 162. *I. patens, var. leucantha,* Ker, *Gen. Irid.* 98.
I. candida, DC. in Red. Lil. t. 426; *Eckl. Top. Verz.* 27. *I.*
anemonæflora, Jacq. Ic. t. 273? *I. conica,* Herb. *Drège.*

VAR. β, **I. aristata** (Thunb. Diss. No. 15, non Ker); outer segments flushed
with red outside.

SOUTH AFRICA: without locality, *Burchell!* Var. β, *Thunberg!*
COAST REGION: near Tulbagh, *Ecklon,* 444! between Slangheuvel, Frenchhoek and Donkerhoek, under 1000 ft., *Drège!* Var. β, Tulbagh, *Thom!*

I. anemonæflora, Jacq., referred by Ker to *Sparaxis,* is probably a garden
form of this with one very large flower.

7. I. lutea (Baker in Journ. Linn. Soc. xvi. 91); habit and
leaves of *I. leucantha,* from which it differs by its flowers, which
are uniformly deep bright yellow. *Handb. Irid.* 162. *I. erecta, Jacq.*
Hort. Schoenbr. t. 18 (*large figure*). *I. erecta, var. lutea,* Ker in Bot.
Mag. t. 846. *I. dubia, Klatt in Linnæa* xxxiv. 644. *I. auran-*
tiaca, Klatt, Ergänz. 63.

South Africa: without locality, *Sieber*, 130 (*ex Klatt*).
Coast Region: Cape Flats at Doornhoogde, *Ecklon and Zeyher*, 86. New Kloof, near Tulbagh, *Ecklon and Zeyher*, 87 (*ex Klatt*).

8. **I. campanulata** (Houtt. Handl. xii. 42, tab. 78, fig. 4); corm globose, $\frac{1}{2}$ in. diam.; tunics fibrous; basal leaves about 4, linear, glabrous, moderately firm in texture, 6–9 in. long; stems simple, slender, terete, a foot long; flowers many, arranged in a dense spike; spathe-valves membranous, tricuspidate, $\frac{1}{4}$–$\frac{1}{3}$ in. long; perianth-tube slender, cylindrical, not longer than the spathe; limb campanulate, $\frac{1}{2}$–$\frac{3}{4}$ in. long, with oblong, concolorous, dark purple-lilac segments; anthers lanceolate, $\frac{1}{6}$ in. long, much exceeding the free filaments; style-branches reaching to the top of the anthers. *Baker in Journ. Linn. Soc.* xvi. 91; *Handb. Irid.* 162; *Klatt, Ergänz.* 62. *I. latifolia, Klatt, Ergänz.* 63.

Coast Region: Clanwilliam Div., *Mader in Herb. MacOwan*, 2182! Worcester Div., *Cooper*, 3611!

9. **I. patens** (Ait. Hort. Kew. i. 59); corm globose, $\frac{1}{2}$–$\frac{3}{4}$ in. diam.; tunics of fine brown matted fibres; produced basal leaves about 4, linear, glabrous, $\frac{1}{2}$–1 ft. long, moderately firm in texture; stems erect, terete, 1–1$\frac{1}{2}$ ft. long, often branched, furnished with 2–3 reduced leaves; flowers several, arranged in a moderately dense, erect spike; outer spathe-valve green, tricuspidate, $\frac{1}{6}$–$\frac{1}{3}$ in. long; perianth-tube slender, cylindrical, equalling or slightly exceeding the spathe; limb campanulate, $\frac{3}{4}$–1 in. long, with pale red, concolorous segments; anthers lanceolate, $\frac{1}{4}$ in. long, exceeding the free filaments; style-branches not reaching to the top of the anthers. *Ker in Bot. Mag. t.* 522; *Red. Lil. t.* 140; *Willd. Enum.* i. 56; *Klatt in Linnæa* xxxiv. 646; *Baker, Handb. Irid.* 163. *I. filiformis, Vent. Hort. Cels. t.* 48; *Red. Lil. t.* 30. *I. aristata, Schneev. Ic. t.* 32. *I. flaccida, Salisb. Prodr.* 35. *I. coccinea, Thunb. Fl. Cap.* i. 241, *in part. I. densiflora, Klatt, Ergänz.* 63.

South Africa: without locality, *Thunberg! Oldenburg! Burchell! Drège*, 8369! *Thom!*

10. **I. speciosa** (Andr. Bot. Rep. t. 186); corm small, globose; tunics of matted fibre; basal leaves about 6, linear, glabrous, moderately firm in texture; stems slender, terete, usually simple; flowers few, arranged in a short, erect spike; spathe-valves membranous, tricuspidate, $\frac{1}{6}$–$\frac{1}{4}$ in. long; perianth-tube cylindrical, equalling or rather exceeding the spathe; limb campanulate, $\frac{1}{2}$–$\frac{3}{4}$ in. long, with oblong, concolorous, dark crimson segments; anthers lanceolate, $\frac{1}{4}$ in. long, exceeding the free filaments; style-branches bright red, reaching nearly to the top of the anthers. *Willd. Enum.* i. 56; *Baker in Journ. Linn. Soc.* xvi. 91; *Handb. Irid.* 163; *Klatt, Ergänz.* 63. *I. crateroides, Ker in Bot. Mag. t.* 594; *Gen. Irid.* 102. *I. coccinea, Thunb. Fl. Cap.* i. 241, *ex parte. I. patens, var. kermesina, Regel Gartenfl. t.* 356. *I. pulcherrima, Eckl. Top. Verz.* 24.

South Africa: without locality, *Thunberg! Masson! Roxburgh!*

COAST REGION: near Tulbagh, *Ecklon,* 432! *MacOwan,* 2480! *MacOwan and Bolus, Herb. Norm.,* 265! Worcester Div., *Cooper,* 1671! 1694! French Hoek Kloof, *Drège!*

11. I. maculata (Linn. Sp. Plant. ed. 2, 1664); corm globose, $\frac{3}{4}$–1 in. diam.; tunics of strong parallel fibres; produced basal leaves about 4, linear, glabrous, strongly ribbed, $\frac{1}{2}$–1 ft. long, $\frac{1}{8}$–$\frac{1}{4}$ in. broad, acuminate; stems slender, terete, 1–2 ft. long, simple or branched; flowers many, arranged in dense, erect spikes; outer spathe-valve oblong, tricuspidate, membranous, $\frac{1}{4}$–$\frac{1}{2}$ in. long; perianth-tube slender, usually twice as long as the spathe; limb campanulate, yellow, $\frac{3}{4}$–1 in. long, with a dark purple or black blotch at the throat and oblong, obtuse segments; anthers yellow, lanceolate, $\frac{1}{4}$–$\frac{1}{3}$ in. long, much exceeding the free filaments; style-branches $\frac{1}{6}$ in. long, shorter than the anthers. *Ait. Hort. Kew.* i. 60; *Thunb. Diss.* No. 19, *ex parte; Jacq. Hort. Schoenbr. t.* 21; *Baker, Handb. Irid.* 163. *I. conica, Salisb. Prodr.* 36; *Ker in Bot. Mag. t.* 539; *Red. Lil. t.* 138; *Ker, Gen. Irid.* 99; *Klatt in Linnæa* xxxiv. 644. *I. Milleri, Bery. Pl. Cap.* 8 (*Miller Ic.* 104, *tab.* 156, *fig.* 1). *I. capitata, Andr. Bot. Rep. t.* 50. *I. abbreviata, Houtt. Handl.* xii. 41, *t.* 78, *fig.* 3. *I. dubia, Vent. Choix, t.* 10. *I. fusco-citrina, Red. Lil. t.* 86; *Klatt, Ergänz.* 63. *I. crocea, Eckl. Top. Verz.* 25. *I. alboflavens, Eckl. Top. Verz.* 27.

VAR. β, **ochroleuca** (Ker in Bot. Mag. t. 1285); flowers sulphur-yellow with a large brown blotch at the throat. *I. ochroleuca, G. Don in Sweet Hort. Brit. edit.* 2, 502.

VAR. γ, **I. nigro-albida** (Klatt, Ergänz. 62); flowers pure white, with a large black blotch at the throat. *I. capitata, var., Andr. Bot. Rep. t.* 159.

VAR. δ, **ornata** (Baker); flowers flushed with bright red or purple outside.

COAST REGION : Clanwilliam Div., *Mader in Herb. MacOwan,* 2235! Malmesbury Div.; near Berg River, *Thunberg!* near Paardeberg, *Ecklon,* 445! Groene Kloof, *MacOwan,* 2914! and *MacOwan and Bolus, Herb. Norm.,* 941! Tulbagh, *Pappe!* Worcester Div., *Cooper,* 3610! vicinity of Cape Town, *Ecklon,* 433! *Zeyher,* 1600! *MacOwan,* 2277! *Bolus,* 3320! Camps Bay, *MacOwan and Bolus, Herb. Norm.,* 263! Rondebosch, *Drège!*

12. I. columellaris (Ker in Bot. Mag. t. 630); habit and leaves of *I. maculata,* from which it only differs by its deeply-coloured flowers, which in the type are bright mauve-purple, with a blue throat. *Ker. Gen. Irid.* 102; *Ait. Hort. Kew. edit.* 2, i. 88; *Baker, Handb. Irid.* 163. *Morphixia columellaris, Klatt, Ergänz.* 50. *Ixia maculata, Thunb. Diss. No.* 19, *ex parte; Jacq. Hort. Schoenbr. tt.* 19, 20, 22. *M. angustifolia, Klatt, loc. cit.*

SOUTH AFRICA: without locality, *Thunberg! Villet! Roxburgh!*
Under this and the last are included a crowd of named garden varieties, of which about thirty are figured by Jacquin. *Watsonia (Beilia) campanulata,* Klatt, Ergänz. 20, founded on Andr. Bot. Rep. t. 196, and *W. racemosa,* Klatt, Ergänz. 20 on Andr. Bot. Rep. t. 256, drawn with style-branches slightly emarginate at the tip, should both, I suspect, be referred here.

13. I. ovata (Klatt, Ergänz. 62); habit entirely of *I. maculata,* from which it differs by its bright red flowers, with a purple-black

throat. *Baker, Handb. Irid.* 164. *I. capitata, var. ovata, Andr.
Bot. Rep. t.* 23. *I. fulgens, Pappe MSS.*

VAR. β, I. **stellata** (Klatt, Ergänz. 62); throat of perianth yellow. *I. capitata, var. stellata, Andr. Bot. Rep. t.* 232.

SOUTH AFRICA: without locality, *Pappe!*

14. I. **viridiflora** (Lam. Encyc. iii. 340); corm depresso-globose, ½ in. diam.; tunics fibrous; leaves narrow, linear, firm in texture, strongly ribbed, the lower a foot or more long; stem long, slender, simple, terete; flowers numerous, arranged in a long, lax, erect spike; outer spathe-valve oblong, pale green, ⅓ in. long; perianth-tube slender, cylindrical, as long as or a little longer than the spathe; limb campanulate, ¾–1 in. long, pale green with a black throat, the segments oblong; anthers linear, yellow, ⅓ in. long, exceeding the free black filaments; style branches not reaching to the top of the anthers. *Lam. Ill.* i. 112; *Lodd. Bot. Cab. t.* 1548. *I. viridis, Thunb. Fl. Cap. edit. Schult.* 62. *I. maculata, Thunb. Diss. No.* 19, *et Ker, Gen. Irid.* 99, *ex parte. I. spicata, var. viridi-nigra, Andr. Bot. Rep. t.* 29. *I. pulchra, Salisb. in Trans. Hort. Soc.* i. 320. *I. maculata, var. viridis, Jacq. Hort. Schoenbr. t.* 23; *Ker in Bot. Mag. t.* 549. *I. spectabilis, Salisb. Prodr.* 35. *I. prasina, Soland. MSS.*

VAR. 6, I. **cana** (Eckl. Top. Verz. 26); limb pale blue with a black throat. *I. maculata, var. amethystina, Ker in Bot. Mag. t.* 579.

VAR. γ, **cœsia** (Ker in Bot. Reg. t. 580); limb pale lilac, with a small, greenish eye.

SOUTH AFRICA: without locality, *Forster! Masson! Oldenburg! Thom!* Var. β, *Ecklon,* 440!

COAST REGION: Clanwilliam Div., *Zeyher!* Tulbagh Div.; between Breede River and the Bokkeveld, *Drège!* near Tulbagh (Roode Sand), *Thunberg! Ecklon,* 439! *MacOwan,* 2692! *MacOwan and Bolus, Herb. Norm.,* 542!

15. I. **monadelpha** (Delaroche, Descr. 22); corm globose, ½ in. diam., with fibrous tunics; leaves linear, glabrous, moderately firm in texture, ⅛–⅓ in. broad; stems slender, terete, simple or branched; flowers few, arranged in a short spike; outer spathe-valve oblong, pale green, tricuspidate, ¼–⅓ in. long; perianth-tube cylindrical, as long as or twice as long as the spathe; limb campanulate, ½–¾ in. long, very variable in colour, in the type lilac, with a greenish or blue throat; filaments dark-blue, 1/12 in. long, united in a column; anthers linear, ¼–⅓ in. long; style-branches much shorter than the anthers. *Burm. Prodr.* 1; *Ker in Bot. Mag. t.* 607; *Gen. Irid.* 101; *Baker in Journ. Linn. Soc.* xvi. 92; *Handb. Irid.* 164. *Morphixia monadelpha, Klatt, Ergänz.* 50. *I. columnaris, Salisb. Prodr.* 36.

The following named forms differ from the type only in flower-colouring, viz.: *M. purpurea,* Klatt, Ergänz. 49 (*Ixia columnaris,* Andr. Bot. Rep. t. 203), limb concolorous, claret-red, with narrow segments; *M. latifolia,* Klatt, Ergänz. 51 (Andr. Bot. Rep. t. 213; *Galaxia ixiæflora vel ramosa,* Red. Lil.

t. 41), blue with a brownish throat; *M. versicolor*, Klatt, Ergänz. 51 (Andr.
Bot. Rep. t. 211), pale yellow, with a large black spot with radiating reddish
lines; *M. curta*, Klatt, Ergänz. 50 (*Ixia curta*, Andr. Bot. Rep. t. 564; *I.
monadelpha, var. curta*, Ker, Bot. Mag. t. 1378), limb fulvous, with a reddish-
green throat, and large flowers with very broad segments. *M. grandiflora*, Klatt,
Ergänz. 49 (Bot. Rep. t. 250; *Ixia grandiflora*, Pers. Syn. i. 48) is a large-
flowered form with a lilac limb with an obscure blue throat.

SOUTH AFRICA: without locality, *Villett! Mund! Drège*, 8373!
COAST REGION: near Groeno Kloof, *Ecklon*, 442! Hout Bay, *MacOwan*,
2481! and *MacOwan and Bolus, Herb. Norm.*, 266!

16. I. linearis (Thunb. Diss. No. 11); corm globose, ½ in. diam.;
basal leaves 2–3, slender, subterete, firm in texture, ½ lin. diam.,
reaching a foot or more in length; stem slender, terete; flowers
3–6, arranged in a lax simple spike; spathe-valves green, ¼–⅛ in.
long; perianth-tube funnel-shaped, ¼ in. long; limb lilac, ½ in.
long, with oblong segments; anthers lanceolate, ⅛ in. long, equalling
the free filaments; style branches shorter than the anthers. *Prodr.*
9; *Fl. Cap.* i. 231; *Linn. fil. Suppl.* 92; *Baker, Handb. Irid.* 165.
Morphixia linearis, Ker, Gen. Irid. 106; *Baker in Journ. Linn.
Soc.* xvi. 97. *Ixia capillaris, Herb. Thunberg! I. capillaris, var.
gracillima, Ker in Bot. Mag. t.* 570. *I. tenuifolia, Vahl, Enum.* ii.
62? *Hyalis gracilis, Salisb. in Trans. Hort. Soc.* i. 317.

SOUTH AFRICA: without locality, *Masson!*
COAST REGION: Stellenbosch, *Thunberg!* near the mouth of Bot River,
MacOwan, 2657, and *MacOwan and Bolus, Herb. Norm.*, 262! George Div.; at
Wolve Drift, Malgat River, *Burchell*, 6118!

17 I. scariosa (Thunb. Fl. Cap. i. 243); corm globose, ½–¾ in.
diam.; tunics of fine, matted fibres; basal leaves 2–3, short, ensiform, rigid in texture, ¼–⅓ in. broad; stem very slender, branched,
about a foot long; flowers 3–6 to a lax spike; spathe-valves
oblong, thin in texture, pale green with a brown tip, ¼–⅓ in. long;
perianth-tube funnel-shaped, ¼–½ in. long; limb reddish or lilac,
½ in. long, with oblong segments; anthers lanceolate, ⅛ in. long,
shorter than the free filaments; style-branches reaching the top of
the anthers. *Baker, Handb. Irid.* 165. *Morphixia capillaris, aulica
and incarnata, Ker, Gen. Irid.* 106-7. *M. capillaris and aulica,
Baker in Journ. Linn. Soc.* xvi. 97. *Ixia rapunculoides, Red.
Lil. t.* 431. *Hyalis latifolia, marginifolia and aulica, Salisb. in
Trans. Hort. Soc.* i. 317–318. *I. incarnata, Jacq. Ic. t.* 282. *I.
lancea, Jacq. Ic. t.* 281. *I. striata, Vahl, Enum.* ii. 65. *I. aulica,
Ait. Hort. Kew.* i. 57. *I. phlogiflora, Red. Lil. t.* 432.

VAR. β, **longifolia** (Baker, Handb. Irid. 165); leaves linear, thinner in
texture, a foot or more long.

I. fucata, Ker, Bot. Mag. t. 1379, is probably a garden hybrid, of which this
is one of the parents.

SOUTH AFRICA: without locality, *Thunberg! Masson!*
COAST REGION: Karoo, near Gawritz River, *Ecklon and Zeyher*, 94! Var. β,
Warm Bokkeveld, near Ceres, *Bolus*, 2621! Uniondale Div.; mountains near
Avontuur, *Bolus*, 2487!

CENTRAL REGION: Tulbagh Div.; on the Wind-heuvel, Koedoes Mts., *Burchell*, 1285!

18. **I. brevifolia** (Baker, Handb. Irid. 165); corm not seen; stem very slender, hairy, 6–9 in. long, with two sheathing rudimentary leaves, the upper with a short, free, erect, hairy, narrow linear tip; flowers 2–4 in a dense spike; spathe-valves oblong, $\frac{1}{4}$ in. long, turning brown; perianth-tube a little longer than the spathe, funnel-shaped in the upper half; segments oblanceolate-oblong, pale lilac, as long as the tube; stamens half as long as the segments.

EASTERN REGION: Griqualand East; on the Zuurberg Range, 5000 ft., *Tyson*, 1872!

19. **I. odorata** (Ker, Gen. Irid. 101); leaves linear, $\frac{1}{4}$–$\frac{1}{3}$ in. broad; stem slender, terete, distantly branched; flowers fragrant, arranged in a short spike; spathe-valves oblong, in. long; perianth-tube funnel-shaped, $\frac{1}{2}$ in. long; limb bright yellow, $\frac{1}{2}$ in. long, with oblong segments; anthers $\frac{1}{6}$ in. long, equalling the free filaments. *Baker, Handb. Irid.* 165. *Morphixia odorata, Baker in Journ. Linn. Soc.* xvi. 97. *I. erecta, var., Jacq. Hort. Schoenbr. t.* 18.; *Ker in Bot. Mag. t.* 1173.

SOUTH AFRICA: without locality.

Known to me only from the figures cited.

20. **I. trichorhiza** (Baker, Handb. Irid. 165); corm globose, $\frac{1}{3}$ in. diam.; tunics of copious fine fibres, prolonged as fine bristles an inch or more above its neck; leaves 3–4, narrow linear, small, erect, all distantly superposed; stem slender, terete, simple, 4–6 in. long; flowers 2, laxly spicate; spathe-valves ovate, brown, membranous, $\frac{1}{4}$–$\frac{1}{3}$ in. long, entire or slightly lacerated; perianth-tube funnel-shaped, $\frac{1}{6}$ in. long; limb concolorous, bright lilac, $\frac{1}{2}$ in. long, with oblong segments; anthers lanceolate, $\frac{1}{6}$ in. long, equalling the free filaments; style protruded from the tube; branches short, spreading. *Morphixia trichorhiza, Baker in Journ. Bot.* 1876, 237.

COAST REGION: Stockenstrom Div.; on the Katberg, *Scully*, 145!

EASTERN REGION: Griqualand East; Zuurberg Range, 4000 ft., *Tyson*, 1567! Natal; near Durban, *Sutherland!*

21. **I. Cooperi** (Baker, Handb. Irid. 166); corm globose; leaves terete, firm in texture, the lower a foot or more long; stems simple, slender, terete, a foot or more long, bearing 2–3 reduced superposed leaves; flowers several, arranged in a lax secund spike; spathe-valves oblong, pale green, tricuspidate, $\frac{1}{6}$ in. long, brown at the very tip; perianth whitish; tube cylindrical, 2 in. long, slightly dilated in the upper third; segments of the limb whitish, patent, oblanceolate, obtuse, $\frac{1}{2}$ in. long; stamens much exserted from the perianth-tube; anthers lanceolate, $\frac{1}{6}$ in. long, deeply sagittate, equalling the free filaments; style exserted from the tube; branches

falcate, $\frac{1}{6}$ in. long. *Morphixia Cooperi, Baker in Journ. Bot.* 1876.
237 ; *Journ. Linn. Soc.* xvi. 98. *Tritonia Cooperi, Klatt,
Ergänz.* 24.

COAST REGION : Worcester Div., *Cooper,* 1628 ! 1683 !

22. I. paniculata (Delaroche, Descr. 26, t. 1); corm globose,
$\frac{1}{2}$–$\frac{3}{4}$ in. diam. ; tunics brown, membranous; produced basal leaves
2–3, linear, glabrous, moderately firm in texture, $\frac{1}{2}$–$1\frac{1}{2}$ ft. long;
stems slender, terete, erect, 1–3 ft. long, often branched ; flowers
many, arranged in lax erect spikes ; outer spathe-valve oblong, pale,
tricuspidate, $\frac{1}{4}$–$\frac{1}{3}$ in. long; perianth-tube straight, cylindrical,
$2\frac{1}{2}$–3 in. long, gradually dilated in the upper third ; limb cream-
white, $\frac{3}{4}$–1 in. long; segments oblanceolate-oblong, obtuse, often
tinged with red, with a concolorous or blackish base ; anthers
exserted wholly or partially from the perianth-tube, $\frac{1}{6}$ in. long,
about equalling the filaments; style often included in the tube ;
branches short, falcate. *Baker, Handb. Irid.* 166. *Ixia longiflora,
Berg. Cap.* 7 ; *Ait. Hort. Kew.* i. 58 ; *Curt. Bot. Mag. t.* 256 ; *Red.
Lil. t.* 34. *Gladiolus longiflorus, Thunb. Diss.* No. 22. *Tritonia
longiflora, Ker in Konig and Sims' Ann.* i. 228 ; *Bot. Mag. t.* 1502 ;
Gen. Irid. 115. *Hyalis longiflora, Salisb. in Trans. Hort. Soc.* i.
318. *Morphixia paniculata, Baker in Journ. Linn. Soc.* xvi. 97.
Tritonia paniculata, Klatt, Ergänz. 24.

VAR. β, **tenuiflora** (Baker) ; perianth-tube $1\frac{1}{2}$–2 in. long ; segments shorter,
concolorous at the throat. *Tritonia tenuiflora, Ker in Bot. Mag. sub t.* 1275 ;
Klatt, Ergänz. 24. *Ixia tenuiflora, Vahl, Enum.* ii. 66. *Gladiolus longiflorus,
Jacq. Ic. t.* 263. *Tritonia concolor, Sweet, Hort. Brit. edit.* 2, 502.
VAR. γ, **rochensis** (Baker) ; perianth-tube much shorter than in the type and
dilated into a broader funnel at the apex ; segments oblong, concolorous, $\frac{3}{4}$–1 in.
long ; filaments longer ($\frac{1}{3}$ in. long), and anthers distinctly exserted from the tube.
Tritonia rochensis, Ker in Bot. Mag. t. 1503 ; *Gen. Irid.* 114; *Klatt,
Ergänz.* 24.

SOUTH AFRICA : without locality, *Burchell !* *Zeyher,* 1620 ! Var. γ, *Drège,*
8409 !
COAST REGION : Clanwilliam Div. ; Lange Kloof, *Thunberg !* between Berg
Valley and Lange Valley, *Drège !* sandy places, St. Helena Bay and Zwartland,
Thunberg ! near Cape Town, *MacOwan and Bolus, Herb. Norm.,* 476 ! Var. β,
Caledon Div. ; near Genadendal, *Drège,* 8375 ! Var. γ, Tulbagh, *Thom !*
Riversdale Div. ; at the foot of the Lange Bergen, between Vet River and
Krombeks River, *Burchell,* 7183 ! Uniondale Div. ; mountains near Avontuur,
Bolus, 2483 !

23. I. nervosa (Baker, Handb. Irid. 166); corm small, globose ; basal
leaves 4–5, rigid, linear, 1–$1\frac{1}{2}$ ft. long, $\frac{1}{4}$–$\frac{1}{3}$ in. broad, with 5 strong ribs ;
stem simple, terete, 2–3 ft. long, with several superposed reduced
leaves, flowers in a lax erect spike $\frac{1}{2}$ ft. long; spathe-valves oblong,
brown rigid in texture, reflexing, $\frac{1}{4}$–$\frac{1}{3}$ in. long ; perianth white ; tube
cylindrical, $1\frac{1}{2}$ in. long, slightly dilated in the upper half ; segments
oblanceolate, obtuse, patent, $\frac{1}{4}$ in. long ; anthers $\frac{1}{8}$ in. long, exceeding
the free filaments ; style protruded from the tube, the short branches

reaching to the top of the anthers. *Morphixia nervosa, Baker in Journ. Bot.* 1876, 237 ; *Journ. Linn. Soc.* xvi. 98.

COAST REGION: Clanwilliam Div., *Mader in Herb. Bolus,* 2175 ! Piquetberg Div.; Cardouw, *Zeyher,* 1632 !

XVII. STREPTANTHERA, Sweet.

Perianth rotate, with a short funnel-shaped tube, and spreading, equal, obovate, obtuse segments. *Stamens* inserted at the throat of the perianth-tube ; filaments short, flattened, valvate ; anthers linear, erect. *Ovary* 3-celled ; ovules crowded in the cells ; style filiform, reaching to the top of the filaments, with 3 short, spreading, clavate, entire branches. *Capsule* small, subglobose, membranous. *Seeds* small, subglobose.

Rootstock a corm with fibrous tunics ; leaves short, arranged in a fan-shaped rosette ; flowers 2–3 to a spike ; spathe-valves large, membranous, much dotted and sheathed with brown.

DISTRIB : An endemic Cape genus, differing from *Sparaxis* by its short perianth-tube and divaricating stamens.

Flowers white in the outer half	(1) **elegans.**
Flowers coppery-yellow in the outer half	(2) **cuprea.**

1. **S. elegans** (Sweet, Brit. Flow. Gard. t. 209) ; corm ovoid, the size of a blackbird's egg, with brown fibrous coats and a long neck ; leaves 6–8 in a fan-shaped basal rosette, lanceolate, glabrous, moderately firm in texture, 3–4 in. long, $\frac{1}{4}$–$\frac{1}{3}$ in. broad ; stems forked at the base, a little longer than the leaves, 1–2-flowered ; spathe $\frac{1}{2}$ in. long, toothed at the top, wrapped round the tube and ovary ; perianth-tube $\frac{1}{4}$ inch long ; limb $\frac{3}{4}$–1 in., white with a yellow eye bordered with dark purple at the top, the segments much imbricated, obovate, very obtuse ; stamens half as long as the perianth-segments. *Lodd. Bot. Cab. t.* 1359 ; *Baker in Journ. Linn. Soc.* xvi. 92 ; *Handb. Irid.* 160 ; *Klatt, Ergänz.* 56.

COAST REGION : Tulbagh, *Thom !*

2. **S. cuprea** (Sweet, Hort. Brit. edit. 2, 501) ; habit, corms and leaves as in *S. elegans;* peduncle shorter than the leaves, 2–4-flowered ; spathe $\frac{1}{2}$–$\frac{3}{4}$ in. long ; perianth 1 in. long, copper-yellow, with a purple throat bordered with black and a yellow spot in the centre ; stamens less than half as long as the limb, the anthers equalling the connivent filaments. *Sweet, Brit. Flow. Gard.* ser. ii. *t.* 122 ; *Paxt. Mag.* i. 8, *cum icone; Baker in Journ. Linn. Soc.* xvi. 93 ; *Handb. Irid.* 160 ; *Klatt, Ergänz.* 56. *Sparaxis cuprea, Klatt in Linnæa* xxxv. 378.

SOUTH AFRICA : without locality.

Known only from the plant figured, which flowered in Mr. Colville's nursery in 1838.

XVIII. DIERAMA, K. Koch.

Perianth infundibuliform, with a short cylindrical tube dilated at the throat and oblong, obtuse, ascending, subequal segments.

Stamens inserted at the throat of the perianth-tube ; filaments short, free ; anthers linear, sagittate, erect. *Ovary* oblong, 3-celled ; ovules crowded in the cells ; style longer than the perianth-tube, with 3 short, entire, clavate, spreading branches. *Capsule* small, membranous, obtuse, loculicidally 3-valved. *Seeds* small, globose or angled by pressure.

Tall herbs, with large corms ; long, rigid, distichous, linear leaves ; flowers in panicled spikes, and long-pointed membranous spathe-valves.

DISTRIB. Only one species, in a broad sense, which extends its range to the mountains of Tropical Africa.

Perianth-limb ½–1 in. long   ...   .   (1) **pendula.**
Perianth-limb 1¼–1½ in. long   ...   .   (2) **pulcherrima.**

1. **D. pendula** (Baker in Journ. Linn. Soc. xvi. 99) ; corm large, globose, with cylindrical root-fibres, and dry coats of parallel fibres, splitting up into long bristles at the top ; basal leaves about half-a-dozen in a rosette, linear, very rigid in texture, acuminate, strongly nerved, 1½–2 ft. long, ¼ in. broad ; stems erect, terete, very slender, 3–4 ft. long, including the inflorescence ; spikes several, very slender, cernuous, with the flowers crowded at the top ; spathe-valves lanceolate-acuminate, very thin in texture, white more or less brown dotted or brownish, not lacerated, the outer about ¾ in. long ; flowers varying in colour from white to pale or dark mauve-purple ; perianth-tube ⅓–½ in. long ; limb ¾–1 in. long ; stamens half as long as the limb, the anthers much longer than the filaments ; style-branches $\frac{1}{12}$ in. long. *Handb. Irid.* 159. *Ixia pendula, Linn. fil. Suppl.* 91 ; *Thunb. Diss.* No. 16 ; *Prodr.* 9 ; *Fl. Cap.* i. 236. *Sparaxis pendula, Ker in Bot. Reg.* t. 1360 ; *Gen. Irid.* 92. *Watsonia palustris, Pers. Syn.* i. 43. *D. ensifolium, K. Koch and Bouché in Walp, Ann.* vi. 43 ; *Klatt in Linnæa* xxxii. 751. *D. cupuliflora, Klatt in Decken Bot. Ost-Afrika* 73, *t.* 3 ; *Ergänz.* 54. *D. ignea, Klatt, Ergänz.* 54.

VAR. β, **pumila** (Baker) ; leaves very narrow, whole flower not more than ½ in. long, white or purple like the type.

COAST REGION : Humansdorp Div. ; near the Kromme River, *Thunberg !* Alexandria Div. ; Zuurberg Range, *Drège !* Albany Div. ; along the rivulet at Grahamstown, *Burchell,* 3529 ! Stockenstrom Div. ; Katberg, *Drège !*
CENTRAL REGION : Somerset Div. ; Boschberg, 4500 ft., *MacOwan.*
KALAHARI REGION : Transvaal ; Saddleback Range, near Barberton, 4000–4500 ft., *Galpin,* 527 ! Orange Free State, *Cooper,* 876 !
EASTERN REGION : Kaffraria ; *Hutton !* Tembuland ; Bazeia Mts. 2000-3000 ft., *Baur,* 529 ! Griqualand East ; near Kokstad, *Tyson,* 1558 ! and *MacOwan and Bolus, Herb. Norm.,* 1201 ! Natal ; *Plant,* 65 ! *Krauss,* 265 ! *Gerrard,* 1830 ! on the Drakensberg, *Cooper,* 3177 ! on the Coast, *Wood.* 311 ! Inanda, *Wood,* 1005 ! 1359 ! Var. β, Natal ; *Buchanan ! Gueinzius ! Sanderson,* 278 ! Pieter Maritzburg, 2000–3000 ft., *Sutherland !*

2. **D. pulcherrima** (Baker in Journ. Linn. Soc. xvi. 99) ; habit entirely that of *D. pendula,* but more robust, with broader leaves, stems sometimes 5–6 ft. long, spathes 1–1¼ in. long, and much larger flowers, typically bright blood-purple, with a funnel-shaped

limb 1¼–1½ in. long.   *Handb. Irid.* 160; *Klatt, Ergänz.* 54.
Sparaxis pulcherrima, Hook. fil in Bot. Mag. t. 5555.   *S. atropurpurea, Hort.*

COAST REGION: Uitenhage, *Harvey,* 311! Albany Div.; *Cooper,* 1527!
King Williamstown Div., *Backhouse!*
KALAHARI REGION: Transvaal, *Sanderson!*
EASTERN REGION: Kaffraria; on mountains, *Mrs. Barber,* 33! Tembuland,
Bazeia Mts., 2500–3000 ft., *Baur,* 430!

XIX. LAPEYROUSIA, Pourr.

Perianth with a long or short subcylindrical tube slightly
dilated towards the throat; segments subequal, spreading, oblong-
lanceolate. *Stamens* inserted at the throat of the perianth-tube,
unilateral, close, arcuate; filaments short; anthers lanceolate, sagittate at the base, basifixed. *Ovary* 3-celled; ovules crowded, superposed; style filiform, with bifid branches. *Fruit* a small oblong or
globose membranous capsule, with loculicidal dehiscence. *Seeds*
small, globose or angled by pressure.

Rootstock an ovoid corm flattened at the base with matted tunics; produced
basal leaves 1–2 or several, distichous; inflorescence various; spathe-valves
mostly herbaceous in texture; flowers small, red, violet, yellow or white.

DISTRIB. Several species also occur in Angola, Abyssinia and Central Africa.

Subgenus SOPHRONIA. Leaves and flowers congested in a dense sessile
rosette.
  Point of central leaves lanceolate   ...   .   (1) fasciculata.
  Point of central leaves linear:
    Perianth-tube ¾–1 in. long ...   .   (2) cæspitosa.
    Perianth-tube 1–1½ in. long   ...   .   (3) galaxioides.
Subgenus OVIEDA. Produced basal leaf usually one, spreading; stems
elongated, usually branched.
  Spathe-valves small (⅓–½ in. long):
    Flowers one or few in a lax raceme   ...   ...   (4) fistulosa.
    Flowers very numerous, densely corymbose:
      Perianth-limb about as long as the tube:
        Perianth-segments lilac or white   ...   (5) corymbosa.
        Perianth-segments light yellow, with a
          violet spot at the base ...   .   (6) purpureo-lutea.
      Perianth-limb shorter than the tube ...   ...   (7) micrantha.
    Flowers fewer, arranged in panicled spikes:
      Perianth-tube ¼ in. long   ...   .   (8) divaricata.
      Perianth-tube ⅝–¾ in. long   ...   .   (9) Pappei.
      Perianth-tube 1½–2 in. long   ...   .   (10) Fabricii.
  Spathe-valves large:
    Inflorescence a simple spike:
      Leaf flat, linear ...   .   (11) fissifolia.
      Leaf subterete   ...   .   (12) montana.
    Flowers in panicled spikes:
      Perianth-tube about as long as the spathe:
        Segments of the limb oblong   ...   ...   (13) Barklyi.
        Segments of the limb oblanceolate:
          Segments ¼ in. long ...   .   (14) silenoides.
          Segments ½–¾ in. long   ...   ...   (15) Burchellii.
      Perianth-tube rather longer than the spathe ...   (16) delagoensis.
      Perianth-tube much longer than the spathe:
        Outer spathe-valve ⅓–½ in. long ...   ...   (17) anceps.
        Outer spathe-valve 1–1¼ in. long   ...   (18) macrospatha.

Subgenus ANOMATHECA. Basal leaves 2–6 in a distichous rosette.
Leaves narrow linear, firm in texture :
  Spikes numerous, few-flowered :
    Perianth-tube ½–⅔ in. long ... ... (19) **Sandersoni.**
    Perianth-tube 1¼ in. long ... ... (20) **Bainesii.**
  Spikes 2–3, many-flowered ... ... (21) **leptostachya.**
Leaves linear, thin :
  Perianth-limb ⅓–½ in. long ... ... (22) **cruenta.**
  Perianth limb 1 in. long ... ... (23) **grandiflora.**
Leaves lorate, obtuse ... ... (24) **juncea.**

1. L. fasciculata (Ker in Konig and Sims' Ann. i. 238); corm
ovoid, ⅓ in. diam., with thick, brown, cancellate tunics ending in
short bristles and a neck 1–1½ in. long below the surface of the
soil ; leaves and flowers congested into a dense, globose, sessile rosette ;
leaves with a dilated, ovate, scariose base ½ in. long, the outer one
with a plane linear point 3–4 in. long ; the inner with a crisped
lanceolate point 1–1½ in. long ; spathe-valves ovate, scariose, under
an inch long ; perianth-tube whitish, cylindrical, about 1 in. long ;
segments spreading, lanceolate, ¼–⅓ in. long ; stamens shorter than
the perianth-segments. *Ker, Gen. Irid.* 109 ; *Baker, Handb. Irid.*
174. *Galaxia plicata, Jacq. Ic. t.* 292; *Collect. Suppl.* 30. *Ovieda
fasciculata, Spreng. Syst. Veg.* i. 147 ; *Klatt in Linnœa* xxxii. 779
& xxxv. 381. *Ixia heterophylla, Willd. Sp. Plant.* i. 159. *Meristostigma heterophyllum, Dietr. Synops.* i. 161.

WESTERN REGION : Little Namaqualand, *Bolus!*

2. L. cæspitosa (Baker, Handb. Irid. 174); corm globose, ½ in.
diam., with thick, cancellate, brown tunics ending in short bristles
and a neck 1 in. long below the surface of the soil; flowers and
leaves congested into a dense sessile rosette ; leaves with an ovate,
scariose base ½ in. long, the outer with a spreading, linear, strongly
ribbed blade of firm texture 4–6 in. long, the inner with short, free
points of the same kind, ½–1½ in. long ; outer spathe-valve ovate,
scariose, ½ in. long ; perianth whitish ; tube slender, ¾–1 in. long ;
segments spreading, lanceolate, ¼–⅓ in. long ; anthers lanceolate,
1/12 in. long, sagittate at the base, equalling the filaments; style-
branches very short. *Sophronia cæspitosa, Lichten. in Roem. et
Schultes Syst. Veg.* i. 482 ; *Ker, Gen. Irid.* 10.

CENTRAL REGION : Frazerburg Div. ; mountains near Zak River Poort,
Lichtenstein. Between Great Reed River and Stink Fontein, *Burchell,* 1394!
Prince Albert Div.; on the Zwarte Bergen near Klaarstroom, 2000–3000 ft.,
Drège, 2188! Murraysburg Div.; between Middelkop and Murraysburg, 4000
ft., *Tyson,* 276!

3. L. galaxioides (Baker, Handb. Irid. 174) ; corm globose, ½ in.
diam., with brown reticulated tunics ending in bristles at the top
and a neck 1–1½ in. long below the surface of the soil; leaves and
flowers congested into a dense, globose, sessile rosette ; leaves with
an ovate, scariose, pale base, ½–1 in. long ; outer with a linear tip
4–6 in. long of firm texture with about 5 strong ribs ; inner with tips
of the same kind 1–2 in. long ; outer spathe-valve ovate-lanceolate,
herbaceous, ½–¾ in. long ; perianth-tube slender, 1–1½ in. long ;

limb spreading, violet or whitish, $\frac{1}{3}$–$\frac{1}{2}$ in. long, with lanceolate acute segments; stamens half as long as the perianth-segments.

SOUTH AFRICA: without locality, *Hutton! Bowker!*
KALAHARI REGION: Transvaal, *McLea in Herb. Bolus,* 5790! Orange Free State; Diamond Fields at Klip Drift, *Mrs. Barber,* 31!
Flowers fragrant.

4. **L. fistulosa** (Baker in Journ. Linn. Soc. xvi. 155); corm globose, $\frac{1}{3}$ in. diam., with thick tunics of brown reticulated fibres; basal leaves 2, thin, oblong, obtuse, spreading, 1–2 in. long, $\frac{1}{2}$–$\frac{1}{3}$ in. broad; stem slender, terete, fragile, leafless, 4–12 in. long; flowers solitary, terminal, or few, arranged in a very lax raceme; spathe-valves oblong, green, $\frac{1}{4}$ in. long; perianth lilac; tube slender, 1 in. long; segments spreading, oblanceolate, $\frac{1}{4}$–$\frac{1}{3}$ in. long; stamens shorter than the perianth-segments; anthers lanceolate, deeply sagittate at the base. *Handb. Irid.* 169. *Ovieda fistulosa, Spreng. MSS.; Klatt in Linnæa* xxxii. 781.

CENTRAL REGION: Clanwilliam Div., *Zeyher!* Tulbagh Div.; on the Wind Heuvel, Koedoes Mts., *Burchell,* 1282!
WESTERN REGION: Little Namaqualand; near Concordia, *Bolus,* 695! Spurs of the Kamies Bergen, 3000–4000 ft., *Drège!*

5. **L. corymbosa** (Ker in Konig and Sims' Ann. i. 238); corm ovoid, $\frac{1}{2}$–$\frac{3}{4}$ in. diam., with thick cancellate tunics of firm texture, lacerated at the base, ending at the top in a ring of bristles; produced basal leaf 1, spreading, falcate, ensiform, 4–6 in. long, $\frac{1}{4}$–$\frac{1}{2}$ in. broad, moderately firm in texture, with 5–7 strong ribs; stem short, ancipitous; inflorescence a dense corymbose panicle, with a few flowers at the tip of the numerous branchlets and a short ancipitous peduncle with 1–2 much-reduced leaves; spathe-valves oblong, obtuse, $\frac{1}{6}$–$\frac{1}{4}$ in. long, green with a brown tip; perianth bright or pale violet; tube funnel-shaped, $\frac{1}{4}$–$\frac{1}{3}$ in. long; segments oblong, of about the same length; stamens sometimes reaching to the tip of the perianth-segments; anthers lanceolate-sagittate, $\frac{1}{8}$ in. long. *Ker in Bot. Mag. t.* 595; *Gen. Irid.* 108; *Baker, Handb. Irid.* 169. *Ixia corymbosa, Linn. Sp. Plant. ed.* 2, i. 51; *Thunb. Diss. No.* 10; *Prodr.* 9; *Fl. Cap.* i. 229; *Houtt. Handl.* xi. 27, *t.* 77, *fig.* 1; *Jacq. Ic. t.* 288. *Ovieda corymbosa, Spreng. Syst.* i. 147; *Klatt in Linnæa* xxxii. 780. *Peyrousia corymbosa, Sweet, Hort. Brit. edit.* 2, 499. *Meristostigma corymbosum, Diefr. Synops.* i. 161. *L. fastigiata, Ker in Konig and Sims' Ann.* i. 238. *Ixia fastigiata, Lam. Ency.* iii. 337. (*Pluken. Almag.* 87, *t.* 275, *fig.* 1.)

VAR. β, **L. azurea** (Eckl. Top. Verz. 31); perianth-limb larger, $\frac{1}{4}$ in. long.
SOUTH AFRICA: without locality, *Zeyher,* 1594!
COAST REGION: Clanwilliam Div., *Mader in Herb. MacOwan,* 2206! Malmesbury Div.; Zwartland, *Thunberg!* sandy flats near Noordhoek, Hout Bay, *MacOwan,* 2284! and *MacOwan and Bolus, Herb. Norm.,* 268! Lion Mountain, *Bolus,* 2819! Simons Bay, *MacGillivray,* 480! Great Drakenstein and foot of the Paarl Mts., below 1000 ft., *Drège,* 8510! Var. β, Malmesbury Div.; Groene Kloof, *MacOwan,* 2284b! and *MacOwan and Bolus, Herb. Norm.,* 269! between Paarl and Pont, *Drège,* 8509a! Tulbagh, *Ecklon.*

6. L. purpureo-lutea (Baker in Journ. Linn. Soc. xvi. 155);
corm, leaves, and inflorescence exactly as in *L. corymbosa;* outer
spathe-valve ovate, acute, herbaceous, $\frac{1}{4}-\frac{1}{8}$ in. long; perianth-tube
$\frac{1}{2}$ in. long, funnel-shaped towards the top; segments of the same
length, oblong, pale yellow, with a small bright violet spot on the
face at the base; stamens half as long as the perianth-segments.
Handb. Irid. 169.   *Ovieda purpureo lutea, Klatt in Linnæa* xxxii.
780.

COAST REGION: Tulbagh, *Thom!*

7. L. micrantha (Baker in Journ Linn. Soc. xvi. 156); corm
conic, $\frac{1}{2}$ in. diam., with thick, blackish, cancellate tunics ending in a
ring of short bristles; basal leaf 1, spreading, falcate, ensiform, 6–9 in.
long, $\frac{1}{4}-\frac{1}{3}$ in. broad, moderately firm in texture; peduncle $\frac{1}{2}-1$ ft. long
including the inflorescence, strongly ancipitous, with a few much-
reduced leaves; flowers arranged in a dense corymbose panicle, 1–2
at the tip of a branchlet; spathe-valves oblong, obtuse, firm in
texture, green with a brown tip, $\frac{1}{8}-\frac{1}{6}$ in. long; perianth dark lilac;
tube cylindrical, $\frac{1}{4}-\frac{1}{3}$ in. long; segments spreading or reflexing, oblanceolate, $\frac{1}{6}$ in. long; stamens half as long as the perianth-segments.
Handb. Irid. 169.   *Ovieda micrantha, E. Meyer in Herb. Drège;
Klatt in Linnæa* xxxii. 781.   *L. manuleæflora, Eckl. Top.
Verz.* 31.

COAST REGION: mountains near Tulbagh, *Bolus,* 5250! *Zeyher,* 1593! Dutoits Kloof, *Drège!* Riversdale Div.; between Little Vet River and Kampsche
Berg, *Burchell,* 6883!

8. L. divaricata (Baker in Journ. Bot. 1876, 337); corm globose,
$\frac{1}{2}$ in. diam., with thick brown tunics; produced basal leaf linear,
6–9 in. long, firm in texture, strongly ribbed; stem $\frac{1}{2}-1$ ft. long including the inflorescence, subterete, copiously branched from near
the base; branches arcuate, ascending; flowers 6–10 in a spike,
which is short and dense in the flowering, but elongated in the
fruiting stage; outer spathe-valve ovate-navicular, with a recurved
tip, $\frac{1}{4}-\frac{1}{3}$ in. long, green with a red edge; perianth-tube slender, $\frac{1}{4}$ in.
long; limb $\frac{1}{3}-\frac{1}{2}$ in. long, with oblanceolate segments; stamens half
as long as the perianth-segments.   *Handb. Irid.* 170.   *Gladiolus
setifolius, Linn. fil. Suppl.* 96; *Thunb. Diss. No.* 19; *Prodr.* 8; *Fl.
Cap.* i. 202; *Ker, Gen. Irid.* 137.

SOUTH AFRICA: without locality, *Thunberg!* *Thom!*
COAST REGION: Clanwilliam Div., *Mader in Hb. MacOwan,* 2187 partly!
WESTERN REGION: Little Namaqualand; between Port Nolloth and the
interior mountains, 200 ft., *Bolus,* 6575! 6577!

9. L. Pappei (Baker, Handb. Irid. 170); corm not seen; produced
leaf 1 from near the base of the stem, lorate, spreading, moderately
firm in texture, 2–3 in. long, $\frac{1}{4}-\frac{1}{3}$ in. broad; stem slender, subterete,
4–6 in. long including inflorescence; branches 2–3; spikes 4–8-
flowered, lax, erect; outer spathe-valve ovate, green, $\frac{1}{6}$ in. long;
perianth lilac; tube $\frac{5}{8}-\frac{3}{4}$ in. long, slightly dilated towards the throat;
segments oblong, $\frac{1}{4}$ in. long; stamens half as long as the perianth-

segments; anthers oblong, shorter than the filaments; style equalling or overtopping the stamens.

COAST REGION: Tulbagh Div.; Great Winter Hoek, *Pappe!*

10. **L. Fabricii** (Ker in Bot. Mag. sub t. 1246); corm globose, with brown, cancellate tunics ending in short bristles and a long neck; basal leaves 2–3, spreading, linear, 2–6 in. long, $\frac{1}{4}$–$\frac{1}{3}$ in. broad, moderately firm in texture, with 7–9 strong ribs; stems $\frac{1}{2}$–1 ft. long including the inflorescence, copiously and diffusely branched from the base, slightly ancipitous; flowers 2–4 at the tip of a branch in a lax spike; outer spathe-valve oblong, obtuse, green, herbaceous, $\frac{1}{4}$ in. long; perianth lilac or white; tube slender, 1$\frac{1}{2}$–2 in. long; segments patent, oblanceolate, $\frac{1}{4}$–$\frac{1}{3}$ in. long; stamens half as long as the perianth-segments. *Ker, Gen. Irid.* 111; *Baker in Journ. Linn. Soc.* xvi. 156; *Handb. Irid.* 170. *Gladiolus Fabricii, Thunb. Prodr.* 186; *Fl. Cap.* i. 200. *Peyrousia Fabricii, Sweet, Hort. Brit.* edit. 2, 499. *Ovieda Fabricii, Spreng. Syst.* i. 147; *Klatt in Linnæa* xxxii. 778. *Meristostigma Fabricii, Dietr. Syn.* i. 161. *Gladiolus anceps, Linn. Herb. ex parte! Diasia iridifolia, Eckl. Top. Verz.* 31, *non DC.*

SOUTH AFRICA: without locality, *Thunberg!*
COAST REGION: Cape Flats, *Pappe! Zeyher! Bolus,* 2818! *MacOwan* 2517! and *MacOwan and Bolus, Herb. Norm.,* 267! New Kloof, near Tulbagh, under 1000 ft., *Drège,* 8507a!

11. **L. fissifolia** (Ker in Konig and Sims' Ann. i. 238); corm ovoid, with thick, blackish, cancellate tunics ending at the top in short bristles; produced basal leaf 1, lanceolate, spreading, 1–4 in. long, firm in texture, prominently ribbed; stem very short, with 2–3 reduced leaves like bracts, sometimes none; flowers in a simple erect spike 2–3 in. long, dense at the top, lax below; outer spathe-valves ovate, amplexicaul, herbaceous in texture, lower $\frac{1}{2}$–$\frac{3}{4}$ in. long, upper shorter and more obtuse; perianth with a slender tube 1–1$\frac{1}{2}$ in. long; segments oblanceolate or oblong, $\frac{1}{3}$ in. long, dark or light violet or whitish; stamens about half as long as the perianth-segments; anthers $\frac{1}{8}$ in. long; style-branches reaching to the top of the stamens. *Ker in Bot. Mag. t.* 1246; *Gen. Irid.* 109; *Baker in Journ. Linn. Soc.* xvi. 155; *Handb. Irid.* 171. *Gladiolus fissifolius, Jacq. Ic. t.* 268; *Collect.* iv. 464. *Ovieda fissifolia, Spreng. Syst.* i. 147. *Gladiolus bracteatus, Thunb. Prodr.* 186; *Fl. Cap.* i. 199. *Lapeyrousia bracteata, Ker in Bot. Mag. sub t.* 1246; *Gen. Irid.* 110. *Peyrousia bracteata, Sweet, Hort. Brit.* edit. 2, 499. *Ovieda bracteata, Spreng. Syst.* i. 147. *Meristostigma fissifolium and bracteatum, Dietr. Synops.* i. 161. *Moræa ovata, Thunb. Fl. Cap.* i. 280.

SOUTH AFRICA: without locality, *Thunberg!*
COAST REGION: Clanwilliam Div., *Mader,* 100! near Olifants River, *Drège,* 8508! Hex River Kloof, near Worcester, 1000–2000 ft., *Drège!* Kluitjes Kraal, near Ceres Road, 850 ft., *MacOwan, Herb. Aust. Afr.,* 1541! Caledon Div.; between Houw Hoek and Caledon, 600 ft., *MacOwan and Bolus, Herb. Norm.,* 805! Zonder Einde River, near Appels Kraal, *Zeyher,* 4025!

CENTRAL REGION: Tulbagh Div.; between Great Doorn River and Little Doorn River, *Burchell*, 1213!

WESTERN REGION: Little Namaqualand, up to between 1000 and 2000 ft.; region near the Groen and Zwartdoorn Rivers, *Drège*, 2641a! 2641b! near Mierenkasteel, *Drège*, 2642!

12. **L. montana** (Klatt, Ergänz. 25); corm ovoid; produced leaf 1, subterete, 4 in. long, ⅛ lin. diam.; stem simple, terete, 2 in. long, flowers in a simple spike; outer spatho-valve ovate, amplexicaul; foliaceous, 1½ in. long; perianth-tube cylindrical, 1½ in. long; segments elliptic, acute, pale violet, blotched with yellow at the base, ⅓ in. long; anthers equalling the filaments. *Baker, Handb. Irid.* 171.

CENTRAL REGION: Calvinia Div.; on the Hantam Mountains, *Meyer.*

May be a variety of *L. fissifolia*, of which a form gathered by Drège on the mountains north of the Olifants River, alt. 1000–2000 ft., has subterete leaves, but much smaller spathe-valves.

13. **L. Barklyi** (Baker, Handb. Irid. 171); corm not seen; produced leaves 3–4, subterete or narrow linear, ½ ft. long, 1/12–⅛ in. broad, with a dilated lanceolate base, moderately firm in texture, strongly ribbed; stem ¼ ft. long including the inflorescence, branched low down, ancipitous, the lower branches divaricate; flowers 3–5 in a lax spike with a triquetrous axis; outer spathe-valve lanceolate, herbaceous, acuminate, ¾–1 in. long; perianth lilac; tube slender, ¾ in. long, dilated at the top; limb of the same length with oblong-unguiculate segments; stamens half as long as the segments; anthers ⅙ in. long, equalling the filaments.

WESTERN REGION: Little Namaqualand; without precise locality, *Barkly!* near Abbevlakte, 600 ft., *Bolus*, 6576!

14. **L. silenoides** (Ker in Bot. Mag. sub t. 1246); corm ovoid, ½ in. diam., with thick, dark brown tunics; basal leaves linear, 1½–2 in. long; stem ½ ft. long including the inflorescence, ancipitous, branched from the base, with many reduced leaves; spikes few-flowered, very lax; outer spathe-valves lanceolate, green, 1 in. long; perianth-tube whitish, subcylindrical, 1–1¼ in. long; segments oblanceolate, bright red, spreading, ½ in. long; stamens as long as the perianth-segments; style-branches overtopping the anthers. *Ker, Gen. Irid.* 111; *Baker in Journ. Linn. Soc.* xvi. 156; *Handb. Irid.* 171. *Gladiolus silenoides, Jacq. Ic. t.* 270; *Collect.* iv. 468. *Ovieda silenoides, Spreng. Syst.* i. 147. *Meristostigma silenoides, Dietr. Synops.* i. 161.

SOUTH AFRICA: without locality.

Known to me only from Jacquin's figure.

15. **L. Burchellii** (Baker, Handb. Irid. 171); corm globose, with reticulated brown tunics; basal leaves subterete or narrow linear, 6–9 in. long, firm in texture, strongly ribbed; stems ½–1 ft. long including the inflorescence, subterete, diffusely and copiously branched from near the base; flowers 4–8 to a branch, arranged in a very lax

spike with a stiff subterete axis; outer spathe-valves lanceolate, acute, green, moderately firm in texture, $\frac{3}{4}$–1 in. long; perianth pale lilac; tube slender, 1–1$\frac{1}{4}$ in. long; segments of the limb lanceolate, $\frac{1}{2}$–$\frac{3}{4}$ in. long; stamens half as long as the perianth-segments.

KALAHARI REGION: Bechuanaland; Chooi Desert, near Giraffe Station, *Burchell*, 2341! 2350!

Very near *L. silenoides*, from which it differs by its longer leaves and perianth-segments.

16. **L. delagoensis** (Baker, Handb. Irid. 171); corm not seen; produced basal leaf single, spreading, linear, 1–2 ft. long; stem branched from the base; branches of panicle 5–6, strongly ancipitous, 4–6 in. long, subtended at the base by reduced linear leaves; spikes lax; outer spathe-valves lanceolate, green, 1–1$\frac{1}{4}$ in. long; perianth-tube slender, 1$\frac{1}{2}$–1$\frac{3}{4}$ in. long; segments half as long as the tube, linear-lanceolate, reddish; stamens half as long as the segments.

EASTERN REGION: Delagoa Bay; sandy places, Lourenço Marquez, under 50 ft., *Bolus*, 7618!

17. **L. anceps** (Ker in Konig and Sims' Ann. i. 238); corm ovoid; tunics brown, cancellate, plicate at the base; produced basal leaf one, linear, $\frac{1}{2}$ ft. long, moderately firm in texture, strongly ribbed, $\frac{1}{4}$–$\frac{1}{3}$ in. broad, the other leaves much shorter, lanceolate, amplexicaul; stems $\frac{1}{2}$–1 ft. long including the inflorescence, with several arcuate branches, very ancipitous; flowers 2–5, in lax spikes with an angled flexuose rachis; outer spathe-valve ovate, herbaceous, pale green, $\frac{1}{3}$–$\frac{1}{2}$ in. long; perianth lilac or white, with a slender tube 1–1$\frac{1}{2}$ in. long; segments of the limb oblanceolate, $\frac{1}{3}$–$\frac{1}{2}$ in. long, the three upper larger than the three lower; stamens about half as long as the perianth-segments. *Ker, Gen. Irid.* 110; *Sweet, Brit. Flow. Gard. t.* 143; *Baker, Handb. Irid.* 172. *Gladiolus anceps, Linn. fil. Suppl.* 94, *ex parte; Thunb. Diss. No.* 17, *t.* 2, *fig.* 3; *Prodr.* 8; *Fl. Cap.* i. 198; *Jacq. Ic. t.* 269. *Ovieda anceps, Spreng. Syst.* i. 147; *Klatt in Linnæa* xxxii. 778. *Meristostigma anceps, Dietr. Syst.* i. 161. *Ixia Fabricii, Delaroche, Descr.* 18. *L. compressa, Pourr. Mém. Acad. Toul.* iii. 80, *t.* 6. *Ixia pyramidalis, Lam. Ency.* iii. 334. *Gladiolus denticulatus, Lam. Ill.* i. 118. *Witsenia pyramidalis, Pers. Syn.* i. 42.

VAR. β, **L. aculeata** (Sweet, Hort. Brit. edit. 2, 396); outer spathe-valve denticulate down the keel; perianth-limb $\frac{1}{4}$–$\frac{3}{4}$ in. long, with broader segments. *Klatt, Ergänz.* 26. *Gladiolus anceps, Herb. Linn., ex parte. Ovieda aculeata, Klatt in Linnæa* xxxii. 777. *Peyrousia aculeata, Sweet, Brit. Flow. Gard.* ser. 2, t. 39.

SOUTH AFRICA: without locality, *Burchell!* Var. β, *Zeyher*, 1617!
COAST REGION: Tulbagh, *Thom!* Malmesbury Div.; Zwartland and region of Saldanha Bay (including Var. β), *Thunberg!* Var. β, Clanwilliam Div., *Mader in Herb. MacOwan*, 2136! Piquetberg, 1000–2000 ft., *Drège!*
WESTERN REGION: Little Namaqualand; near Ookiep, *Morris in Herb. Bolus*, 5791!

18. **L. macrospatha** (Baker); corm small; outer coats crustaceous, brown; produced leaves 1–2, linear, firm, glabrous, falcate, 1$\frac{1}{2}$–3 in. long; inflorescence branched from near the ground; spikes lax, few-

flowered; rachis narrowly winged; outer spathe-valve lanceolate, green, herbaceous, 1–1½ in. long; perianth lilac; tube much longer than the spathe; segments oblanceolate, ¾–⅞ in. long; stamens one quarter the length of the limb; capsule oblong, ⅓ in. long. *L. cærulea, Bolus in Herb. Norm. Austr. Afric.* No. 697, *non Schinz.*

WESTERN REGION: Little Namaqualand; between Port Nolloth and Eleven-mile Station, 200 ft., *Bolus, Herb. Norm.*, 697!

19. L. Sandersoni (Baker, Handb. Irid. 169); corm globose, ½–¾ in. diam.; tunics thick, brown, cancellate; produced basal leaves 2, narrow linear, firm in texture, erect, strongly ribbed, acuminate, 1–1½ ft. long; stems 1–1½ ft. long including the inflorescence, ancipitous; flowers in a lax ample corymbose panicle, 1–2 at the tip of each branch; spathe-valves oblong or oblong-lanceolate, ⅙–¼ in. long, herbaceous or subscariose; perianth white or lilac; tube slender, ½–¾ in. long, rather widened towards the throat; limb spreading, ⅓ in. long with oblanceolate segments; stamens more than half as long as the perianth-segments. *L. Bainesii, var. breviflora, Baker in Journ. Linn. Soc.* xvi. 156.

KALAHARI REGION: Transvaal; without precise locality. Zoutpansberg Div.; summit of Rhenosterpoort, *Nelson*, 402! *Sanderson! Todd*, 20! 21!
Also in Tropical Africa.

Habit and leaves of *L. Bainesii*, from which it differs by its shorter perianth-tube.

20. L. Bainesii (Baker in Journ. Bot. 1876, 338); corm not seen; produced basal leaves 2, erect, linear, firm in texture, 1–1½ ft. long, ¼–⅓ in. broad, acuminate, strongly ribbed; stems ancipitous, 1½–2 ft. long including the inflorescence; flowers in a dense corymbose panicle, 1–2 at the tip of each branchlet; spathe-valves oblong-lanceolate, acute, greenish, herbaceous, ¼–⅓ in. long; perianth whitish; tube slender, 1½ in. long, slightly dilated at the top; segments of the limb oblanceolate, ⅓ in. long; anthers lanceolate, ⅛–⅙ in. long, equalling the filaments. *Baker in Journ. Linn. Soc.* xvi. 156; *Handb. Irid.* 170.

KALAHARI REGION: Transvaal, *Todd*, 19! Bechuanaland; Sirorume River, *Holub!*
Also in Tropical Africa.

21. L. leptostachya (Baker, Handb. Irid. 170); corm not seen; produced leaves 3–4 in a close rosette, linear, glabrous, erect, moderately firm in texture, 6–9 in. long, ⅙ in. broad; stem 1½ ft. long including the inflorescence, slightly ancipitous; spikes 2–3, very lax, the end one 10–12-flowered; outer spathe-valve ovate or ovate-lanceolate, chartaceous, blue-black, ¼–½ in. long; perianth pale lilac; tube cylindrical, broader upwards, ¾ in. long; segments lanceolate, half as long as the tube; stamens half as long as the perianth-segments.

KALAHARI REGION: Transvaal; Lydenberg Div., McMac, near Pilgrims Rest, *Mudd!* Not Natal, as given in the Handb. Irid.

22. L. cruenta (Baker, Handb. Irid. 173); corm ovoid, with finely reticulated, fibrous, brown tunics; produced leaves about 6 in a distichous tuft, linear, thin in texture, erect, $\frac{1}{2}$–1 ft. long, $\frac{1}{4}$–$\frac{1}{2}$ in. broad; stems slender, terete, 1–2 ft. long, simple or branched; flowers 6–12, arranged in a lax secund spike; outer spathe-valve ovate or ovate-lanceolate, acute, herbaceous, $\frac{1}{6}$–$\frac{1}{3}$ in. long; perianth-tube slender, 1–1$\frac{1}{2}$ in. long; limb bright red, $\frac{1}{2}$ in. long, with oblong segments, the three lower with a dark spot at the base; stamens less than half as long as the perianth-segments; capsule small, globose. *Anomatheca cruenta, Lindl. in Bot. Reg. t.* 1369; *Lodd. Cab. t.* 1857; *Paxt. Mag.* i. 103; *Klatt in Linnæa* xxxii. 775.

SOUTH AFRICA: without locality, *Drège*, 8406!
COAST REGION : Howisons Poort, near Grahamstown, *Hutton!* near Grahamstown, 2000 ft., *MacOwan*, 378!
CENTRAL REGION : Somerset Div., *Bowker!*
KALAHARI REGION : Transvaal; damp shady places in wooded ravines around Barberton, 3000–3500 ft., *Galpin*, 1015!
EASTERN REGION : Kaffraria, *Cooper*, 457! Natal; Inanda, *Wood*, 406! Durban, *Wood*, 650! Nottingham, *Buchanan!* and without special locality, *Gerrard*, 553! *Cooper*, 1278! *Plant*, 28! *Sanderson*, 180! *Krauss*, 392!

23. L. grandiflora (Baker in Bot. Mag. t. 6924); corm globose, $\frac{1}{2}$ in. diam. ; tunics brown, finely fibrous ; leaves 4–8 in a distichous basal tuft, linear, erect, a foot or more long, $\frac{1}{4}$–$\frac{1}{2}$ in. broad; stem simple, terete, as long as the leaves; flowers 4–6 in a lax spike; outer spathe-valve lanceolate, green, $\frac{1}{2}$ in. long, much exceeding the inner; perianth-tube pale, subcylindrical, 1 in. long; segments oblanceolate, bright scarlet, as long as the tube, the three lower with a darker blotch at the base; stamens as long as the segments. *Handb. Irid.* 173. *Anomatheca grandiflora, Baker in Journ. Bot.* 1876, 337.

EASTERN REGION : Delagoa Bay, *Mrs. Monteiro.* Flowered at Kew, Oct., 1886.
Also Highlands of the Zambesi country, collected by Dr. Meller, Sir John Kirk, and Mr. Buchanan.

24. L. juncea (Pourr. in *Mém. Acad. Toul.* iii. 79); corm ovoid, with fibrous tunics; produced basal leaves 4–6 in a distichous tuft, lorate, 6–8 in. long, $\frac{1}{2}$–$\frac{3}{4}$ in. broad, moderately firm in texture; stems 1–2 ft. long including the inflorescence, subterete, branched from below the middle; flowers 4–6 to a branch, arranged in a lax spike; outer spathe-valve oblong, obtuse, green, $\frac{1}{6}$–$\frac{1}{4}$ in. long; perianth pale red; tube slender, straight, $\frac{1}{2}$–$\frac{3}{4}$ in. long, dilated at the throat; segments oblanceolate-oblong, obtuse, $\frac{1}{3}$–$\frac{1}{2}$ in. long, the 3 lower spotted at the throat; stamens half as long as the perianth-segments; capsule globose. *Ker, Bot. Mag. t.* 606; *Baker, Handb. Irid.* 173. *Gladiolus junceus, Linn. fil. Suppl.* 94; *Thunb. Diss. No.* 18; *Red. Lil. t.* 141. *Anomatheca juncea, Ker in Konig and Sims' Ann.* i. 227; *Gen. Irid.* 112. *Ixia spicata, Burm. Prod. Cap.* 1. *I..emarginata, Vahl, Enum.* ii. 70; *Lam. Encyc.* iii. 342; *Ill.* i. 112. *I. Gauleri, Schrad. Journ.*

iv. **B.** 67. *Gladiolus polystachius, Andr. Bot. Rep. t.* 66. *G. marmoratus, Lam. Encyc.* ii. 727. *G. paniculatus, Pers. Syn.* i. 45. *G. excisus, Jacq. Hort. Schoenbr. t.* 491. *G. amabilis, Salisb. Prodr.* 41.

COAST REGION: Uniondale Div.; Lange Kloof, *Thunberg!*

XX. MICRANTHUS, Pers.

Perianth with a short, curved, subcylindrical tube and subequal, spreading, oblanceolate, obtuse segments. *Stamens* inserted at the throat of the perianth-tube, arcuate, unilateral, close; filaments filiform; anthers lanceolate-sagittate, versatile. *Ovary* 3-celled; ovules 2 in a cell, erect, collateral; style filiform; branches short, filiform, bifid. *Capsule* oblong, membranous, loculicidally 3-valved. *Seeds* 1–2 in a cell, lanceolate, erect.

Rootstock a corm with thick, reticulated tunics; leaves sheathing the stem, superposed; flowers small, red, arranged in dense, distichous spikes; outer spathe-valve oblong-lanceolate, with a rigid centre and broad, hyaline border.

DISTRIB. Endemic.

Leaves narrowed gradually to the point ... ...   (1) **plantagineus.**
Leaves lorate, obtuse, tipped with a mucro ... ...   (2) **fistulosus.**

1. **M. plantagineus** (Eckl. Top. Verz. 43); corm ovoid, $\frac{1}{2}$ in. diam.; tunics thick, reticulated, ending in bristles at the top; stems $\frac{1}{2}$–1 ft. long, simple or branched; produced leaves about 3, superposed, linear, acute, firm in texture, strongly ribbed, the lower $\frac{1}{2}$–1 ft. long, the upper much shorter, loosely sheathing the stem to the base of the inflorescence; spikes 3–6 in. long, the lower flowers often abortive or replaced by bubillæ; spathes $\frac{1}{4}$ in. long; perianth-tube as long as the spathe; segments $\frac{1}{6}$–$\frac{1}{4}$ in. long; stamens about as long as the perianth-segments; anthers $\frac{1}{8}$ in. long; style-branches reaching to the top of the filaments. *Baker, Handb. Irid.* 179. *Watsonia plantaginea, Ker in Bot. Mag. t.* 553; *Gen. Irid.* 123; *Baker in Journ. Linn. Soc.* xvi. 159. *Ixia plantaginea, Ait. Hort. Kew.* i. 59; *Red. Lil. t.* 198. *Gladiolus plantagineus, Pers. Syn.* i. 46. *Gladiolus alopecuroides, Linn. Sp. Plant. edit.* 2, 54, *ex parte. M. alopecuroides, Eckl. Top. Verz.* 43. *Ixia triticea, Burm. Prodr.* 1. *M. triticeus, Klatt, Ergänz.* 21. *Phalangium spicatum, Houtt. Handl.* xii. 115, *t.* 80, *fig.* 2. *Watsonia compacta, Lodd. Bot. Cab. t.* 1577. *W. triticea, Spreng. Syst.* i. 150.

VAR. β, **juncea** (Baker in Journ. Linn. Soc. xvi. 159); leaves subterete.

COAST REGION: Cape Div.; near Cape Town, *Thunberg! Bolus,* 2829! Lions Rump, *Burchell,* 110! near Rondebosch, *Burchell,* 179! Muizenberg, *MacOwan and Bolus, Herb. Norm.,* 270! Paarl Div.; Klein Drakenstein Mts., under 500 ft., *Drège!* Worcester Div., *Cooper,* 1621! Dutoits Kloof, 3000–4000 ft., *Drège!* Riversdale Div.; near Zoetmelks River, *Burchell,* 6744! 6770! 6823! on the Lange Bergen, near Kampsche Berg, *Burchell,* 7043! Uitenhage Div.; Galgebosch, *MacOwan,* 1935! Var. β, Malmesbury Div.; Groene Kloof and vicinity, under 1000 ft., *Zeyher,* 1611! Piquetberg Div., *Zeyher,* 1610! Paarl Div.; Klein Drakenstein Mts., under 1000 ft., *Drège,* 8444! Worcester Div., *Cooper,* 1679! Simons Bay, *Milne,* 193! MacGillivray, 480! Uitenhage Div.; between Van Stadens Berg and Bethelsdorp, *Drège,* 8445!

2. **M. fistulosus** (Eckl. Top. Verz. 44); corm $\frac{1}{2}$–1 in. diam., the tunics ending at the top in long bristles; leaves 3–4, superposed, sheathing the stem up to the base of the spike, lorate, falcate, glabrous, moderately firm in texture, obtuse, with a conspicuous mucro, finely veined, the lowest 3–6 in. long, the upper shorter; stems simple, $\frac{1}{2}$–1 ft. long; spike 2–5 in. long, dense, many lower flowers often abortive; spathe and flowers exactly like those of the other species. *Baker, Handb. Irid.* 179. *Gladiolus alopecuroides, Linn. Sp. Plant. edit.* 2, 54, *ex parte. G. spicatus, Linn. Sp. Plant. edit.* 2, 53, *ex Ker, non Linn. Herb. Watsonia spicata, Ker ın Konig and Sims' Ann.* i. 229; *Gen. Irid.* 123. *Ixia spicata, Willd. Sp. Plant.* i. 200. *M. spicatus, Klatt, Ergänz.* 21. *Ixia cepacea, Red. Lil. t.* 96. *I. fistulosa, Ker, Bot. Mag. t.* 523. *Gladiolus fistulosus, Jacq. Hort. Schoenbr. t.* 16. *G. tubulosus, Burm. Prodr.* 2.

COAST REGION: Devils Mountain, *MacOwan,* 2509! *Ecklon,* 524! Flats near Rondebosch, *Burchell,* 198! between Cape Town and Simons Bay, *Burchell,* 8555! near Tulbagh, *Burchell,* 1034! Hottentots Holland Mts., *Thunberg!*

XXI. FREESIA, Klatt.

Perianth with a long, funnel-shaped tube; segments of limb subequal, oblong. *Stamens* inserted below the throat of the tube, close, arcuate; anthers linear. *Ovary* ovoid, 3-celled; ovules crowded in the cells; style filiform, arcuate; branches short, slender, bifid. *Capsule* loculicidal, 3-valved. *Seeds* turgid.

DISTRIB. Endemic, monotypic. Differs from *Tritonia* only by its bifid style-branches.

1. **F. refracta** (Klatt in Linnæa xxxiv. 673); corm ovoid, with thick, reticulated, fibrous tunics; stem terete, flexuose, distantly branched, with a few much-reduced leaves; basal leaves about 6, linear, firm in texture, about $\frac{1}{2}$ ft. long, $\frac{1}{4}$–$\frac{1}{2}$ in. broad; flowers very fragrant, resupinate, arranged in lax secund spikes with a flexuose rachis; spathe-valves oblong-lanceolate, acute, scariose, not hiding the ovary; perianth greenish-yellow or bright yellow, 1–1$\frac{1}{2}$ in. long; tube $\frac{1}{2}$ in. diam. at the throat, constricted suddenly below the middle; limb distinctly bilabiate, its segments ovate-oblong, $\frac{1}{3}$–$\frac{1}{2}$ in. long, those of both the two lips much imbricated, the central one of the upper lip broader and more obtuse than the side ones, and not reflexing in the fully-expanded flower; anthers lanceolate, just exserted from the perianth-tube, much shorter than the filiform filaments. *Klatt, Ergänz.* 26; *Baker, Handb. Irid.* 167. *Gladiolus refractus, Jacq. Ic. t.* 241; *Coll. Suppl.* 26; *Red. Lil. t.* 419. *Tritonia refracta, Ker in Konig and Sims' Ann.* i. 228; *Bot. Reg. t.* 135; *Gen. Irid.* 119. *Gladiolus resupinatus, Pers. Syn.* i. 45. *G. Sparrmanni, Thunb. Fl. Cap. edit. Schult.* 49.

VAR. β, **F. odorata** (Klatt in Linnæa xxxiv. 672); leaves broader and less rigid than in the type; inflorescence less branched, and flowers fewer in a spike; spathe-valves broader and more obtuse, covering the ovary, distinctly toothed at the apex; flowers bright yellow, with a suddenly-constricted tube, but the limb less distinctly bilabiate than in the type, with all the segments more oblong and obtuse. *Baker, Handb. Irid.* 167. *Tritonia odorata, Lodd. Bot. Cab. t.* 1820.

VAR. γ, F. alba (Baker, Handb. Irid. 167); leaves, inflorescence and spathe-valves as in β; flowers larger, pure white, with the tube narrowed more gradually downwards, the limb very indistinctly bilabiate, and all the segments obtuse and nearly equal.

SOUTH AFRICA; without locality, *Sparrmann in Herb. Thunberg!*
COAST REGION: Swellendam Div. ; along the Buffeljagts River, *Zeyher,* 4027 ! Riversdale Div. ; Hooge Kraal, near Zoetmelks River, *Drège !* Var. β, Bathurst Div.; between Blaauw Krans and Kowi Poort, *Burchell,* 3652! and between Blaauw Krans and Kaffir Drift Military Post, *Burchell,* 3711 ! Albany Div. ; near Grahamstown, *Galpin,* 207 ! *Zeyher !* Var. γ, Caledon Div., *MacOwan,* 2482!

The above are the three principal types, which I have characterized mainly from detailed notes kindly furnished by Prof. M. Foster. *F. xanthospila, Klatt in Linnæa* xxxiv. 673 (*Gladiolus xanthospilus, Red. Lil. t.* 124), differs mainly from *alba* by its suddenly-constricted perianth-tube, and *F. Leichtlinii, Klatt in Regel Gartenfl. t.* 808, by the same character and its large pale-yellowish flowers.

XXII. WATSONIA, Miller.

Perianth with a long curved tube, cylindrical in the lower half, funnel-shaped in the upper half; segments subequal, oblong, spreading. *Stamens* unilateral, arcuate, contiguous, inserted below the throat of the perianth-tube; filaments filiform ; anthers linear-oblong, versatile. *Ovary* globose, 3-celled ; ovules many, superposed ; style filiform ; style-branches short, subulate, bifid. *Fruit* an oblong loculicidal capsule. *Seeds* globose or angled by pressure.

Rootstock a tunicated corm ; leaves usually rigid in texture, ensiform; flowers large, bright red, rarely white, arranged in simple or branched spikes ; spathe-valves oblong-lanceolate, entire, rigid in texture, brown or herbaceous.

DISTRIB. One species in Madagascar.

Subgenus WATSONIA PROPER. Flowers large ; upper part of tube cylindrical or narrowly funnel-shaped.
Perianth-segments ⅓–½ in. long ... ... (1) aletroides.
Perianth-segments ¼–½ the length of the tube ... (2) angusta.
Perianth-segments ¾–1 in. long :
  Stems tall, often branched :
    Spikes lax, 12–20-flowered ... ... (3) Meriana.
    Spikes dense, 30–50-flowered ... ... (4) densiflora.
  Stems shorter, simple, spikes few-flowered :
    Perianth-tube 1½–2 in. long ... ... (5) coccinea.
    Perianth-tube 1¼–1½ in. long :
      Perianth-tube ¼ in. diam. at throat ... (6) humilis.
      Perianth-tube ⅛ in. diam. at throat ... (7) strictiflora.

Subgenus NEUBERIA. Flowers large ; perianth-tube broadly funnel-shaped in the upper part.
Funnel-shaped upper part of the tube large :
  Stem short and mostly simple ... ... (8) brevifolia.
  Stem long and branched ... ... (9) rosea.
Funnel-shaped upper part of the tube very small ... (10) marginata.

Subgenus BEILIA. Flowers small ; perianth-tube cylindrical, dilated a little at the summit.

Lower leaves erect, 6-9 in. long:
 Spathe-valves ⅓–½ in. long ... ... **(11) punctata.**
 Spathe-valves ¼ in. long:
  Tube as long as spathe ... ... **(12) minuta.**
  Tube much longer than spathe ... ... **(13) juncifolia.**
Lower leaves spreading, 1-3 in. long ... ... **(14) lapeyrousioides.**
Doubtful species ... ... **(15) retusa.**

1. **W. aletroides** (Ker in Bot Mag. t. 533); corm globose, 1 in. diam.; tunics of reticulated fibres, ending in long bristles at the top; stems 1-2 ft. long, including inflorescence, simple or branched, with a few much-reduced sheathing leaves; basal leaves 4-6, ensiform, rigid in texture, 6-12 in. long, ¼–½ in. broad; flowers 6-12, in a very lax spike, the lower ones drooping; spathe-valves oblong-lanceolate, ½-1 in. long, not rigid; perianth bright scarlet or pale pink; tube curved, 18-21 lin. long, dilated below the middle, ¼ in. diam. at throat; segments oblong, ⅓–½ in. long; style sometimes protruded beyond the segments; stamens inserted at the bottom of the dilated part of the tube; anthers ¼ in. long. *Ker, Gen. Irid.* 128; *Ait. Hort. Kew. edit.* 2, i. 96; *Klatt in Linnæa* xxxii. 742; *Baker, Handb. Irid.* 174. *Antholyza alethroides, Burm. Prodr.* 1. *Gladiolus aletroides, Vahl, Symb.* ii. 96. *Gladiolus Merianus, Thunb. Diss. Glad. No.* 12; *Prodr.* 7. *Gladiolus tubulosus, Jacq. Ic. t.* 229; *Coll.* iv. 153. *Watsonia tubulosa, Pers. Syn.* i. 42. *Antholyza tubulosa, Andr. Bot. Rep. t.* 174.

SOUTH AFRICA: without locality, *Thom,* 361! *Harvey,* 495! 841!
COAST REGION: Zwartberg, near Caledon, under 1000 ft., *Zeyher,* 4031! *Drège! MacOwan and Bolus, Herb. Norm.,* 942! Riversdale Div.; Vals River, *Thunberg!*

2. **W. angusta** (Ker in Konig and Sims' Ann. i. 230); corm globose, 1-1½ in. diam.; tunics fibrous; stems terete, reaching a height of 3-4 ft., usually branched; leaves few, distant, long-sheathing; basal leaves 4-6, ensiform, rigid in texture, 1-2 ft. long, ¾-1 in. broad; flowers in a lax spike 6-9 in. long; outer spathe-valve oblong-lanceolate, brown, rigid in texture, ¾-1 in. long; perianth bright scarlet, with a curved tube 18-21 lin. long, dilated in the upper half, ¼ in. diam. at the throat; segments oblanceolate, oblong, cuspidate, ⅝–¾ in. long; style reaching to the tip of the perianth-segments; anthers linear, ⅓ in. long. *Sweet, Hort. Brit. edit.* 2, 500; *Klatt in Linnæa* xxxii. 739; *Baker in Journ. Linn. Soc.* xvi. 157; *Handb. Irid.* 175. *W. fulgida, Salisb. in Trans. Hort. Soc.* i. 323; *Klatt loc. cit. Antholyza fulgens, Andr. Bot. Rep. t.* 192. *W. iridifolia, var. fulgens, Ker in Bot. Mag. t.* 600. *Gladiolus Merianus, var. Jacq. Ic. t.* 231. *W. atrosanguinea, Klatt in Linnæa* xxxii. 738. *Gladiolus marginatus* γ, *Herb. Thunb.*

SOUTH AFRICA: without locality, *Thunberg!*
COAST REGION: Riversdale Div.; between Little Vet River and Kampscheberg, *Burchell,* 6858! Caledon Div.; upper part of the great mountain at Baviaans Kloof, near Genadendal, *Burchell,* 7698! Uitenhage Div.; near Van Stadens River, *MacOwan,* 2053! channel of Zwartkops River, *Zeyher,* 646!
EASTERN REGION: Tembuland; margin of Bazeia River, *Baur,* 359!

3. W. Meriana (Miller, Gard. Dict. edit. viii. No. 1; Ic. ii. 184, t. 276); corm globose, 1–1½ in. diam.; tunics of reticulated fibres, ending at the top in a ring of bristles; stems 3–4 ft. long including the inflorescence, usually branched; basal leaves 3–4, ensiform, rigid in texture, 1–2 ft. long, ½–¾ in. broad; flowers 12–20, arranged in lax spikes 6–9 in. long, usually bright rose-red, rarely scarlet or white; spathe-valves oblong-lanceolate, brownish, rigid in texture, ¾–1 in. long; perianth-tube curved, 1½–2 in. long, cylindrical in the upper half; throat ½ ½ in. diam.; segments oblong-cuspidate, about ¾ in. long; style not reaching the tip of the perianth-segments; anthers ⅓ in. long. *Ker in Konig and Sims' Ann.* i. 230; *Baker, Handb. Irid.* 175. *Antholyza Meriana, Linn. Sp. Plant. edit.* 2, 54; *Ourt. in Bot. Mag. t.* 418. *Gladiolus Merianus, Red. Lil. t.* 11; *Jacq. Ic. t.* 230. *G. marginatus, Thunb., ex parte. Meriana flore rubello, Trew, Ehret.* 11, *t.* 40. *W. litura, Klatt, Ergänz.* 19.

VAR. β, **W. iridifolia** (Ker in Bot. Mag. sub t. 600); leaves broader than in the type; flowers closer and more numerous, white or pinkish. *Gladiolus iridifolius, Jacq. Ic. t.* 234.

VAR. γ, **W. roseo-alba** (Ker in Bot. Mag. t. 537); perianth-tube more slender than in the type and segments narrower. *Gladiolus roseo-albus, Jacq. Hort. Schoenbr.* i. 7, *t.* 13. *W. Ludwigii, Eckl. MSS.*

VAR. δ, **W. dubia** (Eckl. Top. Verz. 36); leaves linear; spathes-valves 1½-2 in. long; perianth-tube 2 in. long, ¼ in. diam. at throat; segments narrower than in the type, oblong-spathulate, ¾ in. long. *Klatt in Linnæa* xxxii. 741.

VAR. ε, **platypetata** (Baker); flowers many in a long spike, bright rose-red, with shorter, broader, rather imbricated segments.

SOUTH AFRICA: without locality, *Thunberg !*
COAST REGION: Paarl Mountains, 1000–2000 ft., *Drège !* Worcester Div., *Cooper,* 1626! Algoa Bay, *Cooper,* 3179! Albany Div.; near Grahamstown, *MacOwan,* 94! Var. β, Piquetberg, 1500–3000 ft., *Drège,* 8439! Var. δ, New Kloof, near Tulbagh, *Drège !*
KALAHARI REGION: Var. γ, Transvaal, *Sanderson !*
EASTERN REGION: Natal; on a stony hill near Byrne, 4550 ft., *Wood,* 1874! Kaffraria, *Cooper,* 67! 456! 3180!

4. W. densiflora (Baker in Journ. Bot. 1876, 336); corm globose, 1 in. diam.; tunics of parallel strands of matted fibres; stems simple, terete, erect, 2–3 ft. long including the inflorescence; basal leaves 4–6, rigidly coriaceous, ensiform, 2–3 ft. long, ½–¾ in. broad; flowers 40–50, arranged in a dense spike 1 ft. long, all imbricated; outer spathe-valve oblong-lanceolate, acute, brown, rigid, 1 in. long; perianth bright rose-red, rarely white; tube curved, 1¼–1½ in. long, narrowly funnel-shaped in the upper half, ¼ in. diam. at the throat; segments of the limb oblong, cuspidate, ¾ in. long; style not reaching to the tip of the perianth-segments; anthers ⅓ in. long, just protruded from the perianth-tube. *Bot. Mag. t.* 6400; *Handb. Irid.* 176; *Klatt, Ergänz.* 18.

KALAHARI REGION: Transvaal; Saddleback Range, near Barberton, 3500–5000 ft., *Galpin,* 813! Orange Free State, *Cooper,* 886!
EASTERN REGION: Natal; Inanda, *Wood,* 538! Rodeborough, 4000 ft., *Miss*

102   IRIDEÆ (Baker).   [*Watsonia.*

Armstrong! without precise locality, *Plant,* 29! *Cooper,* 3186! Pondoland; between Umtata R. and St. John's R., 1000-2000 ft., *Drège,* 4536!

Mr. J. M. Wood sends a fine form with pure white flowers, which flowered for the first time at Kew in September, 1891.

5. W. coccinea (Herbert MSS. in Herb. Kew); corm depresso-globose, $1\frac{1}{4}$ in. diam.; tunics thick, of reticulated fibres, ending at the top in bristles; stem simple, about a foot long including the inflorescence; basal leaves linear, firm in texture, 6–9 in. long; flowers 4–6 in a very lax spike; spathe-valves oblong-lanceolate, brownish, rigid, the lower 1–$1\frac{1}{2}$ in. long; perianth bright crimson; tube curved, $1\frac{1}{2}$–2 in. long, dilated in the upper half, $\frac{1}{3}$ in. diam at the throat; segments oblong-spathulate, cuspidate, an inch long, the lowest a little longer and more spreading than the others; styles reaching to the tip of the perianth-segments. *Klatt, Ergänz.* 18; *Baker, Handb. Irid.* 175. *W. pellucida, Eckl. Top. Verz.* 36. *W. Meriana, Ker in Bot. Mag. t.* 1194.

SOUTH AFRICA : without locality, *Thom!*
COAST REGION: Constantia, *Zeyher!* between Palmiet River and Houw Hoek, *Grey!* Stellenbosch and Hottentots Holland, *Ecklon.*

A plant received from the Glasnevin Botanic Garden, July, 1887, differs by its smaller bright scarlet flowers.

6. W. humilis (Miller, Gard. Dict. edit. viii. No. 2 ; Ic. ii. 198, t. 297, fig. 2); corm globose, 1 in. diam.; tunics thick, of reticulated fibres prolonged into bristles round the neck; stem usually simple, about a foot long including the inflorescence; basal leaves linear, firm in texture, 6–9 in. long, $\frac{1}{4}$–$\frac{1}{2}$ in. broad; flowers 4–6, arranged in a very lax spike, bright rose-red; spathe-valves oblong-lanceolate or lanceolate, firm in texture, the lower sometimes $1\frac{1}{2}$–2 in. long; perianth with a curved tube $1\frac{1}{4}$–$1\frac{1}{2}$ in. long, funnel-shaped in the upper half, $\frac{1}{3}$ in. diam. at the throat; segments oblong, $\frac{3}{4}$ in. long, $\frac{1}{3}$ in. broad; styles falling short of the tip of the perianth-segments; anthers linear oblong, $\frac{1}{4}$ in. long. *Ker in Bot. Mag. t.* 631 *and* 1193; *Gen. Irid.* 127 ; *Baker, Handb. Irid.* 176. *Antholyza caryophyllacea, Houtt. Handl.* xii. 63, t. 79, *fig.* 3. *Gladiolus marginatus, Thunb. Diss. No.* 20, *ex parte. G. laccatus, Jacq. Ic. t.* 232; *Red. Lil. t.* 343. *Neuberia laccata, Eckl. Unio Itin. No.* 560. *W. laccata, Ker in Bot. Mag. sub t.* 631; *Eckl. Top. Verz.* 36. *W. maculata, Klatt, Ergänz.* 18.

SOUTH AFRICA: without locality, *Thunberg!*
COAST REGION: near Cape Town, *Bolus,* 3800! Cape Flats, *Zeyher,* 1634! Malmesbury Div.; Groene Kloof, *Zeyher,* 1635! near Tulbagh, *Ecklon,* 560! between Paarl and Pont, *Drège,* 8441! Swellendam Div.; near Karmelks River, under 1000 ft., *Drège!* mountain ridges along the lower part of Zonder Einde River, *Zeyher,* 4030! near George, *Burchell,* 6066! 6091!
EASTERN REGION: Tembuland; Bazeia Mountain, 4000 ft., *Baur,* 499! Natal, *Cooper,* 3187!

7. W. strictiflora (Ker in Bot. Mag. t. 1406); corm globose, 1 in. diam.; tunics thick, fibrous; stems simple, $1\frac{1}{2}$–2 ft. long, including the inflorescence; basal leaves linear, firm in texture, 6–12 in. long,

½ in. broad; flowers few, arranged in a very lax spike; spathe-valves
oblong-lanceolate, acute, brown, firm in texture, ½–¾ in. long;
perianth bright rose-red; tube slightly curved, 1½ in. long, subcylindrical in the upper half, ⅓ in. diam. at the throat; segments
oblong-spathulate, patent, ¾ in. long; stamens and style reaching to
the tip of the perianth-segments. *Baker in Journ. Linn. Soc.* xvi.
158; *Handb. Irid.* 176; *Klatt, Ergänz.* 18; *Eckl. Top. Verz.* 36.

COAST REGION: Caledon, *Zeyher!* Houw Hoek, *Ecklon!*

8. **W. brevifolia** (Ker in Bot. Mag. t. 601); corm globose, 1 in.
diam.; tunics of reticulated fibres, produced into a ring of bristles at
the top; stems 1–1¼ in. long including the inflorescence, usually
simple; basal leaves about 4, linear, rigid in texture, 6–9 in. long;
spikes lax, 6–8 flowered; spathe-valves oblong-lanceolate, brown,
rigid, ½–¾ in. long; perianth bright rose-red; tube curved, ¾–1 in.
long, dilated into a broad funnel in the upper half; segments oblong,
½–⅝ in. long; style-branches reaching to the tip of the perianth-
segments; anthers just protruded from the perianth-tube. *Gen.
Irid.* 125; *Ait. Hort. Kew. edit.* 2, i. 95; *Baker, Handb. Irid.* 176.
W. hyacinthoides, Pers. Syn. i. 43. *Antholyza spicata, Andr. Bot.
Rep. t.* 56.

SOUTH AFRICA: without locality, *Thom!*
COAST REGION: Swellendam Div.; near Groot Vaderbosch, *Zeyher,* 4029!

Bears the same relation to *W. rosea* that *W. humilis* bears to *W. Meriana.*

9. **W. rosea** (Ker in Konig and Sims' Ann. i. 230); corm large,
globose; tunics thick, fibrous; basal leaves ensiform, rigid in texture, reaching a length of 2–3 ft. and a breadth of 1–2 inches; stems
4–6 ft. long including the inflorescence, usually much branched;
spikes lax or rather dense, the end one ½–1 ft. long, the others
shorter; spathe-valves oblong-lanceolate, acute, brown, rigid, ¾ in.
long; perianth bright rose-red; tube 1–1¼ in. long, dilated into a
broad funnel in the upper third; segments oblong-spathulate, cuspidate, as long as the tube; stamens and style reaching halfway up
the perianth-segments. *Bot. Mag. t.* 1072; *Gen. Irid.* 125; *Ait.
Hort. Kew. edit.* 2, i. 94; *Baker, Handb. Irid.* 177. *Neuberia rosea,
Eckl. Top. Verz.* 37. *Gladiolus pyramidatus, Andr. Bot. Rep. t.* 335.
G. iridifolius var., Jacq. Ic. t. 235. *W. striata, Klatt, Ergänz.* 18,
Gladiolus sceptrum, Forster MSS.

SOUTH AFRICA: without locality, *Grey! Thom,* 955! *Forster!*
COAST REGION: Table Mountain, *MacOwan,* 2556! *MacGillivray,* 482!
Simons Bay, *Milne!* between Simonsberg and Ban Hoek, *Rogers!* Paarl Moun-
tains, 1000–2000 ft., *Drège!* Stockenstrom Div.; Katberg, *Hutton! Mrs.
Barber,* 3!
KALAHARI REGION: Transvaal; Pilgrims Rest, *Roe in Herb. Bolus,* 2655!
Zoutpansberg Div., Houtbosch Berg, *Rehmann,* 5778! 5779!

Habit and leaf of *W. Meriana,* from which it differs in the shorter tube and
larger segments of its perianth.

10. **W. marginata** (Ker in Bot. Mag. t. 608); corm large, globose;

basal leaves ensiform, very rigid in texture, with a thick pale brown border, 1–1½ ft. long, 1–1½ in. broad; stems reaching a length of 4–5 ft. including the inflorescence, much branched, the dense end-spike 6–9 in. long, the others smaller and stiffly erect; outer spathe-valve oblong-lanceolate, ¾ in. long, paler and less rigid than in the preceding species, rather lacerated in the edge; flowers fragrant, bright rose-red; perianth-tube ¾ in. long, dilated into a small funnel at the very top and spreading oblong segments ¾–1 in. long; stamens and style reaching halfway up the perianth-segments. *Gen. Irid.* 125; *Ait. Hort. Kew. edit.* 2, i. 94; *Baker, Handb. Irid.* 177. *Neuberia marginata, Eckl. Top. Verz.* 37. *Ixia marginata, Ait. Hort. Kew.* i. 59. *I. sceptrum, Hort. Gladiolus marginatus, Linn. fil. Suppl.* 95; *Thunb. Diss. No.* 20, *ex parte. G. glumaceus, Thunb. Prodr.* 186; *Fl. Cap.* i. 204. *Ixia cartilaginea, Lam. Encyc.* iii. 340; *Ill.* i. 112.

VAR. β, minor (Ker in Bot. Mag. t. 1530); flower smaller, the expanded limb about an inch in diameter.

SOUTH AFRICA: without locality, *Thunberg!* Grey! Var. β, *Zeyher,* 1636!
COAST REGION: Heaths by the Breede River near Darling Bridge, *Bolus,* 5252!

This very distinct species is the *Gladiolus marginatus* of the Linnean and Smithian herbaria, but Thunberg also included under the name several of the other species as here treated.

11. **W. punctata** (Ker in Konig and Sims' Ann. i. 229); corm globose, ½ in. diam.; tunics thick, cancellate in the upper half, produced into fine bristles at the neck; leaves 3–4, superposed, narrow linear or subterete, rigid in texture, the lower 6–9 in. long; stems simple, terete, ½–1 ft. long including inflorescence; spike dense, 2–3 in. long; spathe-valves oblong-lanceolate, acute, brownish-green, rigid in texture, ⅓–½ in. long; perianth dark red or dark violet, with a subcylindrical tube ¾–1 in. long and spreading oblong segments ½ in. long; stamens and style reaching halfway up the perianth-segments. *Gen. Irid.* 124; *Klatt, Ergänz.* 19; *Baker, Handb. Irid.* 177. *Gladiolus spicatus, Linn. Sp. Plant. edit.* 2, i. 53; *Thunb. Fl. Cap.* i. 193. *Beilia spicata, Eckl. Top. Verz.* 43. *W. rubens, Ker, Gen. Irid.* 124. *Gladiolus rubens, Vahl, Enum.* ii. 98.

VAR. β, triticea (Baker, Handb. Irid. 177); spike shorter and denser; flowers smaller; leaves terete. *Gladiolus triticeus, Thunb. Fl. Cap.* i. 194. *Beilia triticea, Eckl. Top. Verz.* 43. *G. subulatus, Vahl, Enum.* ii. 97. *W. subulata, Klatt, Ergänz.* 19. *W. filifolia, E. Meyer in Herb. Drège. W. sanguinea, Forster MSS.*

VAR. γ, longicollis (Baker, Handb. Irid. 177); perianth with a slender tube 1–1½ in. long and a limb ½ in. long.

SOUTH AFRICA: without locality, *Thunberg!* Var. β, *Thunberg!* Zeyher, 1608! *Forster!*
COAST REGION: Paarl Mountains, under 100 ft., *Drège!* Hottentots Holland Mts., *Thunberg!* Bontebok Flats, between Cape Agulhas and Potberg, *Drège,* 3492! Var. β, Cape Flats near Constantia, and on Table Mountain, *Ecklon,* 5! Worcester Div.; *Cooper,* 1615! Dutoits Kloof, *Drège!* Var. γ, *Zeyher!*

12. **W. minuta** (Klatt, Ergänz. 19); corm globose, ⅓ in. diam.;

tunics thick, cancellate; stem slender, simple, terete, under a foot long; leaves 3–4, superposed, linear, rigid in texture, the lowest 4–6 in. long, $\frac{1}{12}$–$\frac{1}{8}$ in. broad; flowers dark red, arranged in a dense spike 1–1$\frac{1}{2}$ in. long; spathe-valves oblong-lanceolate, brown, rigid, $\frac{1}{3}$–$\frac{1}{6}$ in. long; perianth-tube cylindrical, not exserted from the spathe; segments narrow, oblong, $\frac{1}{4}$ in. long; stamens and style falling short of the perianth-segments. *Baker, Handb. Irid.* 178. *W. punctata, var. Zeyheri, Baker in Journ. Linn. Soc.* xvi. 159.

COAST REGION: Piquetberg Div., *Zeyher*, 1609! Scarcely more than a variety of *W. punctata.*

13. **W. juncifolia** (Baker, Handb. Irid. 178); corm globose, $\frac{1}{2}$ in. diam.; tunics of fine parallel fibres, prolonged more than an inch above its neck; stem slender, terete, simple, 1$\frac{1}{2}$ ft. long; basal leaves usually 2, terete, firm in texture, a foot or more long; flowers in a lax simple erect spike 4–5 in. long; spathe-valves oblong, rigid in texture, $\frac{1}{8}$ in. long, green with a brown tip; perianth bright lilac; tube cylindrical, under an inch long, limb spreading, $\frac{1}{4}$–$\frac{1}{3}$ in. long, with narrow oblong segments; anthers lanceolate, $\frac{1}{6}$ in. long, equalling the free filaments; style-branches falling short of the top of the anthers. *Morphixia juncifolia, Baker in Journ. Bot.* 1876, 238; *Journ. Linn. Soc.* xvi. 98. *Anomatheca calamifolia, Klatt, Ergänz.* 21. *Ixia Zeyheri, Baker, Handb. Irid.* 166.

COAST REGION: Piquetberg Div., *Zeyher*, 1619!

14. **W. lapeyrousioides** (Baker, Handb. Irid. 178); corm ovoid, $\frac{1}{3}$ in. diam.; tunics of reticulated fibres in the upper half; stem simple, 4–6 in. long, with 1–2 lanceolate, erect, small sheathing leaves; lower leaves 2, superposed; lanceolate, falcate, spreading, moderately firm in texture, 1–3 in. long, $\frac{1}{6}$–$\frac{1}{3}$ in. broad; flowers 3–6 in a very lax spike; spathe-valves oblong, $\frac{1}{4}$ in. long; perianth dark red; tube cylindrical, an inch long; segments oblong, $\frac{1}{4}$ in. long; stamens and style-branches reaching nearly to the tip of the perianth-segments. *Anomatheca juncea, Herb. Drège, non Ker.*

COAST REGION: Caledon Div.; Mountains between Villiersdorp and French Hoek, 2300 ft., *Bolus*, 5251! Worcester Div.; Dutoits Kloof, 2000–4000 ft., *Drège! Bolus*, 5497!

15. **W. retusa** (Klatt, Ergänz. 20); stem slender, terete, simple; leaves linear, like those of an *Ixia;* flowers 6–8 in a lax spike with a flexuose rachis; spathe-valves like those of an *Ixia* in shape and texture, $\frac{1}{8}$ in. long; perianth-tube curved, cylindrical, twice as long as the spathe, dilated into a small funnel at the top; segments bright rose-red, oblong, $\frac{1}{2}$ in. long; anthers very small, oblong, with very short filaments; style-branches protruded from the perianth-tube, bifid at the tip. *Baker, Handb. Irid.* 178. *Ixia polystachia, Ker in Bot. Mag. t.* 629.

SOUTH AFRICA: without locality.

Known only from the figure cited. The habit is so entirely that of an *Ixia*, that I fear the styles may have been drawn bifid by some mistake of the artist. In Gen. Irid. Ker refers it to *Tritonia scillaris.*

XXIII. **BABIANA**, Ker.

Perianth with a long slender tube, funnel-shaped at the top; segments usually oblong or oblong-unguiculate, nearly equal. *Stamens* unilateral, contiguous, inserted near the throat of the tube, usually shorter than the segments. *Ovary* 3-celled; ovules many, superposed; style-branches short, simple, usually flattened at the tip. *Capsule* oblong, membranous, loculicidally 3-valved. *Seeds* small, globose or angled by pressure.

Rootstock a tunicated corm; leaves plicate, various in shape, strongly ribbed, hairy, often distinctly petioled and oblique at the base; flowers various in colour, arranged in simple or panicled spikes; spathe-valves various in texture, usually lanceolate, firm, pilose, finely ribbed.

DISTRIB. Besides the Cape species, only one, discovered by Prof. Bayley Balfour in Socotra, is known.

Subgenus EUBABIANA. Segments of the perianth rather unequal, oblong, spathulate or unguiculate.
Perianth with a long tube:
    Leaves two or three times as long as the peduncle
    and spikes:
        Stamens half as long as the limb ... ... (1) **hypogæa.**
        Stamens one-third as long as the limb ... (2) **Bainesii.**
    Leaves shorter than or a little overtopping the
    flowers:
        Tube much longer than the spathe:
            Perianth-limb ½ inch long:
                Leaves linear ... ... (3) **lineolata.**
                Leaves ensiform ... ... (4) **spathacea.**
            Perianth-limb 1 in. long:
                Leaves linear ... ... (5) **tubiflora.**
                Leaves ensiform ... ... (6) **tubata.**
        Tube equalling or but little longer than the
        spathe:
            Leaves hairy:
                Leaves overtopping flowers ... ... (7) **sambucina.**
                Leaves not overtopping flowers ... (8) **densiflora.**
            Leaves glabrous, very thick and rigid ... (9) **Dregei.**
Perianth with a tube not more than 1–1½ in. long:
    Spikes few-flowered:
        Peduncle very short:
            Leaves linear, spirally curled ... ... (10) **namaquensis.**
            Leaves lanceolate ... ... (11) **pygmæa.**
            Leaves oblong ... ... (12) **Sprengelii.**
            Leaves deltoid:
                Perianth-tube an inch long ... ... (13) **flabellifolia.**
                Perianth-tube 1½ in. long ... (14) **cuneifolia.**
        Peduncle elongated ... ... (15) **mononeura.**
    Spikes 1–3, many-flowered:
        Leaves linear, glabrous or slightly hairy:
            Leaves straight ... ... (16) **fimbriata.**
            Leaves spirally curled ... ... (17) **spiralis.**
        Leaves ensiform, glabrous or slightly hairy:
            Leaves not much overtopping the inflor-
            escence ... ... (18) **mucronata.**
            Leaves much longer than the inflorescence (19) **occidentalis.**
        Leaves ensiform, hairy:
            Perianth-tube equalling the spathe ... (20) **plicata.**
            Perianth-tube longer than the spathe ... (21) **disticha.**

Subgenus ACASTE.　Segments of the perianth oblanceolate-oblong, subequal, not unguiculate :
　　Spathe-valves firm in texture, hairy, closely ribbed :
　　　　Spathe and perianth-limb an inch long ...　　... (22) **stricta.**
　　　　Spathe and perianth-limb 1½ in. long　...　　... (23) **macrantha.**
　　Spathe-valves brown, membranous, glabrous　　... (24) **secunda.**

Subgenus ANTHOLYZOIDES.　Perianth with lingulate, unguiculate upper segments, longer than the others (connects *Babiana* and *Antholyza*).
　　Upper segment about 2 in. long　　...　　...　　... (25) **ringens.**
　　Upper segment under an inch long　...　　...　　... (26) **Thunbergii.**

1. B. hypogæa (Burchell, Travels, ii. 589) ; corm large, globose ; tunics of fine brown matted fibres ; leaves several, more than twice as long as the inflorescence, linear or subulate, pilose, above a foot long ; flowers in 1–3 dense spikes on a very short peduncle ; outer spathe-valve oblong-lanceolate, membranous, glabrous, 1½–2 in. long ; perianth bright lilac ; tube cylindrical, 2–2½ in. long, dilated into a funnel at the top ; limb 1½ in. long, with subequal, acute, oblong-spathulate segments ; stamens half as long as the segments ; anthers lanceolate, ⅓ in. long ; style-branches ⅙ in. long, flattened at the tip. *Baker in Journ. Linn. Soc.* xvi. 165 ; *Handb. Irid.* 180. *Antholyza hypogæa, Klatt, Ergänz.* 11.

KALAHARI REGION : Bechuanaland ; Pellat Plains, near Takun, *Burchell*, 2241 ! Diamond Fields around Kimberley, *Mrs. Barber*, 3 ! 8 !

2. B. Bainesii (Baker in Journ. Bot. 1876, 335) ; corm globose, 1 in. diam. ; tunics of fine drab reticulated fibres ; leaves 6–8, linear, a foot long, plicate, finely pilose, acuminate, more than twice as long as the inflorescence ; flowers few, arranged in 1–2 congested erect spikes on a very short peduncle ; spathe-valve brown, chartaceous, lanceolate, obscurely pilose, 1½–2 in. long ; perianth dark bright lilac ; tube filiform, 2 in. long, dilated into a small funnel at the top ; limb 2 in. long, with oblong or oblong-lanceolate segments narrowed into a long claw ; stamens one-third as long as the segments ; style-branches ⅙ in. long, flattened at the tip. *Journ. Linn. Soc.* xvi. 165 ; *Handb. Irid.* 180.

CENTRAL REGION : near Murraysburg, 4000 ft., *Tyson*, 312 !
KALAHARI REGION : Transvaal, in grassy places, *Wood*, 3653 ! *Roe in Herb. Bolus*, 2654 ! near Johannesburg, *Mrs. Saunders*, 15 ! near Rustenberg, *McLea in Herb. Bolus*, 2654 ! Olifants River, *Nelson*, 396 ! Hooge Veldt, *Rehmann*, 6574 ! South African Gold Fields, *Baines !* Orange Free State ; near Modder River, *Hutton !*

3. B. lineolata (Klatt, Ergänz. 13) ; corm small, globose ; tunics of brown matted fibres ; leaves elongated, narrow linear, subglabrous, with about four plications, 1 lin. broad ; stem nearly a foot long including the inflorescence ; spathe-valves hairy at the base, membranous at the margin, sphacelate at the tip ; perianth-tube cylindrical, twice as long as the spathe ; segments ½ in. long, pale violet, lineolate. *Baker, Handb. Irid.* 180.

COAST REGION : Clanwilliam Div. ; near the Olifants River, *Ecklon and Zeyher*, 132.

I have not seen this, and suspect a misprint in the length of tube as stated, which is said to be 4 lines, whilst the plant is classed by Dr. Klatt next to *B. tubiflora* in his Longitubulosæ.

4. B. spathacea (Ker in Konig and Sims' Ann. i. 234, non Bot. Mag. t. 638); leaves ensiform, nearly a foot long, ½ in. broad, plicate, pilose, petioled, oblique at the base; stem a foot long including the inflorescence; spikes 2–3, dense; outer spathe-valve oblong-lanceolate, scariose, glabrous, ¾–1 in. long; perianth pale lilac; tube slender, 1½–2 in. long; limb ½ in. long, with oblanceolate segments; stamens nearly as long as the segments; style-branches reaching to the tip of the anthers. *Baker, Handb. Irid.* 180. *Gladiolus spathaceus, Linn. fil. Suppl.* 96; *Thunb. Diss. No.* 25; *Prodr.* 9; *Fl. Cap.* i. 208. *B. Ecklonii, Klatt, Ergänz.* 14.

CENTRAL REGION: Calvinia Div.; Bokkeveld and Hantam in dry regions, *Thunberg!*

Described from the type specimens in Thunberg's herbarium, the only ones I have seen. Its affinity is with *B. tubata.*

5. B. tubiflora (Ker in Konig and Sims' Ann. i. 233); corm globose, ½ in. diam.; tunics membranous; basal leaves many, linear, ¾–1 ft. long, ⅛–¼ in. broad, plicate, finely pilose; spikes 1–2, shorter than the leaves, moderately dense, secund, the end one 2–3 in. long; spathe-valves lanceolate, finely ribbed, firm in texture, densely villose, 1½–2 in. long; perianth dull pink; tube slender, an inch longer than the spathe; segments subequal, oblanceolate-unguiculate, under an inch long; stamens not reaching to the tip of the segments. *Bot. Mag. t.* 847 *and* 1019; *Gen. Irid.* 153; *Baker, Handb. Irid.* 180. *Gladiolus tubiflorus, Linn. fil. Suppl.* 96; *Thunb. Diss. No.* 23, *t.* 2, *fig.* 2; *Jacq. Ic. t.* 266. *G. angustifolius, Lam. Ill.* i. 119. *G. inclinatus, Red. Lil. t.* 44. *G. mucronatus, Red. Lil. t.* 142. *Ixia tubulosa, Burm. Prodr.* 1. *Babiana tubulosa, Ker, Gen. Irid.* 154.

COAST REGION: Cape Flats, *Bolus*, 3780! *Pappe!* Malmesbury Div.; Zwartland, *Thunberg!* near Groene Kloof, *Zeyher!* *MacOwan and Bolus, Herb. Norm.*, 544!

6. B. tubata (Sweet, Brit. Flow. Gard. edit. ii. 500); corm globose, 1 in. diam.; tunics membranous; leaves petioled, ensiform, plicate, pilose, moderately firm in texture, a foot or more long, ⅓–½ in. broad at the middle; spikes often several, dense, secund, the end one 3–4 in. long; rachis very pilose; outer spathe-valve oblong or oblong-lanceolate, firm in texture, 1–1½ in. long, densely villose, scariose at the very tip; perianth dull pinkish; tube slender, 1½ in. longer than the spathe, dilated gradually to a throat, ¼–⅓ in. diam.; segments oblong-spathulate, 1–1¼ in. long; stamens not reaching to the tip of the segments; style-branches subulate, ¼ in. long. *Baker, Handb. Irid.* 181. *Gladiolus tubatus, Jacq. Ic. t.* 264. *G. longiflorus, Andr. Bot. Rep. t.* 5. *Babiana tubiflora, var. tubata, Ker in Bot. Mag. t.* 680. *Gladiolus ringens, var., Herb. Thunb.*

SOUTH AFRICA : without locality, *Thunberg! Burchell! Villett! Drège,* 8404!
Harvey, 838 !
COAST REGION: Malmesbury Div.; Groene Kloof, *MacOwan,* 2479!

7. B. sambucina (Ker in Konig and Sims' Ann. i. 234); corm globose, tunicated; basal leaves 5–6, ensiform, petioled, plicate, pilose, overtopping the inflorescence, about ½ in. broad; peduncle including inflorescence ½–1 ft. long; flowers in few or several dense spikes; outer spathe-valve lanceolate, pubescent, 1½–2 in. long; perianth deep lilac; tube 1½–2 in. long, funnel-shaped at the top; segments subequal, oblong-spathulate, 1 in. long; stamens not more than half as long as the segments; style-branches cuneate at the tip. *Bot. Mag. t.* 1019; *Gen. Irid.* 152 ; *Baker, Handb. Irid.* 181. *Gladiolus sambucinus, Jacq. Hort. Schoenbr. t.* 15. *G. ringens,* ε, *Herb. Thunb.*

SOUTH AFRICA : without locality, *Thunberg!*

8. B. densiflora (Klatt, Ergänz. 14); corm ovoid; tunics brown, of thick fibres; stems glabrous, terete, curved, 6–8 in. long; leaves hairy, plicate; outer ensiform, ½ ft. long, ⅓ in. broad ; inner linear ; flowers 8–9 in a dense spike; spathe-valves membranous, cuspidate, unequal; outer 1½ in. long; perianth whitish; tube slender, lilac, 2 in. long; limb 8 lines long with oblong-spathulate segments, the three lower with two purple spots. *Baker, Handb. Irid.* 181.

CENTRAL REGION: Calvinia Div.; on the Hantam Mountains, *Meyer.*

9. B. Dregei (Baker in Journ. Bot. 1876, 336) ; corm not seen ; leaves very thick and rigid in texture, ensiform, plicate, glabrous, 9–12 in. long, ½–1 in. broad at the middle, with thick stramineous ribs and margins ; flowers in 3–4 congested erect spikes on a short common peduncle ; outer spathe-valve oblong-lanceolate, 1½–2 in. long, firm in texture, closely ribbed, glabrous ; perianth-tube cylindrical, 2–2½ in. long, dilated into a funnel at the top; segments oblong-spathulate, subequal, an inch long; stamens half as long as the limb; style-branches flattened at the tip. *Journ. Linn. Soc.* xvi. 165; *Handb. Irid.* 181.

WESTERN REGION : Little Namaqualand ; Uitkomst, 2000–3000 ft., *Drège,* 2628 !

10. B. namaquensis (Baker, Handb. Irid. 181); corm globose, ½ in. diam., with a neck 2 in. long ; leaves several, linear, 1–1½ in. long, spirally curled, very hairy; spikes 2, congested, subsessile on the surface of the soil ; spathe-valves oblong, acute, scariose, ½ in. long ; perianth-tube as long as the spathe ; segments dark purple, obovate-oblong, an inch long; style half as long as the perianth-segments, with declinate branches about 2 lines long.

WESTERN REGION : Little Namaqualand ; in sandy ground near Port Nolloth, *Bolus,* 6579 ! (not 4343 as quoted in the Handb. Irid.).

11. B. pygmæa (Baker in Journ. Linn. Soc. xvi. 165) ; corm ovoid, ½ in. diam. ; tunics brown, membranous; leaves 4–5, with a

petiole 2–3 in. long and a lanceolate, hairy, plicate lamina, 2–3 in.
long, $\frac{1}{3}$–$\frac{1}{2}$ in. broad; flowers 2–6 in an erect spike, on a short hairy
peduncle, reaching nearly to the top of the leaves; spathe-valves
oblong-lanceolate, hairy, $\frac{3}{4}$–1 in. long; perianth pale lilac; tube
funnel-shaped, as long as the spathe; segments unequal, oblong-
unguiculate, 1–1$\frac{1}{4}$ in. long; stamens half as long as the perianth-
limb. *Handb. Irid.* 181. *B. nana, Ker, Gen. Irid.* 154. *Gladiolus
nanus, Andr. Bot. Rep. t.* 137. *Ixia pigmœa, Burm. Prodr.* 1.

SOUTH AFRICA: without locality, *Zeyher!*
COAST REGION: Clanwilliam Div.; between Heeren Logement and Kanagas
Berg, *Drège,* 2625!

12. **B. Sprengelii** (Baker in Journ. Linn. Soc. xvi. 165); corm
globose, $\frac{1}{2}$ in. diam.; tunics brown, membranous; basal leaves 4–5,
with a hairy petiole about 2 in. long and an oblique, oblong, acute,
plicate lamina, 1–1$\frac{1}{2}$ in. long, hairy and moderately firm in texture;
flowers 1–2 in an erect spike reaching to the top of the leaves; outer
spathe-valve oblong-lanceolate, hairy, $\frac{3}{4}$–1 in. long; perianth pale
lilac; tube funnel-shaped, as long as the spathe; segments unequal,
oblong-unguiculate, 1–1$\frac{1}{4}$ in. long; stamens half as long as the
perianth-limb. *Handb. Irid.* 182. *B. pygmœa, Spreng. in Herb.
Zeyher.*

COAST REGION: Saldanha Bay, *Zeyher!*
Perhaps only a variety of *B. pygmœa,* with which Dr. Klatt unites it.

13. **B. flabellifolia** (Harv. ex Klatt in Linnæa xxxv. 380); corm
not seen; leaves 4–5, with a petiole 2–3 in. long, and an oblique,
hairy, deltoid, flabellate lamina about 2 in. long and 1 in. broad at the
truncate dentate tip; flowers 2–3, in an erect spike on a short
peduncle, reaching to the top of the leaves; spathe-valves oblong-
lanceolate, hairy, $\frac{3}{4}$–1 in. long; perianth lilac; tube filiform, as long
as the spathe; segments unequal, oblong-unguiculate, 1 in. long;
stamens more than half as long as the perianth-limb. *Baker in
Journ. Linn. Soc.* xvi. 165; *Handb. Irid.* 182.

WESTERN REGION: Little Namaqualand, *Whitehead!*

14. **B. cuneifolia** (Baker in Journ. Bot. 1876, 335); corm
globose, 1 in. diam.; tunics of fine matted brown fibres; leaves 5–6,
with an underground petiole sometimes 5–6 in. long, and an oblique,
deltoid, very plicate lamina of firm texture, and more or less hairy,
1$\frac{1}{2}$–2 in. long; flowers 5–6 in a congested spike reaching to the
top of the leaves; spathe-valves lanceolate, acuminate, 1$\frac{1}{2}$ in. long;
perianth bright lilac; tube as long as the spathe; limb above an inch
long, with unequal, oblong-unguiculate segments; stamens half as
long as the perianth-limb. *Journ. Linn. Soc.* xvi. 165; *Handb.
Irid.* 182.

SOUTH AFRICA: without locality, *Drège,* 2627!

15. **B. mononeura** (Baker, Handb. Irid. 182); corm not seen;
leaves linear, glabrous, rigidly coriaceous, 8–9 in. long, $\frac{1}{12}$ in. broad,

narrowed gradually from the middle to both ends, with a distinct
strongly-raised midrib and revolute edges; peduncle slender, glabrous,
one-flowered, $\frac{1}{2}$ ft. long, with a small sheathing leaf at the middle;
outer spathe-valve lanceolate, green, glabrous, closely ribbed, an inch
long; perianth-tube as long as the spathe; limb 1–1$\frac{1}{4}$ in. long, with
oblong-unguiculate, unequal segments, the upper one the largest;
stamens more than half as long as the perianth-limb; anther $\frac{1}{4}$ in.
long.

SOUTH AFRICA! without locality, *Thunberg!*

16. **B. fimbriata** (Baker in Journ. Linn. Soc. xvi. 166); corm ovoid,
$\frac{3}{4}$–1 in. diam.; tunics of fine brown matted fibres; leaves about 6,
with a long petiole and a slightly hairy, plicate, linear lamina of
moderately firm texture, 4–6 in. long, $\frac{1}{8}$–$\frac{1}{4}$ in. broad; stem a foot
long including the inflorescence; spikes 1–2, lax, with a slender
nearly glabrous axis; outer spathe-valve oblong-lanceolate, $\frac{1}{2}$ in. long,
subglabrous, green, with a brown scariose cusp; perianth pale lilac;
tube as long as the spathe; limb ringent, under an inch long, with
unequal, oblong-unguiculate segments; stamens nearly as long as the
perianth-limb. *Handb. Irid.* 182. *Antholyza fimbriata, Klatt in
Linnæa* xxxv. 299; *Ergänz.* 11.

WESTERN REGION: Little Namaqualand; between Zwartdoorn River and
Groen River, *Drège,* 2619!

17. **B. spiralis** (Baker, Handb. Irid. 183); corm not seen; basal
leaves 6–8, petioled, narrow linear, plicate, rather hairy, moderately
firm in texture, with a blade 6–9 in. long, $\frac{1}{12}$–$\frac{1}{8}$ in. broad, spirally
twisted towards the tip; stem a foot long including the inflorescence;
spikes 2–3, lax, with a flexuose, pubescent rachis; outer spathe-valve
oblong-lanceolate, $\frac{1}{2}$ in. long, green, finely pubescent, with a brown
scariose cusp; perianth pale lilac; tube as long as the spathe;
segments $\frac{3}{4}$ in. long, unequal, oblong-unguiculate; stamens as long as
the perianth-segments.

SOUTH AFRICA: without locality. Two specimens in the Forsyth herbarium,
purchased by Mr. Bentham in 1835.

18. **B. mucronata** (Ker in Konig and Sims' Ann. i. 234); corm
globose, 1 in. diam.; tunics of fine matted fibres; leaves 5-6, with a
petiole 3–4 in. long, and an ensiform, glabrous, plicate lamina $\frac{1}{2}$–$\frac{3}{4}$ in.
broad; flowers several in a dense spike on a short peduncle, reaching
about to the top of the leaves; outer spathe-valve oblong-lanceolate,
firm in texture, green, closely ribbed, glabrous or slightly hairy,
$\frac{3}{4}$–1 in. long; perianth lilac; tube as long as the spathe; segments
unequal, oblong-unguiculate, an inch long; stamens half as long as
the perianth-limb. *Ker, Gen. Irid.* 149; *Baker, Handb. Irid.* 183.
Gladiolus mucronatus, Jacq. Ic. t. 253; *Collect.* iv. 162.
B. scabrifolia, Brehmer ex Klatt Ergänz, 15.

VAR. β, **longicollis** (Baker, Handb. Irid. 183); perianth-tube longer than the
spathe, 18–21 lines long. *B. spathacea, Ker in Bot. Mag. t.* 638, *excl. syn. Thunb.*

COAST REGION: Clanwilliam Div.; near Holle River, under 1000 ft., *Drège,*
2624! Lange Vallei, *Drège,* 2623!

WESTERN REGION: Little Namaqualand, *MacOwan!*

19. B. occidentalis (Baker, Handb. Irid. 183) ; corm not seen ; leaves 4–5, with a plicate, ensiform, glabrous lamina a foot long, about ½ in. broad at the middle, tapering gradually to both ends ; flowers many, arranged in a panicle of three spikes, with a long peduncle, reaching about half-way up the leaves ; spathe-valves oblong-lanceolate, finely pilose, closely ribbed, an inch long ; perianth bright lilac ; tube filiform, 1 in. long, funnel-shaped at the top ; segments unequal, oblong-unguiculate, 1–1¼ in. long ; stamens half as long as the perianth-limb.

COAST REGION : Clanwilliam Div., *Mader in Herb. MacOwan*, 2169 !
May be a luxuriant variety of *B. mucronata.*

20. B. plicata (Ker in Bot. Mag. t. 576, excl. syn.) ; corm globose, ½ in. diam. ; tunics of brown matted fibres ; leaves 5–6, with a petiole 2–4 in. long, and a hairy, plicate, ensiform lamina ¼–½ in. broad, 3–6 in. long ; flowers fragrant, arranged in a simple or forked spike usually shorter than the leaves ; peduncle and rachis hairy ; outer spathe-valve oblong-lanceolate, hairy, about an inch long ; perianth lilac or red, with a tube equalling the spathe and unequal oblong-unguiculate segments an inch long ; stamens half as long as the perianth-limb ; style-branches flattened at the tip. *Ker in Konig and Sims' Ann.* i. 234 ; *Gen. Irid.* 149 ; *Baker, Handb. Irid.* 183. *Gladiolus fragrans, Jacq. Hort. Schoenbr. t.* 14. *G. plicatus, Linn. Sp. Pl. edit.* 2, 53, *ex parte. G. ringens, Thunb. Fl. Cap.* i. 214, *ex parte. B. cærulescens and B. villosa, Eckl. Top. Verz.* 32. *B. punctata, Klatt, Ergänz.* 14.

VAR. β, **B. maculata** (Klatt, Ergänz. 15) ; dwarf, with a small corm, narrow linear lines, and not more than 2–3 crowded, erect flowers. *B. angustifolia, Eckl. Top. Verz.* 31.
VAR. γ, **Forsteri** (Baker, Handb. Irid. 184) ; lower leaves with a broad half-deltoid lamina developed only on the outer side of the midrib.

SOUTH AFRICA : without locality, *Thunberg!* Var. γ, *Forster!*
COAST REGION : Cape Flats, *Burchell*, 8553 ! *Bolus*, 4750 ! *MacOwan and Bolus, Herb. Norm.*, 273 ! Table Mountain and Devils Mountain, under 1000 ft., *Drège! Burchell*, 8489 ! *MacOwan and Bolus, Herb. Norm.*, 272 ! Camps Bay, *MacOwan*, 2559 ! Between Paardeneiland, Blueberg and Tygerberg, *Drège*, 8387 ! Between Paarl Mts. and Paardeberg, under 1000 ft., *Drège!* Paarl Mts., *Drège*, 8396 ! Laauws Kloof, near Groene Kloof, *Drège*, 8384 ! Var. β, Cape Flats at Doornhoogde, *Zeyher!*

21. B. disticha (Ker in Konig and Sims' Ann. i. 234) ; corm globose, ¾–1 in. diam. ; tunics of fine matted brown fibres ; leaves about 6, with a petiole 2–3 in. long, and an oblique, ensiform, plicate, hairy lamina 4–6 in. long, ½–¾ in. broad ; flowers in a simple spike reaching about to the top of the leaves ; outer spathe-valve oblong-lanceolate, very hairy, closely ribbed, ¾–1 in. long ; perianth pale lilac ; tube slender, ½–¾ in. longer than the spathe, funnel-shaped at the top ; limb 1–1¼ in. long, with unequal oblong-unguiculate segments ; stamens half as long as the perianth-limb ; style-branches ⅙ in. long, cuneate at the tip. *Bot. Mag. t.* 626 ; *Gen. Irid.* 149 ; *Baker, Handb. Irid.* 184. *Gladiolus plicatus, Jacq. Ic. t.* 237.

COAST REGION: Table Mountain, *Ecklon*, 124b! Caledon Div.; Klein River Berg, *Zeyher*, 4003 ! George Div.; Kaymans River, *Burchell*, 5806! Lower Albany, *Hutton!*

WESTERN REGION: Little Namaqualand; between Koper Berg, Silver Fontein, and Kaus, 2000–3000 ft., *Drège*, 8386 ! between Pedros Kloof and Lily Fontein, 3000–4000 ft., *Drège*, 8401 ! without special locality, *Scully*, 26!

Differs from *B. plicata* mainly by its longer perianth-tube, and is scarcely more than a variety.

32. B. stricta (Ker in Konig and Sims' Ann. i. 234) ; corm globose, ⅓ in. diam., with a long neck and tunics of fine matted fibres ; basal leaves about 6, moderately firm in texture, with an ensiform hairy lamina 4–6 in. long ; stem ½–1 ft. long including the inflorescence, which overtops the leaves ; spikes 1–3, moderately dense, many-flowered, with a very hairy axis ; spathe-valves oblong, firm in texture, closely ribbed, very hairy, ½–¾ in. long ; perianth various in colour, red or lilac ; tube funnel-shaped at the top, as long as or a little longer than the spathe ; segments subequal, obtuse, oblanceolate-oblong, ¾–1 in. long ; stamens about half as long as the perianth-segments ; style-branches ⅛ in. long, flattened at the tip. *Bot. Mag. tt. 621 and 637 ; Gen. Irid. 150 ; Baker, Handb. Irid. 184. Gladiolus strictus, Ait. Hort. Kew. i. 63 ; Red. Lil. t. 90. G. plicatus, Linn. Sp. Plant. edit. 2. 53, ex parte ; Thunb. Diss. p. 20 and Prodr. 9, ex parte ; Fl. Cap. i. 211. G. nervosus, Lam. Encyc. ii. 724. B. villosa, Ker in Konig and Sims' Ann. i. 234 ; Bot. Mag. t. 583. Ixia villosa, Ait. Hort. Kew. i. 58. I. punicea, Jacq. Ic. t. 287. I. flabelliformis, Salisb. Prodr. 37. Gladiolus villosus and puniceus, Vahl, Symb. ii. 114–5. B. purpurea, Ker in Bot. Mag. sub t. 1019. Ixia purpurea, Jacq. Ic. t. 286.*

VAR. β, **B. rubro-cyanea** (Ker in Konig and Sims' Ann. i. 234) ; perianth-limb lilac, with a red throat. *Ixia rubro-cyanea, Jacq. Ic. t.* 285 ; *Curt. Bot. Mag. t.* 410 ; *Reich. Exot. t.* 30. *B. rubro-cœrulea, Pritz. Ind. Ic.* 134.

VAR. γ, **B. obtusifolia** (Ker in Konig and Sims' Ann. i. 234) ; dwarfer than the type, with a shorter stem and a few larger, pale lilac flowers, with a more funnel-shaped tube. *Ixia villosa, Jacq. Ic. t.* 284.

VAR. δ, **B. sulphurea** (Ker in Konig and Sims' Ann. i. 234) ; flowers milk-white or sulphur-yellow. *Bot. Mag. t.* 1053. *Gladiolus sulphureus, Jacq. Ic. t.* 239. *G. plicatus, Andr. Bot. Rep. t.* 268.

VAR. ε, **B. angustifolia** (Sweet, Hort. Brit. edit. ii. 499) ; leaves linear. *B. stricta, var., Ker in Bot. Mag. t.* 637. *Ixia villosa, var., Jacq. Frag. t.* 14, *fig.* 3.

VAR. ζ, **B. reflexa** (Eckl. Top. Verz. 33) ; flowers smaller than in the type, the lower of the spike deflexing.

SOUTH AFRICA: without locality, *Thunberg!*
COAST REGION: between Paarl and Pont, below 1000 ft., *Drège*, 8390a! Caledon Div.; near Appels Kraal, by the Zonder Einde River, *Zeyher*, 4001! Var. β, Malmesbury Div.; near Groene Kloof, 500 ft., *MacOwan and Bolus, Herb. Norm.*, 584! 586! Var. δ, Tulbagh Div.; Halfmanshof, near Twenty-four Rivers, 400 ft., *MacOwan and Bolus, Herb. Norm.*, 543!

Besides the varieties cited, *B. undulato-venosa, parviflora, quadripartita* and *multiflora, Klatt, Ergänz.* 16–17, belong here.

23. B. macrantha (MacOwan in Journ. Linn. Soc. xxv. 394) ; corm small, ovoid ; tunics of matted fibres ; basal leaves 4–5, lanceo-

I

late, 2–3 in. long, hairy, petioled; peduncle a span long, simple or forked; flowers 3–4 in a lax spike; outer spathe-valve oblong-lanceolate, 1–1½ in. long, brown at the tip, hairy; perianth-tube ½ in. long; segments obovate, sulphur-yellow tinged with purple at the base, 1½ in. long; stamens half as long as the limb; anthers linear, purple. *Baker, Handb. Irid.* 184.

COAST REGION: Malmesbury Div.; moist, sandy places near Darling, 400 ft., *MacOwan and Bolus, Herb. Norm.,* 811!

Flowered at Kew from plants sent by Prof. MacOwan, June, 1888.

24. B. secunda (Ker, Gen. Irid. 154); corm not seen; leaves about 6, petioled, with an ensiform, densely hairy, plicate lamina 5–6 in. long, ½ in. broad; stem a foot long including the inflorescence, branched from halfway down; spikes many-flowered, dense, secund; spathe-valves oblong, brown, membranous, glabrous, ¼–⅓ in. long, with 2–3 teeth at the top, as in *Ixia;* perianth-tube funnel-shaped, ⅓ in. long; segments subequal, oblong, ¾ in. long; stamens not quite as long as the perianth-segments. *Baker, Handb. Irid.* 184. *Gladiolus secundus, Thunb. Fl. Cap.* i. 215; *Skrivt. Nat. Selsk. Kjobenh.* vi. *t.* 4.

SOUTH AFRICA: without locality, *Thunberg!*
Seen only in the Thunbergian herbarium.

25. B. ringens (Ker in Konig and Sims' Ann. i. 233); corm globose, 1 in. diam.; tunics brown, membranous; stem pilose, 1–1½ ft. long, ending in a long sterile branch; basal leaves many, narrow linear, thick in texture, strongly ribbed, plicate, ½–1 ft. long, glabrous, ¼ in. broad at the middle; flowers 8–12 in a dense secund spike on a short, arcuate branch; outer spathe-valve lanceolate, 1½–1¾ in. long, firm in texture, finely ribbed, pilose; perianth bright red; tube greenish, ½ in. longer than the spathe, openly funnel-shaped at the top; upper segment standing forward, oblong, with a distinct claw, 1½–2 in. long; the others much shorter, lanceolate or oblanceolate-spathulate; stamens much longer than the upper segment; style-branches subulate, ⅙ in. long. *Lodd. Bot Cab. t.* 1006; *Bot. Mag. t.* 6667; *Ker, Gen. Irid.* 152; *Baker, Handb. Irid.* 185. *Antholyza ringens, Linn. Sp. Plant. edit.* 2. i. 54; *Thunb. Diss. No.* 5; *Prodr.* 7; *Fl. Cap.* i. 167 (*Commel. Hort.* i. 81. *t.* 41; *Breyn. Ic.* 21, *t.* 8, *fig.* 1).

SOUTH AFRICA: without locality, *Masson!*
COAST REGION: Cape Flats, under 500 ft.; near Cape Town, *Bolus,* 3744! near Constantia, *Thunberg!* near Duiker Vley, *MacOwan and Bolus, Herb. Norm.,* 271! between Cape Town and Simons Bay, *Burchell,* 8563! and at Wynberg, *Drège,* 8341! Clanwilliam Div., *Zeyher!* Malmesbury Div.; Zwartland, *Thunberg!*

26. B. Thunbergii (Ker in Konig and Sims' Ann. i. 233); corm large, globose, with a long neck; tunics membranous; leaves linear or ensiform, 1–1½ ft. long, ⅙–¾ in. broad, firm in texture, plicate, finely pilose; stem 2–3 ft. long including the inflorescence, which is a lax panicle of 4–8 dense secund spikes 2–3 in. long; spathe-valves oblong-lanceolate, hairy, 1–1¼ in. long; perianth-tube broadly funnel-shaped, ½–¾ in. longer than the spathe; limb reddish; upper segment oblong-spathulate, $\frac{3}{4}$ in. long, the others shorter, lanceolate; stamens reaching to the tip of the upper segment. *Ait. Hort. Kew. edit.* 2, i. 104; *Ker, Gen. Irid.* 152; *Baker, Handb. Irid.* 185. *Antholyza plicata, Linn. fil. Suppl.* 96; *Thunb. Diss. No.* 6; *Prodr.* 7; *Fl. Cap.* i. 169.

SOUTH AFRICA: without locality, *Masson!* *Hort. Fothergill.* in 1788!

COAST REGION: Piquetberg Div.; on sand dunes near the seashore below Verloren Vallei, *Thunberg!*

WESTERN REGION: Little Namaqualand; sandy places by the sea, near Port Nolloth, *MacOwan and Bolus, Herb. Norm.*, 699! and without precise locality, *Scully*, 222!

XXIV. MELASPHŒRULA, Ker.

Perianth slit down nearly to the ovary; segments oblong-lanceolate or lanceolate, very acuminate. *Stamens* short, inserted at the base of the perianth; filaments arcuate, filiform; anthers oblong, versatile. *Ovary* short, 3-lobed, 3-celled; ovules 2–3 in a cell; style short, filiform, its branches entire. *Fruit* a membranous loculicidal capsule, with three deep, acutely-angled lobes. *Seeds* oblong, with a thin brown testa, thickened and spongy at the top.

DISTRIB. Endemic; monotypic.

1. **M. graminea** (Ker in Bot. Mag. t. 615); corm globose, $\frac{1}{2}$ in. diam.; tunics thin, brown; leaves about 6 in a distichous basal rosette, linear, thin in texture, $\frac{1}{2}$–1 ft. long, $\frac{1}{2}$ in. broad; stem slender, 1–2 ft. long including the inflorescence, which is a broad, lax panicle; spikes few-flowered, with a very flexuose, slender rachis; spathe-valves ovate, herbaceous, $\frac{1}{4}$–$\frac{1}{3}$ in. long; perianth yellowish-green, $\frac{1}{3}$–$\frac{1}{2}$ in. long; segments all very acuminate, with purplish-black veins, the lower narrower and more spreading; stamens half as long as the perianth; capsule $\frac{1}{3}$ in. diam., very short, acutely 3-lobed. *Ker in Konig and Sims' Ann.* i. 232; *Baker, Handb. Irid.* 189. *Gladiolus gramineus, Linn. fil. Suppl.* 95; *Thunb. Diss. No.* 26; *Prodr.* 9; *Jacq. Ic. t.* 236; *Andr. Bot. Rep. t.* 62. *Aglœa graminea, Eckl. Top. Verz.* 44. *Gladiolus ramosus, Linn. Sp. Plant. edit.* 2. i. 53. *Phalangium ramosum, Burm. Prodr.* 3, excl. syn. *Diasia iridifolia, DC. in Bull. Phil.* 1803, 151; *Red. Lil. t.* 54. *D. graminifolia, DC. loc. cit.; Red. Lil. t.* 163. *Melasphœrula intermedia, graminea, iridifolia and parviflora, Sweet, Hort. Brit. edit.* 2, 502.

COAST REGION: Table Mountain, *Cooper*, 3206! *Harvey!* *Bolus*, 2827! Clanwilliam Div., *Zeyher!* Paarl Mountains 1000–2000 ft., *Drège!* Malmesbury Div.; near Groene Kloof, *Thunberg! Bolus*, 1485! near Malmesbury, 400 ft., *MacOwan, Herb. Aust. Afr.*, 1544! Tulbagh Div., *Galpin*, 72! Knysna Div.; Ruigte Vallei, under 500 ft., *Drège!*

M. iridifolia, G. Don, merely differs from the type by its shorter leaves of firmer texture, and *M. parviflora, Lodd. Bot. Cab. t.* 1444, by its smaller flowers.

XXV. SPARAXIS, Ker.

Perianth with a short, straight, cylindrical tube, dilated into a funnel in the upper half, and six subequal, oblong, ascending segments. *Stamens* inserted at the throat of the perianth-tube, unilateral, assurgent; filaments short, filiform; anthers lanceolate, contiguous. *Ovary* 3-celled; ovules crowded in the cells; style filiform, protruded from the perianth-tube, with 3 falcate, subulate, entire branches. *Capsule* small, membranous, turbinate, loculicidally 3-valved. *Seeds* globose or angled by pressure.

Rootstock a corm with finely-reticulated tunics; leaves in a fan-like, distichous rosette; flowers few, very showy, arranged in simple or panicled spikes; spathe-valves membranous, lacerated, wrapped round the ovary and perianth-tube.

DISTRIB. Endemic. Only one species in a broad sense, varying indefinitely in the size and colouring of the flowers.

Throat of perianth same colour as segments:
Flowers smaller　...　...　...　...　...　...	(1)	bulbifera.
Flowers larger　...　...　...　...　...　...	(2)	grandiflora.

Throat of perianth bright yellow, with a dark blotch below
each segment ...　...　...　...　...　...　...　(3) tricolor.

1. S. bulbifera (Ker in Konig and Sims' Ann. i. 226); corm globose, $\frac{1}{2}$–$\frac{3}{4}$ in. diam., with fleshy root-fibres, and finely-reticulated, white, membranous tunics; stems erect, terete, simple or forked, $\frac{1}{2}$–$1\frac{1}{2}$ ft. long, with 2–3 reduced sheathing leaves low down, often bearing bulbillæ in their axils; leaves of the basal rosette about 4, lanceolate or linear, $\frac{1}{2}$–1 ft. long, moderately firm in texture, the outer spreading; flowers few to a spike, sometimes solitary, typically yellow; spathes $\frac{1}{3}$ in. long, membranous, whitish, with copious brown dots and lines, wrapped tightly round the ovary and tube, deeply laciniated at the top, the processes brown and whip-like; perianth-tube $\frac{1}{2}$ in. long, dilated into a funnel in the upper half; segments oblong, $\frac{1}{2}$–$\frac{3}{4}$ in. long; anthers $\frac{1}{4}$ in. long, protruded from the perianth-tube; style-branches overtopping the anthers. *Ker, Gen. Irid.* 94; *Klatt in Linnæa* xxxii. 748; *Baker, Handb. Irid.* 197. *Ixia bulbifera, Linn. Sp. Plant. edit.* 2. i. 51; *Burm. Prodr.* 1; *Ker in Bot. Mag. t.* 545; *Red. Lil. t.* 128; *Andr. Bot. Rep. t.* 48. *Ixia bulbifera, var. flava, Thunb. Diss. No.* 17.

VAR. β, S. **violacea** (Eckl. Top. Verz. 27); flowers dark purple. *Klatt, Ergänz.* 55.

SOUTH AFRICA: without locality, *Zeyher! Thom! Burchell!*
COAST REGION: Malmesbury Div.; Zwartland, *Thunberg!* between Groene Kloof and Saldanha Bay, below 500 ft., *Drège,* 8346a! Var. β, Caledon Div., *Zeyher!*

S. albiflora, Eckl. Top. Verz. 28, is a form with the flower whitish inside.

2. S. grandiflora (Ker in Konig and Sims' Ann. i. 225); habit, corm, leaves and spathe just as in *S. bulbifera,* but the flowers larger, with a limb an inch or more long, and oblong segments often $\frac{1}{2}$ in. broad, usually yellow or purple; spathes $\frac{3}{4}$–1 in. long, laciniated half-way down; anthers also larger than in *S. bulbifera. Ker in Bot. Mag. t.* 779; *Bot. Reg. t.* 258; *Gen. Irid.* 94; *Baker, Handb. Irid.* 197. *Ixia grandiflora, Delaroche, Descr.* 23; *Houtt. Handl.* xii. 29, *t.* 77, *fig.* 3; *Ker in Bot. Mag. t.* 541; *Red. Lil. tt.* 139 *and* 362. *Ixia bulbifera, Thunb. Diss. No.* 17, *ex parte. I. aristata, Ait. Hort.*

Kew. i. 57; *Andr. Bot. Rep. t.* 87, *non Thunb. I. monanthos, Delaroche, Descr.* 21. *I. uniflora, Linn. Mant.* 27 ; *Jacq. Ic. t.* 283. *I. holosericea, Jacq. Hort. Schoenbr.* i. 9, *t.* 17. *S. fimbriata, Ker in Konig and Sims' Ann.* i. 226. *I. fimbriata, Lam. Encyc.* iii. 339 ; *Ill.* i. 111. *Belamcanda semiflexuosa, Moench, Meth. Suppl.* 214. *S. lacera, Ker in Konig and Sims' Ann.* i. 226. *S. atropurpurea and miniata, Klatt, Ergänz.* 55. *S. Liliago, Sweet, Hort. Brit. edit.* 2, 501. *Ixia Liliago, Red. Lil. t.* 109. *I. anemonæflora, Red. Lil. t.* 85. *S. stellaris, D. Don in Sweet, Brit. Flow. Gard. ser.* 2, *t.* 383.

COAST REGION; Lion Mountain, *Thunberg ! Drège ! MacOwan and Bolus, Herb. Norm.,* 276! Malmesbury Div.; near Groene Kloof, 300 ft., *Bolus,* 4344! Worcester Div., *Zeyher ! Cooper,* 3178! Tulbagh Div.; Steendaal, near Tulbagh, 900 ft., *MacOwan and Bolus, Herb. Norm.,* 583! *MacOwan,* 2679! near Tulbagh, *Thunberg !*

Exceedingly variable in flower colouring, the principal named forms being *atropurpurea,* dark purple, *anemonæflora,* pale yellow, *Liliago,* white flushed with claret purple outside, and *stellaris,* dark purple, with oblanceolate, acute, narrower segments.

3. S. tricolor (Ker in Konig and Sims' Ann. i. 225) ; differs only from *S. grandiflora* in the colour of the flowers, which are fulvous, dark purple, yellow, more or less flushed with brown purple, or white flushed on the back with claret-purple in the upper half, but have always a bright yellow throat with a dark cuneate, emarginate blotch at the base of each segment. *Ker in Bot. Mag. t.* 1482, *and sub t.* 779; *Ait. Hort. Kew. edit.* 2, i. 85 ; *Ker, Gen. Irid.* 93; *Baker, Handb. Irid.* 197. *Ixia tricolor, Curt. in Bot. Mag. t.* 381; *Schneev. Ic. t.* 39; *Red. Lil. t.* 129. *Streptanthera tricolor and lineata, Klatt, Ergänz.* 56. *Sparaxis versicolor, Sweet, Brit. Flow. Gard. t.* 160. *S. Griffini and blanda, Sweet, Hort. Brit. edit.* 2, 501. *Klatt, Ergänz.* 55. *S. lineata, Sweet, Brit. Flow. Gard. ser.* 2, *t.* 131. *S. meleagris and cana, Eckl. Top. Verz.* 27-28.

COAST REGION : Stellenbosch, *Sanderson,* 942!

The favourite species of cultivators. Figures of a series of garden forms will be found in Van Houtte's *Flore des Serres,* vol. ii. July, 1846.

Doubtful species.

4. S. fragrans (Ker in Konig and Sims' Ann. i. 224) ; bulb small, ovoid ; tunics brown, membranous ; stem simple, slender, terete, under a foot long; leaves about 6 in a basal rosette, and one suprabasal, long, linear, glabrous; flowers 2, in a lax, erect spike ; spathe-valves short, membranous, deeply lacerated ; perianth-tube very short, cylindrical ; segments spreading, orange-yellow, under an inch long, narrow oblong ; stamens half as long as the perianth-limb ; anthers large ; style reaching to the top of the stamens ; branches spreading, $\frac{1}{4}$ in. long. *Gen. Irid.* 93. *Ixia fragrans, Jacq. Ic. t.* 274.

SOUTH AFRICA : without locality.

Known only from Jacquin's figure, drawn about 1790 from a garden specimen.

XXVI. TRITONIA, *Ker.*

Perianth with a short or elongated cylindrical tube, funnel-shaped at the top ; segments of the limb obovate or oblong, subequal or rather unequal. *Stamens* unilateral, contiguous, arcuate, inserted at the base of the funnel-shaped portion of the perianth-tube ; filaments filiform ; anthers usually versatile. *Ovary* 3-celled; ovules superposed ; style filiform ; branches simple, short, spreading. *Fruit* a small oblong, membranous capsule, dehiscing loculicidally. *Seeds* small, globose, or angled by pressure.

Habit of *Ixia.* Rootstock a corm with fibrous reticulated tunics ; basal leaves in a fan-shaped rosette, ensiform or linear ; flowers various in colour, arranged in simple or panicled spikes ; spathe-valves oblong, brown, dentate at the tip.

DISTRIB. One of the species extends to the tropical mountains of Central Africa, and a second is endemic there.

Subgenus TRITONIXIA. Flowers large ; segments subequal, obovate, much imbricated ; spikes secund.
  Flower bright red, or fulvous-yellow :
    Segments obovate-cuneate :
      Outer segments not spotted ...   .   (1) **crocata.**
      Outer segments with a large purple-black
        spot on the claw   ...   .   (2) **deusta.**
    Segments obovate-spathulate, the claw with an
      inflexed hyaline edge   ...   .   (3) **hyalina.**
  Flower white or pink   ...   .   (4) **squalida.**
Subgenus DICHONE. Flowers small ; segments equal, oblong.
  Leaves plane :
    Leaf long   ...   .   (5) **scillaris.**
    Leaf short   ...   .   (6) **trinervata.**
  Leaves much frilled on the edge...   ...   .   (7) **undulata.**
Subgenus TRITONIA PROPER. Flowers large or small ; segments of the limb oblong, the three lower without any callus on the face ; tube dilated at the top into a large funnel.
  Tube much longer than the spathe :
    Flowers green   ...   .   (8) **viridis.**
    Flowers bright pink, with a tube dilated in the
      upper half   ...   .   (9) **Cooperi.**
    Flowers white or pale pink, with a tube dilated
      only at the top :
      Leaf linear, crisped   ...   .   (10) **crispa.**
      Leaf plane, linear-ensiform   ...   .   (11) **pallida.**
      Leaf plane, narrow linear   ...   .   (12) **Bakeri.**
  Tube short (at most an inch long) :
    Perianth-segments longer than the tube :
      Leaf subulate or very narrow :
        Perianth-tube ⅛ in. long ...   .   (13) **ventricosa.**
        Perianth-tube ¼ in. long ...   .   (14) **kamisbergensis.**
      Leaf linear, ¼ in. long :
        Perianth-tube ⅛–¼ in. long   ...   ...   (15) **disticha.**
        Perianth-tube ¼–⅓ in. long   ...   ...   (16) **pauciflora.**
        Perianth-tube ½ in. long ...   .   (17) **Templemanni.**
    Perianth-segments about as long as the tube :
      Flowers white or pale pink :
        Leaves terete   ...   .   (18) **teretifolia.**
      Leaves linear :
        Perianth-tube funnel-shaped only at
        the top   ...   .   (19) **dubia.**

Perianth-tube funnel-shaped from the
    middle :
    Perianth $\frac{1}{4}-\frac{3}{4}$ in. long   ...   ... (20) **Kraussii**.
    Perianth 1–1$\frac{1}{4}$ in. long :
        Leaves $\frac{1}{4}-\frac{1}{2}$ in. broad ...   ... (21) **lineata**.
        Leaves under $\frac{1}{12}$ in. broad   ... (22) **Wilsoni**.
Flowers bright red ...   . (23) **rosea**.
Perianth-segments shorter than the tube :
    Perianth-tube slender to the top ...   . (24) **graminifo ia**.
Perianth-tube funnel-shaped in the upper half :
    Flowers bright red   ...   . (25) **laxifolia**.
    Flowers orange-yellow ...   . (26) **Pottsii**.
    Flowers white ...   . (27) **watsonioides**.

Subgenus MONTBRETIA. Perianth-tube funnel-shaped above the middle ; three lower segments furnished with a spreading callus on the claw.
Callus large, oblong. Flowers bright red :
    Leaves $\frac{1}{12}$ in. broad...   ...   . (28) **Nelsoni**.
    Leaves $\frac{1}{4}-\frac{1}{2}$ in. broad   ...   . (29) **securigera**.
Callus small. Flowers yellow ...   . (30) **flava**.

Subgenus STENOBASIS. Perianth-segments unequal, all distinctly unguiculate.
    The only species   ...   . (31) **unguiculata**.
    Garden hybrid   ...   . (32) **crocosmæflora**.

1. T. crocata (Ker in Konig and Sims' Ann. i. 228) ; corm globose, $\frac{3}{4}$–1 in. diam.; tunics of matted, pale brown, reticulated fibres ; basal leaves 4–6, ensiform, spreading, moderately firm in texture, 4–8 in. long, $\frac{1}{3}-\frac{1}{2}$ in. broad ; stem slender, terete, simple or branched low down, 1–1$\frac{1}{2}$ ft. long including the inflorescence ; flowers 4–10 in a lax secund spike ; spathe-valves oblong, $\frac{1}{3}-\frac{1}{2}$ in. long, green in the lower part, brown and toothed at the tip ; perianth-tube slightly exserted, cylindrical, dilated into a small funnel at the top ; segments obovate-spathulate, $\frac{3}{4}$–1 in. long, much imbricated, bright fulvous-yellow, not distinctly spotted on the claw ; stamens one-third the length of the limb ; anthers purple, lanceolate ; style-branches $\frac{1}{6}$ in. long, overtopping the stamens. *Gen. Irid.* 120 ; *Baker, Handb. Irid.* 190. *Ixia crocata, Linn. Sp. Plant. edit.* 2. 52 ; *Linn. fil. Pl. Rar. Hort. Upsal.* 13. *t.* 7 ; *Thunb. Diss. No.* 20 ; *Prodr.* 10 ; *Fl. Cap.* i. 245 ; *Curt. Bot. Mag. t.* 184 ; *Wendl. Bot. Beobacht.* 5. *t.* 1. *fig.* 4. *Gladiolus crocatus, Pers. Syn.* i. 44. *Tritonixia crocata, Klatt, Ergänz.* 22. *Ixia iridifolia, Delaroche, Descr.* 24 (*Miller, Ic.* 160, *t.* 239, *fig.* 2).

VAR. β. **T. miniata** (Ker in Konig and Sims' Ann. i. 228) ; flowers bright red. *Bot. Mag. t.* 609 ; *Gen. Irid.* 121. *Ixia miniata, Jacq. Hort. Schoenbr. t.* 24. *Tritonixia miniata, Klatt, Ergänz.* 22. *Ixia crocata, Red. Lil. t.* 335.

COAST REGION : Rondebosch, *Drège !* Tulbagh, *Pappe!* mountains between Stellenbosch and Helder Berg, 1000 ft., *MacOwan and Bolus, Herb. Norm.*, 953 ! Swellendam Div.; *Thunberg !* hills on both sides of Buffeljaghts River, *Zeyher*, 3974 ! Mossel Bay Div.; between Little Brak River and Hartenbosch, *Burchell*, 6199 ! and between Zout River and Duyker River, *Burchell*, 6371 !

T. purpurea, Ker in Bot. Mag. sub. t. 1275, *and T. sanguinea, coccinea, and aurantiaca, Eckl. Top. Verz.* 29, are varieties of this, differing in the colour of the flower from the type as indicated by their names.

2. T. deusta (Ker in Konig and Sims' Ann. i. 228) ; differs only from *T. crocata* by the three outer segments of the perianth being

furnished with a purple-black blotch on the claw. *Bot. Mag. t.* 622 ;
Gen. Irid. 121; *Baker, Handb. Irid.* 190.   *Ixia deusta, Ait. Hort.*
Kew. i. 60.   *I. miniata, Red. Lil. t.* 89.   *I. gibba, Salisb. Prodr.*
38.   *I. crocata, var., Thunb. Diss. No.* 20 ; *Andr. Bot. Rep. t.* 134.
Tritonixia deusta, Klatt, Ergänz. 22.

SOUTH AFRICA : without locality, *Thom,* 45 !
COAST REGION : Caledon Div.; Zonder Einde River, *Ecklon and Zeyher,* 103 !
Swellendam Div. ; Breede River, *Gill !* hills on both sides of Buffeljaghts
River, 1000-2000 ft., *Zeyher,* 3974 ! and without precise locality, *Thunberg !*

3. **T. hyalina** (Baker in Journ. Linn. Soc. xvi. 163) ; differs
from *T. crocata* only by the segments of perianth-limb being
spathulately narrowed below the middle into a claw with an inflexed
hyaline margin.   *Handb. Irid.* 191.   *Ixia hyalina, Linn. fil. Suppl.*
91.   *Tritonixia hyalina, Klatt, Ergänz.* 22.   *Tritonia fenestrata,*
Ker in Konig and Sims' Ann. i. 228 ; *Bot. Mag. t.* 704 ; *Gen.*
Irid. 120.   *Ixia fenestrata, Jacq. Ic. t.* 289, *non Thunb.*

SOUTH AFRICA : without locality, *Thunberg ! Thom,* 973 !

4. **T. squalida** (Ker in Bot. Mag. t. 581) ; corm, habit, and leaves
of *T. crocata,* from which it differs by the colour of the flower,
which is white more or less flushed with pink ; the veins are often
pink on a white groundwork, and there is an obscure yellow linear
blotch at the base of the inner or all the segments.   *Ait. Hort. Kew.*
edit. 2. i. 92 ; *Ker, Gen. Irid.* 119 ; *Baker, Handb. Irid.* 191.   *Ixia*
squalida, Ait. Hort. Kew. i. 61.   *Tritonixia squalida, Klatt,*
Ergänz. 22.   *I. similis, Salisb. Prodr.* 38.   *I. crocata, var., Burm.,*
Prodr. 1.

SOUTH AFRICA : without locality, *Thom ! Pappe !*

5. **T. scillaris** (Baker in Journ. Linn. Soc. xvi. 163) ; corm
globose, $\frac{1}{3}-\frac{1}{2}$ in. diam.; tunics of thick parallel fibres ; basal leaves
4-6, ensiform, glabrous, moderately firm in texture, 4-6 in. long,
$\frac{1}{4}-\frac{1}{2}$ in. broad, not crisped ; stem about 1 ft. long including the
inflorescence, simple or branched ; flowers in a lax spike 3-4 in.
long with a flexuose axis ; spathe-valves oblong, moderately firm in
texture, $\frac{1}{8}-\frac{1}{6}$ in. long, deeply 2-3-fid ; perianth pink ; tube cylindrical, rather longer than the spathe ; segments equal, oblong, $\frac{1}{4}-\frac{1}{3}$ in.
long ; stamens one-third the length of the perianth-limb ; anthers $\frac{1}{12}$ in.
long, equalling the filaments ; style not longer than the perianth-
tube; branches falcate, $\frac{1}{12}$ in. long.   *Handb. Irid.* 191.   *Ixia scillaris,*
Linn. Sp. Plant. edit. 2. 52 ; *Burm. Prodr.* 1 ; *Houtt. Handl.* xi.
33, *t.* 77, *fig.* 2 ; *Red. Lil. t.* 127, *non Thunb.*   *I. pentandra, Linn.*
fil. Suppl. 92 ; *Thunb. Diss. No.* 22 ; *Prodr.* 10.   *Agretta pentandra,*
Eckl. Top. Verz. 23.   *Hesperantha pentandra, Herb. Drège.*
I. reflexa, Andr. Bot. Rep. t. 14.   *I. rotata, Ker in Andr. Recens.* 3.
I. retusa, Salisb. Prodr. 35.   *I. polystachya, Jacq. Ic. t.* 275.   *Ker*
in Bot. Mag. t. 629, *non Linn.*

VAR. β. **T. stricta** (Klatt, Ergänz. 23) ; leaves linear, rigid in texture, with
thickened stramineous midrib and margins. *Agretta stricta, Eckl. Top.*
Verz. 23.

SOUTH AFRICA: without locality, *Masson! Forster! Burchell! Pappe!*
COAST REGION: Table Mountain, 250 ft., *MacOwan and Bolus, Herb. Norm.*,
274! *Ecklon*, 9b! 546! Lion Mountain, *Bolus*, 2826! Paarl Mountains,
Drège! Malmesbury Div.; Klipberg, *Drège!* Var. β, on the Zwartberg near
the Hot Springs, Caledon, *Zeyher*, 4008!

6. T. trinervata (Baker, Handb. Irid. 191); corm globose; tunics
thick, fibrous; stem simple, slender, erect, a foot long; leaves
3, one basal rudimentary, one arcuate, lanceolate, firm in texture,
glabrous, 2–3 in. long, $\frac{1}{4}-\frac{1}{3}$ in. broad, with three strong, stramineous
ribs, the uppermost a loose stem-sheath 3–5 in. long; flowers 4–6 in
a lax spike; spathe-valves ovate scariose, strongly cuspidate, $\frac{1}{8}-\frac{1}{6}$ in.
long; perianth white, with a tinge of pink; tube slender, curved,
cylindrical, $\frac{1}{6}-\frac{1}{4}$ in. long; limb campanulate, $\frac{1}{2}$ in. long, with oblong
segments; stamens half as long as the limb; anthers yellow, curved,
equalling the filaments.

COAST REGION: Caledon Div.; near Appels Kraal, by the Zonder Einde
River, *Ecklon and Zeyher*, 242!

Near *T. scillaris*, from which it differs by its fewer and larger flowers, short
leaf, and strongly cuspidate spathe-valves.

7. T. undulata (Baker in Journ. Linn. Soc. xvi. 163); corms
ovoid, cæspitose; tunics of fine fibres; basal leaves 3–5, linear, firm
in texture, glabrous, 2–4 in. long, the margin undulated and plicate
down nearly to the midrib; stem slender, terete, usually simple,
$\frac{1}{2}-1$ ft. long including the inflorescence; spike lax, finally
3–4 in. long, with a very flexuose rachis; spathe-valves oblong,
$\frac{1}{3}-\frac{1}{2}$ in. long, deeply 2–3-fid; perianth pink; tube cylindrical, rather
longer than the spathe, and equal oblong segments $\frac{1}{4}-\frac{1}{3}$ in. long;
stamens one-third the length of the limb; anther oblong, equalling
the filament; style as long as the tube, branches $\frac{1}{8}$ in. long. *Handb.
Irid.* 191. *Ixia undulata, Burm. Prodr.* 1. *I. crispa, Linn. fil.
Suppl.* 91; *Thunb. Diss. No.* 8, *t.* 2, *fig.* 3; *Prodr.* 9; *Fl. Cap.* i.
226; *Ker in Bot. Mag. t.* 599; *Red. Lil. t.* 433. *Dichone crispa,
Salisb. in Trans. Hort. Soc.* i. 320. *Agretta crispa, Eckl. Top. Verz.*
24.

SOUTH AFRICA: without locality, *Thom! Bowie!*
COAST REGION: hills near Tulbagh, and near Piquetberg, *Thunberg!* near
Paarl, *Ecklon*, 10! between Paarl and Klein Drakenstein Mountains, under
1000 ft., *Drège!* near Malmesbury, 400 ft., *MacOwan, Herb. Aust. Afr.*, 1545!
Worcester, *Zeyher!* Caledon Div.; Villiersdorp, *Grey!*

8. T. viridis (Ker in Bot. Mag. t. 1275); corm ovoid; tunics of
parallel strands of fine matted fibres; basal leaves 4–6, linear, thin in
texture, glabrous, $\frac{1}{2}-1$ ft. long, $\frac{1}{3}-\frac{1}{2}$ in. broad, plain or crisped, the
midrib only prominent; stems 1–1$\frac{1}{2}$ ft. long including inflorescence,
simple or branched; spikes secund, many flowered, almost horizontal,
with a very flexuose rachis; spathe-valves oblong-lanceolate, entirely
green; $\frac{1}{4}$ in. long, the outer much the largest; perianth green; tube
1–1$\frac{1}{4}$ in. long, strongly curved at the top, not more than $\frac{1}{12}$ in.
diam. at the throat; segments subequal, oblanceolate, obtuse, $\frac{1}{3}$ in.
long; anthers just exserted from the throat of the perianth-tube.

Gen. Irid. 116; *Klatt in Linnœa* xxxii. 756; *Baker, Handb. Irid.*
192.   *Gladiolus viridis, Ait. Hort. Kew.* iii. 481.   *Ker in Konig and
Sims' Ann.* i. 231.   *Montbretia viridis, Baker in Journ. Linn. Soc.*
xvi. 169.

COAST REGION: Clanwilliam Div.; stony places near Heeren Logement,
Zeyher, 1618!

9. **T. Cooperi** (Baker, Handb. Irid. 192); corm not seen; stems
2–3 ft. long, erect, terete, branched; leaves like those of *Antholyza
nervosa*, ensiform, a foot long, ½ in. broad at the middle, moderately
firm and rigid in texture, with 5 strong, subequal ribs; spikes
very lax, ½ ft. long; spathe-valves oblong-lanceolate, ⅓ in. long, firm
in texture, greenish-brown; perianth bright pink; tube 1½ in. long,
gradually dilated in the upper half, ⅓ in. diam. at the throat;
segments oblong-spathulate, subequal, ½ in. long; stamens just
exserted from the throat of the perianth-tube.

COAST REGION: Worcester Div., *Cooper*, 3182!

10. **T. crispa** (Ker in Konig and Sims' Ann. i. 228); corm ovoid,
¾–1 in. diam.; tunics of fine reticulated fibres; basal leaves 4–6,
firm in texture, linear, much crisped towards the margin, 4–6 in.
long, ¼–½ in. broad; stems slender, terete, simple or branched,
½–1 ft. long including inflorescence; spikes moderately dense,
secund, 4–10-flowered; spathe-valves oblong, obtuse, ⅓–½ in. long,
brown in the upper half; perianth whitish or pale pink; tube
cylindrical, 1½–2 in. long, dilated into a funnel at the top; segments
oblong, obtuse, ½ in. long; stamens reaching more than halfway
up the perianth-limb; anthers lanceolate, purple; style-branches
overtopping the anthers. *Ker in Bot. Mag. t.* 678; *Gen. Irid.* 117;
Baker, Handb. Irid. 192.   *Gladiolus crispus, Thunb. Diss. No.*
7, *t.* 1, *fig.* 2; *Prodr.* 8; *Fl. Cap.* i. 191; *Linn. fil. Suppl.* 94;
Jacq. Ic. t. 267; *Andr. Bot. Rep. t.* 142.   *G. laceratus, Burm.
Prodr. Cap.* 2.   *Montbretia lacerata, Baker in Journ. Linn. Soc.*
xvi. 168.   *Tritonia lacerata, Klatt, Ergänz.* 24.   *Freesia crispa,
Eckl. Top. Verz.* 30.

VAR. β, **T. pectinata** (Ker in Bot. Mag. sub t. 1275); leaves not at all crisped.
Gen. Irid. 117.   *Ixia pectinata, Vahl, Enum.* ii. 62.   *Gladiolus pectinatus,
Soland. MSS.*

Var. γ, **grandiflora** (Baker, Handb. Irid. 192); flowers not more than 2–3 to
a spike, with a tube 2½–3 in. and a limb ¾–1 in. long.

Var. δ, **parviflora** (Baker, Handb. Irid. 192); leaves linear, not crisped;
flowers much smaller than in the type, with a tube about an inch long, and
oblanceolate, obtuse segments ⅓ in. long.

SOUTH AFRICA: without locality, Var. β, *Thunberg! Masson!* Var. δ, *Thom!*
COAST REGION: mountains near Tulbagh, *Thunberg!* Piquetberg, *Thunberg!*
Drège! Paarl Mountains, *Drège!* Hottentots Holland Mountains, 600 ft.,
Bolus, 5254! Worcester Div., *Cooper*, 1610! Var. γ, Lions Rump, *Pappe!*
stony places at the foot of Table Mountain, *MacOwan*, 2389! Var. δ, Winter-
hoek, near Tulbagh, *Pappe!*

11. **T. pallida** (Ker in Bot. Mag. sub t. 1275); corm globose,
¾ in. diam.; tunics of fine matted fibres; stem slender, forked,

1½–2 ft. long; basal leaves 4–6, linear, ensiform, glabrous, plane, 1 ft. long, ½–¾ in. broad; spikes very lax, distichous, 3–4 in. long, 4–6-flowered; spathe-valves oblong, ½ in. long; perianth whitish; tube cylindrical, 2–2½ in. long, dilated into a short throat, ¼–⅓ in. diam.; segments oblong, obtuse, ¾ in. long; anthers just protruded from the throat of the perianth-tube. *Gen. Irid.* 116; *Baker, Handb. Irid.* 192. *Montbretia pallida, Baker in Journ. Linn. Soc.* xvi. 168. *Gladiolus longiflorus, Jacq. Ic. t.* 262, *non Linn. fil.*

SOUTH AFRICA: without locality.

12. T. Bakeri (Klatt, *Ergänz.* 24); corm not seen; basal leaves 4–6, narrow, linear, striated, rigid, erect, glabrous, a foot or more long, 1/12 in. diam.; stem slender, terete, simple or branched, about a foot long including inflorescence, with 1–2 reduced leaves; spikes 1–3, lax, distichous, 2–9-flowered; spathe-valves oblong, ¼–⅓ in. long, brown at the tip; perianth whitish; tube 1½–2 in. long, much dilated at the top; segments oblong, ¾ in. long; stamens reaching high up the limb; style-branches overtopping the anthers. *Baker, Handb. Irid.* 193. *Gladiolus striatus, Soland. MSS., non Jacq. Montbretia striata, Baker in Journ. Linn. Soc.* xvi. 168. *G. longiflorus, Thunb. Diss. No.* 22, *ex parte.*

SOUTH AFRICA: without locality, *Thunberg! Masson! Oldenburg!*
COAST REGION: Caledon Div.; near the Zonder Einde River, *Zeyher*, 4017!

13. T. ventricosa (Baker, Handb. Irid. 193); corm small, ovoid; tunics of fine reticulated fibres; stem slender, simple, ½–1 ft. long; leaves about 3, superposed, glabrous, lowest only produced, narrow linear, erect, as long as the stem; two upper sheathing; flowers 2–4 in a short spike; spathe-valves oblong-navicular, ½ in. long, green, entire; perianth bright red; tube funnel-shaped, ¾ in. long; segments oblong, obtuse, ⅝–¾ in. long; stamens half as long as the segments; anthers lanceolate, equalling the filaments.

COAST REGION: Caledon Div.; Appels Kraal, near the Zonder Einde River, *Zeyher*, 3793! mountains near Genadendal, *Pappe!*

14. T. kamisbergensis (Klatt in Linnæa xxxii. 760); stem terete, forked, 2 ft. long; leaves rigid, linear-subulate, 1 lin. diam., overtopping the stem; flowers purple, distichous; spathe-valves oblong, membranous at the edge, the inner toothed at the tip; perianth-tube ⅓ in. long; segments oblong, subequal, twice as long as the tube; anthers half as long as the filaments; style-branches short, linear. *Baker, Handb Irid.* 193. *Freesia kamisbergensis, Eckl. MSS. Montbretia kamisbergensis, Baker in Journ. Linn. Soc.* xvi. 169.

WESTERN REGION: Little Namaqualand, *Ecklon and Zeyher*, 109.

15 T. disticha (Baker, Handb. Irid. 193); stem branched, 3–4 ft. long, with several reduced leaves; leaves linear, firm in texture, ¼ in. broad, 18–21 in. long; midrib and edges thickened, stramineous; flowers up to 14 in a lax distichous spike; spathe-

valves oblong, unequal, toothed, longer than the tube of the
perianth; perianth rose-red; tube funnel-shaped, $\frac{1}{6}$–$\frac{1}{4}$ in. long;
segments subequal, obovate, twice as long as the tube; anthers twice
as long as the filaments; style-branches linear. *Tritonixia disticha,
Klatt, Ergänz.* 22.

EASTERN REGION : Pondoland ; mountains by St. John's River, 1000–2000 ft.,
Drège, 4549.

16. **T. pauciflora** (Baker, Handb. Irid. 193); corm not seen;
stems 1–3, slender, 4–6 in. long; basal leaves 3–4, linear,
moderately firm, glabrous, 3–4 in. long, $\frac{1}{6}$ in. broad; spikes
1–2-flowered; outer spathe-valve ovate, acute, $\frac{1}{2}$ in. long; perianth-
tube cylindrical, $\frac{1}{4}$–$\frac{1}{3}$ in. long; segments oblong, bright purple,
$\frac{5}{8}$–$\frac{3}{4}$ in. long; anthers lanceolate, $\frac{1}{4}$ in. long, as long as the filaments;
style-branches overtopping the anthers.

WESTERN REGION : Little Namaqualand ; mountains near 'Naries, 3400 ft.,
Bolus, 6622 !

17. **T. Templemanni** (Baker, Handb. Irid. 193); corm middle-sized;
stem about as long as the leaves, bearing 2 small sheathing leaves;
leaves linear, rigid, erect, pale green, glabrous, conspicuously ribbed,
$1\frac{1}{2}$–2 ft. long, $\frac{1}{4}$–$\frac{1}{3}$ in. broad; inflorescence a lax panicle; spikes
few-flowered; spathe-valves ovate, subequal, $\frac{1}{4}$–$\frac{1}{3}$ in. long, green at
the base, red-brown at the tip; perianth-tube funnel-shaped, $\frac{1}{2}$ in.
long; segments obovate, bright red, $\frac{3}{4}$ in long; stamens half as
long as the limb; anthers linear.

COAST REGION : Caledon! Div. ; on the Zwartberg near the Hot Springs,
Templeman!

Discovered by Mr. R. Templeman in 1887. Sent to us by Prof. MacOwan, in
a living state.

18. **T. teretifolia** (Baker, Handb. Irid. 194); corm globose, 1 in.
diam.; tunics of fine, pale brown parallel strands of matted fibres;
stem simple, terete, 1 ft. long, with about 3 reduced sheathing leaves;
leaves rigid, subterete, a foot or more long; spike distichous, very
lax, simple, 6–8-flowered, 4–5 in. long; spathe-valves oblong-lanceo-
late, entirely green, $\frac{1}{3}$–$\frac{1}{2}$ in. long; perianth pale pink, $\frac{1}{2}$–$\frac{5}{8}$ in. long;
segments equal, oblong, as long as the narrowly funnel-shaped tube;
stamens reaching nearly to the tip of the perianth-segments; anthers
large, lanceolate.

EASTERN REGION : Natal; at the foot of Table Mountain, *Krauss,* 430!

19. **T. dubia** (Eckl. ex Klatt in Linnæa xxxii. 761); corm ovoid,
$\frac{1}{2}$ in. diam. ; tunics of fine, matted, reticulated fibres; basal leaves 4–6,
spreading, linear, moderately firm in texture, 3–4 in. long, $\frac{1}{6}$–$\frac{1}{4}$ in.
broad; stems $\frac{1}{2}$–1 ft. long including inflorescence, simple or
branched; spikes lax, distichous, 2–3 in. long; spathe-valves oblong,
$\frac{1}{4}$ in. long, scariose and toothed at the tip; perianth-tube slender, as
long as or twice as long as the spathe, dilated into a small funnel at
the top; segments oblong, equal, pale pink, $\frac{1}{2}$ in. long; stamens

half as long as the limb; anthers as long as the filaments; style-branches not overtopping the anthers. *Baker, Handb. Irid.* 194. *T. Bolusii, Baker in Journ. Bot.* 1876, 337. *Tritonixia Bolusii and conferta, Klatt, Ergänz.* 22.

SOUTH AFRICA: without locality, *Zeyher*, 4014.
COAST REGION: Uitenhage Div.; grassy places near the Zwartkops River, *Ecklon*, 141! near Uitenhage, *Bolus*, 1883! *Harvey*, 141!

20. T. Kraussii (Baker, Handb. Irid. 194); corm globose; tunics of fine reticulated fibres; basal leaves 4–6, linear, firm in texture, 6–9 in. long; stem slender, simple or branched, $1\frac{1}{2}$–2 ft. long including the inflorescence; end-spike distichous, 10–12-flowered, finally 4–5 in. long; spathe-valves oblong, $\frac{1}{4}$–$\frac{1}{2}$ in. long, toothed at the tip, brown and membranous in the upper half; perianth pale pink, $\frac{5}{8}$–$\frac{3}{4}$ in. long; segments oblong, about equalling the broadly funnel-shaped tube; anthers just protruded from the throat of the perianth-tube. *Montbretia dubia, Baker in Journ. Linn. Soc.* xvi. 169 *ex parte*.

EASTERN REGION: Natal, summit of Table Mountain, 2000 ft., *Krauss*, 200! and without precise locality, *Buchanan! Sanderson*, 453!

Scarcely more than a variety of *T. lineata*.

21. T. lineata (Ker in Konig and Sims' Ann. i. 228); corm small, globose; tunics of fine, reticulated fibres; leaves about 6 in a basal rosette, moderately firm in texture, linear; outer $\frac{1}{2}$–1 ft. long, $\frac{1}{4}$–$\frac{1}{2}$ in. broad; stem slender, terete, 1–$1\frac{1}{2}$ ft. long including inflorescence, simple or forked; end-spike lax, distichous, 4–6 in. long; spathe-valves oblong, $\frac{1}{4}$–$\frac{1}{2}$ in. long, distinctly toothed at the tip, brown and scariose in the upper half; perianth white or pale pink, 1–$1\frac{1}{4}$ in. long; tube broadly funnel-shaped in the upper half; segments equal, oblong, about as long as the tube; anthers lanceolate, protruding from the perianth-tube, equalling the filaments; style-branches overtopping the stamens. *Gen. Irid.* 118; *Ait. Hort. Kew. edit.* 2, i. 91; *Baker, Handb. Irid.* 194. *Gladiolus lineatus, Salisb. Prodr.* 40; *Curt. Bot. Mag. t.* 487; *Red. Lil. tt.* 55 *and* 400. *Montbretia lineata, Baker in Journ. Linn. Soc.* xvi. 169. *Ixia reticulata, Thunb. Fl. Cap. edit. Schult.* 60. *Gladiolus venosus, Willd. Enum.* i. 58.

SOUTH AFRICA: without locality, *Thunberg! Harvey*, 904!
COAST REGION: Mossel Bay Div.; between Great Brak River and Little Brak River, *Burchell*, 6164! Bathurst Div.; near the source of Kasuga River, *Burchell*, 3912! Albany Div.; Fish River Heights, *Hutton!* Stockenstrom Div.; Katberg, 3000–4000 ft., *Drège*, 3499c!
CENTRAL REGION: Somerset Div.; Bosch Berg, 3000 ft., *MacOwan*, 1199!
KALAHARI REGION: Basutoland, *Cooper*, 3200!
EASTERN REGION: Natal; Itafamasi, *Wood*, 669! and without precise locality, *Sanderson*, 977! *Gerrard*, 397! Griqualand East; Kokstad, 4700 ft., *Tyson*, 1105! and *MacOwan and Bolus, Herb. Norm.*, 507! Kaffraria, *Cooper*, 1807!

22. T. Wilsoni (Baker in Gard. Chron. 1886. xxvi. 38); corm globose; tunics of fine reticulated fibres; stem slender, 2 ft. long;

leaves about 6, narrow, linear, subquadrangular, under a line broad; 3 lower subbasal, 1–1½ ft. long, the others distant, superposed; spikes 1–2, laxly 4–8-flowered; spathe-valves lanceolate, ½–¾ in. long; perianth white, suffused with red-purple, 1–1¼ in. long; tube broadly funnel-shaped, half as long as the obovate, cuspidate segments, of which the arcuate upper one is the broadest; anthers reaching halfway up the limb, purple; style-arms overtopping the anthers.

COAST REGION: Port Elizabeth, *Wilson!*

Described from a living plant and sketches sent me in June, 1886, from Mr. John Wilson, of Greenside Gardens, St. Andrew's. It was sent home by his brother, Mr. Alexander Wilson, of Port Elizabeth.

23. T. rosea (Klatt in Linnæa xxxii. 760); corm globose, 1 in. diam.; tunics of fine reticulated fibres; leaves linear, firm in texture, 1 ft. long, ½ in. broad; stem 2 ft. long including the inflorescence, branched; spikes lax, distichous, 4–12-flowered, the end one finally 4–6 in. long; spathe-valves oblong, ⅓–½ in. long, distinctly toothed at the tip, scariose in the upper half; perianth bright red; tube ½ in. long, broadly funnel-shaped in the upper half; segments subequal, oblong, as long as the tube; anthers lanceolate, protruded from the perianth-tube; style-branches reaching nearly to the tip of the perianth-segments. *Baker, Handb. Irid.* 194. *Montbretia rosea, Baker in Journ. Linn. Soc.* xvi. 169. *T. securigera, Eckl., non Ker.*

COAST REGION: Stockenstrom Div.; Katberg, *Hutton! Baur,* 851! Fort Beaufort Div.; Winterberg, *Zeyher!* Eastern Frontier, *Mrs. Barber!*

KALAHARI REGION: Orange Free State, *Cooper,* 3201!

EASTERN REGION: Tembuland; Bazeia Mountain, 2500 ft., *Baur,* 431! Natal, *Buchanan!*

24. T. graminifolia (Baker, Handb. Irid. 195); corm not seen; basal leaves about 6, thin, linear, 6–9 in. long, ¼–⅓ in. broad; stem slender, simple, 1-flowered and shorter than the leaves in the only specimen seen; outer spathe-valve thin, green, oblong-lanceolate, an inch long; perianth dull purplish-red; tube slender, an inch long, 1/12 in. diam. at the throat; segments lanceolate, acute, unguiculate, 1/12–⅛ in. broad at the middle, rather shorter than the tube; stamens inserted at the throat of the tube, reaching nearly to the tip of the segments.

KALAHARI REGION: Transvaal; near Lydenburg, *Roe in Herb. Bolus,* 2651!

25. T. laxifolia (Benth. Gen. Plant. iii. 708); corm globose, ¾–1 in. diam.; tunics of matted reticulated fibres; stem slender, 1–2 ft. long including inflorescence, usually branched; basal leaves 4–6, linear, thin in texture, ½–1 ft. long, ¼–⅓ in. broad; spikes lax, distichous, 6–12-flowered, the end one 4–7 in. long; spathe-valves oblong, ¼–½ in. long, brown and toothed at the tip; perianth bright red, 1–1¼ in. long; segments oblong, shorter than the tube, which is broadly funnel-shaped in the upper half; anthers lanceolate, reaching halfway up the segments, one-third the length of the filaments; style-branches clavate, overtopping the stamens. *Baker, Handb. Irid.* 195. *Montbretia laxifolia, Klatt in Linnæa* xxxii. 754 ; *Baker in Trans. Linn. Soc.* xxix. 155, *t.* 101. *fig. A.*

VAR. β, **T. strictifolia** (Benth. loc. cit.); leaves shorter, firmer in texture. *Montbretia strictifolia, Klatt in Linnæa* xxxii. 753.

COAST REGION : Uitenhage Div.; between Zwartkops River and Sunday River, *Ecklon and Zeyher*, 260 ! *Harvey*, 758! Alexandria Div. ; Enon, below 1000 ft., *Drège*, 3489a ! Var. β, near George, *Burchell*, 6090 ! Bathurst Div., between Blaauw Krans and Kaffir Drift Military Post, *Burchell*, 3710 !

CENTRAL REGION : Rhinoster Kop, near Beaufort, *Burke* ! stony hills near Graaff Reinet, 2500 ft., *Bolus*, 597 ! plains at the foot of Bosch Berg, near Somerset East, 2500 ft., *MacOwan and Bolus, Herb. Norm.*, 275 ! Var. β, Graaff Reinet Div.; along the Sunday River, *Burchell*, 2867 !

26. T. Pottsii (Benth. Gen. Plant. iii. 708); corms globose, 1 in. diam., connected by a slender rhizome ; stems branched, 3–4 ft. long including the inflorescence ; leaves 4–6, ensiform, moderately firm in texture, 1–$1\frac{1}{2}$ ft. long, $\frac{1}{2}$–$\frac{3}{4}$ in. broad at the middle; spikes lax, distichous, the end one finally 6–9 in. long ; spathe-valves oblong, greenish-brown, $\frac{1}{6}$–$\frac{1}{4}$ in. long ; perianth bright orange-yellow with a tinge of red, 1–$1\frac{1}{4}$ in. long, the equal oblong segments not more than half as long as the broadly funnel-shaped tube; stamens reaching halfway up the limb; anthers lanceolate, half as long as the filaments ; style-branches overtopping the anthers. *Baker in Bot. Mag. t.* 6722; *Handb. Irid.* 195. *Montbretia Pottsii, Baker in Gard. Chron.* 1877, viii. 424.

KALAHARI REGION : Transvaal ; Lydenburg Div., *Roe in Herb. Bolus*, 2652 !
EASTERN REGION : Natal ; Enon, near Richmond, 2000 ft., *Wood*, 1955! and without precise locality, *Mrs. Saunders* ! Griqualand East ; near Kokstad, 5000 ft., *Tyson*, 1372 ! and near Clydesdale, 2500 ft., *Tyson*, 2109 !

27. T. watsonioides (Baker, Handb. Irid. 195) ; corm not seen ; stem 2 ft. long; produced leaf single, narrow linear, 2–3 ft. long, rigid, with revolute edges and a thick midrib ; spikes dense, simple or branched at the base, 4–6 in. long ; spathe-valves ovate, brown, rigid, $\frac{1}{4}$ in. long; perianth cream-white, an inch long; tube much curved, twice as long as the oblong segments ; stamens reaching the tip of the segments.

KALAHARI REGION : Transvaal ; mountain sides near Barberton, Moodies, and throughout Upper Swazieland, 3500–5000 ft., *Galpin*, 814 ! (not 1890, as quoted in the Handb. Irid.), near Barberton, *Thorncroft in Herb. Wood*, 4113 !

28. T. Nelsoni (Baker, Handb. Irid. 195); corm not seen ; basal leaves 6–8, narrow, linear, moderately firm in texture, 6–9 in. long, $\frac{1}{12}$ in. broad ; stem slender, simple, as long as the leaves, bearing 1–4 flowers in a dense spike ; spathe-valves oblong-lanceolate, $\frac{1}{4}$–$\frac{1}{3}$ in. long, brown and scariose in the upper half; perianth bright red, an inch long, the obovate obtuse segments equalling the tube, which is broadly funnel-shaped in the upper half, the three lower with a spreading oblong callus on the claw ; stamens reaching more than halfway up the perianth-segments.

KALAHARI REGION: Transvaal; Zoutpansberg Div.; mountain summit, Houtbosch Berg, *Nelson*, 440!

29. T. securigera (Ker in Konig and Sims' Ann. i. 228); corm globose, ¾–1 in. diam.; tunics of fine, matted, brown fibres; stem slender, terete, ½–1 ft. long including inflorescence, simple or branched; leaves 6–8 in a distichous rosette, linear, spreading, moderately firm in texture, 3–6 in. long, ¼–⅓ in. broad; flowers 4–8 in a lax distichous spike; spathe-valves oblong, ⅓–½ in. long, faintly toothed at the top, green with a brown margin; perianth bright red, with a tube ¾–1 in. long, broadly funnel-shaped in the upper half; segments broad oblong, ½ in. long, the three lower with a large oblong callus at the throat; stamens reaching halfway up the limb; anthers much shorter than the filaments; style-branches ⅙ in. long, over-topping the stamens. *Ait. Hort. Kew. edit.* 2, i. 91; *Ker, Gen. Irid.* 118; *Baker, Handb. Irid.* 196. *Gladiolus securiger, Ait. Hort. Kew.* i. 65; *Curt. Bot. Mag. t.* 383. *Montbretia securigera, DC. in Red. Lil. t.* 53; *Klatt in Linnæa* xxxii. 752. *Ixia gladiolaris, Lam. Encyc.* iii. 341.

COAST REGION: Grahamstown Flats, 2500 ft., *Galpin*, 386! Oudtshoorn Div.; near Kleine Poort, *Tyson*, 3071!

CENTRAL REGION: Fraserburg Div.; on a rocky hill at Dwaal River Poort, *Burchell*, 1482! Zak River, *Burchell*, 1511! Victoria West Div.; Carnarvon, *Burchell*, 1550! near Murraysburg, *Tyson*, 311!

30. T. flava (Ker in Konig and Sims' Ann. i. 228); corm globose, ½ in. diam.; tunics of matted fibres; stem slender, terete, ½–1 ft. long including inflorescence, simple or branched; basal leaves 4–6, linear, spreading, moderately firm in texture, 3–4 in. long; flowers 3–6 in a lax distichous spike; spathe-valves oblong, ⅓–½ in. long, green with a scariose margin; perianth bright yellow; tube ¾–1 in. long, broadly funnel-shaped in the upper half; segments oblong, ½ in. long, the three lower with a small callus at the throat; stamens reaching halfway up the limb. *Bot. Reg. t.* 747; *Gen. Irid.* 118; *Baker, Handb. Irid.* 196. *Gladiolus flavus, Ait. Hort. Kew.* i. 65. *Montbretia flava, Klatt in Linnæa* xxxii. 753. *Ixia flabellularis, Vahl, Enum.* ii. 67.

SOUTH AFRICA: without locality.

31. T. unguiculata (Baker, Handb. Irid. 196); corm small, globose; tunics of fine matted fibres; stem slender, simple or branched, 1–1½ ft. long; leaves linear, moderately firm, ½ ft. long; spikes very lax, consisting of few erect flowers; spathe-valves oblong, ¼–⅓ in. long, green with a brown tip; perianth-tube subcylindrical, twice as long as the spathe; segments unequal, palé lilac, oblanceolate-unguiculate, ¾–1 in. long; stamens reaching nearly to the tip of the segments.

SOUTH AFRICA: without locality, *Kitching*!

Collected in 1880. Remarkable for the narrow claw of its perianth-segments.

32. Montbretia crocosmæflora (Hort. ; Floral Mag. n.s. t. 472) ;
whole plant 3 ft. high ; produced leaves 6–8 in a basal fan, green,
largest a foot long, under an inch broad ; spikes 1–3, lax, distichous,
the end one 6–9 in. long ; spathe-valves lanceolate, herbaceous,
greenish-brown, ¼–⅓ in. long ; perianth bright reddish-yellow, 1½ in.
long ; segments oblong, subequal, as long as the funnel-shaped tube ;
stigmas reaching to tip of segments ; stamens shorter.

A garden hybrid between *Tritonia Pottsii* and *Crocosmia aurea.* Described
from plants flowered at Kew, Aug., 1889.

XXVII. CROCOSMIA, Planch.

Perianth with a cylindrical tube slightly dilated upwards, and 6
subequal, spreading, oblong segments. *Stamens* inserted at the throat
of the perianth-tube, unilateral, close, arcuate ; filaments long,
filiform ; anthers linear-sagittate, versatile. *Ovary* 3-celled ; ovules
few in a cell, superposed ; style filiform ; branches entire, slightly
dilated at the tip. *Fruit* an inflated, 3-lobed, globose capsule,
dehiscing loculicidally, the valves persistent and chartaceous. *Seeds*
few, large, globose.

DISTRIB. The single species extends to the mountains of Tropical Africa.

1. C. aurea (Planch. in Flore des Serres t. 702) ; corm globose ;
tunics thin, membranous ; stems terete, 2–4 ft. long, branched,
furnished with a few small leaves ; basal leaves about 6 in a distichous rosette, linear or ensiform, thin in texture, 1–1½ ft. long,
⅓–¾ in. broad ; spikes few-flowered, lax, equilateral, with a very
flexuose axis ; spathe-valves greenish-brown, ovate or oblong, acute,
¼–⅓ in. long ; perianth bright dark reddish-yellow ; tube ¾–1 in.
long ; limb 1–1½ in. long, with oblanceolate-oblong segments ;
stamens and style reaching to the tip of the segments ; anthers pale
yellow, ¼–⅓ in. long ; capsule ½ in. diam. ; valves rugose, persistent ;
seeds as large as a pea. *Klatt in Linnæa* xxxii. 764 ; *Baker, Handb.
Irid.* 189. *Tritonia aurea, Pappe MSS.* ; *Hook. in Bot. Mag. t.*
4335 ; *Ergänz.* 24 ; *Bot. Reg.* 1847, t. 61 ; *Baker in Journ. Linn.
Soc.* xvi. 163. *Crocanthus mossambicensis, Klotzsch in Peters, Reise
Mossamb. Bot. t.* 57.

VAR. **maculata** (Baker in Gard. Chron. 1888, iv. 565, fig. 80) has a dark spot at
the base of the perianth-segments. Var. **imperialis**, Hort., has a perianth-limb
3 in. diam., and leaves an inch broad.

KALAHARI REGION : Transvaal ; beside streams in wooded ravines around
Barberton, 3000–3500 ft., *Galpin*, 805 ! without precise locality, *McLea in Herb.
Bolus*, 5792 !

EASTERN REGION : Natal ; Umhloti, *Wood*, 797 ! without precise locality,
Gerrard, 404 ! *Plant*, 76 ! *Cooper*, 3202 ! *Buchanan!* Griqualand East ; Kokstad,
4000 ft., *Tyson*, 1701 ! Pondoland ; between Umtata River and St. John's River,
1000 2000 ft., *Drège*, 4551b ! Tembuland ; Bazeia, *Baur*, 31 !

Also occurs in south-east Tropical Africa.

XXVIII. ACIDANTHERA, Hochst.

Perianth with a cylindrical tube slightly dilated towards the top; segments of the limb subequal, oblong or oblanceolate. *Stamens* unilateral, contiguous, arcuate, inserted at or a little below the throat of the perianth-tube; filaments filiform; anthers linear or lanceolate. *Ovary* 3-celled; ovules many, superposed; style long, filiform; branches short, undivided, flattened, falcate. *Capsule* oblong, membranous, loculicidally 3-valved. *Seeds* globose or discoid and broadly winged.

Rootstock a tunicated corm; leaves narrow linear; flowers few, arranged in a lax simple spike; spathe-valves long, lanceolate, entire, herbaceous.

DISTRIB. Five species in Tropical Africa.

Perianth-tube 2½–3 in. long:
  Segments of the limb oblanceolate:
    Leaves linear-subulate ...   .   (1) **brachystachys.**
    Leaves linear, ¼–⅓ in. broad ...   ...   (2) **graminifolia.**
  Segments of the limb oblong:
    Perianth-tube a little longer than the spathe...   (3) **forsythiana.**
    Perianth-tube twice as long as the spathe ...   (4) **platypetala.**
Perianth-tube 1½–2 in. long, exserted from the spathe:
  Only one basal leaf produced:
    Flowers in a short spike... '   ...   .   (5) **flexuosa.**
    Flowers in a lax spike ...   .   (6) **tubulosa.**
  Two or more basal leaves produced:
    Flowers 1–2 in a spike:
      Segments of limb oblanceolate ...   ...   (7) **rosea.**
      Segments of limb oblong:
        Perianth-tube a little longer than the
          spathe ...   .   (8) **pauciflora.**
        Perianth-tube twice as long as the
          spathe ...   .   (9) **brevicaulis.**
    Flowers 2–6 in a spike:
      Outer spathe-valve green:
        Leaves grass-like ...   .   (10) **Huttoni.**
        Leaves moderately firm ...   ...   (11) **Tysoni.**
      Outer spathe-valve brownish, acuminate...   (12) **capensis.**
Perianth-tube an inch long, not exserted from the
  spathe ..   ..   (13) **brevicollis.**
Perianth-tube very short ...   .   (14) **ixioides.**

1. A. brachystachys (Baker in Journ. Bot. 1876, 338); corm not seen; stem ½ ft. long including inflorescence, simple; basal leaves 4–8, crowded, erect, narrow linear, subterete upwards, thick and rigid in texture, a foot long; flowers 4–6 in a lax spike; spathe-valves lanceolate, acuminate, firm in texture, green, finely pubescent, distinctly ribbed, 1¼–1½ in. long; perianth-tube slender, about 3 in. long; segments of the limb oblanceolate, obtuse, cuspidate, ¾ in. long, nearly white, erect; stamens inserted at the throat of the perianth-tube; anthers lanceolate, ⅙ in. long, equalling the filament. *Journ. Linn. Soc.* xvi. 160; *Handb. Irid.* 185.

COAST REGION: Clanwilliam Div., *Mader in Herb. MacOwan*, 2183!

2. A. graminifolia (Baker in Journ. Bot. 1876, 338); corm globose, 1 in. diam.; tunics thick, fibrous, produced into long bristles above the neck; stem a foot long, with 2–3 reduced sheathing leaves; basal leaves 2, linear, grass-like, 8–9 in. long, $\frac{1}{4}$–$\frac{1}{3}$ in. broad; flowers 2, in a lax spike; spathe-valves green, lanceolate, acute, 2 in. long; perianth-tube 3 in. long; segments of the limb oblanceolate-spathulate, $\frac{3}{4}$–1 in. long, $\frac{1}{8}$ in. broad, white with a tinge of purple; stamens reaching halfway up the limb; anthers lanceolate, $\frac{1}{3}$ in. long, *Journ. Linn. Soc.* xvi. 160; *Handb. Irid.* 186.

COAST REGION : plains between Swellendam and Gauritz River, *Bowie !* in the British Museum Herbarium.

3. A. platypetala (Baker in Journ. Bot. 1876, 339); corm globose, $\frac{3}{4}$ in. diam.; tunics of parallel wiry fibres; stems slender, terete, simple, 1–1$\frac{1}{2}$ ft. long; leaves usually 3, distant, superposed, firm in texture, subterete, strongly ribbed, the lowest a foot long, the upper much shorter; flowers 1–2 in a lax secund spike; outer spathe-valve lanceolate, green, moderately firm, 1–1$\frac{1}{2}$ in. long; perianth-tube curved, 2–3 in. long, funnel-shaped at the apex; limb an inch long, yellowish-white, flushed with dull purple; segments obovate, $\frac{1}{3}$–$\frac{1}{2}$ in. broad; stamens reaching halfway up the limb; capsule oblong, an inch long; seeds large, thin, discoid. *Journ. Linn. Soc.* xvi. 160; *Handb. Irid.* 186. *Gladiolus longicollis, Baker in Journ. Bot.* 1876, 182.

COAST REGION : Uitenhage Div.; Galgebosch, *Zeyher !*
KALAHARI REGION : Transvaal; Sabie River, *Mudd !* Saddleback Range, near Barberton, 3000–5000 ft., *Galpin*, 530 ! Lake Chrissie, *Elliot*, 1603 !
EASTERN REGION :-Natal; Attercliffe, *Sanderson*, 265 ! Inanda, *Wood*, 232 ! 243 ! near Durban, *Wood*, 108 ! without precise locality, *Sanderson*, 16 ! *Gerrard*, 547 ! *Mrs. K. Saunders !* Tembuland; Bazeia Mountain, 3000–4000 ft., *Baur*, 505 !

4. A. forsythiana (Baker, Handb. Irid. 186); corm not seen; stem 1 ft. including inflorescence; stem-leaves 3, superposed, linear, grass-like, $\frac{1}{6}$ in. broad; flowers 4, arranged in a lax erect spike; outer spathe-valve lanceolate, herbaceous, 1$\frac{1}{2}$–2 in. long; perianth-tube $\frac{1}{2}$–$\frac{3}{4}$ in. longer than the spathe; segments of the limb oblong-unguiculate, erect, whitish, 1 in. long; stamens and style reaching halfway up the perianth-limb.

SOUTH AFRICA: without locality. A specimen is in the herbarium of Mr. Forsyth, of the Chelsea Garden, purchased by Mr. Bentham in 1835; and a second in Herb. Bolus, gathered by Dr. Bachmann, probably near Hopefield, Malmesbury Division.

5. A. flexuosa (Baker in Berl. Monat. xix. 15); corm not seen; stem about a foot long including the inflorescence, slender, simple, terete; leaves about 3, distant, superposed, narrow linear, the lowest $\frac{1}{2}$ ft. long; flowers 3–4, in a moderately dense, erect spike; outer spathe-valve green, lanceolate, acute, 1 1$\frac{1}{2}$ in. long; perianth with a slender tube 1$\frac{1}{2}$–2 in. long; segments of the limb oblanceolate, nearly white, an inch long; stamens inserted at the throat of the perianth-tube, more than half as long as the segments; anthers lanceolate, $\frac{1}{4}$ in. long. *Handb. Irid.* 186. *Gladiolus flexuosus, Thunb.*
K 2

Diss. No. 8, *t.* 1, *fig.* 1; *Fl. Cap.* i. 189; *Ker, Gen. Irid.* 146.
Sphœrospora flexuosa, Klatt in Linnœa xxxii. 726.

SOUTH AFRICA: without locality, *Thunberg !*
Doubtfully distinct from *A. tubulosa*.

6. A. tubulosa (Baker in Journ. Linn. Soc. xvi. 160); corm
globose, ½ in. diam.; stems 1-1½ ft. long including inflorescence,
simple, slender, terete; leaves about 3, superposed, distant, linear,
firm in texture, with a thick midrib, the lowest basal a foot long, the
others sheathing the stem with only short, free points; flowers 2-5,
arranged in a very lax spike with a flexuose axis; outer
spathe-valve lanceolate, acute, glabrous, 1½-2 in. long; perianth
with a slender tube about 2 in. long; segments oblanceolate
or oblanceolate-oblong, obtuse, white with a slight tinge of pink,
1-1¼ in. long, ⅙-⅓ in. broad; stamens inserted at the throat of the
tube, more than half as long as the segments; anthers lanceolate, half
as long as the filaments. *Handb. Irid.* 186. *Ixia tubulosa, Houtt.
Handl.* xii. 36, *t.* 78, *fig.* 2. *Gladiolus exscapus, Thunb. Fl. Cap.* i.
175; *Ker, Gen. Irid.* 146. *G. longiflorus, Linn. herb., ex parte !
Acidanthera exscapa, Baker in Berl. Monat.* xix. 15. *Freesia
costata, Eckl.*

SOUTH AFRICA: without locality, *Thunberg !*
COAST REGION: Simons Bay, *Wright,* 264! Devils Mountain, 500 ft.,
MacOwan and Bolus. Herb. Norm., 946 ! Worcester Div.; Dutoits Kloof, 2000-
4000 ft., *Drège !* Drakenstein Mts., near Dutoits Kloof, *MacOwan,* 2506! with-
out precise locality, *Cooper,* 1678 ! Tulbagh Div.; Witsen Berg and Skurfde
Berg, *Zeyher,* 1607a ! near Tulbagh Waterfall, 1200 ft., *Bolus,* 5253 ! Caledon
Div., *Pappe !*

7. A. rosea (Schinz in Bull. Herb. Boiss. ii. 222) ; corm not seen ;
stem very slender, simple, under a foot long, with 2 remote, narrow
linear leaves from the middle, the lower sometimes 6-8 in. long ; spike
1-2-flowered; outer spathe green, herbaceous, lanceolate, 1-1¼ in.
long; perianth reddish-lilac ; tube very slender, half as long again as
the spathe; segments oblanceolate, ¾ in. long; stamens and style
shorter than the perianth-segments.

COAST REGION: Table Mountain (Devils Mountain on label), *Schlechter,* 75 !

8. A. pauciflora (Benth. Gen. Plant. iii. 706): corm globose, ½ in.
diam.; tunics membranous ; produced basal leaves 2-3, linear, glabrous,
grass-like in texture, ½ ft. long ; stem ½-1 ft. long, simple, terete, 1-2-
flowered, with 2-3 reduced sheathing leaves; outer spathe-valve lanceolate, acute, herbaceous, 1¼-1½ in. long ; perianth-tube curved, a little
longer than the spathe, dilated into a small funnel at the top ;
segments of the limb oblong-spathulate, white with a slight reddish
tinge, 1-1¼ in. long ; stamens reaching more than halfway up the
perianth-limb; anthers lanceolate, as long as the filaments. *Handb.
Irid.* 187. *Montbretia pauciflora, Baker in Journ. Bot.* 1876, 336.

CENTRAL REGION: Somerset Div., *Bowker !*

9. A. brevicaulis (Baker); corm not seen ; stem very short,
simple ; produced basal leaves 3, linear, glabrous, 6-8 in. long ;

spike laxly 2-flowered; outer spathe-valve green, lanceolate, an inch
long; perianth bright red-lilac; tube twice as long as the spathe;
segments oblong, an inch long; anthers large, linear, much longer
than the filaments.

KALAHARI REGION : Transvaal; Devils Bridge, Makwongwa Range, near
Barberton, 5000 ft., *Galpin*, 1252 !

10. **A. Huttoni** (Baker in Journ. Bot. 1876, 339); corm not seen;
stem 1–1½ ft. long including inflorescence, with 2–3 reduced long-
sheathing leaves; basal leaves 2–3, erect, linear, grass-like, glabrous,
moderately firm in texture, a foot or more long, ⅙–⅓ in. broad, flat,
with a prominent midrib; flowers 2–4, arranged in a very lax, erect
spike; spathe-valves lanceolate, acute, 1–1½ in. long, herbaceous to
the tip; perianth-tube filiform, 1¼–1½ in. long, dilated into a small
funnel at the very top; segments of the limb oblong, erect, whitish,
¾–1 in. long; anthers linear, ⅓ in. long, longer than the filaments,
which are inserted below the throat of the perianth-tube. *Journ.
Linn. Soc.* xvi. 160; *Handb. Irid.* 187.

COAST REGION : Stockenstrom Div.; Katberg, *Hutton!*

11. **A. Tysoni** (Baker, Handb. Irid. 187); corm small, globose;
stem simple, slender, 1–1½ ft. long, with 3 erect, linear, moderately
firm leaves, the lowest a foot long, ⅛ in. broad; flowers 2–3, very
laxly spicate; outer spathe-valve lanceolate, pale green, 1¼–1½ in.
long; perianth bright pink; tube a little longer than the spathe,
funnel-shaped at the top; segments oblong, 1–1¼ in. long; stamens
more than half as long as the segments; anthers ⅓ in. long; style-
branches subulate, overtopping the anthers.

EASTERN REGION : Griqualand East; near the Cataracts of Mount Currie,
5800 ft., *Tyson*, 1151 ! *MacOwan and Bolus, Herb. Norm.*, 895!

12. **A. capensis** (Benth. Gen. Plant. iii. 706); corm globose, 1 in.
diam.; tunics thick, composed of fine fibres; stems simple, slender,
terete, ½–1 ft. long; produced basal leaves 6–8, linear, erect,
acuminate, glabrous, moderately firm in texture, 6–9 in. long;
flowers 3–6, in a lax, erect spike; outer spathe-valve brownish,
lanceolate, acuminate, chartaceous in texture, 1–1¼ in. long;
perianth-tube slender, 1½ in. long, dilated into a short funnel at the
tip; segments of the limb oblong, whitish, 1 in. long, the three lower
more reflexed at the tip when fully expanded; stamens more than
half as long as the limb; anthers lanceolate, basifixed, ¼ in. long;
style-branches overtopping the anthers. *Baker, Handb. Irid.* 187.
Tritonia capensis, Ker in Konig and Sims' Ann. i. 228; *Bot. Mag.*
tt. 618 *and* 1531; *Gen. Irid.* 116; *Klatt in Linnæa* xxxii. 757.
Montbretia capensis, Baker in Journ. Linn. Soc. xvi. 168. *Gladiolus
roseus, Jacq. Ic. t.* 261. *Tritonia rosea, Ait. Hort. Kew. edit.* 2, i. 91.

COAST REGION : Swellendam Div.; on mountain ridges along the lower part of
Zonder Einde River; *Zeyher*, 4017 ! Riversdale Div.; hills near Zoetmelks River,
Burchell, 6739 !

13. A. brevicollis (Baker in Journ. Bot. 1876, 339); corm
depresso-globose; tunics brown, membranous; stem about a foot
long including the inflorescence, simple, terete, flexuose, with 2–3
reduced sheathing leaves; produced basal leaves 4–5, linear, erect,
thick in texture but not rigid, 1–1½ ft. long, ⅛ in. broad; flowers
3–4, arranged in a lax spike, with a flexuose axis; outer spathe-
valve green, oblong-navicular, ¾–1 in. long; perianth bright purple;
tube cylindrical, as long as the spathe; segments equal, oblong, erect,
¾ in. long; stamens half as long as the segments; anthers lanceolate,
basifixed, not sagittate, equalling the filaments.   *Journ. Linn. Soc.*
xvi. 160; *Handb. Irid.* 187.   *Gladiolus Gueinzii, Kunze in Linnæa*
xx. 14.

COAST REGION: sandy places at the mouth of the Great Fish River,
MacOwan, 1890! sea shore at the mouth of Keiskamma River, *Hutton!*
EASTERN REGION: Natal; sea shore at the mouth of Umgeni River, *Wood,*
55! 1092! without precise locality, *Gueinzius! Cooper,* 3197! *Mrs. Saunders!*

14. A. ixioides (Baker, Handb. Irid. 188); corm not seen; stem
slender, simple, ½ ft. long, with one narrow linear leaf nearly a foot
long from the middle, and a much-reduced one higher up; spike lax,
simple, few-flowered; spathe-valves lanceolate, green, an inch long;
perianth bright lilac; tube cylindrical, ¼ in. long, funnel-shaped at
the throat, and oblong segments; stamens more than half as long as
the limb; style-branches not flattened.

SOUTH AFRICA: without locality, *Elliot,* 1174!

Recedes from *Acidanthera* by its very short perianth-tube.

XXIX. SYNNOTIA, Sweet.

Perianth with a long tube, cylindrical in the lower part, broadly
funnel-shaped at the top; segments of the limb oblong, the upper
outer one the broadest, the others reflexing when the flower is
expanded.   *Stamens* unilateral, contiguous, arcuate, inserted at the
base of the dilated upper part of the tube; filaments filiform; anthers
linear-oblong, basifixed.   *Ovary* 3-celled; ovules many in a cell;
style filiform, its branches simple.   *Fruit* a membranous capsule with
loculicidal dehiscence.   *Seeds* globose, or angled by pressure.

Rootstock, leaves, inflorescence and spathe-valves of *Sparaxis,* from which it
differs by its irregular perianth-lobes.

DISTRIB. Endemic.

Cylindrical portion of the perianth-tube not longer than the
spathe ... ... **(1) bicolor.**
Cylindrical portion of the perianth-tube much longer than the
spathe ... ... **(2) variegata.**

1. S. bicolor (Sweet, Hort. Brit. edit. 2, 501); corm ovoid, ½ in.
diam.; tunics thick, pale, strongly honeycombed; stems ½–1½ ft.
long including inflorescence, simple or branched; leaves about 6 in a
distichous basal rosette, ensiform, thin in texture, 3–6 in. long;
flowers 2–6 in very lax spikes; spathe-valves ovate, membranous,

$\frac{1}{2}$–$\frac{3}{4}$ in. long, green at the base, brown and subscariose upwards, deeply
lacerated ; perianth-tube cylindrical, $\frac{1}{2}$ in. long, dilated into a broad
funnel at the top; limb $\frac{3}{4}$–1 in. long, yellow, more or less flushed
with violet; stamens reaching halfway up the upper segment.
Klatt in Linnœa xxxii. 750; *Baker, Handb. Irid.* 198.   *Gladiolus
bicolor, Thunb. Diss. No.* 16, *t.* 2, *fig.* 1 ; *Prodr.* 8 ; *Fl. Cap.* i. 197 ;
Jacq. Collect. Suppl. 25 ; *Ic. t.* 240.   *Sparaxis bicolor, Ker in Konig
and Sims' Ann.* i. 225.   *S. galeata, Sweet, loc. cit.   Gladiolus
galeatus, Jacq. Ic. t.* 258.

VAR. β, **Roxburghii** (Baker, Handb. Irid. 198); bulb tunics of fine threads;
flower all lilac-purple.

SOUTH AFRICA : without locality.   Var. *β, Roxburgh!*
COAST REGION : Lion Mountain, 800 ft., *MacOwan and Bolus, Herb. Norm.*,
798 ! Paarl Mts., *Drège*, 8347 ! Malmesbury Div.; Groene Kloof, *Thunberg!*
Ecklon and Zeyher, 239 !

2. **S. variegata** (Sweet, Brit. Flow. Gard. t. 150) ; corm ovoid,
$\frac{1}{2}$ in. diam. ; tunics coarsely honeycombed ; leaves just like those of
S. bicolor ; flowers 3–6, arranged in very lax, simple, erect spikes ;
spathe just like that of *S. bicolor ;* perianth dark violet; tube
cylindrical, $1\frac{1}{4}$–$1\frac{1}{2}$ in. long, dilated suddenly into a broad funnel at
the top; limb an inch long ; upper segment $\frac{1}{2}$ in. broad; lower
shorter and narrower, flushed with yellow; stamens reaching halfway up the limb ; anthers oblong, $\frac{1}{6}$ in. long.   *Klatt in Linnœa*
xxxii. 750; *Baker, Handb. Irid.* 198. *Sparaxis luteo-violacea,
Eckl. Top. Verz.* 27.   *S. Wattii, Harv. MSS.*

SOUTH AFRICA: without locality, *Thom !*
COAST REGION : Clanwilliam Div.; *Mader,* 136 ! *MacOwan,* 2138 !

XXX. GLADIOLUS, Linn.

Perianth with a curved funnel-shaped tube ; segments of the limb
usually oblong, obtuse or acute, unequal, the three lower smaller than
the three upper.   *Stamens* unilateral, arcuate, contiguous, inserted
below the throat of the perianth-tube ; filaments long, filiform ;
anthers lanceolate, basifixed.   *Ovary* 3-celled ; ovules numerous,
superposed ; style filiform ; branches short, simple, cuneate at the
tip.   *Capsule* obovoid, loculicidally 3-valved.   *Seeds* discoid and
winged in the Cape species.

Rootstock a tunicated corm; leaves terete, linear or ensiform; flowers comparatively large, very various in colour, often maculate, arranged in lax spikes;
spathe-valves usually large, lanceolate, herbaceous.

DISTRIB. Species about 140, many Tropical African, a few European, Mediterranean and Oriental.

Subgenus I. EUGLADIOLUS.   Spathe-valves large, green, lanceolate; perianth-
segments not distinctly unguiculate.
  A. Leaves terete or linear :
    Perianth-segments acute :
      Perianth-tube $1\frac{1}{2}$–2 in. long :
        Leaves subterete :
          Perianth-segments long and gradually
           pointed   ...   .   (1) **grandis.**

Perianth-segments shortly pointed :
 Perianth pale or slightly
  flushed with dark lilac ... (2) **tristis.**
 Perianth dark lilac ... ... (3) **recurvus.**
Leaves linear :
 Segments with a short cusp ... (4) **angustus.**
 Segments with a long cusp ... (5) **cuspidatus.**
Perianth-tube about an inch long :
 Leaves subterete :
  Flowers horizontal :
   Flowers pink ... ... (6) **hastatus.**
   Flowers blue-lilac ... ... (7) **gracilis.**
   Flowers yellowish ... (8) **tenellus.**
  Flowers suberect ... ... (9) **trichonemifolius.**
 Leaves linear :
  Flowers lilac ... ... (10) **vomerculus.**
  Flowers yellowish ... ... (11) **strictus.**
 Whole flower not above an inch long :
  Leaf with scarcely any free point ... (12) **pubescens.**
  Leaf slender, subterete ... ... (13) **Lambda.**
  Leaf linear, long ... ... (14) **rachidiflorus.**
Perianth-segments obtuse or obscurely cuspidate :
Stem-leaves with only very short, free points :
 Sheaths glabrous :
  Flowers pink or lilac :
   Perianth-tube half as long as
    the segments ... ... (15) **microphyllus.**
   Perianth-tube as long as the
    segments ... ... (16) **brevifolius.**
   Perianth-tube longer than the
    segments ... ... (17) **tabularis.**
   Flower-segments white with a
    red keel ... ... (18) **inandensis.**
  Sheaths pilose ... ... (19) **Woodii.**
Stem-leaves with long, free points :
 Leaves subulate or very narrow :
  Flowers erect or suberect :
   Segments shorter than the tube (20) **tenuis.**
   Segments equalling the tube (21) **debilis.**
   Segments rather longer than
    the tube ... ... (22) **Bolusii.**
   Segments 2–3 times the
    length of the tube :
     Flowers bright lilac ... (23) **biflorus.**
     Flowers pale yellow ... (24) **erectiflorus.**
  Flowers horizontal with a curved
   tube :
   Flowers 1–4 in a spike :
    Upper segments ⅛–½ in.
     broad :
      Flowers white ... (25) **cochleatus.**
      Flowers bright red (26) **Rogersii.**
      Flowers pink ... (27) **Pappei.**
    Upper segments ¼–¾ in.
     broad :
      Corm-tunics of fine
       fibres ... ... (28) **inflatus.**
      Corm-tunics of wiry
       strands ... ... (29) **spathaceus.**
    Flowers many in a spike ... (30) **involutus.**

Leaves linear :
    Leaf-sheaths glabrous :
        Perianth-tube 1½ in. long ...　(31) **hyalinus.**
        Perianth-tube 1 in. long　...　(32) **vittatus.**
        Perianth-tube ⅔ in. long　...　(33) **striatus.**
        Perianth-tube ⅓–½ in. long :
            Segments half as long as
              the tube　...　...　(34) **paludosus.**
            Segments twice the
            length of the tube :
              Produced leaves 2　(35) **Niveni.**
              Produced leaves 3–4 :
                Stamens half as
                  long as limb　(36) **punctatus.**
                Stamens as long
                  as the lower
                  segments　...　(37) **brachyscyphus.**
    Leaf-sheaths hairy　...　...　(38) **villosus.**
B. Leaves ensiform.
Parviflori. Perianth-tube under an inch long.
Spikes equilateral ; flowers very numerous :
    Flowers red :
        Perianth-tube ½ in. long　...　...　...　(39) **crassifolius.**
        Perianth-tube ¾ in. long　...　...　...　(40) **Elliotii.**
    Flowers yellow :
        Stem pubescent　...　...　...　...　(41) **Ludwigii.**
        Stem villose ...　...　...　...　...　(42) **sericeo-villosus.**
Spikes secund ; flowers fewer :
    Flowers yellow　...　...　...　...　...　(43) **ochroleucus.**
    Flowers red :
        Upper segments ½ in. broad :
            Perianth-tube ½ in. long　...　...　(44) **Kirkii.**
            Perianth-tube ¾ in. long　...　...　(45) **Eckloni.**
        Upper segments ¾ in. broad ; two inner
        lower with a large dark blotch :
            Flowers yellow ...　...　,...　...　(46) **purpureo-auratus.**
        Flowers purple :
            Outer spathe-valve 1–1¼ in. long　(47) **Papilio.**
            Outer spathe-valve 1½–2 in. long　(48) **Rehmanni.**
Blandi. Perianth-tube 1–2 in. long ; flowers white or pale red.
Segments obovate, obscurely pointed :
    Sheaths and leaves hairy :　...
        Segments as long as the tube　...　...,　(49) **hirsutus.**
        Segments shorter than the tube　...　...　(50) **salmoneus.**
    Sheaths and leaves glabrous :
        Perianth-tube 1–1¼ in. long　...　...　(51) **scaphochlamys.**
        Perianth-tube 1½–2 in. long　...　...　(52) **floribundus.**
Segments oblong, distinctly pointed :
    Perianth-tube curved :
        Segments ⅓–½ in. broad　...　...　...　(53) **oppositiflorus**
        Segments ½–1 in. broad　...　...　...　(54) **blandus.**
    Perianth-tube nearly straight :
        Segments nearly concolorous　...　...　(55) **Milleri.**
        Segments with a bright red central band...　(56) **undulatus.**
Cardinales. Flowers large, bright red, with a nearly straight tube, and upper
segments not distinctly hooded.
Segments subequal, shorter than the tube ...　...　(57) **Macowani.**
    Upper segments as long as the tube　...　...　(58) **Adlami.**
Segments unequal, longer than the tube :
    Upper segments ¾–1 in. broad ...　...　...　(59) **cardinalis.**

Upper segments obovate, 1–1½ in. broad :
    Lower bracts 1½–2 in. long...   ...   ...   (60) **splendens.**
    Lower bracts 3–6 in. long ...   .   (61) **cruentus.**

Dracocephali.   Flowers large, with a much-curved tube and upper segments hooded.
Flowers dull-coloured :
    Leaves ¾–1 in. broad   ...   .   (62) **dracocephalus.**
    Leaves 1–2 in. broad   ...   .   (63) **platyphyllus.**
Flowers bright red :
    Limb shorter than the tube   ...   .   (64) **psittacinus.**
    Limb as long as the tube :
        Perianth 2–3 in. long   ...   .   (65) **Leichtlini.**
        Perianth 4 in. long ...   .   (66) **Tysoni.**
    Limb longer than the tube   ...   .   (67) **Saundersii.**
Flowers bright yellow ...   .   (68) **aurantiacus.**

Subgenus II. HEBEA. Spathe-valves large, green, oblong-lanceolate. Perianth-segments all with a narrow claw.
Side-segments about ⅓ in. broad :
    Flowers red :
        Leaves with many close equal ribs...   ...   (69) **alatus.**
        Leaves with only a thickened midrib and
            edge   ...   .   (70) **spathulatus.**
    Flowers greenish-yellow ...   .   (71) **orchidiflorus.**
Side-segments about ⅓ in. broad :
    Flowers dull reddish   ...   .   (72) **pulchellus.**
    Flowers yellowish   ...   .   (73) **bicolor.**
Side-segments about ¼ in. broad :
    Stems stout ; flowers few to a spike   ...   ...   (74) **arcuatus.**
    Stems slender ; flowers many to a spike :
        Segments cuspidate :
            Upper segments 1–1¼ in. long   ...   (75) **formosus.**
            Upper segments ¾–1 in. long :
                Tunics of fine parallel fibres   ...   (76) **edulis.**
                Tunics lacerated from the base ...   (77) **Scullyi.**
        Segments not cuspidate :
            Claw of upper segments very narrow   (78) **Dregei.**
            Claw of upper segments not very narrow   (79) **permeabilis.**

Subgenus III. SCHWEIGGERA. Spathe-valves small, brown, rigid. Segments all with a distinct slender claw and small blade.
Perianth-limb ⅓–⅔ in. long   ...   .   (80) **arenarius.**
Perianth-limb 1 in. long   ...   .   (81) **montanus.**

1. G. grandis (Thunb. Fl. Cap. i. 186) ; corm globose ; tunics of thick, parallel, wiry fibres ; stem slender, terete, 1–2 ft. long ; leaves 3, superposed, terete, strongly ribbed, firm in texture, the lowest 1–1½ ft. long ; flowers fragrant, 2–6 in a very lax secund spike ; spathe-valves green, lanceolate, the outer 2–2½ in. long ; perianth 2½–3 in. long, with a curved tube funnel-shaped in the upper third ; segments yellowish-white, more or less tinged with purplish-brown, especially on the keel, oblong, ½–¾ in. broad, narrowed into a long point ; stamens reaching halfway up the limb ; capsule oblong, membranous, 1½ in. long. *Klatt in Linnæa* xxxii. 714 ; *Baker, Handb. Irid.* 202. *G. tristis, Linn. herb.!   G. tristis, var. grandis, Thunb. Diss. No.* 8.   *G. versicolor, Andr. Bot. Rep. t.* 19 ; *Ker in Bot. Mag. t.* 1042 ; *Gen. Irid.* 135.

SOUTH AFRICA : without locality, *Thunberg !*

COAST REGION: Clanwilliam Div.; *Mader in Herb. MacOwan*, 2162! Cape
Div.; Chapman Bay, *MacGillivray*, 484! Table Mountain, 500 ft., *MacOwan
and Bolus, Herb. Norm.*, 809! Caledon Div.; Zwartberg, near the Hot Springs,
MacOwan, 2607! and Palmiet R. Valley, near Grabouw, 700 ft., *Bolus*, 4023!
Worcester Div.; *Cooper*, 1630! 1688! 3191! Swellendam Div.; mountain
ridges along the lower part of Zonder Einde R., *Zeyher*, 1630! 3978! Rivers-
dale Div.; near Zoetmelks River, *Burchell*, 6723! George Div.; near George,
Burchell, 5824! 6056! between George and Malgat R., *Burchell*, 6087! and by
Malgat R., *Burchell*, 6116!

2. G. tristis (Linn. Sp. Plant. edit. 2, i. 53, ex parte); corm globose,
1 in. diam.; tunics of fine parallel strands of matted fibres; stems
slender, simple, 1–2 ft. long; leaves 3, superposed, terete, with 3–5
much-raised, stramineous ribs, the lower 1–1½ ft. long; flowers 3–4 in
a very lax secund spike, fragrant; spathe-valves green, lanceolate,
1½–2 in. long; perianth-tube curved, 1½–2 in. long, funnel-shaped
in the upper third; limb yellowish-white, slightly flushed on the
keel of the segments with purplish-black; segments oblong-spathulate, acute, ⅓–½ in. broad; stamens more than half as long as the
perianth-limb; capsule oblong, membranous, an inch long. *Thunb.
Diss. No. 8, ex parte; Curt. in Bot. Mag. t. 272; Jacq. Ic. t. 243;
Ker in Bot. Mag. t.* 1098; *Gen. Irid.* 136; *Baker, Handb. Irid.*
203. *G. spiralis, Pers. Syn.* i. 43; *Red. Lil. t.* 35.

VAR. β, **G. concolor** (Salisb. Parad. t. 8); flowers almost concolorous, and a
purer white than in the type. *G. tristis, Jacq. Ic. t.* 245.

SOUTH AFRICA: without locality, *Thunberg! Villette! Burchell!*
COAST REGION: Paarl Mountains, 900 ft., *Bolus*, 5569! Uitenhage Div.; at
the foot of Van Stadens Berg, below 1000 ft., *Drège*, 8431b!

3. G. recurvus (Linn. Mant. 28); corm globose, ¾–1 in. diam.;
tunics of parallel wiry fibres; stems slender, simple, 1–2 ft. long;
leaves 3, firm in texture, terete, strongly ribbed, the lowest about
a foot long; flowers very fragrant, 2–6, in a very lax secund spike;
outer spathe-valve green, lanceolate, 1½–2 in. long; perianth-tube
curved, 1½–2 in. long, clavate in the upper third; limb 1–1¼ in.
long, yellowish-white, much flushed with dark lilac; segments
oblong, acute, ½ in. broad; stamens reaching more than halfway up
the limb; capsule oblong, membranous, 1–1¼ in. long. *Ker in Bot.
Mag. t.* 578, *non Thunb.; Baker, Handb. Irid.* 203. *G. punctatus,
Jacq. Ic. t.* 247. *G. tristis, var. punctatus, Thunb. Diss. No.* 8.
G. carinatus, Ait. Hort. Kew. i. 64. *G. ringens, Andr. Bot. Rep. tt.* 27
and 227; *Red. Lil. t.* 123. *G. odorus, Salisb. Prodr.* 40. *G. violaceus,
Pers. Syn.* i. 43. *Watsonia recurva, Pers. Syn.* i. 43. *G. breynianus,
Ker, Gen. Irid.* 135. *G. maculatus, Sweet, Hort. Brit. edit.* 1, 397;
Klatt in Linnæa xxxii. 708.

COAST REGION: near Cape Town, *Bolus*, 4591! *Thunberg! Niven!* Cape
Flats, near Diep River, *MacOwan and Bolus, Herb. Norm.*, 286! Camps Bay,
MacOwan, 2558! Devils Mountain, 500 ft., *Bolus*, 4023! Lion Mountain,
200–500 ft., *MacOwan and Bolus, Herb. Norm.*, 283! *MacOwan*, 2564! foot
of Table Mountain, *Pappe!* between Cape Town and Simons Bay; *Burchell*,
8559! Stellenbosch, *Zeyher!* Somerset Div.; Bosch Berg, 2500–4800 ft.,
MacOwan, 321! Albany Div.; mountain slopes around Grahamstown, 2000–
2500 ft., *Galpin*, 196!

EASTERN REGION : Griqualand East; Zuurberg Range, near Kokstad, 4800 ft., *Tyson,* 1874 !

4. G. angustus (Linn. Sp. Plant. edit. 2, i. 53); corm globose; tunics membranous; stems simple, 2–3 ft. long; leaves 3–4, linear, flat, firm in texture, 1–2 ft. long, $\frac{1}{6}$–$\frac{1}{4}$ in. broad; flowers 2–6, in a very lax spike; spathe-valves green, oblong-lanceolate, acute, the outer 1$\frac{1}{2}$–3 in. long; flowers white, with a tube 1$\frac{1}{2}$–2 in. long, funnel-shaped in the upper third; limb 1–1$\frac{1}{2}$ in. long; segments oblong, acute, the upper concolorous, $\frac{1}{2}$–$\frac{3}{4}$ in. broad, the three lower narrower, with a bright purple, spade-shaped mark on the middle; stamens reaching halfway up the limb. *Hort. Cliff.* 20, *t.* 6; *Jacq. Ic. t.* 252; *Ker in Bot. Mag. t.* 602; *Andr. Bot. Rep. t.* 589; *Red. Lil. t.* 344; *Baker, Handb. Irid.* 204. *G. trimaculatus, Lam. Encyc.* ii. 727; *Ill.* i. 116, *t.* 32; *fig.* 3. *G. cordatus, Thunb. Fl. Cap.* i. 185. *G. macowanianus, Klatt in Trans. South Afric. Phil. Soc.* iii. *pt.* 2, 199.

SOUTH AFRICA : without locality, *Thunberg !*
COAST REGION : Table Mountain, 1400 ft., *Bolus,* 2824! grassy places near Hout Bay, 500 ft., *MacOwan and Bolus,* Herb. Norm., 284! *MacOwan,* 2605 (2065 *ex Klatt*). Stellenbosch Div.; mountains near Somerset West, *MacOwan,* 2680!

5. G. cuspidatus (Jacq. Ic. t. 257); corm globose; tunics of fine, parallel strands of matted fibres; stems simple, 2–3 ft. long; leaves 3–4, linear, rigid in texture, glabrous, the lowest 1$\frac{1}{2}$–2 ft. long, about $\frac{1}{2}$ in. broad; flowers 4–8, in a lax secund spike; spathe-valves green, lanceolate, outer 2–3 in. long; perianth white or pale pink; tube slightly curved, 2–3 in. long, clavate in the upper third; segments oblong, 1$\frac{1}{2}$ in. long, $\frac{1}{3}$–$\frac{1}{2}$ in. broad, narrowed into a long, wavy point, the three lower with a spade-shaped purple blotch; stamens reaching halfway up the limb. *Ker in Bot. Mag. t.* 582; *Gen. Irid.* 139; *Andr. Bot. Rep. t.* 219; *Red. Lil. t.* 136; *Baker, Handb. Irid.* 205. *G. undulatus, Linn. Mant.* 27; *Thunb. Fl. Cap.* i. 206, *ex parte. G. affinis, Pers. Syn.* i. 45.

VAR. β, **G. ventricosus** (Lam. Encyc. ii. 727); flowers pink; point of the segments shorter and less wavy. *G. cuspidatus, Andr. Bot. Rep. t.* 147; *Red. Lil. t.* 36. *G. carneus, Jacq. Ic. t.* 255; *Ker in Bot. Mag. t.* 591, *non Delaroche.*
VAR. γ, **ensifolius** (Baker); whole plant under a foot long; leaves short, rigid, ensiform.

SOUTH AFRICA : without locality, *Thunberg ! Burchell !*
COAST REGION : Rondebosch and between Constantia and Steen Bergen, under 1000 ft., *Drège !* Sand-dunes near Cape Town, *Zeyher,* 1631! Paarl Mountains, below 100 ft., *Drège !* Clanwilliam Div., *Mader in Herb. MacOwan,* 2163! Worcester Div., *Cooper,* 1617! 1680! Stellenbosch Div.; near Somerset West, *MacOwan,* 2691! Var. γ, Caledon Div.; Houw Hoek, and Hottentots Holland Mountains, *Pappe !*

6. G. hastatus (Thunb. Fl. Cap. i. 181); corm globose, $\frac{3}{4}$–1 in. diam.; tunics of parallel strands of matted fibres; stems slender, simple, 1–2 ft. long; leaves 3, superposed, subterete, firm in texture, strongly ribbed; flowers 2–4, in a very lax secund spike; outer spathe-valve green, lanceolate, 1–1$\frac{1}{2}$ in. long; flowers pink; perianth-tube curved, 1–1½ in. long; limb 1–1¼ in. long, with oblong, acute segments ½ in. broad at the middle; stamens reaching halfway up the limb. *G. Thunbergii, Eckl. Top. Verz.* 37; *Baker in Journ. Linn. Soc.* xvi. 173; *Baker, Handb. Irid.* 203.

SOUTH AFRICA: without locality, *Thunberg!*
COAST REGION: Table Mountain, *Burchell*, 8414! *Bolus*, 3883! Devils Mountain, *Burchell*, 8506! Simons Bay, *Wright*, 251! Riversdale Div.; on the Kampsche Berg, *Burchell*, 7069!

7. G. gracilis (*Jacq. Ic. t.* 246); corm globose, ½ in. diam.; tunics of strong, matted fibres; stems very slender, simple, 1–2 ft. long; leaves about 3, superposed, long-sheathing, subterete, rigid in texture, the lowest sometimes a foot or more long; flowers up to 5–6 in a lax secund spike, scentless; outer spathe-valve green, lanceolate, an inch long; perianth pale lilac-blue; tube curved, under an inch long, funnel-shaped in the upper half; limb 1–1½ in. long; the upper segments oblong, shortly cuspidate, ⅓–½ in. broad, the three lower narrower, longer, distinctly unguiculate; stamens reaching halfway up the limb. *Thunb. Fl. Cap.* i. 182; *Ker in Bot. Mag. t.* 562; *Red. Lil. t.* 425; *Ker, Gen. Irid.* 138; *Baker, Handb. Irid.* 203. *G. lœvis, Thunb. Fl. Cap.* i. 178; *Ker, Gen. Irid.* 135. *G. elongatus, Thunb. Fl. Cap.* i. 180; *Ker, Gen. Irid.* 136. *G. pterophyllus, Pers. Syn.* i. 43. *G. setifolius, Eckl. Top. Verz.* 37. *G. spilanthus, Spreng. in herb. Zeyher.*

SOUTH AFRICA: without locality, *Thunberg! Sparrmann! Sieber*, 133!
COAST REGION: Devils Mountain, *Bolus*, 4024! sides of Table Mountain near Camps Bay, 100 ft., *MacOwan*, 2275! *MacOwan and Bolus, Herb. Norm.*, 281! mountains near Simons Town, *Pappe!* Caledon Div.; Klein River Berg, 1000–3000 ft., *Zeyher*, 3986! Uitenhage Div.; Van Stadens Berg, *Pappe!*

8. G. tenellus (Jacq. .*Ic. t.* 248); corm globose, ½ in. diam.; tunics of strong, parallel strands of matted fibres; stem simple, very slender, 1–1½ ft. long; produced leaves about 3, long-sheathing, subterete, the lowest sometimes as long as the stem; flowers 2–5, in a very lax secund spike; outer spathe-valve green, lanceolate, 1–1½ in. long; perianth-tube curved, about an inch long, funnel-shaped at the top; segments oblong, subacute, an inch long, ¼–⅓ in. broad, yellowish-white, tinged with lilac, and the lower ones much spotted at the throat; stamens reaching more than halfway up the limb. *Thunb. Fl. Cap.* i. 179; *Schneev. Ic. t.* 40; *Ker, Gen. Irid.* 137; *Baker, Handb. Irid.* 204. *G. tristis, vars. humilis and luteus, Thunb. Diss. No.* 8.

SOUTH AFRICA: without locality, *Thunberg! Villette! Cooper*, 220!
COAST REGION: at the foot of Devils Mountain, near Mowbray, 300 ft., *Bolus*, 4890! Cape Flats at Doornhoogde, *Zeyher*, 1628! Clanwilliam Div.; between Berg Vallei and Lange Vallei, *Drège!*

9. G. trichonemifolius (Ker in Bot. Mag. t. 1483); corm globose, ½–¾ in. diam.; tunics of parallel strands of matted fibres; stem very slender, terete, ½–1½ ft. long; leaves usually 3, superposed, long-sheathing, slender, terete, rigid in texture, the lower often over-

topping the stem; flowers 1–3, erect; spathe-valves green, lanceolate, the outer 1–2 in. long; perianth yellow; tube nearly straight,
$\frac{3}{4}$–1 in. long, funnel-shaped in the upper half; segments oblong,
acute, subequal, $\frac{3}{4}$–1 in. long, $\frac{1}{4}$–$\frac{1}{3}$ in. broad, the three lower blotched
with purple at the throat; stamens reaching halfway up the limb.
Gen. Irid. 137; *Klatt in Linnæa* xxxii. 707; *Baker, Handb.
Irid.* 204. *G. citrinus, Klatt, Ergänz.* 6. *Ixia spathacea, Soland.
MSS.! G. tenellus, Thunb. herb., ex parte!*

SOUTH AFRICA: Without locality, *Thunberg! Masson! Oldenberg!*
COAST REGION: Piquetberg Range, 1000–4000 ft., *Zeyher,* 1625! Tulbagh
Div.; Witsenberg Flats, *Pappe!* Cape Div.; near Tygerberg, *Harvey!*
Paarl Div.; between Paarl and Paarde Berg, *Drège,* 8457!

10. **G. vomerculus** (Ker, Gen. Irid. 142); corm globose, $\frac{1}{2}$ in.
diam.; tunics of fine, parallel strands of matted fibres; lower
sheaths more or less conspicuously mottled with purple; stem
simple, 1–1$\frac{1}{2}$ ft. long; produced leaves 3–4, linear, $\frac{1}{4}$–$\frac{1}{3}$ in. broad,
strongly ribbed, firm in texture, the lowest about a foot long;
flowers 2–8, in a lax spike; spathe-valves green, lanceolate, 1$\frac{1}{2}$–2 in.
long; perianth lilac, with a curved tube $\frac{3}{4}$–1 in. long; segments
oblong, acute, 1–1$\frac{1}{2}$ in. long, the three lower with a spade-shaped
yellow blotch at the throat; stamens reaching halfway up the
limb. *Baker, Handb. Irid.* 205. *G. hastatus, Ker in Bot. Mag.
t.* 1564; *Klatt in Linnæa* xxxii. 712, *non Thunb. G. tigrinus,
Eckl.*

COAST REGION: Eastern declivity of Table Mountain, *Ecklon,* 158
Caledon Div.; grassy places near the Zonder Einde River, *Zeyher!*

11. **G. strictus** (Jacq. Ic. t. 260); corm globose, with thick tunics
of parallel rigid strands; stem about a foot long; leaves 3, superposed, long-sheathing, glabrous, with free linear points not more than
2–3 in. long; flowers 2–3, yellowish, with purple dots and stripes,
erect; spathe-valves green, lanceolate, 1–1$\frac{1}{2}$ in. long; perianth-
tube nearly straight, an inch long, broadly funnel-shaped in the
upper half; segments acute, as long as the tube, the upper oblong,
$\frac{1}{3}$–$\frac{1}{2}$ in. broad, the others narrower; stamens reaching halfway up
the limb. *Baker, Handb. Irid.* 205.

SOUTH AFRICA: without locality.
Known only from the figure cited.

12. **G. pubescens** (Baker in Journ. Bot. 1876, 333); corm not
seen; stem slender, terete, 1–1$\frac{1}{2}$ ft. long; leaves about 3, superposed, with long, hairy, strongly ribbed sheaths of firm texture and
scarcely any free points; flowers 3–6, in a lax secund spike;
spathe-valves green; outer oblong-lanceolate, very acute, $\frac{1}{2}$–$\frac{3}{4}$ in.
long; perianth pale pink, an inch long, the oblong acute segments
as long as the tube, $\frac{1}{4}$–$\frac{1}{3}$ in. broad at the middle; stamens reaching
halfway up the perianth-limb. *Journ. Linn. Soc.* xvi. 173; *Handb.
Irid.* 204.

COAST REGION: British Kaffraria, *Cooper*, 458!
EASTERN REGION: Natal; bank of Mooi River, *Wood!*

13. G. Lambda (Klatt in Linnæa xxxii. 708); corm globose; tunics thick, with obscure areolæ; stem a foot long; leaves superposed, long-sheathing, terete, $\frac{1}{2}$ lin. diam., the lowest much overtopping the stem; flowers 2, secund; spathe-valves unequal, obovate, purple at the base and tip, the outer $\frac{2}{3}$ in. long; perianth white, an inch long; tube curved, as long as the spathe; upper segments concolorous, subequal, ovate-lanceolate; lower narrower, with two purple blotches at the base; filaments as long as the anthers. *Baker, Handb. Irid.* 204.

SOUTH AFRICA: without locality, *Reynaud* (Berlin Herbarium).

14. G. rachidiflorus (Klatt, Ergänz. 5); stem forked, 2 ft. long; leaves narrow linear, the lower nearly 2 ft. long, 1–6 in. broad; flowers in a dense spike more than $\frac{1}{2}$ ft. long; outer spathe-valve ovate, acute, inner toothed at the tip; perianth reddish, tube $\frac{1}{4}$ in. long; segments ovate, acute, subequal, under $\frac{1}{2}$ in. long; filaments twice as long as the anthers. *Baker, Handb. Irid.* 205.

EASTERN REGION: Port Natal, around the bay, *Drège*, 4537!

I have not seen this, and suspect it may belong to *Tritonia*, to which genus it is referred in Drège's catalogue.

15. G. microphyllus (Baker, Handb. Irid. 206); corm small, globose; tunics of parallel strands of fibres, prolonged at the top into bristles; root-leaves narrow linear, rigid, 4–5 in. long; stem 6–8 in. long, with about 4 glabrous sheaths with short, free, linear tips; spike subsecund, lax, 2–4 in. long; outer spathe-valves oblong, $\frac{1}{3}$–$\frac{1}{2}$ in. long; perianth-tube shorter than the spathe; segments obovate, obtuse, pale red, $\frac{1}{2}$–$\frac{5}{8}$ in. long; stamens nearly as long as the segments.

EASTERN REGION: Griqualand East; Zuurberg Range, near Kokstad, 5500 ft., *Tyson*, 1852!

16. G. brevifolius (Jacq. Ic. t. 249); corm globose; tunics of parallel strands of fine matted fibres; leaf linear, firm in texture, glabrous, $1\frac{1}{2}$–2 ft. long, produced separately from the flowering stem; flowering stem simple, very slender, 1–2 ft. long, with about 3 superposed leaves with long glabrous sheaths and short, free, linear points; flowers 4–12, in a lax secund spike, horizontal, pink or lilac; outer spathe-valve green, lanceolate, $\frac{1}{2}$–$\frac{3}{4}$ in. long; perianth-tube curved, $\frac{3}{4}$ in. long, broadly funnel-shaped in the upper half; limb about as long as the tube; upper segments obovate, $\frac{1}{2}$ in. broad; lower oblong-unguiculate, with purplish, spade-shaped marks at the throat; stamens reaching more than halfway up the limb; capsule oblong, $\frac{1}{2}$ in. long. *Thunb. Fl. Cap.* 1. 177; *Ker, Gen. Irid.* 134; *Baker, Handb. Irid.* 206. *G. aphyllus, Ker, Gen. Irid.* 134. *G. carneus, Andr. Bot. Rep. t.* 240, *non Delaroche*. *G. Orobanche, Red. Lil. t.* 125. *G. hirsutus, vars. aphyllus and* *brevifolius, Ker in Bot. Mag. t. 727. G. festivus, Herb. in. Bot. Reg.* 1844, *Misc.* 89. *G. spilanthus, Klatt in Linnæa* xxxii. 711, *ex parte. G. brevicollis, Klatt, Ergänz.* 5. *G. amœnus, Forst. MSS. G. Andrewsii, Klatt, Ergänz.* 6. *G. jonquilodorus, Eckl. MSS. G. fragrans, Pappe MSS.*

COAST REGION: near Cape Town, *Thunberg! Zeyher,* 1627! *Burchell,* 908! Wynberg, *Burchell,* 877! Devils Mountain, *Bolus,* 4652! *MacOwan and Bolus, Herb. Norm.,* 947! Table Mountain 1000-2000 ft., *Drège!* from the gardens at Cape Town, *Burchell,* 748! Stellenbosch Div.; between Stellenbosch and Cape Flats, *Burchell,* 8362! and between Hottentots Holland and Jonkers Valley, *Burchell,* 8324! Bathurst Div.; between Blauw Krans and Kaffir Drift Military Post, *Burchell,* 3689!

17. G. tabularis (Eckl. Top. Verz. 38); corm globose, 1 in. diam.; tunics of fine, parallel strands of matted fibres, produced at the top into long, scariose, linear blades; stem slender, glabrous, 1 ft. long including inflorescence, with 3-4 superposed, sheathing leaves; free points none or very short; flowers 5-6, in a lax spike, white with a tinge of pink; spathe-valves green, lanceolate, $\frac{1}{2}$-$\frac{3}{4}$ in. long; perianth-tube curved, 1-1$\frac{1}{2}$ in. long, funnel-shaped in the upper half; segments oblong, subequal, obscurely cuspidate, $\frac{3}{4}$ in. long, $\frac{1}{4}$-$\frac{1}{3}$ in. broad; stamens reaching halfway up the limb. *Baker, Handb. Irid.* 207.

COAST REGION: Table Mountain, 2300-3500 ft., among *Restiaceæ,* etc., frequent, *Bolus,* 7057! *Pappe!*

18. G. inandensis (Baker, Handb. Irid. 207); corm globose, $\frac{3}{4}$-1 in. diam.; tunics of fine parallel strands of matted fibres; flowering stem simple, 1-1$\frac{1}{2}$ ft. long, with 3-4 superposed leaves with glabrous sheaths, and short, free, rigid, linear points, or rarely the lowest leaf with a longer point; flowers 6-8, in a lax spike; outer spathe-valve green, oblong-lanceolate, $\frac{1}{2}$ in. long; perianth-tube curved, $\frac{1}{2}$ in. long, broadly funnel-shaped in the upper half; segments white, with a red keel outside, obscurely cuspidate, $\frac{1}{2}$-$\frac{5}{8}$ in. long; upper obovate, $\frac{1}{3}$ in. broad; lower oblong, distinctly unguiculate; stamens reaching halfway up the limb.

EASTERN REGION: Natal; Inanda, *Wood,* 177! 237! without precise locality, *Sanderson!*

A near ally of *G. brevifolius, Jacq.*

19. G. Woodii (Baker, Handb. Irid. 207); corm globose; tunics of parallel strands of matted fibres produced beyond its neck; flowering stem simple, 1-1$\frac{1}{2}$ ft. long, with 3-4 superposed leaves with hairy sheaths and short, free, linear, rigid points; flowers 4-6, in a lax spike, dark red; outer spathe-valve green, oblong-lanceolate, $\frac{1}{2}$-$\frac{3}{4}$ in. long; perianth-tube curved, $\frac{1}{2}$ in. long, broadly funnel-shaped in the upper half; limb $\frac{1}{2}$-$\frac{3}{4}$ in. long; upper segments obovate, obscurely cuspidate, $\frac{1}{3}$ in. broad; lower oblong, distinctly unguiculate; stamens reaching halfway up the perianth-limb; capsule subglobose, $\frac{1}{2}$ in. long and broad.

KALAHARI REGION: Transvaal; Saddleback Range and Devils Kantoor, 4500 ft., *Galpin*, 1024!

EASTERN REGION: Natal; Inanda, *Wood*, 618! and without precise locality, *Buchanan!*

A near ally of *G. brevifolius, Jacq.*

20. G. tenuis (Baker in Journ. Bot. 1876, 335); corm not seen; flowering-stem very slender, terete, $1\frac{1}{2}$–2 ft. long, with 3 superposed leaves with long glabrous sheaths and produced linear points of moderately firm texture, the lowest 1–$1\frac{1}{2}$ ft. long, $\frac{1}{8}$–$\frac{1}{6}$ in. broad; flowers 4–6, in a lax spike, pale pink, all ascending; outer spathe-valves green, oblong-lanceolate, $\frac{3}{4}$–1 in. long; perianth with a straight tube 1 in. long, funnel-shaped in the upper half; segments $\frac{3}{4}$ in. long, obscurely cuspidate, the upper obovate, $\frac{1}{2}$ in. broad, the others oblong, $\frac{1}{4}$–$\frac{1}{3}$ in. broad; stamens reaching more than halfway up the perianth-limb. *Journ. Linn. Soc.* xvi. 174; *Handb. Irid.* 208.

COAST REGION: Swellendam Div.; Craggy Peak on the Lange Bergen near Swellendam, *Burchell*, 7303! 7421!

21. G. debilis (Ker in Bot. Mag. t. 2585); corm globose; tunics of strong wiry fibres; flowering-stem very slender, simple, 1–$1\frac{1}{2}$ ft. long; leaves 3, superposed, with long glabrous sheaths and long, free, subterete points of firm texture, that of the lowest overtopping the stem; flowers 1–3, white, suberect; outer spathe-valve green, lanceolate, $\frac{3}{4}$–1 in. long; perianth-tube straight, $\frac{3}{4}$ in. long, narrowly funnel-shaped up to the top; segments oblong, obtuse, subequal, as long as the tube, two of the inner with a spade-shaped lilac or claret-red blotch at the throat; stamens half as long as the perianth-limb. *Klatt in Linnœa* xxxii. 705; *Baker, Handb. Irid.* 207. *Geissorhiza albens, E. Meyer in herb. Drège.*

COAST REGION: Cape Flats, *Thunberg! Pappe!* Mountains near Simons Town, 1000 ft., *Bolus*, 4949! Worcester Div.; Dutoits Kloof and Drakenstein Mts., 2000–4000 ft., *Drège!* Caledon Div.; Klein River Berg, *Zeyher*, 3988! on the Zwartberg near the Hot Springs, 1000 ft., *MacOwan and Bolus, Herb. Norm.*, 280!

Is in Herb. Thunberg, under *Ixia linearis.*

22. G. Bolusii (Baker, Handb. Irid. 208); corm not seen; flowering-stem slender, simple, $1\frac{1}{2}$–2 ft. long, with 3 long-sheathing, glabrous leaves with long, free, terete points, the lower overtopping the stem; flowers 2–4 in a lax spike, bright pink, all ascending; spathe-valves oblong-navicular, green, $\frac{3}{4}$–1 in. long; perianth-tube funnel-shaped, $\frac{3}{4}$ in. long; segments obovate, 1 in. long, $\frac{1}{3}$–$\frac{1}{2}$ in. broad; stamens reaching halfway up the limb.

COAST REGION: Winterhoek Mountain, near Tulbagh, 3000 ft., *Bolus*, 5244!

May be a variety of *G. inflatus.*

23. G. biflorus (Klatt in Trans. South Afric. Phil. Soc. iii. pt. 2, 197); corm globose, $\frac{1}{2}$ in. diam.; tunics dull brown, lacerated from the base upwards; flowering-stems slender, terete, glabrous, 6–8 in. long, with only one produced, strongly-ribbed, subterete leaf of firm texture,

5–6 in. long, $\frac{1}{2}$ line broad, from near its base, and two rudimentary
ones higher up, the lower with a short free point, the upper entirely
sheathing ; flowers 2–3, bright lilac, laxly spicate, erect; outer
spathe-valve oblong-lanceolate, firm in texture, 1–1$\frac{1}{4}$ in. long ;
perianth-tube erect, $\frac{1}{4}$–$\frac{1}{3}$ in. long, broadly funnel-shaped at the apex ;
segments subequal, oblong, twice as long as the tube ; stamens
suberect, reaching halfway up the limb ; anthers $\frac{1}{4}$ in. long ;
style-arms reaching to the tip of the anthers. *Baker, Handb. Irid.*
207.

COAST REGION : Cape Flats near Wynberg, in sandy places that are often
inundated, under 100 ft., *MacOwan*, 2279, *MacOwan and Bolus, Herb. Norm.*,
279 !

This is upon the very edge of the genus in the direction of *Geissorhiza*.

24. G. erectiflorus (Baker) ; corm small, globose, clothed with
strong flattened strands ; stem very slender, with one long, slender,
subulate leaf from the middle and a small one higher up; flowers
2–3, erect ; outer spathe-valve of the lowest rigid, 1$\frac{1}{2}$ in. long ;
perianth-tube $\frac{1}{2}$ in. long ; segments oblong-spathulate, obtuse,
1$\frac{1}{4}$–1$\frac{1}{2}$ in. long, apparently pale yellow ; stamens shorter than the
segments.

COAST REGION : sandy places near Malmesbury, *MacOwan, Herb. Austr. Afric.*
1548 !

25. G. cochleatus (Sweet, Brit. Flow. Gard. ser. 2, t. 140) ; corm
small, globose ; flowering-stem very slender, $\frac{1}{2}$–1$\frac{1}{2}$ ft. long, with 3–4
superposed leaves with glabrous sheaths, the lower with a rigid
terete blade overtopping the stem ; flowers 1–2 in a lax spike ;
outer spathe-valve oblong, $\frac{3}{4}$–1 in. long ; perianth-tube, curved, $\frac{1}{2}$–$\frac{3}{4}$ in.
long, funnel-shaped in the upper half; limb pure white, $\frac{3}{4}$–1 in.
long ; upper segments oblong-spathulate, $\frac{1}{3}$–$\frac{1}{2}$ in. broad ; lower shorter,
unguiculate, with a purple spade-shaped mark at the throat ; stamens
reaching halfway up the perianth-limb. *Baker, Handb. Irid.* 208.
Waitzia hastulifera, Herb. Lubeck. fide Klatt, Ergänz. 8.

COAST REGION : Lion Mountain, *Drège*, 8447b ! Worcester Div. ; Dutoits
Kloof, *Drège*, 1568 !

26. G. Rogersii (Baker, Handb. Irid. 208) ; flowering-stem slender,
a foot long, with 3–4 superposed leaves with glabrous sheaths, the
lower with a slender subterete blade overtopping the stem ; flowers
3 in a lax secund spike ; spathe-valves green, oblong-lanceolate, $\frac{1}{2}$ in.
long ; perianth bright red ; tube curved, funnel-shaped, $\frac{1}{2}$ in. long ;
three upper segments $\frac{1}{2}$–$\frac{5}{8}$ in. long, obovate or oblong-spathulate,
$\frac{1}{3}$–$\frac{1}{2}$ in. broad ; three lower longer, oblong, $\frac{1}{8}$ in. broad, distinctly
unguiculate, without any distinct blotch on the throat; stamens
reaching halfway up the limb.

COAST REGION : Cape Flats, *Rogers !*

A near ally of *G. inflatus*.

27. G. Pappei (Baker, Handb. Irid. 208) ; corm small, globose ;
stem slender, a foot long ; leaves 3, with glabrous sheaths and rigid,

linear, free tips, the lower $\frac{1}{2}$ ft. long; flowers 2–3 in a lax spike; spathe-valves green, lanceolate, lower outer $1\frac{1}{2}$ in. long; perianth pink; tube slender, slightly curved, ascending, as long as the spathe; limb as long as the tube; segments oblong, subacute, $\frac{1}{3}$–$\frac{1}{2}$ in. broad at the middle; stamens less than half as long as the segments.

COAST REGION :.summit of Table Mountain, *Pappe!*

Intermediate between *G. inflatus* and *gracilis.*

28. G. inflatus (Thunb. Fl. Cap. edit. 1, i. 181); corm small, globose; tunics of fine fibres; produced leaves 3, superposed, with long glabrous sheaths and narrow linear or subterete rigid blades, the lower overtopping the stem; flowers 1–4, pink; spathe-valves green, long-pointed, the lowest $1\frac{1}{2}$–2 in. long; perianth bright pink; tube curved, $\frac{3}{4}$–1 in. long, funnel-shaped in the upper half; limb 1–$1\frac{1}{4}$ in. long; three upper segments obovate or broad oblong, $\frac{1}{2}$–$\frac{3}{4}$ in. broad; three lower oblong-spathulate, rather longer; stamens reaching half-way up the limb. *Ker, Gen. Irid.* 138. *G. ornatus, Klatt in Trans. South Afric. Phil. Soc.* iii. *pt.* 2, 198. *G. ringens, Eckl. non Red. G. bullatus, Thunb. herb.*

SOUTH AFRICA : without locality, *Thunberg!*

COAST REGION : sand dunes near Cape Town, 50 f`.`, *Bolus,* 3741! Cape Flats at Doorn Hoogte, *Zeyher,* 1629 ! near Rondebosch, *MacOwan and Bolus, Herb. Norm.,* 285! Worcester Div., *Cooper,* 3193! Swellendam Div.; Tradouw Mts., *Drège!* near Groot Vadersbosch and Voormans Bosch, *Zeyher,* 3984!

29. G. spathaceus (Pappe, ex Baker, Handb. Irid. 208); corm small, globose; tunics of thick wiry strands; stem very slender, 1–$1\frac{1}{2}$ ft. long; leaves 3, with glabrous sheaths and short, slender, subterete tips; flowers generally solitary, horizontal; spathe-valves green, lanceolate-acuminate, $1\frac{1}{2}$–2 in. long; perianth pink; tube curved, $\frac{1}{2}$ in. long; limb ventricose, above an inch long; upper segments obovate, $\frac{3}{4}$ in. broad; three lower longer, oblong; stamens more than half as long as the upper segments.

COAST REGION : Caledon Division; on the Zwartberg near the Hot Springs, 1000 ft., *MacOwan and Bolus, Herb. Norm.,* 282! *Pappe! MacOwan,* 2167! Klein River Berg and Zwartberg, *Zeyher,* 3985 ! *Ecklon and Zeyher,* 148 !

30. G. involutus (Delaroche, Descr. 28, t. 3); corm globose; tunics with matted, brown, linear fibres; flowering-stem $1\frac{1}{2}$ ft. long, with 3 superposed leaves with long, free, subterete points of firm rigid texture, the lower one equalling or overtopping the stem; sheaths long, glabrous; flowers bright pink, 4–8 in a lax secund spike; outer spathe-valve oblong-lanceolate, 1–$1\frac{1}{2}$ in. long; perianth-tube much curved, $\frac{1}{2}$ in. long; limb 1 in. long; three upper segments obovate, obtuse, $\frac{1}{2}$–$\frac{3}{4}$ in. broad; three lower oblong-unguiculate, with spade-shaped purple blotches at the throat; stamens half the length of the limb. *Ker, Gen. Irid.* 142; *Baker, Handb. Irid.* 209. *G. bimaculatus, Lam. Encyc.* ii. 727 (*Miller Ic.* 158, *t.* 236, *fig.* 1). *G. suaveolens, Zeyher.*

COAST REGION : Knysna Div.; between Groene Vallei and Zwart Vallei, *Burchell,* 5631 ! 5683 !

31. G. hyalinus (Jacq. Ic. t. 242) ; corm globose, $\frac{1}{2}$ in. diam. ; stem a foot long, simple, with 3 superposed leaves with glabrous sheaths and short, free, linear blades ; flowers 2, ascending, bright yellow with red dots ; outer spathe-valve lanceolate, 1–1$\frac{1}{2}$ in. long ; perianth-tube 1$\frac{1}{2}$ in. long, broadly funnel-shaped in the upper half ; upper segment oblong, obscurely cuspidate, $\frac{1}{3}$ in. broad, the others narrower ; stamens reaching halfway up the perianth-limb. *Ker, Gen. Irid.* 137 ; *Baker, Handb. Irid.* 209. *G. strictus, Jacq. Collect.* iv. 170 ; *Klatt in Linnæa* xxxii. 704.

SOUTH AFRICA : without locality.

Known to me only from Jacquin's figure and description.

32. G. vittatus (Hornem. Hort. Hafn. ii. 950) ; corm globose ; stem about a foot long ; leaves 3–4, superposed, linear, the lower with free blades 6–9 in. long, $\frac{1}{4}$–$\frac{1}{3}$ in. broad ; flowers 3–6 in a lax spike, suberect, pink ; spathe-valves oblong-lanceolate, 1–2 in. long ; perianth-tube slightly curved, about an inch long ; limb rather longer than the tube ; segments oblong-spathulate, subobtuse, the top one the longest, the three lower with red or lilac central blotches ; stamens reaching halfway up the limb. *Baker, Handb. Irid.* 210. *G. fasciatus, Roem. et Schult. Syst. Veg.* i. 429. *G. angustus, Thunb. herb. !* *G. undulatus, Schneev. Ic. t.* 19. *G. undulatus, var., Ker in Bot. Mag. t.* 538 ; *Gen. Irid.* 143. *G. vinulus, Klatt in Proc. South Afric. Phil. Soc.* iii. *pt.* 2, 199.

SOUTH AFRICA : in the Karoo, *Thunberg !* Cape Flats near Wynberg, *MacOwan and Bolus, Herb. Norm.,* 287 ! *MacOwan,* 5651 !

33. G. striatus (Jacq. Collect. v. 28) ; corm globose ; tunics with thick wiry strands ; flowering-stem simple, 1 ft. long ; leaves many, long, glabrous, sheaths with free, linear points, 6 in. long ; flowers 2–3, yellowish and pale violet streaked with red ; spathe-valves lanceolate, $\frac{3}{4}$ in. long ; perianth-tube broadly funnel-shaped, $\frac{3}{4}$ in. long ; upper segment the largest, oblong-spathulate, $\frac{1}{2}$ in. broad, the others narrower, reflexing ; anthers violet, just protruded from the throat of the tube. *Baker, Handb. Irid.* 209. *Tritonia striata, Ker in Konig and Sims' Ann.* i. 228 ; *Gen. Irid.* 117. *Antholyza striata, Klatt, Ergänz.* 12.

SOUTH AFRICA : without locality.

Known to me only from Jacquin's description.

34. G. paludosus (Baker in Journ. Bot. 1891, 70) ; corm not seen ; stem simple, a foot long, bearing 2–3 reduced linear leaves ; produced basal leaves 4, rigid, linear, erect, a foot long, $\frac{1}{4}$ in. broad ; spike subsecund, moderately dense, 4–10-flowered, 4–6 in. long ; outer spathe-valve lanceolate, green, $\frac{3}{4}$–1 in. long ; perianth-tube $\frac{1}{2}$ in. long ; segments bright red-purple, obovate, subequal, $\frac{3}{4}$ in. long ; stamens rather shorter than the perianth-segments. *Handb. Irid.* 209.

KALAHARI REGION : Transvaal ; marshy places near Lake Chrissie, *Elliot,* 1588 !

35. G. Niveni (Baker, Handb. Irid. 210); corm small, globose; stem slender, a foot long; produced leaves 2, narrow linear, glabrous, the lower ½ ft. long; flowers 3–4 in a lax spike with a flexuose rachis; outer spathe-valves green, lanceolate, the lower an inch long; perianth bright lilac; tube curved, ½ in. long; segments an inch long, the five lower undulated and reflexed, the upper obovate, obtuse, lateral oblong, acute; three lower oblong-lanceolate, tinged with yellow below the tip; stamens short. *G. ringens, var. undulatus, Andr. Bot. Rep. t.* 275.

SOUTH AFRICA: without locality, *Niven!*

Introduced into cultivation by Niven in the year 1800.

36. G. punctatus (Thunb. Fl. Cap. edit. Schult. 44, non Jacq.); corm not seen; stem 2 ft. long including the inflorescence, mottled with purple towards the base, with 3–4 superposed leaves with free, linear blades, ½–1 ft. long, ¼ in. broad, glabrous, strongly ribbed; flowers 6–10 in a lax secund spike, ½ ft. long; spathe-valves lanceolate, 1–1¼ in. long; perianth-tube curved, funnel-shaped, ⅓ in. long; limb an inch long; upper segments obovate, obtuse, nearly ½ in. broad; lower narrower, oblong-unguiculate; stamens reaching half-way up the perianth-limb. *Baker, Handb. Irid.* 209.

COAST REGION: Paard Island, near Cape Town, *Thunberg!*

37. G. brachyscyphus (Baker, Handb. Irid. 210); corm not seen; stem 1½–2 ft. long, with 2–3 reduced leaves; produced leaves about 4, linear, glabrous, moderately firm, ½ ft. long, ¼–⅓ in. broad; flowers 2–6 in a lax spike with a flexuose rachis; outer spathe-valve lanceolate-navicular, an inch long; perianth bright lilac; tube much curved, ½ in. long; segments oblong, obtuse, 1–1¼ in. long, the three lower shorter than the upper; stamens reaching to the tip of the lower segments.

EASTERN REGION: Griqualand East; damp grassy places near Kokstad, 5000 ft., *Tyson,* 1427!

38. G. villosus (Ker, Gen. Irid. 133); corm globose; tunics of strong, matted, wiry fibres; flowering-stem simple, 1–2 ft. long; sheaths conspicuously hairy, the lowest without any blade; the two upper with a produced linear blade of thick, rigid texture, ½–1 ft. long; flowers 3–4 in a lax secund spike, bright red or lilac; outer spathe-valve oblong-lanceolate, green, ½–1 in. long; perianth with a curved tube, ¾–1 in. long, broadly funnel-shaped in the upper half; limb as long as the tube; segments obscurely cuspidate; upper obovate, ⅓–½ in. broad; lower oblong, unguiculate; stamens reaching halfway up the limb. *Baker, Handb. Irid.* 210. *G. hirsutus, var., Ker in Bot. Mag. t.* 823. *G. villosiusculus, Soland. MSS. G. laccatus, Thunb. Fl. Cap. edit. Schult.* 45, excl. syn. *G. pilosus, Eckl. Top. Verz.* 38; *Klatt in Linnæa* xxxii. 709. *G. puniceus, Lam. Encyc.* ii. 727? *G. Lamarcki, Roem. et Schult. Syst. Veg.* i. 445.

SOUTH AFRICA: without locality, *Thunberg! Masson! Auge!*
COAST REGION: sandy places near Cape Town and Wynberg below 100 ft.,
Bolus, 4724! *MacOwan and Bolus, Herb. Norm.;* 288! Worcester Div.; Dutoits
Kloof, *Drège*, 8433b! Drakenstein Mountains, *Drège*, 8433a! and without precise
locality, *Cooper*, 1667! Caledon Div., *Zeyher*, 3983!

39. G. crassifolius (Baker in Journ. Bot. 1876, 334); corm
large, globose; tunics of fine matted fibres; leaves about 6 in a subbasal distichous rosette, ensiform, rigid in texture, glabrous, with
very thick stramineous ribs, sometimes $1\frac{1}{2}$–2 ft. long, $\frac{1}{2}$–$\frac{3}{4}$ in. broad;
stem simple, 2–3 ft. long including the inflorescence; flowers bright
red, very numerous, in a spike $\frac{3}{4}$–1 ft. long; spathe-valves oblong-
cuspidate or oblong-lanceolate, $\frac{1}{2}$–1 in. long, sometimes brown down
to the very base at the flowering time; perianth about an inch long;
tube curved, funnel-shaped, $\frac{1}{3}$ in. long; three upper segments obovate, obtuse, $\frac{1}{4}$ in. broad, $\frac{1}{2}$ in. long; three lower obovate, unguiculate,
$\frac{1}{6}$ in. broad; stamens reaching nearly to the tip of the perianth-
segments. *Journ. Linn. Soc.* xvi. 175; *Handb. Irid.* 215.

KALAHARI REGION; Transvaal; Pretoria Div.; Apies Poort, *Rehmann*,
4031! Middleburg Div.; Botshabelo Mission station, *Nelson*, 143! Orange
Free State; near Harrismith, *Wood*, 4825! and without precise locality, *Cooper*,
3185! 3199!
EASTERN REGION: Natal; Coast-land, *Sutherland!* near Tugela, 3000–4000 ft.,
Wood, 4407! *Allison!* Polela, *Wood*, 4713! near Camperdown, *Wood*, 3175!
and without precise locality, *Gerrard*, 561! 595! Griqualand East; near Kokstad,
5000 ft., *Tyson*, 1350! Pondoland, *Sutherland!*

40. G. Elliotii (Baker in Journ. Bot. 1891, 70); corm not seen;
stem simple, about a foot long, bearing 1–2 reduced leaves; produced
basal leaves about 4, rigid, ensiform, strongly ribbed, 6–9 in. long,
$\frac{1}{2}$ in. broad; spike dense, distichous, 4–5 in. long; outer spathe-valve
oblong, $\frac{3}{4}$–1 in. long; perianth tube as long as the spathe; segments
oblong, acute, subequal, claret red, an inch long; stamens shorter
than the perianth-segments. *Handb. Irid.* 215.

KALAHARI REGION: Transvaal; Middelburg Division, in marshy places by
Steenkool Spruit, *Elliot*, 1557!

41. G. Ludwigii (Pappe, ex Baker, Handb. Irid. 215); corm
large, globose; tunics of parallel strands of matted fibres; leaves
about 6 in a subbasal distichous rosette, rigid, ensiform, with thick,
stramineous ribs, finely pilose, sometimes 2–3 ft. long, $\frac{1}{2}$–1 in. broad;
stem 2–3 ft. long including the inflorescence, finely pubescent;
spike distichous, often a foot long, with 20–30 flowers; spathe-valve
oblong-lanceolate, acuminate, 1–$1\frac{1}{2}$ in. long, pubescent, scariose
towards the edge and top; flower plain pale yellow; perianth-tube
curved, funnel-shaped, $\frac{3}{4}$ in. long; limb about as long as the tube;
three upper segments oblong-spathulate, $\frac{1}{4}$–$\frac{1}{3}$ in. broad; three lower
narrower, unguiculate; stamens reaching halfway up the limb.
Antholyza hirsuta, Klatt in Linnæa xxxv. 379. *G. sericeo-villosus*,
β *Ludwigii, Baker in Journ. Linn. Soc.* xvi. 175.

VAR. β, **calvatus** (Baker); leaf, stem, rachis of spike, and spathe valves
glabrous. *G. ochroleucus, Bot. Mag. t.* 6291, *non Baker.*

KALAHARI REGION : Var. *B*, Transvaal ; Mountain sides, Umvoti Creek, near Barberton, 3000 ft., *Galpin*, 925 ! Orange Free State, *Cooper*, 1198 !

EASTERN REGION : Natal ; Inanda, *Wood*, 160 ! 601 ! Polela, 4000–5000 ft., *Wood*, 4636 ! Maritzberg Road, 600 ft., *Sanderson*, 454 ! 530 ! Between Umkomanzi River and Umlanzi River, *Drège*, 4541b ! and without precise locality, *Pappe!* Griqualand East ; Enyembi Mountain near Clydesdale, 4500 ft., *MacOwan, Herb. Aust.-Afr.*, 1547 ! Tembuland ; banks of rivers near Bazeia, 200 ft., *Baur*, 144 !

42. **G. sericeo-villosus** (Hook. in Bot. Mag. t. 5427) ; corm large, globose ; leaves about 6 in a subbasal distichous rosette, ensiform, glabrous, strongly ribbed, $1\frac{1}{2}$–2 ft. long, $\frac{1}{2}$–1 in. broad ; stem 3–4 ft. long including the inflorescence, clothed throughout with soft, crisped, white, spreading hairs ; spike distichous, 20–30-flowered, with a flexuose, densely villose axis ; outer spathe-valve oblong-lanceolate, villose, scariose in the upper half ; flower bright yellow ; perianth-tube curved, funnel-shaped, $\frac{1}{2}$–$\frac{3}{4}$ in. long ; limb rather longer than the tube ; upper segments oblong-spathulate, $\frac{1}{4}$ in. broad ; lower narrower, unguiculate ; stamens reaching halfway up the limb. *Baker, Handb. Irid.* 215.

KALAHARI REGION : Orange Free State, *Cooper*, 1198 !

EASTERN REGION : Griqualand East ; Zuurberg Range, near Kokstad, 5000 ft., *Tyson*, 216 !

Differs from *G. Ludwigii* in vestiture alone.

43. **G. ochroleucus** (Baker in Journ. Bot. 1876, 182) corm middle-sized, globose ; leaves 4–6 in a basal rosette, rigid, ensiform, strongly ribbed, the longest 9–12 in. long, about $\frac{1}{2}$ in. broad ; stem $2\frac{1}{2}$–3 ft. long including the inflorescence ; flowers 8–12 in a lax secund spike 6–9 in. long ; spathe-valves green, lanceolate, 1–2 in. long ; flower plain creamy-yellow ; perianth-tube curved, $\frac{3}{4}$ in. long, broadly funnel-shaped in the upper half ; limb an inch long ; 3 upper segments oblong-spathulate, $\frac{1}{2}$ in. broad ; 3 lower oblong-unguiculate, $\frac{1}{4}$ in. broad ; uppermost hood-shaped, not reflexing ; stamens reaching halfway up the limb. *Journ. Linn. Soc.* xvi. 175 ; *Handb. Irid.* 216, *non Bot. Mag. t.* 6291.

EASTERN REGION : Tembuland ; Bazeia Mountain, 2000 ft., *Baur*. 94 !

44. **G. Kirkii** (Baker in Gard. Chron. 1890, viii. 524, non Handb. Irid. 222) ; corm depresso-globose, 1 in. diam., crowned with a ring of fibres ; produced leaves 5–6, linear, glabrous, firm in texture, strongly ribbed, slightly glaucous, 1–$1\frac{1}{2}$ ft. long, $\frac{1}{2}$–$\frac{3}{4}$ in. broad ; stem terete, 3 feet long including the inflorescence ; spike lax, secund, $\frac{1}{2}$–1 ft. long ; spathe-valves green, lanceolate, the outer 1–$1\frac{1}{2}$ in. long ; perianth-tube funnel-shaped, $\frac{1}{2}$ in. long ; segments obovate, obtuse, distinctly cuspidate, pale pink, unspotted, an inch long ; stamens shorter than the segments.

COAST REGION : King Williamstown Div. Described from cultivated plants sent by Sir John Kirk.

Near *G. Eckloni* and *crassifolius*. I find I have used the specific name for two different species.

45. **G. Eckloni** (Lehm. in Ann. Sc. Nat. ser. 2, vi. 107) ; corm large, globose ; tunics of strong, matted, wiry fibres ; produced

leaves 4–6, subbasal, ensiform, glabrous, rigid in texture, with
thick stramineous ribs, not more than a foot long, sometimes 1–1½ in.
broad; upper leaves 3–4, much reduced: stems robust, simple,
1½–3 ft. long including the inflorescence; flowers 6–12, bright red,
copiously and minutely spotted, arranged in a very lax spike;
spathe-valves green, oblong-lanceolate, the lowest 1½–2 in. long;
perianth-tube curved, ¾–1 in. long, broadly funnel-shaped in the
upper half; limb 1 in. long, 3 upper segments oblong-spathulate, ½ in.
broad; 3 lower obovate, ¼ in. broad, distinctly unguiculate; stamens
reaching halfway up the limb. *Klatt in Linnæa* xxxii. 712; *Baker
in Bot. Mag. t.* 6335; *Handb. Irid.* 216. *Neuberia longifolia, Eckl.
Top. Verz.* 37. *G. carneus, Klatt in Linnæa* xxxii. 722, *non Delaroche.*

COAST REGION: Stockenstrom Div.; Katberg, 2000–4000 ft., *Hutton,* 35!
Fort Beaufort Div.; Kunap River, *Murray,* 520! Stutterheim Div.; rocky
places near Kabousie River, 3000 ft., *Murray,* 520!

KALAHARI REGION: Transvaal; at the base of mountains near Barberton,
2800 ft., *Galpin,* 864! Houtbosch, *Rehmann,* 5777! plains near Pretoria, 4100
ft., *McLea in Herb. Bolus,* 5794! Orange Free State; on the Drakensberg,
Cooper, 1040! 3194! 3198! Basutoland, *Bowker,* 1!

EASTERN REGION: Natal; Coast-land, *Sutherland!* Biggarsberg, *Wood,*
4712! upper part of Tugela River, *Allison,* 26! Griqualand East; Mount
Currie, 5200 ft., *Tyson,* 1795! Tembuland; Bazeia Mountain, 2300 ft.,
Baur, 141!

46..**G. purpureo-auratus** (Hook. fil. in Bot. Mag. t. 5944);
corm large, globose; tunics of parallel strands of matted fibres;
leaves ensiform, glabrous, rigid in texture, much shorter than the
stem; stem 3 ft. long including the inflorescence; flowers 10–15 in
a lax secund spike a foot long; spathe-valves green, oblong-lanceolate, 1–1½ in. long; perianth primrose-yellow; tube much curved,
funnel-shaped, under an inch long; upper segments plain, obovate-
spathulate, 1¼–1½ in. long, ¾ in. broad; lower obovate-unguiculate,
the two inner with a spade-shaped red-brown blotch at the throat;
stamens reaching halfway up the limb. *Baker in Journ. Linn. Soc.*
xvi. 175; *Handb. Irid.* 216.

EASTERN REGION: Natal; upper part of Tugela River, *Allison!*

Introduced into cultivation by Mr. Bull. Perhaps only a colour-variety of
G. Papilio.

47. **G. Papilio** (Hook. fil. in Bot. Mag. t. 5565); corm middle-sized,
globose; tunics of parallel strands of fine matted fibres; produced
subbasal leaves about 4, ensiform, glabrous, rigid in texture, 1–1½ ft.
long, ¾–1 in. broad at the middle; stem 2–3 ft. long including the
inflorescence; flowers pale purple, 6–12 in a lax spike; spathe-
valves oblong-navicular, cuspidate, the outer 1–1½ in. long; perianth
horizontal; tube curved, ¼ in. long, broadly funnel-shaped in the
upper half; limb 1¼–1½ in. long; 3 upper segments obovate-
spathulate, ½–¾ in. broad, upper not reflexing; 3 lower oblong-
unguiculate, with a large reddish spade-shaped blotch edged with
yellow at the throat; stamens reaching halfway up the limb. *Baker
in Journ. Linn. Soc.* xvi. 175; *Handb. Irid.* 216.

KALAHARI REGION : Transvaal ; plains near Rustenburg, 3500 ft., *McLea in Herb. Bolus,* 3094 ! Houtbosch, *Rehmann,* 5776 ! Lomatie Valley, near Barberton, 4000 ft., *Galpin,* 1193 ! and without precise locality, *Sanderson !* Orange Free State, *Cooper,* 995 ! 3188 !

EASTERN REGION : Natal; near Durban, *Sanderson,* 367 ! Inanda, *Wood,* 442 ! Clairmont, *Wood,* 1729 ! in a swamp by the Mooi River, *Wood,* 3441 ! without precise locality, *Gerrard,* 647·!

48. G. Rehmanni (Baker, Handb. Irid. 216); corm not seen ; produced subbasal leaves about 4, linear, rigid, glabrous, with thick stramineous ribs, $1\frac{1}{2}$–2 ft. long, $\frac{1}{4}$–$\frac{1}{3}$ in. broad ; stem 2 ft. long including the inflorescence, with about 4 reduced linear leaves ; spike lax, secund, $\frac{1}{2}$ ft. long; outer spathe-valve very large, oblong-lanceolate, firm in texture, $1\frac{1}{2}$–$2\frac{1}{2}$ in. long ; perianth bright red ; tube curved, an inch long, funnel-shaped in the upper half ; limb $\frac{1}{4}$ in. long ; 3 upper segments obovate-spathulate, $\frac{5}{8}$ in. broad ; 3 lower oblong, $\frac{1}{4}$–$\frac{1}{3}$ in. broad, distinctly unguiculate; stamens not reaching halfway up the limb.

KALAHARI REGION : Transvaal; Boschveld between Elands River and Klippan, *Rehmann,* 5096 !

Intermediate between *G. Eckloni* and *Papilio.*

49. G. hirsutus (Jacq. Ic. t. 250); corm middle-sized, globose; leaves 4–6, short, ensiform, rigid in texture, very strongly ribbed, both blade and sheath finely hairy ; stems simple, hairy, 1–2 ft. long including the spike ; flowers 3–6 in a very lax spike ; outer spathe-valve lanceolate, $1\frac{1}{2}$–2 in. long; perianth bright red ; tube curved, $1\frac{1}{2}$ in. long, broadly funnel-shaped in the upper half; segments as long as the tube, obovate, obscurely pointed, the upper about $\frac{3}{4}$ in., the others $\frac{1}{2}$ in. broad ; stamens reaching more than halfway up the limb. *Red. Lil. t.* 273 ; *Ker, Gen. Irid.* 132 ; *Baker, Handb. Irid.* 217. *G. hirsutus, var. roseus, Ker in Bot. Mag. t.* 574. *G. roseus, Andr. Bot. Rep. t.* 11.

SOUTH AFRICA: without locality, *Thunberg ! Burchell !*
COAST REGION : Malmesbûry Div.; Groene Kloof, 300–500 ft., *Zeyher ! Bolus,* 4342 ! *MacOwan and Bolus, Herb. Norm.,* 585 !

G. similis, Eckl. Top. Verz. 40, is said to differ by its smaller, paler, less-spotted flowers.

50. G. salmoneus (Baker, Handb. Irid. 217) ; corm not seen ; stem a foot long; stem-leaves 6–8, rigid, linear, acuminate, finely pubescent, with strong raised ribs, 1–$1\frac{1}{2}$ ft. long, $\frac{1}{3}$ in. broad ; spike moderately dense, distichous, a foot long; outer spathe-valves lanceolate, $1\frac{1}{2}$–3 in. long; perianth-tube slender, curved, 2 in. long ; segments obovate, obtuse, salmon-red, 18–21 lines long; stamens half as long as the perianth-segments.

EASTERN REGION : Griqualand East; mountain slopes around Kokstad, 4800 ft., *Tyson,* 1180 !

Near *G. hirsutus* and *oppositiflorus.*

51. G. scaphochlamys (Baker, Handb. Irid. 217); produced sub-basal leaves 3–4, ensiform, firm and rigid in texture, 1 ft. long, $\frac{1}{2}$ in.

broad ; stem 1½–2 ft. long including the inflorescence, simple ; flowers 6–8 in a lax spike ½ ft. long ; spathe-valves boat-shaped, oblong-lanceolate, firm in texture, 1½–2 in. long ; perianth pinkish-white ; tube curved, 1–1¼ in. long ; segments obovate-spathulate, obtuse, ⅓–½ in. broad ; stamens reaching halfway up the limb.

COAST REGION : Flats near Cape Town, *MacOwan*, 2553 ! Swellendam Div. ; Sparrbosch, under 1000 ft., *Drège*, 8427 ! Voormans Bosch, *Pappe !*

52. G. floribundus (Jacq. Ic. t. 254), corm globose ; tunics of matted fibres ; produced leaves 3–4, ensiform, 1–2 ft. long ; stems 1½–2 ft. or more long including the inflorescence, branched when at all luxuriant ; flowers white with a pink tinge, 4–12 in a very lax distichous spike, all ascending ; outer spathe-valve oblong-lanceolate, 1½–2 in. long ; perianth-tube nearly straight, 1½–2 in. long, funnel-shaped in the upper third ; segments as long as the tube, obovate-spathulate, deltoid at the tip, the upper ¾–1 in. broad ; stamens reaching ⅓ or ½-way up the limb. *Ker in Bot. Mag. t.* 610 ; *Gen. Irid.* 143 ; *Baker, Handb. Irid.* 218. *G! grandiflorus, Andr. Bot. Rep. t.* 118.

COAST REGION : Uitenhage Div.; near Van Stadens River, *Bolus*, 1633 ! *MacOwan*, 2062 ! near Uitenhage, *Burchell*, 4261.

Scarcely more than a variety of *G. blandus*, differing mainly from *G. excelsus* by its more obovate perianth-segments.

53. G. oppositiflorus (Herb. in Bot. Reg. 1842, Misc. 86) ; corm large, globose ; tunics of matted fibres ; produced basal leaves about 4, ensiform, firm in texture, 1–1½ ft. long, ¾–1 in. broad ; stem 3–4 ft. long including the inflorescence, often branched ; flowers up to 30 or 40, arranged in a distichous spike often a foot long ; spathe-valves green, lanceolate, acute, thin in texture, 1–1½ in. long ; perianth white ; tube curved, 1–1¼ in. long, slender up to the top ; limb horizontal, 1½ in. long, with oblong-spathulate acute segments not more than ⅓–½ in. broad at the middle ; stamens half as long as the limb. *Baker, Handb. Irid.* 218 ; *Bot. Mag. t.* 7292.

EASTERN REGION : Transkei, *MacOwan*, 2254 ! Not uncommon about the Kei river, *Flanagan.*

Said to be a native of Madagascar, but not received thence in any of the recent collections. It is often wrongly called *G. floribundus* in gardens.

54 G. blandus (Ait. Hort. Kew i. 64) ; corm globose, middle-sized ; tunics of parallel strands of matted fibres ; produced sub-basal leaves 4, ensiform, firm in texture, glabrous, the outer ½–1 ft. long, ½–¾ in. broad ; stem 1–2 ft. long including inflorescence, sometimes branched ; flowers white with a tinge of red, 4–8 in a lax distichous spike, all ascending ; outer spathe-valves green, lanceolate, 1½–2 in. long ; perianth-tube about 1½ in. long, much dilated and curved at the top ; limb rather longer than the tube, segments oblong-spathulate, narrowed to a point, the top one about ¾ in. and the others about ½ in. broad at the middle ; stamens reaching more

than halfway up the limb. *Ker in Bot. Mag. t.* 625 ; *Gen. Irid.* 140 ;
Baker, Handb. Irid. 217. *G. angustus, Linn. herb. ex parte !*

VAR. β, **G. albidus** (Jacq. Ic. t. 256); flower pure white. *G. blandus, Andr.
Bot. Rep. t.* 99. *G. blandus, var. niveus, Ker in Bot. Mag. t.* 648.

VAR. γ, **G. Mortonius** (Herb. in Bot. Mag. t. 3680) ; flowers suberect; segments
white, with copious, faint, vertical, pink streaks.

VAR. δ, **G. excelsus** (Sweet, Hort. Brit. edit. 2, 501); taller than the type,
with longer leaves and a perianth-tube 2 in. long.

VAR. ε, **G. carneus** (Delaroche, Descr. 00, t. 4) ; more robust than the type, with
more numerous, more spreading pink flowers, with broader, less acute segments.
G. campanulatus, Andr. Bot. Rep. t. 188. *G. blandus, var., Ker in Bot. Mag.
t.* 645.

Var. Hibbertii, Hort., has pink flowers with very distinct, red, spade-shaped
marks on the three lower segments.

SOUTH AFRICA : without locality, *Thunberg ! Oldenburg ! Forster ! Zeyher,*
3980! *Harvey,* 918! Var. γ, *Harvey,* 899! Var. ε, from Hesse's garden at Cape
Town, *Burchell !*

COAST REGION : Devils Mountain, 3736! Dutoits Kloof, *Drège,* 8426b !
Var. δ, Paarl Mountains. under 100 ft., *Drège !* Var. ε, Tulbagh Div. ; Mosterts
Berg, 1700 ft., *Bolus,* 5253! British Kaffraria, *MacOwan,* 1467 ! near Grahams-
town, *Baur,* 1100 !

55. G. Milleri (Ker in Bot. Mag. t. 632) ; corm globose, middle-
sized ; tunics of parallel strands of matted fibres ; stem simple,
1–1½ ft. long ; produced basal leaves about 4, ensiform, shorter than
the stem, ½–¾ in. broad ; flowers 4–5, suberect in a very lax spike ;
outer spathe-valves oblong-lanceolate, moderately firm in texture,
1½ in. long ; perianth milk-white ; tube straight, erect, 1½–2 in.
long ; segments oblong, subacute, the upper one the broadest, ½–⅝ in.
broad at the middle ; stamens reaching ⅓–½-way up the limb.
Gen. Irid. 143 ; *Baker, Handb. Irid.* 218. *Antholyza spicata,
Miller, Gard. Dict. edit.* 8, *No.* 2 (*Ic.* 27, *t.* 40).

COAST REGION : Port Elizabeth, *Cult. Harpur-Crewe !*

Close upon *G. blandus* and *floribundus.*

56. G. undulatus (Jacq. Ic. t. 251); corm globose ; tunics of
parallel strands of fine matted fibres ; produced basal leaves 4–6,
ensiform, firm in texture, glabrous, ½–1 ft. long, ½–¾ in. broad ; stem
simple, 1 ft. long including the spike ; flowers 4–6 in a very lax
spike with a flexuose rachis, all suberect ; outer spathe-valve green,
oblong-lanceolate, 1–1½ in. long ; perianth-tube nearly straight,
1½ in. long ; segments as long as the tube, oblong-spathulate, subacute, milk-white with a red keel, ⅓–½ in. broad ; stamens reaching
more than halfway up the limb. *Red. Lil. t.* 122 ; *Ker in Bot. Mag.
t.* 647; *Gen. Irid.* 142 ; *Baker, Handb. Irid.* 218. *G. angustus,
Herb. Linn. ex parte.* *G. vittatus, Zuccagn. in Roem. Coll.* 121.

SOUTH AFRICA : without locality, *Thunberg ! Bowie !*
COAST REGION : Uitenhage Div., *Pappe !* Bathurst Div. ; between Riet
Fontein and the sea-shore, *Burchell,* 4081 !

57. G. Macowani (Baker, Handb. Irid. 219); corm not seen ; stem
2–3 ft. long ; produced leaves ensiform, more rigid in texture than

in the other *Cardinales*, 1½ ft. long, ½–¾ in. broad ; flowers 6–10 in a lax distichous spike, all suberect ; spathe-valves green, lanceolate, lower 1½–2 in. long; perianth bright scarlet; tube straight, 2 in. long, broadly funnel-shaped in the upper third or quarter ; segments oblong, acute, subequal, 1½ in. long, ½ in. broad ; stamens reaching halfway up the limb. *G. secundus, Soland. in Herb. Banks!*

CENTRAL REGION : Somerset Div. ; on the Zuurberg and Bosch Berg, *MacOwan*, 236! Aliwal North Div. ; on the Witte Bergen, *Cooper*, 3600!

Differs from the other *Cardinales* by the segments being subequal and shorter than the tube.

58. **G. Adlami** (Baker in Gard.Chron. 1889, v. 233); stem 1–1¼ ft. long; leaves ensiform, 1–1½ ft. long, an inch broad at the middle, with distant, strong, unequal, stramineous ribs; flowers 5–6, aggregated in a moderately dense, equilateral, simple, erect spike ; outer spathe-valve oblong-lanceolate, brown, 1½–2 in. long ; perianth greenish-yellow ; tube erect, nearly straight, narrowly funnel-shaped, an inch long ; segments oblong, acute ; upper slightly arcuate, as long as the tube, granulated all over with minute red dots ; five others smaller, minutely cuspidate, the inner plain greenish-yellow, the two others granulated with red towards the tip ; stamens and style a little longer than the perianth ; filaments white ; anthers linear, ½ in. long, granulated with red. *Handb. Irid.* 219.

KALAHARI REGION : Transvaal, *Adlam !*

Described from a plant grown in the Cambridge botanic garden by Mr. R. I. Lynch.

59. **G. cardinalis** (Curt. Bot. Mag. t. 135) ; corm large, globose ; stem 3–4 ft. long; produced leaves 4–6, ensiform, rather thin in texture, glaucous green, reaching 2 ft. or more in length, ¾–1 in. broad ; flowers 12–20 in a spike ½–1 ft. long, all more or less ascending ; spathe-valves green, thin in texture, lanceolate, acute, 1½–3 in. long ; perianth bright scarlet ; tube nearly straight, 1½ in. long, funnel-shaped in the upper half ; upper segments oblong-spathulate, acute, concolorous, 2 in. long, ¾–1 in. broad ; 3 lower shorter and narrower, conspicuously mottled with white at the throat ; stamens reaching more than halfway up the limb ; anthers lanceolate, ⅓ the length of the filaments. *Schneev. Ic. t.* 27; *Red. Lil. t.* 112 ; *Ker, Gen. Irid.* 143 ; *Baker, Handb. Irid.* 219. *G. speciosus, Eckl. Top. Verz.* 41, *non Thunb.*

SOUTH AFRICA : without locality, *Masson !*
COAST REGION : Worcester Div. ; Dutoits Kloof, *Drège !*

60. **G. splendens** (Baker in Journ. Bot. 1876, 333) ; corm not seen ; stem about 2 ft. long; produced leaves ensiform, not very rigid in texture, 1–1½ ft. long, ½ in. broad ; flowers 4–6 in a lax distichous spike, all ascending ; outer spathe-valve green, lanceolate, 1½–2 in. long ; perianth bright scarlet ; tube nearly straight, 1½ in. long, funnel-shaped in the upper third ; upper segments obovate,

minutely cuspidate, 2 in. long, above an inch broad; 3 lower rather
narrower and shorter, with a pale keel through the lower half;
stamens reaching more than halfway up the limb. *Journ. Linn.
Soc.* xvi. 176; *Handb. Irid.* 219.

COAST REGION: George Div.; mountains near Oakhurst, *Dumbleton!*

A near ally of *G. cardinalis.*

61. **G. cruentus** (Moore in Gard. Chron. 1868, 1138); corm large,
globose; stem 2–3 ft. long; produced leaves about 4, ensiform, dark
glaucous green, 1½–2 ft. long, ¾–1 in. broad; spike rather dense,
distichous, 6–10 flowered; bracts very large, lanceolate, the lower
sometimes 3–6 in. long; perianth bright scarlet; tube 1½–2 in. long,
nearly straight, funnel-shaped in the upper half; upper segments
concolorous, obovate-spathulate, obscurely cuspidate, 2–2½ in. long,
1¼–1½ in. broad; 3 lower about 1½ in. long, 1 in. broad, with a large
white blotch at the throat with small red spots; anthers lanceolate,
reaching halfway up the limb. *Hook. fil. in Bot. Mag. t.* 5810;
Baker, Handb. Irid. 219.

EASTERN REGION: Natal, *Hort. Bull.*.

A near ally of *G. cardinalis.*

62. **G. dracocephalus** (Hook. fil. in Bot. Mag. t. 5884); corm
large, depresso-globose; stem simple, about 2 ft. long; produced
leaves ensiform, 1–1½ ft. long, ¾–1 in. broad, moderately firm in
texture; flowers few, arranged in a very lax secund spike; outer
spathe-valve lanceolate, green, 2–3 in. long; perianth-tube much-
curved, greenish, 1½–2 in. long; limb 1½ in. long, yellowish-green,
minutely grained and spotted with dull purple; upper segments
obovate, permanently hooded, ¾–1 in. broad; lower lanceolate,
reflexing; stamens reaching near to the top of the segments; anthers
lanceolate, less than half as long as the filaments. *Baker in Journ.
Linn. Soc.* xvi. 176; *Handb. Irid.* 220.

EASTERN REGION: Natal, at the foot of the Drakensberg, *Cooper*, 3593! upper
part of Tugela River, *Allison!*

Also flowered at Kew, Aug. 1887, from bulbs sent by Mr. Adlam.

63. **G. platyphyllus** (Baker in Gard. Chron. 1893, xiv. 456); corm
large, globose; stem 1½ ft. long, bearing 4–5 leaves with long-
sheathing bases; leaves broadly ensiform, firm, green, the lowest a
foot long above the sheathing base, 2 in. broad at the middle; spike
lax, secund, ½–1 ft. long; outer spathe-valve oblong-navicular, green,
1½–2 in. long; perianth-tube curved, 1½ in. long, ⅓ in. diam. at the
throat; limb with fine red lines on a yellow ground; 3 upper
segments oblong, acute, ⅝–¾ in. broad, standing forward, convex;
3 lower much smaller, oblanceolate, reflexing; stamens ½ in. shorter
than the upper segments.

EASTERN REGION: Natal; Inanda, *Wood*, 422! Bothas Hill, 2000 ft., *Wood*,
4819! and without precise locality, *Gerrard*, 1532! Transkei, *Hort. Leichtlin!*

64. G. psittacinus (Hook. in Bot. Mag. t. 3032); corm very large, depresso-globose ; tunics of parallel strands of matted fibres ; produced leaves about 4, ensiform, rigid in texture, 1–2 ft. long, 1–2 in. broad ; stem 3–4 ft. long including the inflorescence ; spike very lax, reaching a foot or more in length ; spathe-valves green, oblong-lanceolate, 2–3 in. long ; perianth-tube curved, 1½–2 in. long, sub-cylindrical in the upper half ; limb about equalling the tube ; upper segments obovate, dark crimson, hooded, ¾–1 in. broad ; lower segments much smaller, reflexing at the top, red and yellow mixed ; stamens reaching nearly to the tip of the segments ; anthers ½ in. long ; filaments about 1½ in. long ; capsule large, oblong. *Bot. Reg. t.* 1442 ; *Reich. Exot. t.* 116 ; *Baker, Handb. Irid.* 220. *G. natalensis, Reinw. ex Hook. in Bot. Mag. sub t.* 3084 ; *Sweet, Brit. Flow. Gard. ser.* 2, *t.* 281 ; *Lodd. Bot. Cab. t.* 1756. *Watsonia natalensis, Eckl. Top. Verz.* 34.

Var. β, G. Cooperi (Baker in Bot. Mag. t. 6202) ; perianth-tube 2½–3 in. long ; segments more acute.

CENTRAL REGION : Albert Div., *Cooper*, 689 ! Aliwal North Div. ; on the Witte Bergen, 4500–5000 ft., *Drège*, 3502 ! Var. β, Somerset Div., *Bowker !*
KALAHARI REGION : VAR. β, Transvaal ; Lomatie Valley near Barberton, 4000 ft., *Galpin*, 1192 ! Orange Free State ; Riet River, *Burke*, 87 !
EASTERN REGION : Natal, *Cooper*, 3189 ! *Buchanan !* Var. β, *Gueinzius !* Tembuland ; Bazeia Mountain, 3000 ft., *Baur*, 461 !

65. G. Leichtlini (Baker, Handb. Irid. 214) ; corm large, globose ; leafy stem, terete, 2 ft. long ; produced leaves 4, ensiform, bright green, moderately firm, a foot long, under an inch broad ; flowers 6–8 in a moderately dense, secund spike ½ ft. long ; spathe-valves lanceolate, erect, 1–1½ in. long ; perianth bright red ; tube arcuate, 1¼ in. long ; upper segments obovate, permanently connivent, as long as the tube ; 3 lower much smaller, acute, spreading, red at the tip, yellow below it, grained with minute spots of red ; stamens shorter than the upper segments.

KALAHARI REGION : Transvaal, *Adlam !*
Described from plants flowered by Max Leichtlin in 1889.

66. G. Tysoni (Baker, Handb. Irid. 220) ; corm small, globose ; produced leaves about 4, crowded, ensiform, rigidly coriaceous, glabrous, strongly ribbed, very oblique at the base, the outer 6–9 in. long, 1½ in. broad at the middle ; peduncle about a foot long, with several lanceolate bract-leaves ; flowers bright red, 4–6 in a very lax spike ; outer spathe-valves lanceolate, lower 2 in. long ; perianth 4 in. long, with a limb about as long as the curved, narrowly funnel-shaped tube ; upper segment oblong, very convex, ½ in. broad ; lower much smaller.

EASTERN REGION : Griqualand East ; near Fort Donald, 5000 ft., *Tyson*, 1653 !

67. G. Saundersii (Hook. fil. in Bot. Mag. t. 5873) ; corm large, depresso-globose ; produced leaves 4–6, ensiform, rigid in texture, strongly ribbed, 1–2 ft. long, ¾–1 in. broad ; stem 2–3 ft. long including inflorescence; spike very lax, $\frac{1}{2}$ ft. long, 6–8-flowered; spathe-valves green, lanceolate, $1\frac{1}{2}$–2 in. long; perianth-tube curved, 1–$1\frac{1}{4}$ in. long, broadly funnel-shaped in the upper half; limb bright scarlet; 3 upper segments concolorous, oblong spathulate, acute, an inch broad; 3 lower shorter, $\frac{1}{2}$ in. broad, with a great blotch of white spotted with scarlet at the throat; stamens reaching nearly to the tip of the segments; anthers $\frac{1}{2}$ in. long, half the length of the filaments. *Baker in Journ. Linn. Soc.* xvi. 176; *Handb. Irid.* 220.

CENTRAL REGION: Aliwal North Div.; summit of the Witte Bergen, *Cooper*, 605! and Avoca, 4000 ft., *MacOwan, Herb. Aust. Afr.* 1546!
KALAHARI REGION: Transvaal; New Scotland, *MacOwan*, 2253!
EASTERN REGION: Natal; Krans Kop, *McKen*, 17! Inanda, *Wood*, 1216!

Midway between *G. cardinalis* and *psittacinus*.

68. G. aurantiacus (Klatt in Linnæa xxxv. 378); corm large, globose; produced subbasal leaves 4–6, ensiform, moderately firm in texture, 1–$1\frac{1}{2}$ ft. long, $\frac{1}{2}$–$\frac{3}{4}$ in. broad; stem simple, about 3 ft. long, inflorescence included; spike lax, many-flowered, sometimes a foot long; outer spathe-valve green, lanceolate, thin in texture, 1–2 in. long; perianth bright orange-yellow or tinged with red; tube curved, 2 in. long, dilated suddenly at the middle, and cylindrical in the upper half; upper segments obovate, 1–$1\frac{1}{2}$ in. long; 3 lower oblong, shorter; stamens reaching halfway up the segments. *Baker in Journ. Linn. Soc.* xvi. 176; *Handb. Irid.* 221.

EASTERN REGION: Natal; Pietermaritzburg, 2000-3000 ft., *Sutherland!* Camperdown, 2400 ft., *Wood*, 4968! and without precise locality, *Sanderson! Krauss*, 77!
Received alive from Messrs. Dammann, of Naples, in July, 1886.

69. G. alatus (Linn. Sp. Plant. edit. 2, 53); corm small, globose; tunics brown, membranous; basal leaves 3–4, linear, rigid in texture, the lowest the longest, $\frac{1}{2}$–1 ft. long, $\frac{1}{6}$–$\frac{1}{4}$ in. broad, closely and strongly ribbed; stem $\frac{1}{2}$–1 ft. long including the inflorescence; spike usually simple, few-flowered, very lax, with a very flexuose axis; spathe-valves broad, green, oblong-navicular, the outer 1–$1\frac{1}{4}$ in. long; perianth pink; tube $\frac{1}{2}$ in. long, funnel-shaped at the top; upper segment cucullate, obovate, cuneate, with a short claw, $1\frac{1}{4}$–$1\frac{1}{2}$ in. long, $\frac{1}{2}$–$\frac{3}{4}$ in. broad; side ones shorter, suborbicular, not unguiculate; 3 lower deflexed, with a small obovate blade and a long distinct claw; stamens reaching nearly to the tip of the upper segments; anthers lanceolate, $\frac{1}{3}$ in. long. *Thunb. Diss. No.* 15, *ex parte; Andr. Bot. Rep. t.* 8; *Ker in Bot. Mag. t.* 586; *Gen. Irid.* 132; *Baker, Handb. Irid.* 223. *G. speciosus, Thunb. Fl. Cap.* i. 196. *G. papilionaceus, Lichten. in Roem. et Schult. Syst. Veg.* i. 408. *Hebea galeata, Eckl. Top. Verz.* 41.

VAR. β, **G. namaquensis** (Ker in Bot. Mag. t. 592); more robust, with lanceolate leaves sometimes $1\frac{1}{2}$–2 in. broad, 9–10 flowers, and upper perianth-segments an inch broad. *Ker, Gen. Irid.* 132. *G. equitans, Thunb. Fl. Cap.* 192. *G. galeatus, Andr. Bot. Rep. t.* 122.

SOUTH AFRICA: without locality, *Pappe! Thom!* Var. β, *Thunberg!*
COAST REGION: Cape Flats near Duiker Vley, below 100 ft., *MacOwan*, 2268!

MacOwan and Bolus, Herb. Norm., 278! near Paarl, *Drège*, 8447a! Stellenbosch, *Sanderson*, 956! Malmesbury Div.; Groene Kloof and Zwartland, *Thunberg!* Clanwilliam Div., *Mader*, 98!

WESTERN REGION: Var. β, Little Namaqualand; near Ookiep, 3000 ft., *Bolus*, 5793! and without precise locality, *Masson! Scully*, 147!

70. **G. spathulatus** (Baker, Handb. Irid. 223); corm not seen; leaves linear, firm in texture, 1–1½ ft. long, ¼–⅓ in. broad, with thickened stramineous midrib and edges; stem 1 ft. long including inflorescence, slender, terete, simple; flowers 4–5 in a very lax spike with a flexuose rachis; outer spathe-valve green, oblong-navicular, an inch long, ½ in. broad; perianth dull red; tube slender, funnel-shaped, ½ in. long; upper segment 1¼ in. long, ½–¾ in. broad, obovate-spathulate; 2 side ones rather shorter, similar in shape and size; 3 lower deflexed, brownish-purple, obovate, ⅓ in. broad, with long claws; stamens much shorter than the perianth-segments.

KALAHARI REGION: Transvaal; De Beer, near the Nylstrom River, *Nelson*, 295!

Allied to *G. alatus* and *orchidiflorus*.

71. **G. orchidiflorus** (Andr. Bot. Rep. t. 241); corm globose, 1 in. diam.; tunics of parallel strands of matted fibres; basal leaves 3–4, linear, firm in texture, ½–1 ft. long, ⅙–¼ in. broad; stem 1–1½ ft. long including inflorescence, usually simple; flowers 4–6 in a very lax spike with a very flexuose rachis; outer spathe-valve green, oblong-lanceolate, 1–1¼ in. long; perianth greenish-yellow; tube slender, ½ in. long; upper segment curved over, oblong, an inch long, ¼–⅓ in. broad, with a very distinct claw; side ones much shorter, ½ in. broad, with a short claw; 3 lower decurved, with an obovate lamina ¼ in. broad and a long claw; stamens reaching nearly to the tip of the top segments; anthers ⅓ the length of the filament. *Baker, Handb. Irid.* 224. *G. alatus, Thunb. Diss. No.* 15, *ex parte; Jacq. Ic. t.* 259. *G. viperatus, Ker in Bot. Mag. t.* 688; *Gen. Irid.* 131; *Sweet, Brit. Flow. Gard. t.* 156. *G. virescens, Thunb. Fl. Cap.* i. 196.

SOUTH AFRICA: without locality, *Thunberg! Oldenburg! Thom! Burchell!*

COAST REGION: Clanwilliam Div., *Mader*, 99!

WESTERN REGION: Little Namaqualand, near Kook Fontein, 3000 ft., *Bolus*, 6626!

72. **G. pulchellus** (Klatt in Linnæa xxxii. 693); corm small, globose; tunics pale brown, membranous; lowest leaf subterete, a foot or more long, the other 2–3, superposed, much shorter; stem 1 ft. long including inflorescence, simple or branched; flowers 4–8 in a very lax spike; outer spathe-valve oblong-lanceolate, green, ½–¾ in. long; flowers pale pinkish-purple; perianth-tube ½ in. long; upper segment arched, oblong, with a long claw, ¾ in. long, ¼ in. broad; side segments shorter, ovate, acute, ½ in. broad; 3 lower deflexed, obovate, with a long claw, ¼ in. broad; stamens reaching nearly to the tip of the perianth-segments; anthers ¼ the length of

the filaments. *Baker, Handb. Irid.* 224   *Hebea pulchella, Eckl. Herb. Gladiolus virescens β, Herb. Thunb.!*

SOUTH AFRICA: without locality, *Thunberg! Bowie!*
COAST REGION: Caledon Div.; near Appels Kraal, Zonder Einde River, *Zeyher*, 3994! Humansdorp Div.; between Kabeljouw River and Gamtoos River, *Drège*, 8415!

73. G. bicolor (Baker in Journ. Linn. Soc. xvi. 178); corm small,

globose; tunics membranous; lower leaf the longest, subterete, firm in texture, a foot long; 2–3 others superposed, with short, free points; stem simple, 6–9 in. long including inflorescence; flowers 2–3 in a very lax spike; outer spathe-valve green, oblong-navicular, an inch long; perianth dull yellow; tube ½ in. long; upper segment arching, 1–1¼ in. long, obovate, ¼ in. broad, with a long claw; 2 side ones much shorter, suborbicular; 3 lowest deflexed, obovate cuspidate, ¼ in. broad, with very long claws; stamens reaching nearly to the tip of the upper segment; anthers ⅓ the length of the filament. *Handb. Irid.* 224. *Hebea bicolor, Eckl. Top. Verz.* 42. *G. luteus, Klatt in Linnæa* xxxii. 694, *ex parte. G. Templemannii, Klatt in Trans. South Afric. Phil. Soc.* iii. *pt.* 2, 197.

COAST REGION: Caledon Div.; on the Zwartberg, near the Hot Springs, *Templeman in Herb. MacOwan*, 2608! Caledon, *Zeyher! Rogers!*

74. G. arcuatus (Klatt, Ergänz. 4); corm ovoid; tunics with

rhomboid areolæ; produced leaves 2–3, narrow linear, firm in texture, 4–6 in. long, ½ lin. broad; stem ½ ft. long including inflorescence, simple; 'flowers 2–6 in a lax spike with a very flexuose rachis; outer spathe-valve green, lanceolate, ½–¾ in. long; perianth pinkish, with a greenish tube ½ in. long; upper segment oblong-unguiculate, acute, arching, an inch long, ¼ in. broad; the other segments similar in shape, but rather shorter; stamens ⅔ the length of upper segment. *Baker, Handb. Irid.* 224.

WESTERN REGION: Little Namaqualand; near Koper Berg, 2000–3000 ft., *Drège*, 2629c! Spektakel Mountain, near 'Naries, 3600 ft., *MacOwan and Bolus, Herb. Norm.* 700! and without precise locality, *Morris in Herb. Bolus*, 5795!

75. G. formosus (Klatt in Linnæa xxxii. 692); corm ovoid; tunics

of thick, wiry, brown fibres; leaves firm in texture, narrow linear or subterete, the lower a foot or more long; stem 1–2½ ft. long including inflorescence, simple or branched; spikes lax, the end one 6–9 in. long; outer spathe-valve green, oblong-lanceolate, ¾ in. long; flowers bright mauve-purple; perianth-tube ½ in. long, funnel-shaped in the upper half; upper segment 1–1¼ in. long, cucullate, oblong-spathulate, acute, ⅓ in. broad: 2 side ones similar in size and shape, but shorter; 3 lower deflexed, with a small obovate, cuspidate lamina and long claws; stamens much shorter than the perianth-segments. *Baker, Handb. Irid.* 225. *Hebea formosa, Eckl. Top. Verz.* 42.

COAST REGION: Clanwilliam Div., *Mader*, 192! 2163!

76. G. edulis (Burch. ex Ker in Bot. Reg. t. 169); corm ovoid; tunics

of parallel strands of fine matted fibres; produced leaves about 3,

narrow linear, firm in texture, the lowest sometimes overtopping the
stem ; stems slender, 1–2 ft. long including the inflorescence, simple
or branched ; flowers 8–15 in a lax distichous spike ; outer spathe-
valve oblong-lanceolate, acute, subscariose, $\frac{1}{2}$–$\frac{3}{4}$ in. long; perianth
pale lilac or whitish ; tube funnel-shaped, $\frac{1}{2}$ in. long; segments all
with a very distinct cusp ; upper obovate-unguiculate, $\frac{3}{4}$ in. long; 2
side ones oblong, with a long claw; 3 lower lanceolate, with a long
claw; stamens reaching more than halfway up the limb ; capsule
membranous, oblong, $\frac{5}{8}$–$\frac{3}{4}$ in. long. *Baker, Handb. Irid.* 225.
G. permeabilis, Baker in Journ. Linn. Soc. xvi. 178, *ex parte.*

COAST REGION : stony mountains around King Williamstown, *MacOwan and
Bolus, Herb. Norm.,* 840! Alexandria Div.; Zuurberg Range, *Cooper,* 3195!
3196! British Kaffraria, *Cooper,* 1809!
CENTRAL REGION : Graaff Reinet Div. ; on the Sneeuw Bergen, 4000–5000 ft.,
Drège, 8417b! and hills near Graaff Reinet, 2500–4000 ft., *Bolus,* 408! near
Somerset East, 2500 ft., *MacOwan,* 1891!
KALAHARI REGION : Bechuanaland; Pellat Plains, *Burchell,* 2240! Basuto-
land, *Cooper,* 3317! Orange Free State ; Nelsons Kop, *Cooper,* 877!
Differs from *G. permeabilis* by its remarkably cuspidate perianth-segments.

77. G. Scullyi (Baker, Handb. Irid. 224); whole plant 1–1$\frac{1}{2}$ ft.
long ; corm 1–1$\frac{1}{2}$ in. diam.; tunics thick, lacerated from the base and
apex ; stem a few inches long; produced leaves 6–8, firm, linear,
strongly ribbed, $\frac{1}{6}$ in. broad, the lower 8–9 in. long ; inflorescence a
panicle of 3–4 lax spikes 3–4 in. long, with a flexuose axis; outer
spathe-valves green, lanceolate, the lower an inch long ; perianth
reddish, about an inch long ; segments obovate-unguiculate, the upper
$\frac{1}{4}$ in. broad.

WESTERN REGION : Little Namaqualand, *Scully,* 158!
Near *G. edulis.*

78. G. Dregei (Klatt in Linnæa xxxii. 693) ; corm ovoid, $\frac{1}{2}$ in.
diam.; tunics brown, with rhomboid areolæ; produced leaves 3–4,
narrow linear, firm in texture, 4–6 in. long, $\frac{1}{12}$ in. broad ; stem 6–9 in.
long including the inflorescence, simple ; flowers 6–9 in a very lax
spike ; outer spathe-valve green, oblong-lanceolate, $\frac{1}{2}$ in. long ; perianth-
tube $\frac{1}{3}$ in. long ; top segment $\frac{1}{2}$ in. long, arching hemispherically, with
a small obovate lamina and a very long distinct claw ; the other
segments shorter, obovate, distinctly unguiculate ; stamens reaching
nearly to the tip of the perianth-segment. *Baker, Handb. Irid.*
225.

WESTERN REGION : Little Namaqualand; Kamies Bergen between Pedros
Kloof and Lily Fontein, 3000–4000 ft., *Drège,* 2631!

79. G. permeabilis (Delaroche, Descr. 27, t. 2) ; corm ovoid, $\frac{1}{2}$ in.
diam.; tunics of parallel fibres ; produced leaves 3–4, firm in texture,
narrow linear or subterete, $\frac{1}{2}$–1 ft. long ; stems slender, 1–2 ft. long,
inflorescence included, simple or forked ; flowers 6–12 in a lax
distichous spike ; outer spathe-valve lanceolate, greenish, $\frac{1}{2}$–1 in.
long ; perianth-tube curved, funnel-shaped, $\frac{1}{2}$ in. long ; limb with a
pink or lilac tinge; upper segment obovate, indistinctly cuspidate,

$\frac{3}{4}$ in. long, $\frac{1}{8}$ in. broad; 2 side ones quadrate, with a long claw and distinct cusp; 3 lower oblong-unguiculate; stamens reaching more than halfway up the limb. *Baker, Handb. Irid.* 225. *G. dichotomus, Thunb. Diss. No.* 6; *Ker, Gen. Irid.* 147. *Hebea Zeyheri, Eckl. Top. Verz.* 42.

COAST REGION: Swellendam Div.; mountains near the Buffeljaghts River, 1000–2000 ft., *Zeyher,* 3990! Alexandria Div.; Zuurberg Range, 2500–3500 ft., *Drège,* 8422! hetween Saintees Flats and Bushman River, under 500 ft., *Drège,* 4546a! Bathurst Div.; between Riet Fontein and the seashore, *Burchell,* 4115! rocky hills near Grahamstown, *Galpin,* 195! King Williamstown Div.; between Yellowwood River and Zandplaat, 1000–200 J ft., *Drège,* 4546c! Cathcart Div.; Windvogel Berg, 4000–5000 ft., *Drège,* 3501!

EASTERN REGION: Tembuland; mountains near Bazeia, 3000 ft., *Baur,* 528!

KALAHARI REGION: Bechuanaland; Pellat Plains near Takun, *Burchell,* 2251!

80. G. arenarius (Baker in Journ. Linn. Soc. xvi. 178); corm large, globose; tunics thick, of copious wiry fibres produced 3–4 inches past its neck; leaves linear, rigid, 4–6 in. long, $\frac{1}{8}$–$\frac{1}{6}$ in. broad at the middle, with 2–3 strong ribs; stem slender, terete, simple, 1–1$\frac{1}{2}$ ft. long including inflorescence; spike 2–4 in. long, dense upwards; spathe-valves oblong, brownish, rigid, $\frac{1}{4}$ in. long; perianth bright red or yellow; tube slender, about as long as the spathe; limb $\frac{1}{2}$–$\frac{3}{4}$ in. long, all the segments with a small obovate lamina $\frac{1}{8}$–$\frac{1}{6}$ in. broad, and a very distinct claw; stamens reaching nearly to the tip of the segments; anthers small, oblong; capsule ovoid, inflated, 1 in. long. *Handb. Irid.* 226. *G. trinervis and G. montanus, ex parte, Herb. Thunberg! Antholyza orchidiflora, Klatt in Linnæa* xxxii. 733. *G. orchidiflorus, Pers. Syn.* i. 44, *non Andrews. Schweiggera montana, E. Meyer in herb. Drège. Hebea orchidiflora, Eckl. Top. Verz.* 43.

COAST REGION: Table Mountain, 2000–3000 ft., *Thunberg! Drège,* 8342! *Burchell,* 640! sandy places near Cape Town below 100 ft., *Bolus,* 3837! Muizenberg, near Kalk Bay, 750 ft., *Bolus,* 4002! Tulbagh Div.; near Ceres, 1500 ft., *Bolus,* 7455! Breede River, near Darling Bridge, 1000 ft., *Bolus,* 2821! Dutoits Kloof, 2000–4000 ft., *Drège!* Alexandria Div.; sand dunes at the mouth of Bushman River, *Zeyher,* 1612!

81. G. montanus (Linn. fil. Suppl. 95); corm globose; tunics thick, of copious, wiry, long fibres; basal leaves linear, rigid in texture, 1–1$\frac{1}{2}$ ft. long, $\frac{1}{4}$–$\frac{1}{3}$ in. broad at the middle, with about 5 strong stramineous ribs; stems 2–3 ft. long including inflorescence, simple or branched; spikes lax, many-flowered, distichous, sometimes a foot long; spathe-valves oblong, rigid, brownish, $\frac{1}{4}$–$\frac{1}{3}$ in. long; perianth pale red; tube slender, rather longer than the spathe; limb an inch long; segments all with a small obovate lamina $\frac{1}{6}$ in. broad and a very long claw; stamens reaching nearly to the tip of the segments; anthers $\frac{1}{3}$ the length of the filaments; capsule rigid, oblong, $\frac{1}{2}$ in. long. *Thunb. Diss. No.* 1, *t.* 1, *fig.* 4; *Prodr.* 8; *Baker, Handb. Irid.* 226. *Antholyza montana, Ker in Konig and Sims' Ann.* i. 233; *Gen. Irid.* 156; *Lodd. Bot. Cab. t.* 1022. *G. parviflorus, Jacq. Obs.* iv. 2, *t.* 78. *G. tabularis, Pers. Syn.* i. 44. *Hebea tabularis, Eckl. Top. Verz.* 43.

M 2

VAR. β, **nemorosus** (Baker); more robust than the type, with leaves ¼-½ in. broad and much-branched stems 4–5 ft. long. *Antholyza nemorosa, Klatt, Ergänz.* 12. *Schweiggera nemorosa, E. Meyer in herb. Drège.*

VAR. γ, **ramosus** (Baker in Journ. Linn. Soc. xvi. 178); stems slender, branched; leaves narrow linear, with 2–3 ribs; spikes laxer than in the type; perianth-segments with a smaller lamina. *Antholyza ramosa, Klatt in Linnæa* xxxii. 734; *Ergänz.* 13. *Hebea ramosa, Eckl. Top. Verz.* 43.

COAST REGION: summit of Table Mountain, *Thunberg!* Worcester Div.; Drakenstein Mts., near Bains Kloof, 1600 ft., *Bolus,* 4069! Dutoits Kloof, 2000–3000 ft., *Drège,* 1580a! Caledon Div.; Baviaans Kloof and mountains near Genadendal, *Burchell,* 7848! 864! Nieuwe Kloof, Houw Hoek Mts., *Burchell,* 8113! Var. β, between Paarl and Lady Grey Railway Bridge, below 1000 ft., *Drège,* 1579! Clanwilliam Div.; between Lange Vallei and Oliphants River, 1000–1500 ft., *Drège!* on the Witsenberg near Tulbagh, *Burchell,* 8276! Caledon Div.; near Genadendal, *Burchell,* 7921! Var. γ, Worcester Div., *Cooper,* 3314! Caledon Div.; Ganse Kraal, Slang River, *Burchell,* 7559! Baviaans Kloof, near Genadendal, *Burchell,* 7602! between Donker Hoek and Houw Hoek Mts., *Burchell,* 8012! George Div.; on the Postberg near George, *Burchell,* 5949!

Hybrid Gladioli.

The following are the principal cultivated hybrid types which have originated from Cape species, to which Latin names have been given, viz. :—

1. G. GANDAVENSIS, figured in Flore des Serres, tome 2, liv. 3, t. 1, in which the *psittacinus* element predominates. The flower is similar to that of *psittacinus* in size and shape, but the red is pure and bright, instead of being mixed with yellow in minute grains and streaks. *G. brenchleyensis, Hort.,* is closely allied. Most likely they are both hybrids between *psittacinus* and *cardinalis.*

2. G. GANDAVENSIS CITRINUS, Flore des Serres, t. 539. Flower like that of *psittacinus* in size and shape, but bright yellow.

3. G. WILLMOREANUS, Flore des Serres, t. 639. Said to be a hybrid between *oppositiflorus* and *gandavensis.* Flower like *psittacinus* in size and shape, but white, with radiating lines of pink.

4. G. RAMOSUS, figured in Paxton's Mag. vi. p. 99. Said to be a hybrid between *cardinalis* and *oppositiflorus.* The flower resembles that of *blandus,* but is pink, with dark blotches at the base of the three lower segments. *G. spofforthianus,* raised by Dean Herbert by crossing *cardinalis* with *blandus,* is closely allied. These flower earlier and are more hardy than the *gandavensis* hybrids.

5. G. PUDIBUNDUS, figured in Paxton's Mag. ii. p. 197. A type near *G. floribundus, Jacq.,* with erect, pinkish-white flowers, with obovate segments, with a faint blotch at the base of the three lower.

6. G. COLVILLEI, Sweet, Brit. Flow. Gard. t. 155, raised from seeds of *tristis var. concolor* fertilized with the pollen of *cardinalis.* It has bright scarlet flowers with oblong acute segments, with a lanceolate blotch of bright yellow at the base of the three lower ones. *G. insignis* is closely allied. *G. Colvillei albus* (the bride) is a frequent form with white flowers.

The following are rare hybrid Cape types to which Latin names have been given, viz. :—

7. G. MITCHAMIENSIS, a hybrid between *tristis* and *hirsutus,* figured by Dean Herbert in Trans. Hort. Soc. vol. iv. t. 2, as *G. tristi-hirsutus.*

8. G. RIGIDUS, a hybrid between *tristis* and *blandus,* figured in the same plate as the last.

9. G. PROPINQUUS, a hybrid between *floribundus* and *blandus,* very near the last.

10. G. FRAGRANS, a hybrid between *recurvus* and *tristis.*

11 and 12. G. HAYLOCKIANUS and G. DELICATUS, hybrids between *recurvus* and *blandus.*

13. G. HERBERTIANUS, a hybrid between *tristis* and *spofforthianus*.
14. G. ODORATUS, a hybrid between *hirsutus* and *spofforthianus*.
15. and 16. G. CANDIDUS and INCARNATUS, hybrids between *blandus* and *cardinalis*.

In the old series of the *Floral Magazine* there are plates of hybrid Cape *Gladioli* at tt. 36, 77, 123, 171, 184, 222, 266, 315, 363, 364, 405, 419, 463, 464, 507, 508 and 556; and in the new series on tt. 43, 102, 295 and 296.

Of late years several hybrids have been raised, especially by M. Lemoine, of Nancy, between *G. gandavensis* and *G. purpureo-auratus*. These are known in gardens under the name of *G. Lemoinei*. Three of them are figured in the "Garden," July 24th, 1886. Max Leichtlin has lately raised several fine hybrids between *G. Saundersii* and *gandavensis*, which have been sent out in America under the name of *G. Childsi*. M. Lemoine has also lately crossed *Lemoinei* with *Saundersii*, and this is called *G. nanceianus, Hort.*

XXXI. ANTHOLYZA, Linn.

Perianth with a long, curved tube, filiform at the base, cylindrical in the upper half ; segments oblong or lanceolate, generally unequal, the uppermost the longest. *Stamens* unilateral, contiguous, arcuate, inserted low down in the perianth-tube ; anthers lanceolate, sagittate at the base. *Ovary* 3-celled ; ovules superposed in the cells ; style filiform ; branches simple, flattened and cuneate at the tip. *Capsule* membranous, loculicidally 3-valved. *Seeds* turgid or discoid, generally winged.

Rootstock a tunicated corm ; leaves generally linear or ensiform ; inflorescence a simple, rarely branched spike ; spathe-valves lanceolate, entire, herbaceous or rigid in texture ; flowers large, usually bright red.

DISTRIB. Besides the Cape species there are four in Tropical Africa.

Subgenus EUANTHOLYZA. Segments of the limb unequal, the upper lingulate-unguiculate, standing forward, the others shorter, spreading.

Upper part of perianth-tube not saccate at the base :
　　Spathe very long ; flowers few ... 　　... 　　... (1) quadrangularis.
　　Spathe middle-sized ; flowers many :
　　　　Leaves narrow linear 　　... 　　... 　　... (2) caffra.
　　　　Leaves ensiform, ¼–½ in. broad :
　　　　　　Perianth-limb bright red 　　... 　　... (3) intermedia.
　　　　　　Perianth-limb dull violet 　　... 　　... (4) spicata.
　　　　Leaves ensiform, 1–1½ in. broad ... 　　... (5) æthiopica.
　　Spathe small, valves rigid :
　　　　Leaves ½–¾ in. broad 　　... 　　... 　　... (6) fucata.
　　　　Leaves 2–3 in. broad 　　... 　　... 　　... (7) paniculata.
Upper part of perianth-tube obscurely saccate 　　... (8) Cunonia.
Upper part of perianth-tube strongly saccate 　　... (9) saccata.
Subgenus HOMOGLOSSUM. Segments of the limb nearly equal.
Spikes few-flowered, secund :
　　Leaf-sheaths glabrous 　　... 　　... 　　... 　　... (10) revoluta.
　　Leaf-sheaths hairy 　　... 　　... 　　... 　　... (11) Merianella.
Spikes dense, equilateral :
　　Leaves narrow linear or subulate 　　... 　　... (12) Lucidor.
　　Leaves ensiform, with 3–4 strong, distant ribs 　　(13) nervosa.
Spike long, lax... 　　... 　　... 　　... 　　... 　　... (14) laxiflora.

1. A. quadrangularis (Burm. Prod. Cap. 1); corm large, globose,
stoloniferous ; stem slender, terete, simple, 2–3 ft. long; leaves like
those of *Gladiolus tristis*, superposed, narrow linear ; flowers 2–4 in
a very lax spike ; outer spathe-valve lanceolate, herbaceous, 1½–2 in.
long ; perianth variegated, bright red and yellow ; tube curved, 2 in.
long, cylindrical in the upper half, not saccate ; upper segment red,
lingulate, ¾ in. long, the others much smaller, greenish, oblong or
lanceolate, spreading ; anthers purple, lanceolate, reaching to the tip
of the upper segment. *Baker, Handb. Irid.* 231. *Gladiolus quadrangularis, Ker in Bot. Mag. t.·567 ; Gen. Irid.* 131. *Petamenes
quadrangularis, Salisb. in Trans. Hort. Soc.* i. 324. *Anisanthus
quadrangularis, Klatt in Linnœa* xxxii. 727. *Gladiolus abbreviatus,
Andr. Bot. Rep. t.* 166. *Antholyza abbreviata, Pers. Syn.* i. 42.

SOUTH AFRICA : without locality.

2. A. caffra (Ker in Konig and Sims' Ann. i. 232) ; corm large,
globose ; tunics brown, breaking up into fine bristles ; stem slender,
terete, simple, 1½–2 ft. long, with 2–3 much-reduced linear leaves ;
leaves several, basal, narrow linear, firm in texture, 1 ft. long, green ;
flowers 12–20, arranged in a lax spike 4–6 in. long, bright red ;
spathe-valves brown, lanceolate, ½–¾ in. long ; perianth-tube 1–1¼ in.
long, cylindrical in the upper half ; upper segment lingulate-spathu-
late, ¾–1 in. long ; the others much smaller, spreading, oblanceolate
or lanceolate ; stamens reaching to the tip of the upper segment ;
anthers lanceolate, ¼–⅓ in. long ; capsule large, inflated ; seeds
triquetrous, narrowly winged. *Ker, Gen. Irid.* 156 ; *Baker, Handb.
Irid.* 230. *Anisanthus splendens, Sweet, Brit. Flow. Gard. ser.* ii.
t. 84 ; *Klatt in Linnœa* xxxii. 727. *Gladiolus splendens, Herb. in
Bot. Reg.* 1843, *Misc.* 46.

COAST REGION : mountains near George, in damp places, 1000–2000 ft., *Drège*,
3494b ! *Burchell*, 6010 ! Uitenhage, *Harvey*, 306 ! Alexandria Div. ; Zuurberg
Range, *Drège*, 3494d ! Albany Div. ; near Riebeek, *Burchell*, 3487 ! near Grahamstown, *MacOwan*, 15 !
EASTERN REGION : Natal, *Cooper*, 3209 !
This is in Herb. Thunberg as a variety of *A. œthiopica*. Plant's hybrid *Anisanth*,
figured in Bot. Reg. 1842, t. 53, is a hybrid between this and a hybrid *Gladiolus*
between *cardinalis* and *tristis*.

3. A. intermedia (Baker in Journ. Linn. Soc. xvi. 180) ; corm
not seen ; stem terete, simple, about as long as the leaves ; basal
leaves 6–10, linear, 6–9 in. long, moderately firm in texture, ¼–⅓ in.
broad at the middle, with 3 prominent ribs ; flowers many, arranged
in a moderately dense equilateral spike 4–5 in. long ; spathe-valves
lanceolate, red-tinted, about ½ in. long, the inner one rather the
largest ; perianth-tube curved, 1–1¼ in. long, cylindrical in the upper
half ; segments all bright red, lingulate-unguiculate, black on the
claw, the upper one an inch long, the others considerably shorter ;
stamens reaching nearly to the tip of the upper segment. *Handb.
Irid.* 230.

COAST REGION : Uitenhage Div. ; in rocky ground at Van Stadens Hooghte,
MacOwan, 2070 !
Comes in between *A. œthiopica* and *A. caffra*.

4. A. spicata (Brehmer ex Klatt, Ergänz. 11) ; stem simple, terete, 1½-2 ft. long ; basal leaves 6–8, linear, 1 ft. long, ⅓ in. broad, with 3–4 strong ribs ; flowers arranged in a close spike 3–4 in. long ; spathe-valves ovate, acute, reddish, the inner one the largest, an inch long ; perianth-tube slender, as long as the spathe, widened to a broadly cylindrical lurid throat ; all the segments dull violet, the upper ½ in. long, the others subequal, oblong, ⅓ in. long ; anthers one-sixth the length of the filaments. *Baker, Handb. Irid.* 229.

COAST REGION : Worcester Div. : on hills near Dutoits Kloof, *Drège*, 1576.

5. A. æthiopica (Linn. Sp. Plant. edit. 2. i. 54) ; corm globose, large ; tunics brown, membranous ; stem branched, 3–4 ft. long ; basal leaves several, ensiform, forming a fan-shaped rosette, 1–1½ ft. long, 1 in. or more broad, moderately firm in texture, green; flowers red-yellow, many, arranged in a moderately dense spike, which is sometimes 6–9 in. long ; spathe-valves oblong-lanceolate, greenish, ½–¾ in. long; perianth-tube curved, 18–21 lines long, cylindrical in the upper two-thirds ; upper segment red, lingulate, 1–1¼ in. long, the others much shorter, spreading ; stamens reaching to the top of the upper segment ; capsule oblong, ½ in. long ; seeds turgid, not winged. *Thunb. Prodr.* 7 ; *Fl. Cap.* i. 163 ; *Ker in Bot. Mag. t.* 561 ; *Andr. Bot. Rep. t.* 210 ; *Baker, Handb. Irid.* 230. *A. præalta, Red. Lil. t.* 387. *A. ringens, Andr. Bot. Rep. t.* 32. *A. floribunda, Salisb. in Trans. Hort. Soc.* i. 324.

VAR. β, **A. immarginata** (Thunb. Herb.) ; not so tall ; leaf narrower, about ⅓ in. broad ; flower red, with a little dull yellow, the limb as long as in the type, but the tube much shorter, 1–1¼ in. long; spathe shorter. *A. æthiopica, Red. Lil.* t. 110.

VAR. γ, **A. vittigera** (Salisb. in Trans. Hort. Soc. i. 324) ; stature of the type ; perianth-tube bright yellow, with stripes of red. *Bot. Mag. t.* 1172.

VAR. δ, **A. bicolor** (Gasp. in Belg. Hort. ii. 145, cum icone) ; habit dwarfer than in the type ; leaves narrower ; perianth-tube red at the top, bright pale yellow in the lower half. *A. æthiopica, var. minor, Bot. Reg. t.* 1159.

SOUTH AFRICA : without locality, *Masson !* Vars. γ and δ, cultivated.

COAST REGION : Lion Mountain and Camps Bay, *Pappe !* near Cape Town, *MacOwan, Herb. Aust. Afr.*, 1549! George Div.; by rivulets in woods, *Thunberg !* Var. β, Table Mountain, 200 ft., *MacOwan, Herb. Aust. Afr.*, 1550! Devils Mountain, 500 ft., *Bolus*, 4747 ! Swellendam Div. ; Buffeljaghts River, near Sparrbosch, *Drège*, 8340! Mossel Bay Div. ; between Mossel Bay and Zout River, *Burchell*, 6339 ! George Div., *Thunberg !* Knysna Div. ; Melville, in the forest, *Burchell*, 5422 ! Albany Div. ; near Grahamstown, *MacOwan*, 619 !

6. A. fucata (Baker in Journ. Linn. Soc. xvi. 180) ; corm large, globose ; tunics brown, membranous ; basal leaves ensiform, firm in texture, green, 2–2½ ft. long, ½–¾ in. broad ; inflorescence branched, overtopping the leaves ; flowers arranged in a lax, spreading, secund spike 3–4 in. long, with a flexuose rachis ; spathe-valves ovate, under ¼ in. long ; perianth bright red and yellow ; tube curved, 18–21 lin. long, cylindrical in the upper half; upper segment red, lingu-late, ¾ in. long ; the others shorter, oblanceolate or lanceolate, yellow, spreading ; stamens reaching nearly to the tip of the top segment; style protruding beyond it. *Handb. Irid.* 229. *Tritonia fucata; Herbert in Bot. Reg.* 1838, *t.* 35.

SOUTH AFRICA: without locality.

Known only from the figure cited, which was drawn from specimens grown by Dean Herbert at Spofforth in 1837.

7. **A. paniculata** (Klatt in Linnæa xxxv. 379); corm large, globose; stems stout, terete, 3–4 ft. long; fully developed lower leaves lanceolate, very oblique at the base, rigid, but thin in texture, 1½–2 ft. long, 2–3 in. broad, with 9–11 distant prominent ribs; inflorescence copiously panicled; spikes equilateral, close, many-flowered, with a very wavy rachis; spathe-valves rigid, brown, oblong, ¼–⅓ in. long; perianth bright reddish-yellow; tube curved, 1–1¼ in. long, gradually widened to a throat ⅛ in. diam.; upper segment lingulate, ½–¾ in. long; the others smaller, spreading, lanceolate; stamens reaching to the tip of the upper segment. *Baker, Handb. Irid.* 229.

KALAHARI REGION: Transvaal; by the side of streams, Lomatie Valley, near Barberton, 4000 ft., *Galpin*, 1211! McMac, near Pilgrims Rest, *Mudd!*

EASTERN REGION: Natal; upper part of Tugela River, *Allison!* Klip River division; bank of a stream, *Wood*, 3499! and without precise locality, *Gerrard*, 1530!

A very fine and distinct species, remarkable for its palm-like leaves, copiously-panicled spikes, and small rigid spathes.

8. **A. Cunonia** (Linn. Sp. Plant. edit. 2. i. 54); corm small, globose; tunics membranous; stem simple, 1–1½ ft. long, with several reduced leaves with short, free, linear points; produced leaves about 4, linear, moderately firm in texture, ½–1 ft. long; flowers 4–6, bright red, arranged in a very lax spike; spathe-valves herbaceous, green, lanceolate, ¾ in. long; perianth-tube an inch long, faintly saccate at the base of the cylindrical upper half; upper segment spoon-shaped, 1½ in. long; lateral much shorter, obovate, adnate to the upper segment halfway up; basal very small and inserted lower down; stamens reaching to the tip of the upper segment; capsule oblong, membranous, ½–¾ in. long; seeds discoid, with a broad wing. *Miller Ic.* 75, *t.* 113; *Thunb. Diss. No.* 4; *Prodr.* 7; *Fl. Cap.* i. 165; *Red. Lil. t.* 12; *Baker, Handb. Irid.* 231. *Cunonia Antholyza, Miller, Gard. Dict. edit.* viii. *Gladiolus Cunonia, Ker in Konig and Sims' Ann.* i. 230; *Gen. Irid.* 130. *Anisanthus Cunonia, Sweet, Hort. Brit. edit.* 2, 500; *Klatt in Linnæa* xxxii. 728.

COAST REGION: near Cape Town, *Thunberg! Pappe!* Fish Hoek, near Muizenberg, *Bolus*, 4871! Simons Bay, *Wright!* Mossel Bay and Breede River, *Rogers!*

9. **A. saccata** (Baker in Journ. Linn. Soc. xvi. 180); corm ovoid, small; tunics brown, membranous; stems 2–3 ft. long, simple or branched, with several superposed long-sheathing leaves with free linear points; lower leaves narrow linear, moderately firm in texture, bright green, about a foot long, with 3 strong ribs; flowers 4–10 in lax secund spikes; spathe-valves lanceolate, green, herbaceous, ¾–1 in. long; perianth-tube 1–1¼ in. long, conspicuously

saccate at the base of the cylindrical upper part; upper segment
lingulate, an inch long; lateral oblong-lanceolate, much shorter;
lower very small; stamens reaching nearly to the tip of the upper
segment; anthers lanceolate-sagittate, $\frac{1}{3}$ in. long. *Handb. Irid.*
231. *Anisanthus saccatus, Klatt in Linnœa* xxxv. 300.

WESTERN REGION : Clanwilliam Div.; Karoo-like heights near Holle River,
under 100 ft., *Drège*, 2646!

10. **A. revoluta** (Burm. Prodr. Cap. 1); corm globose, 1 in. diam.;
tunics thick, cancellate; stem slender, terete, simple, 1–1$\frac{1}{2}$ ft. long,
with 2–3 reduced leaves with long glabrous sheaths; produced leaf
one, linear-subulate, 1 ft. long, firm in texture; flowers 2–4 in a
very lax secund spike, bright red; spathe-valves lanceolate, green,
herbaceous, 1–2 in. long; perianth-tube curved, 1$\frac{1}{2}$–2 in. long,
cylindrical in the upper half; upper segments oblong, $\frac{3}{4}$–1 in. long;
lower smaller, lanceolate, falcate; stamens much shorter than the
perianth-segments; anthers $\frac{1}{4}$ in. long; style-arms short, cuneate.
Gladiolus Watsonius, Thunb. Diss. No. 10; *Prodr.* 8; *Fl. Cap.* i.
173; *Jacq. Ic. t.* 233; *Bot. Mag. t.* 450; *Red. Lil. t.* 369. *G.
præcox, Andr. Bot. Rep. t.* 38. *G. recurvus, Houtt. Handl.* xii. 49,
t. 79, *fig.* 1, *non Linn. Watsonia revoluta and præcox, Pers. Syn.*
i. 42. *Homoglossum revolutum, Baker in Journ. Linn. Soc.* xvi. 161.
H. præcox, Salisb. in Trans. Hort. Soc. i. 325.

VAR. β, **Gawleri** (Baker); segments of the limb larger, variegated red and
yellow over the lower half of the face. *Homoglossum revolutum, var. Gawleri,
Baker in Journ. Linn. Soc.* xvi. 161. *Gladiolus Gawleri, Klatt, Ergänz.* 7.

COAST REGION : Lion Mountain, *Thunberg! MacOwan and Bolus, Herb.
Norm.,* 289 ! *MacOwan,* 2302 ! near Cape Town, 50 ft., *Bolus,* 3729 ; Malmesbury
Div.; between Eikenboom and Riebeeks Castle, *Drège,* 8448b ! Albany Div.;
mountain sides of Broekhuzens Poort, near Grahamstown, 2000–2500 ft., *Galpin,*
66 !
WESTERN REGION : Little Namaqualand ; hills along the Hartebeest River,
2000–3000 ft., *Zeyher,* 3976!

11. **A. Merianella** (Linn. Syst. Veg. edit. 13, 77); corm globose,
1 in diam.; stem slender, terete, simple, 1–1$\frac{1}{2}$ ft. long; leaves 3–4,
superposed, all with short, narrow linear, strongly-ribbed free
points and long hairy sheaths; flowers 3–6 in a very lax secund
spike, pale pink; outer spathe-valve lanceolate, herbaceous, $\frac{3}{4}$–1$\frac{1}{2}$ in.
long; perianth with a curved tube 18–21 lin. long, cylindrical in the
upper half; segments broad, oblong, subequal, obtuse, $\frac{1}{2}$–$\frac{3}{4}$ in. long;
stamens reaching nearly to the tip of the perianth-segments; anthers
lanceolate, $\frac{1}{8}$ in. long; style-branches very slender. *Gladiolus
Merianellus, Thunb. Diss.* 11; *Prodr.* 7; *Fl. Cap.* i. 172; *Ker,
Gen. Irid.* 133 ; *Baker, Handb. Irid.* 227. *Homoglossum Merianella, Baker in Journ. Linn. Soc.* xvi. 161. *Gladiolus hirsutus, var.
Merianellus, Ker in Bot. Mag. sub t.* 727. *Watsonia humilis, Pers.
Syn.* i. 42, *non Miller. W. pilosa, Klatt in Trans. South Afric.
Phil. Soc.* iii. *pt.* 2, 200.

SOUTH AFRICA : without locality, *Oldenburg! Masson! Harvey,* 63 !
COAST REGION : mountains of the Cape Peninsula, 800–1000 ft., *Thunberg !*

Rogers! Bolus, 4869! *MacOwan and Bolus, Herb. Norm.*, 290! *MacOwan*, 2510, and *Herb. Aust. Afr.*, 1551!

12. **A. Lucidor** (Linn. fil. Suppl. 96); corm large, globose; tunics brown, breaking up into fine fibres; stems simple, terete, 1–1½ ft. long, with several superposed leaves with subulate free points; basal leaves rigid, narrow linear at the base, subulate upwards; flowers many, bright-red, arranged in a dense equilateral spike 3–6 in. long; spathe-valves oblong-navicular, rigid in texture, ½–¾ in. long, the inner the largest; perianth-tube curved, 1–1¼ in. long, cylindrical in the upper two-thirds; segments oblong-unguiculate, subequal, ½–¾ in. long; stamens reaching nearly to the tip of the segments; anthers lanceolate, ⅓ in. long; style-branches short, slender, cuneate at the tip. *Thunb. Diss. No.* 1; *Prodr.* 7; *Fl. Cap.* i. 162; *Ker in Konig and Sims' Ann.* i. 233; *Gen. Irid.* 156. *Gladiolus Lucidor, Baker, Handb Irid.* 227. *Homoglossum Lucidor, Baker in Journ. Linn. Soc.* xvi. 161. *Watsonia lucidior, Eckl. Top. Verz.* 36.

COAST REGION: near Constantia, *Thunberg!* Devils Mountain, *Ecklon*, 88! Paarl Mts., *Drège!* Caledon Div.; Donker Hoek Mts., *Burchell*, 7962! between Donker Hoek and Houw Hoek Mts., *Burchell*, 8013! mountains near Genadendal, *Burchell*, 7697/2! 8638!

I know nothing of *W. pottbergensis* and *tigrina*, Eckl. loc. cit., named only, not described, said to be closely allied to this species.

13. **A. nervosa** (Thunb. Diss. No. 3); corm large, globose; outer tunics breaking up into fine fibres; inner produced past its neck as flattened bristles; stems terete, simple, 1–2 ft. long; basal leaves 4–6, ensiform, ½–1 ft. long, rigid in texture, ½–1 in. broad at the middle, with 4–5 strong ribs; flowers bright-red, arranged in a dense equilateral spike 3–6 in. long; spathe-valves oblong-lanceolate, firm in texture, ⅓–½ in. long; perianth-tube curved, 1–1¼ in. long, cylindrical in the upper two-thirds; limb ½–¾ in. long, all the segments oblanceolate-unguiculate, the lower spreading, the upper straight; stamens reaching nearly to the tip of the perianth-segments: anthers lanceolate, ¼–⅓ in. long; style-branches very short, cuneate at the top; capsule large, inflated; seeds flat, broadly winged. *Thunb. Prodr.* 7; *Fl. Cap.* i. 164; *Ker in Bot. Mag. sub t.* 1172; *Gen. Irid.* 156. *Gladiolus nervosus, Baker, Handb. Irid.* 228.

SOUTH AFRICA: without locality, *Thunberg! Zeyher*, 1633! COAST REGION: sides of Devils Mountain near Rondebosch, 500 ft., *Bolus*, 4528! Muizenberg, 600 ft., *Bolus*, 4870! Worcester Div.; Goudine, *Cooper*, 3181! mountains near Swellendam, *Burchell*, 7371! Riversdale Div.; between Valsche River and Zoetemelks River, *Burchell*, 6599! near Zoetemelks River, *Burchell*, 6642! 6699! Uitenhage Div.; near Wittklip Mountain, *MacOwan*, 1936! Algoa Bay, *Cooper*, 1485!

Connects in the shape of its perianth *Homoglossum* with *Euantholyza* through *A. Lucidor*.

14. **A. laxiflora** (Baker); corm not seen; stem simple, 3 ft. long, bearing 4 superposed, rigid, linear, strongly-ribbed leaves, the lowest 2 ft. long, ⅓ in. broad, the second a foot long, the uppermost short;

spike lax, many-flowered, a foot long; spathe-valves green, lanceolate, the outer reaching 2 in. long; perianth bright-red; tube curved, dilated at the middle where the stamens are inserted, 1½ in. long, ¼ in. diam. at the throat; 3 upper segments of the limb obovate-cuneate, ¾ in. long, 3 lower smaller; stamens reaching nearly to the tip of the upper segments. *Gladiolus antholyzoides, Baker in Journ. Bot.* 1891, 70; *Handb. Irid.* 227.

KALAHARI REGION: Transvaal; damp ground near Pretoria, *Elliot*, 1447!
Near *A. revoluta.*

ORDER CXXXV. AMARYLLIDEÆ.

(By J. G. BAKER.)

Flowers hermaphrodite, regular or nearly so. *Perianth* corolline, with or without a funnel-shaped tube above the ovary; segments 6, biserial. *Stamens* usually 6, one opposite each segment of the perianth, rarely 3 or numerous; filaments filiform, free or united in a cup; anthers long or short, basifixed or versatile, dehiscing usually by introrse longitudinal slits. *Ovary* inferior, 3-celled; placentation axile; style entire; stigmas 3, distinct or confluent; ovules usually numerous and superposed. *Fruit* usually a 3-valved capsule with loculicidal dehiscence, sometimes indehiscent. *Seeds* usually numerous and superposed, with a black membranous or crustaceous testa; albumen fleshy, rarely horny; embryo small, central.

Usually acaulescent herbs, with all the leaves radical, rarely shrubs, with leafy stems; rootstock a tunicated bulb in the largest typical tribe, not so in the others; flowers usually umbellate, sometimes solitary, corymbose or racemose.

DISTRIB. Genera 64; species 650; spread widely in the temperate and tropical regions of both hemispheres. Two tribes out of five restricted to the new world.

Tribe 1. *HYPOXIDEÆ.* *Rootstock* a tunicated corm. *Leaves* all radical, dry, persistent. *Inflorescence* not subtended by a spathe. *Flowers* usually yellow and hairy outside.

I. **Pauridia.**—*Stamens* 3.
II. **Curculigo.**—*Stamens* 6. *Perianth* with a tube. *Fruit* indehiscent.
III. **Hypoxis.**—*Stamens* 6. *Perianth* without a tube. *Fruit* a capsule with circumscissile dehiscence.

Tribe 2. *AMARYLLEÆ.* *Rootstock* a tunicated bulb. *Leaves* all radical, rarely coriaceous. *Flowers* various in colour, never hairy on the outside, usually in an umbel subtended by a spathe. *Stamens* free.

* Anthers basifixed.

IV. **Hessea.**—*Anthers* small, subglobose. *Perianth* cut down nearly or quite to the ovary.
V. **Carpolyza.**—*Anthers* small, subglobose. *Perianth* with a distinct tube.
VI. **Anoiganthus.**—*Anthers* oblong, deeply sagittate. *Flowers* umbellate.
VII. **Gethyllis.**—*Flowers* solitary. *Perianth* hypocrateriform. *Stamens* uniseriate; anthers linear.
VIII. **Apodolirion.**—*Flowers* solitary. *Perianth* funnel-shaped. *Stamens* biseriate; anthers linear.

** *Anthers dorsifixed, versatile.*

* *Fruit indehiscent or bursting irregularly. Seeds one or few bulbiform.*

IX. **Crinum.**—*Segments* of perianth (in the Cape species) broad. *Stamens* declinate. *Leaves* persistent.
X. **Amaryllis.**—*Segments* of perianth broad. *Stamens* declinate. *Leaves* short-lived.
XI. **Ammocharis.**—*Segments* of perianth narrow. *Stamens* erect.

** *Fruit a 3-valved capsule. Seeds many, compressed.*

XII. **Brunsvigia.**—*Perianth* cut down to the ovary. *Style* not swollen at the base. *Capsule* turbinate, acutely angled.
XIII. **Nerine.**—*Perianth* cut down to the ovary. *Style* not swollen at the base. *Capsule* globose, obtusely angled.
XIV. **Strumaria.**—*Perianth* cut down to the base. *Style* swollen and triquetrous towards the base.
XV. **Vallota.**—*Perianth* with a tube rather shorter than the limb.
XVI. **Cyrtanthus.**—*Perianth* with a tube longer than the limb.

*** *Fruit baccate or capsular. Ovules 2 or few, clustered at the middle of the placentas.*

XVII. **Clivia.**—*Bulb* imperfect. *Spathe-valves* several. *Fruit* baccate; ovules few.
XVIII. **Hæmanthus.**—*Bulb* large, tunicated. *Spathe-valves* several. *Fruit* baccate; ovules 2, collateral.
XIX. **Buphane.**—*Fruit* capsular. *Spathe-valves* 2.

Tribe 3. *VELLOZIEÆ. Rootstock* not bulbous. *Leaves* coriaceous, persistent. *Flowers* solitary, more or less glandular outside.

XX. **Vellozia.**—The only genus.

I. PAURIDIA, Harv.

Perianth with a short infundibuliform tube above the ovary; segments oblong, subequal, spreading. *Stamens* 3, inserted in the perianth tube opposite the inner segments; filaments short, filiform; anthers linear, basifixed. *Ovary* clavate, 3-celled; style short, entire; stigmas 3, subulate, falcate, sometimes 1, 2, or all short and sterile. *Capsule* membranous, obconic, crowned with the faded perianth. *Seeds* many, globose, minute.

DISTRIB. Endemic, monotypic.

1. **P. hypoxidoides** (Harv. Gen. S. Afr. Pl. 342); corm globose, ¼ in. diam.; outer tunics rigid, black, cancellate, with a ring of bristles at the top; produced leaves 6–12, linear, falcate, glabrous, 1–2 in. long; peduncles several to a corm, very slender, 1-flowered, about as long as the leaves; perianth-segments yellow, tipped with green, 3–5-nerved, glabrous, ⅙ in. long; ovary glabrous, ⅛ in. long; capsule often oblique, with one or two of the cells abortive. *Baker in Journ. Linn. Soc.* xvii. 126. *Ixia minuta, Linn. fil. Suppl.* 92; *Thunb. Diss. Ixia, No.* 2, *t.* 1, *fig.* 1; *Prodr.* 9; *Fl. Cap.* i. 216. *Galaxia minuta, Ker in Konig and Sims' Ann.* i. 241; *Gen. Irid.* 71. *Romulea minuta, Eckl. Top. Verz.* 19. *Hypoxis triandra, Pappe MSS. H. nana, E. Meyer in herb. Drège.*

COAST REGION: hills and flats near Cape Town, *Thunberg! Burchell*, 8448! *Pappe! Bolus*, 2815! near Rondebosch, *MacOwan and Bolus, Herb. Norm.*, 291! *MacOwan*, 2615! Stellenbosch Div.; between Stellenbosch and Somerset West, below 1000 ft., *Drège!*

II. CURCULIGO, Gaertn.

Perianth cut down to the ovary; segments 6, lanceolate, subequal, spreading. *Stamens* 6, inserted at the base of the perianthsegments; filaments short, filiform; anthers linear, basifixed. *Ovary* 3-celled, narrowed into a long beak resembling a perianth tube; style short, entire; stigmas 3, oblong, erect, adpressed. *Fruit* indehiscent. *Seeds* globose; testa black, crustaceous.

Rootstock tuberous; leaves plicate, marcescent, broad or narrow; flowers yellow, solitary or many, capitate.

DISTRIB. Species 12, extending to Tropical Africa, Tropical Asia, Australia, and Tropical America.

Rootstock a globose corm ... ... (1) **plicata.**
Rootstock a rhizome ... ... (2) **namaquensis.**

1. **C. plicata** (Ait. Hort. Kew. edit. 2, ii. 253); corm depressoglobose, $\frac{1}{2}-\frac{3}{4}$ in. diam.; outer tunics of brown reticulated fibres, crowned with bristles; produced leaves 2–6 to a corm, linear, weak, glabrous, 3–9 in. long; peduncles 1–3 to a corm, hidden like the ovary in the basal sheaths; perianth-limb $\frac{1}{2}$–1 in. long; pale green and glabrous outside, yellow within; segments acute, laxly 5–7-nerved; anthers linear, obtuse, $\frac{1}{3}$ in. long; ovary with a filiform, green, glabrous beak 2–6 in. long; capsule oblong, rostrate, indehiscent, narrowed into a persistent beak. *Roem. et Schultes, Syst. Veg.* vii. 755; *Baker in Journ. Linn. Soc.* xvii. 122. *Hypoxis plicata, Linn. fil. Suppl.* 197; *Thunb. Prodr.* 60. *Fabricia plicata, Thunb. in Fabric. Reise* 29. *Gethyllis plicata, Jacq. Hort. Schoenbr. t.* 80. *Hypoxis luzulæfolia, D.C. in Red. Lil. t.* 260. *Empodium plicatum, Salisb. Gen.* 43. *Forbesia plicata and angustifolia, Eckl. Top. Verz.* 4.

VAR. β, **Barberi** (Baker in Journ. Linn. Soc. xvii. 123); leaves as in the type, but peduncle longer, so that the ovary is protruded from the basal sheaths; beak of ovary much shorter, $\frac{1}{4}$–1 in. long.

VAR. γ, **C. veratrifolia** (Baker in Journ. Linn. Soc. xvii. 123); leaves 1–2, lanceolate, 1–1$\frac{1}{2}$ in. broad at the middle; beak of the ovary 1–4 in. long. *Hypoxis plicata, Jacq. Collect. Suppl.* 55; *Ic. t.* 367. *H. veratrifolia, Willd. Sp. Plant.* ii. 109. *C. plicata β, Ker in Bot. Reg. t.* 345.

SOUTH AFRICA: without locality, *Sieber*, 124! Var. γ, *Zeyher*, 1664!
COAST REGION: hills near Cape Town, *Thunberg! Pappe! Burchell*, 8441! Uniondale Div.; on a rocky hill near Groot River in Lange Kloof, *Burchell*, 4979! Stockenstrom, *Scully*, 198!
CENTRAL REGION: Somerset Div.; shady places at the foot of Bosch Berg near Somerset East, 2500 ft., *MacOwan*, 1874! near Somerset East, *Bolus*, 336! Var. β, Somerset Div., *Miss Bowker!* near Hopetown, *Mrs. Barber!*
EASTERN REGION: Natal; Klip River, 3500–4500 ft., *Sutherland!* Var. β, Inanda, *Wood*, 273!

2. C. namaquensis (Baker); rootstock a tortuose rhizome $\frac{1}{8}$ in. diam. ; leaves solitary, oblong-lanceolate, subpetiolate, thin, narrowed gradually from the middle to both ends, 15–18 in. long; flowers solitary, pedicel 2–3 in. long; tube as long as the ovary; segments 8, lanceolate, an inch long, pale yellow inside ; anthers pale yellow, $\frac{1}{4}$ in. long ; filaments very short; ovary cylindrical, $\frac{3}{4}$ in. long.

WESTERN REGION: Little Namaqualand, near Ookiep, *Bolus*, 6629! in herb. Bolus.

III. HYPOXIS, Linn.

Perianth cut down to the ovary ; segments 6, subequal, spreading, the outer often hairy on the outside. *Stamens* 6, inserted at the base of the perianth-segments; filaments short, filiform; anthers linear, sagittate, basifixed or versatile. *Ovary* 3-celled, rarely beaked ; style short, entire; stigmas 3, distinct or concrete. *Capsule* usually opening with circumscissile dehiscence below the apex. *Seeds* globose; testa black, crustaceous, with a couple of prominences.

Rootstock tuberous, small and monocarpic or large and polycarpic ; leaves plicate, persistent, linear or lanceolate ; flowers generally yellow, solitary, corymbose or racemose, rarely umbellate ; bracts small, acute.

DISTRIB. A genus of 50–60 species, with its headquarters at the Cape, extending to Tropical Africa, Tropical Asia, Australia, and through a great part of America.

Subgenus IANTHE. Whole plant entirely glabrous.
Leaves linear or linear-subulate :
  Expanded perianth $\frac{1}{3}$ in. diam. ... ... (1) **minuta.**
  Expanded perianth $\frac{3}{4}$–1 in. diam. :
    Flowers usually solitary :
      Flowers white : leaves not denticulate... (2) **alba.**
      Flowers yellow : leaves not denticulate (3) **curculigoides.**
      Flowers yellow : leaves denticulate ... (4) **serrata.**
    Flowers 3–6 in an umbel :
      Leaves subterete ... ... (5) **aquatica.**
      Leaves linear ... ... (6) **Scullyi.**
    Expanded perianth 1½–2 in. diam. ... ... (7) **stellata**
Leaves lanceolate :
  Expanded perianth about $\frac{1}{2}$ in. diam. ... ... (8) **ovata.**
  Expanded perianth about 1 in. diam. ... ... (9) **Andrewsii.**

Subgenus EUHYPOXIS. Leaves, peduncles, and outside of the flower more or less hairy.
Flowers white or red, solitary :
  Leaves subterete ... ... (10) **milloides.**
  Leaves linear, flat ... ... (11) **Baurii.**
Flowers yellow, solitary ... ... (12) **Flanagani.**
Flowers yellow, corymbose or racemose :
  Perianth-limb $\frac{1}{4}$–$\frac{1}{2}$ in. diam. :
    Leaves subterete :
      Corm $\frac{1}{4}$–$\frac{1}{3}$ in. diam. ... ... (13) **filiformis.**
      Corm 1–2 in. diam. ... ... (14) **kraussiana.**
    Leaves linear :
      Flowers corymbose :
        Leaves thin ; veins slender :
          Leaves slightly hairy ... ... (15) **angustifolia.**
          Leaves very hairy ... ... (16) **floccosa.**

Leaves firm, strongly veined :
    Leaves glabrous   ...   ... (17) **Zeyheri.**
    Leaves softly pilose ...   ... (18) **Gerrardi.**
    Leaves silky ...   . (19) **argentea.**
Flowers racemose :
    Leaves thin ... ... (20) **Jacquini.**
    Leaves rigid ... ... (21) **Arnottii.**
Leaves lanceolate or oblong-lanceolate :
    Flowers corymbose :
        Leaves membranous :
            Dwarf: leaves hairy ... ... (22) **membranacea.**
            Tall : leaves glabrous... ... (23) **Woodii.**
        Leaves moderately firm :
            Leaves lanceolate ... (24) **parvifolia.**
            Leaves oblong-lanceolate ... (25) **brevifolia.**
    Leaves firm in texture : veins thickened :
        Leaves obscurely bristly on the
            edge only ... ... (26) **setosa.**
        Leaves hairy on the surface ... (27) **villosa.**
    Flowers racemose :
        Leaves about ¼ in. broad ... ... (28) **obtusa.**
        Leaves 1½–2 in. broad ... ... (29) **latifolia.**
Flowers larger (perianth-limb ½–¾ in. long) :
    Leaves subterete ... ... (30) **longifolia.**
    Leaves linear :
        Flowers corymbose ... ... (31) **Ludwigii.**
        Flowers racemose :
            Leaves weak ... ... (32) **acuminata.**
            Leaves rigidly coriaceous ... ... (33) **rigidula.**
Leaves lanceolate or oblong-lanceolate :
    Flowers few, corymbose :
        Leaves glabrous ... ... (34) **colchicifolia.**
        Leaves pilose ... ... (35) **multiceps.**
        Leaves matted with persistent white
            tomentum beneath ... ... (36) **stellipilis.**
    Flowers many, racemose :
        Ovary glabrous ... ... (37) **oligotricha.**
        Ovary villose :
            Lower pedicels very short :
                Leaves lanceolate-acuminate (38) **hemerocallidea.**
                Leaves oblong-lanceolate :
                    Leaves strongly ciliated (39) **costata.**
                    Leaves not ciliated ... (40) **Galpini.**
            Lower pedicels ½–1 in. long ... (41) **Rooperii.**

1. **H. minuta** (Linn. fil. Suppl. 197) ; corm annual, small, ovoid, much flattened out at the base, crowned with a ring of black bristles ; leaves 3–4, linear, suberect, glabrous, 1–2 in. long, moderately firm in texture ; peduncles 1–2, slender, erect, glabrous, simple or deeply forked ; bracts linear or subulate ; perianth-limb ⅛–¼ in. long ; segments oblong-lanceolate, pale yellow inside, the three outer glabrous and greenish on the back ; stamens half as long as the perianth-limb ; anthers lanceolate, basifixed ; filaments very short ; stigmas distinct ; ovary obconic, glabrous, ¹⁄₁₂ in. long and broad ; capsule turbinate. *Thunb. Prodr.* 59 ; *Fl. Cap. edit. Schult.* 303 ; *Roem. et Schultes, Syst. Veg.* vii. 773 ; *Baker in Journ. Linn. Soc.* xvii. 101. *Helonias minuta, Linn. Mant.* ii. 225. *Hypoxis pumila, Lam. Encyc.* iii. 184. *H. triflora, Harv. MSS.*

SOUTH AFRICA: without locality, *Zeyher*, 1665!
COAST REGION: Malmesbury Div.; by the Berg River, *Burke!* sandy
places near Cape Town, *Thunberg!* near Rondebosch, *Pappe!*

2. H. alba (Linn. fil. Suppl. 198); corm globose, $\frac{1}{4}-\frac{1}{3}$ in. diam.;
tunics densely beset with erect black bristles; produced leaves 3–4,
erect, subterete, firm in texture, glabrous, 2–3 in. long; peduncle
slender, erect, usually 1-flowered, with a lanceolate bract near the
base, rarely forked; perianth-limb $\frac{1}{3}-\frac{1}{2}$ in. long, white inside,
glabrous, and tinged with red or green outside; segments oblong-
lanceolate; anthers linear, basifixed; ovary clavate, glabrous, $\frac{1}{6}$ in. long;
capsule green, obconic, membranous. *Thunb. Prodr.* 60; *Fl. Cap.
edit. Schult.* 303; *Jacq. Collect.* iv. 135, *t.* 2, *fig.* 1 ; *Fragm.* 13, *t.* 7,
fig. 4; *Baker in Journ. Linn. Soc.* xvii. 102. *H. affinis and dubia,
Roem. et Schultes, Syst. Veg.* vii. 774–5. *H. obliqua, Eckl. et Zeyher,
Exsic.* 420, *non Jacq. H. crassifolia, Pappe MSS. H. tabularis,
Eckl. Top. Verz.* 10.

VAR. β, **gracilis** (Baker in Journ. Linn. Soc. xvii. 102); leaves very narrow,
and not so firm in texture; perianth-limb $\frac{1}{4}-\frac{1}{3}$ in. long. *H. alba, Lodd. Bot.
Cab. t.* 1074. *H. minor, Eckl. Top. Verz.* 10.
VAR. γ, **Burkei** (Baker, loc. cit.); peduncles usually forked low down; bracts
linear, foliaceous, 1–2 in. long; ovary cylindrical, $\frac{3}{4}$–1 in. long, narrowed into
a distinct beak. A robust aquatic variety.

SOUTH AFRICA: without locality, *Sieber*, 126! *Harvey*, 104! 105! *Zeyher*,
4132! Var. γ, *Thunberg!*
COAST REGION: near Cape Town, *Thunberg!* Cape Flats, *Burchell*, 8570!
Bolus, 2813! *Schlechter*, 624! Paarl Mountains, 1000–2000 ft., *Drège!* between Paarl and Lady Grey Railway Bridge, under 1000 ft., *Drège*, 2395!
Var. β, Caledon Div.; Klein River Berg, 1000–3000 ft., *Zeyher*, 4131! at the
foot of mountains near the mouth of Bot River, 400 ft., *Bolus*, 7478! Uniondale Div.; Lange Kloof, between Wagenbooms River and Apies River, *Burchell*,
4940! Knysna Div.; near Melville, *Burchell*, 5495! Uitenhage, *Harvey*, 135!
Albany Div.; near Grahamstown, 2000 ft., *MacOwan*, 1222! Var. γ, Worcester
Div.; Wagenbooms River, *Burke!*

3. H. curculigoides (Bolus in Hook. Ic. t. 2259 A); corm small,
globose, crowned with a ring of fibres; leaves usually 2, rarely 3,
linear, erect, entire, 2–3 in. long at the flowering time, $\frac{1}{2}$ lin. broad;
peduncle slender, glabrous, 1-flowered, 3–4 in. long; perianth-
segments oblong-lanceolate, $\frac{1}{2}$ in. long, yellow, glabrous and much
tinged with green outside; stamens more than half as long as the
perianth-segments ; ovary clavate, glabrous, $\frac{1}{4}$ in. long.

VAR. β, **H. Schlechteri** (Bolus in Hook. Ic. t. 2259 B); flowers smaller
than in the type, orange-yellow inside, tinged with red outside ; perianth-segments
oblong.

COAST REGION: Cape Flats in heathy places, near Kenilworth, *Schlechter*,
627! and near Wynberg, *MacOwan and Bolus, Herb. Norm.*, 1383! Var. β,
near Kenilworth, *Schlechter*, 628! near Wynberg, *MacOwan and Bolus, Herb.
Norm.*, 1384!

4. H. serrata (Linn. fil. Suppl. 197); corm globose, annual,
$\frac{1}{4}-\frac{1}{3}$ in. diam., crowned with a ring of bristles; produced leaves
6–12, linear, subulate, glabrous, 3–6 in. long, moderately firm in

texture, obscurely denticulate; peduncles 1–5 to a corm, very slender, simple, erect, with a linear clasping bract below the middle; perianth-limb about $\frac{1}{2}$ in. long; segments oblong-lanceolate, pale yellow inside, the three outer glabrous, and pale green with a tinge of red on the back; stamens half as long as the perianth-limb; anthers linear, basifixed; filaments very short; ovary clavate, glabrous, $\frac{1}{4}$–$\frac{1}{3}$ in. long; stigmas distinct, lanceolate; capsule clavate, $\frac{1}{3}$ in. long. *Thunb. Prodr.* 60; *Fl. Cap. edit. Schult.* 304; *Jacq. Ic. t.* 369; *Bot. Mag. t.* 709 (*excl. var.* β); *Roem. et Schultes, Syst. Veg.* vii. 768; *Baker in Journ. Linn. Soc.* xvii. 103. *H. luzulæfolia, Eckl. Top. Verz.* 10, *non DC.*

COAST REGION: Lion Mountain, *Thunberg!* near Hout Bay, 250 ft., *Bolus,* 7195! *MacOwan and Bolus, Herb. Norm.,* 1382! *Ecklon and Zeyher,* 649!
WESTERN REGION: Little Namaqualand, *Scully,* 115!

5. H. aquatica (Linn. fil. Suppl. 197); corm small, globose; leaves 4–6, weak, flaccid, glabrous, subterete from a lanceolate sheathing base, channelled down the face, 1–1$\frac{1}{2}$ ft. long; peduncle $\frac{1}{2}$–1 ft. long; flowers 4–6 in an umbel; pedicels 1–3 in. long; bracts large, linear, foliaceous; perianth-limb about $\frac{1}{2}$ in. long, whitish inside, green outside; segments oblong or oblong-lanceolate; stamens half as long as the perianth-segments; anthers linear, basifixed; filaments very short, ovary clavate, glabrous, $\frac{1}{3}$–$\frac{1}{2}$ in. long; stigmas distinct; capsule $\frac{1}{2}$–$\frac{3}{4}$ in. long. *Willd. Sp. Plant.* ii. 108; *Roem et Schultes, Syst. Veg.* vii. 776; *Baker in Journ. Linn. Soc.* xvii. 102.

SOUTH AFRICA: without locality, *Oldenburg! Drège,* 8515a! 8515b!
COAST REGION: Clanwilliam Div., *Zeyher! Mader,* 172! near Cape Town, *Thunberg! Bolus,* 2814! Cape Flats near Wynberg, *MacOwan and Bolus, Herb. Norm.,* 293! *MacOwan,* 2658.
WESTERN REGION: Little Namaqualand; near Klip Fontein, 3000 ft., *Bolus,* 6583! and without precise locality, *Whitehead!*

6. H. Scullyi (Baker in Journ. Bot. 1889, 2); corm small, globose; leaves linear, membranous, glabrous, $\frac{1}{2}$–1 ft. long, $\frac{1}{8}$–$\frac{1}{4}$ in. broad; flowers 1–3 to a peduncle; pedicels 2–3 in. long, bracteated at the base by reduced linear leaves; perianth-segments yellow, $\frac{1}{2}$ in. long, glabrous outside; stamens more than half as long as the segments; anthers $\frac{1}{6}$ in. long; ovary clavate, glabrous, $\frac{1}{4}$ in. long.

WESTERN REGION: Little Namaqualand, in moist places; near Ookiep, *Bolus,* 6581! between Nababeep and Modder Fontein, *Bolus,* 6582! and without precise locality, *Scully,* 10! *MacOwan and Bolus, Herb. Norm.,* 1381!

7. H. stellata (Linn. fil. Suppl. 197); corm globose, annual, $\frac{1}{2}$–$\frac{3}{4}$ in. diam.; tunics brown, coarsely cancellate; leaves 4–12 to a corm, subterete from a clasping linear base, glabrous, moderately firm in texture, 4–15 in. long; peduncles 1–4, simple, rarely forked, with a lanceolate clasping bract below the middle; perianth-limb $\frac{3}{4}$–1$\frac{1}{2}$ in. long; segments lanceolate, acute, plain white inside, the three outer green and glabrous on the back; anthers linear, basifixed, $\frac{1}{3}$ in. long; filaments very short; ovary clavate, glabrous, about $\frac{1}{2}$ in. long; stigmas lanceolate, joined at the base; capsule clavate, $\frac{1}{2}$–$\frac{3}{4}$ in. long, dehiscing round the middle. *Thunb. Prodr.* 60; *Fl. Cap. edit. Schult.* 304; *Roem. et Schultes, Syst. Veg.* vii. 776; *Baker in Journ. Linn. Soc.* xvii. 101. *Fabricia stellata, Thunb. in Fabric. Reiss Norw.* 27. *Amaryllis capensis, Linn. Sp. Plant. edit.* 2, 420. *H. elata, Roem. et Schultes, Syst. Veg.* vii. 778.

VAR. β, **H. elegans** (Andr.); perianth-segments with a large or small black, blue-black or purplish spot at the base. *H. stellata, Jacq. Ic. t.* 368; *Andr. Bot. Rep. t.* 236; *Bot. Mag. t.* 1223; *Flore des Serres, t.* 1027. *H. stellata, var. elegans, Pers. Syn.* i. 362. *H. tridentata, DC. in Red. Lil. sub t.* 169. *Spiloxene pavonina, Salisb. Gen.* 44. *H. cærulescens, DC. in Red. Lil. sub t.* 169.

VAR. γ, **H. Gawleri** (Baker in Journ. Linn. Soc. xvii. 101); flowers large, pale yellow, with a black spot at the base of each segment. *H. stellata, Bot. Mag. t.* 662; *DC. in Red. Lil. t.* 169; *Andr. Bot. Rep. t.* 101.

VAR. δ, **H. linearis** (Andr. Bot. Rep. t. 171); leaves more slender than in the type; flowers smaller, pale yellow, without any spot at the base of the segments. *Poir. Encyc. Suppl.* iii. 112; *Roem. et Schultes, Syst. Veg.* vii. 769. *Ianthe linearis, Salisb. Gen.* 44. *H. serrata, β, Gawl. in Bot. Mag. t.* 917. *H. juncea, Eckl. Top. Verz.* 10.

SOUTH AFRICA: without locality, *Forster!* Var. β, *Masson! Oldenburg!* Var. γ, *Masson!*

COAST REGION: Malmesbury Div.; Zwartland and Groene Kloof (including vars. β and δ), *Thunberg!* Cape Flats at Dornhoogte, *Zeyher!* between Cape Town and Simous Bay, *Burchell*, 8572! near Rondebosch, *Cooper*, 3236! Var. β, Clanwilliam Div., *Mader*, 144! moist places on the sides of Table Mountain, 250 ft., *MacOwan*, 2380! *MacOwan and Bolus, Herb. Norm.*, 292! Paarl Div.; Achter de Paarl, under 1000 ft., *Drège!* Var. δ, Table Mountain, *Ecklon*, 417! Worcester Div.; Drakenstein Mts., 2000–3000 ft., *Drège!*

WESTERN REGION: Little Namaqualand; near Lily Fontein, 4000–5000 ft., *Drège*, 2657!

H. geniculata, acuminata and *laxa, Eckl. Top. Verz.* 9 (names only), probably belong here, as they are placed in a group of which the flowers are said to be blueish.

8. H. ovata (Linn. fil. Suppl. 197); corm globose, perennial, 1 in. diam., densely coated with black wiry fibres; produced leaves 4–8, lanceolate, falcate, glabrous, moderately firm in texture, 2–3 in long, ½ in. broad; flowers 1–3 to a corm; peduncles slender, erect, glabrous, 1-flowered, 1–3 in. long; perianth-limb yellow, ¼–⅓ in. long; segments oblong-lanceolate, tinged with red outside, glabrous; anthers linear, basifixed, ⅛ in. long; ovary turbinate, glabrous, ⅙–⅛ in. long; stigmas linear, distinct. *Thunb. Prodr.* 60; *Fl. Cap. edit. Schult.* 306; *Ker in Bot. Mag. t.* 1010·; *Roem. et Schultes, Syst. Veg.* vii. 771; *Baker in Journ. Linn. Soc.* xvii. 103. *Ianthe ovata, Salisb. Gen.* 44.

SOUTH AFRICA: without locality, *Thunberg! Oldenburg!*

COAST REGION: Piquetberg, 1800 ft., *Bodkin in Herb. Bolus*, 7559! Worcester Div.; Dutoits Kloof, 1000–2000 ft., *Drège*, 1555! Caledon Div.; on the Zwartberg, near the Hot Springs, *Zeyher*, 4135!

9. H. Andrewsii (Baker in Journ. Linn. Soc. xvii. 104); root-fibres copious, long, slender; produced leaves 5–6, lanceolate, glabrous, spreading, 3–4 in. long; peduncles 3 to a cluster, branched about the middle, 2-3-flowered; bracts large, lanceolate; perianth-limb $\frac{1}{2}$ in. long, bright yellow inside, green and glabrous outside; segments ovate-lanceolate; stamens half as long as the perianth-limb; anthers linear, basifixed, twice as long as the filaments; ovary clavate, glabrous, $\frac{1}{4}-\frac{1}{3}$ in. long. *H. obliqua, Andr. Bot. Rep. t.* 195, *non Jacq.*

SOUTH AFRICA: without locality.

Known only from the figure cited, which was drawn from a plant cultivated by Mr. G. Hibbert at Clapham in June, 1801.

10. H. milloides (Baker in Journ. Linn. Soc. xvii. 105); corm small, globose; leaves 4–6, subulate from a lanceolate clasping base, erect, 2–4 in. long, moderately firm in texture, slightly hairy when young, glabrous when mature; peduncles solitary, erect, 1-flowered, 1–2 in. long; perianth limb $\frac{1}{3}-\frac{1}{2}$ in. long; segments oblong, obtuse, bright red on both surfaces, the outer slightly hairy on the back; stamens very small; ovary turbinate, obconic, densely bristly, $\frac{1}{8}$ in. long.

EASTERN REGION: Griqualand East; top of Mount Currie, 7500 ft., *Tyson*, 1357! Natal; mountains of Klip River, 3500–4000 ft., *Sutherland!* Dargle Farm, *Mrs. Fannin*, 58! and without precise locality, *Krauss*, 24!

Is perhaps only a variety of *H. Baurii*.

11. H. Baurii (Baker in Journ. Bot. 1876, 181); corm ovoid, $\frac{1}{2}$ in. diam., crowned with a ring of bristles; leaves 6–8 to a corm, linear, firm in texture, erect, densely hairy, $1\frac{1}{2}-2$ in. long, $\frac{1}{8}-\frac{1}{6}$ in. broad; peduncles 2–3, simple, erect, slender, densely hairy, 2–3 in. long; perianth-limb $\frac{1}{2}$ in. long; segments oblong, obtuse, bright red, the outer hardly at all hairy on the back; stamens very small; ovary turbinate, $\frac{1}{6}$ in. long, densely hairy; capsule turbinate, densely pilose, $\frac{1}{4}$ in. long. *Journ. Linn. Soc.* xvii. 105; *Gard. Chron.* 1877, viii. 584.

VAR. *β*, **H. platypetala** (Baker, loc. cit.); flowers whitish.

EASTERN REGION: Tembuland; Bazeia Mountains, 3500–4000 ft., *Baur*, 501! Griqualand East; Mount Currie, 5500–6000 ft., *Tyson*, 1571! and *MacOwan and Bolus, Herb. Norm.*, 481! Vaal Bank, near Kokstad, *Haygarth in Herb. Wood*, 4182! Natal; Liddesdale, in marshy places, 5000 ft., *Wood*, 4261! Var. *β*, Griqualand East; on the Zuurberg Range, 4000 ft., *MacOwan and Bolus, Herb. Norm.*, 1212! *Groom in Herb. Wood*, 1976! between Durban and Kokstad, *Groom in Herb. Wood*, 1753! Natal; Liddesdale, on a stony hill, 5000 ft., *Wood*, 4260! Dargle Farm, *Mrs. Fannin!* and without precise locality, *McKenn! Sutherland!*

12. H. Flanagani (Baker); corm very small; leaves 4–6 to a corm, spreading, linear, very hairy, $1-1\frac{1}{2}$ in. long; $\frac{1}{2}-1$ lin. broad; peduncle slender, hairy, 1-flowered, $1-1\frac{1}{2}$ in. long; perianth-segments oblong, yellow, $\frac{1}{4}-\frac{1}{3}$ in. long, greenish and hairy outside; stamens half as long as the perianth-limb; filaments as long as or shorter than the oblong anthers; ovary clavate, hairy, $\frac{1}{8}$ in. long; stigmas connate.

COAST REGION: Komgha, *Flanagan*, 314!

13. H. filiformis (Baker in Journ. Linn. Soc. xvii. 109); corm oblong, $\frac{1}{4}-\frac{1}{3}$ in. diam., with a long neck and dark brown membranous tunics; leaves 6–8, setaceous, strongly ribbed, rigid, 3–4 in. long, under a line in diameter, bright green, loosely hairy; peduncles 1–2, very slender, 2–5 in. long, 1–2 flowered; perianth-limb $\frac{1}{4}-\frac{1}{3}$ in. long; segments oblong-lanceolate, pale yellow, the outer green and hairy on the back; anthers lanceolate, deeply sagittate; filament as long as the anther; ovary turbinate, $\frac{1}{8}-\frac{1}{6}$ in. long, densely hairy; stigmas concrete.

COAST REGION: Queenstown Div., *Cooper*, 462!

KALAHARI REGION: Transvaal; Saddleback Range, near Barberton, 4500–5000 ft., *Galpin*, 1101!

EASTERN REGION: Natal; Mohlamba Range, 5000–6000 ft., *Sutherland!* Inanda, *Wood*, 1030!

14. H. kraussiana (Buching. in Flora 1845, 311, name only); corm ovoid, $1\frac{1}{2}-2$ in. diam., crowned with a dense mass of bristles; tunics blackish; leaves 10–15, subulate from a linear base, rigid in texture, strongly ribbed, glabrous, $1-1\frac{1}{2}$ ft. long, channelled down the face; peduncles 2–3, hairy upwards, 4–9 in. long; flowers 2–5, corymbose; pedicels $\frac{1}{4}-1$ in. long, densely hispid; bracts linear-subulate; perianth-limb $\frac{1}{3}-\frac{1}{2}$ in. long; segments oblong-lanceolate, yellow, the outer densely hairy all over the back; anthers $\frac{1}{8}$ in. long, deeply sagittate; ovary between clavate and turbinate, densely hairy, $\frac{1}{6}$ in. long; stigmas concrete; capsule turbinate, $\frac{1}{4}$ in. long. *Baker in Journ. Linn. Soc.* xvii. 109.

COAST REGION: Uitenhage Div.; Van Stadens River, *MacOwan*, 2123! Van Stadens Berg, *Burchell*, 4742! between Van Stadens Berg and Bethelsdorp, *Drège*, 8534a! Bathurst Div.; between Riet Fontein and the sea-shore, *Burchell*, 4099!

EASTERN REGION: Natal; hills near Pietermaritzberg, *Krauss*, 104! Swazieland; Nottingham Peak, Havelock Concession, 4000 ft., *Saltmarsh in Herb. Galpin*, 984!

15. H. angustifolia (Lam. Encyc. iii. 182); corm oblong, $\frac{1}{3}$ in. diam., with a long neck and brown membranous tunics; leaves 6–12, linear, thin in texture, 4–6 in. long, $\frac{1}{6}-\frac{1}{4}$ in. broad, slightly hairy; peduncles 1–4, slender, arcuate, hairy upwards; flowers often 2, corymbose; pedicels hairy, $\frac{1}{2}-1$ in. long; bracts setaceous; perianth-limb $\frac{1}{4}-\frac{1}{3}$ in. long; segments oblong-lanceolate, pale yellow; outer green and hairy on the back; stamens half as long as the perianth-limb; anthers lanceolate, deeply sagittate; filament short; ovary turbinate, densely pilose, $\frac{1}{6}-\frac{1}{8}$ in. long; stigmas concrete; capsule small, hairy, turbinate. *Roem. et Schultes, Syst. Veg.* vii. 767; *Fisch. et C. A. Mey. Ind. Sem. Petrop.* x. 49; *Baker in Journ. Linn. Soc.* xvii. 111. *H. biflora, Baker in Journ. Bot.* 1876, 181.

VAR. β, **Buchanani** (Baker in Journ. Linn. Soc. xvii. 111); a large shade-grown variety, with longer leaves of very thin texture; pedicels longer, very slender.

KALAHARI REGION: Orange Free State, *Cooper*, 1039!

EASTERN REGION: Tembuland; near Bazeia, 2000–2500 ft., *Baur*, 347! Var.
β, Natal; Iuanda, *Wood*, 426! 771! and without precise locality, *Buchanan!*
Also found in Tropical Africa, Madagascar, and Mauritius.

16. **H. floccosa** (Baker in Kew Bullet. 1894, 357); corm small,
oblong; radical leaves 5–6, linear, not rigid, 2–3 in. long, $\frac{1}{12}$ in.
broad at the middle, densely clothed with short, soft, spreading
hairs; peduncle very slender, shorter than leaves; flowers 1–2;
bracts linear; pedicels very hairy, $\frac{1}{2}$–$\frac{3}{4}$ in. long; perianth yellow,
$\frac{1}{4}$ in. long; segments oblong lanceolate, the outer very hairy outside; stamens much shorter than the perianth; ovary hairy,
clavate.

COAST REGION: hills near Swellendam, 500 ft., *Bolus*, 7469.

17. **H. Zeyheri** (Baker in Journ. Linn. Soc. xvii. 112); corm
oblong, $\frac{3}{4}$–1 in. diam., crowned with a ring of bristles; neck elongated; tunics membranous; leaves 6–10, linear, glabrous, moderately
firm in texture, 4–6 in. long, $\frac{1}{3}$–$\frac{1}{2}$ in. broad; peduncles 1–2, slender,
flexuose, hairy upwards; flowers 2–3, corymbose; pedicels $\frac{1}{2}$–1 in.
long; bracts setaceous; perianth-limb $\frac{1}{4}$–$\frac{1}{3}$ in. long; segments
oblong-lanceolate, yellow, the three outer hairy outside; stamens
half as long as the perianth-limb; anthers lanceolate, deeply sagit-
tate, $\frac{1}{8}$ in. long; ovary turbinate, $\frac{1}{8}$ in. long, densely hairy; stigmas
concrete; capsule $\frac{1}{4}$ in. long.

COAST REGION: Albany, *Williamson!* Uitenhage? *Ecklon and Zeyher*, 7!

18. **H. Gerrardi** (Baker in Journ. Linn. Soc. xvii. 110); corm
oblong, $\frac{1}{2}$–$\frac{3}{4}$ in. diam., with a long neck and brown membranous
tunics; leaves 6–10, linear, rigid in texture, strongly ribbed, 6–15
in. long, $\frac{1}{6}$–$\frac{1}{4}$ in. broad, acuminate, shortly and softly pilose all over;
peduncles 2–4, slender, hairy, 2–6 in. long; flowers 2–4, corymbose;
pedicels densely hairy, $\frac{1}{2}$–1 in. long; bracts small, linear or subulate;
perianth-limb $\frac{1}{4}$–$\frac{1}{3}$ in. long; segments oblong-lanceolate, yellow,
the outer densely pilose all over the back; stamens more than half
as long as the perianth-limb; anthers lanceolate, $\frac{1}{8}$ in. long; ovary
clavate-turbinate, densely pilose, $\frac{1}{6}$ in. long; stigmas concrete;
capsule turbinate, densely pilose, $\frac{1}{4}$–$\frac{1}{3}$ in. long.

EASTERN REGION: Natal; Inanda, *Wood*, 327! and without precise locality,
Gerrard! Buchanan!

19. **H. argentea** (Harv. ex Baker in Journ. Linn. Soc. xvii·
110); corm ovoid, $\frac{1}{2}$–$\frac{3}{4}$ in. diam., with a long neck, and crowned
with a dense ring of bristles; leaves 6–12, linear-subulate, firm in
texture, falcate, acuminate, 4–6 in. long, $\frac{1}{12}$ in. broad at the base,
densely persistently silky all over both surfaces; peduncles 1–2,
slender, densely silky; flowers often 2, corymbose; pedicels $\frac{1}{2}$–1
in. long; bracts setaceous; perianth-limb $\frac{1}{4}$ in. long; segments
oblong-lanceolate, the outer densely silky on the back; stamens
half as long as the perianth-limb; anthers lanceolate, deeply

sagittate; ovary clavate, silky, $\frac{1}{8}-\frac{1}{6}$ in. long; stigmas concrete; capsule turbinate, $\frac{1}{4}$ in. long.

VAR. β, **H. sericea** (Baker in Journ. Linn. Soc. xvii. 111) ; leaves not so firm and thick in texture and less silky. *H. sericea var. Dregei, Baker loc. cit.* 112.

VAR. γ, **flaccida** (Baker,) ; a large form with thinner longer leaves and longer pedicels. *H. sericea var. flaccida, Baker loc. cit.* 112.

SOUTH AFRICA: without locality, *MacOwan*, 50 !
COAST REGION: near Uitenhage, *Burchell*, 4469 ; Var. β, near Swellendam, 500 ft., *Bolus*, 7468 ! Uitenhage, *Zeyher*, 950 ! Stockenstrom Div.; Katberg, 3000–4000 ft., *Drège*, 8525 ! British Kaffraria, *Cooper*, 1811 ! Var. γ, Albany, *Williamson !*
CENTRAL REGION : Var. β, Cave Mountain, 4300 ft., near Graaff Reinet, *Bolus*, 176 ! near Somerset East, *MacOwan*, 1593b ! Fish River, *Burke !* Orange River, *Burke !*
KALAHARI REGION: Transvaal ; plains around Pretoria, 4000 ft., *Bolus*, 176 ! Var. β, Orange Free State ; Bloemfontein, *Rehmann*, 3761 ! Var. γ, Orange Free State ; near Seven Fontein Mission-station, *Burke !* Transvaal ; Apies River, *Burke !* bank of Vaal River, *Burke*, 447 !
EASTERN REGION : Var. β, Griqualand East ; near Kokstad, 4700 ft., *Tyson*, 1095 ! *MacOwan and Bolus, Herb. Norm.*, 480 ! Natal ; Berea, near Durban, *Wood*, 101 ! Inanda, *Wood*, 191 !

20. **H. Jacquini** (Baker in Journ. Linn. Soc. xvii. 112) ; corm oblong, $\frac{1}{2}-\frac{3}{4}$ in. diam., with a long neck and brown membranous tunics; leaves 10–12, linear, $\frac{1}{2}$–1 ft. long, $\frac{1}{6}-\frac{1}{8}$ in. broad, thin in texture, hairy on both sides and on the margin ; peduncles slender, pilose, $\frac{1}{2}$ ft. long; flowers 3–4, racemose ; pedicels all very short ; bracts linear, $\frac{1}{4}-\frac{1}{2}$ in. long ; perianth-limb $\frac{1}{4}$ in. long ; segments oblong-lanceolate, yellow, the outer green and hairy on the back ; stamens half as long as the perianth-limb ; anthers lanceolate, $\frac{1}{12}$ in. long ; ovary clavate, densely pilose, $\frac{1}{3}-\frac{1}{2}$ in. long; stigmas concrete ; capsule clavate, $\frac{1}{2}-\frac{3}{4}$ in. long. *H. villosa, Jacq. Collect Suppl.* 51; *Ic. t.* 370, *non Linn. fil.*

SOUTH AFRICA: without locality, *Hort. Jacquin*.

21. **H. Arnottii** (Baker in Gard. Chron. 1877, viii. 552) ; corm globose, 3–4 in. diam., crowned with a dense mass of bristles; leaves 5–6, linear, erect, falcate, rigid in texture, strongly ribbed, a foot long, $\frac{1}{4}-\frac{1}{3}$ in. broad, clothed with soft, short, spreading hairs ; peduncles 2, slender, arcuate, $\frac{1}{2}$ ft. long; densely pilose upwards ; flowers 6–8, racemose ; pedicels $\frac{1}{4}-\frac{1}{3}$ in. long ; bracts linear-subulate, densely pilose, $\frac{1}{3}-\frac{1}{2}$ in. long ; perianth-limb under $\frac{1}{2}$ in. long ; segments oblong or oblong-lanceolate, the outer green and pilose on the back ; anthers $\frac{1}{8}$ in. long ; filaments very short ; ovary clavate, densely pilose, $\frac{1}{4}$ in. long ; stigmas concrete ; capsule turbinate, $\frac{1}{4}$ in. long. *Baker in Journ. Linn. Soc.* xvii. 112.

CENTRAL REGION: Colesberg Div., *Arnott !* Cultivated at Kew in 1870.

22. **H. membranacea** (Baker in Journ. Linn. Soc. xvii. 106); corm globose, $\frac{1}{4}-\frac{1}{3}$ in. diam., with a long neck ; tunics brown, membranous ; leaves 4–8, lanceolate, membranous, 1–3 in. long, $-\frac{1}{2}$ in. broad, thinly clothed with long soft hairs ; peduncles 1–2,

very slender, flexuose, hairy, deeply forked; bracts minute, linear; pedicels ½–2 in. long; perianth-limb ¼ in. long; segments oblong-lanceolate, whitish-yellow inside, the outer green and rather hairy on the back; stamens as long as the perianth-segments; anthers lanceolate, versatile; filament as long as the anther; ovary turbinate, ⅛ in. long, pilose; capsule small, turbinate. *H. parvula, Baker in Journ. Linn. Soc.* xvii. 113.

EASTERN REGION: Griqualand East; Pumngwan Mountain, near Clydesdale, 3500 ft., *Tyson,* 2880! Natal; Tugela, *Gerrard,* 1835! Itafamasi, *Wood,* 844! 862! and without precise locality, *Sanderson!* Swazieland; mountain sides, Havelock Concession, 4000 ft., *Saltmarsh in Herb. Galpin,* 1049!

23. **H. Woodii** (Baker in Journ. Bot. 1889, 3); corm oblique, oblong, crowned with a ring of slender fibres; leaves about 6, linear or lanceolate, erect, thin in texture, with slender ribs, quite glabrous on both faces and margin, the longest at the flowering time about a foot long, ½–1 in. broad; peduncle slender, slightly hairy, 4–5 in. long; flowers 2–3, corymbose; pedicels ½–1 in. long; bracts linear-setaceous; perianth-limb ¼–⅓ in. long; segments oblong, acute, yellow, the outer hairy on the back; stamens half as long as the perianth-limb; ovary turbinate, hairy, ⅙ in. long; stigmas concrete.

EASTERN REGION: Natal; Inanda, *Wood,* 426a!

24. **H. parvifolia** (Baker); corm oblong, ¼ in. diam.; leaves 3–4, lanceolate, moderately firm, hairy, 1–1½ in. long; peduncles very slender, hairy, 1–2 in. long, 1–2-flowered; segments oblong, bright yellow, hairy outside, ¼ in. long; stamens ⅓ the length of the perianth-segments; ovary turbinate, very hairy, ⅛ in. long.

KALAHARI REGION: Transvaal; summit of Saddleback Range, near Barberton, 5000 ft., *Galpin,* 1059!

25. **H. brevifolia** (Baker); corm very small; leaves about 3, oblong-lanceolate, acute, moderately firm, thinly and softly hairy, 1–1¼ in. long, ¼–⅓ in. broad; peduncle very slender, hairy, always 1-flowered, 2–3 in. long; perianth-segments oblong-lanceolate, yellow, ¼ in. long, hairy outside; stamens ⅓ the length of the perianth-segments; ovary turbinate, hairy, ⅛ in. long.

EASTERN REGION: Natal; Liddesdale, *Wood,* 3940!

26. **H. setosa** (Baker in Journ. Linn. Soc. xvii. 113); corm oblong, 1–1½ in. diam., crowned with a dense ring of strong bristles 1½–2 in. long; leaves 6–8, lanceolate, rigidly coriaceous, strongly ribbed, 4–6 in. long, ½–1 in. broad at the middle, quite glabrous on both surfaces, obscurely bristly on the margin; peduncle shorter than the leaves, hairy upwards; flowers 2–3, corymbose; bracts small, linear; perianth-limb ⅓–½ in. long; segments oblong-lanceolate, yellow, the outer greenish and densely hairy on the back; anthers lanceolate, deeply sagittate, ⅙ in. long; ovary hairy, turbinate, ⅙ in. long; stigmas concrete.

SOUTH AFRICA : without locality, *MacOwan*, 72 ! in Dublin Herbarium.
Flowered at Kew in Sept., 1873.

27. **H. villosa** (Linn. fil. Suppl. 198); corm oblong, 1–1½ in. diam.,
crowned with a dense ring of bristles; leaves a dozen or more,
lanceolate, acuminate, moderately firm in texture, 4–6 in. long, about
½ in. broad, more or less densely clothed with appressed hairs;
peduncles 1–4, slender, finely pilose, 2–3 in. long; flowers 2–5,
corymbose; pedicels ½–1 in. long; bracts small, setaceous; perianth-
limb ⅓–½ in. long; segments oblong or oblong-lanceolate, yellow, the
outer densely hairy on the back; stamens half as long as the perianth-
limb; filament about as long as the lanceolate-sagittate anther; ovary
turbinate, densely hairy, ⅛ in. long; stigmas concrete; capsule
turbinate, densely hairy, ¼ in. long. *Thunb. Prodr.* 60; *Fl. Cap.
edit. Schult.* 305; *Roem. et Schultes, Syst. Veg.* vii. 765; *Baker in
Journ. Linn. Soc.* xvii. 113, *non Jacq. Fabricia villosa, Thunb. in
Fabric. Reise Norw.* 31. *H. tomentosa, Lam. Encyc.* iii. 182. *H.
sobolifera, Jacq. Collect. Suppl.* 53; *Ic. t.* 372; *Red. Lil. t.* 170;
Bot. Mag. t. 711; *Fisch. et Mey. Ind. Sem. Petrop.* x. 51.

VAR. β, **H. scabra** (Lodd. Bot. Cab. t. 970); leaves thinner in texture,
scarcely at all hairy except on the margin.

VAR. γ, **H. obliqua** (Jacq. Collect. Suppl. 54; Ic. t. 371); more robust, with
leaves clothed with dense, short, pale, bristly hairs principally on the margin and
keel beneath; peduncle longer, densely clothed with short, stellate, bristly hairs.
Roem. et Schultes, Syst. Veg. vii. 766, *non Andrews.*

VAR. δ, **H. canescens** (Fisch. et Mey. Ind. Sem. Petrop. x. 50); more robust
than the type, with larger leaves, sometimes an inch broad, densely clothed with
short, soft, often brown, stellate hairs; pedicels longer; perianth-limb longer.
Walp. Rep. i. 847.

VAR. ε, **H. pannosa** (Baker in Gard. Chron. 1874, ii. 130); a large robust form
with flaccid leaves above a foot long and an inch broad, densely clothed on both
surfaces with long, appressed, fine, soft hairs. *H. microsperma, Lallem. in Fisch.
and Mey. Ind. Sem. Petrop.* x. 50.

SOUTH AFRICA : without locality, Var. ε, cultivated at Kew in 1874.
COAST REGION : Swellendam Div., *Thunberg !* Mossel Bay Div.; between
Duyker River and Gauritz River, *Burchell*, 6401! Uitenhage Div.; Zwartkops
River, *Zeyher*, 4138! between Van Stadens Berg and Bethelsdorp, *Drège*, 2192!
Algoa Bay, *Cooper*, 3237! Var. β, Mossel Bay, near the Landing-place, *Burchell*,
6307! Uitenhage Div.; Van Stadens Berg, *Burchell*, 4745! Albany Div.;
Blaauw Krans, *Burchell*, 3632! Var. δ, Alexandria Div.; Zwartwater Poort,
Burchell, 3380–2! Albany Div.; Grahamstown, *Burchell*, 3542! *MacOwan*,
1899!
CENTRAL REGION : near Somerset East, 2500 ft., *MacOwan*, 1593! Var. β,
Somerset Div.; Bosch Berg, 4500 ft., *MacOwan*, 1898! Var. γ, Bosch Berg,
4500 ft., *MacOwan*, 1594a! 1594b!
EASTERN REGION : Var. β, Tembuland, Bazeia Mountains, 2000 ft., *Baur*,
74! Var. δ, Natal; Inanda, *Wood*, 184! Var. ε, Natal; slopes of the Drakens-
berg, *Wood*, 3434!

28. **H. obtusa** (Burch. in Bot. Reg. t. 159); corm globose, 1½–2 in.
diam., densely bristly at the crown; leaves 8–12, linear or lanceolate,
acuminate, firm in texture, strongly ribbed, hairy, especially on the
margins and keel beneath, finally a foot or more long. ½–¾ in. broad;

peduncles slender, hairy, much shorter than the leaves; flowers few
or several, racemed; lowest pedicels $\frac{1}{4}$ in. long; bracts linear;
perianth-limb $\frac{1}{3}$–$\frac{1}{2}$ in. long; segments oblong, yellow, outer densely
hairy on the back; anthers lanceolate, deeply sagittate, $\frac{1}{6}$ in. long;
filaments shorter; ovary turbinate, densely bristly, $\frac{1}{4}$ in. long; stigmas
concrete; capsule turbinate, densely hairy, circumscissile round the
middle. *Roem. et Schultes, Syst. Veg.* vii. 766; *Baker in Trans.
Linn. Soc.* xxix. 156; *Journ. Linn. Soc.* xvii. 114.

KALAHARI REGION : Orange Free State, *Mrs. Barber,* 685! Bechuanaland;
Pellat Plains, near Takun, *Burchell!* Transvaal; Saddleback Range, near
Barberton, 3200–4500 ft., *Galpin,* 412!
EASTERN REGION : Natal; Drakensberg, *Bolus,* 2572!

Also mountains of Equatorial Africa.

29. H. latifolia (Hook. in Bot. Mag. t. 4817); corm globose,
$2\frac{1}{2}$–3 in. diam., crowned with bristles; leaves 6–8, lanceolate, rigidly
coriaceous, strongly ribbed, entirely glabrous, finally $1\frac{1}{2}$–2 ft. long,
$1\frac{1}{2}$–2 in. broad at the middle; peduncles slender, hairy, 3–4 in. long;
raceme 10–12-flowered, $1\frac{1}{2}$–2 in. long; lower pedicels $\frac{1}{4}$ in. long.;
bracts linear; perianth-limb $\frac{1}{2}$ in. long; segments oblong, yellow, the
outer greenish and hairy on the back; anthers lanceolate, $\frac{1}{4}$ in. long;
filament much shorter; ovary turbinate, hairy, $\frac{1}{4}$ in. long; stigmas
concrete. *Baker in Journ. Linn. Soc.* xvii. 115.

EASTERN REGION : Natal, *Garden, Adlam !* Described from plants cultivated
at Kew in 1854 and 1887.

30. H. longifolia (Baker in Bot. Mag. t. 6035); corm oblong,
$1\frac{1}{2}$–2 in. diam.; leaves 8–9, subterete, rigidly coriaceous, deeply
channelled down the face, 1–$1\frac{1}{2}$ ft. long, $\frac{1}{3}$–$\frac{1}{6}$ in. broad, shortly hairy
mainly on the margins and keel beneath; peduncles 2–3, densely
pilose, 6–9 in. long; flowers 2–4, corymbose; pedicels $\frac{1}{2}$–1 in. long,
densely pilose; bracts linear-setaceous; perianth-limb $\frac{5}{8}$–$\frac{3}{4}$ in. long;
segments oblong-lanceolate, yellow, the outer densely pilose on the
back; anthers lanceolate-sagittate, slightly versatile, $\frac{1}{8}$ in. long;
ovary turbinate, densely pilose, $\frac{1}{6}$–$\frac{1}{4}$ in. long; stigmas concrete;
capsule turbinate, densely pilose, $\frac{1}{4}$ in. long. *Baker in Journ. Linn.
Soc.* xvii. 115.

SOUTH AFRICA : without locality, *Thunberg !*
COAST REGION : Humansdorp Div.; Clarkson, *Kitching !* Queenstown, *Cooper,*
3238!
KALAHARI REGION : Orange Free State; Vet River, *Burke!*

31. H. Ludwigii (Baker in Journ. Bot. 1876, 181); corm oblong,
1–$1\frac{1}{2}$ in. diam.; leaves 8–9, linear, subcoriaceous, with thickened
ribs, 1–$1\frac{1}{2}$ ft. long, $\frac{1}{4}$–$\frac{1}{3}$ in. broad, shortly bristly mainly on the
margins and keel beneath; peduncle 9–12 in. long, weak, loosely
hairy; flowers 4–12, corymbose; lower pedicels 1–$1\frac{1}{2}$ in. long; bracts
linear; perianth-limb $\frac{5}{8}$–$\frac{3}{4}$ in. long; segments oblong or oblong-
lanceolate, yellow, the outer densely hairy outside; anthers
lanceolate-sagittate, $\frac{1}{6}$ in. long; filaments much shorter; ovary

turbinate, densely hairy, $\frac{1}{8}-\frac{1}{4}$ in. long; stigmas concrete. *Baker in
Journ. Linn. Soc.* xvii. 116.

EASTERN REGION : Tembuland; Bazeia, *Baur,* 301 !

Originally described from a specimen from Baron Ludwig's Garden.

32. H. acuminata (Baker in Journ. Bot. 1889, 3); corm not
seen; leaves 6–8, erect, linear, not rigid in texture, loosely hairy all
over, 12–15 in. long, $\frac{1}{2}$ in. broad low down, tapering to the acuminate
apex very gradually; peduncles single, weak, villose, 6–8 in. long;
flowers 2–4, subracemose; lower pedicels $\frac{1}{4}-\frac{1}{2}$ in. long; bracts small,
linear-subulate; perianth-limb $\frac{5}{8}-\frac{3}{4}$ in. long; segments oblong-lanceolate, yellow, the outer densely villose on the back; anthers lanceolate-
sagittate, versatile, $\frac{1}{8}$ in. long; ovary obconic, densely villose, $\frac{1}{8}$ in.
long; stigmas concrete; capsule turbinate, villose, $\frac{1}{4}$ in. long.

EASTERN REGION : Natal; Inanda, *Wood,* 1347 !

33. H. rigidula (Baker in Journ. Linn. Soc. xvii. 116); corm
oblong, 1–1$\frac{1}{2}$ in. diam.; leaves 5–6, linear, rigidly coriaceous, strongly
ribbed, erect, shortly pilose, 1–1$\frac{1}{2}$ ft. long, $\frac{1}{4}-\frac{1}{3}$ in. broad; peduncles
2–3, sometimes a ·foot long, slender, shortly pilose; flowers 3–8,
racemose; pedicels all short; bracts linear; perianth-limb $\frac{5}{8}-\frac{3}{4}$ in.
long; segments oblong-lanceolate, yellow, the outer densely pilose
outside; anthers versatile, $\frac{1}{8}$ in. long; filaments short; ovary
turbinate, $\frac{1}{4}$ in. long, clothed with dense, long, whitish, bristly hairs;
stigmas concrete; capsule turbinate, densely villose, $\frac{1}{4}$ in. long,
slitting round the middle, not dividing into valves.

VAR. β, **pilosissima** (Baker, loc. cit. 117); leaves, peduncle, and perianth
shaggy, with denser, longer hairs.

SOUTH AFRICA : without locality, *Zeyher,* 1670 !
COAST REGION : Alexandria Div.; Zuurberg Range, 2000–3000 ft., *Drège,*
2194a! Bathurst Div.; between Blaauw Krans and Kaffir Drift Military Post,
Burchell, 3694! Queenstown Div.; Shiloh, *Baur,* 904! B.itish Kaffraria,
Cooper, 3239 !
CENTRAL REGION : Somerset Div.; Bosch Berg, 4500 ft., *MacOwan,* 1649 !
Albert Div., *Cooper,* 1763 ! Mooyplaats, 4500–5000 ft., *Drège,* 2194d !
KALAHARI REGION : Orange Free State, *Cooper,* 883 ! Basutoland, *Cooper,*
3241 ! Transvaal; Wonderfontein, *Nelson,* 261 ! Var. β, Transvaal; Magalies
Berg, *Burke,* 156 ! Saddleback Range, near Barberton, 2300–4500 ft., *Galpin,*
1099 !
EASTERN REGION : Griqualand East; mountains around Kokstad, *MacOwan
and Bolus, Herb. Norm.,* 1211 ! Natal; Inanda, *Wood,* 407 ! near Gorton,
Wood, 3433 ! Var. β, Natal; Tugela, *Gerrard,* 1826 !

34. H. colchicifolia (Baker in Journ. Bot. 1889, 3); corm globose,
2 in. diam., crowned with bristles; leaves 6–8, oblong-lanceolate or
lanceolate, moderately firm in texture, strongly ribbed, rather glaucous,
quite glabrous on both surfaces and margin, the largest 6–8 in. long,
1$\frac{1}{2}$–2 in. broad at the flowering time; peduncle slender, much shorter
than the leaves; flowers 3–4, corymbose; pedicels $\frac{1}{2}$–1 in. long;
bracts linear; perianth-limb $\frac{1}{2}-\frac{5}{8}$ in. long; segments oblong, yellow,
the outer green and hairy on the back; stamens half as long as the

perianth-segments; anthers lanceolate, $\frac{1}{6}$ in. long; filaments shorter; ovary obconic, hairy, $\frac{1}{8}$ in. long; stigmas concrete.

SOUTH AFRICA: without locality. Described from a plant that flowered with Mr. Bull in 1884.

35. **H. multiceps** (Buching. in Flora 1845, 311, name only); corm globose, 1$\frac{1}{2}$–2 in. diam., crowned with a dense ring of long, wiry bristles; leaves 5–6, oblong or oblong-lanceolate, coriaceous, conspicuously and closely ribbed, 4–6 in. long at the flowering time, 1–1$\frac{1}{2}$ in. broad, clothed over both surfaces and margin with short, stellate hairs; peduncle 4–6 in. long, densely coated with stellate hairs; flowers 2–4, corymbose; pedicels $\frac{1}{2}$–1 in. long; bracts linear; perianth-limb $\frac{3}{4}$ in. long; segments oblong or oblong-lanceolate, yellow, the outer densely villose on the back; anthers lanceolate, $\frac{1}{4}$ in. long; filaments short; ovary obconic, densely villose, $\frac{1}{4}$ in. long; stigmas concrete; capsule turbinate, villose, $\frac{1}{2}$ in. long. *Baker in Journ. Linn. Soc.* xvii. 117.

COAST REGION: Cathcart Div.; between Windvogel Mountain and Zwart Kei River, 3000-4000 ft., *Drège*, 3513d!
KALAHARI REGION: Transvaal; Saddleback Range, 4500 ft., *Galpin*, 1058!
EASTERN REGION: Natal; near Pietermaritzburg, *Krauss*, 248! Iuanda, *Wood*, 1011!

36. **H. stellipilis** (Ker in Bot. Reg. t. 663); corm globose, 1$\frac{1}{2}$–2 in. diam., crowned with a ring of bristles; leaves a dozen or more, lanceolate, acuminate, neither rigid in texture nor conspicuously ribbed, $\frac{1}{2}$–1 ft. long, $\frac{1}{2}$–1 in. broad at the middle, green and glabrous above, covered with persistent white tomentum beneath; peduncles 3–4 to a corm, 4–6 in. long, densely villose; flowers 4–10, subcorymbose; lower pedicels 1–1$\frac{1}{2}$ in. long; bracts large, linear; perianth-limb $\frac{3}{4}$ in. long; segments oblong, the outer villose on the back; anthers lanceolate, $\frac{1}{4}$ in. long; filaments shorter; ovary obconic, $\frac{1}{4}$ in. long, densely clothed with long white hairs; stigmas concrete; capsule turbinate, villose, circumscissile round the middle. *Roem. et Schultes, Syst. Veg.* vii. 767; *Fisch. et Mey. Ind. Sem. Petrop.* x. 51; *Baker in Journ. Linn. Soc.* xvii. 118. *H. lanata, Eckl.*

COAST REGION: Zwartkops River, near Uitenhage, 50–500 ft., *Zeyher*, 4140! *Drège*, 8527! Bedford Div.; Kagaberg, *Elliot*, 685!
CENTRAL REGION: Somerset Div.; Commadagga, *Burchell*, 3303!

37. **H. oligotricha** (Baker in Journ. Bot. 1889, 3); corm not seen; leaves oblong-lanceolate, glabrous, erect, subcoriaceous, strongly ribbed, 15–18 in. long, 1$\frac{1}{2}$–2 in. broad at the middle; peduncles ancipitous, glabrous, 6–8 in. long; flowers 10–15 in a lax raceme 3–4 in. long; lower pedicels $\frac{1}{4}$–$\frac{1}{3}$ in. long; bracts linear, $\frac{1}{2}$–1 in. long; perianth $\frac{1}{2}$–$\frac{3}{4}$ in. long; segments oblong-lanceolate, the outer with only a few scattered adpressed hairs on the back; anthers lanceolate, $\frac{1}{4}$ in. long; ovary globose, nearly glabrous, $\frac{1}{8}$ in. long and broad; stigmas concrete.

EASTERN REGION: Natal coast, *Wood*, 1170!

38. H. hemerocallidea (Fisch. et Mey. Ind. Sem. Petrop, viii. 64, x. 50) ; corm globose, 3–4 in. diam. ; leaves 6–8, lorate, acuminate, subcoriaceous, shortly hairy on the keel and midrib beneath, $1\frac{1}{2}$–2 ft. long, $1\frac{1}{2}$–2 in. broad in the lower half; peduncles 2–3, a foot long, ancipitous, densely pilose ; flowers 6–12 in a lax raceme ; lower pedicels very short ; bracts linear, villose ; perianth-limb $\frac{1}{4}$ in. long ; inner segments oblong, yellow ; outer oblong-lanceolate, densely pilose all over the back ; anthers lanceolate, $\frac{1}{4}$ in. long ; filaments short ; ovary obconic, densely villose, $\frac{1}{4}$ in. long ; stigmas concrete. *Walp. Ann.* i. 847 ; *Baker in Journ. Linn. Soc.* xvii. 119. *H. elata, Hook. fil. in Bot. Mag. t.* 5690, *non Roem. et Schultes.*

KALAHARI REGION : Basutoland, *Cooper,* 3242 !
EASTERN REGION : Tembuland ; near Eutwanazana, *Baur !*

39. H. costata (Baker in Journ. Linn. Soc. xvii. 119) ; corm globose, $1\frac{1}{2}$–2 in. diam. ; leaves 5–6, oblong or oblong-lanceolate, coriaceous, conspicuously and closely ribbed, 5–6 in. long at the flowering time, 1–$1\frac{1}{2}$ in. broad at the middle, densely ciliated with long hairs on the margins and keel beneath, otherwise glabrous ; peduncles 1–2, as long as the leaves, densely hairy ; flowers few, racemose ; lower pedicels very short ; bracts linear, villose ; perianth-limb $\frac{1}{2}$–$\frac{3}{4}$ in. long, the outer segments densely hairy on the back ; anthers lanceolate, $\frac{1}{4}$ in. long ; filaments short ; ovary obconic, densely hairy, $\frac{1}{4}$ in. long ; stigmas concrete.

KALAHARI REGION : Orange Free State ; Nelsons Kop, *Cooper,* 879 !

40. H. Galpini (Baker) ; corm not seen ; leaves lanceolate or oblong-lanceolate, erect, reaching above a foot in length and above an inch in breadth, erect, strongly ribbed, nearly naked ; peduncle very hairy, 6–8 in. long ; raceme many-flowered, 3–4 in. long ; pedicels short, ascending ; perianth-segments $\frac{5}{8}$–$\frac{3}{4}$ in. long, very hairy on the outside ; stamens less than half as long as the perianth-segments ; ovary turbinate, very hairy, $\frac{1}{4}$ in. long.

KALAHARI REGION : Transvaal ; Saddleback Range, near Barberton, 4000–4500 ft., *Galpin,* 1098 !

Very near *H. oligotricha,* except that the flowers are very hairy.

41. H. Rooperii (Moore in Gard. Comp. 1, 65, cum icone) ; corm globose, 2–3 in. diam., crowned with a ring of bristles ; leaves 12–18 to a corm, lanceolate, acuminate, 1–2 ft. long, 1–$1\frac{1}{2}$ in. broad in the lower half, moderately firm in texture, glabrous on the face when mature, shortly hairy on the back and margin ; peduncles 2–6, 9–12 in. long, ancipitous, hairy upwards ; flowers 4–10, racemose ; lower pedicels $\frac{1}{2}$–1 in. long ; bracts linear ; perianth-limb $\frac{3}{4}$ in. long ; segments oblong, yellow, the outer hairy all over the back ; anthers lanceolate, $\frac{1}{4}$ in. long ; filaments shorter ; ovary turbinate, $\frac{1}{4}$ in. long, densely villose ; stigmas concrete ; capsule turbinate, densely villose, $\frac{1}{2}$ in. long, circumscissile round the middle. *Lemaire, Jard. Fleur. t.* 303 ; *Baker in Journ. Linn. Soc.* xvii. 118.

VAR. β, **Forbesii** (Baker, loc. cit.); dwarfer, with leaves hairy only on the edge and keel beneath and a shorter perianth.

COAST REGION: Albany Div.; near Bushman River, below 1000 ft., *Drège*, 8529! Stockenstrom, *Scully*, 117! British Kaffraria, *Cooper*, 154! 3240!

KALAHARI REGION: Transvaal; Potchefstrom district, Mooi River, *Nelson*, 302! Houtbosch, *Rehmann*, 5810! plains around Barberton, 1800-2600 ft., *Galpin*, 1190! and without precise locality, *McLea in Herb. Bolus*, 5801!

EASTERN REGION: Griqualand East; mountains around Kokstad, 4300 ft., *MacOwan and Bolus, Herb. Norm.*, 1210! Natal; Mooi River Valley, 2000-3000 ft., *Sutherland!* Inanda, *Wood*, 373! and without precise locality, *Buchanan! Gerrard*, 1828! Var. β, Delagoa Bay, *Forbes!*

IV. HESSEA, Herb.

Perianth cut down nearly or quite to the ovary; segments subequal, patent, obtuse, 3-nerved in the middle. *Stamens* inserted at the base of the perianth-segments; filaments filiform or flattened towards the base; anthers small, globose, basifixed. *Ovary* globose, 3-celled; ovules few in a cell, superposed; style subulate or strumose towards the base, tricuspidate at the stigmatose apex. *Capsule* globose, membranous, loculicidally 3-valved. *Seeds* 1-3 in a cell, globose, greenish.

Rootstock a bulb with membranous tunics; leaves generally produced after the flowers, filiform, linear or lorate; flowers few or many in an umbel, small, reddish-purple; pedicels elongated; spathe-valves 2, linear or lanceolate.

DISTRIB. Endemic.

Subgenus HESSEA PROPER. *Style* not dilated at the base.
Stamens much shorter than the oblong perianth-segments:
 Style very slender, not swollen ... ... (1) **stellaris**.
 Style stout, rather swollen ... ... (2) **crispa**.
Stamens nearly or quite as long as the oblong-lanceolate perianth-segments:
 Perianth cut down to the ovary:
  Leaves lorate ... ... (3) **dregeana**.
  Leaves filiform ... ... (4) **Rehmanni**.
 Perianth with a short tube above the ovary:
  Perianth ¼-½ in. long ... ... (5) **Zeyheri**.
  Perianth ⅛ in. long ... ... (6) **brachyscypha**.
Subgenus IMHOFIA. *Style* swollen at the base.
Leaves filiform. Style-base much swollen:
 Peduncle straight ... ... (7) **filifolia**.
 Peduncle spiral ... ... (8) **spiralis**.
Leaves lorate. Style-base slightly swollen ... ... (9) **gemmata**.

1. **H. stellaris** (Herb. Amaryll. 289); bulb subglobose, ¾-1 in. diam., with a long neck and brown or pale membranous tunics; leaves 2, produced after the scape, lorate, glabrous, 6-12 in. long; scape lateral, 3-9 in. long, stout, for the genus; flowers 6-30; pedicels 1-2 in. long; spathe-valves lanceolate, green; perianth-segments oblong, crisped, purplish-red, ⅓-½ in. long; stamens half as long as the perianth-segments; ovary globose, ⅛-¼ in. diam.; style slender, not swollen, subulate, not dilated at the base. *Kunth, Enum.* v. 630; *Baker, Handb. Amaryll.* 21. *Amaryllis stellaris*,

Jacq. Hort. Schoenbr. i. 37, *t.* 71 ; *Willd. Sp. Plant.* ii. 61 ; *Ait. Hort. Kew. edit.* 2, ii. 229. *Strumaria stellaris, Gawl. in Bot. Mag. sub t.* 1363 ; *Roem. et Schultes, Syst. Veg.* vii. 790. *H. crispa, Kunth, Enum.* v. 632, *excl. syn. Amaryllis pulchella, Spreng. in herb. Zeyher. Periphanes stellaris, Salisb. Gen.* 118.

SOUTH AFRICA : without locality, *Thunberg! Zeyher!*
COAST REGION : Near Cape Town ; on the Lions Rump, *Burchell,* 8444! and on sand-dunes, *Bolus;* 4012! Cape Flats, near Claremont, *MacOwan,* 2567! and near Wynberg, *MacOwan and Bolus, Herb. Norm.,* 297!

2. **H. crispa** (Kunth, Enum. v. 632, ex parte); bulb globose, 1 in. diam., with tunics produced an inch above its neck ; leaves 2, produced after the scape, lorate, glabrous, $\frac{1}{2}$ ft. long ; scape lateral, 4–6 in. long; flowers 6–8 ; pedicels $1\frac{1}{2}$–2 in. long; spathe-valves green, reflexing, lanceolate acuminate ; perianth-segments oblong, much crisped, $\frac{1}{3}$ in. long ; stamens half as long as the perianth-segments ; ovary depresso-globose, $\frac{1}{8}$ in. diam. ; ovules 4–6 in a cell ; style cylindrical, as long as the stamens, slightly swollen, minutely tricuspidate at the apex. *Baker, Handb. Amaryllid.* 22. *Amaryllis crispa, Jacq. Hort. Schoenbr.* i. 37, *t.* 72 ; *Willd. Sp. Plant.* ii. 61 ; *Ait. Hort. Kew. edit.* 2, ii. 229. *Strumaria crispa, Ker in Bot. Mag. t.* 1363. *Imhofia crispa, Herb. Amaryll.* 290. *Amaryllis cinnamomea, L'Herit. Sert.* 16, *t.* 17. *Imhofia cinnamomea, Roem. Amaryllid.* 28. *Periphanes crispa, Salisb. Gen.* 118.

SOUTH AFRICA: without locality, *Herb. Gouan! Herb. Gay!*

3. **H. dregeana** (Kunth, Enum. v. 633); bulb globose, with several pale membranous tunics produced 1–$1\frac{1}{2}$ in. above its neck ; leaves 2, lorate, glabrous, enveloped at the base in a long green funnel shaped sheath, like that of a *Strumaria ;* peduncle 3–4 in. long; flowers 20–30 ; pedicels $1\frac{1}{2}$ in. long, all ascending; spathe-valves 2, green, lanceolate, reflexing ; perianth cut down to the ovary ; segments oblanceolate-oblong, $\frac{1}{4}$ in. long, under $\frac{1}{12}$ in. broad; stamens as long as the perianth-segments ; ovary globose, $\frac{1}{12}$ in. diam. ; ovules 2–3 in a cell : style subulate, as long as the stamens, slightly thickened downwards, minutely tricuspidate at the tip. *Baker, Handb. Amaryllid.* 22.

COAST REGION : Clanwilliam Div. ; between Berg Vallei and Lange Vallei, *Drège,* 252b!

4. **H. Rehmanni** (Baker, Handb. Amaryllid. 22); bulb globose, under 1 in. diam. ; outer tunics brown, produced an inch over its apex ; leaves very slender, subterete, glabrous, shorter than the peduncle ; peduncle slender, subterete, 4–8 in. long ; flowers 8–12 ; pedicels strongly angled, $\frac{1}{4}$–$\frac{3}{4}$ in. long ; spathe-valves 2, ovate or ovate-lanceolate, small, tinged with red ; perianth cut down to the ovary ; segments whitish, oblanceolate, much crisped, $\frac{1}{6}$ in. long ; stamens as long as the perianth-segments ; filaments white, filiform ; anthers purplish, globose; ovary globose, $\frac{1}{12}$ in. diam. ; style rather shorter than the stamens, distinctly tricuspidate at the apex.

KALAHARI REGION : Transvaal; Hooge Veld, Donkershock, *Rehmann*, 6549 ! near Johannesberg, *Miss Saunders*, 13 !

5. H. Zeyheri (Baker, Handb. Amaryllid. 22) ; bulb globose, with pale membranous tunics produced $1\frac{1}{2}$–2 in. above its neck ; leaves 2, contemporary with the flowers, linear, glabrous, $\frac{1}{2}$–1 ft. long, $\frac{1}{6}$ in. broad ; peduncle stout, 4–8 in. long ; flowers 20–30 ; pedicels $1\frac{1}{2}$–$2\frac{1}{2}$ in. long ; spathe-valves lanceolate, greenish, reflexing ; perianth-segments lanccolate, $\frac{1}{4}$–$\frac{1}{3}$ in. long, $\frac{1}{12}$ in. broad, joined in a very short tube at the base ; stamens as long as the perianth-segments, filaments connate at the base ; ovary globose, $\frac{1}{12}$ in. diam. ; style as long as the stamens, slender, minutely tricuspidate at the tip.

WESTERN REGION : Little Namaqualand ; Hardeveld, *Zeyher*, 1661 !

6. H. brachyscypha (Baker) ; bulb globose, $\frac{1}{2}$ in. diam., with thick, pale, outer tunics produced 2 in. above its neck ; leaves linear, glabrous ; peduncle slender, $1\frac{1}{2}$–2 in. long ; umbel 6–9 flowered ; spathe-valves linear, reflexed, tinged red, $\frac{1}{2}$ in. long ; pedicels stiffly erect, $\frac{3}{4}$–1 in. long ; perianth bright red, $\frac{1}{6}$ in. long, with a distinct tube and oblong segments ; stamens as long as the perianth ; anthers uniform ; ovary globose, $\frac{1}{8}$ in. diam.; style overtopping the anthers.

COAST REGION : Malmesbury Div. ; Schaapplaats near Hopefield, *Bachmann* !

7. H. filifolia (Benth. Gen. Plant. iii. 721) ; bulb globose, $\frac{1}{2}$ in. diam., with membranous tunics produced an inch above its neck ; leaves 2–5, produced after the flowers, filiform, very slender, glabrous, weak, sometimes 6–8 in. long ; peduncle very slender, 3–6 in. long, not twisted spirally ; flowers 4–12 ; pedicels slender, 1–2 in. long ; perianth-limb $\frac{1}{4}$ in. long, cut down to the base, whitish or red-purple ; segments oblong-lanceolate, subacute, 3-nerved in the centre ; stamens rather shorter than the perianth-segments ; ovary green, globose, $\frac{1}{12}$ in. diam. ; style as long as the stamens, subulate from an ovoid base, minutely tricuspidate at the tip ; capsule depresso-globose, membranous ; seeds 1–2 in a cell. *Baker, Handb. Amaryllid.* 22. *Imhofia filifolia, Herb. Ama·yll.* 290, *t.* 29, *fig.* 8; *Kunth, Enum.* v. 626. *Strumaria filifolia, Jacq. Ic.* ii. 14, *t.* 361 ; *Bot. Reg. t.* 440. *Leucojum strumosum, Thunb. Prodr.* 58 ; *Ait. Hort. Kew.* i. 407, *t.* 5. *Crinum tenellum, Linn. fil. Suppl.* 194.

SOUTH AFRICA : without locality. *Sparrmann ! Masson ! Oldenburg !*
COAST REGION : near Cape Town, *Thunberg !* Cape sand-dunes, *Bolus*, 4013 ! Cape Flats near Rondebosch, *MacOwan and Bolus, Herb. Norm.*, 294 ! Mac-Owan, 2655 ! Tulbagh Div. ; between Tulbagh and the Drostdy, *Burchell*, 1030 ! Worcester Div., Hex River Kloof, 1000–2000 ft., *Drège !*

8. H. spiralis (Baker, Handb. Amaryllid. 22, non Berg.) ; bulb globose, $\frac{1}{2}$ in. diam., with membranous tunics produced an inch above its neck ; leaves produced after the flowers, filiform, glabrous ; peduncle 2–4 in. long, very slender, spirally twisted ; flowers 2–8 ; pedicels slender, 1–2 in. long ; perianth-limb $\frac{1}{6}$ in. long, pale reddish ; segments oblong-lanceolate ; stamens nearly as long as the perianth segments ;

ovary green, globose ; style as long as the stamens, cylindrical from
an ovoid base ; capsule depresso-globose, membranous, deeply 3-lobed.
Strumaria spiralis, Zeyher, exsic. non Ait.

WESTERN REGION : Little Namaqualand ; Hardeveld, *Zeyher*, 1662 !

9. **H. gemmata** (Benth. Gen. Plant. iii. 721) ; bulb globose, 1 in.
diam., with pale brown membranous tunics, produced 1½–2 in. above
its neck ; leaves 2–3, developed after the flowers, lorate, flaccid,
½ ft. or more long, ¼–½ in. broad, pilose towards the margin ; peduncle
stout, ½–1 ft. long ; flowers 10–20 ; pedicels 1–3 in. long ; spathe-
valves 2, lanceolate, greenish, reflexing ; perianth-limb ¼–⅓ in. long,
cut down to the base ; segments oblanceolate-oblong, obtuse,
3-nerved in the centre ; stamens nearly as long as the perianth-
segments ; ovary green, globose, ₁/₁₂ in. diam. ; style as long as the
stamens, with a slightly dilated conical base ; capsule depresso-globose,
membranous, deeply 3-lobed, ¼ in. diam. *Baker, Handb. Amaryllid.*
23. *Strumaria gemmata, Ker in Bot. Mag. t.* 1620 ; *Poir. Encyc.
Suppl.* v. 259. *Imhofia` gemmata, Herb. Amaryll.* 291 ; *Kunth,
Enum.* v. 628. *I. burchelliana, Herb. Amaryllid.* 290, *t.* 29, *fig.* 5.
Kunth, Enum. v. 627. *I. bergiana, Kunth, loc. cit. Hessea
burchelliana, Benth. Gen. Plant. loc. cit.*

COAST REGION : Uniondale Div. ; Lange Kloof, *Burchell*, 4954 ! 4967 ! 4978 !
Uitenhage Div. ; Grass Ridge, between Coega R. and Sunday R., *Zeyher*,
4107 !
CENTRAL REGION : Graaff Reinet Div. ; Sneeuw Berg, 5000 ft., *Bolus*, 1817 !
Colesburg Div. ; near Hondeblats River, *Burchell*, 2702 ! between Riet Fontein
and Plettenbergs Beacon, *Burchell*, 2727 ! and without precise locality, *Bolus*,
1817 ! Albert Div., *Cooper*, 575 ! *Drège*, 3517 ! Hopetown Div., *Burchell*, 2683.

V. CARPOLYZA, Salisb.

Perianth with a short funnel-shaped tube above the ovary ; segments linear-oblong, subequal, patent, laxly 3-nerved in the middle.
Stamens inserted in the perianth-tube ; filaments short, filiform ;
anthers subglobose, minute, basifixed. *Ovary* globose, 3-celled ;
ovules few in a cell, superposed ; style filiform, tricuspidate at the
stigmatose apex. *Capsule* globose, membranous, loculicidally 3-valved.
Seeds few in a cell, subglobose, greenish.

DISTRIB. Monotypic, endemic.

1. **C. spiralis** (Salisb. Parad. t. 63) ; bulb ovoid, ¼–½ in. diam., with
a long neck ; tunics brown, membranous ; leaves 4–6, contemporary
with the flowers, subulate, very slender, 2–4 in. long ; peduncle filiform, spirally twisted, 2–4 in. long ; flowers 1–3, rarely 4–6 ; pedicels
1–2 in. long ; spathe-valves 2, small, linear, membranous ; perianth
pinkish-white ; tube ⅛–⅙ in. long ; segments ½ in. long ; stamens emerg-
ing but little from the perianth-tube ; ovary globose, ₁/₁₂ in. diam. ; capsule ¼ in. diam. *Salisb. Gen.* 119 ; *Herb. Amaryll.* 292, *t.* 29, *fig.*
9 ; *Baker, Handb. Amaryllid.* 23. *Hessea spiralis, Berg. in Linnæa*
1826, 252. *Strumaria spiralis, Ait. Hort. Kew. edit.* 2, ii. 213 ;

Gawl. in Bot. Mag. t. 1383. *Hæmanthus spiralis, Thunb. Prodr.* 58 ;
Fl. Cap. edit. Schult. 296. *Crinum spirale, Andr. Bot. Rep. t.* 92.
C. tenellum, Jacq. Ic. ii. 14, *t.* 363. *Amaryllis spiralis, L'Herit. Sert.
Angl.* 10, *t.* 13.

COAST REGION : vicinity of Cape Town ; Lion Mountain, *Thunberg ! Burchell*, 837–2 ! *MacOwan and Bolus, Herb. Norm.*, 295 ! and 1385 ! *MacOwan,*
2572, *Drège !* Sea Point, 200 ft., *Bolus*, 3730 !

VI. ANOIGANTHUS, Baker.

Perianth with a short infundibuliform tube above the ovary and 6
oblong-lanceolate, subequal, ascending segments. *Stamens* biserial, 3
inserted in the perianth-tube and 3 at its throat ; filaments filiform ;
anthers oblong, erect, deeply sagittate. *Ovary* oblong, 3-celled ;
ovules numerous, superposed ; style filiform, with 3 short, falcate,
clavate, stigmatose forks. *Capsule* globose, membranous, loculicidally
3-valved. *Seeds* numerous, flat.

DISTRIB. Endemic ; monotypic.

1. **A. breviflorus** (Baker in Journ. Bot. 1878, 76) ; bulb ovoid, 1 in.
diam., with a short neck and brown membranous tunics ; leaves 3–4,
contemporary with the flowers, lorate, obtuse, erect, glabrous, $\frac{1}{2}$–1 ft.
long, $\frac{1}{4}$–$\frac{1}{2}$ in. broad ; peduncles 3–12 in. long ; flowers 2–10, umbellate ;
pedicels erect, 1–2 in. long ; spathe-valves 2, lanceolate ; perianth
yellow or milk-white ; tube $\frac{1}{4}$–$\frac{1}{3}$ in. long ; segments $\frac{1}{2}$–$\frac{3}{4}$ in. long,
with about 5 ribs ; stamens half as long as the perianth-segments ;
ovary green, glabrous, oblong ; style overtopping the anthers ;
capsule $\frac{1}{2}$ in. long. *Bot. Mag. t.* 7072 ; *Handb. Amaryllid.* 27.
Cyrtanthus breviflorus, Harv. Thes. Cap. t. 139.

VAR. β, *minor* (Baker, Handb. Amaryllid. 28) ; dwarf, 1–2-flowered, with
short pedicels and narrow perianth-segments. *A. luteus, Baker in Journ. Bot.*
1878, 77. *Cyrtanthus luteus, Baker in* Journ. Bot. 1876, 66.

COAST REGION : Stockenstrom Div. ; Katberg, 4000–5000 ft., *Drège! Scully*,
150 ! Fort Beaufort Div., *Cooper*, 255 !
CENTRAL REGION : Somerset Div. ; Bosch Berg, *MacOwan*, 2133 ! Graaff
Reinet Div. ; Koudevelds Berg, 5000 ft., *Bolus*, 2577 !
EASTERN REGION : Tembuland ; mountains near Bazeia, 2500–3000 ft., *Baur*,
248 ! Griqualand East ; near Kokstad, 4700 ft., *Tyson*, 1548 ! Natal ; Inanda,
Wood, 175 ! Weston, *Rehmann*, 7358 ! Table Mountain, *Krauss*, 255 ! and without
precise locality, *Plant*, 106 ! *Sanderson*, 692 ! Swazieland ; Havelock Concession, 4000 ft., *Saltmarshe in Herb. Galpin*, 1097 ! Var. β, Tembuland ; mountains near Bazeia, 2500–3000 ft., *Baur*, 248 ! Natal ; Inanda, *Wood*, 230 !
and without precise locality, *Buchanan !*

VII. GETHYLLIS, Linn.

Perianth hypocrateriform, with a long, slender, cylindrical tube and
6 equal, spreading, oblong acute segments. *Stamens* 6 or many, inserted in a single row at the throat of the perianth-tube ; filaments
short, sometimes bearing 2–5 anthers ; anthers linear, basifixed, twisting up spirally when the flower expands. *Ovary* 3-celled, hidden by

the spathe and amongst the sheaths of the bulb-neck; ovules numerous, superposed; style long, filiform; stigma capitate. *Fruit* clavate, indehiscent, succulent. *Seeds* globose, immersed in pulp; testa loose, hyaline.

Acaulescent herbs, with the habit of *Crocus;* leaves generally contemporary with the fruit, produced after the flowers, which are delicate in texture, whitish, and fugitive; spathe monophyllous, membranous, amplexicaul, clasping the ovary and lower part of the perianth-tube; berry yellowish, fragrant, edible.

DISTRIB. Endemic.

Stamens 6 :
   Style falling short of the tip of the perianth-
     segments:
       Leaves glabrous, spirally twisted all down ...   (1) **spiralis.**
       Leaves glabrous, rolled back spirally
         towards the tip ...   .   (2) **verticillata.**
       Leaves densely hispid ...   .   (3) **villosa.**
   Style exserted, declinate ...   .   (4) **longistyla.**
Stamens numerous:
   Leaves linear, glabrous, spirally twisted:
     Stamens 9–12 ...   .   (5) **afra.**
     Stamens in 6 bundles, several in each bundle (6) **britteniana.**
   Leaves linear, hairy   ...   .   (7) **ciliaris.**
   Leaves lanceolate, very much crisped   ...   ...   (8) **undulata.**
   Leaves lorate, glabrous, twisted ...   .   (9) **latifolia.**

1. G. spiralis (Linn. fil. Suppl. 198); bulb globose, 1–1¼ in. diam.; tunics produced 1–2 in above its top; leaves 4–6, produced after the flower, linear-subulate, glabrous, very much spirally twisted all the way down, 4–6 in. long; perianth-tube 2–3 in. long; limb 1–1½ in. long, whitish, tinted red on the outside; segments oblong-lanceolate, ¼–⅓ in. broad at the middle; stamens 6, about ½ in. long; filament equalling the anther; style straight, rather overtopping the anthers; fruit clavate, 2–3 in. long, under ½ in. diam., narrowed to the base. *Thunb. Diss.* 15; *Prodr.* 59; *Fl. Cap. edit. Schult.* 302; *Bot. Mag. t.* 1088; *Herb. Amaryllid.* 185; *Baker in Journ. Bot.* 1885, 225; *Handb. Amaryllid.* 24. *Papiria spiralis, Thunb. in Act. Lund.* i. 2, 111.

SOUTH AFRICA : without locality, *Masson! Harvey! Burchell!*
COAST REGION: Cape Flats, *Rogers!* Worcester Div., *Cooper,* 1644! 1710! Uniondale Div.; Lange Kloof, *Thunberg!*

Flowers at the Cape in December and fruits about February. *G. rosea, Eckl. Top. Verz.* 4, is a small form with a red tinted perianth-limb.

2. G. verticillata (R. Br. Prodr. 290); bulb small, ovoid, with a spotted cylindrical neck 1½–2 in. long; leaves 4–5, narrow linear, glabrous, not spirally twisted, but rolled up like a watch-spring towards the tip, 4–6 in. long; perianth pure white; tube slender, 3–4 in. long; segments lanceolate, acute, reflexing, an inch long; stamens 6; filaments filiform, as long as the anthers; style a little overtopping the stamens; fruit yellow, clavate, contemporary with the leaves, 1½ in. long, ⅓ in. diam. *Roem. et Schultes, Syst.* vii. 781;

Herb. Amaryllid. 186, *t.* 25, *fig.* 6 ; *Kunth, Enum.* v. 697 ; *Baker in Journ. Bot.* 1885, 226, *t.* 259, *fig.* 2 ; *Handb. Amaryllid.* 24.

COAST REGION : Piquetberg, *Masson!*

3. **G. villosa** (Linn. fil. Suppl. 198) ; bulb small, ovoid ; tunics produced an inch or more above its apex ; leaves 5–10, developed after the flowers, linear, 2–3 in. long, ⅛ in. broad, twisted spirally, clothed with dense ascending or deflexed whitish and bristly hairs ; perianth-tube pilose, 2–4 in. long ; limb tinged with pink, about an inch long ; segments lanceolate, reflexing, ⅛ in. broad ; stamens 6, one-third the length of the segments ; anthers about as long as the filaments ; style straight, reaching to the top of the stamens ; fruit clavate, yellowish, contemporary with the leaves, 2 in. long, ½ in. diam. *Thunb. Diss.* 15 ; *Prodr.* 59 ; *Fl. Cap. edit. Schult.* 303 ; *Willd. Sp. Plant.* ii. 104 ; *Herb. Amaryllid.* 186 ; *Kunth, Enum.* v. 697 ; *Baker in Journ. Bot.* 1885, 226 ; *Handb. Amaryllid.* 24. *Papiria villosa, Thunb. in Act. Lund.* i. 2, 111, *cum icone.*

SOUTH AFRICA : without locality, *Masson!*
COAST REGION : Mossel Bay Div. ; Attaquas Kloof, near Ganse Kraal, *Thunberg!*

4. **G. longistyla** (Bolus in Journ. Linn. Soc. xviii. 396) ; bulb ovoid or subglobose ; inner tunics pale reddish, produced some distance above its apex ; leaves 12–18, produced after the flowers, linear from a dilated base, acuminate, ciliated, 3–4 in. long, ¼ in. broad, covered with linear, white, lacerated, centrally-affixed scales ; perianth-tube 2 in. long ; segments oblong-lanceolate, acuminate, about an inch long, ¼ in. broad ; stamens 6, about ½ in. long ; anther longer than the filament ; style exserted, ½–⅝ in. beyond the segments, stout, subangular, thicker gradually towards the base. *Baker in Journ. Bot.* 1885, 226 ; *Handb. Amaryllid.* 24.

CENTRAL REGION : Graaff Reinet Div. ; Sneeuw Berg Range, 4000 ft., *Tyson in Herb. Bolus,* 842 !

5. **G. afra** (Linn. Sp. Plant. 633) ; bulb globose, 1½–2 in. diam., with brown membranous tunics produced 2–3 in. over its apex ; leaves 12–20, as long as the flower, linear, twisted, strongly ribbed, glabrous ; perianth-tube 3–4 in. long ; limb whitish, 1½–2 in. long ; segments oblong or oblanceolate-oblong, acute, varying from ⅜ to ¾ in. broad ; stamens 9–12, about 1½ in. long ; anthers equalling the filaments ; style straight, rather overtopping the anthers ; fruit yellowish, clavate, edible, with an agreeable scent. *Lindl. Bot. Reg. t.* 1016 ; *Roem. et Schultes, Syst.* vii. 780 ; *Herb. Amaryllid.* 185 ; *Kunth, Enum.* v. 696 ; *Baker in Journ. Bot.* 1885, 226 ; *Handb. Amaryllid.* 24.

COAST REGION : Clanwilliam Div. ; Lange Vallei and Berg Vallei, *Zeyher,* 1663 ! Riversdale Div. ; near Heidelberg, *Burchell,* 7209 !

6. **G. britteniana** (Baker in Journ. Bot. 1885, 227, t. 260) ; bulb globose, 2–3 in. diam., with a thick spotted neck 1½ in. long ; leaves 12–15, linear, glabrous, firm in texture, spirally twisted,

4–6 in. long, $\frac{1}{12}$–$\frac{1}{8}$ in. broad; perianth pure white; tube stout, 2–3 in. long, its base hidden by the sheathing membranous bract; segments oblong-lanceolate, 2 in. long, $\frac{3}{4}$ in. broad; stamens very numerous, arranged in 6 clusters; filament about as long as the anther. *Handb. Amaryllid.* 24. *G. bivaginata, Masson MSS.*

CENTRAL REGION : Karoo, *Masson!*

I have described this from Masson's three sketches, and named it after Mr. Britten, to whom we are indebted for seeking out and rendering available for use the series of drawings by Masson on which I have mainly relied in characterizing the species of this genus. The same plant is in Herb. Bolus, but the locality is doubtful.

7. G. ciliaris (Linn. fil. Suppl. 198); bulb globose, $1\frac{1}{2}$ in. diam., with the tunics produced beyond its apex in a cylindrical sheath 4–5 in. long; leaves 20 or more, produced after the flowers, linear, spirally twisted, conspicuously ciliated; perianth-tube 2–3 in. long; limb whitish, $\frac{1}{2}$ in. long; segments oblong-lanceolate, $\frac{1}{3}$–$\frac{1}{2}$ in. broad; anthers numerous, linear, $\frac{1}{4}$–$\frac{1}{3}$ in. long; filaments very short, confluent; style straight, not longer than the stamens; fruit clavate, yellow, 2–3 in. long, above $\frac{1}{2}$ in. diam. *Thunb. Diss.* 15; *Prodr.* 59; *Fl. Cap. edit. Schult.* 302; *Jacq. Hort. Schoenbr.* i. 41, *t.* 79; *Herb. Amaryllid.* 185; *Kunth, Enum.* v. 696; *Baker in Journ. Bot.* 1885, 227; *Handb. Amaryllid.* 25. *Papiria ciliaris, Thunb. in Act. Lund.* i. 2, 111. *G. polyanthera, Solander MSS.*

SOUTH AFRICA : without locality, *Masson!*
COAST REGION : near Cape Town, in sandy places, *Thunberg!*

Jacquin describes this species as having 6 filaments, with 3 anthers to each. There is a single flower in Thunberg's herbarium marked *G. cuspidata*, with 6 filaments, with 2 anthers to about two of them, and one only to the others, and the same plant or a near ally is in Herb. Bolus, No. 4383, gathered in Basutoland in 1880. Solander's full original description will be found in a paper by Mr. Britten in Journ. Bot. 1884, p. 148.

8. G. undulata (Herb. Amaryll. 186, t. 25, fig. 5); bulb globose, 2–3 in. diam., with a neck 2–3 in. long; leaves 12–20, lanceolate, spreading, villose, much undulated, 5–6 in. long, $\frac{1}{3}$–$\frac{1}{2}$ in. broad; perianth pure white; tube stout, 2–3 in. long; segments oblong-lanceolate, $1\frac{1}{2}$ in. long; stamens numerous; filaments short, filiform; style not much overtopping the stamens. *Kunth, Enum.* v. 697; *Baker in Journ. Bot.* 1885, 227; *Handb. Amaryllid.* 25.

COAST REGION : Zeekoe Vallei, March, 1794, *Masson!*

9. G. latifolia (Masson ex. Baker in Journ. Bot. 1885, 228, t. 259, fig. 1); bulb globose, $1\frac{1}{2}$ in. diam. with a cylindrical neck 8–9 in. long; leaves 10–12, spreading, lorate, twisted, glabrous, 4–5 in. long, $\frac{1}{2}$ in. broad; perianth-tube stout, 2–3 in. long; segments pinkish, oblong-lanceolate, acute, 2 in. long, $\frac{1}{2}$ in. broad; stamens about 20, all distinct, $\frac{1}{4}$–$\frac{1}{2}$ in. long; filaments filiform, as long as the anthers. *Handb. Amaryllid.* 25.

WESTERN REGION : Little Namaqualand ; Meerhofs Kasteel, Sept., 1793, *Masson!*
Described from Masson's drawing.

VIII. APODOLIRION, Baker.

Perianth infundibuliform, with a long cylindrical tube rather dilated at the apex, and 6 subequal, ascending, oblong-lanceolate or lanceolate segments. *Stamens* biseriate, 3 inserted at the throat, and 3 below the throat of the perianth-tube; filaments filiform, about as long as the linear, rather recurving, basifixed anthers. *Ovary* 3-celled, hidden down amongst the sheaths of the bulb-neck; ovules numerous, superposed; style filiform, obscurely lobed at the stigmatose apex. *Fruit* and *Seeds* unknown.

Bulbous herbs, with the habit of *Crocus*; leaves usually not produced till after the flowers, linear or lanceolate; flowers white or red.

DISTRIB. Endemic.

Perianth-tube 1–1½ in. long :
   Anthers ½ in. long ... ... (1) **lanceolatum.**
   Anthers ¼ in. long ... ... (2) **Ettæ.**
Perianth-tube 2–4 in. long :
   Perianth-segments lanceolate ... ... (3) **Buchanani.**
   Perianth-segments oblong :
      Leaves plane ... ... (4) **Bolusii.**
      Leaves crisped ... ... (5) **Macowani.**
Perianth-tube 6–8 in. long ... ... (6) **Mackenii.**

1. **A. lanceolatum** (Benth. Gen. Plant. iii. 722); bulb globose, ½ in. diam., with brown membranous tunics produced 1–1½ in. over its apex; leaf solitary, lanceolate, contemporary with the flower, 2 in. long, ⅛ in. broad, narrowed from the middle to both ends, moderately firm in texture, crisped at the edge; perianth-tube an inch long; limb whitish, ¾–1 in. long; segments oblanceolate, acute, ⅛ in. broad; anthers linear, ¼ in. long; ovary hidden in the produced tunics of the bulb. *Baker, Handb. Amaryllid.* 26. *Gethyllis lanceolata, Linn. fil. Suppl.* 198; *Thunb. Diss.* 15; *Prodr.* 59; *Fl. Cap. edit. Schult.* 303. *Papiria lanceolata, Thunb. in Act. Lund.* i. 2, 112, *cum icone.*

SOUTH COAST : without locality, *Harvey !*
COAST REGION : Swellendam Div.; Buffeljaghts River, *Thunberg !*

2. **A. Ettæ** (Baker, Handb. Amaryllid. 26); bulb globose, ½ in. diam. ; outer tunics brown, membranous, produced an inch above its neck ; leaves linear, glabrous; peduncle and spathe hidden in the neck of the bulb; perianth-tube cylindrical, 3 in. long; limb 1–1¼ in. long, white, tinged with red; segments ⅙ in. broad; anthers in two superposed rows opposite the lower part of the segments, ¼ in. long.

KALAHARI REGION : Transvaal; mountain tops near Barberton, 4000-5000 ft., *Galpin,* 436 !
EASTERN REGION : Natal, on a grassy hill at Imbumbulu, Umlaas Location, *Wood,* 3193 !

3. **A. Buchanani** (Baker in Journ. Bot. 1878, 75) ; bulb globose, 1 in. diam, with the white membranous tunics produced 1–1½ in.

above its apex ; peduncle about ½ in. long; leaves unknown ; spathe
of one membranous valve an inch long, wrapped tightly round the
perianth-tube ; perianth-tube very slender, about 2 in. long ; limb
whitish, 1½ in. long ; segments oblanceolate, acute, ¼ in. broad above
the middle ; longer stamens, reaching nearly halfway up the
perianth-limb ; anthers linear, ¼ in. long. *Handb. Amaryllid.* 26 ;
Hook. Ic. t. 1388. *Cyphonema Buchanani, Baker in Journ. Bot.*
1876, 66.

EASTERN REGION : Natal; hills near Pietermaritzburg, *Krauss,* 449 ! and
without precise locality, *Buchanan !*

4. A. Bolusii (Baker in Journ. Bot. 1878, 75) ; bulb not seen ;
leaves 3, contemporary with the flower, linear, erect, glabrous, rather
twisted, with a green blade 3–4 in. long, ⅛ in. broad, dilated at the
base into a clasping hyaline membrane 2 in. long ; perianth-tube
2–3 in. long ; limb whitish, 1½–2 in. long; segments obovate-oblong,
subacute, permanently imbricated, ½ in. broad ; upper stamens reaching
almost halfway up the perianth-limb ; anthers linear, ¼–¾ in. long ;
ovary oblong, ½ in. long. *Handb. Amaryllid.* 26.

CENTRAL REGION : Cave Mountain near Graaff Reinet, 4300 ft., *Bolus,* 717 !

5. A. Macowani (Baker, Handb. Amaryllid. 26) ; bulb globose or
ovoid, 1 in. diam., with membranous whitish tunics produced
1–1½ in. over its top ; leaves produced after the flowers, only seen
in a young state, linear, glabrous, firm in texture, very much
crisped ; perianth-tube 3–4 in. long ; limb whitish, 1–1½ in. long ;
segments oblanceolate-oblong, subacute, ¼–⅓ in. broad above the
middle ; longer stamens reaching ⅓ up the perianth-limb ; anthers
linear, ⅛ in. long ; ovary hidden in the basal sheaths.

COAST REGION : Uitenhage Div. ; sandy places by Coega River, *MacOwan,*
1928 ! Port Elizabeth, *Holland in Herb. MacOwan,* 1928 !

6. A. Mackenii (Baker in Journ. Bot. 1878, 75) ; bulb globose,
½ in. diam., with white membranous tunics produced 1½ in. above its
apex ; leaves unknown; perianth-tube reddish, 6–8 in. long; limb
1¾–2 in. long, bright red ; segments oblanceolate-unguiculate, acute,
¼–⅓ in. broad above the middle ; anthers linear-oblong, ⅛ in. long.
Handb. Amaryllid. 26.

EASTERN REGION : Natal ; Great Noodsberg, 2500 ft., *McKen,* 1 ! *Wood,* 110 !
4147 ! Lorenço Marquez ; Lobombo Mountains, *Mrs. Saunders !*

IX. CRINUM, Linn.

Perianth (in the Cape species) infundibuliform, with a cylindrical
tube above the ovary, which is generally as long as the limb, and
6 equal, connivent, oblong segments. *Stamens* inserted at the throat
of the perianth-tube ; filaments long, filiform, declinate ; anthers
linear, dorsifixed, versatile. *Ovary* oblong, 3-celled ; ovules few in
a cell, superposed, sessile or immersed in the thick placenta ; style

long, filiform; stigma capitate. *Fruit* a globose capsule, bursting irregularly. *Seeds* few, large, bulbiform.

Bulb large, tunicated; leaves few or many, linear or lorate, persistent; flowers umbellate, white or tinged, or striped with red; spathe of two large deltoid valves.

DISTRIB. Species about 60, spread through the tropical and warm-temperate regions of both the Old and New worlds.

Perianth-tube short :
  Perianth-segments acute :
    Leaf linear ... ... (1) **lineare.**
    Leaf lorate, acuminate ... ... (2) **variabile.**
  Perianth-segments obtuse ... ... (3) **campanulatum.**
Perianth-tube 3–4 in. long :
  Perianth-segments obtuse ... ... (4) **imbricatum.**
  Perianth-segments subacute... ... ... (5) **Moorei.**
Perianth-segments acute :
  Perianth-segments white, with a distinct keel
    of red :
      Leaves ciliated ... ... (6) **forbesianum.**
      Leaves scabrous on the margin ... ... (7) **longifolium.**
  Perianth-segments pink, without a distinct keel
    of red ... ... (8) **Macowani.**

1. **C. lineare** (Linn. fil. Suppl. 195); bulb small, ovoid; leaves linear, 1½–2 ft. long, about ½ in. broad, glaucous green, weak in texture, channelled down the face, glabrous and entire on the margin; scape slender, subterete, about a foot long; flowers 5–6 in an umbel; pedicels ½–¾ in. long; spathe-valves small, lanceolate-deltoid; perianth-tube curved, slender, 1½–2½ in. long; segments 2–3 in. long, oblanceolate-oblong, acute, about ½ in. broad, tinged with red in the centre; stamens much shorter than the perianth-segments; anthers ⅓ in. long; ovary oblong, ½ in. long; style nearly straight, reaching to the top of the perianth-segments. *Thunb. Prodr.* 59; *Fl. Cap. edit. Schult.* 301; *Kunth, Enum.* v. 582; *Baker, Handb. Amaryllid.* 92. *C. revolutum, Herb. in Bot. Mag. sub t.* 2121; *Amaryllid.* 267. *Amaryllis revoluta, L'Herit. Sert. Angl. t.* 14; *Gawl. in Bot. Mag. t.* 915. *A. revoluta var. gracilior, Bot. Reg. t.* 623. *Crinum algoense, Herb. Amaryllid.* 272; *Kunth, Enum. loc. cit.*

COAST REGION : Uitenhage Div.; between Van Stadens River and Zwartkops River, *Thunberg!*

2. **C. variabile** (Herb. Amaryllid. 268, t. 44, fig. 23); bulb ovoid, 3–4 in. diam., without any distinct neck; leaves 10–12 to a bulb, linear, green, weak in texture, glabrous, entire on the margin, the outer reaching a length of 2 feet and more, and a breadth of 2 inches; scape erect, compressed, 1–1½ ft. long, ½ in. thick at the base; flowers 10–12 to an umbel; pedicels ½–1 in. long; spathe-valves deltoid, 2–3 in. long; perianth with a curved greenish tube about 1½ in. long; segments oblong, acute, 2½–3½ in. long, ½ in. broad, tinged with red down the keel; filaments red, an inch shorter than the perianth-segments; anthers ½ in. long; ovary oblong, ½ in.

long; style reaching to the tip of the perianth-segments. *Kunth,
Enum.* v. 578; *Baker, Handb. Amaryllid.* 92. *Amaryllis variabilis,
Jacq. Hort. Schoenbr.* iv. 14, *t.* 429 ; *Roem. et Schultes, Syst. Veg.* vii.
866. *A. revoluta, var. robustior, Gawl. in Bot. Reg. t.* 615. *Crinum
variabile, var. roseum, Herb. in Bot. Reg.* 1844, *t.* 9. *C. crassifolium,
Herb. App.* 23.

SOUTH AFRICA: without locality, *Thunberg!*
COAST REGION : British Kaffraria, *Cooper,* 3311 !

3. C. campanulatum (Herb. in Bot. Mag. sub t. 2121); bulb
small, ovoid; leaves linear, reaching a length of 3–4 ft., ½–1 in.
broad, 1–1½ in. broad at the base, weak in texture, deeply channelled
down the face, scabrous on the margin ; peduncle slender, a foot or
more long ; flowers 6–8 to an umbel ; pedicels ½–1 in. long ; spathe-
valves lanceolate-deltoid, reddish-brown, 1½–2 in. long ; perianth
with a slender, curved, cylindrical tube 1½–3 in. long ; limb campanulate, about as long as the tube ; segments oblong, obtuse, ½–¾ in.
broad, permanently connivent, bright rose-red ; filaments much
shorter than the perianth-segments ; anthers linear-oblong, ¼ in.
long ; style reaching to the tip of the perianth-segments. *Herb.
Amaryllid.* 270 ; *Kunth, Enum.* v. 580 ; *Baker, Handb. Amaryllid.*
92. *C. aquaticum, Herb. App.* 23 ; *Bot. Mag. t.* 2352; *Roem. et
Schultes, Syst. Veg.* vii. 866. *C. caffrum, Herb. in Sweet, Hort. Brit.
edit.* 3, 678. *Hæmanthus hydrophilus, Thunb. ex Roem. et Schultes,
Syst.* vii. 892. *Kunth, Enum.* v. 600.

COAST REGION: Bathurst Div.; between Kaffir Drift and Blaauw Krans,
Burchell, 3785! and between Port Alfred and Kaffir Drift, *Burchell,* 3838!
Albany, in ponds, *Zeyher!*

4. C. imbricatum (Baker in Gard. Chron. 1881, xvi. 760);
bulb very large, globose ; leaves lorate, very thin in texture, reaching
a length of at least 3 ft., 4 in. broad in the middle, narrowed to 2 in.
near the base, glabrous, the distinctly-marked veins connected by
numerous distinct cross-bars; peduncle a foot or more long, as thick
as a man's finger ; flowers 5–6 to an umbel ; pedicels about an inch
long ; spathe-valves membranous, 3–4 in. long ; perianth white, with
a slender cylindrical tube about 3 in. long ; limb campanulate, about
as long as the tube, with oblong, obtuse, permanently imbricated
segments 1¼–1½ in. broad ; filaments declinate, about an inch shorter
than the perianth-segments; anthers linear-oblong, ¾ in. long ; ovary
narrow-oblong, ½ in. long; style reaching to the tip of the perianth-
segments. *Baker, Handb. Amaryllid.* 92.

EASTERN REGION : Transkei ; Butterworth Forest, *MacOwan,* 2027 !

Nearly allied to the well-known tropical African *C. giganteum, Andr. Bot. Rep.*
t. 169.

5. C. Moorei (Hook. fil. in Bot. Mag. t. 6113); bulb ovoid,
½ ft. diam , copiously stoloniferous, with a neck often a foot or
more long ; leaves 12–15 to a bulb, spreading, spaced out, lorate,
bright green, thin in texture, 2–3 ft. long, 3–4 in. broad at the middle,
narrowed gradually to the point and to 2 in. near the base, not at

all ciliated on the margin ; peduncle green, moderately stout, 2–3 ft.
long; flowers 6–10 to an umbel; pedicels 1½–3 in. long; spathe-
valves very large, thin, greenish or red-tinted; perianth with a
greenish tube about 3 in. long, and a funnel-shaped pinkish limb of
about the same length, with oblong, subacute, connivent segments
1–1½ in. broad at the middle; filaments pink, declinate, an inch
shorter than the perianth-segments; anthers linear-oblong, ½ in.
long ; ovary oblong, ½ in. long ; style very slender, declinate, as long
as the perianth-segments. *Garden* 1881, 260, *with a coloured figure ;
Baker in Gard. Chron.* 1881, xvi. 760 ; *Handb. Amaryllid.* 93. *C.
makoyanum, Carrière in Rev. Hort.* 1877, 417, *fig.* 75. *C. Colensoi,
Mackenii and natalense, Hort.*

EASTERN REGION : Natal; Inanda, *Wood*, 1164 ! and without precise locality,
Mrs. K. Saunders.

Our first knowledge of it was from a sketch sent by Bishop Colenso in 1858.
C. Schmidti, Regel Gartenflora, vol. xxxi. (1882) 34, *t.* 1072, is probably a white-
flowered variety of this species.

6. C. forbesianum (Herb. Amaryllid. 267) ; bulb globose, 6–8 in.
diam., without any distinct neck ; leaves not fully developed till after
the flowers, 10–12 to a bulb, spreading, lorate, obtuse, closely veined,
glaucous, 3–4 ft. long, 3–4 in. broad, with a conspicuously ciliated
margin ; peduncle stout, compressed, pale green, 9–12 in. long, an
inch thick ; flowers 30–40 to an umbel, faintly scented ; pedicels
stout, ½–¾ in. long; spathe-valves deltoid, red-tinted, 3–4 in. long ;
perianth-tube about 3 in. long, nearly straight in the central flowers
of the umbel, curved in the outer ones ; limb funnel-shaped, 4–4½
in. long ; the oblong, acute, connivent segments ¾–1 in. broad, white,
distinctly banded with red down the back ; filaments bright red,
nearly as long as the perianth-segments ; anthers about ½ in. long ;
ovary oblong, ¾ in. long ; style reaching to the tip of the perianth-
segments. *Kunth, Enum.* v. 577 ; *Baker in Bot. Mag. t.* 6545 ;
Handb. Amaryllid. 93. *Amaryllis Forbesii, Lindl. in Trans. Hort.
Soc.* vi. 87.

EASTERN REGION : Delagoa Bay, *Forbes, Monteiro*, 53 !

7. C. longifolium (Thunb. Prodr. 59) ; bulb ovoid, 3–4 in. diam.,
with only a short neck ; leaves a dozen or more to a bulb, lorate
acuminate, narrowed gradually to the apex, glaucous, firmer in
texture than in *zeylanicum* and *Moorei*, suberect in the lower half,
the outer reaching a length of 2–3 ft. and a breadth of 2–3 in. ;
distinctly scabrous on the margins; peduncle a foot or more long,
but little compressed ; flowers 6–12 or more to an umbel; pedicels
1–2 in. long ; spathe-valves about 3 in. long ; perianth generally more
or less tinged with red, rarely pure white, with a curved cylindrical
tube 3–4 in. long, and a limb of about the same length, with oblong
acute segments ¾–1 in. broad ; stamens nearly as long as the perianth-
segments ; filaments bright red ; anthers linear-oblong, ⅓–½ in. long ;
style reaching to the tip of the perianth-segments. *Thunb. Fl. Cap.*

edit. Schult. 302, *non Roxb.*; *Baker, Handb. Amaryllid.* 93. *Amaryllis longifolia, Linn. Sp. Plant.* 421 ; *Ait. Hort. Kew.* i. 419 ; *Jacq. Ic. t.* 362 ; *Red. Lil. t.* 347 ; *Gawl. in Bot. Mag. t.* 661 ; *Bot. Reg. tt.* 303 *and* 546. *C. capense, Herb. Amaryllid.* 269 ; *Kunth, Enum.* v. 579. *Amaryllis capensis, Miller, Gard. Dict. edit.* 8, *No.* 12. *A. bulbisperma, Burm. Prodr. Cap.* 9. *Crinum riparium, Herb. App.* 23.

COAST REGION: near Cape Town, between the foot of Lion Mountain and the sea-shore, *Thunberg!* Cathcart Div.; between Windvogel Berg and Zwart Kei 3000–4000 ft., *Drège*, 4519b !

CENTRAL REGION : Colesberg Div., *Shaw!*

WESTERN REGION : Little Namaqualand; near the mouth of the Orange R. below 600 ft., *Drège*, 2653 !

KALAHARI REGION : Hopetown Div.; inundated banks of Orange River, *Burchell*, 2662! Bechuanaland; between Hamapery, near Kuruman, and Kosi Fontein, *Burchell*, 2525 !

EASTERN REGION : Transkei; Gekau, below 4000 ft., *Drège*, 4519a! Natal; Klip River, 3500–4500 ft., *Sutherland!*

It is the most hardy in English gardens, and the most extensively cultivated of any species of the genus, and is one of the parents of numerous hybrids, of which the following have received Latin names :—*C. Goweni*, a hybrid with *zeylanicum*; *C. Mitchamiæ*, with *pedunculatum*; *C. Herberti*, with *scabrum*; *C. Puseyæ*, with *latifolium*; *C. Wallichii*, with *careyanum*; *C. Szymouri*, with *lineare*; *C. Roxburghii*, with *defixum*; *C. altaclaræ*, with *erubescens*; *C. Shepherdi*, with *cruentum*; and *C. Powellii*, with *Moorei*. *C. Lesemanni, Beck. in Wein. Illust. Gart. Zeit.* 1876, 125, t. 1, is identical with *C. Powellii*.

8. C. Macowani (Baker in Gard. Chron. 1878, ix. 298); bulb globose, reaching a diameter of 9–10 inches, with a neck 6–9 in. long; leaves 12–15 to a bulb, spreading, lorate, bright green, thin in texture, 2–3 ft. long, 3–4 in. broad in the middle, slightly scabrous on the margin ; peduncles often 2–3 to a bulb, green, compressed, 2–3 ft. long, an inch thick at the top ; flowers 10–15 to an umbel ; spathe-valves very large ; pedicels 1–2 in. long ; perianth-tube curved, cylindrical, greenish, 3–4 in. long ; limb funnel-shaped, of about equal length, with oblong, very acute pinkish, segments 1–1½ in. broad at the middle ; stamens red, declinate, a little shorter than the perianth-segments ; anthers about ½ in. long ; ovary oblong, ½ in. long ; style not longer than the stamens. *Bot. Mag. t.* 6381 ; *Handb. Amaryllid.* 94.

EASTERN REGION : Transkei; near Butterworth, *MacOwan and Bolus, Herb. Norm.*, 508! Griqualand East, *MacOwan*, 2122 ! Natal; Upper Tugela, *Wood*, 5614 !

Both this and *C. Moorei* are closely allied to the Asiatic *C. latifolium, Linn. C. zeylanicum, Linn.*, which differs by its shorter pedicels and perianth-segments with a distinct keel of red down the centre, may perhaps also occur within the Cape area.

X. AMARYLLIS, Linn.

Perianth infundibuliform, with a short funnel-shaped tube and 6 subequal large, oblong, acute, ascending segments. *Stamens* inserted at the throat of the perianth-tube ; filaments long, filiform, declinate ; anthers linear, versatile. *Ovary* oblong, 3-celled ; ovules many, superposed; style filiform, declinate; stigma capitate. *Capsule* globose, bursting irregularly. *Seeds* few, globose, bulbiform.

The name *Amaryllis* is universally given in gardens to the species and numerous hybrids of the large American genus *Hippeastrum*, which differs from *Amaryllis* in its fruit and seeds.

DISTRIB. Endemic; monotypic.

1. A. Belladonna (Linn. Sp. Plant. edit. 2, 421); bulb the size of a swan's egg ; tunics fibroso-membranous ; leaves 7–9, produced after the flowers, lorate, bifarious, dull green, 1–1½ ft. long, ½–¾ in. broad ; scape a foot or more long, solid, compressed ; flowers 6–12 in an umbel, produced in spring in a wild state ; pedicels 1–1½ in. long ; spathe-valves 2, large, deltoid ; perianth rose-red or whitish, 3–3½ in. long ; tube under ½ in. long ; segments permanently connivent, oblong, acute, ½–¾ in. broad above the middle ; stamens rather shorter than the limb ; ovary green, oblong ; style as long as the limb ; capsule globose, membranous, about 1 in. diam. *L'Herit. Sert. Angl.* 12 ; *Ait. Hort. Kew.* i. 417 ; *Bot. Mag. t.* 733 ; *Red. Lil. t.* 180 ; *Herb. Amaryll.* 275 ; *Salisb. Gen.* 117 ; *Kunth, Enum.* v. 601 ; *Baker, Handb. Amaryllid.* 95. *A. rosea, Lam. Encyc.* i. 122. *Callicore rosea, Link. Handb.* i. 193. *A. pallida, Red. Lil. t.* 479. *A. pudica, Gawl. in Journ. Sci. & Arts* ii. 348. *Coburgia Belladonna and pudica, Herb. in Bot. Mag. sub t.* 2113. *C. pallida, Herb. in Trans. Hort. Soc.* iv. 181.

VAR. β, **A. blanda** (Ker in Bot. Mag. t. 1450); leaves longer and broader, reaching 2–3 ft., sheathing each other at the base; pedicels 2–3 in. long; flowers larger, lighter coloured than in the type, and opening wider, with an infundibuliform tube ½–¾ in. long, and segments 3–4 in. long, 1–1¼ in. broad. *Herb. Amaryll.* 277, t. 36, *fig.* 10; *Kunth, Enum.* v. 601. *Coburgia blanda, Herb. in Bot. Mag. sub t.* 2113.

SOUTH AFRICA : without locality, *Thunberg!*
COAST REGION : near Cape Town, on the sides of mountains, *MacOwan,* 2561 ! Camps Bay, *Burchell,* 857 ! Cape Flats near Wynberg, *MacOwan and Bolus, Herb. Norm.,* 296 ! near Bredasdorp, *Kitching!*

A. pallida, Red., figured also in Bot. Reg. t. 714, is a form with whitish flowers. *Hæmanthus longifolius, Hort. Nymph.,* mentioned in *Kunth, Enum.* v. 600, is a spurious species made up from a mixture of *Amaryllis* with *Crinum variabile.*

XI. **AMMOCHARIS,** Herb.

Perianth infundibuliform, with a short cylindrical tube above the ovary, and 6 equal, ascending, oblanceolate unguiculate segments. *Stamens* inserted at the throat of the perianth-tube, about as long as the segments ; filaments filiform, equally divergent ; anthers linear-oblong, dorsifixed, versatile. *Ovary* ampullæform, narrowed into a distinct neck, 3-celled ; ovules numerous, superposed, sessile ; style filiform, a little longer than the perianth-segments ; stigma capitate. *Fruit* unknown.

Rootstock a large tunicated bulb ; leaves many, bifarious, produced before the densely umbellate flowers.

HAB. Endemic.

1. A. falcata (Herb. App. 17; Amaryll. 241); bulb ovoid,
reaching a diameter of 6–9 in., not produced into a neck; tunics
very numerous, brown, membranous; leaves 6–12, bifarious, lorate,
spreading, produced before the flowers, in summer or autumn,
reaching a length of 1–2 ft., about an inch broad, green, glabrous;
peduncle stout, lateral, ancipitous, ½–1 ft. long; flowers 20–40
in an umbel, fragrant, bright red, produced in December in the
wild plant; pedicels 1–1½ in. long; spathe-valves 2, large, deltoid,
perianth-tube ½ in. long; segments 1½–2 in. long, narrowed gradually
into a filiform claw, reflexing at the top; ovary ⅓ in. long. *Baker,
Handb. Amaryllid.* 96. *Amaryllis falcata, L'Herit. Sert. Angl.* 13;
Ait. Hort. Kew. i. 418. *Hæmanthus falcatus, Thunb. Prodr.* 58;
Fl. Cap. edit. Schult. 297. *Crinum falcatum, Jacq. Hort. Vind.*
iii. 34, *t.* 60. *A. coranica, Herb. App.* 17; *Amaryllid.* 241, *t.* 44,
fig. 22; *Kunth, Enum.* v. 613. *Amaryllis coranica, Burchell in
Bot. Reg. t.* 139 *and t.* 1219 (*var. pallida*). *Palinetes falcata and
coranica, Salisb. Gen.* 116.

COAST REGION: near Cape Town, *Thunberg!* Uitenhage Div.; Coega River,
MacOwan, 1850! Cathcart Div.; Blesbok Flats, near Windvogel Mountain,
3000 ft., *Drège,* 8544!
CENTRAL REGION: Beaufort West Div.; Nieuwveld Mountains, 3000–5000 ft.,
Drège, 8543! Somerset Div.; Bosch Berg, 2800 ft., *MacOwan,* 1815! meadows
near Somerset East, *MacOwan,* 1850! Albert Div.; by the Orange River,
Burke!
KALAHARI REGION: Griqualand West; Klip Fontein, *Burchell,* 2638! Orange
Free State; by the Caledon River, *Burke!* Bechuanaland; between Hama-
pery, near Kuruman and Kosi Fontein, *Burchell,* 2532!
EASTERN REGION: Natal; Pietermaritzburg and Mooi River Valley, 2000–
3000 ft., *Sutherland!*

2. A. coccinea (Pax in Engl. Jahrb. x. 3); bulb ovoid-
spherical, 8 in. diam.; leaves lorate, subglaucescent, striated, an
inch broad, crisped at the edge; peduncle ½ ft. or more long;
umbel ½ ft. or more in diameter; spathe-valves large, ovate-
lanceolate; pedicels above an inch long; flowers bright crimson,
produced in December; perianth-tube above ½ in. long; segments
oblong-lanceolate, three times the length of the tube.

KALAHARI REGION: Griqualand West, in sandy soil at Kimberley Boshof,
3500 ft., *Marloth,* 784.

XII. BRUNSVIGIA, Heist.

Perianth suberect or decurved, cut down to a campanulate or
short cylindrical tube above the ovary; segments subequal, falcate,
oblanceolate or oblanceolate-oblong. *Stamens* more or less declinate,
inserted in the perianth-tube; filaments filiform, three shorter;
anthers oblong, dorsifixed, versatile. *Ovary* turbinate, 3-celled;
ovules numerous, superposed; style filiform, declinate; stigma
capitate. *Capsule* large, turbinate, acutely triquetrous, narrowed
gradually into the long ancipitous pedicel, loculicidally 3-valved.
Seeds subglobose, with a conspicuous funiculus.

Rootstock a large tunicated bulb; leaves produced after the flowers, lorate or lingulate; peduncle robust, solid; flowers bright red, umbellate; spathe 2-valved.

DISTRIB. Endemic.

Leaves lorate, 1–2 in. broad:
    Perianth-tube cylindrical, ¼ in. long ... ... (1) **Josephinæ.**
    Perianth-tube campanulate, very short :
    Perianth about 2 in. long ... ... (2) **grandiflora.**
    Perianth 1–1½ in. long :
        Stamens shorter than the perianth-segments (3) **slateriana.**
        Stamens as long as the perianth-segments ... (4) **minor.**
Leaves lingulate or oblong :
    Perianth 2–2½ in. long :
        Leaf 2–2½ in. broad ... ... (5) **gigantea.**
        Leaf 3–4 in. broad :
        Ovary turbinate ... ... (6) **Cooperi.**
        Ovary subglobose... ... ... (7) **sphærocarpa.**
    Perianth 1–1¼ in. long :
        Umbel many-flowered :
        Leaf 2–3 in. broad ... ... (8) **striata.**
        Leaf 4–5 in. broad ... ... (9) **natalensis.**
        Umbel 3–5 flowered ... ... (10) **radula.**
Imperfectly known ... ... (11) **radulosa.**

1. B. Josephinæ (Gawl. in Bot. Reg. t. 192–3); bulb ½ ft. diam.; tunics brown, membranous; leaves 8–10, lorate, glaucous-green or greenish, suberect, 1½–2 in. broad, thick in texture, closely ribbed, reaching a length of 2, or even 3 ft.; peduncle 1½ ft. long, subterete, an inch or more in diameter; flowers usually 20–30, sometimes 50–60 in an umbel; pedicels 6–12 in. long; spathe-valves small, deltoid, cuspidate; perianth bright red, 2½–3 in. long, with a curved cylindrical tube about ½ in. long, and lanceolate segments; stamens as long as the perianth-segments; style finally a little exserted beyond them; anthers linear-oblong, ¼ in. long; ovary oblong or turbinate, ¾–1 in. long; capsule turbinate, smaller and less strongly angled than in *B. gigantea. Kunth, Enum.* v. 607; *Baker, Handb. Amaryllid.* 97. *Amaryllis Josephinæ, Red. Lil. tt.* 370–372. *A. josephiniana, Herb. Amaryllid.* 278. *A. griffiniana, Roem. Amaryll.* 113. *B. grandiflora and glauca, Salisb. Gen.* 117.

COAST REGION : Uitenhage Div.; between Galgebosch and Melk River, *Burchell,* 4757 ! and without precise locality, *Cooper,* 1580 !
CENTRAL REGION : Albert Div., *Cooper,* 1396 !
KALAHARI REGION : Orange Free State, *Cooper,* 3228 !
EASTERN REGION : Natal ; Zuurberg, *Wood,* 914 ! near Durban, *Wood,* 259 ! Tugela Valley, *Allison,* 20 ! and without precise locality, *Cooper,* 1003 !

2. B. grandiflora (Lindl. Bot. Reg. t. 1335) ; bulb large, ovoid ; leaves lorate, obtuse, pale green, suberect, subscabrous on the margin ; peduncle slightly glaucous, compressed, 1½ ft. long ; flowers, about 30 in an umbel ; pedicels 3–6 in. long ; spathe-valves deltoid, cuspidate ; perianth about 2 in. long, pale red ; tube scarcely any ; segments oblong-lanceolate ; stamens scarcely as long as the perianth-

segments; ovary turbinate, ½ in. long ; style declinate. *Kunth, Enum.*
v. 609 ; *Baker, Handb. Amaryllid.* 97.

SOUTH AFRICA : without locality.

Described from the figure cited, which was drawn from a plant that flowered
in London with Mr. Tate in 1829. Cooper's 1530 from the division of Albany,
and a specimen of old date received from Prof. MacOwan, perhaps belong here ;
but in both cases the leaf is absent. Herbert's *var. banksiana* is no doubt
founded on a mis-matching of leaf and flower.

3. B. slateriana (Benth. Gen. Plant. iii. 727) ; bulb globose, 4–5
in. diam. ; tunics membranous, pale brown ; leaves 6, lorate, suberect,
glaucescent, 1 in. broad when young, not developed till after the
flowers ; peduncle stout, compressed, ½ ft. long, above ½ in. diam. ;
flowers 15–20 in an umbel ; pedicels 3–4 in. long ; spathe-
valves lanceolate-deltoid, shorter than the pedicels ; perianth-limb
horizontal, bright rose-red, ½ in. long, cut down nearly to the
ovary ; segments lanceolate, ¼ in. broad ; stamens and style declinate,
shorter than the perianth-segments ; ovary turbinate, ¼ in. long.
Baker, Handb. Amaryllid. 97. *Ammocharis slateriana, Kunth,
Enum.* v. 613. *Amaryllis banksiana, Lindl. in Bot. Reg.* 1842,
t. 11.

SOUTH AFRICA : without locality.

Known only from the figure cited, which was drawn from a plant flowered by
J. H. Slater, Esq., at Newick Park, Uckfield, in 1849.

4. B. minor (Lindl. in Bot. Reg. t. 954) ; bulb ovoid, about
2 in. diam., with pale brown tunics ; leaves 3–4, produced after the
flowers, lorate, about half a foot long, above an inch broad ; peduncle
6–9 in. long, ½–¾ in. diam. ; flowers 12–40 to an umbel, pale red ;
pedicels 3–6 in. long ; spathe-valves ovate, tinged red, under 2 in.
long ; perianth 1¼–1½ in. long, with a very short tube and
lanceolate acute segments ; stamens declinate, as long as the
perianth-segments ; anthers oblong, ⅛ in. long ; ovary turbinate,
⅓ in. long ; capsule turbinate, 1½ in. long. *Herb. Amaryllid.*
281 ; *Kunth, Enum.* v. 610; *Baker, Handb. Amaryllid.* 97. *B.
humilis, Ecklon Exsic.*

COAST REGION : Uniondale Div. ; Lange Kloof, *Burchell*, 4944! Uitenhage
Div., *Cooper*, 1587 !
CENTRAL REGION : Somerset Div. ; Bosch Berg, *MacOwan*, 507 !
WESTERN REGION : Little Namaqualand ; near Meerhofs, *Zeyher*, 4110 !

A plant from Klip Fontein, in Griqualand West, *Burchell*, 2628 ! and two
gathered by Drège (8546b, at Wupperthal, in the province of Clanwilliam ; and
8546a, for which no locality is given), and a plant sketched by Masson about
1795, of which the leaves are unknown, are either this or a near ally.

5. B. gigantea (Heist. Monogr. cum icone); bulb the size of a
child's head ; tunics brown, membranous ; leaves about 4, spreading,
lingulate, 3–5 in. broad, obtuse, closely ribbed, moderately firm in
texture, usually not more than 6–9 in. long, but sometimes, in
cultivation, reaching 12–15 in. ; scape red or green, 8–12 in. long,

as thick as a man's finger; flowers 20–30 in an umbel; pedicels stout, strongly ribbed, 4–6 in. long; spathe-valves deltoid, much shorter than the pedicels; perianth bright red, decurved, 2–2½ in. long, with a campanulate tube ¼ in. long, and lanceolate segments; stamens about as long as the perianth; anthers oblong, ⅛ in. long; ovary turbinate, ¾ in. long in the flowering stage; style finally exceeding the stamens; capsule turbinate, acutely triquetrous, 2–3 in. long, 1–1¼ in. diam. *Baker, Handb. Amaryllid.* 98. *B. multiflora, Ait. Hort. Kew. edit.* 2, ii. 230, *Guwl. in Bot. Mag. l.* 1019; *Herb. Amaryllid.* 280, *t.* 36, *fig* 1; *Kunth, Enum.* v. 606. *Amaryllis orientalis, Linn. Sp. Plant.* 422; *Jacq. Hort. Schoenbr.* i. 38, *t.* 74. *Hæmanthus orientalis, Thunb. Prodr.* 59; *Fl. Cap. edit. Schult.* 298. *B. orientalis, Eckl. Top. Verz.* 7. *Coburgia multiflora, Herb. in Bot. Mag. sub t.* 2213. *B. rubricaulis, Roem. Amaryllid.* 57.

CoAST REGION: Malmesbury Div.; Zwartland, *Thunberg!* Cape Div.; sand-dunes near Constantia and Zeekoe Vallei, *Thunberg!* at the foot of Muizenberg, near Kalk Bay, 100 ft., *Bolus,* 4526! Caledon Div.; near Genadendal, *Burchell,* 7930!

CENTRAL REGION : Colesberg, *Shaw!*

I have not seen *B. albiflora, Eckl. Top. Verz.* 7 (name only), founded on an unlocalized specimen gathered by Brehm about 1795.

6. **B. Cooperi** (Baker in Saund. Ref. Bot. t. 330); bulb ovoid, 3–4 in. diam. ; tunics pale brown, membranous; leaves 4–6, lingulate, spreading almost horizontally, 3½–4 in. broad, 9–12 in. long, thick in texture, closely veined, minutely granulated, the margin thickened and scabrous; peduncle terete, an inch thick, 1½ ft. long; flowers 12–16 in an umbel, bright red; pedicels 3–6 in. long; spathe-valves ovate, 3 in. long; perianth 2–2½ in. long, with a very short tube and oblong-lanceolate segments ⅓–½ in. broad at the middle; stamens as long as the perianth-segments; anthers linear-oblong, ¼ in. long; ovary turbinate, ¾ in. long; style finally a little exserted; capsule turbinate, acutely angled, 1½–2 in. long. *Baker, Handb. Amaryllid.* 98.

CENTRAL REGION: Graaff Reinet Div. ? ; plain on the northern side of the Sneeuw Berg Range, 4500 ft., *Bolus,* 1816 ! Aliwal North Div.; Witte Bergen, 5000-6000 ft., *Drège,* 3518 !

KALAHARI REGION : Orange Free State, *Cooper,* 881 !

7. **B. sphærocarpa** (Baker); leaves lingulate, obtuse, smooth, glabrous, a foot long, 3 in. broad at the middle; margin entire; pedicels 8–9 in. long; perianth bright red, 2½–2¾ in. long, with a short tube and lanceolate segments ¼ in. broad; stamens as long as the perianth-segments; anthers oblong, ⅛ in. long; ovary subglobose, ½–¾ in. diam., not at all acutely angled.

EASTERN REGION : Griqualand East, alt. 6000 ft., *Tyson,* 1268!

Very near *B. Cooperi.*

8. **B. striata** (Ait. Hort. Kew. edit. 2, ii. 231); bulb ovoid, the size of a walnut; leaves 3–6, produced after the flowers, oblong or

lingulate, spreading or suberect, 4–6 in. long, 2½ in. broad, subcoriaceous, with an entire thickened scabrous margin and close strong
ribs; peduncle 6–8 in. long, ½ in. diam.; flowers 10–20 in a dense
umbel, ½ ft. diam.; pedicels slender, 1–3 in. long; spathe-valves
ovate, 1–1½ in. long; perianth bright red or rose-red, erect, 1–1¼ in.
long, with a cylindrical, straight tube about ¼ in. long and oblong-
lanceolate, acute segments ¼ in. broad; stamens distinctly exserted;
ovary turbinate, under ½ in. long; style finally exserted ½–¾ in.;
capsule turbinate, triquetrous. *Herb. Amaryll.* 281 ; *Kunth, Enum.*
v. 609; *Baker, Handb. Amaryllid.* 98.   *Amaryllis striata, Jacq.*
Hort. Schoenbr. i. 36, *t.* 70.   *A. nervosa, Poir. Encyc. Suppl.* i
321.

SOUTH AFRICA: without locality, *Masson ! Villette !*
Sketched by Masson in 1788.

9. B. natalensis (Baker); bulb large; leaves oblong, subcoriaceous, scabrous and glaucous on both sides, 6–7 in. long, 4–4¼ in.
broad, not ciliated on the margin; peduncle stout, much longer than
the leaves; umbel densely many-flowered; spathe-valves large, ovate;
pedicels 4–5 in. long; perianth-segments lanceolate, deep pink,
united in a short cup at the base, above an inch long; stamens
shortly exserted; anthers small, oblong; ovary turbinate.

EASTERN REGION: Natal; South Downs, Weenen County, 5000 ft., *Wood,*
4421!

10. B. Radula (Ait. Hort. Kew. edit. 2, ii. 230); bulb globose,
larger than a hazel nut; leaves 2, developed after the flowers, lingulate, obtuse, spreading, 2–3 in. long, 1–1¼ in. broad, covered all over
the surface with raised, rough papillæ; peduncle ancipitous, not
more than 2–3 in. long; flowers 3–5 to an umbel, nearly horizontal,
inodorous; pedicels 1–1½ in. long; spathe-valves ovate, reddish,
½ in. long; perianth ¾–1 in. long, cut down very nearly to the ovary;
segments lanceolate; stamens declinate, equalling the perianth-
segments; ovary turbinate; style as long as the stamens. *Herb.*
Amaryllid. 281; *Kunth, Enum.* v. 610; *Baker, Handb. Amaryllid.*
98. *Amaryllis Radula, Jacq. Hort. Schoenbr.* i. 35, *t.* 68. *Coburgia*
Radula, Herb. in Trans. Hort. Soc. iv. 181.

WESTERN REGION: Little Namaqualand; Karroo, near Olifants River,
Masson !
Sketched by Masson in 1790.

11. B. radulosa (Herb. Amaryllid. 281, *t.* 22, fig. 2); leaves
spreading, lingulate, obtuse, 8–9 in. long, 3–3½ in. broad at the
middle, thick in texture, closely ribbed, rough all over the surfaces,
the margin thickened and scabrous; flowers and fruit unknown.
Kunth, Enum. v. 610; *Baker, Handb. Amaryllid.* 98.

CENTRAL REGION: Colesberg Div.; near the Hondeblats River, *Burchell,*
2703–3!

XIII. NERINE, Herb.

Perianth infundibuliform, cut down nearly or quite to the ovary, erect or rather decurved; segments equal, oblanceolate, falcate, more or less crisped. *Stamens* inserted at the base of the perianth-segments; filaments filiform, thickened at the base, suberect or declinate, 3 shorter; anthers oblong, dorsifixed, versatile. *Ovary* globose, 3-lobed; ovules few in a cell, superposed; style filiform, straight or declinate, obscurely tricuspidate at the stigmatose apex. *Capsule* globose, deeply 3-lobed, membranous, loculicidally 3-valved. *Seeds* 1 or few in a cell, globose.

Bulb tunicated, not produced into a neck; leaves usually lorate, produced with or a little after the flowers; scape slender or robust; umbels few-or many-flowered; spathe-valves 2, lanceolate; flowers pale or deep red.

DISTRIB. Endemic.

Peduncle long, slender. Perianth-limb, stamens and
   style nearly erect:
  Segments hardly at all crisped:
    Leaves green, suberect ... ... (1) **sarniensis.**
    Leaves glaucous, falcate ... ... (2) **curvifolia.**
  Segments distinctly crisped ... ... (3) **Moorei.**
Peduncle long, slender. Perianth-limb, stamens and
   style declinate:
  Filaments not distinctly appendiculate at the base:
    Umbel centripetal:
      Leaves 4–6, linear or lorate:
        Stamens long:
          Perianth-segments distinctly crisped (4) **flexuosa.**
          Perianth-segments hardly at all
           crisped:
            Perianth-segments ⅓ in. broad (5) **angustifolia.**
            Perianth-segments ¼ in. broad (6) **pudica.**
        Stamens short ... ... (7) **brachystemon.**
      Leaves 6–10, filiform... ... ... (8) **filifolia.**
    Umbel centrifugal:
      Perianth ½–¾ in. long, much crisped ... (9) **undulata.**
      Perianth 1–1½ in. long ... ... (10) **humilis.**
  Filaments distinctly appendiculate at the base:
    Process at the back of the filaments ... ... (11) **appendiculata.**
    Process between the filaments ... ... (12) **pancratioides.**
Peduncle short, stout:
  Leaves linear, contemporary with the flowers ... (13) **lucida.**
  Leaves lorate, contemporary with the flowers ... (14) **duparquetiana.**
  Leaves lingulate, produced after the flowers ... (15) **marginata.**

1. **N. sarniensis** (Herb. App. 19); bulb ovoid, 1½–2 in. diam.; tunics pale brown; leaves about 6, bright green, suberect, not curved laterally, developed after the flowers, linear, finally about a foot long, ½–¾ in. broad; peduncle slender, rather compressed, 1–1½ ft. long; umbel 10–20-flowered, centripetal; pedicels 1–2 in. long; spathe-valves ovate-lanceolate, 1½–2 in. long; perianth erect, 1¼–1½ in. long, bright crimson; segments oblanceolate, ¼–½ in. broad, equally falcate, slightly crisped; filaments suberect, bright red, ½ in. longer than the segments; anthers oblong, ⅙ in. long; ovary green,

globose, $\frac{1}{8}$ in. diam. ; style straight, nearly 2 in. long. *Kunth, Enum.* v. 617 ; *Baker, Handb. Amaryllid.* 99. *Amaryllis sarniensis, Linn. Sp. Plant.* 293 ; *Curt. Bot. Mag. t.* 294 ; *Red. Lil. t.* 33 ; *Jacq. Hort. Schoenbr.* i. 34, *t.* 66. *A. dubia, Houtt. Handl.* xii. 181, *t.* 84, *fig.* 1. *Hæmanthus sarniensis, Thunb. Prodr.* 58 ; *Fl. Cap. edit. Schult.* 298. *A. Jacquini, Tratt. Gartenpfl.* 43. *N. Jacquini, Roem. Amaryllid.* 105. *Imhofia sarniensis, Salisb. Gen.* 118. (*Lilium sarniense, Dougl. Descr. Guern. Lil. tt.* 1–2.)

VAR. β, N. Plantii (Hort.) ; differs from the type by the colour of the flower being a duller crimson, the segments more distinctly unguiculate, and the peduncle longer.

VAR. γ, N. venusta (Herb. App. 19) ; leaves green ; flowers bright scarlet, produced earlier than in any of the other varieties. *Ker in Bot. Mag. t.* 1090. *Imhofia venusta, Salisb. Gen.* 118.

VAR. δ, N. rosea (Herb. App. 19) ; leaves darker green than in the type ; flowers rose-red ; seeds oblong. *Bot. Mag. t.* 2124. *Imhofia rosea, Salisb. Gen.* 118.

VAR. ε, N. corusca (Herb. App. 19) ; bulb tunics not chaffy ; leaves broader than in the type, bright green, with distinct cross-bars between the main veins ; flowers large, bright scarlet, resembling those of *N. curvifolia. Herb. Amaryllid.* 283 ; *Kunth, Enum.* v. 617. *Amaryllis corusca, Gawl. in Bot. Mag. sub t.* 1430. *A. humilis* β, *Gawl. in Bot. Mag. t.* 1089. *Imhofia corusca, Salisb. Gen.* 118.

COAST REGION : Table Mountain, *Thunberg! Rogers? Ecklon!*

Commonly cultivated in European gardens under the name of Guernsey lily.

2. N. curvifolia (Herb. App. 19) ; bulb ovoid, $1\frac{1}{2}$–2 in. diam. ; tunics pale brown ; leaves 6, developed after the flowers, lorate, obtuse, curved laterally, thicker in texture than in *N. sarniensis,* glaucous, closely veined, a foot long, $\frac{1}{2}$–$\frac{2}{3}$ in. broad ; peduncle slender, glaucous, 1–$1\frac{1}{2}$ ft. long ; umbel 8–12-flowered, centripetal ; pedicels 1–$1\frac{1}{2}$ in. long ; spathe-valves about as long as the pedicels ; perianth erect, bright scarlet, $1\frac{1}{4}$–$1\frac{1}{2}$ in. long ; segments oblan-ceolate, $\frac{1}{4}$ in. broad, equally falcate, but little crisped ; stamens suberect, about as long as the perianth-limb ; ovary globose, $\frac{1}{8}$ in. diam. ; style suberect, finally 2 in. long. *Herb. Amaryllid.* 283, *t.* 36, *fig.* 4, & *t.* 45, *fig.* 3 ; *Kunth, Enum.* v. 616 ; *Baker, Handb. Amaryllid.* 100. *Amaryllis curvifolia, Jacq. Hort. Schoenbr.* i. 33, *t.* 64 ; *Gawl. in Bot. Mag. t.* 725 ; *Red. Lil. t.* 274. *Imhofia glauca, Salisb. Gen.* 118.

VAR. β, N. Fothergillii (Roem. Amaryllid. 104) ; more robust in all its parts than in the type ; leaf broader ; flowers in an umbel more numerous, between crimson and scarlet. *Amaryllis Fothergillii, Andr. Bot. Rep. t.* 163.

SOUTH AFRICA : without locality.

Common in European gardens, where it flowers from the middle to the end of September.

3. N. Moorei (Leichtl. in Gard. Chron. 1886, xxvi. 681) ; leaves produced a little after the flowers, 9–12 in. long, $\frac{1}{2}$–$\frac{3}{4}$ in. broad, much curved, slightly twisted, blunt-pointed, bright green ; peduncle stout, compressed, shorter than the leaves ; flowers 6–9 in a centripetal umbel ; pedicels $\frac{3}{4}$–1 in. long ; spathe-valves lanceolate, longer than the pedicels; perianth erect, bright scarlet; segments oblanceolate, crisped, 1¼ in. long, nearly ¼ in. broad; filaments erect, longer than the perianth-segments; anthers oblong; ovary ⅛ in. diam.; style straight, 1½ in. long. *Baker, Handb. Amaryllid.* 100.

ORIGIN DOUBTFUL.

Described from a plant sent to Leichtlin by Mr. F. W. Moore, of the Glasnevin Garden. Is it a hybrid between *curvifolia* and *flexuosa*?

4. N. flexuosa (Herb. App. 19); bulb subglobose, 1½ in. diam.; leaves 4–6, contemporary with the flowers, linear-lorate, arcuate, bright green, ½–¾ in. broad, sometimes rough with pustules on the surfaces; peduncle slender, subterete, slightly glaucous, flexuose, sometimes 2–3 ft. long; flowers 10–20 in a centripetal umbel; pedicels slender, 1–2 in. long; spathe-valves lanceolate, as long as the pedicels; perianth-limb generally pale pink, 1–1¼ in. long, cut down very nearly to the ovary; segments crisped, oblanceolate, ⅙ in. broad; stamens declinate, the 3 longer ones rather shorter than the perianth-segments; anthers oblong, claret-red, ⅛–⅙ in. long; ovary globose-trigonous, ⅙ in. diam.; style declinate, as long as the stamens. *Herb. Amaryllid.* 283; *Kunth, Enum.* v. 619; *Baker, Handb. Amaryllid.* 100. *Amaryllis flexuosa, Jacq. Hort. Schoenbr.* i. 35, *t.* 67; *Willd. Sp. Plant.* ii. 60; *Ait. Hort. Kew. edit.* 2, ii. 229; *Ker in Bot. Reg. t.* 172.

VAR. β, **Sandersoni** (Baker); peduncle and pedicels more robust; leaves an inch broad; perianth-segments 1¼ in. long, less crisped, and more distinctly united in a cup at the base.
VAR. γ, **N. pulchella** (Herb. App. 19); leaves glaucous, firmer in texture than in the type; peduncle not flexuose; perianth-segments pale pink, with a rose-red keel; stamens and style whitish. *Herb. in Bot. Mag. t.* 2407; *Kunth, Enum.* v. 620.

CENTRAL REGION: Somerset Div.; lower part of Bruintjes Hoogte, *Burchell,* 2989! and on mountain tops, 4400 ft., *Bolus,* 2202! Var. γ, Somerset Div.; Bosch Berg, *MacOwan,* 1549!
KALAHARI REGION: Var. β, Transvaal, *Sanderson!*

5. N. angustifolia (Baker); bulb ovoid, 1½–2 in. diam.; leaves 3–4, contemporary with the flowers, linear, glaucous green, a foot long, ⅛–¼ in. broad; peduncle stiffly erect, moderately stout, 1½–2 ft. long; umbel many-flowered, centripetal; spathe-valves small, ovate-lanceolate; pedicels very pubescent, 1½–2 in. long; perianth-segments oblanceolate, 1–1¼ in. long, pink, scarcely at all crisped; stamens rather shorter than the perianth-segments; ovary pubescent, obtusely angled; style as long as the stamens. *N. pulchella, var. angustifolia, Baker in Saund. Ref. Bot. t.* 329. *N. flexuosa, var. angustifolia, Baker, Handb. Amaryllid.* 101.

CENTRAL REGION: Somerset Div.; summit of Bosch Berg, 4500 ft., *MacOwan,* 1889!
KALAHARI REGION: Orange Free State, *Cooper,* 3221!
EASTERN REGION: Natal; Polela District, *Adlam!* and Griffins Hill, Estcourt, *Rehmann,* 7320!

6. **N. pudica** (Hook. fil. in Bot. Mag. t. 5901); bulb globose, about 1 in. diam.; leaves 4–6, glaucous, contemporary with the flowers, suberect, 8–9 in. long, $\frac{1}{4}$–$\frac{1}{3}$ in. broad; peduncle slender, subterete, 1–1$\frac{1}{2}$ ft. long; umbel centripetal, 4–6-flowered; pedicels slender, 1–1$\frac{1}{2}$ in. long; spathe-valves lanceolate, 1$\frac{1}{2}$ in. long; perianth-tube short; limb erect or deflexed, 1$\frac{1}{4}$–1$\frac{1}{2}$ in. long; segments oblanceolate, thin in texture, scarcely at all crisped, white with a pink keel upwards; stamens declinate, the three longest a little shorter than the perianth-segments; anthers oblong, $\frac{1}{8}$ in. long; ovary globose, $\frac{1}{8}$ in. diam.; style declinate, reaching to the tip of the perianth-segments; seeds globose. *Flore des Serres, t.* 2464; *Baker, Handb. Amaryllid.* 101.

VAR. β, **N. Elwesii** (Hort. Leichtlin); leaves much broader, bright green, more persistent, distinctly veined, with a prominent midrib; flowers more compact; pedicels stouter; perianth-segments pale rose with a darker keel, thicker and more wavy.

ORIGIN DOUBTFUL. History not clearly known. First described from a plant that flowered at Kew in 1868.

7. **N. brachystemon** (Baker); leaves about 5, subterete, glabrous, a foot long, $\frac{1}{2}$ lin. diam.; peduncle slender, 1$\frac{1}{2}$ ft. long; umbel 10–20-flowered, centripetal; pedicels stiffly erect, pubescent, 1$\frac{1}{2}$–2 in. long; spathe-valves as long as the pedicels; perianth bright red; tube none; segments linear, rather crisped, $\frac{3}{4}$–$\frac{7}{8}$ in. long; stamens not more than half as long as the segments; anthers small, oblong; ovary $\frac{1}{6}$ in. diam., globose, deeply 3-lobed.

EASTERN REGION: Griqualand East, near Kokstad, 5000 ft., *Tyson,* 1269!

8. **N. filifolia** (Baker in Bot. Mag. t. 6547); bulb globose, under 1 in. diam., with pale tunics produced a short distance over its neck; leaves 6–10, contemporary with the flowers, grass-green, subulate, very slender, suberect, 6–8 in. long; peduncle slender, terete, green, finely glandular-pubescent, a foot long; flowers 8–10 in a centripetal umbel; pedicels slender, 1–1$\frac{1}{2}$ in. long; spathe-valves lanceolate, green, under an inch long; perianth deflexed, bright red, 1 in. long, cut down to the ovary; segments oblanceolate, unguiculate, $\frac{1}{12}$–$\frac{1}{8}$ in. broad, crisped; stamens declinate, shorter than the perianth-segments; anthers oblong, red; ovary globose, green, $\frac{1}{6}$ in. diam.; style equalling the longer stamens.

KALAHARI REGION: Orange Free State. Sent to Kew by Mr. Ayres in 1879.

9. **N. undulata** (Herb. App. 19); bulb ovoid, $\frac{3}{4}$–1 in. diam.; tunics pale, membranous; leaves 4–6, linear, contemporary with the flowers, bright green, 1–1$\frac{1}{2}$ ft. long, $\frac{1}{4}$–$\frac{1}{2}$ in. broad; peduncle slender, 1–1$\frac{1}{2}$ ft. long; flowers 8–12 in a centrifugal umbel; pedicels slender, 1–1$\frac{1}{2}$ in. long; spathe-valves lanceolate, as long as the pedicels; perianth-limb pale pink, 8–9 lin. long, cut down very nearly to the ovary; segments oblanceolate, much crisped, $\frac{1}{12}$ in. broad; stamens declinate, about as long as the perianth-segments; ovary globose-trigonous, $\frac{1}{8}$–$\frac{1}{6}$ in. diam.; style declinate, as long as the

perianth-limb; capsule ¼ in. diam., deeply lobed; cells 1-seeded.
Herb. Amaryllid. 283, *t.* 45, *fig.* 2 ; *Kunth, Enum.* v. 621 ; *Baker,*
Handb. Amaryllid. 102. *Amaryllis undulata, Linn. Syst. Nat. edit.*
12, 237 ; *Hill, Hort. Kew.* 352, *cum icone* ; *L'Herit. Sert.* 16 ;
Bot. Mag. t. 369 ; *Red. Lil. t.* 115 ; *Jacq. Hort. Vind.* iii. *t.* 13 ;
Tratt. Arch. t. 393. *Hæmanthus undulatus, Thunb. Fl. Cap. edit.*
Schult. 297. *N. crispa, Hort.*

Vᴀʀ. *β*, major (Tratt. Arch. t. 394); perianth-limb larger and less crisped.
N. aucta, Roem. Amaryll. 107. *Amaryllis aucta, Tratt. Thes.* 9, *t.* 45.

Cᴏᴀsᴛ Rᴇɢɪᴏɴ: hills below Table Mountain, and at Wynberg, *Thunberg!*
Albany Div., *Cooper,* 1532 ! 3220 ! near Dassies Klip, between Port Elizabeth
and Grahamstown, *Bolus,* 2686 !
Kᴀʟᴀʜᴀʀɪ Rᴇɢɪᴏɴ : Orange Free State, *Cooper,* 2235 !

10. N. humilis (Herb. App. 19); bulb ovoid, 1–1½ in. diam. ;
tunics membranous ; leaves about 6, contemporary with the flowers
in November, linear, bright green, suberect, channelled down the
face, about a foot long, about ¼ in. broad; peduncle slender, subterete,
slightly glaucous, ½–1½ ft. long; flowers 10–20 in a centrifugal
umbel ; pedicels slender, 1–1½ in. long; spathe-valves lanceolate,
green, about as long as the pedicels ; perianth-limb bright pink or
rose-red, cut down very nearly to the ovary, 1–1¼ in. long ; segments
oblanceolate, acute, crisped; stamens declinate, the three longer
about equalling the perianth-segments ; anthers oblong, purple, ⅛ in.
long ; ovary globose-trigonous, ⅛–⅙ in. diam. ; style declinate, as long
as the perianth-limb. *Herb. Amaryllid.* 283 ; *Kunth, Enum.* v. 621 ;
Baker, Handb. Amaryllid. 102. *Amaryllis humilis, Jacq. Hort.*
Schoenbr. i. 36, *t.* 69 ; *Bot. Mag. t.* 726 ; *Red. Lil. t.* 450.

Sᴏᴜᴛʜ Aғʀɪᴄᴀ : without locality, *Thunberg!*
Cᴏᴀsᴛ Rᴇɢɪᴏɴ : Table Mountain, 1000–2000 ft., *Drège !* New Kloof, near
Tulbagh, *Drège !* Piquetberg, 1000–2000 ft., *Drège,* 2654 ! Albany Div.,
Zeyher !

Differs from *flexuosa* by the centrifugal expansion of its inflorescence, by its
dwarfer habit, and narrower channelled leaves.

11. N. appendiculata (Baker in Gard. Chron. 1894, xvi. 336) ;
bulb ovoid, middle-sized ; leaves 3, contemporary with the flowers,
linear, green, glabrous, a foot long, deeply channelled down the face ;
peduncle stout, terete, about 2 feet long ; flowers 10–15 in a centripetal umbel ; spathe-valves small, membranous, tinged with red ;
pedicels very pubescent ; perianth red, an inch long ; segments linear,
distinctly keeled, crisped in the upper half ; filaments each appendiculate at the base outside with a lacerated membranous strap-shaped
process ⅙ in. long ; anthers small, oblong; ovary ¼ in. diam., deeply
3-lobed ; style as long as the stamens.

Eᴀsᴛᴇʀɴ Rᴇɢɪᴏɴ: Natal, *Wood !*
Described from a cultivated specimen sent by Mr. Jas. O'Brien in 1894.

12. N. pancratioides (Baker in Gard. Chron. 1891, x. 576);
leaves above a foot long, bright green, subterete in the lower part,

nearly flat towards the top, $\frac{1}{16}$–$\frac{1}{12}$ in. diam. ; peduncle robust, terete, 2 ft. long ; umbel 10–20-flowered, centripetal ; spathe-valves lanceolate, an inch long ; pedicels 1–1½ in. long, densely pubescent ; perianth pure white, ¾–1 in. long ; segments oblanceolate, ¼ in. broad, not crisped ; stamens less than half as long as the perianth ; filaments filiform, alternating with large square bifid scales ; anthers small, black, oblong ; style very short.

EASTERN REGION : Natal ; in a ravine on a hill near Greytown, 4000–5000 ft., *Wood*, 4311 ! and Weenen County, at Waterfall No. 7, 5000–6000 ft., *Evans*, 410 !

Originally described from a plant flowered Nov. 1891 by Mr. R. A. Todd, of Foots Cray.

13. **N. lucida** (Herb. Amaryllid. 283, t. 36, fig. 3) ; bulb 3–4 in. diam., with the membranous tunics produced round a neck 2–3 in. above its apex ; leaves 6–8, contemporary with the flowers, linear, bright green, spreading, flaccid, a foot or more in length, ½–¾ in. broad, scabrous on the edge ; peduncle stout, ancipitous, 3–8 in. long, ¼–⅓ in. diam. ; flowers 20–40 in a centripetal umbel ; peduncles stiff, moderately stout, 3–4 in. long ; spathe-valves ovate-lanceolate, much shorter than the pedicels ; perianth-limb 1½–2 in. long, pale or bright red, cut down to a cup ⅓ in. long ; segments oblanceolate-unguiculate, ¼ in. broad, hardly at all crisped ; stamens and style declinate, nearly as long as the perianth-segments ; ovary globose-trigonous, ¼ in. diam. ; capsule depresso-globose, ⅓–½ in. diam. *Kunth, Enum.* v. 620 ; *Baker, Handb. Amaryllid.* 102. *Amaryllis lucida, Burchell, Travels*, i. 536. *Brunsvigia lucida, Herb. App.* 16 ; *Roem. et Schultes, Syst. Veg.* vii. 847. *Amaryllis laticoma, Ker in Bot. Reg.* t. 497.

KALAHARI REGION : Griqualand West ; between Griqua Town and Witte Water, *Burchell*, 1969 ! Orange Free State ; Sand River, *Burke !* Transvaal, *Todd*, 22 ! Bechuanaland ; Pellat Plains near Kuruman, *Burchell*, Bulb No. 68 !

14. **N. duparquetiana** (Baker) ; leaves lorate, contemporary with the flowers, 8–12 in. long, straight or arcuate ; peduncle as thick as a man's finger ; flowers 20 to an umbel ; pedicels 4 in. long ; perianth 2 in. long ; tube very short, funnel-shaped ; segments linear-lanceolate, acuminate, not undulated, white with a many-nerved carmine keel ; stamens nearly as long as the segments ; ovary as broad as long, strongly angled ; ovules 2 in each cell ; style papillose. *Imhofia duparquetiana, Baill. in Bull. Linn. Soc. Paris* 1132.

KALAHARI REGION : Kalahari, *Duparquet*, 120, 122, 196, 197 (Herb. Mus. Paris).

15. **N. marginata** (Herb. Amaryllid. 283) ; bulb ovoid, 2–3 in. diam., with imbricated, dark brown, membranous tunics and very strong root-fibres ; leaves 4, produced after the flowers, spreading, lingulate, obtuse, 6–9 in. long, 2–2½ in. broad, with a reddish crisped cartilaginous margin ; peduncle stout, compressed, ½ ft. long ; flowers

12–20 in a centripetal umbel; pedicels 1½–2 in. long; spathe-valves deltoid, shorter than the pedicels; perianth erect, bright scarlet, cut down to the ovary; segments lanceolate, but little crisped, 1¼ in. long, ¼ in. broad; stamens suberect, a little longer than the perianth-segments; style suberect, a little longer than the stamens.   *Kunth, Enum.* v. 615; *Baker, Handb. Amaryllid.* 102.   *Amaryllis marginata, Jacq. Hort. Schoenbr.* i. 34, *t.* 65.   *Brunsvigia marginata, Ait. Hort. Kew. edit.* 2, ii. 230; *Gawl. in Bot. Mag. sub t.* 1443.   *Imhofia marginata, Herb. App.* 18.   *Elisena marginata, Roem. Amaryllid.* 63.

CENTRAL REGION: Tulbagh Div.; Cold Bokkeveld, *MacOwan*, 2994! and *MacOwan, Herb. Aust. Afr.*, 1552!

Hybrid Nerines.

The following hybrid Nerines were raised long ago by Dean Herbert, viz. :—

1. N. MITCHAMIÆ, Herb. Amaryll. 283, t. 45, fig. 1, and N. VERSICOLOR, between *curvifolia* and *undulata.*
2. N. HAYLOCKI, between *curvifolia* and *pulchella.*
3. N. PULCHELLA UNDULATA.
4. N. SPOFFORTHIÆ, between *venusta* and *undulata.*
5. N. PULCHELLA HUMILIS.
6. N. HUMILIS UNDULATA.
7. N. CURVIFOLIA VENUSTA.

These, I believe, have all died out, but the following are in existence in cultivation at the present time, viz.:—

8. N. AMABILIS, between *pudica* and *humilis.*
9. N. CAMI, between *curvifolia* and *undulata.*
10. N. ATROSANGUINEA, between *Plantii* and *flexuosa.*
11. N. CINNABARINA, between *Fothergillii* and *flexuosa.*
12. N. O'BRIENI (of which CARMINATA and CÆRULEA are forms), between *pudica* and *Plantii.*
13. N. ERUBESCENS, between *flexuosa* and *undulata.*
14. N. ELEGANS, between *flexuosa* and *rosea.*
15. N. MEADOWBANKI, between *sarniensis* and *Fothergillii.*
16. N. MANSELLII, Hort. O'Brien., between *flexuosa* and *Fothergillii.*
17. N. ROSEO-CRISPA, Hort. Cam., between *undulata* and *flexuosa.*
18. N. EXCELLENS, Moore in Florist 1882, t. 567, between *flexuosa* and *humilis* major.
19. N. STRICKLANDI, O'Brien in Gard. Chron. 1894, xvi. 690, between *curvifolia* and *pudica.*

XIV. STRUMARIA, Jacq.

Perianth infundibuliform, cut down to the ovary, with 6 equal spreading oblanceolate segments.   *Stamens* inserted at the base of the perianth-segments; filaments filiform, usually united to one another and to the dilated style towards the base; anthers oblong, versatile.   *Ovary* globose, 3-celled; ovules few in a cell, superposed; style swollen and triquetrous in the lower half, tricuspidate at the stigmatose apex.   *Capsule* small, globose, membranous, loculicidally 3-valved.   *Seeds* one or few in a cell, globose, bulbiform.

Rootstock a tunicated bulb; leaves lorate, glabrous, contemporary with the flowers; scape slender, terete, solid; flowers umbellate, small, whitish or reddish; spathe-valves 2, small, lanceolate.

DISTRIB. Endemic.

Perianth-segments plane; filaments connate with one another,
 and with the style towards the base:
  Leaves short, lorate, $\frac{1}{3}$–$\frac{1}{2}$ in. broad:
    Spathe-valves shorter than the pedicels; flowers
    inodorous ...   . (1) **truncata.**
    Spathe-valves as long as the pedicels; flowers scented (2) **linguæfolia.**
  Leaves longer and narrower, $\frac{1}{8}$–$\frac{1}{4}$ in. broad:
    Leaves enclosed at the base in a funnel-shaped
    truncate sheath ...   . (3) **angustifolia.**
    Leaves without a funnel-shaped basal sheath   ... (4) **rubella.**
  Perianth-segments crisped; filaments free to the base   ... (5) **undulata.**

1. S. truncata (Jacq. Ic. ii. t. 357); bulb globose, 1 in. diam., with pale brown tunics produced an inch over its top; leaves 3–6, erect, lorate, obtuse, about $\frac{1}{2}$ ft. long, $\frac{1}{3}$–$\frac{1}{2}$ in. broad, enclosed at the base in a funnel-shaped, truncate, membranous, brown sheath, about 2 in. long, $\frac{1}{3}$–$\frac{1}{2}$ in. diam. at the throat; peduncle slender, about a foot long; flowers 6–15 in an umbel, inodorous; pedicels 1–1$\frac{1}{2}$ in. long; spathe-valves cuspidate, greenish, shorter than the pedicels; perianth-segments oblanceolate, plane, pinkish, $\frac{1}{3}$–$\frac{1}{2}$ in. long; stamens distinctly exserted, joined to one another and to the style towards the base; anthers $\frac{1}{24}$ in. long; style as long as the stamens, dilated towards the base. *Willd. Sp. Plant.* ii. 31; *Ait. Hort. Kew. edit.* 2, ii. 212; *Herb. Amaryllid.* 288, *t.* 29, *fig.* 11; *Kunth, Enum.* v. 623; *Baker, Handb. Amaryllid.* 104. *Hœmanthus vaginatus, Thunb. Fl. Cap. edit. Schult.* 297. *Hessea vaginata, Herb. Amaryllid.* 289; *Kunth, Enum.* v. 634. *Hymenetron truncatum, Salisb. Gen.* 128.

WESTERN REGION: Little Namaqualand; Hardeveld, *Thunberg!* *Zeyher,* 1660! Rhenoster Fonteiu, *Masson!*

2. S. linguæfolia (Jacq. Ic. ii. t. 356); bulb globose, 1–1$\frac{1}{4}$ in. diam., with pale brown membranous tunics produced an inch above its neck; leaves 4–8, lorate, obtuse, erect, glabrous, 4–5 in. long, $\frac{1}{3}$ in. broad, enclosed at the base in a loose funnel-shaped scariose sheath; peduncle slender, terete, about a foot long; flowers 10–15 in an umbel, sweet-scented; pedicels slender, about an inch long; spathe-valves as long as the pedicels; perianth-segments oblanceolate, plane, white, $\frac{1}{2}$ in. long; filaments as long as the perianth-segments, united and connate with the style towards the base; anthers oblong; style as long as the stamens, triquetrous, and swollen towards the base; capsule globose, $\frac{1}{6}$ in. diam., with about 3 seeds in a cell. *Willd. Sp. Plant.* ii. 31; *Herb. Amaryllid.* 288, *t.* 29, *fig.* 10; *Kunth, Enum.* v. 624. *S. baueriana, Herb. Amaryllid.* 394, *t.* 45, *fig.* 5. *Hymenetron linguæfolium, Salisb. Gen.* 128.

WESTERN REGION: Little Namaqualand, *Whitehead!*

Doubtfully distinct from *S. truncata.*

3. S. angustifolia (Jacq. Ic. ii. t. 359); bulb ovoid, $\frac{1}{2}$ in. diam., without any neck; leaves 2, narrow, lorate, suberect, glabrous, 5–6 in. long, $\frac{1}{8}$–$\frac{1}{6}$ in. broad, enclosed in a reddish-brown cylindrical sheath at the base, 1$\frac{1}{2}$–2 in. long; peduncle slender, about a foot long; flowers 8–10 in an umbel, sweet-scented; pedicels 1–1$\frac{1}{2}$ in. long; spathe-valves lanceolate, $\frac{1}{2}$–$\frac{3}{4}$ in. long; perianth-segments oblanceolate, $\frac{1}{2}$ in. long, white with a pinkish tinge, plane; stamens exserted, united to the style towards the base; style swollen, acutely triquetrous towards the base; capsule globose; seeds usually solitary. *Willd. Sp. Plant.* ii. 32; *Ait. Hort. Kew.* 2, ii. 212; *Herb. Amaryllid.* 287, *t.* 29, *fig.* 14; *Kunth, Enum.* v. 623; *Baker, Handb. Amaryllid.* 104. *Pugionella angustifolia, Salisb. Gen.* 128.

SOUTH AFRICA: without locality.
Known to me only from Jacquin's figure.

4. S. rubella (Jacq. Ic. ii. t. 358); bulb ovoid, under an inch in diam.; leaves 3–4, narrow lorate, suberect, 6–9 in. long, $\frac{1}{4}$ in. broad, not enclosed in a produced funnel-shaped sheath at the base; peduncle slender, terete, 1–1$\frac{1}{2}$ ft. long; flowers 6–10 in an umbel, scentless; pedicels about an inch long; spathe-valves lanceolate, purplish; perianth-segments pinkish, plane, lanceolate, $\frac{1}{3}$–$\frac{1}{2}$ in. long; stamens exserted; filaments united to one another, and the style towards the base; style clavate-triquetrous in the lower third. *Willd. Sp. Plant.* ii. 31; *Ait. Hort. Kew.* 2, ii. 212; *Herb. Amaryllid.* 288, *t.* 29, *fig.* 12; *Kunth, Enum.* v. 624; *Baker, Handb. Amaryllid.* 104. *Stylago rubella, Salisb. Gen.* 127.

SOUTH AFRICA: without locality.
Known to me only from Jacquin's figure.

5. S. undulata (Jacq. Ic. ii. t. 360); bulb ovoid, 1–1$\frac{1}{2}$ in. diam.; leaves 3, narrow lorate, suberect, glabrous, 6–10 in. long, $\frac{1}{3}$–$\frac{1}{2}$ in. broad, enclosed at the base in a loose, red-brown, funnel-shaped, truncate sheath 2 in. long; peduncle slender, terete, 1$\frac{1}{2}$ ft. long; flowers about 6 in an umbel, scentless; pedicels 1–1$\frac{1}{2}$ in. long; spathe-valves lanceolate, purplish, as long as the pedicels; perianth-segments pinkish, crisped, lanceolate, $\frac{1}{2}$ in. long; stamens exserted; filaments filiform, free to the base; style as long as the stamens, clavate-triquetrous below the middle. *Willd. Sp. Plant.* ii. 32; *Herb. Amaryllid.* 288, *t.* 29, *fig.* 13; *Kunth, Enum.* v. 624; *Baker, Handb. Amaryllid.* 104. *Eudolon undulatum, Salisb. Gen.* 127.

SOUTH AFRICA: without locality.
Known to me only from Jacquin's figure.

XV. VALLOTA, Herb.

Perianth erect, with an infundibuliform tube, and 6 equal, ascending, oblong, connivent segments, with a pulvinate callus at the base. *Stamens* inserted in the upper part of the perianth-tube; filaments filiform, elongated; anthers oblong, dorsifixed, versatile. *Ovary* oblong, 3-celled; ovules numerous, superposed; style filiform, obscurely tricuspidate at the stigmatose apex. *Capsule* ovoid-oblong, obtusely angled, loculicidally 3-valved. *Seeds* black, compressed, produced into a wing at the base.

DISTRIB. Endemic; monotypic.

1. **V. purpurea** (Herb. App. 29); bulb ovoid, with brown membranous tunics; leaves 6–18, subdistichous, developed with the flowers, lorate, green, finally 1½–2 ft. long, 1–1¼ in. broad, dying down in autumn; peduncle 2–3 ft. long, slightly compressed, sub-terete, hollow, glaucous; flowers 6–9 to an umbel, bright scarlet; pedicels 1–2 in. long; spathe-valves oblong-lanceolate, 2–3 in. long; perianth erect, with a greenish tube 1¼–1½ in. long, ¾ in. diam. at the throat, the divisions about 2 in. long and 1 in. broad; stamens less than an inch shorter than the perianth-segments; anthers oblong, ⅙ in. long; ovary oblong-triquetrous, ⅓–½ in. long. *Herb. Amaryllid.* 134, 414, *t.* 31, *fig.* 15, *t.* 1, *fig.* 52; *Kunth, Enum.* v. 531; *Baker, Handb. Amaryllid.* 53. *Amaryllis purpurea, Ait. Hort. Kew.* i. 417; *Gawl. in Bot. Mag. t.* 1430. *A. speciosa, L'Herit. Sert.* 12. *Crinum speciosum, Linn. fil. Suppl.* 195; *Thunb. Prodr.* 59. *Cyrtanthus purpureus, Herb. in Bot. Mag. sub t.* 2113. *Amaryllis elata, Jacq. Hort. Schoenbr.* i. 32, *t.* 62. *Vallota elata, Roem. Amaryllid.* 110.

Var. **magnifica**, Hort., is a form with large (4 in. long) bright scarlet flowers, with a white eye. Introduced into cultivation by Masson in 1774. There is a white-flowered form. *V. elata*, Roem., is a form with smaller cherry-red flowers and shorter anthers.

COAST REGION : Hartequas Kloof, *Thunberg!* on the Post Berg, near George, *Burchell*, 5949–2 !

XVI. CYRTANTHUS, Ait.

Perianth infundibuliform, with a narrow tube two or three times as long as the segments, dilated gradually upwards to the throat; segments oblong, subequal. *Stamens* inserted in the perianth-tube; filaments filiform, sometimes very short; anthers oblong, dorsifixed, versatile. *Ovary* 3-celled; ovules numerous, crowded, superposed; style long, filiform, indistinctly or distinctly 3-lobed at the stigmatose apex. *Capsule* oblong, loculicidally 3-valved. *Seeds* flattened, numerous; testa black.

Rootstock a tunicated bulb; leaves persistent, linear or lorate; peduncle fistulose; flowers umbellate, pendulous or erect, usually red or white with green stripes; spathe-valves 2–4, greenish, lanceolate.

DISTRIB. In addition to the Cape species there is one in Tropical Africa.

Subgenus CYRTANTHUS PROPER. *Leaves* lorate. *Flowers* pendulous, many in an umbel.
Leaves 1½–2 in. broad   ...   .   (1) **obliquus.**

Leaves about an inch broad:
Perianth 2–2½ in. long ... ... (2) carneus.
Perianth 1¼–1½ in. long ... ... (3) Elliotii.
Leaves ½–¾ in. broad ... ... (4) Huttoni.

Subgenus MONELLA. *Leaves* linear, 1/12–¼ in. broad. *Flowers* many in an umbel, usually pendulous.
Flowers red:
Perianth ¾–1¼ in. long:
Segments nearly as long as the tube ... ... (5) brachyscyphus.
Segments short ... ... (6) parviflorus.
Perianth usually 1½, rarely 2, in. long:
Perianth-tube very slender in the lower half:
Leaves straight ... ... (7) collinus.
Leaves spirally twisted ... ... (8) spiralis.
Perianth-tube not so slender in the lower half:
Perianth-segments ⅓ as long as the tube:
Flowers drooping ... ... (9) pallidus.
Flowers suberect ... ... (10) rectiflorus.
Perianth-segments about ⅙ as long as the tube:
Perianth-tube about an inch long ... (11) Macowani.
Perianth-tube 1½ in. long ... ... (12) O'Brieni.
Perianth usually 2, rarely 3, in. long:
Perianth-segments linear-oblong; tube ⅛ in. diam. at its throat ... ... (13) odorus.
Perianth-segments oblong; tube ⅙–¼ in. diam. at its throat:
Perianth-segments spreading:
Flowers bright red ... ... (14) angustifolius.
Flowers red striped with yellow ... (15) striatus.
Perianth-segments connivent ... ... (16) Tuckii.
Flowers yellow ... ... (17) lutescens.
Flowers white ... ... (18) Mackenii.

Subgenus GASTRONEMA. *Flowers* 1 or few in an umbel, erect or slightly decurved. *Perianth-tube* more dilated in the upper half, with a throat ½–1 in. diam.
Flowers white, with 6 red or green stripes:
Leaves linear:
Perianth 1½–2 in. long:
Leaves straight ... ... (19) uniflorus.
Leaves spirally twisted ... ... (20) helictus.
Perianth 3–4 in. long:
Segments ½ in. long ... ... (21) vittatus.
Segments 1 in. long ... ... (22) smithianus.
Flowers bright red:
Leaves lanceolate, petioled ... ... (23) sanguineus.
Leaves linear ... ... (24) Galpini.

1. **C. obliquus** (Ait. Hort. Kew. i. 414); bulb ovoid, 3–4 in. diam.; tunics brown, membranous; leaves 10–12, erect-falcate, produced after the flowers, distichous, lorate, obtuse, 1½–2 ft. long, 1–2 in. broad, twisted, glaucous, with a smooth reddish margin; peduncle subterete, mottled, 1–2 ft. long, ½–¾ in. diam.; flowers 6–12 in an umbel, very drooping, inodorous; pedicels ½–1 in. long; spathe-valves 4, lanceolate, reflexing; perianth curved, 2–3 in. long, yellow at the base, bright red upwards tipped with green; tube twice as long as the segments, dilated gradually to a throat ⅓–½ in. diam.; segments oblong, much imbricated, ¾–1 in. long, ⅓–½ in.

broad; stamens biseriate, inserted low down in the perianth-tube; filaments $\frac{3}{4}$–1 in. long; anthers small, oblong, reaching about half-way up the segments; ovary green, oblong, $\frac{1}{4}$ in. diam.; style reaching to the tip of the segments; stigma capitate; capsule oblong-triquetrous, an inch long. *Jacq. Hort. Schoenbr.* i. 39, *t.* 75; *Andr. Bot. Rep. t.* 265; *Gawl. in Bot. Mag. t.* 1133; *Red. Lil. t.* 381; *Lodd. Bot. Cab. t.* 947; *Herb. Amaryllid.* 128; *Kunth, Enum.* v. 534; *Baker, Handb. Amaryllid.* 54. *Crinum obliquum, Linn. fil. Suppl.* 195; *Thunb. Prodr.* 59; *Fl. Cap. edit. Schult.* 301. *Amaryllis Umbella, L'Herit. Sert.* 15, *t.* 16. *Timmia obliqua, Gmel. Syst. Nat.* ii. 538. *C. varius, Roem. Amaryllid.* 48.

COAST REGION: Humansdorp Div.; hills near Kabeljouw River, *Thunberg!* Bathurst Div.; Riet Fontein, between Port Alfred and Theopolis, *Burchell,* 3970! and between Riet Fontein and the source of Kasuga River, *Burchell,* 4154!

CENTRAL REGION: Somerset Div.; Commadagga Mts., *Burchell!* and Bosch Berg, 2300 ft., *MacOwan!*

2. **C. carneus** (Lindl. in Bot. Reg. t. 1462); bulb ovoid, 2–3 in. diam.; leaves 8–10, lorate, obtuse, produced after the flowers, glaucous, twisted, $1\frac{1}{2}$ ft. long, about 1 in. broad; peduncle subterete, glaucous, twisted, a foot or more long; flowers 8–10 in an umbel, very cernuous; pedicels $\frac{1}{4}$–$\frac{1}{2}$ in. long; spathe-valves 3-4, lanceolate, greenish, $1\frac{1}{2}$–2 in. long; perianth 2–$2\frac{1}{2}$ in. long, rather curved, bright red; tube twice as long as the segments, dilated gradually from the base to a throat $\frac{1}{3}$ in. diam.; segments oblong, much imbricated, $\frac{1}{3}$ in. broad; stamens inserted low down in the perianth-tube; filaments an inch long; anthers oblong, reaching halfway up the segments; ovary oblong, $\frac{1}{6}$ in. diam.; style reaching to the tip of the perianth-segments; stigma capitate. *Herb. Amaryllid.* 129; *Kunth, Enum.* v. 535; *Baker, Handb. Amaryllid.* 54.

SOUTH AFRICA: without locality, *Thom!*
COAST REGION: on the Zwartberg, near Caledon, *Templeman in Herb. Mac-Owan,* 2619! Alexandria Div.; between Hoffmans Kloof and Drie Fontein, 1000–2000 ft., *Drège!*

3. **C. Elliotii** (Baker); leaves lorate, flaccid, above a foot long, an inch broad; peduncle stout, above a foot long; flowers bright red, many in an umbel, cernuous; pedicels 1–2 in. long; spathe-valves oblong, brownish, 2 in. long; perianth $1\frac{1}{4}$–$1\frac{1}{2}$ in. long; tube slender at the base, $\frac{1}{6}$ in. diam. at the throat; segments oblong, $\frac{1}{2}$ in. long.

COAST REGION: Stockenstrom Div.; bank of stream on the south slope of Katberg, *Elliot,* 149!

4. **C. Huttoni** (Baker, Handb. Amaryllid. 55); leaves lorate, contemporary with the flowers, a foot long, $\frac{1}{2}$–$\frac{3}{4}$ in. broad, narrowed to an obtuse point; peduncle above a foot long, $\frac{1}{6}$–$\frac{1}{4}$ in. diam.; flowers 6–8 in an umbel; pedicels slender, erect, $\frac{3}{4}$–1 in. long; spathe-valves 2, oblong-lanceolate, whitish, membranous, $1\frac{1}{2}$ in. long; perianth about an inch long, pale red; tube much curved, widening from a

subcylindrical base to a throat $\frac{1}{8}$–$\frac{1}{6}$ in. diam.; segments oblong, half
as long as the tube; stamens biseriate; filaments very short; anthers
oblong, yellow, the three lower scarcely, the three upper a little
exserted from the perianth-tube; style reaching halfway up the
segments; stigma distinctly tricuspidate.

SOUTH AFRICA: without locality, *Hutton!*

Described from a single specimen in the Kew herbarium, that flowered at Kew
in May, 1864.

5. **C. brachyscyphus** (Baker, Handb. Amaryllid. 55); bulb ovoid;
leaves about 3, contemporary with the flowers, linear, 9–10 in. long,
$\frac{1}{4}$ in. broad; scape slender, terete, as long as the leaves; umbel 6–8-
flowered; pedicels $\frac{1}{2}$–$\frac{3}{4}$ in. long; spathe-valves lanceolate, an inch
long; perianth pale red, under an inch long, with a funnel-shaped
tube $\frac{1}{6}$ in. diam. at the throat, a little longer than the oblong-
lanceolate segments; stamens as long as the segments; anthers
oblong, $\frac{1}{12}$ in. long.

EASTERN REGION: Pondoland, *Bachmann*, 295!

Described from a plant flowered in July, 1886, by Mr. J. H. Tillett, of
Sprowston, near Norwich, who introduced it from Pondoland.

6. **C. parviflorus** (Baker in Gard. Chron. 1891, ix. 104); bulb
ovoid, under $\frac{3}{4}$ in. diam.; leaves linear, flaccid, above a foot long;
peduncle slender, $\frac{1}{2}$–1 ft. long; flowers 6–8 in an umbel; pedicels $\frac{1}{4}$–1
in. long; spathe-valves lanceolate, 1–1$\frac{1}{2}$ in. long; perianth bright
red, 1–1$\frac{1}{4}$ in. long; tube slightly curved, slender at the base, $\frac{1}{6}$ in.
diam. at throat; segments broad oblong, $\frac{1}{6}$ in. long.

COAST REGION: Port Elizabeth, *Hort. Dawson Paul! Hort. Strickland!*
Stockenstrom Div.; garden at Fort Seymour, *Scully*, 59! Komgha, *Flanagan*,
67!

KALAHARI REGION: Transvaal; Saddleback Range, near Barberton, 4000–
4500 ft., *Galpin*, 477!

EASTERN REGION: Griqualand East; near Fort Donald, 4500 ft., *Tyson*,
1745!

7. **C. collinus** (Gawl. in Bot. Reg. t. 162); bulb ovoid,
1$\frac{1}{2}$ in. diam.; leaves 3, produced with the flowers, linear, glaucous, weak in texture, not twisted, 6–9 in. long, $\frac{1}{6}$–$\frac{1}{4}$ in. broad;
peduncle slender, subterete, glaucous, about a foot long; flowers
6–10 in an umbel, bright scarlet, inodorous; pedicels slender,
suberect, $\frac{1}{2}$–1 in. long; spathe-valves 2, lanceolate, about an inch
long; perianth bright red, 1$\frac{1}{2}$–2 in. long; tube curved, very slender
in the lower half, dilated gradually to a throat $\frac{1}{4}$ in. diam.; segments
oblong, obtuse, $\frac{1}{3}$ in. long; stamens almost uniseriate at the throat
of the perianth-tube; filaments under $\frac{1}{12}$ in. long; anthers small,
oblong; style reaching halfway up the perianth-segments; stigma
tricuspidate. *Roem. et Schultes, Syst. Veg.* vii. 898; *Kunth, Enum.*
v. 536; *Baker, Handb. Amaryllid.* 56. *Monella glauca, Herb. App.*
29. *M. collina, Salisb. Gen.* 139.

COAST REGION: Caledon Div.; Baviaans Kloof, near Genadendal, *Burchell*,
7783! Alexandria Div.; Zuurberg Range, *Cooper*, 3223!

8. C. spiralis (Burch. ex Gawl. in Bot. Reg. t. 167, excl. syn.) ;
bulb ovoid, 1½ in. diam. ; leaves 2–3, produced after the flowers,
linear, several times spirally twisted, weak in texture, glaucous green,
6–9 in. long, ¼–⅓ in. broad at the middle ; peduncle slender, reddish,
terete, glaucous, 9–12 in. long ; flowers 4–6 in an umbel, inodorous ;
pedicels slender, ascending, ½–1 in. long ; spathe-valves 2, greenish,
lanceolate, an inch long ; perianth bright crimson-red, 1½–2 in. long,
tube curved, dilated from a subcylindrical base to a throat ¼ in.
diam. ; segments oblong, patent, ⅓–½ in. long ; stamens a little
exserted from the perianth-tube ; filaments ¼–⅓ in. long ; anthers
oblong, yellow ; ovary oblong, under ¹⁄₁₂ in. diam. ; style overtopping
the stamens ; stigma tricuspidate. *Reich. Fl. Exot. t. 42 ; Herb.
Amaryllid.* 129 ; *Kunth, Enum.* v. 537 ; *Baker, Handb. Amaryllid.*
55. *Monella spiralis, Herb. App.* 29 ; *Salisb. Gen.* 139.

COAST REGION : Port Elizabeth Div. ; near Bethelsdorp, *Zeyher*, 347 !
Burchell !

9. C. pallidus (Sims in Bot. Mag. t. 2471) ; bulb ovoid, 1½ in.
diam., produced into a short neck ; leaves 5, produced after the flower,
linear, green, not twisted, ¼ in. broad ; peduncle subterete, purplish,
½ ft. long ; flowers 4–5 in an umbel, pale red ; pedicels slender,
½–1 in. long ; spathe-valves 2, lanceolate, greenish ; perianth 1½ in.
long ; tube dilated gradually to a throat ¼ in. diam. ; segments
oblong, ½ in. long ; stamens and style exserted from the perianth-
tube. *Roem. et Schultes, Syst. Veg.* vii. 897 ; *Herb. Amaryllid.* 129 ;
Kunth, Enum. v. 536 ; *Baker, Handb. Amaryllid.* 56.

SOUTH AFRICA : without locality.

Introduced into cultivation by Villette in 1822. Known to me only from the
figure cited.

10. C. rectiflorus (Baker) ; bulb ovoid, 1½ in. diam. ; leaves 2,
contemporary with the flowers, linear, flaccid, straight, green, ¼–⅓ in.
broad at the middle ; peduncle slender, 1–1½ ft. long ; flowers 8–10
in an umbel, red ; pedicels slender, ascending, ¾–1 in. long ; spathe-
valves 2, lanceolate, not exceeding the pedicels ; perianth straight,
erect, scarcely over an inch long, the tube dilated gradually from
the base to a throat ⅙ in. diam. ; segments oblong, half as long as
the tube ; stamens distinctly biseriate ; filaments very short ; anthers
minute, oblong, the 3 lower placed one-third of the way down the
perianth-tube, the 3 upper at its throat ; style reaching to the tip
of the segments ; stigma obscurely tricuspidate.

COAST REGION : King Williamstown Div. ; mountains near Perie, *Tidmarsh
in Herb. MacOwan !*

11. C. Macowani (Baker in Gard. Chron. 1875, iv. 98) ; bulb ovoid,
1 in. diam. ; tunics brown ; leaves 1–3, contemporary with the
flowers, linear, green, straight, erect, ½–1 ft. long, ⅛–¼ in. broad ;
peduncle slender, terete, purplish, about a foot long ; flowers 6–8 in
an umbel, bright scarlet, inodorous, cernuous ; pedicels ⅝–¾ in. long ;
spathe-valves 2, lanceolate, greenish, 1–1½ in. long ; perianth 1¼–1½

in. long; tube curved, gradually dilated from the base to a throat $\frac{1}{8}$ in. diam.; segments ovate, spreading horizontally, acute, $\frac{1}{6}-\frac{1}{4}$ in. long; stamens distinctly biserial; filaments very short; anthers oblong, yellow, the 3 lower a short distance down the tube and the 3 upper at its throat; ovary oblong, $\frac{1}{6}-\frac{1}{8}$ in. diam.; style just exserted from the perianth-tube; stigma tricuspidate. *Baker, Handb. Amaryllid.* 56; *Regel, Gartenfl.* xxviii. 1, *t.* 960.

COAST REGION : Fort Beaufort Div. ; Winterberg, *Bowkor !*
CENTRAL REGION : Somerset Div.; summit of Bosch Berg, 4700 ft., *Mac-Owan,* 1006 ! Graaff Reinet Div., Koudevelds Berg, 5000 ft., *Bolus,* 1309 !
Introduced into cultivation in 1875.

12. C. O'Brieni (Baker in Gard. Chron. 1894, xv. 716); bulb ovoid, middle-sized; leaves linear, contemporary with the flowers, a foot long, $\frac{1}{4}-\frac{1}{3}$ in. broad, flaccid, bright green; scape terete, longer than the leaves; flowers 7–8 in an umbel; pedicels $\frac{1}{4}-\frac{1}{2}$ in. long; spathe-valves lanceolate, scariose, $1\frac{1}{2}$ in. long; perianth pale bright scarlet; tube curved, above $1\frac{1}{2}$ in. long, $\frac{1}{6}$ in. diam. at the throat, narrowed gradually to the base; lobes ovate, $\frac{1}{6}$ in. long; stamens and style included in the perianth-tube.

EASTERN REGION : Natal; Drakensberg, *Hort. J. O'Brien !* crevices of rocks, valley of Buffalo River, 5000–6000 ft., *Wood,* 4812 ! and the same or a very nearly allied form from Griqualand East ; Vaal Bank, near Kokstad, *Haygarth in Herb. Wood,* 4203 !
Intermediate between *C. angustifolius* and *C. Macowani.*

13. C. odorus (Gawl. in Bot. Reg. t. 503); bulb ovoid, brown, $1\frac{1}{2}$ in. diam.; leaves 2–3, linear, as long as the peduncle, straight, green, $\frac{1}{12}$ in. broad; peduncle $\frac{1}{2}$–1 ft. long; flowers 4–10 in an umbel, drooping, bright red, fragrant; pedicels slender, ascending, 1–$1\frac{1}{2}$ in. long; spathe-valves 2, lanceolate, greenish, $1\frac{1}{2}$–2 in. long; perianth 2–$2\frac{1}{4}$ in. long; tube curved, slender, dilated gradually from the base to a throat $\frac{1}{8}$ in. diam.; segments linear-oblong, $\frac{1}{3}-\frac{1}{2}$ in. long; stamens exserted from the perianth-tube; filaments $\frac{1}{8}-\frac{1}{6}$ in. long; anthers oblong, yellow; ovary oblong, $\frac{1}{8}$ in. diam.; style sometimes protruded beyond the tip of the perianth-segments; stigma tricuspidate. *Herb. Amaryllid.* 129 ; *Roem. et Schultes, Syst. Veg.* vii. 895 ; *Kunth, Enum.* v. 539 ; *Baker, Handb. Amaryllid.* 56. *Monella odora, Herb. App.* 29. *Eusipho odora, Salisb. Gen.* 139.

SOUTH AFRICA : without locality, *Villette !*
EASTERN REGION : Natal ; near Pietermaritzburg, *Krauss,* 266 ! and without precise locality, *Gerrard,* 548 !

14. C. angustifolius (Ait. Hort. Kew. i. 414); bulb ovoid, $1\frac{1}{2}$–2 in. diam.; tunics brown, membranous; leaves 2–3, usually contemporary with the flowers, linear, flaccid, green, straight, 1–$1\frac{1}{2}$ ft. long, $\frac{1}{4}-\frac{1}{3}$ in. broad at the middle ; peduncle terete, 1–$1\frac{1}{2}$ ft. long ; flowers 4–10 in an umbel, bright red, inodorous; pedicels slender, ascending, 1–$1\frac{1}{2}$ in. long; spathe-valves 2, lanceolate, greenish, $1\frac{1}{2}$–2 in. long; perianth $1\frac{1}{2}$–2 in. long; tube curved, dilated gradually from

the base to a throat $\frac{1}{6}$ in. diam.; segments oblong, patent, $\frac{1}{3}$–$\frac{1}{2}$ in.
long; stamens exserted from the perianth-tube; filaments $\frac{1}{3}$–$\frac{1}{4}$ in.
long; anthers oblong, yellow; ovary oblong, $\frac{1}{8}$–$\frac{1}{6}$ in. diam.; style
sometimes protruded beyond the perianth-segments; stigma tricus-
pidate. *Curt. Bot. Mag. t.* 271; *Lodd. Bot. Cab. t.* 368; *Red. Lil.
t.* 388; *Herb. Amaryllid.* 129; *Kunth, Enum.* v. 538; *Baker,
Handb. Amaryllid.* 57. *Crinum angustifolium, Linn. fil. Suppl.*
195; *Thunb. Prodr.* 59; *Fl. Cap. edit. Schult.* 300. *Amaryllis
cylindracea, L'Herit. Sert. Angl.* 15. *Timmia angustifolia, Gmel.
Syst. Nat.* ii. 538. *Monella angustifolia, Herb. App.* 29. *Eusipho
angustifolius, Salisb. Gen.* 139.

VAR. β, **grandiflorus** (Baker, Handb. Amaryllid. 57); peduncle stouter;
perianth 2½–3 in. long, with a throat ¼ in. diam.
VAR. γ, **C. ventricosus** (Willd. Sp. Plant. ii. 49); perianth-tube more dilated,
ventricose in the upper half. *Herb. Amaryllid.* 129; *Kunth, Enum,* v. 537.
C. angustifolius, Jacq. Hort. Schoenbr. i. 40, *t.* 76. *Monella ventricosa, Herb.
App.* 29.

SOUTH AFRICA: without locality, *Thunberg! Masson! Oldenburg!*
COAST REGION: near Kasteel Poort on Table Mountain, 2500 ft., *MacOwan
and Bolus, Herb. Norm.,* 502! King Williamstown Div., near Perie Mountains,
Tidmarsh, 528! Queenstown Div, *Cooper,* 3272! Var. β, British Kaffraria,
Cooper, 1805! Var. γ, Worcester Div.; Dutoits Kloof, *Drège,* 1520! Bathurst
Div.; between Blauw Krans and the source of Kasuga River, *Burchell,* 3903!
CENTRAL REGION: Cave Mountain near Graaff Reinet, 3800 ft., *Bolus,* 179!
KALAHARI REGION: Basutoland, *Cooper,* 3224! Transvaal, *Sanderson!*
EASTERN REGION: Natal; Pietermaritzburg, 2000–3000 ft., *Sutherland!*
Inanda, *Wood,* 469! Var. β, Tembuland; Bazeia, 2000 ft., *Baur,* 532! Gri-
qualand East, near Kokstad, 5000 ft., *Tyson,* 1555! Natal; Mooi River
Valley, 2000–3000 ft., *Sutherland!* Var. γ, Natal; Pietermaritzburg, 2000–
3000 ft., *Sutherland!*

15. **C. striatus** (Herb. in Bot. Mag. t. 2534); bulb ovoid, 1½ in.
diam.; leaves 2, green, linear, straight, contemporary with the
flowers, a foot long, $\frac{1}{4}$–$\frac{1}{3}$ in. broad at the middle, narrowed to both
ends; peduncle slender, terete, reddish, under a foot long; flowers
2–3 in an umbel, pendulous; pedicels slender, $\frac{3}{4}$–1 in. long; spathe-
valves 2, lanceolate, greenish; perianth 2½ in. long, bright red, with
yellow ribs; tube curved, gradually dilated to a throat $\frac{1}{8}$ in. diam.;
segments oblong, $\frac{1}{3}$–$\frac{1}{2}$ in. long; stamens exserted from the perianth-
tube; filaments ¼ in. long; style overtopping the stamens; stigma
tricuspidate. *Herb. Amaryllid.* 129; *Roem. et Schultes, Syst. Veg.*
vii. 898; *Kunth, Enum.* v. 538.

SOUTH AFRICA: without locality.

Known to me only from the figure cited. Probably a mere variety of
C. angustifolius.

16. **C. Tuckii** (Baker in Journ. Bot. 1876, 183); bulb ovoid, 1½ in.
diam., with tunics produced some distance above its top; leaves 2,
contemporary with the flowers, linear, erect, straight, green, 1–1½ ft.
long, $\frac{1}{4}$–$\frac{1}{3}$ in. broad; peduncle 1–1½ ft. long, $\frac{1}{4}$–$\frac{1}{3}$ in. diam.; flowers
10–12 in an umbel; pedicels slender, cernuous or suberect, 1–2½ in.
long; spathe-valves 2, lanceolate, green, 3–3½ in. long; perianth

$1\frac{1}{2}$–2 in. long, yellowish at the base, passing gradually upwards into deep blood-red; tube curved, dilated gradually from the base to a throat $\frac{1}{3}$ in. diam.; segments oblong, $\frac{1}{4}$–$\frac{1}{3}$ in. long, persistently connivent; stamens biseriate, the three lower not exserted from the perianth-tube; filaments as long as the anthers; ovary oblong, $\frac{1}{8}$ in. diam.; style just overtopping the anthers; stigma tricuspidate. *Baker, Handb. Amaryllid.* 57; *Gard. Chron.* 1892, xii. 155, *fig.* 28.

CENTRAL REGION: Somerset Div.; summit of Bosch Berg, 4500 ft., *MacOwan*, 2133!

17. **C. lutescens** (Herb. Amaryllid. 129, t. 33, fig. 14); bulb globose, about 1 in. diam.; leaves 2–4, produced with or after the flowers, linear, green, straight, flaccid, a foot long, $\frac{1}{4}$ in. broad; peduncle slender, $\frac{1}{2}$–1 ft. long; flowers 2–3 to an umbel; pedicels slender, erect, $\frac{1}{4}$–$\frac{3}{4}$ in. long; spathe-valves 2, lanceolate, greenish, $\frac{3}{4}$–1 in. long; perianth suberect, yellowish, 2 in. long; tube slightly curved, very slender in the lower half, dilated gradually to a throat $\frac{1}{6}$ in. diam.; segments oblong, $\frac{1}{4}$–$\frac{1}{3}$ in. long; stamens distinctly biseriate; filaments very short; anthers minute, oblong, the three upper placed at the throat of the perianth-tube, the three lower a short space below; style exserted from the perianth-tube; stigma tricuspidate. *Kunth, Enum.* v. 539; *Baker, Handb. Amaryllid.* 58. *C. alboluteus, Burch. MSS. Monella ochroleuca, Herb. App.* 29.

VAR. **Cooperi** (Baker, Handb. Amaryllid. 58); peduncle more robust; umbel sometimes 6–10-flowered; pedicels and spathe-valves longer; perianth-segments ovate, not more than $\frac{1}{8}$–$\frac{1}{6}$ in. long, more spreading. *C. lutescens, Hook. in Bot. Mag. t.* 5374.

COAST REGION: Riversdale Div.; on or near the Lange Bergen, *Burchell*, 7144! Var. *β*, mountains near King Williamstown, *Mrs. Barber!* Komgha, *Flanagan*, 294! 300! Stockenstrom Div.; southern spurs of Katberg, *Mrs. Barber! Hutton!* British Kaffraria, *Cooper*, 1806! 3225! *Mrs. Hutton!*

KALAHARI REGION: Transvaal; mountain tops near Barberton, 5000–5500 ft., *Galpin*, 654! and without precise locality, *Sanderson!*

EASTERN REGION: Var. *β*, Tembuland; Bazeia, in low moist places, 2000 ft., *Baur*, 242!

18. **C. Mackenii** (Hook. fil. in Gard. Chron. 1869, 641, with figure); bulb ovoid, 1–$1\frac{1}{2}$ in. diam.; tunics brown, membranous; leaves 2–6, contemporary with the flower, linear, green, suberect, straight, 9–12 in. long, $\frac{1}{4}$–$\frac{1}{3}$ in. broad; peduncle about a foot long, subterete, slightly glaucous, red-brown towards the base; flowers 4–10 in an umbel, pure white; pedicels slender, erect, $\frac{1}{2}$–1 in. long; spathe-valves 2, lanceolate, greenish, 1–$1\frac{1}{2}$ in. long; perianth suberect, 2 in. long; tube slightly curved, dilated gradually from the base to a throat $\frac{1}{6}$ in. diam.; segments oblong, ascending, $\frac{1}{4}$–$\frac{1}{3}$ in. long; stamens distinctly biseriate; filaments very short; anthers minute, oblong, the 3 upper placed opposite the throat of the tube, the 3 lower a little below it; ovary oblong, $\frac{1}{9}$–$\frac{1}{6}$ in. diam.; style exserted from the perianth-tube; stigma tricuspidate. *Baker in Saund. Ref. Bot. t.* 355.

EASTERN REGION: Natal; Ifafa, *Tyson*, 2103! and without precise locality, *Gerrard*, 1855! *Sutherland! Wood*, 1336! *Adlam*, 152!

19. C. uniflorus (Gawl. in Bot. Reg. t. 168) ; bulb globose, $\frac{3}{4}$–1 in. diam.; tunics pale, membranous, produced about an inch over its top; leaves 1–2, contemporary with the flowers, linear, straight, erect, $\frac{1}{16}$–$\frac{1}{8}$ in. broad; peduncle slender, 3–9 in. long; umbel 1–3-flowered; pedicel short if the flower is solitary, reaching 1–1$\frac{1}{2}$ in. if they are more than one ; spathe-valves 2, lanceolate, green, 1–1$\frac{1}{4}$ in. long ; perianth 1$\frac{1}{2}$–2 in. long, erect or slightly decurved, white with red-brown or green stripes ; tube dilated gradually from the base to a throat $\frac{1}{3}$–$\frac{1}{2}$ in. diam. ; segments oblong, $\frac{1}{2}$–$\frac{3}{4}$ in. long, with about 5 fine distinct ribs in the middle, which are decurrent down the tube ; stamens biseriate; filaments about $\frac{1}{4}$ in. long, erect or incurved ; anthers oblong, yellow; ovary green, oblong, $\frac{1}{12}$–$\frac{1}{8}$ in. diam.; style overtopping the anthers, with 3 falcate stigmatose forks $\frac{1}{12}$ in. long; capsule oblong-trigonous, under an inch long ; seeds discoid, black, $\frac{1}{3}$ in. diam. *Roem. et Schultes, Syst. Veg.* vii. 899; *Kunth, Enum.* v. 540 ; *Baker, Handb. Amaryllid.* 58. *Amaryllis clavata, L'Herit. Sert.* 11 ; *Willd. Sp. Plant.* ii. 52. *Gastronema clavatum, Herb. in Bot. Mag. t.* 2291 ; *Amaryllid.* 132 ; *App.* 30. *Amaryllis Pumilio, Ait. Hort. Kew* i.,415. *A. humilis, Link. Enum.* i. 313. *Cyrtanthus Pumilio, Roem. et Schultes, Syst. Veg.* vii. 900.

SOUTH AFRICA: without locality, *Sparrman! Drège,* 2190a!
COAST REGION: Uitenhage Div.; Van Stadens Berg, *MacOwan,* 1938! near Uitenhage, *Burchell,* 4265! between Bethelsdorp and Uitenhage, *Burchell,* 4399–3! and near the Zwartkops River, *Ecklon,* 606!

20. C. helictus (Lehm. Delect. Sem. Hort. Hamburg. 1839, 7) ; bulb globose, 1–1$\frac{1}{2}$ in. diam., with tunics produced an inch over the top ; leaves 2–4, contemporary with the flowers, linear, spirally twisted, 6–9 in. long, $\frac{1}{12}$–$\frac{1}{8}$ in. broad ; peduncle slender, 3–9 in. long; flowers 1–3 in an umbel; pedicel $\frac{1}{4}$–$\frac{1}{2}$ in. long if there is only one flower, reaching 1–1$\frac{1}{4}$ in. if there are 2–3 ; spathe-valves 2, lanceolate, 1–2 in. long ; perianth white, 1$\frac{1}{2}$–2$\frac{1}{2}$ in. long ; tube straight or slightly curved, dilated gradually from the base to a throat $\frac{1}{2}$ in. diam.; segments oblong, cuspidate, $\frac{1}{2}$–$\frac{3}{4}$ in. long, with five fine distinct ribs down the middle ; stamens and style as in *C. uniflorus;* ovary green, oblong, $\frac{1}{8}$–$\frac{1}{6}$ in. diam. *Kunth, Enum.* v. 541 ; *Baker, Handb. Amaryllid.* 58. *Cyphonema loddigesianum, Herb. in Bot. Mag. sub tt.* 3710 *and* 3747. *Gastronema spirale, Zeyher Exsic.*

CENTRAL REGION: Somerset Div.; near the Great and Little Fish Rivers, *Drège,* 2191a! *MacOwan,* 1580! and *Burke!* near Graaff Reinet, 2700 ft., *Bolus,* 1308 !

21. C. vittatus (Desf. ex Red. Lil. t. 182); bulb globose, 1$\frac{1}{4}$ in. diam. ; leaves 5–6, contemporary with the flowers, linear, suberect ; peduncle terete, 6–9 in. long; flowers 5–6 in an umbel, white striped with red-brown ; pedicels very short; spathe-valves 2, lanceolate, greenish, 1$\frac{1}{2}$ in. long ; perianth about 3 in. long ; tube slender in the lower half, dilated gradually to a throat $\frac{1}{2}$ in. diam.; segments oblong, $\frac{1}{2}$ in. long ; stamens exserted from the throat of the tube ; ovary oblong, green ; style overtopping the stamens. *Poir.*

Encyc. Suppl. ii. 438; *Roem. et Schultes, Syst. Veg.* vii. 899; *Kunth, Enum.* v. 541; *Baker, Handb. Amaryllid.* 58.

SOUTH AFRICA: without locality.

Known only from the figure cited.

22. **C. smithianus** (Herb. in Bot. Mag. sub t. 3779); bulb ovoid, 1½ in. diam.; leaves 2–4, contemporary with the flowers, linear, spirally twisted, rather glaucous, 6–9 in. long, ¼ in. broad; peduncle slender, terete, ½ ft. long; flowers 1 2 in an umbel; pedicels 1–1½ in. long, spathe-valves 2, lanceolate; perianth 3½–4 in. long, white streaked with green or red-brown; tube slightly curved, dilated gradually from the base to a throat an inch in diameter; segments oblong, an inch long, with 5–7 fine distinct ribs in the centre, decurrent down the tube; stamens biseriate; filaments filiform, ½ in. long; anthers yellow, oblong; ovary oblong, green, ¼ in. diam.; style much overtopping the anthers; stigmas 3, spreading. *Kunth, Enum.* v. 541; *Baker, Handb. Amaryllid.* 58.

EASTERN REGION : Lorenço Marquez ; Lombobo Mountains, *Mrs. K. Saunders !*

Described from a specimen sent by Prof. Balfour from the Edinburgh Botanic Garden in May, 1876.

23. **C. sanguineus** (Hook. in Bot. Mag. t. 5218); bulb ovoid, 2 in. diam.; tunics brown, scariose; leaves 3–4, contemporary with the flowers, lanceolate, acute, bright green, a foot long, ½–¾ in. broad at the middle, narrowed to the apex, and with a distinct petiole 3–4 in. long; peduncle slender, terete, pale green, 6–9 in. long; flowers 1–3 in an umbel, scentless; pedicels ½–2 in. long; spathe-valves 2–4, lanceolate, green, 2–3 in. long; perianth 3–4½ in. long, bright red; tube suberect or decidedly curved, sub-cylindrical in the lower half, dilated in the upper half to a throat an inch in diameter; segments ovate, falcate or rather revolute, 1–2 in. long; stamens uniseriate, exserted a little from the throat of the perianth-tube; filaments arcuate, incurved, ½ in. long; anthers small, oblong; ovary oblong-trigonous, ⅛–⅙ in. diam.; style overtopping the anthers; stigmas 3, falcate, ¼ in. long. *Baker, Handb. Amaryllid.* 59. *Gastronema sanguineum, Lindl. in Journ. Hort. Soc.* iii. 315, *with woodcut.*

EASTERN REGION : Natal, *Sanderson,* 551 ! *Plant !*

Introduced into cultivation by Backhouse in 1860. Also Galla country, East Tropical Africa, *Wakefield !*

C. hybridus, N. E. Brown in Gard. Chron. 1885, xxiv. 391, is a garden hybrid between this species and *Vallota purpurea.*

24. **C. Galpini** (Baker in Kew Bullet. 1892, 83); bulb ovoid, ¾ in. diam.; leaves linear, glabrous; peduncle very slender, 3 in. long, bearing a single erect flower; spathe-valves lanceolate, membranous, an inch long; pedicel ¼ in. long; perianth bright red, under 2 in. long; tube broadly funnel-shaped in the upper two-thirds, cylindrical at the base; segments ovate, ½ in. long; stamens inserted in two rows above the middle of the tube; filaments ¼ in. long; anthers small, oblong; style overtopping the anthers.

KALAHARI REGION: Transvaal; rocky hill sides, Berea Ridge near Barberton, 3000 ft., *Galpin*, 409!

Near *C. sanguineus.*

XVII. CLIVIA, Lindl.

Perianth infundibuliform, curved or straight; tube infundibuliform, short; segments connivent, oblanceolate or oblanceolate-oblong, obtuse, much longer than the tube, the three outer narrower than the three inner. *Stamens* inserted at the throat of the perianth-tube; filaments filiform, about as long as the segments; anthers oblong, dorsifixed, versatile. *Ovary* globose, 3-celled: ovules 5–6 in a cell, fascicled at the centre of the placenta; style long, very slender, tricuspidate at the stigmatose apex. *Fruit* a bright red pulpy berry. *Seeds* one or few, large, globose, bulbiform.

Bulb imperfect, consisting only of the dilated bases of the leaves; root-fibres very thick; leaves numerous, lorate or oblanceolate, distichous, persistent; peduncle stout, solid, ancipitous; flowers bright red-yellow, scentless, many in an umbel; spathe-valves several, unequal, green, lanceolate, membranous, imbricated.

DISTRIB. Endemic.

Subgenus EUCLIVIA. *Perianth* curved.

Leaf not narrowed to the apex   ...   .   (1) nobilis.
Leaf narrowed to the apex...   ...   .   (2) Gardeni.

Subgenus IMANTOPHYLLUM. *Perianth* straight.

The only species   ...   .   (3) miniata.

1. **C. nobilis** (Lindl. in Bot. Reg. t. 1182); leaves about a dozen, lorate, very obtuse, bright green, not narrowed below the middle, $1-1\frac{1}{2}$ ft. long, $1\frac{1}{4}-1\frac{1}{2}$ in. broad, firm in texture, scabrous on the margin; peduncle ancipitous, about a foot long; flowers often 40–60 in an umbel; pedicels $\frac{1}{2}-1$ in. long; spathe-valves several, unequal, lanceolate, greenish; perianth reddish-yellow, curved; tube infundibuliform, $\frac{1}{4}-\frac{1}{3}$ in. long; segments oblanceolate, tipped with green, $\frac{3}{4}-1\frac{1}{4}$ in. long, $\frac{1}{4}-\frac{1}{3}$ in. broad; stamens a little exserted; anthers oblong, yellow, $\frac{1}{12}$ in. long; ovary globose, $\frac{1}{8}$ in. diam.; style finally a little longer than the stamens. *Roem. et Schultes, Syst. Veg.* vii. 892. *Herb. Amaryllid.* 230, *t.* 36, *fig.* 6, *and t.* 44. *fig.* 29; *Kunth, Enum.* v. 585; *Baker, Handb. Amaryllid.* 61. *Imantophyllum Aitoni, Hook. in Bot. Mag. t.* 2856.

COAST REGION: Bathurst Div.; near Kaffir Drift, *Burchell*, 3873! and between Riet Fontein and the source of Kasuga River, *Burchell*, 4131! Eastern Frontier, *Mrs. Barber!*
KALAHARI REGION: Transvaal; Saddleback Range, near Barberton, 4000–4650 ft., *Galpin*, 1102!
EASTERN REGION: Natal; near Murchison, *Wood*, 3066! Zululand; Eshowe, *Mrs. Saunders*, 8!

Introduced into cultivation in 1828.

2. **C. Gardeni** (Hook. in Bot. Mag. t. 4895); leaves 10–12, lorate, distichous, bright green, $1\frac{1}{2}-2\frac{1}{2}$ ft. long, $1-1\frac{1}{2}$ in. broad,

narrowed gradually to the apex, but hardly at all below the middle;
peduncle ancipitous, 1–1½ ft. long; flowers 12–20 in an umbel;
pedicels 1–1½ in. long; spathe-valves several, linear or lanceolate,
greenish, 1½ in. long; perianth reddish-yellow, curved; tube infundibuliform, ⅓–½ in. long; segments all oblanceolate, obtuse, 1¼ in.
long, ¼–⅓ in. broad, tipped with green; stamens as long as the
perianth-segments; anthers oblong, yellow, ⅛ in. long; ovary
globose, ⅛ in. diam.; style finally much exserted; fruit ovoid, bright
red, an inch long. *Baker, Handb. Amaryllid. 62.*

KALAHARI REGION: Transvaal, *Hort. Nelson!*
EASTERN REGION : Natal, *Garden!*

Introduced into cultivation by Captain Garden in 1855.

3. C. miniata (Regel, Gartenflora, 1864, 131, t. 434); leaves
16–20, oblanceolate, suberect, bright green, 1½–2 ft. long, 1½–2 in.
broad, narrowed to the apex and gradually to the base; peduncle
stout, ancipitous, 1–1½ ft. long; flowers 12–20 in an umbel;
pedicels 1–2 in. long; spathe-valves several, linear or lanceolate,
greenish, 1½–2 in. long; perianth erect, bright scarlet with a yellow
throat; tube infundibuliform, ⅓–½ in. long; limb 2–2½ in. long;
outer segments oblanceolate, inner oblanceolate-oblong, ½ in. broad;
stamens shorter than the perianth-segments; anthers yellow, ⅛ in.
long; ovary globose, ¼ in. diam.; style reaching to the top of the
perianth-segments. *Imantophyllum miniatum, Hook. in Bot. Mag.*
tt. 4783; *Flore des Serres, tt.* 949, 950; *Baker, Handb. Amaryllid.* 62.
Himantophyllum miniatum, Groenland in Rev. Hort. 1859, 125,
*tt.*29-30; *Flore des Serres, tt.* 2373-4. *Vallota? miniata, Lindl. in Gard.*
Chron. 1854, 119. *Cyrtanthus Smithii, Krauss, in Flora,* 1845, 311.

EASTERN REGION : Natal; mountains near Pietermaritzburg, 2800 ft., *Mac-*
Owan and Bolus, Herb. Norm., 804! and without precise locality, *Buchanan!*
Cooper, 3219!

Introduced into cultivation by Backhouse in 1854. *Var. Lindeni, André in*
Ill. Hort. 1878, *t.* 343, is a robust large-flowered variety with the inner segments
of the limb an inch broad.

Between *C. nobilis* and *miniata* a fine hybrid has been raised, which is now
common in European gardens under the name of *Imantophyllum cyrtanthiflorum.*
A full account of it, with figures, will be found in Van Houtte's Flore des Serres,
t. 1877, and by Groenland in the Revue Horticole for 1859, vol. viii. p. 258,
fig. 65. It has a curved perianth, with the inner segments of the limb twice as
broad as the outer, and exserted stamens.

XVIII. HÆMANTHUS, Linn.

Perianth straight, erect, with a short subcylindrical tube above the
ovary; segments longer than the tube, linear or lanceolate, equal,
spreading or permanently ascending. *Stamens* inserted at the throat
of the perianth-tube; filaments filiform, often longer than the
perianth-segments; anthers oblong, dorsifixed, versatile. *Ovary*
globose, 3–celled; ovules solitary or in collateral pairs; style filiform,
erect; stigma minutely tricuspidate. *Fruit* globose, indehiscent,
baccate. *Seeds* globose, often solitary; testa pale, membranous.

Rootstock a tunicated bulb; leaves broad and obtuse, various in texture, often thick and fleshy; peduncle stout, solid, compressed; flowers numerous, usually red or white, in dense capitate umbels, surrounded by a whorl of erect or spreading membranous bracts.

DISTRIB. Several species in Tropical Africa; one in Socotra. Two of the four subgenera are endemic at the Cape.

Subgenus I. NERISSA (Salisb.). *Leaves* membranous. *Spathe-valves* and *perianth-segments* spreading.

The only Cape species ... ... (1) **Katherinæ.**

Subgenus II. GYAXIS (Salisb.) *Leaves* membranous. *Spathe-valves* and *perianth-segments* persistently ascending.

Leafy stem short, leaves distinctly petioled ... ... (2) **puniceus.**
Leafy stem elongated, leaves not petioled :
    Bracts brown, flowers greenish ... ... (3) **natalensis.**
    Bracts green, flowers bright red ... ... (4) **magnificus.**

Subgenus III. MELICHO (Salisb.). *Bulb* with thick bifarious tunics. *Leaves* thick, fleshy. *Bracts* and *perianth-segments* spreading.

Umbel few-flowered :
    Leaves glabrous on the margin ... ... (5) **Pumilio.**
    Leaves ciliated on the margin ... ... (6) **lanceæfolius.**
Umbel many-flowered :
    Leaves oblong or obovate, hairy ... ... (7) **carneus.**
    Leaves lorate or lingulate, glabrous :
      Leaves 1½–2 in. broad :
       Perianth ⅜–½ in. long ... ... (8) **amarylloides.**
       Perianth ¼ in. long ... ... (9) **montanus.**
      Leaves 4–5 in. broad... ... ... (10) **candidus.**

Subgenus IV. DIACLES (Salisb.). *Bulb* with thick bifarious tunics. *Leaves* thick, fleshy. *Bracts* and *perianth-segments* permanently ascending.

Bracts white, veined with green; flowers pure white :
    Peduncle produced; umbel not compressed :
      Pedicels very short :
       Leaves lingulate :
        Leaves ciliated ... ... (11) **albiflos.**
        Leaves not ciliated ... ... (12) **albo-maculatus.**
       Leaves suborbicular ... ... (13) **Baurii.**
      Pedicels ½–¾ in. long ... ... (14) **Arnottii.**
    Peduncle very short; umbel compressed :
      Leaves orbicular ... ... (15) **deformis.**
      Leaves round-oblong ... ... (16) **Mackenii.**
Bracts reddish, ¾–1 in. long, falling short of the flowers :
    Leaves suborbicular :
      Peduncle glabrous ... ... (17) **Cooperi.**
      Peduncle densely hairy ... ... (18) **hirsutus.**
    Leaves oblong ... ... (19) **incarnatus.**
    Leaves lorate :
      Leaves glabrous on the margin :
       Leaves ½ in. broad, much undulated ... (20) **undulatus.**
       Leaves 2–3 in. broad :
        Pedicels ¼ in. long ... ... (21) **concolor.**
        Pedicels as long as the flowers ... (22) **sanguineus.**
      Leaves ciliated on the margin ... ... (23) **humilis.**
Bracts oblong, reddish, 1½–2 in. long, equalling or overtopping the flowers :
    Leaves suborbicular :
      Leaves with scabrous margins; bracts shorter
       than the flowers ... ... (24) **rotundifolius.**
      Leaves with smooth margins; bracts over-
       topping the flowers ... ... (25) **callosus.**

Leaves lingulate, 4–8 in. broad :
  Mature leaf, glabrous on the face and edge :
    Leaves barred on the back   ...   ...   (26)   **moschatus.**
    Leaves not barred on the back ...   ...   (27)   **coccineus.**
  Mature leaf ciliated   ...   .   (28)   **tigrinus.**
Leaves lorate, 1½–2 in. broad :
  Leaves glabrous...   ...   .   (29)   **hyalocarpus.**
  Leaves ciliated only   ...   .   (30)   **crassipes.**
  Leaves usually both ciliated and hairy on the
    surface   ...   .   (31)   **pubescens.**

1. **H. Katharinæ** (Baker in Gard. Chron. 1877, vii. 656); bulb
globose, 2–3 in. diam.; leafy stem, short; leaves 3–5, contemporary
with the flowers, oblong, membranous, bright green, 9–12 in. long,
4–6 in. broad, narrowed at the base into a distinct spotted petiole
4–5 in. long; vertical veins 8–10 on each side of the distinct midrib,
the central ones $\frac{1}{4}$–$\frac{1}{3}$ in. apart; peduncle lateral, distinct from the
leafy stem, spotted low down, about a foot long, $\frac{1}{2}$–1 in. diam. at the
base; umbel globose, 4–6, or in cultivation 8–9 in. diam.; pedicels
slender, 1–2 in. long; spathe-valves 5–6, lanceolate, very thin,
reflexing, fugacious, 1½–2 in. long; perianth bright red; tube
cylindrical, $\frac{1}{2}$–$\frac{3}{4}$ in. long; segments lanceolate, reflexing, $\frac{3}{4}$–1 in.
long; filaments ascending, bright red, 1–1½ in. long; anthers oblong,
yellow, $\frac{1}{12}$ in. long; ovary green, globose, $\frac{1}{8}$–$\frac{1}{6}$ in. diam.; style erect,
2 in. long; berry bright red, the size of a small cherry, $\frac{1}{2}$ in. diam.
Hook. fil. in Bot. Mag. t. 6778; *Baker, Handb. Amaryllid.* 64.

EASTERN REGION : Natal; Inanda, *Wood,* 403 ! and without precise locality,
Mrs. K. Saunders ! Sanderson !

KALAHARI REGION : Transvaal; stony ground at the base of hills, Queen's
River Valley, near Barberton, 2000 ft., *Galpin,* 711 !

Introduced into cultivation by Mr. Keit in 1877. Proves to be the finest and
most easily cultivated of all the species of its subgenus.

2. **H. puniceus** (Linn. Sp. Plant. 325) ; bulb subglobose, oblique,
2–3 in. diam.; scale-leaves orbicular ; leafy stem not produced ;
leaves 2–4, oblong, membranous, bright green, much undulated, 6–9
in. long, 2–3 in. broad, narrowed at the base into a distinct petiole 3–6
in. long, fully developed by the time the flowers appear; main veins
about 6 on each side of the distinct midrib, the central ones $\frac{1}{6}$–$\frac{1}{4}$ in.
apart ; peduncle lateral, $\frac{1}{2}$–1 ft. long; umbel globose, capitate, about
3 in. diam.; pedicels erect, $\frac{1}{2}$–1 in. long; bracts 6–8, oblong, imbricated, permanently ascending, 1½–2 in. long; flowers inodorous;
perianth pale scarlet, rarely white, 1 in. long; tube cylindrical,
$\frac{1}{4}$–$\frac{1}{3}$ in. long; segments lanceolate, 3–nerved; filaments bright red,
an inch long; anthers yellow, oblong, $\frac{1}{12}$ in. long; style straight, red,
$\frac{1}{4}$ in. long; berry bright red, globose, the size of a small cherry. *Bot.
Mag. t.* 1315 ; *Ait. Hort. Kew.* i. 404 ; *Thunb. Prodr.* 59 ; *Fl. Cap.
edit. Schult.* 299 ; *Red. Lil. t.* 320 ; *Herb. Amaryll.* 233 ; *Kunth,
Enum.* v. 588 ; *Baker, Handb. Amaryllid.* 65. *H. rodoutoanus,
Roem. Amaryllid.* 38. *Gyaxis puniceus, Salisb. Gen.* 131.

VAR. *β,* **membranaceus** (Baker); habit weaker; petiole longer in proportion
to the lamina ; peduncle more slender ; pedicels shorter, so that the bracts sometimes overtop the flowers.

SOUTH AFRICA: without locality, *Auge!*
COAST REGION: Humansdorp Div.; Kromme River, *Thunberg!* Port Eliza-
beth Div.; between Drostdy Farm and the Leadmine, near Glassen Point,
Burchell, 4472! and Cragga Kamma, *Zeyher,* 573!
CENTRAL REGION: Somerset Div.; Bruintjes Hoogte, *Burchell,* 3019! 3072!
and Bosch Berg, 3000 ft., *MacOwan,* 1890!
EASTERN REGION: Tembuland; Bazeia, in forests, 2000–2500 ft., *Baur,* 729!
Var. β, Natal; Umymyati, *Wood,* 1219! and without precise locality, *Peddie!*
Gerrard, 735! *Cooper,* 3230!

Introduced into cultivation before the end of the seventeenth century.

3. **H. natalensis** (Pappe ex Hook. in Bot. Mag. t. 5378); bulb
globose, oblique, 2–3 in. diam.; scale-leaves about 4, round-ovate,
pale green, much dotted and tipped with red-brown; leafy stem
produced, about a foot long; leaves 8–9, oblong, membranous, bright
green, not petioled, but narrowed to a clasping base, above a foot long,
4–5 in. broad, the outer tipped and spotted on the back with
red-brown; peduncle produced from the axis of one of the scale-
leaves, distinct from the leafy stem, compressed, sulcate, about a foot
long, ¾–1 in. thick; umbel very dense, globose, capitate, 3–4 in.
diam.; pedicels 1–1½ in. long; bracts about 6, oblong, bright red, imbricated, permanently ascending, 1½–2 in. long; perianth greenish, 1 in.
long; tube cylindrical, ¼ in. long; segments linear, ¾ in. long; filaments yellowish, above an inch long; anthers minute, oblong; ovary
globose, ⅓ in. diam; style overtopping the anthers; berry subglobose,
red, usually 1-seeded, ⅓ in. diam. *Baker, Handb. Amaryllid.* 66.

EASTERN REGION: Natal, *Buchanan!* Griqualand East; moist places near
Kokstad, 5000 ft., *Tyson,* 1843!

Described from living plants in the Kew collection.

4. **H. magnificus** (Herb. in Bot. Reg. 1841, Misc. 71, No. 153);
bulb globose, 3–4 in. diam.; scale-leaves 6–8, orbicular; leafy stem
reaching a length of 2 feet, not developed till after the peduncles,
spotted with red-brown; produced leaves 6–8, oblong, membranous,
bright green, undulated, 12–15 in. long, 4–5 in. broad, not petioled,
narrowed to a clasping base; main veins 8–10 on each side of the
distinct midrib, tranverse veinlets close and very oblique; peduncle
lateral, very stout, about a foot long; umbel globose, very dense,
sometimes 5–6 in. diam.; pedicels slender, ½–1 in. long; bracts 6–8,
ovate, ascending, imbricated, bright green, about 2 in. long, plain
or spotted with red-brown; perianth bright scarlet, 1 in. long;
tube cylindrical; segments ascending, twice as long as the tube;
filaments bright scarlet, an inch long; anthers oblong, yellow; style
overtopping the anthers; berry bright red, the size of a small cherry.
Kunth, Enum. v. 589. *Baker, Handb. Amaryllid.* 66. *H. puniceus,*
var. magnifica, Herb. in Bot. Mag. sub t. 3870. *H. Rouperi, Floral*
Mag. 1875, *t.* 148.

VAR. β, **H. insignis** (Hook. in Bot. Mag. t. 4745); bracts plain green, much
overtopping the flowers, 5–6 in. long.

VAR. γ, **Gumbletoni** (Baker; leafy stem about a foot long; leaves orbicular, unspotted, 5-6 in. long and broad; peduncle 6-8 in. long; bracts and flowers of the type.

VAR. δ, **H. superbus** (Hort.); leafy stem not produced; leaves 5-6, rosulate, contemporary with the flowers, oblong, narrowed from the middle to the base and apex; peduncle, bracts, and flowers of the typical form.

SOUTH AFRICA: without locality.

EASTERN REGION: Delagoa Bay, *Monteiro!* Var. β, Introduced from Natal by the Rev. Mr. Rooper. Var. γ, Described from a specimen and sketch of Mr. Gumbleton's. Var. δ, Described from living plants grown at Kew, and by Messrs. Kreluge and Sir C. W. Strickland.

Mr. Krelage tells me that under similar treatment *superbus* flowers in March, *puniceus* not till July or August.

5. **H. Pumilio** (Jacq. Hort. Schoenbr. i. 32, t. 61); bulb sub-globose, about 1 in. diam.; leaves 2, developed after the flowers, suberect, lorate, subacute, glabrous, 4-5 in. long, about $\frac{1}{2}$ in. broad, green, barred and spotted on the back near the base with red-brown; peduncle slender, about 2 in. long, spotted; flowers 5-6 in an umbel; pedicels $\frac{1}{2}-\frac{3}{4}$ in. long; bracts oblong-lanceolate, $\frac{1}{2}-\frac{3}{4}$ in. long; perianth whitish, about $\frac{1}{2}$ in. long; tube short; segments spreading; stamens and style exserted. *Willd. Sp. Plant.* ii. 27; *Ait. Hort. Kew. edit.* 2, ii. 209; *Poir. Encyc. Suppl.* iii. 34; *Herb. Amaryll.* 234; *Kunth, Enum.* v. 591; *Baker, Handb. Amaryllid.* 67. *Melicho Pumilio, Salisb. Gen.* 130.

SOUTH AFRICA: without locality.

Known to me only from Jacquin's figure.

6. **H. lanceæfolius** (Jacq. Hort. Schoenbr. i. 31, t. 60); bulb ovoid, compressed, 1-1$\frac{1}{2}$ in. diam.; tunics thick, bifarious; leaves 2, rarely 3, not developed till after the flowers, lanceolate, acute, 6-8 in. long, 2 in. broad, pale green, glabrous on the surfaces, ciliated on the margin with minute deflexed hairs, dotted and barred with red, usually spreading; peduncle compressed, green, glabrous, slender, 4-5 in. long; flowers 6-8 in an umbel; pedicels $\frac{1}{2}-\frac{3}{4}$ in. long; bracts lanceolate, about $\frac{1}{2}$ in. long; perianth white slightly tinged with pink, under $\frac{1}{2}$ in. long; tube short; segments linear, spreading; stamens and style exceeding the perianth-segments; anthers oblong, minute, yellow. *Willd. Sp. Plant.* ii. 27; *Ait. Hort. Kew. edit.* 2, ii. 209; *Poir. Encyc. Suppl.* iii. 33; *Herb. Amaryllid.* 234; *Kunth, Enum.* v. 590; *Baker, Handb. Amaryllid.* 67. *Melicho lanceæfolius, Salisb. Gen.* 130.

SOUTH AFRICA: without locality.

Known to me only from Jacquin's figure.

7. **H. carneus** (Gawl. in Bot. Reg. t. 509); bulb compressed, 2-3 in. diam.; tunics thick, imbricated, green, bifarious; leaves 2, developed after the flowers, pale green, oblong or obovate, 4-6 in. long, 2-4 in. broad, narrowed to the base, softly hairy, especially towards the margin; peduncle slender, $\frac{1}{2}$-1 ft. long, usually pilose and mottled with purple; flowers 40-50 or more in a dense globose

umbel 2–3 in. diam.; pedicels slender, $\frac{1}{2}$–1 in. long; bracts 5–6, oblong, usually reflexing, about an inch long; perianth pink, rarely white, $\frac{1}{2}$–$\frac{3}{4}$ in. long; tube $\frac{1}{12}$–$\frac{1}{4}$ in. long; segments lanceolate; stamens generally exserted; anthers minute, oblong, yellow; style overtopping the stamens. *Hook. in Bot. Mag. t.* 3373; *Herb. Amaryllid.* 234; *Kunth, Enum.* v. 590; *Baker, Handb. Amaryllid.* 67. *Melicho carneus, Salisb. Gen.* 130. *H. brevifolius, Herb. Amaryllid.* 234, *t.* 30, *fig.* 3; *Kunth, Enum.* v. 591.

VAR. β, H. **strigosus** (Herb. Amaryllid. 234, t. 30, fig. 2); leaves hairless on the surface when mature. *Kunth, Enum.* v. 590.

SOUTH AFRICA : Var. β, a sketch by *Masson* in the British Museum.
CENTRAL REGION : Somerset Div. ; Bosch Berg, 3500 ft., *MacOwan*, 1578! mountains near Graaff Reinet, 4300 ft., *Bolus*, 764 ! *Bowker*, 33 !
KALAHARI REGION : Orange Free State ; Rhinoster Kop, *Burke!* and Caledon River, *Burke*, 444 !
EASTERN REGION : Transkei ; between Kei River and Gekau, 1000–2000 ft., *Drège*, 4517 ! Tembuland ; Equtyeni, Gatberg country, 4000 ft., *Baur*, 1169 !

8. H. **amarylloides** (Jacq. Hort. Schoenbr. iv. 5, t. 408); bulb ovoid, not compressed, 1$\frac{1}{2}$–2 in. diam. ; tunics thick, bifarious; leaves 2, developed after the flowers, lingulate, obtuse, bright green, glabrous on both edge and margin, 6–12 in. long, 1$\frac{1}{2}$–2 in. broad ; peduncle slender, glabrous, $\frac{1}{2}$–1 ft. long; flowers 50 or more in a dense globose umbel 2–3 in. diam.; pedicels slender, $\frac{1}{2}$–1 in. long ; bracts 5–6, oblong or oblong-lanceolate, spreading, about an inch long ; perianth pinkish or white, $\frac{3}{8}$–$\frac{1}{2}$ in. long ; tube short ; segments lanceolate, spreading ; stamens and style exceeding the perianth-segments. *Poir. Encyc. Suppl.* iii. 33 ; *Kunth, Enum.* v. 589 ; *Baker, Handb. Amaryllid.* 68. *H. amaryllidioides, Herb. Amaryllid.* 233. *Melicho amarylloides, Salisb. Gen.* 130.

COAST REGION : Bedford Div.; hills between Bedford and Esterhuyzens Poort, *MacOwan*, 2248 !
CENTRAL REGION : Somerset Div. ; Brak Fontein, near Somerset East, *MacOwan.*
KALAHARI REGION : Transvaal ; Mooi River, *Burke!*
EASTERN REGION : Transkei ; between Gekau and Bashee River, 1000–2000 ft., *Drège*, 4524 ! Tembuland ; Geigeira, near Bazeia, 2000 ft., *Baur*, 568 !

9. H. **montanus** (Baker); bulb middle-sized ; leaves 2, contemporary with the flowers, lorate, obtuse, glabrous, 8–9 in. long, 1$\frac{1}{2}$ in. broad at the middle ; peduncle slender, glabrous, $\frac{1}{2}$ ft. long; expanded head 2 in. diam. ; bracts many, lanceolate, reflexing ; pedicels $\frac{1}{2}$–$\frac{3}{4}$ in. long; flowers white ; perianth-tube subcylindrical, $\frac{1}{6}$ in. long ; segments lanceolate, spreading, $\frac{1}{2}$ in. long ; stamens and style not longer than the segments.

EASTERN REGION : Natal; valley of the Buffalo River near Charleston, 5000–6000 ft., *Wood*, 4810 !

Very near *H. amarylloides, Jacq.*

10. H. **candidus** (Bull. Cat. New Plants 1894, 3); bulb large, globose ; leaves 2, contemporary with the flowers, fleshy, lingulate, a foot or more long, 4–5 in. broad, hairy on both sides; peduncle as

long as the leaves, very hairy; expanded head 5 in. diam.; bracts
many, lanceolate, reflexed at the flowering-time, $1\frac{1}{2}$ in. long; pedicels
about an inch long; flowers white; perianth-tube cylindrical, $\frac{1}{4}$ in.
long; segments linear, $\frac{5}{8}$ in. long; stamens and style longer than the
segments.

EASTERN REGION: Natal; Tugela Valley, *Allison!*

Sent to us by Mr. F. Boyle, April, 1891.

11. **H. albiflos** (Jacq. Hort. Schoenbr. i. 31, t. 59); bulb compressed, 2–3 in. in the longer diameter; tunics thick, green, bifarious;
leaves 2–4, contemporary with the flowers, lingulate, suberect, obtuse,
6–8 in. long, $2\frac{1}{2}$–$3\frac{1}{2}$ in. broad above the middle, narrowed gradually
to the base, green and glabrous on the surfaces, more or less densely
ciliated on the margin; peduncle pale green, glabrous, 6–9 in. long,
$\frac{1}{3}$ in. diam.; umbel dense, globose, 2 in. diam.; pedicels about $\frac{1}{4}$ in.
long; bracts 5–6, oblong, obtuse, permanently ascending, white with
distinct green veins; flowers scentless, pure white; perianth $\frac{3}{4}$ in.
long; tube subcylindrical, $\frac{1}{4}$ in. long; segments linear, $\frac{1}{2}$–$\frac{5}{8}$ in. long;
filaments white, $\frac{3}{4}$ in. long; anthers oblong, yellow, very small;
ovary oblong, green; style overtopping the anthers; berry globose,
bright red, $\frac{1}{2}$ in. diam. *Willd. Sp. Plant.* ii. 27; *Poir. Encyc. Suppl.*
iii. 32; *Red. Lil. t.* 398; *Lodd. Bot. Cab. t.* 602; *Gawl. in Bot.
Mag. t.* 1239; *Baker, Handb. Amaryllid.* 68. *H. intermedius, Roem.
Amaryllid.* 36. *H. virescens, vars. intermedius and albiflos, Herb.
Amaryllid.* 235; *Kunth, Enum.* v. 592. *Diacles ciliaris, Salisb.
Gen.* 130.

VAR. β, **pubescens** (Herb.); leaves hairy also on the face. *H. pubescens, Gawl.
in Bot. Reg. t.* 382; *Lodd. Bot. Cab. t.* 702, *non Linn. fil. H. virescens, var.
pubescens, Herb. Amaryllid.* 235. *Kunth, Enum.* v. 592, *excl. syn. Diacles
pubescens, Salisb. Gen.* 130.

VAR. γ, **brachyphyllus** (Baker); leaves oblong, about 3 in. long, 2 in. broad,
finely ciliated on the margin, glabrous on the surfaces; peduncle glabrous, a
little shorter than the leaves.

VAR. δ, **Burchellii** (Baker); leaves oblong, 3–4 in. long, 2–$2\frac{1}{4}$ in. broad, hairy
on the surface, densely ciliated on the margin; peduncle hairy, a little longer
than the leaves.

SOUTH AFRICA: without locality, *Villette!*

COAST REGION: Var. δ, Uitenhage Div., *Zeyher,* 967!

CENTRAL REGION: stony places and among shrubs near Graaff Reinet, 2000 ft.,
Bolus, 85! Var. γ, Somerset Div.; mountain above the spring of Commadagga,
Burchell, 3348! Var. δ, Graaff Reinet Div.; between Milk River and Platte
River, *Burchell,* 2955!

Frequent in gardens. Described from a plant that flowered at Kew in
November, 1877.

H. Clarkei, Hort., is a hybrid raised by Colonel Trevor Clarke in 1891, between
albiflos and a variety of *coccineus.*

12. **H. albomaculatus** (Baker in Gard. Chron. 1878, x. 202);
bulb 2 in. diam.; tunics thick, green, bifarious; leaves 2, contemporary with the flowers, lorate, obtuse, suberect, above a foot long,
2–$2\frac{1}{2}$ in. broad above the middle, glabrous when mature, spotted
with white, very obscurely ciliated in an early stage; peduncle

glabrous, green, slightly compressed, 4–6 in. long, $\frac{1}{2}$–$\frac{3}{4}$ in. diam ; umbel dense, globose, 2 in. diam.; bracts 6–7, oblong, imbricated, white with green veins, about 1$\frac{1}{2}$ in. long; pedicels $\frac{1}{8}$–$\frac{1}{4}$ in. long; perianth white, an inch long; tube cylindrical, $\frac{1}{3}$ in. long; segments lanceolate, twice as long as the tube; filaments white, an inch long; anthers oblong, pale yellow; style as long as the stamens. *Baker, Handb. Amaryllid. 69.*

EASTERN REGION: Pondoland, *Bachmann,* 293! Natal; Mount Moreland, *Wood,* 1006! near Durban, *Wood,* 1989!

Described from plants that flowered with Messrs. Low at Clapton, November, 1875, and Messrs. Henderson, Maida Vale, December, 1877.

13. H. Baurii (Baker in Bot. Mag. t. 6875); bulb compressed, oblong, 3–4 in. in the long diameter; tunics thick, green; leaves 2, suborbicular, truncate at the apex, $\frac{1}{2}$ ft. long, 7–8 in. broad, glabrous on the surface, densely ciliated on the margin with minute spreading whitish hairs; peduncle under 2 in. long, $\frac{1}{2}$ in. diam., glabrous; umbel dense, about 2 in. diam. ; bracts 5–6, oblong, white, membranous, imbricated, very obtuse, 2 in. long, overtopping the flowers; pedicels about $\frac{1}{4}$ in. long; perianth pure white, above an inch long; tube infundibuliform, $\frac{1}{3}$ in. long; segments lanceolate, $\frac{3}{4}$ in. long; stamens as long as the perianth-segments; anthers oblong, bright yellow, $\frac{1}{12}$ in. long ; ovary green, oblong; style just exserted. *Baker, Handb. Amaryllid. 69.*

EASTERN REGION: Griqualand East; Shawbury, close to the Tsitsa Falls, 1500 ft., *Baur,* 231!

Flowered at Kew, Nov., 1885. Included amongst the drawings of Natal plants exhibited by Miss Large at the Indian and Colonial Exhibition of 1886.

14. H. Arnottii (Baker in Gard. Chron. 1878, x. 492); bulb globose, 3 in. diam.; tunics thick, green, bifarious; leaves 2, produced after the flowers, round-oblong, spreading, very obtuse, 4–5 in. long, 3–4 in. broad, dull green on the face, very pale green on the back, glabrous except on and near the margin, ciliated with long, soft, whitish hairs; peduncle 4–5 in. long, rather compressed, purplish red, glabrous; umbel 1$\frac{1}{2}$–2 in. diam. ; pedicels $\frac{1}{2}$–$\frac{3}{4}$ in. long; bracts 5–6, oblong-lanceolate, ascending, whitish with conspicuous green veining, 1$\frac{1}{4}$–1$\frac{1}{2}$ in. long; perianth white, $\frac{5}{8}$ in. long; tube $\frac{1}{8}$ in. long; segments lanceolate, twice as long as the tube; filaments distinctly exserted ; anthers minute, oblong, yellow; ovary green, globose; style a little overtopping the anthers. *Baker, Handb. Amaryllid. 68.*

CENTRAL REGION : Colesberg Div., *Arnot!*

Described from a plant that flowered at Kew in June, 1878, the leaves not being fully developed till September.

15. H. deformis (Hook. fil. in Bot. Mag. t. 5903); bulb slightly compressed, 3–4 in. diam.; tunics thick, green, bifarious; leaves 4, distichous, spreading, contemporary with the flowers, orbicular, pilose, 3½–4 in. long and broad, not mottled nor specially ciliated ; peduncle scarcely any; umbel compressed, 1½ in. in the long diameter ; bracts 5–6, subequal, ascending, obovate-oblong, obtuse, white, 1¼–1½ in. long; pedicels ¼ in. long; perianth white, nearly an inch long; tube ¼ in. long ; segments linear, ⅝ in. long; stamens longer than the perianth-segments ; anthers oblong, yellow ; style as long as the stamens. *Baker, Handb. Amaryllid.* 69.

EASTERN REGION : Natal, *McKen!*

Described from a specimen that flowered at Kew in March, 1871.

16. **H. Mackenii** (Baker, Handb. Amaryllid. 69); bulb 3–4 in. diam.; tunics thick, green, bifarious; leaves 2, contemporary with the flowers, round-oblong, spreading, 6–8 in. long, 4–5 in. broad, glabrous on the face, clothed on the margin and under surface with soft, whitish, spreading hairs ; peduncle scarcely any; umbel few-flowered, compressed, about 2 in. in the long diameter ; bracts oblong, the outer greenish, densely pubescent, the inner white, veined with green ; pedicels very short ; perianth white, nearly an inch long ; tube cylindrical, ¼ in. long; segments linear; filaments white, ⅞ in. long ; anthers oblong, yellow ; style as long as the stamens.

EASTERN REGION : Natal, *McKen!*

Described from a plant that flowered with Mr. Bull, Nov., 1870.

17. **H. Cooperi** (Baker, Handb. Amaryllid. 70) ; bulb compressed, 4 in. diam. ; leaves 2, produced in October, spreading orbicular, 4–5 in. long and broad, shortly ciliated on the margin ; scape compressed, glabrous, as long as the leaves, ½ in. diam. ; umbel dense, globose, 3 in. diam., produced in July ; pedicels ¼ in. long ; bracts 5–6, ovate, red, an inch long, ascending ; perianth deep flesh-coloured, an inch long; segments lanceolate, twice as long as the tube ; stamens distinctly exserted ; anthers minute, oblong, yellow ; style overtopping the stamens.

SOUTH AFRICA : without locality.

Described from a sketch of a plant from Mr. Cooper, which flowered at Kew in 1863.

18. **H. hirsutus** (Baker in Gard. Chron. 1878, ix. 756); bulb compressed, 3 in. diam. ; tunics thick, bifarious, brownish ; leaves 2, contemporary with the flowers, round-oblong, 5–6 in. long at the flowering-time, 3½–4 in. broad, suberect, very obtuse, shortly hairy all over both sides, dark green on the face, paler green on the back, the hairs at and near the edge longer than those of the surface ; peduncle a little longer than the leaves, ¼ in. diam., greenish in the upper, reddish in the lower, densely clothed with spreading, soft, whitish hairs ; umbel globose, 3–4 in. diam. ; pedicels about ½ in. long; bracts 6–8, oblong, membranous, bright red, ascending or spreading ; perianth white or pink, an inch long ; segments oblanceolate, obtuse, twice as long as the tube ; filaments white, an inch

long; anthers minute, oblong, pale yellow; ovary green, turbinate; style as long as the stamens. *Baker, Handb. Amaryllid.* 70.

KALAHARI REGION : Transvaal; near Barberton, 4000–4500 ft., *Galpin*, 1183!
EASTERN REGION: Natal; amongst rocks near Mooi River, *Wood*, 3443! and without precise locality, *Cooper*, 3231!
Described from a specimen sent from the Transvaal by Mr. C. Mudd, which flowered with Messrs. Veitch in April, 1878.

19. H. incarnatus (Burchell ex Herb. Amaryllid. 237, t. 31, fig. 1); bulb compressed, 2–3 in. diam.; tunics thick, bifarious; leaves 2, oblong, spreading, 7–8 in. long, $3\frac{1}{2}$–4 in. broad at the middle, narrowed gradually to a base $1\frac{1}{4}$–$1\frac{1}{2}$ in. broad, quite glabrous on both surfaces and margin; peduncle reddish, glabrous, 3–4 in. long; umbel dense, globose; pedicels $\frac{1}{4}$–$\frac{1}{2}$ in. long; bracts 4–5, oblong, ascending, pale red, an inch long; perianth pale red, $\frac{3}{4}$ in. long; tube very short; segments lanceolate; stamens slightly exserted; anthers small, oblong; style as long as the stamens. *Kunth, Enum.* v. 598; *Baker, Handb. Amaryllid.* 70.

COAST REGION : Port Elizabeth Div.; Kragga Kamma, *Burchell*, 4556!

20. H. undulatus (Herb. Amaryllid. 233, t. 30, fig. 1); bulb ovoid, $\frac{5}{8}$ in. diam., narrowed into a neck $\frac{1}{2}$ in. long; leaves 2, suberect, lorate, glabrous, much undulated, 5–6 in. long, $\frac{1}{2}$ in. broad above the middle, narrowed gradually to $\frac{1}{8}$ in. at the base; peduncle slender, glabrous, spotted, 2–3 in. long; umbel dense, about an inch in diameter; pedicels $\frac{1}{4}$–$\frac{1}{3}$ in. long; bracts 4, oblong, ascending, bright red; perianth about $\frac{1}{2}$ in. long; tube very short; segments lanceolate; stamens much exserted, just reaching to the top of the bracts. *Kunth, Enum.* v. 589; *Baker, Handb. Amaryllid.* 70.

COAST REGION : Clanwilliam Div.; Rhenoster Fontein, near Lange Vallei, July, 1793, *Masson!*

21. H. concolor (Herb. Amaryllid. 238, t. 31, fig. 2); bulb oblong, compressed; tunics thick, bifarious; leaves 2, suberect, lorate, about a foot long, $1\frac{1}{2}$–2 in. broad, quite glabrous on both surfaces and margin; peduncle slender, glabrous, rather shorter than the leaves; umbel globose, $1\frac{1}{2}$ in. diam.; pedicels $\frac{1}{4}$ in. long; bracts 4–6, oblong or lanceolate, ascending, reddish, an inch long; perianth bright red, $\frac{3}{4}$ in. long; tube very short; segments lanceolate, above $\frac{1}{2}$ in. long; filaments bright red, $\frac{1}{4}$ in. longer than the perianth-segments; anthers oblong, yellow; style red, overtopping the anthers. *Kunth, Enum.* v. 600; *Baker, Handb. Amaryllid.* 71.

SOUTH AFRICA : without locality, *Burchell*, bulb No. 276!
Very near *H. sanguineus, Jacq*.

22. H. sanguineus (Jacq. Hort. Schoenbr. iv. 4, t. 407); bulb oblong, compressed, 3 in. diam.; tunics thick, whitish; leaves 2, suberect, lorate, about a foot long, 2–3 in. broad above the middle, green, glabrous, unspotted; peduncle produced before the leaves, slender, compressed, glabrous, unspotted, 8–9 in. long; umbel dense, globose; pedicels as long as the flowers; bracts 6–8, lanceolate, acute, reddish, ascending, 1–1½ in. long; perianth ¾ in. long; tube whitish; segments lanceolate, obtuse, blood-red, ½ in. long; filaments exserted; anthers oblong, yellow; style overtopping the anthers. *Poir. Encyc. Suppl.* iii. 32; *Roem. et Schultes, Syst. Veg.* vii. 884; *Herb. Amaryllid.* 235, *t.* 31, *fig.* 5; *Kunth, Enum.* v. 595; *Baker, Handb. Amaryllid.* 71.   *Melicho sanguineus, Salisb. Gen.* 130.

SOUTH AFRICA : without locality.

Known only from Jacquin's figure. The obscure *H. hookerianus, Herb. Amaryll.* 404, *t.* 46, *fig.* 7, may perhaps be the same species, but its leaves are not known.

23. **H.** humilis (Jacq. Hort. Schoenbr. iv. 6, t. 411); bulb ovoid, slightly compressed, 1 in. diam. ; leaves 2, suberect, lorate, produced after the flowers, light green, unspotted, subacute, ½ ft. long, 1½ in. broad above the middle, glabrous on the face, shortly ciliated with reflexed hairs on the margin; peduncle much compressed, 2 in. long; umbel compressed, with about 20 flowers; pedicels nearly as long as the flowers; bracts about 6, subequal, lanceolate, pale red, under an inch long; perianth whitish, ⅓ in. long; tube very short; segments linear; filaments exserted, white; anthers oblong, yellow; style as long as the stamens; berry white; seed solitary, subglobose, brown, shining. *Poir. Encyc. Suppl.* iii. 34; *Roem. et Schultes, Syst. Veg.* vii. 888; *Herb. Amaryllid.* 234; *Kunth, Enum.* v. 591; *Baker, Handb. Amaryllid.* 71.   *Melicho humilis, Salisb. Gen.* 130.

SOUTH AFRICA: without locality.

Known to me only from Jacquin's figure.

24. **H** rotundifolius (Gawl. in Bot. Mag. t. 1618); bulb compressed, 3–4 in. diam. ; tunics thick, bifarious; leaves 2, spreading, suborbicular, produced after the flowers, 4–5 in. long, about 4 in. broad, scaberulous above, the margins scabrous and purplish coloured; peduncle 3–6 in. long, moderately stout, red; umbel dense, compressed, 1½–2 in. diam. ; bracts about 6, oblong, obtuse, ascending, bright red, 1–1¼ in. long, much shorter than the flowers; pedicels ¼–½ in. long; perianth red, ¾ in. long; segments twice as long as the subcylindrical tube; stamens distinctly exserted; anthers small, yellow, oblong; style as long as the stamens. *Roem. et Schultes, Syst. Veg.* vii. 891; *Herb. Amaryllid.* 235, *t.* 31, *fig.* 8; *Kunth, Enum.* v. 594; *Baker, Handb. Amaryllid.* 70. *H. orbicularis, Donn, Hort. Cant. edit.* ii. 38.   *H. lambertianus, Roem. et Schultes, Syst. Veg.* vii. 891.

SOUTH AFRICA : without locality.

25. **H.** callosus (Burchell ex Kunth, Enum. v. 594); bulb compressed, 3–4 in. diam. ; tunics thick, bifarious; leaves 2, spreading, orbicular-oblong, glabrous and smooth on both sides, 9–10 in. long, 6–7 in. broad, margins smooth, not ciliated; peduncle stout, compressed, dotted with red, about 3 in. long; umbel dense, 1½–2 in. diam. ; bracts about 6, oblong, erect, thick, rigid, glabrous, bright red, 1½–2 in. long, and much overtopping the flowers; pedicels 3–4

lines long; perianth ¾ in. long, the tube 1½ line long, the segments linear, obtuse, pink; stamens exserted; anthers small, yellow. *Baker, Handb. Amaryllid.* p. 71.

SOUTH AFRICA: without locality, *Burchell*, bulb No. 129.

The specimen is from a bulb collected by Burchell, and grown in his garden at Fulham, in the years 1817-1819.

26. H. moschatus (Jacq. Hort. Schoenbr. iv. 6, t. 410); bulb compressed, 3 in. diam.; tunics thick, bifarious, brown or reddish at the top; leaves 2, produced after the flowers, lingulate, 1½ ft. long, 4-4½ in. broad at the middle, narrowed to 2 in. at the base, obscurely pubescent when young, glabrous when mature, marked on the back with interrupted transverse bars of darker green on a pale green ground; peduncle glabrous, much compressed, ½ ft. long, pale green blotched with dark green; umbel dense, globose, 2-3 in. diam.; pedicels ¼-½ in. long; bracts 6-8, oblong, subacute, bright red, ascending, about 2 in. long; flowers with a strong foxy or musky scent; perianth nearly an inch long, light scarlet; segments three times as long as the tube; stamens distinctly exserted; anthers oblong, bright yellow; style as long as the stamens. *Poir. Encyc. Suppl.* ii. 33; *Roem. et Schultes, Syst. Veg.* vii. 890; *Herb. Amaryllid.* 236, *t.* 31, *fig.* 7; *Kunth, Enum.* v. 596; *Baker, Handb. Amaryllid.* 72.

SOUTH AFRICA: without locality.

Described after Jacquin, and a plant sent in 1878 by Colonel Trevor Clarke.

27. H. coccineus (Linn. Sp. Plant. 325); bulb compressed, 3 in. diam.; tunics many, thick, bifarious, brown at the apex; leaves 2, fully developed in English gardens in December or January, lingulate, suberect, 1½-2 ft. long, 6-8 in. broad above the middle, narrowed gradually to 3-4 in. at the base, green, not maculate, glabrous both upon the surfaces and margin; peduncle 6-9 in. long, compressed, green, mottled with minute dots of purplish-brown; umbel dense, globose, 2-3 in. diam.; pedicels ¼-½ in. long; bracts 6-8, oblong, obtuse, imbricated, bright red, permanently ascending, 2-2½ in. long; perianth bright red, about an inch long; tube short; segments linear; stamens exserted; anthers oblong, bright yellow; style as long as the stamens. *Thunb. Prodr.* 59; *Fl. Cap. edit. Schult.* 300; *Gawl. in Bot. Mag. t.* 1075; *Red. Lil. t.* 39; *Lodd. Bot. Cab. t.* 240; *Herb. Amaryllid.* 236; *Kunth, Enum.* v. 596; *Baker, Handb. Amaryllid.* 71.

VAR. β, H. coarctatus (Jacq. Hort. Schoenbr. i. 30, t. 57); fully developed leaves about a foot long, 3-4 in. broad; bracts shorter. *Bot. Reg. t.* 181; *Willd. Sp. Plant.* ii. 25; *Poir. Encyc. Suppl.* iii. 32; *Ait. Hort. Kew. edit.* 2, ii. 207.

VAR. γ, H. carinatus (Linn. Sp. Plant. edit. 2, 413); leaves above a foot long, much narrower and more deeply channelled down the face than in the type. *Lam. Encyc.* iii. 102; *Willd. Sp. Plant.* ii. 27; *Ait. Hort. Kew. edit.* 2, ii. 209.

SOUTH AFRICA: without locality, *Villette!*
COAST REGION: near Cape Town, *Thunberg! Burchell*, 892!

28. H. tigrinus (Jacq. Hort. Schoenbr. i. 29, t. 56); bulb sub-globose, compressed; tunics thick, bifarious; leaves 2, falcate, pro-duced after the flowers, lingulate, 9–12 in. long, 3–4 in. broad at the middle, much spotted on the lower part of the back, glabrous on the surfaces, minutely ciliated; peduncle ½ ft. long, green, much spotted with red; umbel dense, globose, 1½–2 in. diam.; bracts 6–8, oblong, obtuse, bright red, permanently erect, 1½–2 in. long, over-topping the flowers; pedicels ¼–⅓ in. long; perianth pale red, ¾–⅞ in. long; tube very short; stamens distinctly assorted; anthers small, oblong, yellow; style as long as the stamens. *Bot. Mag. t.* 1705; *Willd. Sp. Plant.* ii. 25; *Ait. Hort. Kew. edit.* 2, ii. 207; *Poir. Encyc. Suppl.* iii. 32; *Roem. et Schultes, Syst. Veg.* vii. 889; *Herb. Amaryllid.* 237, *t.* 31, *fig.* 3; *Kunth, Enum.* v. 598; *Salisb. Gen.* 131; *Baker, Handb. Amaryllid.* 72.

COAST REGION: Uitenhage Div., *Zeyher! Cooper,* 1588!

29. H. hyalocarpus (Jacq. Hort. Schoenbr. iv. 5, t. 409); bulb ovoid, compressed, 2 in. diam.; tunics thick, bifarious; leaves 2, produced after the flowers, lorate, suberect, about a foot long, 2 in. broad, glabrous on both surfaces and margin, not maculate; peduncle 4–6 in. long, green, mottled with dark red; umbel dense, globose, 2 in. diam.; pedicels ¼ in. long; bracts about 6, oblong, bright red, ascending, overtopping the flowers; perianth an inch long, red; tube short; segments linear; stamens slightly exserted; anthers small, oblong; style as long as the stamens; berry globose, whitish, ⅛ in. diam.; seed solitary, brown. *Poir. Encyc. Suppl.* iii. 33; *Roem. et Schultes, Syst. Veg.* vii. 887; *Herb. Amaryllid.* 236, *t.* 31, *fig.* 9; *Kunth, Enum.* v. 595; *Salisb. Gen.* 131; *Baker, Handb. Amaryllid.* 72.

SOUTH AFRICA: without locality.

Known to me with certainty only from Jacquin's figure. *H. zebrinus, Herb. Amaryllid.* 237, of which the flowers are not known, differs by its narrower leaves, densely maculate on the back, and sometimes also slightly on the face.

30. H. crassipes (Jacq. Hort. Schoenbr. iv. 7, t. 412); bulb ovoid, 1½ in. diam.; tunics thick, bifarious; leaves 2, produced after the flowers, lorate, suberect, about ½ ft. long, 1½–2 in. broad, glabrous on the surfaces, softly ciliated, mottled with red on the back in the lower half; peduncle erect, compressed, 3 in. long, green, mottled with red; flowers inodorous; umbel dense, globose; pedicels short; bracts 5–6, oblong or oblong-lanceolate, reddish, ascending, 1½ in. long; perianth pale red, ¾–⅞ in. long; tube short; segments linear; filaments exserted; anthers small, yellow, oblong; style as long as the stamens. *Poir. Encyc. Suppl.* iii. 33; *Roem. et. Schultes, Syst. Veg.* vii. 889; *Herb. Amaryllid.* 237, *t.* 31, *fig.* 10; *Kunth, Enum.* v. 599; *Baker, Handb. Amaryllid.* 72.

SOUTH AFRICA: without locality.

Known to me only from Jacquin's figure.

31. H. pubescens (Linn. fil. Suppl. 193); bulb compressed, 2–3 in. diam.; tunics thick, bifarious; leaves 2, produced after the flowers, lorate, suberect, 6–9 in. long, $1\frac{1}{2}$–2 in. broad above the middle, narrowed to $\frac{1}{4}$–$\frac{1}{3}$ in. at the base, minutely ciliated, glabrous or finely pilose on the face and back, blotched with red on the back towards the base; peduncle compressed, dark red, 3–4 in. long; umbel dense, globose, 1–$1\frac{1}{2}$ in. diam.; pedicels very short; bracts usually 4, rarely 5–6, oblong, red, ascending, $1\frac{1}{2}$–2 in. long, overtopping the flowers; perianth bright red, $\frac{3}{4}$–$\frac{7}{8}$ in. long; tube short; segments linear; stamens distinctly exserted; anthers yellow, oblong; style overtopping the stamens. *Thunb. Prodr.* 59; *Fl. Cap. edit. Schult.* 299; *Baker, Handb. Amaryllid.* 73. *H. quadrivalvis, Jacq. Hort. Schoenbr.* i. 30, t. 58; *Bot. Mag. t.* 1523; *Ait. Hort. Kew. edit.* ii. 208; *Herb. Amaryllid.* 235, *t.* 31, *fig.* 4; *Kunth, Enum.* v. 594; *Salisb. Gen.* 131.

SOUTH AFRICA: without locality, *Masson !*
COAST REGION: Malmesbury Div.; Groene Kloof, *Thunberg !* Worcester Div.; Dutoits Kloof, *Drège,* 8552 !

I have seen the type specimens of *H. pubescens* both in the Linnean and Thunbergian herbaria.

XIX. BUPHANE, Herb.

Perianth erect, hypocrateriform, with a short campanulate or subcylindrical tube and equal spreading linear or lanceolate segments. *Stamens* inserted at the throat of the perianth-tube; filaments straight, filiform; anthers oblong, dorsifixed, versatile. *Ovary* turbinate, 3-celled; ovules 2–4, attached to the middle of the placenta; style simple, obscurely 3-lobed at the stigmatose apex. *Capsule* membranous, turbinate, loculicidally 3-valved. *Seeds* solitary, globose, bulbiform.

Rootstock a large tunicated bulb; leaves ensiform or lingulate, coriaceous, peduncle short, stout, solid; flowers aggregated in dense umbels; spathe of two opposite deltoid valves.

DISTRIB. One of the two Cape species extends its range to the mountains of Equatorial Africa.

Perianth-tube short; leaves linear, not ciliated ... (1) **longipedicellata.**
Perianth-tube subcylindrical; leaves ensiform, not
  ciliated ... ... (2) **disticha.**
Perianth-tube campanulate; leaves lingulate, ciliated
  with scales ... . (3) **ciliaris.**

1. B longipedicellata (Pax in Engl. Jahrb. x. 4); bulb not seen; leaves linear, erecto-patent, glaucous, glabrous, 4–5 in. long, crisped at the edge; peduncle stout, as long as the leaves; umbel 8 in. or more in diameter; spathes deciduous at the flowering-time; pedicels 5 in. long; perianth-tube very short; segments linear, recurved, rose-red, vittate, $1\frac{1}{2}$ in. long; stamens as long as the perianth-segments; capsule turbinate; seeds 1–2 in a cell.

KALAHARI REGION: Griqualand West; in sandy soil at Barkly West, 3700 ft., *Marloth,* 974 !

2. B. disticha (Herb. in Bot. Mag. sub t. 2578); bulb subglobose, 6–8 in. diam., with many hundreds of tunics, the outer brown and firm in texture; leaves 8–16, ensiform, distichous, erect, coriaceous, glaucous, closely ribbed, narrowed gradually to the point, finally 1–1½ ft. long, 1–1¼ in. broad, hairy on the margin, not ciliated, often much undulated; peduncle stout, ancipitous, glaucous, ½–1 ft. long; flowers very numerous, bright red, arranged in a dense umbel; pedicels 2–4 in. long; spathe-valves deltoid, 2–3 in. long; perianth-tube infundibuliform or subcylindrical, ¼–½ in. long; segments of the limb linear, ¾–1 in. long; stamens equalling or a little exceeding the perianth-segments; anthers yellow, oblong, $\frac{1}{12}$–⅛ in. long; ovary turbinate, green, ⅛ in. diam.; style red, rather curved, slightly exceeding the stamens; capsule turbinate, ¾ in. long, ½ in. diam. *Herb. Amaryll.* 239; *Kunth, Enum.* v. 603; *Baker, Handb. Amaryllid.* 73. *Amaryllis disticha, Linn. fil. Suppl.* 195; *Paterson's Travels,* 51, *cum icone. B. toxicaria, Herb. loc. cit. Hæmanthus toxicarius, Thunb. Prodr.* 59; *Fl. Cap. edit. Schult.* 298; *Jacq. Fragm. tt.* 39 & 41, *fig.* 1; *Gawl. in Bot. Mag. t.* 1217. *Brunsvigia toxicaria, Gawl. in Bot. Reg. t.* 567. *Amaryllis toxicaria, Dietr. Syn.* 1181. *Hæmanthus sinuatus, Thunb. ex Kunth, Enum.* v. 600.

COAST REGION: Uitenhage Div.; Sand Fontein near Coega River, *Burke!* and grassy heights between Uitenhage and Enon, under 1000 ft., *Drège,* 4523a! Albany Div.; between Assegai Bosch and Rautenbach's Drift, *Burchell,* 4194!

CENTRAL REGION: Interior Region, *Thunberg!* Somerset Div.; Bruintjes Hoogte, 3000 ft., *MacOwan,* 2134! and Fish River, *Burke!* Victoria West Div.; Nieuwveld, between Brak River and the Uitvlugt 3000–4000 ft., *Drège,* 4523b!

KALAHARI REGION: Griqualand West Div.; Griqua Town, *Burchell,* 1881/2! Bechuanaland; Pellat Plains near Takun, *Burchell,* 2247/3!

EASTERN REGION: Natal; Inanda, *Wood,* 1401! Clairmont, *Wood,* 1584! and on the Drakensberg, *Bolus,* 2834!

Extends to Angola, *Welwitsch!* and the mountains round Lake Nyassa, *Thomson!* and Lake Tanganyika, *Cameron!*

3. **B. ciliaris** (Herb. in Bot. Mag. sub t. 2573); bulb globose, 3–4 in. diam.; tunics brown, truncate at the top; leaves 4–6, spreading, lingulate, rigidly coriaceous, ½–1 ft. long, 2–4 in. broad, closely ribbed, spotted with red beneath, densely ciliated with persistent, ascending, brown or stramineous flattened bristles ⅛–⅓ in. long; peduncle solid, ancipitous, 4–6 in. long, ½ in. diam.; flowers 50–100 in an umbel; pedicels rigid, triquetrous, 3–4 in. long; spathe-valves deltoid, much shorter than the pedicels; perianth with a small campanulate tube; segments lanceolate, spreading or reflexing, dull purple, ¼–⅓ in. long; stamens shorter than the perianth-segments; anthers minute, oblong; ovary turbinate, ⅛ in. diam.; style slightly declinate; capsule turbinate, triquetrous, ¼–⅓ in. diam. *Herb. Amaryllid.* 240, *t.* 36, *fig.* 8; *Kunth, Enum.* v. 604; *Baker, Handb. Amaryllid.* 74. *Amaryllis ciliaris, Linn. Sp. Plant. ed.* 2, 422; *Linn. fil. Suppl.* 195. *Hæmanthus ciliaris, Linn. Sp. Plant. ed.* 2, 413; *Thunb. Prodr.* 59; *Fl. Cap. edit. Schult.* 299; *Jacq. Fragm. t.* 40, *fig.* 1 & *t.* 41, *fig.* 2. *Brunsvigia ciliaris, Gawl. in Bot. Reg. sub t.* 193; *Lindl. in Bot. Reg. t.* 1153. *Coburgia ciliaris, Herb. in Trans. Hort. Soc.* iv. 181. *Crossyne ciliaris, Salisb. Gen.* 117.

VAR. β, B. **guttata** (Herb. Amaryllid. 240, t. 22, fig. 1); leaf narrrower, the bristles subulate, ¼–½ in. long, variously directed. *Kunth, Enum.* v. 605.

SOUTH AFRICA : without locality, *Villette !* Var. β, *Thunberg !* and a sketch by *Masson* in the British Museum !

COAST REGION: hills near Cape Town, *Thunberg !* near Tulbagh, *Burchell,* 1038! Caledon Div.; near Genadendal, *Burchell,* 7929 !

XX. VELLOZIA, Vand.

Perianth cut down to the ovary ; segments equal, lanceolate or oblong, spreading, marcescent. *Stamens* 6 in all the Cape species (sometimes indefinite) ; filaments very short ; anthers linear, basi-fixed. *Ovary* 3-celled ; ovules numerous, superposed ; style columnar ; stigma capitate, 3-lobed. *Capsule* globose or turbinate, coriaceous, laterally 3-lobed, finally dehiscing loculicidally. *Seeds* numerous, angled or compressed; testa black; albumen firm in texture ; embryo minute, remote from the hilum.

Stems generally thick and woody, surrounded by many fibrous sheaths; leaves linear or subterete, coriaceous, persistent ; peduncles slender, elongated, 1-flowered ; flowers white or tinged with blue or purple.

DISTRIB. The headquarters of the genus is in Central Brazil. In Africa it extends to Abyssinia and Madagascar. Species about 50.

Stems thick and woody, with numerous fibrous coats :
  Ovary echinate :
    Leaves smooth on the edge ...   . (1) **retinervis.**
    Leaves bristle-ciliated ...   . (2) **clavata.**
  Ovary viscose, not echinate :
    Perianth not glandular outside   ...   ... (3) **equisetoides.**
    Perianth glandular outside   ...   . (4) **viscosa.**
  Ovary densely villose   ...   . (5) **villosa.**
  Herbaceous, with shorter, slender stems   ...   ... (6) **elegans.**
  Acaulescent ; flowers very small   ...   . (7) **humilis.**

1. **V. retinervis** (Baker); fruticose, reaching a height of many feet, with branches 1½–2 in. diam., clothed with many sheaths of wiry, brown, parallel fibres with clathrate interspaces ; leaves linear or linear-subulate, ½–1 ft. long, $\frac{1}{12}$–¼ in. broad, closely and finely ribbed, smooth upon the margins and keel ; peduncles 1–3-nate, stiffly erect, 3–5 in. long, not at all viscid, smooth except the very top ; perianth-limb 1½–1¾ in. long ; segments purplish-white, lanceolate, neither glandular nor hairy outside ; anthers linear, ¾ in. long ; ovary turbi-nate, ⅓ in. long, densely echinate with subulate, drab, wiry, ascending bristles ; style cylindrical, as long as the anthers. *Xerophyta retinervis, Baker in Journ. Bot.* 1875, 233 ; *Gard. Chron.* 1876, vi. 836, *fig.* 153 ; *Regel, Gartenfl.* xxvi. 161, *t.* 903.

KALAHARI REGION : Transvaal ; Magalies Berg, *Burke !* *Zeyher,* 1672 ! plains near Pretoria, 4100 ft., *Bolus,* 5802 ! hills near Pretoria, *Rehmann,* 4317 ! near Barberton, *Galpin,* 438 !

Introduced into cultivation by Haage and Schmidt in 1876.

2. **V. clavata** (Baker) ; fruticose ; leaves linear-subulate, glabrous, coriaceous, acuminate, 3–4 in. long, closely ribbed, margined with short ascending bristles ; peduncles 1–2-nate, 4–6 in. long, smooth except towards the tip, where they are black and bristly ; perianth-

limb 1½–2 in. long; segments lanceolate, whitish, the outer neither
gland-dotted nor hairy on the back; anthers and style ¾ in. long;
ovary clavate, ½ in. long, black, echinate with laxly-disposed, short,
ascending, drab bristles. *Xerophyta clavata, Baker in Journ. Bot.*
1875, 233.

EASTERN REGION : Natal; without precise locality, *Gerrard*, 1824 !

3. **V. equisetoides** (Baker); frúticose, with branches an inch
thick, coated with many sheaths of parallel, brown, wiry strands
of matted fibres; leaves linear, glabrous, rigidly coriaceous, acuminate, ½–1 ft. long, ¼–⅓ in. broad, closely and finely ribbed, smooth on
the margins and keel; peduncle 4–6 in. long, rough upwards, with
sessile black glands; perianth-limb an inch long; segments lanceolate, whitish, the outer not gland-dotted on the back; anthers linear,
⅓–½ in. long; ovary clavate, ⅓ in. long, black, glandular-scabrous;
capsule subglobose. *Xerophyta equisetoides, Baker in Journ. Bot.*
1875, 233.

KALAHARI REGION : South African Gold Fields, *Baines !*

Possibly this was collected in Matabeleland, and not south of the Tropic.

Also gathered in the Zambesi country by Meller and Buchanan, where it is
called Chiapiji, and reaches a height of 5 ft.

4. **V. viscosa** (Baker); fruticose; leaves linear, rigidly coriaceous,
acuminate, 6–9 in. long, ⅛ in. broad, closely and strongly ribbed,
scabrous on the edges and prominent keel; peduncle 8–9 in. long,
clothed with sessile black glands down to the base; perianth-limb
1 in. long; segments lanceolate, rose-red, the outer dotted with large
black glands all over the back; anthers linear, half as long as the
perianth-segments; ovary clavate, ⅓ in. long, black and viscous all
over; capsule oblong-turbinate. *Xerophyta viscosa, Baker in Journ.*
Bot. 1875, 235.

KALAHARI REGION : Transvaal; Zoutpansberg Div., Hout Bosch, *Rehmann*,
5790! 5791!
EASTERN REGION : Pondoland, *Sutherland !* Griqualand East; Mount Currie,
7000 ft. ? *MacOwan and Bolus, Herb. Norm.*, 896! Natal; on the Drakensberg near Tugela Falls, *Wood*, 3439!

5. **V. villosa** (Baker in Journ. Bot. 1889, 3); fruticose, with
woody branches an inch thick, coated with many sheaths of parallel,
slender, wiry fibres; leaves linear-subulate, falcate, rigidly coriaceous,
4–6 in. long, densely clothed throughout with short spreading
whitish hairs; peduncle 3–4 in. long, villose; perianth-limb 1¼ in.
long; segments lanceolate, whitish, the outer villose on the back;
anthers linear, ½ in. long; ovary turbinate, ½ in. long, densely
villose; style cylindrical, equalling the anthers.

KALAHARI REGION. Transvaal ; Zoutpansberg Div., Hout Bosch, *Rehmann*,
5792 !

6. **V. elegans** (Oliver in Bot. Mag. t. 5803); herbaceous; stems
tufted, slender, reaching sometimes a length of half a foot, with

leaves laxly disposed all down, the lower spreading, the upper
ascending ; leaves linear, glabrous, bright green, subcoriaceous, 4–6
in. long, $\frac{1}{4}$–$\frac{1}{3}$ in. broad, smooth upon the margin and midrib beneath ;
peduncles slender, 1–3-nate, glandular, glabrous, 4–5 in. long ;
perianth limb $\frac{1}{2}$–$\frac{3}{4}$ in. long ; segments oblong-lanceolate, white, the
outer glabrous on the back ; anthers lanceolate, $\frac{1}{6}$ in. long ; ovary
turbinate, green, glabrous, $\frac{1}{6}$ in. long ; stigma oblong, overtopping
the anthers ; capsule oblong, $\frac{1}{2}$ in. long. *Talbotia elegans, Balf. in
Proc. Bot. Soc. Edinen.* ix. 192. *Xerophyta elegans, Baker in
Journ. Bot.* 1875, 234. *Hypoxis barbacenioides, Harv. MSS.*

VAR. β, **minor** (Baker) ; dwarfer, with falcate leaves 1½–2 in. long ; perianth-
limb ¼–⅓ in. long. *Xerophyta minuta, Baker loc. cit.*

EASTERN REGION : Natal ; Inanda, *Wood,* 1114 ! Fields Hill, *Sanderson,* 598 !
and without precise locality, *Gerrard,* 1555 ! *Cooper,* 2563 ! Var. β, Natal,
Gueinzius !

7. **V. humilis** (Baker in Journ. Bot. 1889, 4) ; herbaceous ;
acaulescent, with a tuft of strong wiry root-fibres ; rosettes of leaves
densely cæspitose, surrounded by dense mass of sheaths composed
of parallel, drab, wiry strands of matted fibres $\frac{1}{2}$–$\frac{3}{4}$ in. long ; leaves
linear, falcate, glabrous, rigidly coriaceous, strongly ribbed, 1–2 in.
long ; peduncles very slender, 1–2 in. long, rough with glands ;
perianth-limb $\frac{1}{4}$ in. long, segments oblong-lanceolate, the outer
green and glabrous on the back ; anthers linear, nearly as long as
the perianth-limb ; ovary obconic, $\frac{1}{12}$ in. long and broad.

KALAHARI REGION : Transvaal ; Bosch Veldt between Elands River and
Klippan, *Rehmann,* 5138 ! Apies River, *Burke,* 122 ! and without precise
locality, *Greenstock !*

Also Mozambique, *Sir John Kirk !*

ORDER CXXXVI. **DIOSCOREACEÆ.**

(By J. G. BAKER.)

Flowers typically diœcious (universally so in the Cape genera).
Perianth regular, with 6 biseriate subequal segments. *Male flowers :*
Perianth campanulate, infundibuliform or oblong ; tube short ; segments ovate, oblong or oblanceolate. *Stamens* 6, inserted in the tube
of the perianth or the 3 opposite the inner segments absent or
imperfect ; filaments short, incurved ; anthers 2-celled. *Pistil* rudimentary. *Female flowers : Perianth* similar, but smaller. *Staminodia*
minute. *Ovary* inferior, linear-oblong, acutely triquetrous, 3-celled ;
ovules 2 in a cell, superposed ; style short ; stigmas 3, recurved,
entire or bifid. *Fruit* in the Cape genera an acutely-angled triquetrous
coriaceous capsule, with loculicidal dehiscence. *Seeds* flat, usually
winged ; embryo small, surrounded by fleshy albumen.

Rootstock tuberous ; stems twining ; leaves usually alternate, petioled, simple
or compound ; venation reticulated ; flowers small, greenish, spicate or racemose,
sometimes panicled ; bracts and bracteoles minute.

DISTRIB. Cosmopolitan, mainly tropical or subtropical. Genera 8. Species under 200.

I. **Dioscorea.** *Tuber* fleshy, subterranean. *Seed* winged all round or (§ *Helmia*) at the base only.

II. **Testudinaria.** *Tuber* large, firm, half-exposed. *Seed* winged only at the apex.

I. DIOSCOREA, Linn.

Flowers diœcious. *Male perianth* campanulate, with a short tube and 6 subequal oblanceolate segments. *Stamens* usually 6, inserted in the perianth-tube, sometimes only 3, the three inner abortive or entirely absent; filaments short, incurved; anthers oblong or globose. *Style* rudimentary. *Female perianth* smaller, with minute staminodes usually present. *Ovary* linear-triquetrous, 3-celled; ovules 2 in a cell, superposed; style very short; stigmas 3, falcate, entire or bifid. *Capsule* oblong or obovate, rigidly coriaceous, acutely triquetrous. *Seeds* discoid, winged all round or (§ *Helmia*) at the base only.

Rootstock a hypogæous tuber; stems twining; leaves simple or compound, usually alternate; flowers small, greenish-yellow, spicate, racemose or panicled.

The sections to which species belong, which have not been seen in fruit, cannot be determined.

DISTRIB. Cosmopolitan. Species about 150.

Leaves simple, not lobed :
  Leaves shallowly cordate :
    Leaves twice as long as broad, 3-nerved   ... (1) **Burchellii.**
    Leaves as long as broad, 5-nerved ... ... (2) **malifolia.**
  Leaves deeply cordate :
    Leaves acute ; basal sinus broad ... ... (3) **Mundii.**
    Leaves obtuse ; basal lobes touching ... (4) **Rehmanni.**
Leaves simple, lobed :
  Male flowers subspicate :
    Leaves palmate ...   . (5) **multiloba.**
    Leaves cordate-ovate with a large central lobe :
      Stamens 6 ...   . (6) **diversifolia.**
      Stamens 3 ...   . (7) **rupicola.**
  Male flowers racemose...   ...   . (8) **undatiloba.**
Leaves compound :
  Leaflets 3 ; flowers panicled :
    Male flowers longer, not crowded ... ... (9) **dregeana.**
  Leaflets 5 ; flowers panicled :
    Leaflets sessile :
      Outer leaflets large, deeply lobed ... ... (10) **microcuspis.**
      Outer leaflets small, entire ... ... (11) **Forbesii.**
    Leaflets stalked :
      Pedicels very short :
        Bracts much longer than the flowers ... (12) **crinita.**
        Bracts as long as the flowers ... ... (13) **retusa.**
      Pedicels ¼–⅓ in. long   ...   . (14) **Tysoni.**

1. **D. Burchellii** (Baker in Journ. Bot. 1889, 1) ; stems slender, very twining, glabrous ; leaves alternate ; petiole ¼–½ in. long ; blade simple, ovate-lanceolate, shallowly cordate at the base, always entire, mucronate, 1–2 in. long, moderately firm in texture, bright green, triplinerved ; male flowers in lax subspicate racemes 1–2 in. long

with a slender straight glabrous axis ; pedicels very short, subtended
by a minute ovate-lanceolate bract; perianth campanulate, $\frac{1}{12}$ in.
long ; tube very short; segments oblong, obtuse ; stamens 6 ; filaments incurved, longer than the globose anthers ; rudimentary pistil
globose ; female flowers and fruit unknown.

COAST REGION : George Div. ; in a forest near Touw River, *Burchell*, 5728!
Stockenstrom Div. ; Katberg, *Hutton!* British Kaffraria, *Mrs. Barber*, 34 !

2. D. (Helmia) malifolia (Baker in Journ. Bot. 1889, 1) ; stems
slender, twining, glabrous; leaves alternate ; petiole $\frac{1}{2}$–1 in. long ;
blade simple, entire, broad ovate, truncate or slightly cordate at the
base, 1–2 in. long and broad, moderately firm in texture, green and
glabrous on both surfaces, 5-nerved from base to apex, minutely
mucronate ; male flowers in copious lax fascicled racemes 2–3 in.
long with a flexuose or straight glabrous rachis ; pedicels short,
ascending ; bracts minute ovate, acuminate ; perianth glabrous,
$\frac{1}{16}$–$\frac{1}{12}$ in. long ; tube short; segments oblong, obtuse ; fertile stamens
6, much shorter than the perianth-segments ; female flowers in lax
racemes 4–6 in. long ; ovary cylindrical-triquetrous, glabrous, $\frac{1}{3}$ in.
long ; capsule obovate-triquetrous, emarginate, an inch long ; seeds
with a large basal wing.

COAST REGION : Koingha, *Flanagan*, 97 !
KALAHARI REGION : Transvaal; near Burberton, *Galpin*, 764 ! 1189 !
EASTERN REGION : Tembuland; Morley, 1000–2000 ft., *Drège*, 4500 ! Natal;
Inanda, *Wood*, 753 ! Northdene, near Durban, 500 ft., *MacOwan and Bolus*,
Herb. Norm., 1035 ! Port Natal, *Sanderson*, 232 ! and without precise locality,
Cooper, 3247 ! *Gerrard*, 444 !

3. D. (Helmia) Mundii (Baker in Journ. Bot. 1889, 1) ; stems very
slender, twining, glabrous ; leaves alternate ; petiole $\frac{1}{2}$–1 in. long ;
blade simple, entire, ovate, deeply cordate at the base, 2–3 in. long,
mucronate, membranous, glabrous, triplinerved ; male flowers in lax
fascicled simple racemes 3–4 in. long with a straight glabrous axis ;
pedicels short, subtended by a pair of minute ovate acuminate bracts ;
perianth infundibuliform, glabrous, $\frac{1}{12}$ in. long ; segments oblanceolate, obtuse ; fertile stamens 6, nearly as long as the perianth-
segments ; female flowers also in lax simple racemes ; capsule oblong-
triquetrous, glabrous, $1\frac{1}{4}$ in. long, $\frac{5}{8}$ in. broad ; seed winged at the
base only. *Testudinaria nemorum, Mund Exsic.*

SOUTH AFRICA : without locality, *Mund!*
COAST REGION : Knysna Div. ; Koratra, below 1000 ft., *Drège*, 8559a !

4. D. Rehmanni (Baker) ; stems slender, twining, glabrous ;
leaves cordate-ovate, obtuse, moderately firm, green and glabrous on
both surfaces, those of the main stem about 2 in. long and broad ;
basal lobes rounded, touching or slightly imbricate ; petiole short ;
male flowers in lax axillary simple racemes 2–3 in. long; pedicels
spreading or erecto-patent, $\frac{1}{6}$ in. long ; bracts lanceolate, minute,
persistent ; perianth of male flowers $\frac{1}{6}$ in. long ; tube funnel-shaped ;
segments oblong-lanceolate, as long as the tube ; stamens shorter than
the perianth-segments ; female flowers and fruit unknown.

KALAHARI REGION: Transvaal ; Zoutpansberg Div.; Hout Bosch, *Rehmann*, 5783 !

5. D. multiloba (Kunth, Enum. v. 376); stems very twining, glabrous; leaves alternate; petiole ½–1 in. long; blade simple, firm in texture, green and glabrous on both surfaces, cordate-deltoid when fully developed, 3–4 in. long and broad, with 5–7 deltoid lobes, reaching about halfway down to the petiole, the central one prominently triplinerved from base to apex, conspicuously mucronate, sometimes shallowly 3- or 5-lobed with a large rounded central lobe, or even cordate-ovate or cordate-orbicular and quite undivided; male flowers in lax spikes 1–2 in. long with a very flexuose, strongly-angled, glabrous rachis, each subtended by a small deltoid cuspidate bract; perianth campanulate, glabrous, ⅛ in. diam. when expanded; tube short; segments ovate, acute; fertile stamens 6; anthers elliptical; rudimentary style very short, triquetrous; female flowers deflexed, arranged in lax spikes with a flexuose rachis; perianth campanulate, ⅙ in. diam.; ovary linear-oblong, glabrous, ⅓ in. long; capsule obovate-oblong, ¾–1 in. long; seeds with an orbicular nucleus, with a broad membranous wing all round.

EASTERN REGION: Natal; Inanda, *Wood*, 329 ! 825 ! and without precise locality, *Gueinzius ! Gerrard*, 772 ! Also *Drège*, 4496 ! collected probably in Pondoland.

6. D. diversifolia (Griseb. in Endl. and Mart. Fl. Bras. iii. 41); stems glabrous, twining, very slender; leaves alternate; petiole 1–2 in. long; blade membranous, glabrous, cordate-ovate, reaching a length and breadth of 3–5 in., shallowly or deeply 3–5-lobed at the base when fully developed, but often quite entire, acute or obtuse, conspicuously mucronate, the large central segment triplinerved from base to apex; male flowers in 2–3-nate very lax spikes 1–3 in. long with a slender, straight, glabrous rachis, each subtended by a minute ovate bract; perianth campanulate, ⅙ in. diam. when expanded; tube short; segments ovate, acute; fertile stamens 6, much shorter than the perianth; rudimentary style very minute; female flowers in lax simple racemes; pedicels short, patent; capsule obovate, nearly truncate at the apex, ¾–⅞ in. broad; seeds with an orbicular nucleus, with a broad membranous wing all round. *Kunth, Enum.* v. 375.

EASTERN REGION: Pondoland; between Umtentu River and Umzimkulu River, under 500 ft., *Drège*, 4497 ! Natal; Groenberg, *Wood*, 892 ! and without precise locality, *Gerrard*, 1920 !

Described by Grisebach by mistake as a Brazilian plant.

7. D. rupicola (Kunth, Enum. v. 378); stems very slender, very twining, glabrous; leaves alternate; petiole ¾–1 in. long; blade simple, cordate-hastate, 2–3 in. long and broad, membranous, bright green, shallowly 5-lobed with a large deltoid, mucronate, central segment triplinerved from base to apex, and two small orbicular lobes on each side of the spreading base; male flowers in lax subspicate racemes 2–3 in. long with a flexuose glabrous axis; pedicels very short, subtended at the base by a pair of minute bracts; perianth campanulate, $\frac{1}{8}-\frac{1}{6}$ in. diam. when expanded; segments ovate, acute; fertile stamens 3; filaments 2-3 times as long as the minute subglobose anthers; rudimentary pistil absent; female spikes lax, solitary, simple, 4-4$\frac{1}{2}$ in. long; bracts ovate acuminate, $\frac{1}{4}$ as long as the glabrous ovary; perianth with 3 minute staminodia; fruit unknown. *Testudinaria rupicola, Ecklon MSS.*

SOUTH AFRICA : without locality, a specimen from Berlin Botanic Garden!
EASTERN REGION : Griqualand East; at the foot of Mount Currie, 5300 ft., *Tyson,* 1433! *MacOwan and Bolus, Herb. Norm.,* 468! Natal; Inanda, *Wood,* 1167!

8. D. undatiloba (Baker in Journ. Bot. 1889, 2); stems very slender, very twining, glabrous; leaves alternate; petiole about an inch long; blade cordate-deltoid, membranous, glabrous, bright green, 2-3 in. broad, palmately 7-lobed more than halfway down to the petiole, the central lobe the largest, conspicuously mucronate, repand-pinnatifid, triplinerved from base to apex, the two lobes on each side of it also repand, obliquely truncate at the apex, the 4 others shorter and entire; male flowers in lax simple racemes 1-2 in. long; pedicels nearly as long as the flower, subtended by an ovate-lanceolate bract; perianth campanulate, $\frac{1}{8}$ in. long; segments ovate, acute; female flowers in lax spikes 3-4 in. long; ovary clavate, glabrous, $\frac{1}{4}$ in. long; perianth campanulate; segments ovate-lanceolate; fruit not seen.

EASTERN REGION : Natal; without precise locality, *Gerrard,* 1617!

9. D. (Helmia) dregeana (Baker); stems stout, very twining, pubescent; leaves alternate; fully developed petiole 3-4 in. long; blade digitately trifoliolate; leaflets 3-8 in. long, thin but firm in texture, green, pubescent beneath, the end one obovate cuspidate, triplinerved from base to apex, the side ones obliquely ovate, much produced on the lower side; male flowers in ample panicles in sessile globose clusters spaced out on the slender, spreading, pubescent branches; perianth campanulate, $\frac{1}{16}$ in. long; segments ovate; bracts ovate, very hairy, about as long as the clusters; fertile stamens 6; female flowers in ample panicles, with spreading spicate branches; perianth-segments very small, ovate; ovary linear-oblong, densely pubescent, $\frac{1}{4}$ in. long; capsule oblong, deflexed, 1$\frac{1}{2}$-2 in. long; seeds with a semiorbicular apical wing about the breadth of the nucleus. *Helmia dregeana, Kunth, Enum.* v. 437.

EASTERN REGION : Pondoland; between St. John's River and Umtsikaba River, 1000-2000 ft., *Drège,* 4502b! Griqualand East; Malowe Woods, 4000 ft., *Tyson,* 1584! and *MacOwan and Bolus, Herb. Norm.,* 1204! Natal; Inanda, *Wood,* 417! 429! Durban, *Sanderson,* 461! *Rehmann,* 8575! and without precise locality, *Gerrard,* 1455!

A near ally of the Tropical Asiatic *D. dæmona,* Roxb.

10. D. microcuspis (Baker); stems slender, twining, glabrous; leaves membranous, glabrous, digitately 5-foliolate; leaflets sessile, oblong, about 3 in. long, usually acute, with a minute cusp, the end one rarely retuse and emarginate, the outer ones with an oblong

obtuse lobe on the lower side; petiole shorter than the blade; flowers of both sexes unknown; fruit in lax, simple, shortly peduncled, axillary racemes; pedicels very short, deflexed; capsule oblong, acutely angled, retuse, glabrous, an inch long; seeds oblong, narrowly winged along the margin as well as at the base.

EASTERN REGION: Natal; Adendorf's farm, Newcastle, *Rehmann*, 7042!

11. **D. Forbesii** (Baker in Journ. Bot. 1889, 2); stems slender, very twining, shortly pubescent; leaves alternate; petiole ½–1 in. long; blade digitately compound; leaflets 5, sessile, oblanceolate-oblong, obtuse, distinctly mucronate, moderately firm in texture, 1–2 in. long, finely pubescent; male flowers in shortly-peduncled geminate spikes with a slender very hairy rachis; perianth campanulate, densely villose, $\frac{1}{12}$ in. long, subtended by an ovate-lanceolate acuminate bract a little longer than the flower; fertile stamens 6; filaments very short; anthers globose; rudimentary style very short; female flowers and capsule not seen.

EASTERN REGION: Delagoa Bay, *Forbes!*

12. **D. crinita** (Hook. fil. in Bot. Mag. t. 6804); stems slender, twining, pubescent; leaves alternate; petiole 1–2 in. long; blade compound; leaflets 5, distinctly stalked, obovate, membranous, conspicuously mucronate, 1½–3 in. long, green on both surfaces, thinly pilose; male flowers in 2–4-nate, shortly peduncled, axillary racemes with a very slender, densely pilose rachis; pedicels very short; perianth campanulate, very hairy, $\frac{1}{12}$ in. long, subtended on the outside by an ovate-lanceolate, very acuminate, hairy, membranous bract two or three times as long as the flower; fertile stamens 3; filaments very short; female flowers in shortly peduncled dense spikes; ovary oblong, densely villose; perianth-segments very small; capsule not seen.

EASTERN REGION: Natal; Berea, near Durban, 150 ft., *Wood!* Inanda, *Wood*, 1618! and without precise locality, *Gerrard*, 445! 512!

Introduced into cultivation by Mrs. Steane in 1883!

13. **D. retusa** (Masters in Gard. Chron. 1870, 1149, with woodcut); stems slender, very twining, finely pubescent; leaves alternate; petiole 2 in. long; blade compound; leaflets 5, distinctly stalked, obovate, retuse, mucronate, 1½–2 in. long; moderately firm in texture, green, glabrous; male flowers in 3–4-nate, shortly-peduncled racemes with a slender hairy rachis; pedicels very short; perianth oblong, $\frac{1}{12}$ in. long, subtended on the outside by a large ovate acuminate, membranous bract, usually shorter than the flower, and 2 minute broad bracteoles; perianth-segments oblong, connivent; fertile stamens 3; filaments very short; staminodia 3; rudimentary style 3-lobed; female flower and fruit not seen.

SOUTH AFRICA: without locality, *Zeyher*, 1846!
KALAHARI REGION: Orange Free State, *Cooper*, 907; Transvaal; Magalies River, *Burke*, 266!

14. D. Tysoni (Baker in Journ. Bot. 1889, 2); stems slender, glabrous, sarmentose ; petiole slender, 1½-2 in. long; leaflets 5, petiolulate, obovate-oblong, distinctly mucronate, glabrous, membranous, 1-2 in. long ; racemes 2-4-nate, lax, simple, nearly sessile ; rachis pubescent; pedicels ⅛-¼ in. long, not bracteate at the base ; male perianth ⅛ in. long, campanulate, with a short tube and oblong segments.

EASTERN REGION : Griqualand East, near Fort Donald, *Tyson*, 1646 !

II. TESTUDINARIA, Salisb.

Flowers diœcious. *Male perianth* campanulate, with a short tube, and 6 subequal, spreading, oblanceolate segments. *Stamens* 6, inserted in the perianth-tube; filaments short, incurved; anthers oblong. *Style* rudimentary. *Female perianth* smaller, with minute staminodes. *Ovary* 3-celled ; ovules 2 in a cell, superposed; style very short ; stigmas 3, recurved, 2-lobed. *Capsule* rigid, acutely triquetrous. *Seeds* flat, with a broad apical wing.

Rootstock a large spherical half-epigæous tuber ; stems very slender, twining ; leaves simple, petioled, alternate ; flowers of both sexes in simple racemes.

DISTRIB. Endemic.

Leaf cordate-orbicular, not lobed   ...   .   (1)   **elephantipes.**
Leaf cordate-deltoid, shallowly 3-lobed   ...   ...   (2)   **sylvatica.**

1. T. Elephantipes (Burch. Trav. ii. 147); rootstock often 2-3 ft. diam., immersed in the lower half only, firm in texture and greyish-brown on the outer surface, and deeply areolated ; stems slender, glabrous, much branched, twining, the branchlets twining by the slender tips ; leaves alternate, shortly petioled, suborbicular, rather broader than long, 1-2 in. broad, entire, moderately firm in texture, bright green or glaucous, conspicuously mucronate ; male flowers in lax shortly-peduncled racemes 2-3 in. long ; pedicels erecto-patent, equalling or exceeding the flowers, with a lanceolate bract at the base, and often a bracteole at the middle ; perianth ⅛ in. long ; segments spreading widely ; stamens nearly as long as the perianth-segments ; filaments filiform ; anthers oblong or globose, pale yellow, shorter than the filaments ; female flowers also in simple racemes ; pedicels finally deflexed, ¼-⅓ in. long ; capsule obovate-oblong, under an inch long ; seed with a membranous wing ½ in. long, ¼ in. broad. *Lindl. Bot. Reg. t. 921; Kunth, Enum.* v. 441. *Tamus Elephantipes, L'Herit. Sert.* 29 ; *Ait. Hort. Kew.* iii. 401 ; *edit.* 2. v. 386; *Bot. Mag. t.* 1347. *Dioscorea Elephantopus, Spreng. Syst.* iv. *cur. post.* 143. *Testudinaria montana, Burch. l.c.* 148 ; *Lindl. in Bot. Reg. sub t.* 921. *Dioscorea montana, Spreng. Syst.* iv. *cur. post.* 143.

COAST REGION : Uitenhage Div., *Ecklon,* 788 ! *Zeyher,* 583 ! Alexandria Div. ; Hoffmanns Kloof, between Enon and the Zuur Berg Range, 1000 2000 ft., *Drège,* 8561!

CENTRAL REGION : Alexandria Div. ; on the rocks of Zwartwater Poort, 3390! Somerset Div. ; Bosch Berg, 3500 ft., *MacOwan and Bolus, Herb. Norm.,* 1214! Mountains on the S.W. side of Graaff Reinet, *Burchell,* 2912! Mountain sides near Graaff Reinet, 3000-4000 ft., *Bolus,* 288! *Drège,* 8560c ! *Bowker,* 23 !

KALAHARI REGION : Orange Free State, *Cooper,* 905 ! 906 !

2. T. sylvatica (Kunth, Enum. v. 443); rootstock and habit as in the other species; leaves deltoid, deeply cordate, usually shallowly 3-lobed, 2–3 in. long and broad; apical mucro $\frac{1}{8}$–$\frac{1}{6}$ in. long; male racemes simple, shortly peduncled, 2–3 in. long; pedicels $\frac{1}{8}$–$\frac{1}{6}$ in. long, erecto-patent or patent with a small deltoid bract at the base and usually a bracteole at the middle; expanded flower $\frac{1}{8}$ in. diam.; segments oblanceolate, obtuse; stamens much shorter than the perianth-segments; female flowers in lax simple racemes; pedicels shorter than those of the male flowers; ovary linear-oblong, $\frac{1}{4}$–$\frac{1}{3}$ in. long; capsule obovate, under an inch long. *Dioscorea sylvatica, Ecklon. Exsic. D. hederifolia, Griseb. in Endl. and Mart. Fl. Bras.* iii. 42.

COAST REGION : Humansdorp Div.; north side of Kromme River, *Burchell*, 4855 ! Uitenhage, *Zeyher*, 892 ! Port Elizabeth Div.; at Cragga Kamma, *Burchell*, 4519 ! Albany Div.; Howisons Poort, near Grahamstown, *Zeyher*, 4152 ! British Kaffraria, *Cooper*, 64 !
CENTRAL REGION : Somerset, *Bowker !*
KALAHARI REGION : Transvaal; hillside near Edwin Bray Battery, Barberton, 2200 ft., *Galpin*, 1185 !
EASTERN REGION : Transkei; rocky banks of Bashee River, *Drège*, 4499a ! Tembuland; between Bashee River and Morley, 1000–2000 ft., *Drège*, 4499b ! Natal; Inanda, *Wood*, 420 ! 780 ! 1329 ! and without precise locality, *Gerrard*, 508 ! *Cooper*, 1186 ! 3243 ! *Sanderson !*

This was supposed by Grisebach by mistake to be Brazilian.

ORDER CXXXVII. **LILIACEÆ**.

(By J. G. BAKER.)

Flowers usually hermaphrodite and regular, rarely unisexual or slightly irregular. *Perianth* inferior, corolline, with 6 biseriate subequal segments, either free to the base or united in a tube. *Stamens* 6, placed opposite the perianth-segments; filaments usually filiform; anthers didymous, basifixed or dorsifixed, dehiscing by introrse, extrorse or lateral slits, rarely by terminal pores. *Ovary* superior, 3-celled; ovules collateral or superposed, rarely solitary, usually anatropous, rarely orthotropous; style entire or trifid; stigmas rarely capitate. *Fruit* capsular or baccate. *Seeds* with copious fleshy or horny albumen, enclosing a straight or curved embryo.

Erect herbs, or sometimes climbers or erect shrubs; rootstock often bulbous; leaves various in shape and texture, thick and fleshy in the *Aloineæ*, often all radical; inflorescence umbellate, spicate, racemose or panicled; bracts various.

DISTRIB. Cosmopolitan. Genera 187. Species above 2000.

Series I. ASPARAGACEÆ. *Fruit* baccate. *Anthers* with introrse dehiscence.

Tribe 1. *SMILACEÆ. Stems* woody, climbing. *Leaves* normal, with reticulated veining. *Ovules* orthotropous.

I. **Smilax.**—*Flowers* diœcious, umbellate.

Tribe 2. *ASPARAGEÆ. Stems* generally woody, climbing or erect. *Leaves* small and bract-like, with solitary or fascicled cladodia in their axils. *Perianth-segments* distinct.

II. Asparagus.—*Stamens* free.

Tribe 3. *LUZURIAGEÆ. Stems* woody, erect. *Leaves* normal. *Perianth*-segments united into a tube at the base.

III. Behnia.—*Veins* of leaves lax, conspicuous. *Flowers* cymose. *Perianth* oblong.

IV. Dracæna.—*Veins* of leaves close, inconspicuous. *Flowers* racemose. *Perianth* cylindrical.

Series II. LILIACEÆ VERÆ. *Fruit* a loculicidal capsule. *Anthers* with introrse dehiscence.

Tribe 4. *HEMEROCALLEÆ. Rootstock* not bulbous. *Leaves* not fleshy. *Inflorescence* racemose. *Perianth*-segments united into a tube at the base.

V. Kniphofia.—*Perianth* cylindrical.
VI. Notosceptrum.—*Perianth* short, campanulate.

Tribe 5. *ALOINEÆ. Rootstock* not bulbous. *Leaves* thick, fleshy, toothed. *Inflorescence* racemose. *Perianth*-segments united into a tube at the base.

VII. Gasteria.—*Perianth* with a ventricose tube and short segments.
VIII. Aloe.—*Perianth* cylindrical, with a short tube and long connivent segments.
IX. Apicra.—*Perianth* cylindrical, with a long tube and short equal spreading segments.
X. Haworthia.—*Perianth* with an oblong tube and bilabiate limb.

Tribe 6. *ASPHODELEÆ. Leaves* linear or ensiform, entire, not thick and fleshy. *Inflorescence* racemose. *Perianth* cut down to the base.

* *Rootstock not bulbous. Anthers dorsifixed, versatile.*

XI. Bulbinella.—*Flowers* yellow or white. *Filaments* glabrous. *Ovules* 2 in each cell.
XII. Bulbine.—*Flowers* bright yellow. *Filaments* bearded. *Ovules* more than 2 in each cell.

** *Rootstock bulbous or tuberous. Anthers dorsifixed, versatile.*

XIII. Bowiea.—*Rootstock* bulbous. *Stems* scandent. *Seeds* naked.
XIV. Schizobasis.—*Rootstock* bulbous. *Stems* erect. *Seeds* naked.
XV. Eriospermum.—*Roo'stock* tuberous. *Stems* erect. *Seeds* woolly.

*** *Rootstock not bulbous. Anthers basifixed, erect.*

XVI. Anthericum.—*Perianth* not twisted. *Capsule* obtusely angled. *Seeds* triquetrous.
XVII. Chlorophytum.—*Perianth* not twisted. *Capsule* acutely angled. *Seeds* flattened.
XVIII. Cæsia.—*Perianth* twisting as it fades. *Peduncle* long; flowers many, racemose.
XIX. Nanolirion.—*Perianth* twisting as it fades. *Peduncle* very short, 1–3-flowered.

Tribe 7. *ALLIEÆ. Rootstock* generally bulbous. *Flowers* in an umbel, which is subtended by a spathe of 1–2 or more membranous bracts.

* *Rootstock not bulbous. Perianth gamophyllous.*

XX. Agapanthus.—*Periant h*-tube short. *Corona* 0.
XXI. Tulbaghia.—*Perianth*-tube oblong-cylindrical. *Corona* annular or formed of 6 distinct scales.

** *Rootstock bulbous. Perianth gamophyllous.*

XXII. Allium.—The only genus.

*** *Rootstock bulbous. Perianth gamophyllous. Scape very short.*

XXIII. Massonia.—*Perianth*-limb regular.

XXIV. Daubenya.—*Perianth*-limb irregular.

Tribe 8. *SCILLEÆ.* *Rootstock* bulbous. *Peduncle* naked; inflorescence racemose or spicate, not subtended by one common spathe.

* *Perianth gamophyllous. Seeds not crowded nor compressed.*

XXV. Whiteheadia.—*Perianth*-tube short. *Filaments* connate in a ring at the base.

XXVI. Polyxena.—*Perianth*-tube cylindrical; segments equal. *Filaments* free.

XXVII. Lachenalia.—*Perianth*-tube cylindrical or oblong; segments unequal, usually of two different lengths. *Filaments* free.

** *Perianth gamophyllous. Seeds crowded, angled or discoid.*

XXVIII. Drimia.—*Perianth*-tube campanulate; segments linear, reflexing.

XXIX. Rhadamanthus.—*Perianth*-tube campanulate; segments erect or spreading; anthers connivent.

XXX. Litanthus.—*Perianth*-tube oblong; segments very short.

XXXI. Dipcadi.—*Perianth*-tube oblong; segments of two kinds; the outer caudate.

XXXII. Galtonia.—*Perianth*-tube oblong; segments subequal, spreading.

*** *Perianth polyphyllous. Seeds crowded, angled or discoid.*

XXXIII. Albuca.—Inner *perianth*-segments connivent.

XXXIV. Urginea.—*Perianth*-segments all spreading.

**** *Perianth gamophyllous. Seeds not crowded, turgid.*

XXXV. Veltheimia.—*Perianth*-tube cylindrical; segments short, ovate.

XXXVI. Hyacinthus.—*Perianth*-tube oblong; segments as long as the tube.

***** *Perianth polyphyllous. Seeds not crowded nor compressed.*

XXXVII. Drimiopsis.—*Perianth*-segments cucullate, connivent.

XXXVIII. Eucomis.—*Perianth*-segments flat, spreading. *Raceme* surmounted by a coma of leaf-like empty bracts.

XXXIX. Scilla.—*Perianth*-segments spreading, 1-nerved. *Racemes* not comose. *Flowers* usually blue or mauve-purple.

XL. Ornithogalum.—*Perianth*-segments spreading, keeled with more than one nerve. *Racemes* not comose. *Flowers* usually white or yellow, never blue.

Series III. COLCHICACEÆ. *Fruit* capsular. *Anthers* dehiscing extrorsely.

Tribe 9. *ANGUILLARIEÆ.* *Rootstock* a bulb or corm. *Perianth*-segments narrowed into a distinct claw.

* *Capsule dehiscing septicidally.*

XLI. Androcymbium.—*Peduncle* very short; inflorescence capitate. *Perianth*-segments distinct

XLII. Wurmbea.—*Flowers* spicate. *Perianth*-segments united at the base.

XLIII. Bæometra.—*Flowers* racemose. *Perianth*-segments distinct.

XLIV. Dipidax.—*Flowers* spicate. *Perianth*-segments distinct.

** *Capsule dehiscing loculicidally.*

XLV. Ornithoglossum.—The only genus.

Tribe 10. *UVULARIEÆ.* *Rootstock* not bulbous. *Capsule* (in the Cape genera) dehiscing loculicidally.

* *Anthers dehiscing by longitudinal slits.*

XLVI. Gloriosa. — *Stems* climbing. *Perianth*-segments free, unguiculate, spreading.

XLVII. Sandersonia. —*Stems* erect. *Perianth* gamophyllous; segments short.

XLVIII. Littonia.—*Stems* scandent. *Perianth*-segments free, connivent, not unguiculate.

** *Anthers dehiscing by terminal pores.*

XLIX. Walleria. – The only genus.

I. SMILAX, Linn.

Flowers diœcious. *Perianth*-segments distinct, subequal. *Male flower*: *Stamens* 6, inserted at the base of the perianth-segments; filaments filiform; anthers erect, dehiscing introrsely. Rudimentary *pistil* none. *Female flower* : *Staminodia* 6, filiform. *Ovary* sessile, ovoid, 3-celled; ovules 1–2 in a cell, pendulous, orthotropous; style 0 or short; stigmas 3. *Berry* globose, often by abortion 1–2-seeded. *Seed* globose; testa thin, adpressed; albumen horny; embryo minute.

Rootstock often large; stems usually woody and sarmentose, often prickly; leaves alternate, with a pair of spiral tendrils at the base of the petiole; venation reticulated; flowers of both sexes in umbels, small, greenish.

DISTRIB. Cosmopolitan. Species nearly 200.

1. S. kraussiana (Meisn. in Flora, 1845, 312); stems slender, woody, glabrous, many-ribbed, often armed with minute prickles; leaves alternate; petiole ¼–½ in. long, with a pair of firm, spirally-curled tendrils at the base; blade oblong, moderately firm in texture, 3–4 in. long, 5-nerved, rounded or obscurely cordate at the base, acute or obtuse with a cusp; flowers in copious dense axillary umbels; pedicels short, furnished with a pair of persistent ovate-amplexicaul obtuse bracteoles about the middle; pedicels rather longer than the flowers; perianth ⅛ in. long; segments obtuse, recurved from about the middle; stamens shorter than the perianth; anthers oblong, much shorter than the filaments; berry globose, the size of a pea; pedicel ⅓ in. long, articulated at the apex. *Kunth, Enum.* v. 242 ; *A.D.C. Monog. Phanerog.* i. 171 ; *Baker in Trans. Linn. Soc.* xxix. 162, *t.* 106. *S. morsaniana, Kunth, Enum.* v. 241. *S. mossambicensis, Garcke in Peters' Reise Mozamb. Bot.* 518.

KALAHARI REGION : Transvaal; in woods around Barberton, 3000–4500 ft., *Galpin*, 728!

EASTERN REGION : Natal; in woods near Durban, *Cooper*, 2035! Inanda, *Wood*, 594! Port Natal, under 500 ft., *Drège*, 4503! and without precise locality, *Krauss*, 316! *Gerrard*, 293! *Gueinzius* ! *Sanderson* ! *Plant* !

II. ASPARAGUS, Linn.

Flowers in all the Cape species hermaphrodite. *Perianth* poly-phyllous, campanulate; segments subequal, spreading. *Stamens* 6, inserted at the base of the perianth-segments; filaments short, fili-form or slightly flattened; anthers dorsifixed, versatile, dehiscing introrsely. *Ovary* sessile, globose, 3-celled; ovules 2 or few in a cell; style short, filiform; stigma tricuspidate. *Berry* globose, pulpy. *Seeds* often solitary, globose, or angled by pressure; testa black; albumen horny.

Stems erect or scandent, often woody; leaves small, scariose, often produced into spines at the base; cladodia solitary or fascicled in the axils of the leaves, usually subterete, rarely flattened; flowers small, whitish, usually axillary, rarely racemose; pedicels articulated, so that the flowers fall easily.

DISTRIB. Species about 100, spread through the temperate and tropical regions of the Old World.

1. DECLINATI. Spur of the leaves of the main stem only small and deltoid, not spreading and pungent. Cladodia terete. Pedicels axillary, mostly 1–2-nate.

Cladodia usually absent	(1) **denudatus.**
Cladodia few in a cluster :	
Stems woody :	
Perianth minute	(2) **Eckloni.**
Perianth $\frac{1}{12}$–$\frac{1}{8}$ in. long	(3) **exuvialis.**
Stems very weak, sarmentose	(4) **crispus.**
Stems not woody, straight, erect	(5) **virgatus.**
Cladodia many in a cluster :	
Cladodia very short	(6) **nodosus.**
Cladodia filiform, $\frac{1}{4}$–$\frac{1}{2}$ in. long :	
Cladodia 6–12 in a cluster	(7) **plumosus.**
Cladodia often 20 or more in a cluster ...	(8) **consanguineus.**
Cladodia more robust, $\frac{1}{4}$–$\frac{1}{2}$ in. long :	
Pedicel as long as the flower	(9) **declinatus.**
Pedicel much longer than the flower	(10) **Macowani.**

2. UMBELLATI. Spur of the leaves of the main stem not large and pungent. Cladodia terete. Flowers many in a cluster.

Cladodia short :	
Stems pubescent	(11) **multiflorus.**
Stems glabrous	(12) **Burkei.**
Cladodia long	(13) **subulatus.**

3. THUNBERGIANI. Spur of the leaves of the main stem produced into a distinct pungent spine. Cladodia subulate. Branchlets not fascicled. Flowers mostly 1–2-nate.

Cladodia minute, elliptic	(14) **stellatus.**
Cladodia rigid, cylindrical	(15) **microrhaphis.**
Cladodia robust, subulate, $\frac{1}{8}$–$\frac{1}{4}$ in. long	(16) **thunbergianus.**
Cladodia rigid, subulate, $\frac{1}{4}$–$\frac{1}{2}$ in. long	(17) **minutiflorus.**
Cladodia filiform, $\frac{1}{4}$–$\frac{1}{2}$ in. long	(18) **Cooperi.**

4. CAPENSES. Spur of the leaves of the main stem produced into a pungent spine. Branchlets fascicled, with 1–4-nate flowers from the tip only. Cladodia terete.

Pedicels very short :	
Branches not ending in spines :	
Cladodia pubescent, $\frac{1}{12}$–$\frac{1}{6}$ in. long	(19) **capensis.**
Cladodia glabrous, $\frac{1}{4}$–$\frac{1}{4}$ in. long	(20) **densus.**
Branches ending in spines	(21) **Nelsoni.**
Pedicels as long as the flowers :	
Branches not ending in spines	(22) **Burchellii.**
Branches ending in spines	(23) **stipulaceus.**

5. AFRICANI. Spur of the leaves of main stem produced into a pungent spine. Cladodia subulate. Flowers generally more than two in a cluster.

Cladodia $\frac{1}{4}$–$\frac{1}{2}$ in. long :	
Main stem slender ; prickles small	(24) **asiaticus.**
Main stem robust and woody ; prickles large ...	(25) **africanus.**
Cladodia $\frac{1}{2}$–1 in. or more long :	
Cladodia arcuate	(26) **retrofractus.**
Cladodia straight, ascending	(27) **laricinus.**

6. STRIATI. Cladodia flattened (phyllocladia). Flowers axillary, solitary, geminate or umbellate.

Stems woody, suberect ; phyllocladia rigid in texture :	
Flowers generally solitary	(28) **erectus.**
Flowers umbellate	(29) **striatus.**

Stems slender, sarmentose; phyllocladia not rigid:
Pedicels long ... ... (30) scandens.
Pedicels short ... ... (31) ramosissimus.

7. Racemosi. Cladodia subulate. Flowers in racemes.

Lower pedicels $\frac{1}{8}$–$\frac{1}{6}$ in. long ... ... (32) racemosus.
Lower pedicels $\frac{1}{4}$ in. long ... ... (33) Saundersiæ.

8. Falcati. Cladodia flattened (phyllocladia). Flowers in racemes.
Phyllocladia generally solitary:
Phyllocladia $\frac{1}{4}$–1 in. long ... ... (34) sarmentosus.
Phyllocladia 1–2 in. long ... ... (35) oxyacanthus.
Phyllocladia 1–4-nate, 1–2 in. long:
Phyllocladia $\frac{1}{6}$–$\frac{1}{4}$ in. broad ... ... (36) falcatus.
Phyllocladia $\frac{1}{16}$ in. broad ... ... (37) Sprengeri.
Phyllocladia 3–6-nate:
Phyllocladia $\frac{1}{4}$ lin. broad ... ... (38) myriocladus.
Phyllocladia $\frac{1}{2}$–1 lin broad ... ... (39) æthiopicus.

9. Myrsiphylli. Phyllocladia solitary, ovate or lanceolate. Flowers axillary, greenish, with segments reflexing from above the base.
Stems and branchlets very slender, voluble:
Phyllocladia lanceolate ... ... (40) Kraussii.
Phyllocladia ovate-lanceolate ... ... (41) volubilis.
Phyllocladia ovate ... ... (42) medeoloides.
Stems stiff, suberect; branchlets stout, acutely angled:
Phyllocladia ovate ... ... (43) undulatus.
Phyllocladia linear ... ... (44) juniperoides.

1. A. denudatus (Baker in Journ. Linn. Soc. xiv. 606); stems sarmentose, glabrous, strong in texture, very copiously compound; main stem subterete, $\frac{1}{8}$ in. diam.; spur of the leaves of the main stems small, deltoid; final branchlets slender, angled, straight or arcuate, ascending; cladodia entirely absent in the typical form; flowers 1–4-nate, produced almost solely from the tip of the branchlets; pedicels $\frac{1}{12}$ in. long, articulated at the middle; perianth campanulate, $\frac{1}{12}$ in. long; segments oblong; stamens nearly as long as the perianth; anthers oblong; ovules 5–6 in a cell; berry small, globose, 1-seeded. *Asparagopsis denudata, Kunth, Enum.* v. 82.

Var. β, **fragilis** (Baker, loc. cit.); branchlets with a few subulate 2–3-nate cladodia $\frac{1}{12}$–$\frac{1}{8}$ in. long; internodes $\frac{1}{8}$–$\frac{1}{4}$ in. long. *Asparagus fragilis, Burch. MSS.*

Central Region: Queenstown Div.; Stormberg 5000–6000 ft., *Drège*, 3533! Var. β, Graaff Reinet Div.; Sneeuw Berg Range, 3500 ft., *Bolus*, 791! Colesberg Div.; Carolus Poort, *Burchell*, 2758!
Kalahari Region: Orange Free State, *Zeyher, Aspar. No.* 29!

2. A. Eckloni (Baker in Journ. Linn. Soc. xiv. 615); stems woody, sarmentose, glabrous, with curved, slender, pale, moderately rigid branches; main leaves not produced into a spur at the base; upper internodes $\frac{1}{4}$–$\frac{1}{3}$ in. long; cladodia 1–6-nate, subulate, straight, very slender, $\frac{1}{4}$–$\frac{1}{2}$ in. long; flowers axillary, usually 1–2-nate, rarely 3–4-nate; pedicels cernuous, articulated at the middle, $\frac{1}{12}$ in. long; perianth very small.

South Africa: without locality, *Ecklon and Zeyher!* (as "*A. stipularis?*") (Dublin herbarium).

3. A. exuvialis (Burchell, Travels i. 432) ; stems slender, woody, glabrous, very zigzag, with the white epidermis peeling off ; main leaves produced only in a small deltoid spur at the base ; branches very zigzag, with deflexed branchlets ; cladodia very slender, subulate, 2–6 in a cluster, $\frac{1}{4}$–$\frac{1}{2}$ in. long ; flowers 1–2-nate, axillary ; pedicels $\frac{1}{12}$–$\frac{1}{8}$ in. long, articulated below the middle, spreading or cernuous ; perianth campanulate, $\frac{1}{12}$–$\frac{1}{8}$ in. long ; segments oblong, obtuse ; stamens $\frac{2}{3}$ the length of the perianth-segments ; berry small, globose. *Bresl. Diss. No. 40 ; Roem. et Schultes, Syst. Veg. vii. 337 , Kunth, Enum. v. 74 ; Baker in Journ. Linn. Soc.* xiv. 608.

COAST REGION : Malmesbury Div. ; near Hopefield, *Bachmann,* 822 ! Uitenhage Div. ; near the Zwartkops River, *Zeyher,* 607 ! 4157 ! (a dwarfed, suberect variety with crowded branches).

KALAHARI REGION : Griqualand West ; along the Vaal River, *Burchell,* 1768 ! and Griqua Town, *Burchell,* 1854 ! Bechuanaland ; Kosi Fontein, *Burchell,* 2566 !

4. A. crispus (Lam. Encyc. i. 295) ; main stem very slender, very weak, glabrous, sarmentose, very flexuose ; leaves of the main stem with a minute, adpressed, deltoid spur ; branchlets very slender, curved, deflexed ; final internodes $\frac{1}{16}$–$\frac{1}{12}$ in. long ; cladodia ascending, 2–3-nate, subulate or rather flattened, $\frac{1}{6}$–$\frac{1}{4}$ in. long ; flowers axillary, solitary ; pedicels $\frac{1}{4}$–$\frac{1}{3}$ in. long, articulated near the tip ; perianth campanulate, $\frac{1}{6}$ in. long ; segments lanceolate, reflexing from halfway down ; stamens half as long as the perianth-segments ; anthers minute, globose ; ovary ovoid, narrowed into a short style ; berry subglobose, $\frac{1}{3}$ in. diam., 5–6-seeded. *Poir. Encyc. Suppl.* i. 482 ; *Bresl. Diss. No.* 22 ; *Roem. et Schultes, Syst. Veg.* vii. 326 ; *Kunth, Enum.* v. 73 ; *Baker in Journ. Linn. Soc.* xiv. 607. *A. decumbens, Jacq. Hort. Schoenbr.* i. 51, t. 97 ; *Bresl. Diss. No.* 21. *Asparagopsis decumbens, Kunth, Enum.* v. 77. *Asparagus flexuosus, Thunb. Prodr.* 66 ; *Fl. Cap. edit.* 2, ii. 332 ; *Kunth, Enum.* v. 74. *Medeola triphylla, Thibaud MSS. A. sinensis, Hort.*

COAST REGION : Clanwilliam Div., *Zeyher !* Malmesbury Div. ; near Hopefield, *Bachmann,* 814 ! Table Mountain. *Ecklon,* 93b ! *Bolus,* 4753 ! Simons Bay, *Wright !* George Div. ; Outeniqua, *Thunberg !* Alexandria Div. ; Zuurberg Range, *Cooper,* 3252 ! British Kaffraria, *Cooper,* 3254 !

5. A. virgatus (Baker in Saund. Ref. Bot. t. 214) ; stems fascicled, erect, very much branched, 3–6 ft. long ; main stem stiffly erect, simple in the lower half ; branches slender, angled ; largest leaves with only a minute deltoid spur ; final branchlets very slender, straight, ascending, with internodes $\frac{1}{6}$–$\frac{1}{3}$ in. long ; cladodia 1–3-nate, terete, ascending, channelled down the face, mucronate, moderately firm in texture, $\frac{1}{4}$–$\frac{1}{3}$ in. long ; pedicels axillary, 1–2-nate, $\frac{1}{4}$–$\frac{1}{3}$ in. long, articulated below the middle ; perianth campanulate ; segments linear-oblong, spreading from the base ; stamens nearly as long as the perianth-segments ; anthers $\frac{1}{3}$–$\frac{1}{4}$ the length of the filaments ; berry globose, bright scarlet, $\frac{1}{6}$ in. diam., 1-seeded. *Baker in Journ. Linn. Soc.* xiv. 606.

VAR. β, **capillaris** (Baker, loc. cit.) ; cladodia rather longer, more slender and more copiously developed.

s 2

COAST REGION : Bedford Div. ; in a kloof near Bedford, *Elliot*, 691 ! Stocken-strom Div. ; Katberg, *Mrs. Barber!* British Kaffraria, *Cooper*, 202 partly !
CENTRAL REGION : Somerset Div. ; Bosch Berg, 3000 ft., *Burchell*, 3197 ! *MacOwan*, 1918 !
KALAHARI REGION : Transvaal ; damp grassy hollows around Barberton, 2800 ft., *Galpin*, 1155 !
EASTERN REGION : Pondoland ; between Umtata River and St. John's River, 1000-2000 ft., *Drège*, 4483b ! Natal ; woods near Umlaas (Umlazi ?) River, *Krauss*, 356 ! Port Natal, *Drège*, 4483c ! Inanda, *Wood*, 1373 ! and on the Drakensberg, *Cooper*, 2248 !

Also occurs in British Central Africa.

6. A. nodosus (Solander in herb. Banks.); stems woody, suberect, pale, densely pubescent ; main leaves not produced into distinct spines at the base ; branches straight ; branchlets erecto-patent ; cladodia subulate, $\frac{1}{12}-\frac{1}{8}$ in. long, moderately robust, in clusters of 6-10 at the axillary nodes ; flowers 2-4-nate ; pedicels very short ; perianth very small ; stamens about as long as the perianth-segments ; anthers oblong. *Baker in Journ. Linn. Soc.* xiv. 608.

SOUTH AFRICA : without locality, *Masson!* in the British Museum Herbarium.

7. A. plumosus (Baker in Journ. Linn. Soc. xiv. 613); stems terete, green, glabrous, sarmentose ; main leaves not produced into distinct prickles ; branches decompound, spreading horizontally, with the branchlets and cladodia in one plane ; cladodia very slender, filiform, bright green, 6-12 in a cluster, $\frac{1}{3}-\frac{1}{2}$ in. long ; pedicels 1-2-nate, shorter than the flowers, articulated at the middle ; perianth campanulate, $\frac{1}{8}$ in. long ; segments oblanceolate-oblong, spreading widely ; stamens $\frac{2}{3}$ the length of the perianth ; berry globose, bright black, 1-seeded, $\frac{1}{4}$ in. diam.

COAST REGION : British Kaffraria, *Cooper*, 202 partly !
EASTERN REGION : Pondoland, *Bachmann*, 261 ! Natal ; Port Natal, *Drège*, 4482 ! Inanda, *Wood*, 1649 ! and without precise locality, *Gerrard*, 227 ! *Gerrard and McKen*, 754 !

The finest of all the species for decorative purposes, and now widely spread in European gardens.

8. A. consanguineus (Baker in Journ. Linn. Soc. xiv. 615); stems slender, terete, glabrous, green, sarmentose ; lower leaves with only a small deltoid spur ; branches pinnate, spreading or deflexed, flexuose, with very slender deflexed branchlets ; cladodia filiform, bright green, flexuose, about $\frac{1}{4}$ in. long, 10-30 in a cluster ; pedicels 1-2-nate, cernuous, about as long as the cladodia, articulated at the apex ; perianth campanulate, $\frac{1}{8}$ in. long ; segments oblong, obtuse, spreading falcately ; stamens nearly as long as the perianth ; ovules 5-6 in a cell. *Asparagopsis consanguinea, Kunth, Enum.* v. 76. *Asparagus scandens, Herb. Drège, non Thunb. A. declinatus, Eckl. and Zeyh. Exsic., non Linn.*

SOUTH AFRICA : without locality, *Thunberg ! Mund !*
COAST REGION : Piquetberg Div. ; between Krom River and Pieters Fontein on the Piquet Berg, below 1000 ft., *Drège !*

CENTRAL REGION : Calvinia Div. ; limestone hills and karoo-like depressions near Bitter Fontein, 3000-4000 ft., *Zeyher*, 1675 !

WESTERN REGION : Little Namaqualand ; stony places near Klip Fontein, 3000 ft., *Bolus*, 6589 !

9. **A. declinatus** (Linn. Sp. Plant. 313) ; stems sarmentose, slender, woody, terete, glabrous ; main leaves produced into a hard spreading deltoid spur ; branches decompound, with numerous slender straight spreading or deflexed branchlets ; cladodia slender, subulate, bright green, $\frac{1}{3}$-$\frac{1}{2}$ in. long, 6-10 in the lateral clusters, up to 20 in those at the tip and base of the branchlets ; pedicels 2-3-nate, $\frac{1}{8}$-$\frac{1}{6}$ in. long, articulated at the middle ; perianth campanulate, $\frac{1}{12}$ in. long ; segments obtuse ; stamens nearly as long as the perianth ; berry small, globose, 1-seeded. *Miller, Dict. edit.* vi. *No.* 7 ; *Lam. Encyc.* i. 294 ; *Bresl. Diss. No.* 32 ; *Thunb. Prodr.* 66 ; *Fl. Cap. edit. Schult.* 333 ; *Kunth, Enum.* v. 59 ; *Baker in Journ. Linn. Soc.* xiv. 609. *Asparagopsis setacea, Kunth, Enum.* v. 82.

COAST REGION : George Div ; Outeniqua, in woods, *Thunberg !* Knysna Div. ; Koratra, under 1000 ft., *Drège*, 8584b ! Uitenhage Div., *Zeyher, Asparag. No.* 3 ! Alexandria Div. ; Enon, in Olyvenhout Kloof and Olifants Kloof, under 1000 ft., *Drège*, 8584c !

CENTRAL REGION : Somerset Div. ; in woods on the sides of Bosch Berg, 3000–4000 ft., *MacOwan*, 1917 !

EASTERN REGION : Pondoland, *Bachmann*, 262 ! Natal, *Cooper*, 1127 !

10. **A. Macowani** (Baker in Journ. Linn. Soc. xiv. 609, ex parte) ; stems slender, erect, glabrous ; largest leaves not produced into a spur at the base ; branchlets straight, slender, ascending, angled ; cladodia very slender, densely fascicled, often 20 or more in a cluster, subulate, bright green, curved, moderately firm in texture, about $\frac{1}{2}$ in. long ; pedicels 1-2-nate, axillary, $\frac{1}{4}$ in. long, articulated near the base ; perianth campanulate, $\frac{1}{12}$ in. long ; segments oblanceolate-oblong, obtuse ; stamens nearly as long as the perianth-segments ; anthers minute, globose.

COAST REGION : Albany Div., *Zeyher*, 879 !

Allied to *A. declinatus.*

11. **A. multiflorus** (Baker in Journ. Linn. Soc. xiv. 610) ; stem woody, terete, pale green, slender, finely pubescent ; main leaves with only a small deltoid basal spur ; branchlets straight, slender, rigid, woody, deflexed ; cladodia subulate, rigid, mucronate, 6-12 in a cluster, about $\frac{1}{4}$ in. long ; flowers densely clustered ; pedicels $\frac{1}{4}$ in. long, articulated below the middle ; perianth campanulate, $\frac{1}{8}$ in. long ; segments oblanceolate-oblong ; stamens $\frac{2}{3}$ the length of the perianth-segments ; anthers minute, subglobose.

CENTRAL REGION : Somerset Div. ? *Bowker !*

12. **A. Burkei** (Baker in Journ. Linn. Soc. xiv. 607) ; stem woody, slender, pale green, glabrous, rather zigzag ; branchlets long, slender, straight, woody ; main leaves only produced at the base

into a small deltoid spur; cladodia slender, subulate, $\frac{1}{4}$–$\frac{1}{2}$ in. long,
3–12 in a cluster; flowers 2–6 in a cluster; pedicels $\frac{1}{12}$–$\frac{1}{8}$ in. long,
articulated at the middle; perianth campanulate, $\frac{1}{12}$ in. long; segments oblanceolate, obtuse; stamens nearly as long as the perianth;
berry small, globose.

CENTRAL REGION: Prince Albert Div. ; Gamka River, *Burke!*
KALAHARI REGION: Transvaal; Crocodile River, *Burke!*

13. A. subulatus (Thunb. Prodr. 66); main stem slender, terete,
woody, pale green, glabrous, very flexuose; main leaves produced at
the base into a deltoid pungent spur; branchlets slender, firm in
texture, very zigzag; cladodia 2–3-nate, subulate, rigid, $\frac{3}{4}$–1 in. long;
flowers many in an umbel; pedicels $\frac{1}{6}$ in. long, articulated below
the middle; perianth campanulate, $\frac{1}{8}$–$\frac{1}{6}$ in. long; segments
oblanceolate-oblong, reflexing from below the middle; stamens
much shorter than the perianth-segments; anthers minute, oblong.
Fl. Cap. edit. Schult. 333; *Willd. Sp. Plant.* ii. 154; *Roem. et
Schultes, Syst. Veg.* vii. 326; *Kunth, Enum.* v. 74; *Baker in Journ.
Linn. Soc.* xiv. 612.

SOUTH AFRICA: without locality, *Thunberg! Gill!*
COAST REGION: Uitenhage, *Zeyher,* 250! *Bolus,* 1644!
CENTRAL REGION: Bruintjes Hoogte, upper part, *Burchell,* 3063!
EASTERN REGION: Natal, *Gerrard,* 228!

14. A. stellatus (Baker in Journ. Linn. Soc. xiv. 612); main stem
suberect, woody, terete, glabrous; prickles woody, pungent, patent,
$\frac{1}{8}$–$\frac{1}{4}$ in. long; branches crowded, decompound, erecto-patent, straight,
prickly, internodes very short; cladodia 5–10 in a cluster, stellate,
elliptical, distinctly mucronate, $\frac{1}{16}$ in. long; flowers not seen.

CENTRAL REGION: Albert Div., *Cooper,* 622! Aliwal North Div. ; on the
Witte Bergen, 5000–6000 ft., *Drège,* 8589!

15. A. microrhaphis (Baker in Journ. Linn. Soc. xiv. 612); main
stem woody, suberect, multisulcate, glabrous, $\frac{1}{6}$ in. diam.; prickles
woody, pungent, patent, $\frac{1}{6}$ in. long; branches woody, pale, patent,
decompound, but little flexuose; internodes very short; cladodia
4–10 in a cluster, terete, rigid, mucronate, $\frac{1}{8}$–$\frac{1}{6}$ in. long; flowers
axillary, 1–2-nate; pedicels spreading, $\frac{1}{8}$–$\frac{1}{6}$ in. long, articulated
below the middle; perianth campanulate, $\frac{1}{12}$ in. long; segments
oblong, obtuse; stamens nearly as long as the perianth-segments;
ovules 7–9 in a cell. *Asparagus microrhaphis, Kunth, Enum.* v. 83.

COAST REGION: Queenstown Div. ; Table Mountain, 6000–7000 ft., *Drège,*
3534!
CENTRAL REGION: Colesberg Div. ; Naauw Poort, *Burchell,* 2781!

16. A. thunbergianus (Schult. fil. Syst. Veg. vii. 333); stems pale
or dark brown, woody, terete, glabrous; prickles spreading, woody,
pungent, uncinate, $\frac{1}{8}$–$\frac{1}{4}$ in. long; branchlets slightly flexuose,
copiously pinnate; internodes very short; cladodia 10–15 in a
cluster, stout, terete, subulate, ascending, mucronate, $\frac{1}{8}$–$\frac{1}{4}$ in. long;

pedicels 1–2-nate, $\frac{1}{8}$–$\frac{1}{6}$ in. long, articulated below the middle; perianth campanulate, $\frac{1}{12}$ in. long; segments oblong, obtuse, spreading horizontally ; stamens nearly as long as the perianth-segments; ovules 5–6 in a cell; berry globose, $\frac{1}{4}$ in. diam. *Baker in Journ. Linn. Soc.* xiv. 615. *Asparagopsis Thunbergii, Kunth, Enum.* v. 85. *A. Dregei, Kunth, Enum.* v. 84. *Asparagus albus, Thunb. Prodr.* 66; *Fl. Cap. edit. Schult.* 333, *non Linn. A. nitidus, Forster herb.*

SOUTH AFRICA : without locality, *Forster ! Thunberg !*
COAST REGION : Malmesbury Div. ; Saldanha Bay, *Grey !* Cape Div. ; Lions Rump, *Burchell,* 8453 ! Table Mountain and Devils Mountain, below 1000 ft., *Drège !* Cape Flats, *Burke !* Stellenbosch Div. ; between Hottentots Holland and Jonkers Valley, *Burchell,* 8318 !

17. **A. minutiflorus** (Baker in Journ. Linn. Soc. xiv. 616); stems slender, woody, terete, multisulcate, pubescent, slightly flexuose ; prickles pungent, subulate, $\frac{1}{8}$–$\frac{1}{6}$ in. long ; branches short, crowded, stiff, slender, erecto-patent ; cladodia 3–10 in a cluster, slender, subulate, mucronate, rigid, $\frac{1}{4}$–$\frac{1}{3}$ in. long ; flowers 1–2 in a cluster ; pedicels $\frac{1}{12}$ in. long, articulated at the middle ; perianth campanu-late, $\frac{1}{12}$ in. long ; segments oblanceolate, obtuse, spreading widely ; stamens nearly as long as the perianth ; ovules about 3. in a cell. *Asparagopsis minutiflora, Kunth, Enum.* v. 89.

EASTERN REGION : Cafferland, *Gill !* Delagoa Bay, *Forbes !*

18. **A. Cooperi** (Baker in Gard. Chron. 1874, i. 818); stems terete, glabrous, green, wide-climbing; prickles woody, spreading, pungent, $\frac{1}{6}$ in. long ; branches spreading, slender, copiously decom-pound, with patent branchlets ; cladodia very slender, subulate, often a dozen or more in a cluster, $\frac{1}{4}$–$\frac{1}{3}$ in. long ; flowers axillary, 1–3-nate ; pedicels $\frac{1}{6}$ in. long, articulated below the middle ; perianth campanu-late, $\frac{1}{12}$ in. long ; segments oblanceolate-oblong, spreading horizon-tally ; stamens nearly as long as the perianth-segments ; berry globose, $\frac{1}{6}$ in. diam., 1-seeded. *Baker in Journ. Linn. Soc.* xiv. 613.

CENTRAL REGION : Somerset Div. ; in woods on the sides of Bosch Berg, 2300–4000 ft., *MacOwan,* 1810 !

Originally described from specimens in the garden of Mr. Wilson Saunders in 1871, collected by Mr. Thos. Cooper.

19. **A. capensis** (Linn. Sp. Plant. 314); stems woody, copiously branched, suberect, terete ; prickles large, spreading, pungent, one at the base of each fascicle of final branchlets ; branches ascending, woody, rather flexuose ; branchlets in dense clusters, $\frac{3}{4}$–1 in. long ; cladodia subulate, densely clustered, ascending, pale green, pubescent, distinctly mucronate, $\frac{1}{12}$–$\frac{1}{8}$ in. long ; flowers produced only from the tip of the branchlets, usually solitary, subsessile ; perianth campanu-late, $\frac{1}{8}$ in. long ; segments oblong, obtuse ; ovules 4–5 in a cell ; berry small, 1-seeded. *Miller, Dict. edit.* viii. *No.* 9 ; *Jacq. Hort. Schoenbr.* iii. 8, *t.* 266 ; *Thunb. Prodr.* 66; *Fl. Cap. edit. Schult,*

834 ; *Bresl. Diss.* 21 ; *Baker in Journ. Linn. Soc.* xiv. 617, *non
Linn. herb. A. retrofractus, Wendl. Hort. Herren, t.* 22, *non Linn.
Asparagopsis passerinoides, Kunth, Enum.* v. 90.

SOUTH AFRICA : without locality, *Masson !* *Oldenburg !*
COAST REGION : near Cape Town, *Thunberg !* Table Mountain, 1000–2000 ft.,
Drège ! Lion Mountain, *Burchell,* 8451 ! Paarl Div. ; hills by the Berg River,
near Onderste Pont, below 500 ft., *Drège !* Tulbagh Div. ; on the Wind-heuvel,
Koedoes Mountains, *Burchell,* 1288/1 ! Uitenhage Div., *Cooper,* 1573 ! *Zeyher,*
46 !
CENTRAL REGION : Graaff Reinet Div. ; Sneeuw Berg Range, 4000–5000 ft.,
Drège, 8591 !
WESTERN REGION : Little Namaqualand, between Kaus, Natvoet, and Doorn
Poort, 1000–2000 ft., *Drège,* 8593 !

The plant that represents *A. capensis* in the Linnean herbarium is *A. asiaticus.*

20. **A. densus** (Soland. in Herb. Banks.) ; main stem woody,
sulcate, pubescent ; spines pungent, produced from the base of the
clusters of final branchlets, which are densely fascicled ; cladodia 4–6
in a cluster, ascending, subulate, glabrous, strongly angled, $\frac{1}{4}$–$\frac{1}{2}$ in.
long ; flowers 2–4 in a cluster, from the tip of the final branchlets ;
pedicels very short ; perianth campanulate, $\frac{1}{12}$ in. long ; segments
oblong, obtuse ; stamens nearly as long as the perianth ; anthers very
minute. *Baker in Journ. Linn. Soc.* xiv. 616.

SOUTH AFRICA : without locality, *Auge !* in the British Museum Herbarium.

21. **A. Nelsoni** (Baker in Journ Linn. Soc. xiv. 617) ; main stem
suberect, woody, densely pubescent ; branches woody, erecto-patent,
pungent at the tip ; final branchlets in dense clusters, with a spine at
the base ; cladodia 3–6-nate, subulate, ascending, glabrous, pale
green, $\frac{1}{8}$–$\frac{1}{6}$ in. long ; flowers produced only from the tips of the
branchlets, nearly sessile ; perianth small, campanulate.

SOUTH AFRICA : without locality, *Nelson !* in the British Museum Herbarium.

Combines the branching of *A. stipulaceus* and the subsessile flowers of *A. capensis.*

22. **A Burchellii** (Baker in Journ. Linn. Soc. xiv. 618) ; stems
woody, much-branched, terete, pale, glabrous ; branches woody, very
zigzag ; prickles large, pungent, produced only from the base of the
clusters of final branchlets, which are in dense bundles, and $\frac{1}{2}$–1 in.
long ; cladodia generally 4–8-nate, subulate, ascending, distinctly
mucronate, strongly angled, glabrous, $\frac{1}{12}$–$\frac{1}{8}$ in. long ; flowers produced
only from the tip of the branchlets, generally solitary ; pedicels as
long as the cladodia ; perianth campanulate, $\frac{1}{12}$ in. long ; segments
oblong, obtuse ; stamens nearly as long as the perianth ; berry small,
1-seeded.

COAST REGION : Uitenhage Div.,*Cooper,* 1574 ! *Zeyher,* 10 !
CENTRAL REGION : Somerset Div. ; by the Blyde River, *Burchell,* 2962 !
2962/1 ! On stony hills near Graaff Reinet, 2500 ft., *Bolus,* 411 !

23. **A. stipulaceus** (Lam. Encyc. i. 297) ; a much-branched, sub-

erect shrub, with slender, glabrous, white, terete main stems; branchlets woody, rather zigzag, ascending, ending in pungent spines; final
branchlets very slender, about $\frac{1}{2}$ in. long, arranged in dense fascicles
with a small pungent spine at the base; cladodia 3–6 in a cluster,
subulate, glabrous, very slender, fragile, deciduous, $\frac{1}{6}$–$\frac{1}{4}$ in. long;
flowers produced from the tip of the branchlets only, usually
1–2-nate; pedicels as long as the flowers; perianth campanulate,
$\frac{1}{12}$ in. long; segments oblong, obtuse; ovules 3–4 in a cell; berry
small, 1-seeded. *Bresl. Diss. No. 5; Roem. et Schultes, Syst. Veg.*
vii. 333; *Baker in Journ. Linn. Soc.* xiv. 617. *A. suaveolens,*
Burchell, Travels ii. 226. *Asparagopsis stipulacea, triacantha, and*
Zeyheri, Kunth, Enum. v. 91–92.

VAR. β, **A. spinescens** (Steud. ex Roem. et Schultes, Syst. Veg. vii. 334);
cladodia more robust, strongly angled, $\frac{1}{4}$–$\frac{1}{2}$ in. long. *Asparagopsis spinescens,*
Kunth, Enum. v. 93.

SOUTH AFRICA: without locality, *Sonnerat! Mund!*
COAST REGION: Uitenhage Div.; near the Zwartkops River, *Zeyher,* 4163!
Fort Beaufort Div.; Koonap, 400–500 ft., *Baur,* 1079!
CENTRAL REGION: Somerset Div.; near Somerset East, *MacOwan,* 1774!
Colesberg Div., *Shaw!*
KALAHARI REGION: Griqualand West; around Griqua Town, *Burchell,* 1956!
and Barkly, near the Vaal River, *Nelson,* 176!
EASTERN REGION: Natal; mountain ranges 2000–3000 ft., 30–60 miles from
the sea, *Sutherland!* and without precise locality, *Gerrard,* 542! Delagoa Bay,
Bolus!

"This karroid Asparagus is very mischievous to the Angora goat-hair. The
withered shoots fasten themselves to the long flank-hair, break off, and form an
inextricable snare. Often the goat, when stretching out his hind foot to scratch,
catches it in the tangle and fetters himself effectually."—*MacOwan.*

24. **A. asiaticus** (Linn. Sp. Plant. 313); main stem slender,
subterete, green, glabrous, sarmentose; main prickles small; branches
long, slender; cladodia subulate, rigid, ascending, $\frac{1}{4}$–$\frac{1}{2}$ in. long, 3–12
in a cluster; flowers axillary, generally 1–4-nate; pedicels $\frac{1}{4}$ in. long,
articulated below the middle; perianth campanulate, $\frac{1}{8}$ in. long;
segments oblong, obtuse, spreading horizontally; stamens nearly as
long as the perianth-segments; anthers oblong, $\frac{1}{3}$ the length of the
filament; ovules 5–6 in a cell; berry globose, $\frac{1}{6}$ in. diam., 1-seeded.
Linn. Mant. 366; *Miller, Dict. edit.* viii. *No.* 8; *Bresl. Diss. No.* 12;
Willd. Sp. Plant. ii. 153; *Baker in Journ. Linn. Soc.* xiv. 618.
Asparagopsis Willdenowii, Kunth, Enum. v. 86. *A. scoparia,*
Kunth, Enum. v. 80. *Asparagus rivalis, Burchell, Trav.* i. 400.
A. mitis and scaberulus, A. Rich, Fl. Abyss. ii. 319–320. *A. capen-*
sis, Linn. Herb.

CENTRAL REGION: Colesberg, *Shaw!* Aliwal North Div.; banks of the
Orange River, near Aliwal North, 4300 ft.. *Drège,* 4481b!
KALAHARI REGION: Bechuanaland; Maadji Mountain, *Burchell,* 2376! Kosi
Fontein, *Burchell,* 2587!
EASTERN REGION: Natal, *Gerrard,* 70! Delagoa Bay, *Bolus!*

Also Tropical Africa and Tropical Asia.

25. **A. africanus** (Lam. Encyc. i. 295); stems woody, terete, much-

branched, sarmentose ; main prickles large, spreading, pungent ; branchlets slender, woody, glabrous or pubescent, slightly zigzag ; cladodia densely clustered, rigid, subulate, about $\frac{1}{2}$ in. long; flowers generally umbellate, sometimes only 1–2-nate ; pedicels $\frac{1}{6}$–$\frac{1}{4}$ in. long, articulated below the middle; perianth campanulate, $\frac{1}{12}$–$\frac{1}{6}$ in. long ; segments oblong, obtuse ; stamens nearly as long as the perianth-segments ; anthers oblong, $\frac{1}{3}$ the length of the filaments; ovules 4–6 in a cell ; berry globose, ^{1} in. diam., 1-seeded.   *Bresl. Diss. No.* 9 ; *Roem. et Schultes, Syst. Veg.* vii. 331 ; *Baker in Journ. Linn. Soc.* xiv. 619.   *Asparagopsis Lamarckii, Kunth, Enum.* v. 87.   *A. juniperina, Kunth, Enum.* v. 85.   *A. niveniana, Kunth, Enum.* v. 88.

VAR. β, **A. dependens** (Thunb. Prodr. 66) ; branches deflexed, often pubescent ; cladodia shorter and stouter than in the type.   *Fl. Cap. edit. Schult.* 333 ; *Willd. Sp Plant.* ii. 154; *Bresl. Diss. No.* 7 ; *Kunth, Enum.* v. 74.   *A. rubicundus, Berg. Cap.* 88.

VAR. γ, **A. microphyllus** (Burch. MSS.) ; cladodia short, stout, pale green, curved.

VAR. δ, **concinnus** (Baker in Journ. Linn. Soc. xiv. 619) ; stems slender ; branches ascending ; cladodia rigid, ascending, $\frac{1}{4}$ in. long.

VAR. ε, **Wrigh ii** (Baker, loc. cit.) ; cladodia very stout, angled, with a groove down each face; berries larger.

SOUTH AFRICA : without locality, Var. β, *Thunberg! Zeyher,* 4158! COAST REGION : Malmesbury Div. ; near Hopefield, *Bachmann,* 815 ! 816 ! 821 !   Cape Div. ; Kloof between the Lion's Head and Table Mountain, *Burchell,* 243 !   Cape Flats near Rondebosch, *Burchell,* 719 !   Camps Bay, *Burchell,* 389 ! Paarl Div. ; Simons Berg, near the waterfall, 2000 ft., *Drège,* 8577 !   Swellendam Div. ; at Groot Vadersbosch, *Burchell,* 7221 !   Uitenhage Div. ; Zwartkops River, under 100 ft., *Drège,* 8576 !   Uitenhage, *Zeyher,* 178 !   Alexandria Div. ; between Rautenbachs Drift and Addo Drift, *Burchell,* 4269 !   Albany Div., *Cooper,* 3250 !   British Kaffraria, *Cooper,* 191 !   Var. β, Uitenhage, *Zeyher,* 236 !   Fort Beaufort Div., *Cooper,* 523 !   British Kaffraria, *Cooper,* 404 !   Var. ε, Simons Bay, *Wright,* 454 ! CENTRAL REGION : near Graaff Reinet, 2500 ft., *Bolus,* 132 !   Somerset Div. ; by the Blyde River, *Burchell,* 2984 !   Var. δ, Somerset Div. ; lower part of Bruintjes Hoogte, *Burchell,* 3020 ! KALAHARI REGION : Hopetown Div. ; between "Puff-adder Halt and Bare Station," *Burchell,* 2677 !   Transvaal ; Berea Hill, near Barberton, *Galpin,* 622 !   Bechuanaland ; at Hamapery, near Kuruman, *Burchell,* 2443 !   Var. γ, Bechuanaland ; Kosi Fontein, *Burchell,* 2572 ! EASTERN REGION : Pondoland, *Bachmann,* 258 !   Natal ; near Umlazi River, *Krauss,* 413 !   Inanda, *Wood,* 989 ; Colenso, *Wood,* 4089 ! and without precise locality, *Gerrard,* 126 !

Also Angola.

26. A. retrofractus (Linn. Sp. Plant. 313) ; stems much-branched, slender, woody, terete ; main prickles small, pungent, uncinate ; branchlets slender, woody, very zigzag ; cladodia densely clustered, subulate, curved, ascending, $\frac{3}{4}$–1 in. long ; flowers axillary, umbellate ; pedicels $\frac{1}{4}$ in. long, articulated below the middle ; perianth campanulate, $\frac{1}{8}$ in. long ; segments oblong, obtuse ; stamens nearly as long as the perianth ; anthers oblong, $\frac{1}{3}$–$\frac{1}{2}$ the length of the filament ; ovules 5–6 in a cell ; berry small, globose, 1-seeded. *Miller, Dict. edit.* viii. *No.* 5 ; *Lam. Encyc.* i. 295 ; *Ait. Hort. Kew.*

edit. 2, ii. 274; *Bresl. Diss. No.* 10; *Roem. et Schultes, Syst. Veg.* vii. 330; *Baker in Journ. Linn. Soc.* xiv. 621, *non Forst.* *A. declinatus, Sieber, Herb. Cap. No.* 84, *non Linn. Asparagopsis retrofracta, Kunth, Enum.* v. 88.

SOUTH AFRICA: without locality, *Sieber*, 84! *Zeyher*, 1674!
COAST REGION: Tulbagh Div.; Mitchell's Pass, *Dickson in Herb. Bolus*, 5576! Albany Div.; in a wooded Kloof west of Grahamstown, *Burchell*, 3599!
CENTRAL REGION: Somerset Div.; upper part of Bruintjes Hoogte, *Burchell*, 3060!

27. A. laricinus (Burchell, Trav. i. 537); stems woody, subterete, much-branched; main prickles large, pungent, spreading; branches slender, ascending, pale, woody, zigzag; cladodia subulate, rigid, densely clustered, ascending, $\frac{3}{4}$–1 in. long; flowers axillary, many in a cluster; pedicels $\frac{1}{6}$–$\frac{1}{4}$ in. long, articulated below the middle; perianth campanulate, $\frac{1}{12}$–$\frac{1}{8}$ in. long; segments oblong, obtuse; stamens nearly as long as the perianth; anthers oblong, $\frac{1}{3}$–$\frac{1}{4}$ the length of the filament; ovules 5–6 in a cell; berry globose, 1-seeded, $\frac{1}{6}$ in. diam. *Bresl. Diss. No.* 40; *Roem. et Schultes, Syst. Veg.* vii. 337; *Kunth, Enum.* v. 75; *Baker in Journ. Linn. Soc.* xiv. 620.

SOUTH AFRICA: without locality, *Zeyher, Asparagus No.* 21!
COAST REGION: Malmesbury Div.; near Hopefield, *Bachmann*, 1937! Hex River Berg, *Rehmann*, 2676!
CENTRAL REGION: Somerset Div.? *Bowker!*
KALAHARI REGION: Griqualand West; between Griqua Town and Witte Water, *Burchell*, 1971! Bechuanaland; Pellat Plains, between "Pintado Fontein and Thermometer Fontein," *Burchell*, 2227!

28. A. erectus (Thunb. Prodr. 65); stems woody, suberect, glabrous; prickles indistinct; branchlets slender, ascending, terete, not zigzag; upper internodes $\frac{1}{4}$–$\frac{1}{2}$ in. long; phyllocladia solitary, linear, acuminate, rigid in texture, $\frac{3}{4}$–1 in. long, with about 5 distinct ribs; flowers usually solitary, rarely 2–4-nate; pedicels $\frac{1}{6}$ in. long, articulated above the middle; perianth $\frac{1}{6}$ in. long; segments obtuse, falcate from above the base; stamens $\frac{2}{3}$ the length of the perianth; berry globose, $\frac{1}{4}$ in. diam. *Fl. Cap. edit. Schult.* 332; *Baker in Journ. Linn. Soc.* xiv. 621. *Dracæna erecta, Linn. fil. Suppl.* 204; *Willd. Sp. Plant.* ii. 158; *Roem. et Schultes, Syst. Veg.* vii. 346. *Myrsiphyllum erectum, Schlechtend. ex Kunth, Enum.* v. 109.

SOUTH AFRICA: without locality, *Thunberg!*
Confused in Thunberg's herbarium with a form of *A. striatus*.

29. A. striatus (Thunb. Prodr. 65); main stem stout, woody, suberect, green, sulcate, glabrous; prickles small, brown, uncinate, deflexed; branchlets ascending or spreading, slightly zigzag; phyllocladia solitary at the sides, 2–3-nate at the tip of the branchlets, lanceolate, very rigid in texture, pungent, $\frac{3}{4}$–1 in. long, with 5–7 distinct ribs; flowers axillary, generally umbellate; pedicels $\frac{1}{4}$ in. long, articulated below the middle; perianth campanulate, $\frac{1}{6}$–$\frac{1}{5}$ in.

long; segments obtuse, falcate from above the base; stamens ⅔ the
length of the perianth; anthers very minute; berry globose, ¼ in.
diam. *Fl. Cap. edit.˙Schult.* 332; *Baker in Journ. Linn Soc.*
xiv. 621. *Dracœna striata, Linn. fil. Suppl.* 204; *Willd. Sp.
Plant.* ii. 157. *D. stricta, Roem. et Schultes, Syst. Veg.* vii. 346.
Myrsiphyllum striatum, Schlechtend. ex Kunth, Enum. v. 110.

VAR. β, **linearifolius** (Baker in Journ. Linn. Soc. xiv. 622); phyllocladia
linear, about ₁⁄₁₂ in. broad, 5-ribbed. *Asparagus erectus, herb. Thunberg. ex
parte.*

VAR. γ, **Dregei** (Baker, loc. cit.); branchlets spreading, zigzag; phyllocladia
subterete, 3-nerved.

VAR. δ, **Zeyheri** (Baker); branches very zigzag, with numerous short deflexed
final branchlets; phyllocladia subterete, 3-nerved, ½ in. long, ¼ lin. broad.

SOUTH AFRICA: without locality, *Thunberg !* Var. β, *Thunberg !*
COAST REGION: Queenstown Div., *Cooper*, 2464! Var. β, Uitenhage Division,
by the Zwartkops River, *Drège*, 8564a! Uitenhage, *Zeyher*, 480! Var. γ, Cath-
cart Div.; between Windvogel Mountain and Zwart Kei River, *Drège*, 8565 !
Var. δ, Uitenhage Div., *Zeyher, Asparagus No. 6*!
CENTRAL REGION: Somerset Div.; rocky places at the foot of Bosch Berg,
3000 ft., *MacOwan*, 1802! and by the Little Fish River at "Otter Station,"
Burchell, 3271/1! Graaff Reinet Div.; by the Sunday River near Graaff Reinet,
Drège, 8564c! mountains on the S.W. side of Graaff Reinet, *Burchell*, 2900!
Var. β, Carnarvon Div.; near Carnarvon at the northern exit of Karee Bergen
Poort, *Burchell*, 1570! Graaff Reinet Div.; Cave Mountain, 3300 ft., near
Graaff Reinet, *Bolus*, 516!

30. **A. scandens** (Thunb. Prodr. 66); stems slender, sarmentose;
leaves not produced at the base into distinct prickles; branchlets
very slender, glabrous, acutely angled, not zigzag; internodes very
short; phyllocladia ternate, linear, spreading, curved, moderately
firm in texture, distinctly costate, bright green, ¼–½ in. long; flowers
axillary, generally solitary; pedicels ¼–½ in. long, articulated above
the middle; perianth campanulate, ⅛ in. long; segments oblong,
obtuse; stamens nearly as long as the perianth; anthers oblong,
⅓–¼ the length of the filament; berry scarlet, 1-seeded. *Fl. Cap.
edit. Schult.* 334; *Willd. Sp. Plant.* ii. 152; *Bresl. Diss. No. 5*;
Baker in Saund. Ref. Bot. t. 21; *Journ. Linn. Soc.* xiv. 622. *A.
pectinatus, Red. Lil. t.* 407. *Dracœna volubilis, Linn. fil. Suppl.* 204.
Asparagopsis scandens, Kunth, Enum. v. 78.

VAR. β, **deflexus** (Baker in Journ. Linn. Soc. xiv. 622); branchlets very
zigzag; phyllocladia firmer in texture, only ⅛–¼ in. long; flowers smaller.

COAST REGION: vicinity of Cape Town, *Burchell*, 434! Worcester Div. ;
Drakenstein Mountains, 1000–2000 ft., *Drège*, 8570 ! Worcester, *Zeyher !* George
Div.; forest near Touw River, *Burchell*, 5729! Outeniqua, *Thunberg !* Knysna
Div.; woody ravine near Keurbooms River, *Burchell*, 5149! Stockenstrom
Div., *Scully*, 267!
CENTRAL REGION: Somerset Div.; woods on Bosch Berg, 4000 ft., *MacOwan*,
1771.

31. **A. ramosissimus** (Baker in Journ. Linn. Soc. xiv. 622);
stems sarmentose, very much branched, with very slender, short,
ascending or spreading final branchlets; main leaves not produced

at the base into a distinct prickle ; phyllocladia in clusters of 3 to 8, linear, subpatent, falcate, 1-nerved, $\frac{1}{4}$–$\frac{1}{3}$ in. long ; flowers solitary from the tips of the branchlets ; pedicels cernuous, $\frac{1}{8}$–$\frac{1}{6}$ in. long, articulated above the middle ; perianth campanulate, $\frac{1}{12}$ in. long ; segments oblong, obtuse, spreading horizontally ; stamens but little shorter than the perianth-segments ; anthers oblong ; style distinctly tricuspidate at the tip.

SOUTH AFRICA : without locality.

Described from a living plant in the garden of the late Mr. Wilson Saunders at Reigate, introduced by Mr. Thos. Cooper.

32. A. racemosus (Willd. Sp. Plant. ii. 152) *var.* **tetragonus,** (Baker in Journ. Linn. Soc. xiv. 624) ; main stems woody, sarmentose, terete, stramineous ; prickles large, pungent ; branchlets copious, slender, glabrous, acutely angled, spreading or ascending, not zigzag ; cladodia tetragonous, rigid, mucronate, bright green, 4–8-nate, $\frac{1}{3}$–$\frac{1}{2}$ in. long ; racemes copious, produced from the woody branches, spreading, 2–4 in. long ; pedicels 1–2-nate, the lower $\frac{1}{8}$–$\frac{1}{6}$ in. long, articulated about the middle ; bracts minute, deltoid, scariose ; perianth campanulate, $\frac{1}{12}$ in. long ; segments oblanceolate-oblong, obtuse ; stamens nearly as long as the perianth ; anthers minute, subglobose ; ovules 4–6 in a cell ; berry globose, $\frac{1}{4}$ in. diam., pulpy, 1-seeded. *Asparagus tetragonus, Bresl. Diss. No.* 15. *Asparagus stachyoides, Sprengel ex Baker in Journ. Linn. Soc.* xiv. 624. *Asparagopsis krebsiana and subquadrangularis, Kunth, Enum.* v. 93–94.

VAR. β, **decipiens** (Baker) ; racemes more elongated and laxer, with cladodia produced from some of the nodes instead of flowers.

SOUTH AFRICA : without locality, *Drège,* 3489b !
COAST REGION : Uitenhage Div. ; Coega Valley, *Cooper,* 3255 ! Albany Div., *Cooper,* 1511 ! Stockenstrom, *Scully,* 44 !
CENTRAL REGION : Somerset Div. ; between Little Fish River and Commadagga, *Burchell,* 3281 ! and at the foot of Bosch Berg, 2500–2800 ft., *MacOwan,* 1773 ! 1808 ! Graaff Reinet Div. ; by the Sunday River near Monkey Ford, *Burchell,* 2889 ! Var. β, Alexandria Div. ; on the rocks of Zwartwater Poort, *Burchell,* 3403 !
EASTERN REGION : Pondoland, *Bachmann,* 257 ! Griqualand East ; Shawbury, 1800 ft., *Baur,* 228 !

The type of the species is spread through Tropical Africa and Tropical Asia.

33. A. Saundersiæ (Baker in Journ. Bot. 1889, 42) : stems slender, sarmentose, woody, terete, stramineous, glabrous ; prickles small, spreading, pungent ; branchlets spreading or ascending, very slender, acutely angled, not zigzag ; cladodia 3–4-nate, subterete, rigid, ascending, about $\frac{1}{2}$ in. long, rounded on the back, flat or rather channelled down the face ; racemes spreading from the main woody branches, lax, about an inch long ; pedicels very slender, solitary, articulated above the middle, the lower $\frac{1}{4}$ in. long ; bracts minute, ovate white ; perianth $\frac{1}{12}$ in. long ; segments oblanceolate-oblong, obtuse, spreading horizontally ; stamens nearly as long as the perianth; anthers globose, very minute; ovary pedicellate; style very short.

EASTERN REGION : Natal, *Mrs. K. Saunders !*

34. **A sarmentosus** (Linn. Sp. Plant. 314) ; stems woody, multi-sulcate, suberect or twining; prickles deflexed, pungent ; branchlets numerous, slender, square, not zigzag, spreading or ascending, inter-nodes very short ; phyllocladia linear, rigid, mucronate, $\frac{1}{2}$–$\frac{3}{4}$ in. long, solitary unless at the tip of the branchlets, flat on both faces, bright green, furnished with a distinct midrib ; racemes 1–3 in. long, spreading from the woody branches ; lower pedicels $\frac{1}{12}$–$\frac{1}{6}$ in. long, articulated about the middle ; bracts minute, scariose ; perianth $\frac{1}{12}$ in. long ; segments oblong, obtuse, spreading widely ; stamens nearly as long as the perianth ; anthers minute, globose ; berry globose, 1-seeded, $\frac{1}{4}$ in. diam. *Miller, Dict. edit.* viii. *No.* 10 ; *Willd. Sp. Plant.* ii. 155 ; *Ait. Hort. Kew. edit.* 2, ii. 275 ; *Bresl. Diss. No.* 18 ; *Baker in Journ. Linn. Soc.* xiv. 625. *Asparagopsis sarmentosa, Kunth, Enum.* v. 97.

VAR. β, **comatus** (Baker, loc. cit.) ; stems twining ; racemes lengthened out with phyllocladia produced from some of the nodes, especially the upper ones. *A. sarmentosus, Red. Lil. t.* 460.—*A. aculeatus maximus sarmentosus zeylanicus, Herm. Hort. Acad. Lug. Bat.* 63, *cum icone.*

VAR. γ, **densiflorus** (Baker, loc. cit.) ; stems suberect; branchlets short, crowded, ascending ; phyllocladia linear, rigid, ascending, about $\frac{1}{4}$ in. long. *Asparagopsis densiflora, Kunth, Enum.* v. 96.

VAR. δ, **Kunthii** (Baker, loc. cit.) ; stems twining ; phyllocladia not so close as in the other forms and larger (about an inch long). *Asparagopsis lancea, Kunth, Enum.* v. 104, *excl. syn.*

SOUTH AFRICA : without locality, Var. δ, *Grey !*
COAST REGION : Var. β, Albany Div., *Cooper,* 3251 ! Var. γ, near the Zwart-kops River, *Ecklon and Zeyher,* 1061 ! Alexandria Div.; between Rautenbachs Drift and Addo Drift, *Burchell,* 4208 ! Stutterheim Div. ; Kabousie Mountain, 3500 ft., *Murray in Herb. MacOwan !* British Kaffraria, *Cooper,* 192 ! 402 ! Var. δ, Knysna Div. ; " Kaatjes Kraal," in the forest and by the rivulet, *Burchell,* 5232 !
CENTRAL REGION : Alexandria Div.; Zwartwater Poort, *Burchell,* 3383 ! Somerset Div. ; Commadagga, *Burchell,* 3302 ! Var. β, Somerset Div.; sides of Bosch Berg, 3500 ft., *MacOwan,* 1772 ! Var. γ, Somerset Div. ; upper part of Bruintjes Hoogte, *Burchell,* 3061 ! and without precise locality, *Bowker !*
KALAHARI REGION : Orange Free State, *Cooper,* 3603 !
EASTERN REGION : Natal ; Inanda, *Wood,* 1626 ! and without precise locality, *Sutherland !* Var. β, Griqualand East ; mountains near Kokstad, 4300 ft., *Tyson,* 3158 !

35. **A. oxyacanthus** (Baker in Journ. Linn. Soc. xiv. 625) ; stems woody, terete, straight, densely pubescent ; prickles pungent, deflexed ; phyllocladia produced from the stout woody stems, solitary, linear, mucronate, erecto-patent, rigid in texture, bright green, 1-nerved, 1–1$\frac{1}{2}$ in. long, $\frac{1}{6}$–$\frac{1}{5}$ in. broad ; racemes produced from the woody stems, dense, 1–1$\frac{1}{2}$ in. long; pedicels very short, articulated near the tip, bracts lanceolate, white, scariose, $\frac{1}{4}$ in. long ; perianth campanulate, $\frac{1}{12}$ in. long ; segments oblong, obtuse ; stamens nearly as long as the perianth ; anthers minute, globose.

CENTRAL REGION : Albany, *Bowie !* Somerset, *Bowker !*

36. A. falcatus (Linn. Sp. Plant. 449); stems stout, terete, woody, wide-climbing; prickles large, spreading, pungent; branchlets slender, woody terete, stramineous, flexuose; phyllocladia 1–3-nate at the side, up to 6–8 in a cluster at the tip of the branchlets, lanceolate, 1–2 in. long, $\frac{1}{6}$–$\frac{1}{4}$ in. broad, bright green, 1-nerved, moderately firm in texture; racemes 1–2 in. long, 1–3-nate; pedicels 1–2-nate at the sides, clustered at the tips, articulated at the middle, $\frac{1}{6}$–$\frac{1}{4}$ in. long; perianth campanulate, $\frac{1}{12}$–$\frac{1}{6}$ in. long; segments oblong, obtuse, spreading widely; stamens shorter than the perianth; anthers minute; berry globose, 1-seeded, $\frac{1}{4}$ in. diam. *Bresl. Diss. No.* 2; *Kunth, Enum.* v. 71; *Baker in Journ. Linn. Soc.* xiv. 626. *A. æthiopicus, var. ternifolius, Baker in Saund. Ref. Bot. t.* 261; *Gard. Chron.* 1872, 1588, *fig.* 338.

EASTERN REGION: Kaffraria, *Cooper*, 3253! Natal; near Durban, *Wood*, 227! and without precise locality, *Gerrard*, 746! *Gerrard and McKen*, 753!

Also Tropical Asia and Tropical Africa.

37. A. Sprengeri (Regel in Act. Hort. Petrop. xi. 302); stems woody, terete, wide-climbing; prickles small, hooked; phyllocladia 1–4 nate, flat, linear, glabrous, bright green, straight or slightly curved, 1–1$\frac{1}{2}$ in. long, $\frac{1}{16}$ in. broad; racemes $\frac{3}{4}$–1 in. long; pedicels $\frac{1}{12}$ in. long, articulated at the middle; bracts as long as the pedicels; perianth $\frac{1}{12}$ in. long; segments oblong; stamens shorter than the perianth; anthers minute; berry $\frac{1}{6}$ in. diam.

EASTERN REGION: Natal. Described from cultivated specimens.

38. A. myriocladus (Baker in Journ. Bot. 1889, 43); main stem woody, terete, suberect, many-ribbed; prickles slender, pungent, deflexed; branchlets short, crowded, very slender, strongly angled, simple, ascending; phyllocladia 3–8-nate, linear, rigid, mucronate, $\frac{1}{4}$ in. long; racemes lax, about an inch long, produced from the woody main stems; pedicels $\frac{1}{12}$–$\frac{1}{8}$ in. long, articulated at the middle; bracts ovate-lanceolate; perianth $\frac{1}{12}$ in. long; segments obtuse; stamens nearly as long as the perianth; anthers minute, globose.

EASTERN REGION: Natal; Inanda, *Wood*, 355!

39. A. æthiopicus (Linn. Mant. 63); stems woody, terete, climbing, many-ribbed; prickles large, spreading, pungent; branchlets short, simple, spreading, strongly angled; phyllocladia linear, rigid, 3–6-nate, ascending, mucronate, about $\frac{1}{2}$ in. long; racemes 2–3 in. long, produced from the main stems, spreading; pedicels $\frac{1}{12}$–$\frac{1}{8}$ in. long; articulated at the middle; bracts minute, lanceolate; perianth campanulate, $\frac{1}{12}$ in. long; segments oblong, obtuse; stamens nearly as long as the perianth; anthers minute, globose; berry 1-seeded, $\frac{1}{4}$ in. diam. *Willd. Sp. Plant.* ii. 153; *Roem. et Schultes, Syst. Veg.* vii. 335; *Baker in Journ. Linn. Soc.* xiv. 626. *A. lanceus, Thunb. Prodr.* 66; *Fl. Cap. edit. Schult.* 334. *A. falcatus, Linn. Herb. Asparagopsis æthiopica, Kunth, Enum.* v. 95.

VAR. β, **natalensis** (Baker); woody stems not ribbed; phyllocladia larger, less rigid, darker green, 3–6-nate at the sides, densely clustered at the tips of the branchlets; flowers laxly panicled.

SOUTH AFRICA: without locality, *Thunberg! Villette! Nelson! Hort. Upsal.!*
EASTERN REGION: Var. β, Natal; Inanda, *Wood,* 1351!

Also in Tropical Africa.

40. A. Krausii (Baker in Journ. Linn. Soc. xiv. 628); stems very slender, climbing indefinitely; prickles none; branchlets very slender, subterete; phyllocladia solitary unless at the tip of the branchlets, lanceolate, moderately firm in texture, $\frac{1}{2}$–2 in. long, $\frac{1}{6}$–$\frac{1}{4}$ in. broad, finely and equally 7–11-nerved; flowers axillary, generally solitary; pedicels cernuous, $\frac{1}{6}$–$\frac{1}{4}$ in. long, articulated above the middle; perianth campanulate, greenish, $\frac{1}{8}$ in. long; segments linear-oblong, reflexing above the base; stamens nearly as long as the perianth; anthers minute; ovules 5–6 in a cell. *Myrsiphyllum krausianum and gramineum, Kunth, Enum.* v. 107–8.

SOUTH AFRICA: without locality, *Mund! Hutton!*
COAST REGION: Malmesbury Div.; near Hopefield, *Bachmann,* 820! 1835! 2235! Cape Div.; near Constantia, *Krauss,* 1333! Worcester Div.; valley of the Breede River, near Bains Kloof, *Bolus,* 2845! Bathurst Div.; near Kaffir Drift, *Burchell,* 3861!
CENTRAL REGION: Somerset Div.; between Little Fish River and Commadagga, *Burchell,* 3282! and on the mountain above the spring of Commadagga, *Burchell,* 3348!

41. A. volubilis (Thunb. Prodr. 66); stems very slender, green, glabrous, many-ribbed, climbing indefinitely; prickles none; branchlets slender, square; phyllocladia solitary, ovate-lanceolate, rigid in texture, bright green, acute, $\frac{1}{2}$–1 in. long, with a distinct midrib and thickened margin; pedicels axillary, 1–2-nate, $\frac{1}{8}$ in. long; perianth campanulate, greenish, $\frac{1}{6}$–$\frac{1}{8}$ in. long; segments linear-oblong, reflexing above the base; stamens nearly as long as the perianth; anthers minute. *Baker in Journ. Linn. Soc.* xiv. 628.

SOUTH AFRICA: without locality, *Thunberg! Masson!*
CENTRAL REGION: Fraserburg Div.; between Kuilen Berg and Great Reed River, *Burchell,* 1361!

I do not cite the synonym of Linn. fil. Suppl. 204, because the specimen in the Smithian herbarium is not this species, but *A. scandens.*

42. A. medeoloides (Thunb. Prodr. 66); stems very slender, much-branched, wide-climbing; prickles none; branchlets slender, quadrangular; phyllocladia solitary, ovate, cordate or rounded at the base, moderately firm in texture, many-nerved, $\frac{1}{2}$–2 in. long; pedicels axillary, generally geminate and cernuous, $\frac{1}{4}$–$\frac{1}{2}$ in. long, articulated near the apex; perianth greenish, campanulate, $\frac{1}{8}$ in. long; segments linear-oblong, reflexing above the base; stamens nearly as long as the perianth; filaments rather flattened; anthers minute, $\frac{1}{3}$ the length of the filament; berry often 3-seeded. *Thunb. Fl. Cap. edit. Schult.* 333; *Baker in Journ. Linn. Soc.* xiv. 627. *Dracæna medeoloides, Linn. fil. Suppl.* 203. *Medeola asparagoides, Linn. Sp. Plant.*

484; *Mant. alt.* 370; *Red. Lil. t.* 442. *Myrsiphyllum asparagoides*, *Willd. in Ges. Naturf. Fr. Berl. Mag.* ii. 25; *Kunth, Enum.* v. 105; *Hook. in Bot. Mag. t.* 5584. *Ruscus volubilis*, *Thunb. Prodr.* 13; *Fl. Cap. edit. Schult.* 38; *Kunth, Enum.* v. 276.

VAR. β, **angustifolius** (Baker); phyllocladia ¾–1 in. long, twice as long as broad. *Myrsiphyllum angustifolium, Willd. in Ges. Naturf. Fr. Berl. Mag.* ii. 25; *Kunth, Enum.* v. 106. *Medeola angustifolia, Miller, Dict. edit.* viii. *No.* 2.

VAR. γ, **falciformis** (Baker); phyllocladia ovate-lanceolate, 1½–2 in. long, thinner than in the type. *Myrsiphyllum falciforme, Kunth, Enum.* v. 107; *Baker in Saund. Ref. Bot. t.* 47.

SOUTH AFRICA: without locality, *Thunberg!* Var. β, *Hort. Fothergill.* year 1779!

COAST REGION: Knysna Div.; near Goukamma River, *Burchell*, 5570! Uitenhage, *Zeyher*, 972! Bathurst Div.; at the mouth of the Kowie River, *MacOwan*, 1920! near Port Alfred, *Burchell*, 3801! and at Kaffir Drift Military Post, *Burchell*, 3769! Albany Div.; *Cooper*, 14! *Hutton!* Queenstown Div.; Shiloh, 3500–4000 ft., *Baur*, 1143! and without precise locality, *Cooper*, 328! British Kaffraria, *Cooper*, 188!

CENTRAL REGION: Tulbagh Div.; by the Ongeluks River, *Burchell*, 1230! Somerset Div.; at Commadagga, *Burchell*, 3310! near Somerset East, 2500 ft., *MacOwan*, 1802! 1862! Graaff Reinet Div.; along the Sunday River, near Monkey Ford, *Burchell*, 2886!

WESTERN REGION: Little Namaqualand, Silver Fontein near Ookiep, 2000–3000 ft., *Drège*, 2704b! Var. γ, Little Namaqualand near Mieren Kasteel, 1000–2000 ft., *Drège*, 2704a!

KALAHARI REGION: Transvaal; Houtbosch, *Rehmann*, 5785! and Umvoti Creek, Barberton, *Galpin*, 857!

EASTERN REGION: Pondoland, *Sutherland!* Griqualand East; around Clydesdale, 3000 ft., *Tyson*, 2544! Natal; Inanda, *Wood!* and without precise locality, *Cooper*, 3257! Var. β, Natal, *Gerrard*, 1523!

43. **A. undulatus** (Thunb. Prodr. 66); stem suberect, a foot or more long, simple towards the base, copiously branched from the middle upwards; prickles none; branchlets stout, acutely angled; phyllocladia ovate, solitary unless at the tip of the branchlets, ½–1 in. long, rigid in texture, often complicate, with a distinct midrib and close unequal ribs, scabrous on the margin and stronger ribs beneath; pedicels axillary, 1–3-nate, cernuous, ⅙–¼ in. long, articulated near the tip; perianth campanulate, greenish, ⅙ in. long; segments linear-oblong, falcate from above the base; stamens nearly as long as the perianth; anthers minute. *Fl. Cap. edit. Schult.* 332; *Baker in Journ. Linn. Soc.* xiv. 628. *Dracæna undulata, Linn. fil. Suppl.* 203; *Roem. et Schultes, Syst. Veg.* vii. 346. *Myrsiphyllum undulatum, Schlechtend. ex Kunth, Enum.* v. 109.

VAR. β, **rugosus** (Baker, loc. cit.); dwarf form, with phyllocladia only ¼-⅓ in. long, very scabrous on the edge and ribs.

SOUTH AFRICA: without locality! *Thunberg! Oldenburg! Harvey!* β, *Oldenburg.*

COAST REGION: Malmesbury Div.; near Hopefield, *Bachmann*, 812! 813! near Cape Town, *Bolus*, 3690!

44. **A. juniperoides** (Engl. in Jahrb. x. 3); stem erect, sulcate, 8–16 in. long; branches short; internodes not above $\frac{1}{12}$–⅙ in. long; leaves scariose, deltoid-acuminate, ⅕–¼ in. long; phyllocladia linear,

flat, very acute, glabrous, $\frac{1}{2}$–1 in. long; flowers solitary, hermaphrodite, shortly pedicellate; perianth $\frac{1}{4}$ in. long; segments linear-oblong; stamens nearly as long as the perianth; berry subglobose, 3-seeded.

WESTERN REGION: Great Namaqualand; near Aus, *Marloth*, 1538!

III. BEHNIA, Didrichs.

Flowers hermaphrodite. *Perianth* infundibuliform, gamophyllous; tube long; segments subequal, ovate, spreading. *Stamens* 6, inserted at the middle of the perianth-tube; filaments short, thick; anthers oblong, dorsifixed, versatile, dehiscing introrsely. *Ovary* contracted at the base into a short gynophore, 3-celled; ovules few in a cell; style short; stigmas 3, spreading. *Berry* globose, fleshy. *Seeds* subglobose; testa black; albumen horny.

DISTRIB. Endemic; monotypic.

1. **B. reticulata** (Didrichs. in Viddensk. Meddel. Kjöben. 1854, 183); stems glabrous, slender, woody, terete, unarmed, copiously branched, sarmentose; branchlets flexuose; leaves alternate, sessile, ovate, acute, 2–3 in. long, firm in texture, green, glossy, broadly rounded at the base, furnished with a distinct midrib and 5–7 strongly marked vertical veins on each side of it, connected by close distinct cross-veinlets; flowers in simple or slightly compound cymes from the axils of the upper leaves; bracts persistent, scariose, ovate-lanceolate; pedicels $\frac{1}{3}$ in. long, articulated at the apex; perianth greenish, $\frac{1}{3}$ in. long; segments $\frac{1}{4}$ as long as the tube; stamens and style not exserted beyond the tip of the perianth-segments; berry $\frac{1}{3}$–$\frac{1}{2}$ in. diam. *Ruscus reticulatus, Thunb. Prodr.* 13; *Fl. Cap. edit. Schult.* 38; *Willd. Sp. Plant.* iv. 876; *Kunth, Enum.* v. 276. *Dictyopsis Thunbergii, Harv. in Bot. Mag. t.* 5638; *Cape Gen. edit.* 2, 406. *Hylonome reticulata, Baker in Journ. Linn. Soc.* xiv. 561, *vix Webb.*

SOUTH AFRICA : without locality, *Thunberg!*
COAST REGION: Uitenhage, *Zeyher*, 542! Port Elizabeth Div.; Cragga Kamma, *Burchell*, 4554! Alexandria Div.; Zuurberg Range, *Cooper*, 3258! Bathurst Div.; near Port Alfred, *Burchell*, 3800! Albany Div.; *Hutton!* Stutterheim Div.; Kabousie, 3000–3500 ft., *Murray in Herb. MacOwan*, 1809! British Kaffraria, *Cooper*, 189!
CENTRAL REGION : Somerset Div., *Zeyher! Bowker!*
KALAHARI REGION: Transvaal; woods around Barberton, *Galpin*, 517! Houtbosch Berg, *Nelson*, 494!
EASTERN REGION : Natal; coast-land, up to 1000 ft., *Sutherland!* Inanda, *Wood*, 204! and without precise locality, *Cooper*, 1266! 3256!

IV. DRACÆNA, Vand.

Flowers hermaphrodite. *Perianth* cylindrical, gamophyllous; tube cylindrical or campanulate; segments linear, spreading or reflexing.

Stamens 6, inserted at the throat of the perianth-tube; filaments elongated, filiform or slightly flattened; anthers oblong, dorsifixed, versatile, dehiscing introrsely. *Ovary* sessile, ovoid, 3-celled; ovules solitary, erect; style long, filiform; stigma capitate. *Berry* globose, often by abortion 1-2-seeded. *Seeds* globose or by pressure angled; testa thick, often pulpy; albumen horny; embryo small.

Stems woody, simple or branched, usually erect and stout, marked with the scars of the fallen leaves; leaves sessile and ensiform, or petioled and elliptic, closely veined; flowers racemose, white or greenish, often panicled; pedicels articulated; bracts small, scariose, persistent.

DISTRIB. Species about 50, spread through the tropical and subtropical zones of the Old World.

1. **D. hookeriana** (K. Koch, Wochenschr. iv. 394, excl. patr.); stem 2-10 ft. long, usually simple; leaves crowded towards its apex, sessile, ensiform, 2-3 ft. long, 1-1½ in. broad, slightly narrowed to the base, moderately firm in texture, pale green, with a distinct whitish border, recurving from about the middle, tapering to a long point; midrib hidden before it reaches the apex; racemes in an ample erect panicle 1½-2 ft. long; branches ascending, lower compound; pedicels 3-4-nate, articulated at the apex; bracts white, scariose, ovate, persistent; perianth greenish, about an inch long; segments about as long as tube; stamens as long as the perianth-segments; filaments rather flattened; anthers oblong, pale yellow, $\frac{1}{12}$ in. long; berry 1-3-seeded, yellow. *Baker in Journ. Linn. Soc.* xiv. 527. *D. Rumphii, Regel, Revis.* 41, *excl. syn. Cordyline Rumphii, Hook. in Bot. Mag. t.* 4279, *excl. syn. Sansevieria paniculata, Schinz in Durand and Schinz, Conspect. Fl. Afric.* 141.

VAR. **D. latifolia** (Regel, Rev. 41); leaves 2-3 in. broad at the middle, much narrowed to the base. *Baker in Saund. Ref. Bot. t.* 353.

D. latifolia var. schmidtiana, Regel, Gartenfl. t. 1023, is a form with broad variegated leaves.

COAST REGION: Uitenhage Div.; near Strand Fontein and Matjes Fontein, and by Van Stadens River, *Drège*, 4494a! Albany Div.; in a wooded Kloof west of Grahamstown, *Burchell*, 3564! Bathurst Div.; Port Alfred, *Schönland*, 290!

EASTERN REGION: Pondoland, *Bachmann*, 250! 251!

V. **KNIPHOFIA**, Moench.

Perianth cylindrical or funnel-shape; tube long; segments sub-equal, small, ovate or oblong. *Stamens* 6, hypogynous, as long as the perianth or longer, the three opposite the inner segments longest; filaments filiform, slightly declinate; anthers oblong, dorsifixed, versatile, dehiscing introrsely. *Ovary* sessile, ovoid, 3-celled; ovules many, superposed; style long, filiform; stigma subentire, minute, capitate. *Seeds* triquetrous, acutely angled; testa thin, brown-black, punctate; albumen fleshy.

Rootstock neither bulbous nor tuberous; leaves all radical, persistent, narrowed gradually to the apex; peduncle long, naked; flowers many, arranged in a subspicate raceme, generally bright red or yellow; pedicels short, articulated at the apex, shorter than the white scariose persistent bracts.

DISTRIB. Also twelve species in Tropical Africa, and two in Madagascar.

- Perianth $\frac{1}{6}$–$\frac{1}{3}$ in. long:
  - Perianth subcylindrical; stamens scarcely exserted:
    - Perianth $\frac{1}{6}$ in. long ... ... (1) **Buchanani.**
    - Perianth $\frac{1}{4}$–$\frac{1}{3}$ in. long:
      - Racemes short:
        - Bracts ovate ... ... (2) **parviflora.**
        - Bracts oblong-lanceolate ... ... (3) **breviflora.**
      - Racemes long ... ... (4) **modesta.**
  - Perianth funnel-shaped; stamens much exserted ... (5) **infundibularis.**
- Perianth $\frac{1}{2}$–$\frac{3}{4}$ in. long:
  - Perianth subcylindrical:
    - Leaves $\frac{1}{12}$–$\frac{1}{8}$ in. broad:
      - Flowers pale yellow ... ... (6) **gracilis.**
      - Flowers bright scarlet ... ... (7) **Evansii.**
    - Leaves $\frac{1}{3}$–$\frac{1}{2}$ in. broad ... ... (8) **citrina.**
    - Leaves $\frac{3}{4}$–1 in. broad low down:
      - Stamens shortly exserted ... ... (9) **Tuckii.**
      - Stamens much exserted ... ... (10) **Tysoni.**
    - Leaves 1$\frac{1}{2}$–2 in. broad low down ... ... (11) **foliosa.**
  - Perianth funnel-shaped:
    - Leaves $\frac{1}{8}$–$\frac{1}{6}$ in. broad; raceme lax ... ... (12) **pauciflora.**
    - Leaves $\frac{1}{2}$ in. broad; raceme dense ... ... (13) **pumila.**
    - Leaves 1–1$\frac{1}{2}$ in. broad; raceme dense ... (14) **ensifolia.**
- Perianth an inch or more long:
  - Acaulescent; leaves linear:
    - Leaves $\frac{1}{16}$–$\frac{1}{12}$ in. broad:
      - Flowers yellow ... ... (15) **triangularis.**
      - Flowers bright scarlet ... ... (16) **Nelsoni.**
    - Leaves $\frac{1}{8}$–$\frac{1}{6}$ in. broad:
      - Leaves scabrous on the edge ... ... (17) **Macowani.**
      - Leaves smooth on the edge ... ... (18) **Galpini.**
    - Leaves $\frac{1}{4}$–$\frac{1}{2}$ in. broad:
      - Racemes lax ... ... (19) **laxiflora.**
      - Racemes moderately lax ... ... (20) **natalensis.**
      - Racemes dense:
        - Stamens not exserted:
          - Leaves under a foot long ... (21) **Baurii.**
          - Leaves 1$\frac{1}{2}$ ft. long ... ... (22) **porphyrantha.**
          - Leaves 3 ft. long ... ... (23) **decaphlebia.**
        - Stamens exserted ... . (24) **linearifolia.**
  - Acaulescent; leaves short lanceolate ... ... (25) **drepanophylla.**
  - Acaulescent; leaves ensiform-acuminate:
    - Stamens much exserted ... ... (26) **sarmentosa.**
    - Stamens finally just exserted:
      - Leaves dull green:
        - Leaves $\frac{3}{4}$–1 in. broad low down ... (27) **aloides.**
        - Leaves 1$\frac{1}{2}$ in. broad low down ... (28) **Rooperi.**
      - Leaves bright green:
        - Pedicels very short ... ... (29) **longicollis.**
        - Pedicels $\frac{1}{6}$–$\frac{1}{4}$ in. long ... ... (30) **Burchellii.**
  - Caulescent; leaves ensiform-acuminate:
    - Leaves 2–3 in. broad ... ... (31) **caulescens.**
    - Leaves 5–6 in. broad ... ... (32) **Northiæ.**

1. **K. Buchanani** (Baker in Journ. Bot. 1885, 276); leaves linear,

moderately firm, 2–3 ft. long, $\frac{1}{8}$–$\frac{1}{6}$ in. broad, with about 5 veins between the thickened midrib and smooth margin; peduncle slender, 3 ft. long; raceme dense, 2–3 in. long, $\frac{1}{2}$ in. diam.; pedicels very short; bracts ovate-lanceolate, $\frac{1}{8}$ in. long; flowers all whitish; perianth subcylindrical, $\frac{1}{6}$ in. long, $\frac{1}{12}$ in. diam. at the throat; stamens finally just exserted.

EASTERN REGION: Natal; Fields Hill, 1500 ft., *Wood*, 1972! near Boston, 3000–4000 ft., *Wood*, 4572! and without precise locality, *Buchanan!*

2. K. parviflora (Kunth, Enum. iv. 553); leaves few, linear, subrigid, about a foot long, $\frac{1}{8}$ in. broad, smooth on the margin; peduncle slender, a little overtopping the leaves; raceme dense, 2–3 in. long; pedicels very short; bracts ovate, acute, $\frac{1}{8}$ in. long; flowers yellow; perianth cylindrical, $\frac{1}{3}$ in. long, not constricted above the ovary, $\frac{1}{8}$ in. diam. at the throat; segments very short, obtuse; stamens finally just exserted; ovules 4–5 in a cell. *Baker in Journ. Linn. Soc.* xi. 361; *Journ. Bot.* 1885, 277.

EASTERN REGION: Tembuland; Bazeia Mountains, 3500–4000 ft., *Baur*, 617! Pondoland; between St. John's River and Umtsikaba River, 1000–2000 ft., *Drège*, 4528.

3. K. breviflora (Harv. ex Baker in Journ. Linn. Soc. xi. 361); leaves linear, not rigid, 1–1$\frac{1}{2}$ ft. long, $\frac{1}{12}$ in. broad, with about 5 strong ribs and a scabrous margin; peduncle slender, terete, as long as the leaves; raceme dense, 1–1$\frac{1}{2}$ in. long; pedicels very short; bracts oblong-lanceolate, $\frac{1}{8}$ in. long; flowers all yellow; perianth cylindrical, $\frac{1}{4}$ in. long, $\frac{1}{12}$ in. diam. at the throat; segments ovate, obtuse, $\frac{1}{2}$ lin. long; stamens as long as the perianth; ovules 4–6 in a cell.

KALAHARI REGION: Orange Free State, *Cooper*, 1029! 3294! EASTERN REGION: Natal; Van Reenens Pass, 5000–6000 ft., *Wood*, 5606!

4. K. modesta (Baker in Journ. Bot. 1889, 43); leaves linear, rigid, 1$\frac{1}{2}$ ft. long, $\frac{1}{12}$–$\frac{1}{8}$ in. broad, acutely keeled on the back; peduncle slender, as long as the leaves; empty bracts lanceolate; raceme dense, subspicate, subsecund, 3–6 in. long; pedicels very short; bracts lanceolate, $\frac{1}{4}$–$\frac{1}{3}$ in. long; perianth cylindrical, white, $\frac{1}{3}$ in. long; segments ovate; anthers oblong, finally a little exserted. *Bot. Mag. t.* 7293.

EASTERN REGION: Griqualand East; Mount Currie, near Kokstad, 6000 ft., *Tyson*, 1418! Natal; hill-side near Bothas, *Wood*, 4409! Glencoe, 4000–5000 ft., *Wood*, 4826!

5. K. infundibularis (Baker in Journ. Bot. 1885, 277); leaves linear, a foot or more long, $\frac{1}{6}$ in. broad low down, smooth on the margin, with 4–5 fine veins on each side of the raised midrib; peduncle slender, terete; raceme dense, 2 in. long, 1 in. diam.; pedicels very short; bracts lanceolate, acute, $\frac{1}{6}$ in. long; flowers all yellow; perianth funnel-shaped, $\frac{1}{3}$ in. long, $\frac{1}{2}$ lin. diam. at the base,

dilated suddenly above the base, about $\frac{1}{6}$ in. diam. at the throat ;
segments as broad as long ; stamens exserted $\frac{1}{6}-\frac{1}{4}$ in.

SOUTH AFRICA : without locality.

Described from a specimen at Kew from the herbarium of Bishop Goodenough,
dried from Kew Gardens about 1780.

6. K. gracilis (Harv. ex Baker in Journ. Linn. Soc. xi. 362) ;
leaves linear, $1\frac{1}{2}-2$ ft. long, $\frac{1}{6}$ in. broad, with 5-6 veins between the
midrib and smooth margin ; peduncle as long as the leaves ; raceme
dense, 2-3 in. long ; pedicels very short ; bracts ovate, obtuse,
$\frac{1}{8}-\frac{1}{6}$ in. long ; perianth pale yellow, $\frac{1}{2}-\frac{5}{8}$ in. long ; tube very slender ;
throat dilated ; segments oblong, $\frac{1}{12}$ in. long ; style and longer
stamens exserted.

EASTERN REGION : Zululand, *Gerrard and McKen*, 2140 !

7. K. Evansii (Baker) ; relics of old leaves fibrous ; produced
leaves linear, subcoriaceous, $1\frac{1}{2}$ ft. long, $\frac{1}{12}$ in. broad low down,
with few nerves, thickened entire edges and back not acutely keeled ;
scape moderately stout, $1\frac{1}{2}-2$ ft. long ; raceme very dense, oblong,
2-3 in. long ; pedicels $\frac{1}{8}-\frac{1}{6}$ in. long, the lower cernuous ; bracts
ovate, white, scariose, not longer than the pedicels ; perianth
cylindrical, $\frac{1}{2}$ in. long, bright scarlet, drying purplish-black ; lobes
short, ovate ; stamens and style included.

EASTERN REGION : Natal; on the Drakensberg, 6000-7000 ft., *Evans*, 353 !

Near *K. gracilis, Harv.*

8. K. citrina (Baker in Gard. Chron. 1893, xiv. 552) ; leaves
many to a stem, linear, moderately firm, green, $1\frac{1}{2}-2$ ft. long, $\frac{1}{3}$ in.
broad low down, triquetrous on the back and acutely channelled
down the face, slightly scabrous on the edge ; peduncle slender,
shorter than the leaves ; raceme oblong, dense, 2-3 in. long ;
pedicels very short ; bracts oblong-lanceolate, $\frac{1}{6}$ in. long ; perianth
subcylindrical, pale yellow, 9-10 lines long, $\frac{1}{6}$ in. diam. at the throat ;
lobes small, ovate ; stamens and style much exserted.

COAST REGION : Albany Div.; mountains north of Grahamstown, *Hort.
Leichtlin. !*

Introduced in 1893.

9. K. Tuckii (Baker in Gard. Chron. 1893, xiii. 68) ; leaves
ensiform, bright green, $1-1\frac{1}{2}$ ft. long, $\frac{3}{4}$ in. broad low down, tapering
gradually to the apex ; edges distinctly serrated ; peduncle
moderately stout, shorter than the leaves ; raceme very dense,
oblong-cylindrical, 5-6 in. long, $1\frac{1}{2}$ in. diam.; flowers all deflexed,
yellow, tinged with bright red when young ; pedicels very short ;
bracts oblong-lanceolate, scariose, $\frac{1}{8}$ in. long ; perianth subcylindrical,
$\frac{5}{8}$ in. long ; tube $\frac{1}{12}$ in. diam. low down, $\frac{1}{6}$ in. at the throat ; lobes
short, ovate, obtuse ; stamens shortly exserted.

CENTRAL REGION : Colesberg Div., *Tuck !*

Described from a living specimen sent by Mr. Tuck to Mr. Max Leichtlin in
1892.

10. **K. Tysoni** (Baker in Journ. Bot. 1889, 43); leaves linear, 3–4 ft. long, $\frac{3}{4}$ in. broad at the base, tapering gradually to a long point, acutely keeled on the back; peduncle as long as the leaves; raceme very dense, $\frac{1}{2}$ ft. long, $2\frac{1}{2}$ in. diam.; pedicels very short; bracts oblong, obtuse, $\frac{1}{6}$ in. long; perianth cylindrical, red-yellow, $\frac{3}{4}$ in. long, $\frac{1}{8}$ in. diam. at the throat; segments suborbicular; stamens exserted $\frac{1}{4}$–$\frac{1}{3}$ in.

EASTERN REGION: Griqualand East; on the Zuurberg, 4000 ft., *Tyson*, 1700!

Between *K. pumila* and *K. sarmentosa*.

11. **K. foliosa** (Hochst. in Flora 1844, 31); acaulescent; leaves densely tufted, ensiform, acuminate, acutely keeled, 2–3 feet long, $1\frac{1}{2}$–2 in. broad above the clasping base; peduncle very stout, as long as the leaves; raceme dense, oblong; pedicels very short; bracts oblong-lanceolate, the lower $\frac{1}{4}$ in. long; perianth yellow, cylindrical, $\frac{3}{4}$–$\frac{7}{8}$ in. long; lobes small, ovate, obtuse; stamens much exserted. *Baker in Journ. Bot.* 1874, 4; *Bot. Mag. t.* 6742. *K. quartiniana, A. Rich. Fl. Abyss.* ii. 323; *Baker in Journ. Linn. Soc.* xi. 362.

KALAHARI REGION: Transvaal.

Flowered at Kew, May, 1895, from seeds sent by Mr. R. Loveday from Pretoria. Also Abyssinia.

12. **K. pauciflora** (Baker in Journ. Bot. 1885, 280); leaves few, linear, rigid, firm, 1–$1\frac{1}{2}$ ft. long, $\frac{1}{8}$–$\frac{1}{6}$ in. broad, with 10–15 veins and a thickened smooth margin; peduncle slender, terete, $1\frac{1}{2}$–2 ft. long; raceme lax, 2–3 in. long; pedicels $\frac{1}{8}$–$\frac{1}{6}$ in. long; bracts lanceolate, $\frac{1}{8}$–$\frac{1}{6}$ in. long; flowers pale yellow; perianth narrowly funnel-shaped, $\frac{3}{4}$ in., long, $\frac{1}{6}$ in. diam. at the throat; stamens shortly exserted. *Bot. Mag. t.* 7269.

EASTERN REGION: Natal; Clairmont, *Wood*, 1096! and without precise locality, *Sanderson*, 416!

Mr. Dewar has crossed this species with *K. Macowani*. The hybrid first flowered at Kew, Oct., 1893.

13. **K. pumila** (Kunth, Enum. iv. 552); leaves linear, $1\frac{1}{2}$–2 ft. long, $\frac{1}{2}$ in. broad low down, glaucous, acutely keeled, with 10–12 veins on each side of the midrib; peduncle moderately stout, as long as the leaves; raceme very dense, 3–4 in. long, under 2 in. diam.; pedicels very short; bracts oblong-lanceolate, acute, $\frac{1}{4}$–$\frac{1}{3}$ in. long; perianth narrowly funnel-shaped, $\frac{5}{8}$–$\frac{3}{4}$ in. long; $\frac{1}{2}$ lin. diam. at the base, suddenly dilated above the base, $\frac{1}{6}$ in. diam. at the throat; segments ovate, obtuse, $\frac{1}{12}$ in. diam.; stamens and style exserted $\frac{1}{4}$–$\frac{1}{3}$ in.; capsule subglobose, $\frac{1}{6}$ in. long. *Baker in Journ. Linn. Soc.* xi. 363; *Journ. Bot.* 1885, 277. *Veltheimia pumila, Willd. Sp. Plant.* ii. 182. *Aletris pumila, Ait. Hort. Kew.* i. 464. *Tritoma pumila, Gawl. in Bot. Mag. t.* 764. *Tritomantho pumila, Link, Enum.* i. 333; *Roem. et Schultes, Syst. Veg.* vii. 631.

KALAHARI REGION: Bechuanaland; Kosi Fontein, *Burchell*, 2554!
CENTRAL REGION: Colesberg, *Shaw*!

14. K. ensifolia (Baker in Journ. Bot. 1885, 278); leaves ensiform, very acuminate, moderately firm in texture, 3 ft. long, 1–1½ in. broad at the base, scabrous on the margin, with about 20 veins on each side of the midrib; peduncle moderately stout; raceme dense; pedicels very short; bracts oblong-lanceolate, ¼ in. long; flowers yellow; perianth funnel-shaped, ⅝–¾ in. long, ⅛ in. broad above the ovary, above ⅙ in. at the throat; segments ovate, obtuse, $\frac{1}{12}$ in. long; stamens and style much exserted.

KALAHARI REGION: Transvaal; Matebe River, *Holub*, 1530!

15. K. triangularis (Kunth, Enum. iv. 551); leaves rather rigid, narrow linear, erect, a foot long, $\frac{1}{16}$–$\frac{1}{12}$ in. broad, subtriquetrous, smooth on the margin; peduncle slender, terete, 1–1½ ft. long; raceme dense, 1–1½ in. long; pedicels very short; bracts lanceolate, ¼ in. long; flowers all yellow; perianth cylindrical, an inch long, $\frac{1}{12}$ in. diam. at the throat; segments ovate-oblong, obtuse, longer than broad; stamens and style not exserted. *Baker in Journ. Linn. Soc.* xi. 362; *Journ. Bot.* 1885, 278.

CENTRAL REGION: Aliwal North Div.; on the Witte Bergen, 6000–7000 ft., *Drège*, 3524!

16. K. Nelsoni (Masters in Gard. Chron. 1892, xi. 554, fig. 83); old leaves persistent as weak fibres; produced leaves narrow linear, 1½–2 ft. long, $\frac{1}{16}$ in. broad, with a thick midrib and recurved serrulate edges; peduncle as long as the leaves; raceme dense, oblong, 2–3 in. long, all the flowers deflexed when expanded; pedicels very short; bracts lanceolate, ¼ in. long; perianth cylindrical, bright scarlet, with sometimes a tinge of orange, 1¼ in. long, $\frac{1}{12}$ in. diam. throughout; lobes small, oblong; stamens irregularly biserial, much shorter than the perianth.

KALAHARI REGION: Orange Free State, *Nelson!*

17. K. Macowani (Baker in Journ. Bot. 1874, 3); leaves linear, erect, rigid, 1½–2 ft. long, ⅛–⅙ in. broad, with 3–5 veins in each side of the midrib, and a thickened scabrous margin; peduncle slender, 1½–2 ft. long; raceme very dense, 2–4 in. long, under 2 in. diam; pedicels very short; bracts oblong-lanceolate, acute, ¼–⅓ in. long; many upper flowers bright dark red; perianth cylindrical, an inch long, ⅛ in. diam. at the throat; segments ovate, obtuse, $\frac{1}{12}$ in. long; stamens not exserted. *Bot. Mag. t.* 6167; *Carrière in Rev. Hort.* 1879, 390, *with coloured figure*; *Baker in Journ. Bot.* 1885, 278. *Tritoma maroccana and rigidissima, Hort.*

COAST REGION: Swellendam Div.; on the Zwartberg near Swellendam, *MacOwan!*

CENTRAL REGION: Somerset Div.; grassy slopes of Bosch Berg, 4500–4800 ft., *MacOwan*, 1536! Graaff Reinet Div.; summit of Tandjes Berg, near Graaff Reinet, 5000 ft., *Bolus*, 761!

K. corallina, Hort. (*Tritoma corallina* (in text *coralloides*), *Rev. Hort. Belg.* 1893, 25, with coloured figure), raised by Deleuil of Marseilles, is a hybrid between this species and *K. Uvaria.*

18. K. Galpini (Baker); leaves linear, firm in texture, 2–3 ft. long, $\frac{1}{8}$ in. broad, with a stout midrib, 3–5 veins, and smooth edges; peduncle slender, as long as the leaves; raceme dense, oblong, 2–3 in. long, with all the flowers deflexed; pedicels very short; bracts lanceolate, $\frac{1}{4}$–$\frac{1}{3}$ in. long; perianth cylindrical, slender, flame-red, an inch long; lobes very small; stamens not exserted.

KALAHARI REGION: Transvaal; mountain slopes, Upper Moodies, near Barberton, 4500 ft., *Galpin*, 1208!

19. K. laxiflora (Kunth, Enum. iv. 552); leaves flat, linear, rigid, 3–5 ft. or more long, $\frac{1}{3}$ in. broad, scabrous on the margin; peduncle terete; raceme lax, 6–8 in. long; pedicels very short; bracts ovate, obtuse, $\frac{1}{4}$–$\frac{1}{3}$ in. long; perianth cylindrical, 14 lines long; segments ovate, obtuse; stamens not exserted; ovules about 10 in a cell.

EASTERN REGION : Pondoland or Natal; between Umtentu River and Umzimkulu River, below 500 ft., *Drège*, 4527!

20. K. natalensis (Baker in Journ. Bot. 1885, 278); leaves linear, $1\frac{1}{2}$–2 ft. long, $\frac{1}{4}$–$\frac{1}{3}$ in. broad, firm, with 10–12 distinct veins between the raised midrib and thickened margin; peduncle terete, 2–3 ft. long; raceme less dense than in *K. Uvaria*, 6–8 in. long, 2 in. diam.; pedicels very short; bracts oblong, $\frac{1}{6}$ in. long; flowers mostly yellow; perianth subcylindrical, about an inch long, $\frac{1}{8}$ in. diam. at the throat; segments ovate, $\frac{1}{12}$ in. long; stamens as long as the perianth; style finally exserted.

VAR. β, condensata (Baker); scape and leaves shorter; raceme dense.
VAR. γ, angustifolia (Baker); leaves not more than $\frac{1}{8}$ in. broad, few-nerved; racemes lax.

SOUTH AFRICA : without locality. Var. β, *Hort. Leichtlin.* !
EASTERN REGION: Griqualand East; Mount Currie, 6000 ft., *Tyson*, 1768! Natal; Inanda, *Wood*, 636! Botanic Garden, Durban, *Wood*, 3435! Zululand; hills above Tugela River, *Wood*, 3871! and without precise locality, *Adlam*! Var. γ, Tabamhlope Mountain, 6000–7000 ft., *Evans*, 411

Introduced to Kew in 1883 by Miss North. Has been hybridized by Leichtlin with *K. corallina* and *K. aloides, var. Saundersii.*

21. K. Baurii (Baker); leaves linear, rigid, flat, under a foot long, $\frac{1}{4}$ in. broad low down, with about 20 ribs and a thickened scabrous margin; peduncle a foot long; raceme dense, $1\frac{1}{2}$–2 in. long; pedicels very short; bracts oblong, $\frac{1}{4}$–$\frac{1}{3}$ in. long; flowers pale yellow; perianth cylindrical, $1\frac{1}{4}$ in. long, $\frac{1}{12}$ in. diam. at the throat; segments ovate, $\frac{1}{12}$ in. long; stamens as long as the perianth; style exserted.

EASTERN REGION: Tembuland; Bazeia, 2000–3000 ft., *Baur*, 755!

22. K. porphyrantha (Baker in Journ. Bot. 1874, 4); leaves linear, moderately firm in texture, $1\frac{1}{2}$ ft. long, $\frac{1}{4}$–$\frac{1}{3}$ in. broad low down, sharply keeled, smooth on the edge, with 10–12 fine in-

distinct veins on each side of the midrib; peduncle slender, not
longer than the leaves; raceme very dense, 2–3 in. long, under 2 in.
diam.; pedicels very short; bracts oblong-lanceolate, $\frac{1}{4}$–$\frac{1}{3}$ in. long;
flowers pale yellow, or tinged with purple; perianth cylindrical,
1–1$\frac{1}{4}$ in. long, $\frac{1}{6}$ in. diam. at the throat; segments broad ovate, as
broad as long; stamens and style not exserted. *Journ. Bot.* 1885,
279.

KALAHARI REGION: Orange Free State, *Cooper*, 3207! 3208!

23. K. decaphlebia (Baker in Engl. Jahrb. xv. Beibl. 35, 6);
leaves many, linear, 3 ft. long, $\frac{1}{4}$ in. broad low down, tapering
gradually to a long point, with about 5 strong ribs on each side
between the prominent midrib and recurved smooth margin;
peduncle slender, nearly as long as the leaves; raceme dense, oblong,
3–4 in. long; flowers bright yellow, the lower only deflexed;
pedicels very short; bracts oblong, $\frac{1}{4}$ in. long, white, with a brown
keel; perianth an inch long, $\frac{1}{12}$ in. diam. at the base, $\frac{1}{6}$ in. at the
throat; lobes suborbicular; stamens not exserted; style finally just
exserted.

EASTERN REGION: Pondoland, *Bachmann*, 280!

24. K. linearifolia (Baker in Engl. Jahrb. xv. Beibl. 35, 5);
leaves many, linear, moderately firm, strongly and closely ribbed,
above 2 ft. long, $\frac{1}{3}$–$\frac{1}{2}$ in. broad, tapering gradually to a long point,
obscurely serrulate; peduncle long, slender; raceme very dense;
axis 2–3 in. long; upper flowers ascending, lower deflexed; pedicels
very short; bracts oblong, obtuse, $\frac{1}{4}$ in. long, white with a brown
keel; perianth an inch long, $\frac{1}{8}$ in. diam. low down, $\frac{1}{8}$ in. at the
throat; lobes small, ovate; stamens exserted, the longer $\frac{1}{4}$ in.
beyond the tip of the perianth.

EASTERN REGION: Pondoland, *Bachmann*, 279!

25. K. drepanophylla (Baker in Engl. Jahrb. xv. Beibl. 35, 5);
root-fibres cylindrical, fleshy; leaves few, lanceolate, falcate, thin,
finely veined, $\frac{1}{2}$ ft. long, $\frac{1}{2}$–1 in. broad low down, obscurely serrated;
peduncle a foot long; raceme dense, very short; all the flowers
deflexed; pedicels very short; bracts small, oblong; perianth cylindrical, bright yellow, 1$\frac{1}{2}$ in. long; segments oblong; stamens as long
as the perianth; style finally exserted.

EASTERN REGION: Pondoland, *Bachmann*, 281!

26. K. sarmentosa (Kunth, Enum. iv. 552); leaves ensiform-
acuminate, 2–3 ft. long, $\frac{3}{4}$–1 in. broad low down, glaucous green,
acutely keeled, with about a dozen ribs on each side of the midrib;
peduncle stout, as long as the leaves; raceme dense, cylindrical,
$\frac{1}{2}$–1 ft. long, 2 in. diam.; pedicels very short; bracts lanceolate,
$\frac{1}{3}$–$\frac{1}{2}$ in. long; perianth cylindrical, $\frac{3}{4}$–1 in. long, under $\frac{1}{12}$ in. diam.
above the ovary, $\frac{1}{6}$ in. at the throat; segments broad ovate, obtuse,

$\frac{1}{12}$ in. long; longer stamens and style finally exserted $\frac{1}{3}$–$\frac{1}{2}$ in.
Baker in Journ. Linn. Soc. xi. 362 ; *Journ. Bot.* 1885, 279.
Aletris sarmentosa, Andr. Bot. Rep. t. 54. *Veltheimia sarmen-*
tosa, Willd. Enum. 380. *V. repens, Andr. Recens.* 18. *V. media,*
Donn. Hort. Cant. edit. ii. 131. *Tritoma media, Gawl. in Bot.*
Mag. t. 744 ; *Red. Lil. t.* 161 ; *Ait. Hort. Kew. edit.* 2, ii. 290.
Tritomanthe media, Link, Enum. i. 333 ; *Roem. et Schultes, Syst.*
Veg. vii. 630.

Grown Australia without locality.

27. K. alooides (Moench, Meth. 632); leaves ensiform-acuminate,
slightly glaucous, moderately firm in texture, 2–3 ft. long, $\frac{3}{4}$–1 in.
broad low down, scabrous on the margin, acutely keeled, with 30–40
close vertical veins; peduncle as long as the leaves, $\frac{1}{3}$–$\frac{1}{2}$ in. diam.
low down; raceme dense, often $\frac{1}{2}$ ft. long, 2$\frac{1}{2}$–3 in. diam.; pedicels
$\frac{1}{12}$–$\frac{1}{8}$ in. long; bracts ovate, obtuse, $\frac{1}{4}$ in. long; upper flowers bright
scarlet, lower yellow; perianth cylindrical, 1$\frac{1}{4}$–1$\frac{1}{2}$ in. long, $\frac{1}{8}$ in.
diam. at the throat; segments ovate, obtuse, $\frac{1}{12}$ in. long and broad;
style and sometimes also the stamens just exserted in the lower
flowers; ovules 12–15 in a cell; capsule ovoid, $\frac{1}{4}$–$\frac{1}{3}$ in. long. *Kunth,*
Enum. iv. 551 ; *Flore des Serres, t.* 1393; *Baker in Journ. Linn.*
Soc. xi. 364 ; *Journ. Bot.* 1885, 279. *K. Uvaria, Hook. in Bot.*
Mag. t. 4816. *Aloe Uvaria, Linn. Sp. Plant.* 323. *Aletris*
Uvaria, Linn. Mant. alt. 368. *Tritoma Uvaria, Gawl. in Bot. Mag.*
t. 758; *Red. Lil. t.* 291. *Veltheimia Uvaria, Willd. Sp. Plant.* ii.
182; *Jacq. Fragm.* 7, *t.* 4, *fig.* 1. *V. speciosa, Roth, Nov. Sp.*
190.

VAR. **maxima** (Baker in Bot. Mag. t. 6553); taller than the type, with leaves
4–5 ft. long, an inch broad at the base; raceme longer; flowers larger. *Tritoma*
grandiflora, Hort. T. Saundersi, Carrière in Rev. Hort. 1882, 504, *with coloured*
figure.

VAR. **nobilis** (Baker); a still more robust form, with peduncle, including the
1–1$\frac{1}{2}$ ft. raceme, sometimes 6–7 ft. long; flowers 1$\frac{1}{2}$ in. long. *Tritoma nobilis,*
Guillon in Rev. Hort. 1882, 24; 1885, 252, *with coloured figure.*

VAR. **serotina** (Hort.); a late-flowering form, with slender perianth 1$\frac{1}{4}$ in.
long, and distinctly exserted stamens.

K. præcox, Baker in Saund. Ref. Bot. t. 169, is a plant imported by Mr. T.
Cooper, that on one occasion flowered in summer, and at other times in autumn.
K. Sandersoni, Hort., is a form with green leaves. Other garden varieties are
carnosa, glaucescens, and *refulgens.*

COAST REGION : Cape Peninsula, between Rondebosch and Hout Bay, below
1000 ft., *Drège!* Uitenhage, *Zeyher,* 113 ! Bedford Div. *Burke!*
CENTRAL REGION : Somerset Div.; Bosch Berg, 4000 ft., *MacOwan,* 1908 !
Graaff Reinet Div. ; Sneeuw Berg Range, 3800–6000 ft., *Bolus,* 217 ! *Drège!*
KALAHARI REGION : Orange Free State, *Cooper,* 3599 ! Basutoland, *Cooper,*
3234 ! Transvaal; Saddleback Range, near Barberton, 4000–4600 ft., *Galpin,*
1230 !
EASTERN REGION : Natal ; Inanda, *Wood,* 1330 !

28. K. Rooperi (Lemaire, Jard. Fleur. t. 362) ; leaves ensiform-
acuminate, 4 ft. long, 1$\frac{1}{2}$ in. broad low down, dull green, acutely

keeled, scabrous on the margin; peduncle stout, stiffly erect, as long
as the leaves ; raceme very dense, 4–6 in. long, $2\frac{1}{2}$ in. diam. ; pedicels
very short; bracts obtuse, $\frac{1}{4}$–$\frac{1}{3}$ in. long ; perianth cylindrical, $1\frac{1}{4}$–$1\frac{1}{2}$ in.
long, $\frac{1}{6}$ in. diam. at the throat ; segments ovate, obtuse, $\frac{1}{12}$ in. long;
stamens finally just exserted.   *Baker in Journ. Linn. Soc.* xi. 363;
Bot. Mag. t. 6116 ; *Journ. Bot.* 1885, 280.   *Tritoma Rooperi,
Moore in Gard. Comp.* i. 116, *with figure.*

COAST REGION : British Kaffraria, *Rooper !*

29. **K. longicollis** (Baker in Gard. Chron. 1893, xiii. 682) ; leaves
ensiform, bright green, acutely keeled, 2 feet long, an inch broad at
the base, tapering gradually to the point, smooth on the edges ;
peduncle terete, slender, $1\frac{1}{2}$ ft. long ; raceme dense, short, oblong ;
pedicels very short ; bracts ovate, scariose, $\frac{1}{3}$ in. long ; perianth subcylindrical, $1\frac{1}{4}$–$1\frac{1}{2}$ in. long, lemon-yellow, tinged orange-yellow when
young, $\frac{1}{6}$ in. diam. at the throat, constricted above the ovary ; lobes
small, ovate ; stamens and style finally just shortly exserted.

EASTERN REGION : Natal, *Hort. Leichtlin,* 1893 !

30. **K. Burchellii** (Kunth, Enum. iv. 552) ; leaves ensiform-
acuminate, bright green, moderately firm in texture, 2–3 ft. long,
$\frac{1}{2}$–$\frac{3}{4}$ in. broad low down, acutely keeled, smooth on the margin, with
15–20 ribs on each side of the midrib ; peduncle stout, 3 ft. long ;
raceme dense, $\frac{1}{2}$–1 ft. long, 2 in. diam. ; pedicels $\frac{1}{6}$–$\frac{1}{4}$ in. long ;
bracts oblong-lanceolate, $\frac{1}{4}$ in. long ; perianth subcylindrical, $1\frac{1}{4}$–$1\frac{1}{2}$ in.
long, $\frac{1}{6}$ in. diam. at the throat, bright yellow, much tinged with red
when young ; segments ovate, as long as broad ; stamens finally
just exserted. *Baker in Journ. Linn. Soc.* xi. 363, *excl. syn. ; Journ.
Bot.* 1885, 280.   *Tritoma Burchellii, Herb. in Bot. Reg. t.* 1745.

SOUTH AFRICA : without locality, *Thunberg ! Burchell !*

31. **K. caulescens** (Baker ex Hook. fil. in Bot. Mag. t. 5946) ;
caulescent, with a thick stem $\frac{1}{2}$–1 ft. long below the rosette of leaves ;
leaves ensiform-acuminate, 2–3 ft. long, 2–3 in. broad low down,
very glaucous, not acutely keeled, with a distinctly serrulate margin ;
peduncle stout, 3 ft. long ; raceme very dense, 4–6 in. long, 2–$2\frac{1}{2}$ in.
diam. ; pedicels $\frac{1}{6}$–$\frac{1}{4}$ in. long ; bracts obtuse ; perianth an inch long,
$\frac{1}{6}$ in. diam. at the throat ; segments broad ovate, obtuse, $\frac{1}{12}$ in. long ;
stamens and style exserted half an inch.   *Journ. Bot.* 1885, 281.
Tritoma caulescens, Rev. Hort. 1887, 132, *fig.* 28.

CENTRAL REGION : Wodehouse Div. ; on the Storm Bergen, *Cooper!*
Only known from cultivated plants, introduced by Mr. T. Cooper about 1860.

32. **K. Northiæ** (Baker in Journ. Bot. 1889, 43) ; stem shortly
produced, 2–3 in. diam. ; leaves 30–40 in a dense rosette, lanceolate-acuminate, broadly channelled down the face, not acutely
keeled on the back, 4–5 ft. long, 5–6 in. broad low down, tapering
gradually to a long point, the margin distinctly serrulate, the inner

leaves of the rosette growing gradually narrower and shorter;
peduncle shorter than the leaves, above 1 in. diam.; raceme very
dense, above a foot long, 3 in. diam.; bracts ovate, dark brown,
scariose, the upper ¼ in. long; pedicels short; all the lower flowers
yellow, only the upper red towards the tip; perianth cylindrical, an
inch long; segments small, ovate; stamens finally much exserted.
Gard. Chron. 1891, x. 67; *Bot. Mag. t.* 7412.

COAST REGION: Albany Div.; near Grahamstown.

Described from a painting by Miss North and a living plant in the Cactus-house
at Kew, which she brought home. I am informed by Mr. Tidmarsh that it was
discovered in a wild state by Mr. W. Dugmore. It was flowered by Mr. W. E.
Gumbleton in County Cork in July, 1891, and has done so several times at Kew.

VI. NOTOSCEPTRUM, Benth.

Perianth campanulate, divided down nearly to the base; segments
oblong, the three outer 3-nerved, the three inner 1-nerved in the keel.
Stamens 6, hypogynous; filaments rather flattened, as long as the
perianth; anthers small, oblong, versatile, dorsifixed, dehiscing introrsely. *Ovary* ovoid, 3-celled; ovules several in a cell, superposed;
style declinate, longer than the stamens; stigma entire. *Fruit*
unknown.

DISTRIB. Two species in Angola.

Perianth-tube as long as the segments; leaves dry, rigid   (1) natalense.
Perianth-tube almost obsolete; leaves rather fleshy ...   (2) alooides.

1. **N. natalense** (Baker); leaves rigid, lanceolate, 1–1½ ft. long,
1 in. broad at the middle, narrowed gradually to the base and apex,
with a thickened serrulate margin; peduncle stout, as long as the
leaves, with many small empty bracts in the upper part; spike dense,
cylindrical, ½ ft. long, ⅝ in. diam.; bracts ovate, scariose, persistent,
shorter than the flowers; perianth oblong, ⅙ in. long; segments
oblong, obtuse, rather shorter than the campanulate tube; stamens
finally a little exserted.

EASTERN REGION: Natal; Klip River County, *Mrs. K. Saunders in Herb.
Wood,* 3895!

Nearly allied to the two Angolan species collected by Welwitsch.

2. **N. alooides** (Benth. in Gen. Plant. iii. 775); leaves several,
lorate, 2 ft. long, rather fleshy, reflexing towards the apex; peduncle
naked except a few empty bracts, stout, terete, woody, 2–3 ft. long;
flowers in a dense cylindrical spike about a foot long, an inch in
diameter; bracts ovate-cuspidate, scariose, nearly as long as the
flowers; perianth ¼–⅓ in. long; tube very short; outer segments
ovate-oblong, brownish-yellow; inner oblong, bright yellow, with a
1-nerved brown keel; filaments linear-subulate, ⅓ in. long; anthers
1/16 in. long; style finally ¼ in. longer than the perianth. *Urginea
alooides, Bolus in Journ. Linn. Soc.* xviii. 395.

KALAHARI REGION: Transvaal; Lydenburg Div., on the summit of the Devils Kop, near McMac, 5000–6000 ft., *McLea in Herb. Bolus*, 3011!

I leave this where Mr. Bentham has placed it, but feel satisfied that when fully known it will prove to be distinct generically from the true *Notosceptra*.

VII. GASTERIA, Duval.

Perianth tubular, curved; tube dilated into a ball in the lower half, cylindrical in the upper; segments small, ovate, subequal. *Stamens* 6, hypogynous, slightly declinate, about as long as the perianth; filaments filiform; anthers oblong, versatile, dehiscing down the face. *Ovary* sessile, oblong-trigonous; ovules numerous, superposed; style filiform; stigma capitate, slightly 3-lobed. *Capsule* chartaceous, loculicidally 3-valved. *Seeds* compressed, winged; testa brown; albumen fleshy; embryo axile.

Caudex none or short; leafy stem short or elongated; leaves thick, fleshy, distichous or multifarious, usually spotted with white; flowers laxly racemose, red, white and green; pedicels articulated at the tip; bracts small, persistent.

DISTRIB. Endemic.

Leaves regularly distichous.

VERRUCOSÆ. Leaf-stem short; tubercles raised.
    Tubercles much raised ... ... (1) **verrucosa.**
    Tubercles slightly raised :
        Leaf lorate, 6–8 in. long; tubercles scattered :
            Leaf ½ in. thick in centre... ... ... (2) **radulosa.**
            Leaf ¼ in. thick ... ... (3) **subverrucosa.**
        Leaf lingulate; tubercles obscurely banded ... (4) **repens.**

LINGUÆ. Leaf-stem short; tubercles (unless on the margin of the leaf) immersed.

Leaves thick, glossy, lorate; leaf-stem 3–5 in. long ... (5) **nigricans.**
Leaves thin, rather glossy, lorate; leaf-stem 1½–2 in. long :
    Spots scattered ... ... (6) **elongata.**
    Spots aggregated in transverse bands ... ... (7) **transvaalensis.**
Leaves rather glossy, lingulate :
    Leaf ½ in. thick in middle ... ... (8) **brevifolia.**
    Leaf ⅛ in. thick in middle ... ... (9) **obtusifolia.**
Leaves thin, lorate, dull green ... ... (10) **disticha.**
Leaves thicker, dull green :
    Leaf lorate, 6–8 in. long ... ... (11) **sulcata.**
    Leaf lingulate, 3–4 in. long ... ... (12) **mollis.**
BICOLORES. Leaf-stem long; leaves glossy, thick, scarcely at all spotted.
    The only species... ... ... (13) **bicolor.**
PLANIFOLIÆ. Leaf-stem long; leaves ensiform, much spotted.
    The only species... ... ... (14) **planifolia.**
Leaves arranged in a distichous rosette, which is more or less twisted spirally.
SUBLINGUÆ. Leafy stem short.
    Spots small, immersed :
        Leaves lorate, 4–6 in. long :
            Dull green ... ... (15) **excavata.**
            Smooth and glossy ... ... (16) **dicta.**
        Leaves ensiform, about 1 ft. long ... ... (17) **retata.**

Spots larger, white, confluent :
    Spots rather raised   ...   .   (18) **cheilophylla.**
    Spots immersed :
       Leaves about 1 in. bròad :
          Margin of all the leaves single   ...   (19) **pallescens.**
          Margin of inner leaves double   ...   (20) **porphyrophylla.**
       Leaves 1½ in. broad   ...   .   (21) **variolosa.**

MACULATÆ.   Leafy stem elongated.

Leaves lorate :
    Spots indistinct ...   .   (22) **Zeyheri.**
    Spots small, distinct :
       Leaves 1–1¼ in. broad   ...   .   (23) **colubrina.**
       Leaves 1½–2 in. broad :
          Leaves 4–6 in. long ...   .   (24) **spiralis.**
          Leaves 9–15 in. long ...   .   (25) **picta.**
   Leaves ensiform or lorate-ensiform ; spots large,
      confluent :
       Leaves 4–6 in. long   ...   .   (26) **maculata.**
       Leaves a foot or more long   ...   ...   (27) **pulchra.**
Leaves arranged in a dense multifarious rosette.

CARINATÆ.   Spots raised.

Leaves lanceolate ; tubercles large, white   ...   ...   (28) **carinata.**
Leaves lorate ; tubercles small, white   ...   .   (29) **subcarinata.**
Leaves deltoid ; tubercles concolorous   ...   .   (30) **decipiens.**

NITIDÆ.   Spots immersed ; perianth ¾–1 in. long.

Leaves lanceolate-deltoid, 2–3 in. long ...   .   (31) **parvifolia.**
Leaves lanceolate, 3–4 in. long ...   .   (32) **gracilis.**
Leaves lanceolate, 4–6 in. long :
   Leafy stem short :
      Spots not very distinct ...   .   (33) **obtusa.**
      Spots very distinct   ...   .   (34) **lætepunctata.**
   Leafy stem elongated...   ...   .   (35) **marmorata.**
Leaves lanceolate or ensiform, ¼–1 ft. long :
   Leaves ensiform, 1¼–1½ in. broad low down   ...   (36) **trigona.**
   Leaves lanceolate, 2–3 in. broad low down :
      Spots small, white, numerous :
         Edges tubercled   ...   .   (37) **glabra.**
         Edges not distinctly tubercled   ...   ...   (38) **nitida.**
      Spots few, obscure, greenish-white   ...   ...   (39) **excelsa.**
      Spots few, whitish, and many brown ...   ...   (40) **fuscopunctata.**

GRANDIFLORÆ.   Spots immersed ; perianth 1½–2 in. long.

Leaves ensiform   ...   .   (41) **acinacifolia.**
Leaves lanceolate :
    Spots distinct   ...   .   (42) **candicans.**
    Spots indistinct ...   .   (43) **Croucheri.**
Hybrids between *Gasteria* and *Aloe*   ...   . { (44) **Peacockii.**
                                                    { (45) **pethamensis.**
Hybrid between *Gasteria* and *Haworthia*   ...   ...   (46) **Bayfieldii.**

1. G. verrucosa (Haw. Syn. 89); leafy stem 1½–2 in. long;
leaves 10–12, distichous, lanceolate, spreading, 6–9 in. long, 1–1½ in.
broad, dull green, edges thickened on both sides, face concave in
the lower half ; apex deltoid-cuspidate ; spots pure white, crowded,
scattered, prominently raised ; clasping base with a reddish-white
cartilaginous edge ; peduncle usually simple in the type ; raceme lax,
a foot long ; pedicels ⅓–½ in. long ; bracts lanceolate, shorter than the

pedicels; perianth $\frac{3}{4}-\frac{7}{8}$ in. long; ball oblong, $\frac{1}{6}$ in. diam. *Baker in Journ. Linn. Soc.* xviii. 184. *Aloe verrucosa, Miller, Gard. Dict. edit.* viii. *No.* 20; *Willd. Sp. Plant.* ii. 189; *Bot. Mag. t.* 837; *Salm-Dyck, Aloe sect.* 29, *t.* 25; *Kunth, Enum.* iv. 543. *A. carinata, DC., Plantes Grasses, t.* 63. *A. disticha, Linn. Sp. Plant. edit.* 2, 459, *ex parte. A. acuminata, Lam. Encyc.* i. 90, *excl. syn.*

VAR. β, **latifolia** (Haw. Revis. 47); more robust, with leaves sometimes a foot long, and 6–8 racemes forming a deltoid panicle.

VAR. γ, **G. intermedia** (Haw. Syn. 89); leaves more ensiform, 6–9 in. long, 1-1$\frac{1}{4}$ in. broad low down; tubercles not so white; inflorescence usually simple. *Aloe intermedia, Haw. in Trans. Linn. Soc.* vii. 12; *Salm-Dyck, Aloe sect.* 29, *fig.* 24; *Kunth, Enum.* iv. 542. *A. Lingua, Ker in Bot. Mag. t.* 1322, *excl. syn.*

VAR. δ, **scaberrima** (Baker in Journ. Linn. Soc. xviii. 184); leaves more lorate, incurved with most of the tubercles, greenish-white, nearly the same colour as the leaf. *Aloe scaberrima, Salm-Dyck, Hort.* 322; *Aloe sect.* 29, *fig.* 26; *Kunth, Enum.* iv. 543. *G. intermedia, var. asperrima, Haw. in Phil. Mag.* 1827, 356.

SOUTH AFRICA: without locality.

Cultivated by Boerhaave and Miller, 1720–1730. There is a specimen in the herbarium of Bishop Goodenough, dried probably from Kew about 1780.

2. G. radulosa (Baker in Journ. Bot. 1889, 43); leafy stem 1–1$\frac{1}{2}$ in. long; leaves about 6, distichous, lorate-ensiform, 6–8 in. long, 1$\frac{1}{2}$ in. broad, flexible in texture, face flat, dull green, $\frac{1}{6}$ in. thick in the middle, rounded, and with a cusp at the apex, dentate on the edge upwards, spots on the face crowded, small, whitish, slightly raised; flowers unknown.

SOUTH AFRICA: without locality.

Described from a living plant in the Kew collection in 1885, received from Berlin. Allied to *G. subverrucosa.*

3. G. subverrucosa (Haw. in Phil. Mag. 1827, 355); leafy stem 1$\frac{1}{2}$–2 in. long; leaves 8–10, lorate, distichous, $\frac{1}{2}$ ft. long, 1–1$\frac{1}{4}$ in. broad, not so thick as in *verrucosa*, rounded, and with a mucro and continuously hairy at the apex, edges not thickened but beaded with raised tubercles, spots of the face and back small, irregularly scattered and slightly raised; peduncle simple, 1 ft. long; raceme a foot long; pedicels $\frac{1}{4}-\frac{1}{3}$ in. long; bracts lanceolate, shorter than the pedicels; perianth $\frac{7}{8}$ in. long; tube oblong, $\frac{1}{6}$ in. diam. *Baker in Journ. Linn. Soc.* xviii. 184. *Aloe subverrucosa, Salm-Dyck, Obs.* 67; *Roem. et Schultes, Syst. Veg.* vii. 671; *Kunth, Enum.* iv. 544.

VAR. β, **parvipunctata** (Haw. loc. cit.); leaves longer, more ensiform; spots smaller, less crowded in the upper part of the leaf.

VAR. γ, **marginata** (Baker, loc. cit.); pale hairy margin of leaf continued all the way down.

COAST REGION: living plants of the type and var. γ, introduced from Algoa Bay by *T. Cooper!*

Introduced about 1820. Intermediate between *G. verrucosa* and *G. disticha.*

4. G. repens (Haw. Revis. 48); leafy stem about an inch long;

leaves 8–10, distichous, spreading, lingulate, $2\frac{1}{2}$–3 in. long, 8–9 lines broad, deltoid-cuspidate at the apex; spots whitish, small, rather raised, aggregated into irregular transverse bands; flowers unknown. *Baker in Journ. Linn. Soc.* xviii. 185. *Aloe repens, Roem. et Schultes, Syst. Veg.* vii. 674; *Kunth, Enum.* iv. 542.

SOUTH AFRICA: without locality.

Described from a drawing of a plant grown at Kew in 1821. Perhaps a hybrid between *G. verrucosa var. intermedia* and *G. carinata.*

5. G. nigricans (Haw. Syn. 86), leafy stem 4–5 in. long; leaves 10–20, distichous, crowded, lingulate, rigidly coriaceous, erecto-patent, 4–8 in. long, $1\frac{1}{2}$–2 in. broad, rounded and with a cusp at the horny apex, not tubercled and never thickened on the margin below it, $\frac{1}{4}$–$\frac{1}{3}$ in. thick in the middle, dark green and glossy on the surface, with copious, scattered, immersed whitish spots, face rather turgid in the lower half, back rounded; peduncle a foot or more long, usually simple; raceme lax, 1–$1\frac{1}{2}$ ft. long; lower pedicels $\frac{1}{4}$–$\frac{1}{3}$ in. long; bracts lanceolate, as long as the pedicels; perianth $\frac{5}{8}$–$\frac{3}{4}$ in. long, ball oblong, $\frac{1}{6}$ in. diam. *Baker in Journ. Linn. Soc.* xviii. 185. *Aloe nigricans, Haw. in Trans. Linn. Soc.* vii. 13; *Salm-Dyck, Aloe, sect.* xxix. *fig.* 7; *Kunth, Enum.* iv. 534. *A. obliqua, Jacq. Hort. Schoenbr. t.* 418, *excl. syn. A. Lingua var. crassifolia, Ait. Hort. Kew.* i. 469; *Ker in Bot. Mag. t.* 838.

VAR. β, G. crassifolia (Haw. in Phil. Mag. 1827, 351); leaves narrower and thicker, with the margin slightly tubercled.

VAR. γ, marmorata (Haw. in Phil. Mag. 1827, 350); leaves shorter and narrower than in the type, marbled with white and reddish spots. *Aloe nigricans var. marmorata, Salm-Dyck, Obs.* 64.

VAR. δ, G. fasciata (Haw. in Phil. Mag. 1827, 349); leafy stem not so short; leaves longer than in the type, lorate, very smooth, with the spots aggregated into indistinct transverse bands. *Aloe nigricans var. fasciata, Salm-Dyck, Obs.* 64. *A. vittata, Roem. et Schultes, Syst. Veg.* vii. 662; *Salm-Dyck, Aloe, sect.* xxix. *fig.* 6.

VAR. ε, polyspila (Baker in Journ. Linn. Soc. xviii. 185); leafy stem 3 in. long; leaves 10–12, lorate, $\frac{1}{2}$ ft. long, 1–$1\frac{1}{4}$ in. broad, not tubercled on the edge, the blotches smaller, more crowded and greener than in the type.

VAR. ζ, A. guttata (Salm-Dyck, Hort. 332); leafy stem longer than in the type; leaves not so thick, lorate, with smaller spots and the margins tubercled. *Salm-Dyck, Aloe, sect.* xxix. *fig.* 9; *Kunth, Enum.* iv. 547. *Gasteria subnigricans var. glabrior, Haw. in Phil. Mag.* 1827, 352.

VAR. η, G. subnigricans (Haw. in Phil. Mag. 1827, 351); leaves lorate, strongly tubercled on the margin, and rough on the surfaces through the tubercles being slightly raised. *Aloe subnigricans, Spreng. Syst.* ii. 71; *Salm-Dyck, Aloe, sect.* xxix. *fig.* 10; *Kunth, Enum.* iv. 547.

VAR. θ, platyphylla (Baker in Journ. Linn. Soc. xviii. 186); leafy stem 2–3 in. long; leaves lingulate, $2\frac{1}{4}$–3 in. broad, with the margins tubercled below the apex; inflorescence branched, with a stout peduncle and 3–4 racemes $1\frac{1}{2}$ ft. long. *G. latifolia, Hort. vix Haworth.*

COAST REGION: living plants of the type and var. ε, introduced from Algoa Bay by *T. Cooper!* Hills near Port Elizabeth, under 500 ft., *Zeyher,* 1482!

Introduced into cultivation at the beginning of the eighteenth century. Vars. *fasciata, guttata,* and *subnigricans* first described from the Vienna collections about 1815.

6. G. elongata (Baker) ; leafy stem 1–1½ in. long; leaves 6–8, lorate, 6–10 in. long, 1–1¼ in. broad, flexible, rather glossy, flat on the face, rounded on the back, ¼ in. thick in the middle ; apex deltoid-cuspidate ; edges not thickened ; upper half of the leaf denticulate on the margin ; spots copious, immersed, scattered, whitish ; flowers not seen.

SOUTH AFRICA : without locality.

Described from a living plant obtained from Berlin, in the Kew collection 1885. Intermediate between *G. nigricans* and *G. disticha.*

7. G. transvaalensis (Hort. De Smet. ex Baker in Journ. Bot. 1889, 44) ; rosette distichous or slightly oblique ; leafy stem short ; leaves about 8, lorate, dark green, rather glossy, 4–5 in. long, an inch broad, ¼ in. thick in the centre, face not excavated, border not thickened, but toothed towards the white, horny, deltoid-cuspidate apex ; spots greenish-white, immersed, aggregated into transverse bands ; flowers unknown.

KALAHARI REGION : Transvaal.

Described from a living plant in the Kew collection in 1885. Allied to *G. nigricans.*

8. G. brevifolia (Haw. Syn. 89) ; leafy stem 1–1½ in. long ; leaves 10–12, densely crowded, lingulate, 3–4 in. long, 2 in. broad, very thick, rather glossy, but not so rigid and smooth as in *G. nigricans,* rounded, and with a cusp and continuously horny at the tip, tubercled, but never doubled on the margin, spots small, white, immersed, the clasping base with a reddish-white horny border ; peduncle usually simple ; raceme 1½–2 ft. long ; pedicels ¼–⅓ in. long ; bracts minute, lanceolate ; perianth ¾–⅞ in. long, ball oblong, ¼ in. diam. *Haw. in Phil. Mag.* 1827, 351 ; *Baker in Journ. Linn. Soc.* xviii. 186. *Aloe brachyphylla, Salm-Dyck, Hort. Dykensis* 332 ; *Aloe, sect.* xxix. *fig.* 8 ; *Kunth, Enum.* iv. 535.

SOUTH AFRICA : without locality.

Cultivated in England by Mr. Hill before 1809.

9. G. obtusifolia (Haw. in Phil. Mag. 1827, 351) ; leafy stem 1½–2 in. long ; leaves 12–16, distichous, lingulate, densely crowded, 4–6 in. long, 2–2½ in. broad, broadly rounded and with a minute cusp at the apex, tubercled on the edge below it, ⅙ in. thick in the middle, never doubled on the margin, quite smooth on the surfaces, with copious, small, immersed whitish spots ; peduncle simple, above a foot long ; raceme 1½ ft. long ; pedicels ⅓–½ in. long ; bracts minute, lanceolate ; perianth ⅞–1 in. long, ball oblong, ¼–⅓ in. diam. *Baker in Journ. Linn. Soc.* xviii. 186. *Aloe obtusifolia, Salm-*

Dyck, Obs. 62; *Aloe, sect.* xxix. *fig.* 37; *Kunth, Enum.* iv. 548.
A. Lingua var. brevifolia, Salm-Dyck, Cat. 17.

SOUTH AFRICA : without locality.

First described from the Vienna collections in 1815.

10. G. disticha (Haw. in Phil. Mag. 1827, 352); leafy stem
1–1½ in. long; leaves 10–12, crowded, lorate, distichous, 4–6 in.
long, 1½ in. broad, ⅙ in. thick in the middle, flat on the face and dull
green with copious, small, immersed, scattered white spots, deltoid-
cuspidate and horny at the apex, denticulate below it, border
never doubled in the type, clasping base with a white horny edge;
peduncle simple or branched; racemes 1–1½ ft. long; pedicels
⅓–½ in. long ; bracts minute, lanceolate; perianth ¾–⅞ in. long, ball
oblong, ¼ in. diam. *Baker in Journ. Linn. Soc.* xviii. 186.
G. denticulata, Haw. Suppl. 50. *Aloe disticha, a, Linn. Sp. Plant.*
321 ; *Roem. et Schultes, Syst. Veg.* vii. 666 ; *Kunth, Enum.* iv. 546.
A. Lingua, Thunb. Diss. No. 14, *ex parte*; *Salm-Dyck, Aloe, sect.*
xxix. *fig.* 33. *A. linguiformis, Miller, Gard. Dict. edit.* viii. *No.* 13,
ex parte; *DC. Plant. Grasses, t.* 68.

VAR. β, **minor** (Baker in Journ. Linn. Soc. xviii. 187); leaves fewer, not
more than 2–3 in. long, 1 in. broad ; peduncle simple ; raceme of fewer flowers.

VAR. γ, **G. conspurcata** (Haw. in Phil. Mag. 1827, 353) ; leaves longer, sometimes a foot long ; spots denser, smaller ; tubercles of the edge less raised ;
flowers rather larger. *Aloe conspurcata, Salm-Dyck, Obs.* 59; *Aloe, sect.* xxix.
fig. 31 ; *Kunth, Enum.* iv. 546.

VAR. δ, G. **angustifolia** (Haw. Syn. 88) ; leaf thicker than in the type, with
the face slightly concave. *Aloe angustifolia, Salm-Dyck, Obs.* 57 : *Aloe, sect.*
xxix. *fig.* 30. *A. Lingua var. angustifolia, Haw. in Trans. Linn. Soc.* vii. 13.

VAR. ε, G. **angulata** (Haw. in Phil. Mag. 1827, 354) ; leaves thicker than in
the type, with one of the borders often doubled; spots sometimes aggregated
into irregular bands; perianth an inch long. *G. longifolia, Haw. Syn.* 88. *Aloe
angulata, Willd. in Ges. Naturf. Fr. Berl. Mag.* v. 276; *Salm-Dyck, Aloe, sect.*
xxix. *fig.* 29; *Kunth, Enum.* iv. 546. *A. Lingua var. longifolia, Haw. in
Trans. Linn. Soc.* vii. 13.

SOUTH AFRICA : without locality.

Cultivated in Europe from the beginning of the eighteenth century.

11. G. sulcata (Haw. in Phil. Mag. 1827, 354); leafy stem 1½–2 in.
long; leaves 10–12 in a distichous or slightly twisted rosette,
crowded, lorate, 6–8 in. long, 1½–1¾ in. broad, rounded at the tip and
with a small cusp, usually doubled at the borders and conspicuously
tubercled, thicker in texture than in *A. disticha*, dull pale green, with
copious, minute, scattered, immersed, sometimes indistinct spots,
horny border of the base very narrow ; peduncle simple, a foot or
more long; raceme simple, a foot or more long; pedicels ⅜–½ in.
long ; bracts small, lanceolate; perianth ¾ in. long, tube oblong,
¼ in. diam. *Baker in Journ. Linn. Soc.* xviii. 187. *Aloe sulcata,
Salm-Dyck, Obs.* 54 ; *Aloe, sect.* xxix. *fig.* 32 ; *Kunth, Enum.* iv. 545.
A. Lingua var. angulata, Haw. in Trans. Linn. Soc. vii. 13.

SOUTH AFRICA: without locality.

Introduced into cultivation before 1804.

12. G. mollis (Haw. Revis. 46, 203); leafy stem 1–1½ ft. long; leaves 6–8, distichous, crowded, spreading, lingulate, 3–4 in. long, 1¼–1½ in. broad, ¼–⅓ in. thick, dull green, with small immersed spots, apex deltoid-cuspidate, borders tubercled, base narrowly margined; peduncle simple, about a foot long; raceme simple, a foot long; lower pedicels ⅜–½ in. long; bracts small, lanceolate; perianth ⅞ in. long, ball oblong, ¼ in. diam. *Haw. in Phil. Mag.* 1827, 351; *Baker in Journ. Linn. Soc.* xviii. 187. *Aloe mollis, Roem. et Schultes, Syst. Veg.* vii. 665; *Salm-Dyck, Aloe, sect.* xxix. *fig.* 38; *Kunth, Enum.* iv. 548.

SOUTH AFRICA: without locality.

Introduced into cultivation by Bowie in 1819.

13. G. bicolor (Haw. in Phil. Mag. 1826, 275); leafy stem 4–5 in. long; leaves 12–20, distichous, lorate, 8–9 in. long, 1¼–1½ in. broad at the middle, deltoid-cuspidate at the apex, very rigid, very thick, a slightly glaucous green, slightly convex on the face in the lower half, more so on the back, with no spots except a few on the back towards the base, edges of the clasping base bright red; peduncle simple, above a foot long; raceme lax, above a foot long; pedicels ¼–⅓ in. long; bracts small, lanceolate; perianth ⅝ in. long, much curved, ball oblong, ⅛ in. diam. *Baker in Journ. Linn. Soc.* xviii. 188. *Aloe bicolor, Roem. et Schultes, Syst. Veg.* vii. 682; *Salm-Dyck, Aloe, sect.* xxix. *fig.* 5; *Kunth, Enum.* iv. 537.

SOUTH AFRICA: without locality.

Introduced into cultivation about 1825. A very distinct species.

14. G. planifolia (Baker in Journ. Linn. Soc. xviii. 188); shortly caulescent; leafy part of stem 6–10 in. long; leaves 6–10 on a side, distichous, lax, ensiform, rigid, erecto-patent, 6–8 in. long, ¾–1 in. broad low down, ⅙ in. thick in the centre, glossy, dark green, with copious, confluent, white immersed blotches, like those of *G. maculata;* tip deltoid-cuspidate, horny, entire, margin below it neither thickened nor tubercled; peduncle simple, 1–1½ ft. long; raceme a foot long; pedicels ¼–½ in. long; bracts small, lanceolate; perianth ¾ in. long, ball subglobose, nearly white, ⅓ in. diam. *Aloe planifolia, Baker in Saund. Ref. Bot. t.* 162.

COAST REGION: Living plants, introduced from Algoa Bay about 1860 by *T. Cooper!*

15. G. excavata (Haw. in Phil. Mag. 1827, 354); leafy stem 1½–2 in. long; leaves 12–16 in a more or less irregular distichous rosette, crowded, spreading, lorate, 4–6 in. long, 1¼–1½ in. broad, pale dull green with small, immersed, indistinct whitish spots, deltoid-cuspidate at the apex, borders tubercled, one or both doubled; peduncle simple, a foot or more long; raceme a foot or more long; lower pedicels ⅓–½ in. long; bracts small, lanceolate; perianth an inch long, ball oblong, ¼ in. diam. *Baker in Journ. Linn. Soc.* xviii. 188. *Aloe excavata, Willd. in Ges. Naturf. Fr. Berl. Mag.* v. 276; *Salm-Dyck,*

Aloe, sect. xxix. *fig.* 22 ; *Kunth, Enum.* iv. 545. *A. obscura, Willd. in Ges. Naturf. Fr. Berl. Mag.* v. 275.—*A. africana, etc., Miller, Ic. t.* 19.

SOUTH AFRICA : without locality.

Introduced into cultivation at the beginning of the eighteenth century. Salm-Dyck distinguishes two varieties, *unilateralis* and *multifaria.*

16. G. dicta (N. E. Brown in Gard. Chron. 1876, vi. 68, fig. 18) ; leafy stem 2 in. long ; leaves 12–14 in a slightly spiral rosette, lorate, 4–5 in. long, 1½ in. broad, firm in texture, very smooth, dull green, with one side often doubled, deltoid-cuspidate at the apex, toothed and tubercled below it, spots on face small, whitish and immersed ; peduncle simple or branched ; racemes 1–1½ ft. long ; pedicels ¼–⅓ in. long ; bracts small, lanceolate ; perianth ¾ in. long, ball oblong, ¼ in. diam. *Baker in Journ. Linn. Soc.* xviii. 189. *G. sub-nigricans var. torta, Hort.*

SOUTH AFRICA : without locality.

Described from living plants in the Kew collection. History unknown.

17. G. retata (Haw. in Phil. Mag. 1827, 350) ; leafy stem about 3 in. long ; leaves 10–12, crowded, spirally distichous, ensiform, outer spreading, inner ascending, a foot long, 1¼–1½ in. broad, ⅓–½ in. thick low down, one border doubled, surface smooth, spots small and immersed, apex deltoid-cuspidate and toothed and borders tubercled ; peduncle branched ; racemes a foot long ; lower pedicels ¼–⅓ in. long ; bracts small, lanceolate ; perianth ¾ in. long, ball oblong, ¼ in. diam. *Baker in Journ. Linn. Soc.* xviii. 189. *Aloe dictyodes, Roem. et Schultes, Syst. Veg.* vii. 663 ; *Salm-Dyck, Aloe, sect.* xxix. *fig.* 4 ; *Kunth, Enum.* iv. 539.

SOUTH AFRICA : without locality.

Introduced into cultivation by Bowie in 1826.

18. G. cheilophylla (Baker in Journ. Linn. Soc. xviii. 189) ; leafy stem about 2 in. long ; leaves 14–18 in a much twisted distichous rosette, ensiform, erecto-patent, 9–12 in. long, 1–1½ in. broad low down, narrowed to a deltoid-cuspidate apex, sometimes doubled on one edge, dark green, with copious, confluent, slightly raised white spots, tip with an entire white horny margin, spots of the edge more raised than those of the face ; flowers not seen. *G. undata, Hort. non Haw.*

SOUTH AFRICA : without locality.

Described from living plants in the Kew and Peacock collections. History not known. Leaf in shape and marking most like *G. pulchra,* but the leafy stem much shorter.

19. G. pallescens (Baker in Journ. Linn. Soc. xviii. 190) ; leafy stem about 2 in. long ; leaves 8–10, crowded, ensiform, outer spreading, inner erecto-patent, ½ ft. long, an inch broad, ⅓–½ in. thick at the base, ¼ in. in the centre, not doubled on the border, apex with a toothed horny border, edge not tubercled, surfaces with very copious, confluent, greenish-white immersed spots ; peduncle

simple, a foot or more long; raceme a foot long; lower pedicels
¼–⅓ in. long; bracts lanceolate, as long as the pedicels; perianth
¾ in. long, ball oblong, ¼ in. diam.

COAST REGION : living plant introduced about the year 1860 from Algoa Bay
by *T. Cooper*

20. G. porphyrophylla (Baker in Journ. Linn. Soc. xviii. 190);
leafy stem 1½–2 in. long; leaves 8–10, crowded, erecto-patent, lorate,
7–8 in. long, an inch broad, dull purple, very smooth, doubled on
one border, furnished with copious, immersed, confluent whitish spots,
apex deltoid-cuspidate with a toothed horny border, edges below
it not tubercled; peduncle simple, a foot long; raceme a foot long;
lower pedicels ¼–⅓ in. long; bracts lanceolate, as long as the
pedicels; perianth ¾ in. long, ball oblong, ¼ in. diam.

SOUTH AFRICA : without locality.

Described from a living plant in the Kew collection, received from Mr. Tisley
in 1873. Closely allied to *G. pallescens.*

21. G. variolosa (Baker in Saund. Ref. Bot. t. 347); leafy stem
1½–2 in. long; leaves 15–18 in a spirally distichous rosette, crowded,
ensiform, 8-9 in. long, 1½ in. broad low down, narrowed gradually to
a deltoid-cuspidate tip, smooth, dull green, firm in texture, concave
on the face, not tubercled on the edge, one border sometimes doubled,
spots immersed, copious, large, whitish and confluent; peduncle simple,
a foot or more long; raceme 1½ ft. long; lower pedicels ⅓–½ in. long;
bracts lanceolate; perianth ¾ in. long, ball oblong, ¼ in. diam.
Baker in Journ. Linn. Soc. xviii. 190.

COAST REGION : living plant introduced about the year 1860 from Algoa Bay
by *T. Cooper!*

22. G. Zeyheri (Baker in Journ. Linn. Soc. xviii. 190); leafy
stem ½ ft. long; leaves 16–20, laxly disposed in a twisted distichous
rosette, all erecto-patent, lorate, 8–9 in. long, about an inch broad,
smooth, firm in texture, flat on the face, apex deltoid-cuspidate,
tubercled and not doubled on the edge, spots obscure and immersed,
clasping base much dilated and vertically plicate; peduncle simple,
a foot long; raceme a foot or more long; lower pedicels ¼–⅓ in.
long; bracts lanceolate; perianth ¾ in. long, ball oblong, ¼ in. diam.
Aloe Zeyheri, Salm-Dyck, Aloe, sect. xxix. *fig.* 3 *bis.*

SOUTH AFRICA : without locality.

Introduced into cultivation by Zeyher.

23. G. colubrina (N. E. Brown in Gard. Chron. 1877 viii. 38);
leafy stem 4–6 in. long; leaves 8–10, arranged in a spirally distichous or almost multifarious rosette, lorate, erecto-patent, 9–14 in.
long, 1¼ in. broad low down, ⅓ in. thick, smooth, bright green or purple,
face concave, spots small, whitish, immersed, one border often
doubled, apex deltoid-cuspidate, margins not tubercled; peduncle
stout, branched, 4 ft. long including inflorescence; pedicels ¼–⅓ in.

long ; bracts small, lanceolate ; perianth $\frac{5}{8}-\frac{3}{4}$ in. long, ball oblong, $\frac{1}{4}$ in. diam.   *Baker in Journ. Linn. Soc.* xviii. 190.

COAST REGION : Uitenhage Div., *Bolus !*

Introduced into cultivation about 1870.

24. G. spiralis (Baker in Journ. Linn. Soc. xviii. 189) ; leafy stem 4–6 in. long ; leaves 16–28, crowded, arranged in a spirally twisted rosette, lorate, erecto-patent, glossy, dark green, 4–6 in. long, $1\frac{1}{2}$ in. broad, apex deltoid-cuspidate, border neither doubled nor tubercled, face flat, back rounded, centre $\frac{1}{4}$ in. thick, spots on the surfaces copious, whitish, immersed, clasping base furnished with a red horny border ; peduncle simple, a foot long ; raceme 1–$1\frac{1}{2}$ ft. long ; lower pedicels $\frac{1}{4}-\frac{1}{3}$ in. long ; bracts lanceolate, as long as the pedicels ; perianth $\frac{3}{4}$ in. long, ball oblong, $\frac{1}{4}$ in. diam.

VAR. tortulata (Baker, loc. cit.) ; leaf 5–6 in. long, more narrowed upwards, and sometimes doubled on one border.

SOUTH AFRICA : without locality.

First described from living plants in the Kew collection in 1872. Also sent home alive from the Cape by Sir H. Barkly in 1879. May be only a variety of *G. nigricans.*

25. G. picta (Haw. in Phil. Mag. 1827, 349) ; shortly caulescent ; leafy part of stem 6–8 in. long ; leaves 12–20, laxly arranged in a spirally distichous rosette, lorate-ensiform, all erecto-patent, $\frac{1}{2}$–1 ft. long, $1\frac{1}{2}$–2 in. broad, $\frac{1}{2}$ in. thick at the base, dark glossy green, with copious, immersed, distinct white spots, flat on the face, one border often doubled, apex deltoid-cuspidate, margin below it not tubercled, dilated base with a broad red horny border ; peduncle often branched ; racemes 1–$1\frac{1}{2}$ ft. long ; lower pedicels $\frac{1}{4}-\frac{1}{3}$ in. long ; bracts lanceolate, as long as the pedicels ; perianth $\frac{5}{8}-\frac{3}{4}$ in. long, ball oblong, $\frac{1}{4}-\frac{1}{3}$ in. diam.   *Baker in Journ. Linn. Soc.* xviii. 191.   *Aloe bowieana, Roem. et Schultes, Syst. Veg.* vii. 662 ; *Salm-Dyck, Aloe,* sect. xxix. *fig.* 3 ; *Kunth, Enum.* iv. 536.

VAR. β, G. formosa (Haw. in Phil. Mag. 1827, 350) ; leaves narrower ; spots purer white.

SOUTH AFRICA : without locality.

Introduced into cultivation by Bowie about 1827. Intermediate between *G. nigricans* and *G. maculata.*

26. G. maculata (Haw. in Phil. Mag. 1827, 349) ; shortly caulescent ; leafy part of stem often $\frac{1}{2}$ ft. long, leaves 12–20, laxly disposed in a spirally twisted rosette, lorate, ensiform, all erecto-patent, firm in texture, 4–6 in. long, about an inch broad, $\frac{1}{6}$ in. thick in the centre, smooth and tinted with purple, immersed spots copious, large, white and confluent, one border doubled, edge not tubercled, clasping base with a red horny border ; peduncle 1–$1\frac{1}{2}$ ft. long, usually simple ; racemes a foot or more long ; lower pedicels $\frac{1}{3}-\frac{1}{2}$ in. long ; bracts lanceolate ; perianth $\frac{3}{4}$ in. long, ball oblong, $\frac{1}{4}$ in. diam.   *Baker in Journ. Linn. Soc.* xviii. 191.   *Aloe*

maculata, Thunb. Diss. No. 10, *ex parte; Salm-Dyck, Aloe, sect.*
xxix. *fig.* 1; *Kunth, Enum.* iv. 536. *A. obliqua, Haw. in Trans.*
Linn. Soc. vii. 14. *A. maculata, var. obliqua, Ait. Hort. Kew.* i.
469. *A. Lingua, Ker in Bot. Mag. t.* 979, *excl. syn. G. obliqua,*
Haw. Syn. 85.

VAR. β, **fallax** (Haw. in Phil. Mag. 1827, 349); smaller in all its parts;
leaves narrower, with more numerous white blotches.

COAST REGION: George Div.; Outeniqua, *Thunberg.*

Introduced into cultivation early in the eighteenth century.

27. G. pulchra (Haw. Syn. 86); shortly caulescent; leafy part
of stem ½ ft. or more long; leaves 12–20, laxly arranged in a spirally
distichous rosette, all erecto-patent, ensiform, 1–1½ ft. long, an inch
broad low down, narrowed gradually to the tip, smooth, often tinted
with purple, furnished with copious, large, immersed, white confluent
spots, one border doubled, edge not tubercled; peduncle branched,
3 ft. long including the inflorescence; racemes 1–1½ ft. long;
lower pedicels ⅓–½ in. long; bracts small, lanceolate; perianth
¾ in. long, ball oblong, ¼ in. diam. *Haw. in Phil. Mag.* 1827, 349;
Baker in Journ. Linn. Soc. xviii. 191. *Aloe pulchra, Jacq. in Hort.*
Schoenbr. t. 419; *Haw. in Trans. Linn. Soc.* vii. 14; *Salm-Dyck,*
Aloe, sect. xxix. *fig.* 2; *Kunth, Enum.* iv. 536. *A. obliqua, DC.*
Plantes Grasses, t. 91. *A. maculata, Ker in Bot. Mag. t.* 765.—
A. foliis linguiformibus variegatis, Miller, Ic. t. 292.

SOUTH AFRICA: without locality.

Introduced into cultivation early in the eighteenth century.   Closely allied to
G. maculata.

28. G. carinata (Haw. Syn. 87); leafy stem 1–1½ in. long;
leaves 12–20, arranged in a dense multifarious rosette, lanceolate,
5–6 in. long, 1½–2 in. broad low down, and ⅓ in. thick, narrowed
gradually to the deltoid denticulate horny apex, concave on the face,
obliquely keeled down the back, dull green, rough, with copious,
raised, white, scattered papillæ, the margin and keel prominently
papillose; peduncle simple, 1½ ft. long; raceme 1–1½ ft. long;
lower pedicels ⅓–½ in. long; bracts small, lanceolate; perianth ⅞–1 in.
long, ball oblong, ⅙ in. diam. *Baker in Journ. Linn. Soc.* xviii. 192.
Aloe carinata, Miller, Gard. Dict. edit. viii. *No.* 21; *Haw. in Trans.*
Linn. Soc. vii. 13; *Bot. Mag. t.* 1331, *ex parte; Salm-Dyck, Aloe,*
sect. xxix. *fig.* 20; *Kunth, Enum.* iv. 541.—*Aloe africana sessilis*
foliis carinatis verrucosis, Dill. Hort. Elth. t. 18.

VAR. β, **G. strigata** (Haw. in Phil.. Mag. 1827, 357); more robust; leaves
8–10 in. long, less rough; tubercles sometimes confluent. *Aloe carinata var.*
lævior, Salm-Dyck, loc. cit.

VAR. γ, **G. parva** (Haw. in Phil. Mag. 1827, 356); a small, very proliferous
form, with few tubercles.

SOUTH AFRICA: without locality, *Zeyher,* 4180!

Introduced into cultivation early in the eighteenth century.

29. G. subcarinata (Haw. in Phil. Mag. 1827, 358); leafy stem 1–1½ in. long; leaves 10–15, arranged in a dense multifarious rosette, spreading, lorate, 4–6 in. long, 1½ in. broad at the base, ¼ in. thick in the centre, deltoid-cuspidate at the apex, doubled on one of the borders, dull green, rough, with small raised whitish spots, the edges strongly tubercled; peduncle simple, 1–1½ ft. long; raceme 1–1½ ft. long; lower pedicels ⅓–½ in. long; bracts small, lanceolate; perianth ¾–⅞ in. long, ball oblong, ¼ in. diam. *Baker in Journ. Linn. Soc.* xviii. 192. *Aloe subcarinata, Salm-Dyck, Obs.* 51 ; *Aloe, sect.* xxix. *fig.* 21 ; *Kunth, Enum.* iv. 541. *A. pseudo-angulata, Salm-Dyck, Cat.* 16.

VAR. β, **G. undata** (Haw. in Phil. Mag. 1827, 358) ; leaves smoother, brighter, narrower, and shorter. *Aloe undata, Roem, et Schultes, Syst. Veg.* vii. 677 ; *Kunth, Enum.* iv. 541. *A. subcarinata, var. minor, Salm-Dyck, Aloe, loc. cit.*

SOUTH AFRICA : without locality.

First described from the Vienna collections in 1815. Probably a garden hybrid between *G. carinata* and *G. disticha. Var. striata, Salm-Dyck,* is a form with the leaves striped with bands of white.

30. G. decipiens (Haw. in Phil. Mag. 1827, 357); leafy stem 1–1½ in. long; leaves 12–20, arranged in a dense multifarious rosette, deltoid, 2–3 in. long, 1¼–1½ in. broad and ⅓–½ in. thick low down, narrowed gradually to a deltoid cusp, concave on the face, obliquely keeled in the back, dull dark green, rough with copious concolorous raised spots; peduncle simple, a foot long; raceme a foot long; lower pedicels ¼–⅓ in. long; bracts lanceolate; perianth ⅞–1 in. long, ball oblong, ¼ in. diam. *Baker in Journ. Linn. Soc.* xviii. 192. *Haworthia nigricans, Haw. in Phil. Mag.* 1824, 301. *Aloe decipiens, Roem. et Schultes, Syst. Veg.* vii. 674 ; *Salm-Dyck, Aloe, sect.* xxix. *fig.* 16 ; *Kunth, Enum.* iv. 539.

SOUTH AFRICA : without locality.

Introduced into cultivation by Bowie in 1824. Very different from every other *Gasteria* in its habit, which resembles that of *Haworthia viscosa.*

31. G. parvifolia (Baker in Journ. Linn. Soc. xviii. 193); acaulescent; leaves 10–12, arranged in a dense multifarious rosette, lanceolate-deltoid, 2–2½ in. long, 1–1¼ in. broad, and ⅓–½ in. thick low down, dull green, changing to purple when old, face very concave, back obliquely keeled, spots on the face copious, small, whitish, immersed, forming irregular bands, borders and keel tubercled; flowers not seen.

SOUTH AFRICA : without locality, living plants introduced into cultivation about 1860 by *T. Cooper!*

32. G. gracilis (Baker in Journ. Linn. Soc. xviii. 193); acaulescent; leaves 9–10, arranged in a dense multifarious rosette, lanceolate, 3–4 in. long, ¾–1 in. broad low down, outer recurved, narrowed from the base to the continuously horny tip, back obliquely keeled, spots on the face very numerous, small and immersed, those on the edges and keel raised; flowers unknown.

EASTERN REGION : Natal, living plants introduced into cultivation about 1860 by *T. Cooper.*

33. G. obtusa (Haw. in Phil. Mag. 1827, 359) ; leafy stem 1½–2 in. long ; leaves 12–18, arranged in a dense multifarious rosette, outer spreading, inner ascending, lanceolate, about half a foot long, 1½ in. broad and ½ in. thick low down, concave on the face, obliquely keeled down the back, dull green, with copious, scattered, immersed, small whitish spots, margins tubercled ; peduncle simple, a foot or more long ; raceme a foot or more long ; pedicels ¼–⅓ in. long ; bracts small, lanceolate ; perianth under an inch long, ball oblong, ¼ in. diam. *Baker in Journ. Linn. Soc.* xviii. 194. *Aloe obtusa, Roem. et Schultes, Syst. Veg.* vii. 679. *A. trigona, Salm-Dyck, Aloe, sect.* xxix. *fig.* 18 ; *Kunth, Enum.* iv. 540. *A. trigona, var. obtusa, Salm-Dyck, Obs.* 46.

SOUTH AFRICA : without locality.
Introduced into cultivation about 1820.

34. G. lætepunctata (Haw. in Phil. Mag. 1827, 357) ; leafy stem 1½–2 in. long ; leaves 15–20, arranged in a dense multifarious rosette, lanceolate, 4–6 in. long, 1½–2 in. broad and ½ in. thick low down, narrowed gradually to the apex, concave on the face, obliquely keeled down the back, bright green or tinged with purple, smooth, with copious, small, pure white, immersed spots and tubercled edges ; peduncle a foot or more long, simple or forked ; racemes a foot long ; pedicels ¼–⅓ in. long ; bracts small, lanceolate ; perianth ⅞ in. long, ball oblong, ¼ in. diam. *Baker in Journ. Linn. Soc.* xviii. 193. *Aloe lætepunctata, Roem. et Schultes, Syst. Veg.* vii. 676 ; *Kunth, Enum.* iv. 537.

SOUTH AFRICA : without locality.
Introduced into cultivation by Bowie in 1825.

35. G. marmorata (Baker in Journ. Linn. Soc. xviii. 194) ; shortly caulescent ; leafy part of stem 6–10 in. long ; leaves 20–30, arranged in a multifarious rosette, lorate-lanceolate, 5–6 in. long, 1¼–1½ in. broad low down, rounded and with a horny cusp at the apex, very smooth, dull green, with obscure, large, immersed, confluent spots, one border conspicuously doubled, edge not tubercled ; peduncle branched ; raceme a foot long ; pedicels ¼–⅓ in. long ; bracts small, lanceolate ; perianth ⅞ in. long, ball oblong, ¼ in. diam.

SOUTH AFRICA : without locality.
Described from living plants in the Peacock and Kew collections. Allied to *G. maculata* and *G. picta,* but truly multifarious.

36. G. trigona (Haw. in Phil. Mag. 1827, 359) ; leafy stem about 2 in. long ; leaves 12–15, arranged in a dense multifarious rosette, lanceolate, all except the outer ascending, 6–8 in. long, 1¼–1½ in. broad and ⅓–½ in. thick low down, narrowed gradually to the apex, face concave, back obliquely keeled, green, very

smooth, with copious, small, white, immersed spots, edges tubercled; peduncle a foot or more long, simple or forked; racemes
a foot or more long; pedicels $\frac{1}{4}$–$\frac{1}{3}$ in. long; bracts lanceolate;
perianth $\frac{7}{8}$ in. long, tube oblong, $\frac{1}{4}$ in. diam. *Baker in Journ. Linn.
Soc.* xviii. 194. *A. trigona var. elongata, Salm-Dyck, Obs.* 45. *A.
elongata, Salm-Dyck, Aloe,* sect. xxix. *fig.* 15; *Kunth, Enum.* iv.
539.

SOUTH AFRICA: without locality.

First raised in Europe from seeds sent home by Herr Swellingreben in 1790.
Closely allied to *G. obtusa* and *G. glabra.*

37. G. glabra (Haw. Syn. 87); leafy stem $1\frac{1}{2}$–2 in. long; leaves
15–18, arranged in a dense multifarious rosette, lanceolate, 6–9 in.
long, 2–3 in. broad and $\frac{1}{2}$–$\frac{3}{4}$ in. thick low down, face concave,
back obliquely keeled, smooth, dark green, with copious, small,
immersed, whitish spots, edges tubercled; peduncle simple, a foot
long; raceme simple, a foot or more long; pedicels $\frac{1}{4}$–$\frac{1}{3}$ in. long;
bracts small, lanceolate; perianth $\frac{7}{8}$–1 in. long, ball oblong, $\frac{1}{4}$ in.
diam. *Haw. in Phil. Mag.* 1827, 358; *Baker in Journ. Linn. Soc.*
xviii. 194. *Aloe glabra, Salm-Dyck, Obs.* 48; *Aloe,* sect. xxix. *fig.*
19; *Roem. et Schultes, Syst. Veg.* vii. 677; *Kunth, Enum.* iv. 540.
A. carinata var. subglabra, Haw. in. Trans. Linn. Soc. vii. 14.
A. carinata, Ker. in Bot. Mag. t. 1331, *ex parte.*

VAR. β, **major** (Kunth, loc. cit.); more robust, with leaves a foot or more
long.

SOUTH AFRICA: without locality.

Introduced into cultivation by Masson in 1796.

38. G. nitida (Haw. in Phil. Mag. 1827, 359); leafy stem $1\frac{1}{2}$–2 in.
long; leaves 12–15, arranged in a dense multifarious rosette, lanceo-
late, 8–9 in. long, 2–$2\frac{1}{2}$ in. broad and $\frac{1}{2}$–$\frac{3}{4}$ in. thick low down,
face concave, back obliquely keeled, smooth, bright green,
with copious, small, immersed, whitish spots, edges and keel
scarcely at all rugose; peduncle simple, a foot or more long; raceme
simple, a foot or more long; pedicels $\frac{1}{4}$–$\frac{1}{3}$ in. long; bracts lanceolate,
about as long as the pedicels; perianth $\frac{7}{8}$–1 in. long, tube oblong,
$\frac{1}{4}$ in. diam. *Baker in Journ. Linn. Soc.* xviii. 195. *Aloe nitida,
Salm-Dyck, Cat.* 13; *Ker in Bot. Mag. t.* 2304; *Salm-Dyck, Aloe,*
sect. xxix. *fig.* 17; *Kunth, Enum.* iv. 539.

VAR. **grandipunctata** (Salm-Dyck, Obs. 47); spots larger, arranged in
irregular bands.

SOUTH AFRICA: without locality.

Introduced into the Dutch gardens about 1790.

39. G. excelsa (Baker in Journ. Linn. Soc. xviii. 195); leafy stem
2–3 in. long; leaves 10–30, arranged in a dense multifarious
rosette, outer spreading, inner erecto-patent, lanceolate, 8–12 in.
long, $2\frac{1}{2}$–3 in. broad and $\frac{1}{2}$–$\frac{3}{4}$ in. thick low down, face concave, back obliquely keeled, dull green, with a few faint im-

mersed spots of greenish-white, edges smooth or only very
obscurely papillose; peduncle about a foot long; racemes 8–10,
forming a large deltoid panicle, end one 2 ft. long; ·pedicels
$\frac{1}{4}$–$\frac{1}{3}$ in. long; bracts lanceolate, as long as the pedicels; perianth
$\frac{7}{8}$ in. long, tube oblong, $\frac{1}{4}$ in. diam.

COAST REGION: King Williamstown Div.; near the Chalumna River, living
plants introduced into cultivation about 1860 by *T. Cooper!*

40. **G. fuscopunctata** (Baker in Journ. Linn. Soc. xviii. 195);
leafy stem 2–3 in. long; leaves 12–20, arranged in a dense multi-
farious rosette, stiffly erecto-patent, lanceolate, $\frac{1}{2}$–1 ft. long, 2$\frac{1}{2}$–3 in
broad and $\frac{3}{4}$–1 in. thick low down, face slightly concave,
back obliquely keeled, dull green, when old reddish-brown, with a
few whitish immersed and many brown star-like often confluent
spots, smooth on the margin; peduncle simple, 1$\frac{1}{2}$ ft. long; raceme
simple, 1$\frac{1}{2}$–2 ft. long; pedicels $\frac{1}{4}$–$\frac{1}{3}$ in. long; bracts lanceolate;
perianth 1 in. long, tube oblong, $\frac{1}{4}$–$\frac{1}{3}$ in. diam.

SOUTH AFRICA: without locality, living plants introduced about 1860 by
T. Cooper!

41. **G. acinacifolia** (Haw. Suppl. 49); leafy stem 1$\frac{1}{2}$–2 in. long;
leaves 10–20, arranged in a dense multifarious rosette, stiffly erecto-
patent, ensiform, about a foot long, 1$\frac{1}{2}$–2 in. broad and $\frac{1}{2}$–$\frac{3}{4}$ in. thick
low down, narrowed gradually from the base to the apex, slightly
concave on the face, unequally keeled on the back, smooth, shining,
dark green, with copious, small, immersed, greenish-white spots,
margins slightly rugose; peduncle 3–4 ft. long including the
inflorescence; racemes 1–1$\frac{1}{2}$ ft. long; pedicels $\frac{1}{3}$–$\frac{1}{2}$ in. long; bracts
lanceolate; perianth 1$\frac{1}{2}$–2 in. long, tube oblong-cylindrical, $\frac{1}{4}$ in.
diam. *Baker in Journ. Linn. Soc.* xviii. 196. *Aloe acinacifolia,*
Jacq. Eclog. 49, *t.* 31; *Ker in Bot. Mag. t.* 2369; *Salm-Dyck, Aloe,*
sect. xxix. *fig.* 11; *Kunth, Enum.* iv. 537.

VAR. β, **G. venusta** (Haw. in Phil. Mag. 1827, 360); leaves narrower, more
glossy, with whiter spots aggregated into obscure bands.
VAR. γ, **G. ensifolia** (Haw. in Phil. Mag. 1825, 282); leaves shorter and less
acuminate. *Aloe ensifolia, Roem. et Schultes, Syst. Veg.* vii. 681; *Salm-Dyck,
Aloe, sect.* xxix. *fig.* 12; *Kunth, Enum.* iv. 538.
VAR. δ, **G. pluripuncta** (Haw. in Phil. Mag. 1827, 360); an immature form
with subdistichous lorate leaves of thinner texture, not keeled on the back.
VAR. ε, **G. nitens** (Haw. Suppl. 48); leaves longer, sometimes 1$\frac{1}{2}$ ft. long, less
acuminate, very smooth, light green, with copious confluent spots. *Aloe nitens,
Roem. et Schultes, Syst. Veg.* vii. 680; *Kunth, Enum.* iv. 538.

SOUTH AFRICA: without locality.
Cultivated at Vienna before 1810.

42. **G. candicans** (Haw. in Phil. Mag. 1827, 361); leafy stem
1$\frac{1}{2}$–2 in. long; leaves 12–20, arranged in a dense multifarious rosette,
lanceolate, 6–15 in. long, 2$\frac{1}{2}$–3 in. broad and $\frac{1}{2}$–$\frac{3}{4}$ in. thick low
down, face concave, back obliquely keeled, bright green, smooth,
with copious, immersed, minute whitish spots, margins and
keel not tubercled; peduncle 3–4 ft. long including the inflores-

cence; racemes 5–6, forming a deltoid panicle; pedicels $\frac{1}{3}$–$\frac{1}{2}$ in.
long; bracts lanceolate; perianth 1$\frac{1}{2}$–2 in. long, with an oblong-
cylindrical tube $\frac{1}{4}$ in. diam. *Baker in Journ. Linn. Soc.* xviii. 196.
Aloe candicans, Roem. et Schultes, Syst. Veg. vii. 681; *Salm-Dyck,
Aloe, sect.* xxix. *fig.* 13; *Kunth, Enum.* iv. 538. *G. linita, Haw.
loc. cit.*

SOUTH AFRICA: without locality.

Introduced into cultivation at Kew about 1820, probably by Bowie.

43. G. Croucheri (Baker in Journ. Linn. Soc. xviii. 196); leafy
stem 1$\frac{1}{2}$–2 in. long; leaves 12–18, arranged in a dense multifarious
rosette, lanceolate, 12–15 in. long, 3–3$\frac{1}{2}$ in. broad and $\frac{3}{4}$–1 in. thick
low down, narrowed gradually to a deltoid-cuspidate tip, face con-
cave, back obliquely keeled, smooth, dark green, with copious,
small, immersed, greenish-white spots, edges smooth or obscurely
papillose; peduncle 3–4 ft. long including the inflorescence;
racemes several, forming a deltoid panicle; pedicels $\frac{1}{3}$–$\frac{1}{2}$ in. long;
bracts lanceolate; perianth 1$\frac{3}{4}$–2 in. long, tube oblong-cylindrical,
$\frac{1}{4}$ in. diam. *G. natalensis, Baker in Journ. Linn. Soc.* xviii. 187.
Aloe Croucheri, Hook. fil. in Bot. Mag. t. 5812.

EASTERN REGION: Natal, living plants introduced about 1862 by *T. Cooper!*
and *M. J. McKen!*

G. natalensis is the juvenile state of *G. Croucheri.*

44. G. Peacockii (Baker in Journ. Linn. Soc. xviii. 195); leafy
stem 2–3 in. long; leaves 12–15, arranged in a dense multifarious
rosette, lanceolate, 9–12 in. long, 2 in. broad low down, and $\frac{1}{2}$–$\frac{3}{4}$ in.
thick, gradually narrowed to a deltoid-cuspidate apex, face concave,
back obliquely keeled, green, with a few immersed scattered whitish
spots, principally on the back, margin rugose with raised white
tubercles; flowers unknown.

A garden hybrid between *G. acinacifolia var. ensifolia* and *Aloe heteracantha*,
raised by Pfersdorff in Paris about 1875. A similar hybrid, of which the
parentage is not known, had long been cultivated by Mr. Bullen at the Glasgow
Botanic Garden, of which he sent me flowers in February, 1883. This has an
ensiform leaf about a foot long, and 1$\frac{1}{4}$–1$\frac{1}{2}$ in. broad low down, a branched inflorescence, a cylindrical pale red perianth under an inch long, with segments
twice as long as the tube and included stamens.

45. G. pethamensis (Baker in Journ. Linn. Soc. xviii. 193); leafy
stem 1$\frac{1}{2}$–2 in. long; leaves 16–20, crowded, multifarious, lanceolate,
3–4 in. long, $\frac{3}{4}$–1 in. broad low down, dark green, tinged with purple
when old, narrowed gradually to the tip, $\frac{1}{4}$–$\frac{1}{3}$ in. thick, face concave,
back obliquely keeled, surfaces furnished with small scattered white
raised spots, margins and keel prominently tubercled; flowers
unknown.

A garden hybrid between *G. carinata* and *Aloe variegata*, raised at Petham in
Kent in 1840 by Mr. Ricketts. Its history is given in Gard. Chron. 1841, 183.

46. G. Bayfieldii (Baker in Journ. Linn. Soc. xviii. 197); leafy
stem about 2 in. long; leaves 12–15, arranged in a dense multi-

farious rosette, lanceolate, 3–4 in. long, about an inch broad low down, gradually narrowed to a deltoid-cuspidate apex, face slightly concave, obliquely keeled on the back, green, with copious nearly immersed distinct white spots, margin strongly tubercled; peduncle simple, slender, about a foot long ; raceme ½–1 ft. long; pedicels ¼ in. long; bracts lanceolate, as long as the pedicels; perianth subcylindrical, ⅝ in. long, ⅛ in. diam. in the lower half ; tips of the segments spreading. *Aloe Bayfieldii, Salm-Dyck, Aloe, sect.* xxix. *fig.* 14 ; *Kunth, Enum.* iv. 538.

SOUTH AFRICA ? without locality.

First described from the collection of Mr. Hitchin, of Norwich, in 1828.

G. apicroides, Baker, in Journ. Linn. Soc. xviii. 197 ; is an allied form, with 20–30 leaves on a stem about half a foot long.

SOUTH AFRICA : without locality, living plants introduced about 1862 by T. *Cooper !*

G. squarrosa, Baker, in Journ. Linn. Soc. xviii. 197 ; another allied form with a long leafy stem and 20–30 squarrose leaves, with a concave face and much less distinct spots.

SOUTH AFRICA : without locality.

Both of these were described from living plants at Kew about 1878.

VIII. **ALOE**, Linn.

Perianth subcylindrical, straight or slightly recurved ; tube campanulate or cylindrical ; segments elongated, much imbricated, spreading only at the tip. *Stamens* 6, hypogynous, equalling or exceeding the perianth ; filaments subulate ; anthers small, oblong, dorsifixed, versatile, dehiscing introrsely. *Ovary* sessile, oblong-trigonous ; ovules many, superposed ; style filiform ; stigma small, capitate. *Capsule* oblong-trigonous, coriaceous, loculicidally 3-valved. *Seeds* triquetrous or flattened, often winged ; testa thin, brownish-black ; albumen fleshy ; embryo straight.

Caudex sometimes none, sometimes produced, simple or branched; leaves fleshy, usually multifarious and toothed on the margin, generally aggregated in a dense rosette; flowers usually bright red or bright yellow, racemose or umbellate ; pedicels solitary, short or elongated; bracts persistent, scariose.

DISTRIB. Species about 100; many Tropical African. Extends also to Madagascar, Socotra, India, China, Arabia, and the Mediterranean region.

Subgenus 1. EUALOE. Perianth straight (rarely slightly curved); leaves multifarious in the mature plant, except in *A. Cooperi.*

Group 1. *Acaules.* Stem none, or very short below the dense rosette of leaves.

Stamens and style about as long as the perianth :

Leaves narrowly linear ; teeth very minute :				
Raceme short, subcapitate	...	...	...	(1) **minima.**
Raceme long, lax	...	...	...	(2) **kniphofioides.**
Leaves linear ; teeth very minute :				
Perianth 1¼–1½ in. long :				
Leaves distichous	...	...	...	(3) **Cooperi.**
Leaves multifarious	..	...	...	(4) **micracantha.**
Perianth ½–¾ in. long :				
Perianth straight	...	...	...	(5) **Kraussii.**
Perianth curved ...	...	...	...	(6) **myriacantha.**

Leaves lanceolate or oblong-lanceolate:
Leaves with a long pellucid awn ... ... (7) **aristata.**
Leaves thin without an awn... ... ... (8) **Boylei.**
Leaves thick without an awn:
Perianth $1\frac{1}{4}$–$1\frac{1}{2}$ in. long:
Leaves copiously tubercled on the back ... ... (9) **humilis.**
Leaves smooth or slightly tubercled on the back ... ... (10) **pratensis.**
Perianth $1\frac{3}{4}$–2 in. long ... ... (11) **virens.**
Stamens and style much exserted:
Leaves linear ... ... (12) **Bowiea.**
Leaves lanceolate ... ... (13) **longistyla.**
Leaves ensiform ... ... (14) **Ecklonis.**

Group 2. *Brevicaules.* Old plants with a simple stem a few inches long below the dense rosette of leaves; leaves not spotted.

Leaves margined with prickles:
Leaves without raised ribs on their face:
Leaves very glaucous, 4–6 in. long, not distinctly lineate:
Prickles not confluent ... ... (15) **brevifolia.**
Prickles confluent in a horny border ... (16) **Serra.**
Leaves with very distinct vertical lines ... (17) **lineata.**
Leaves with 1–2 raised ribs on their face ... (18) **heteracantha.**
Leaves entire ... ... (19) **striata.**

Group 3. *Maculatæ.* Old plants with a short stem below the dense rosette of very much spotted leaves.

Leaves lanceolate or oblong-lanceolate:
Prickles minute ... ... (20) **serrulata.**
Prickles larger; racemes dense, capitate:
Leaves $1\frac{1}{2}$–2 in. broad: ... ... (21) **Saponaria.**
Leaves $2\frac{1}{2}$–4 in. broad:
Prickles $\frac{1}{8}$–$\frac{1}{6}$ in. long:
Leaf $\frac{1}{4}$ in. thick in the middle ... (22) **latifolia.**
Leaf $\frac{1}{6}$ in. thick in the middle ... (23) **leptophylla.**
Prickles $\frac{1}{4}$ in. long ... . (24) **macracantha.**
Prickles larger; racemes elongated:
Lower pedicels 1–$1\frac{1}{2}$ in. long ... (25) **obscura.**
Lower pedicels about $\frac{1}{2}$ in. long:
Leaf $2\frac{1}{2}$–3 in. broad:
Prickles large ... ... (26) **grandidentata.**
Prickles smaller ... ... (27) **Greenii.**
Leaf $1\frac{1}{3}$–2 in. broad:
Perianth much constricted ... ... (28) **gasterioides.**
Perianth but little constricted ... ... (29) **tricolor.**
Leaves ensiform:
Rosette of leaves dense:
Perianth $\frac{5}{8}$ in. long; tube short ... ... (30) **microstigma.**
Perianth an inch long; tube long... ... (31) **Monteiroi.**
Rosette of leaves lax:
Leaves 6–8 in. long, an inch broad... ... (32) **consobrina.**
Leaves 1-$1\frac{1}{2}$ ft. long, $1\frac{1}{2}$–2 in. broad ... (33) **spicata.**

Group 4. *Laxifoliæ.* Leaves of the mature plant laxly disposed on an elongated stem, not spotted.

Leaves linear, $\frac{1}{2}$–1 in. broad; stems sarmentose:
Perianth $\frac{1}{3}$–$\frac{1}{2}$ in. long ... ... (34) **tenuior.**

Perianth 1–1¼ in. long:
  Perianth-segments very short:
    Internodes obscurely striped with green   (35) **ciliaris.**
    Internodes conspicuously striped with
      green ... ...   (36) **striatula.**
  Perianth-segments much longer than the
    tube ... ...   (37) **gracilis.**
Leaves lanceolate; stems sarmentose... ... ...   (38) **Macowani.**
Leaves lanceolate; stems suberect ... ...   (39) **aurantiaca.**
Leaves ovate-lanceolate; stems suberect:
  Leaves 3–4 in. long, 1½–2 in. broad ... ...   (40) **distans.**
  Leaves 6–8 in. long, 2–3 in. broad:
    Prickles smaller, more distant ... ...   (41) **mitriformis.**
    Prickles larger, closer ... ...   (42) **albispina.**
  Leaves 9–12 in. long, 3–4 in. broad ... ...   (43) **nobilis.**

Group 5. *Veræ.* Leaves of the mature plant aggregated in a dense rosette at the end of an elongated often-branched cylindrical stem a few inches thick, not spotted, except sometimes a little when young.
Leaves lanceolate, 1–1½ ft. long:
  Leaves green, rather lax ... ...   (44) **Brownii.**
  Leaves dull green, slightly glaucous ... ...   (45) **sigmoidea.**
  Leaves very glaucous ... ...   (46) **cœsia.**
Leaves ensiform:
  Stamens about as long as the perianth:
    Leaves glaucous:
      Stem short; teeth small:
        Bracts pale reddish; leaves slightly
          glaucous ... ...   (47) **succotrina.**
        Bracts purple; leaves very glau-
          cous ... ...   (48) **purpurascens.**
      Stem slender, elongated; teeth larger .   (49) **arborescens.**
    Leaves bright green, very acuminate ...   (50) **pluridens.**
  Stamens much longer than the perianth:
    Pedicels much shorter than the bracts ..   (51) **speciosa.**
    Lower pedicels about as long as the bracts:
      Leaves straight:
        Leaves green, 1½ in. broad ...   (52) **longiflora.**
        Leaves slightly glaucous, 1½–2 in.
          broad ... ...   (53) **chloroleuca.**
        Leaves very glaucous, 2–3 in. broad:
          Leaves 1½ ft. long ... ...   (54) **platylepis.**
          Leaves a foot long ... ...   (55) **fulgens.**
      Leaves sickle-shaped ... ...   (56) **drepanophylla.**
    Pedicels much longer than the bracts ...   (57) **salmdyckiana.**
Flowers unknown ... ...   (58) **nitens.**

Group 6. *Dichotomæ.* Mature plant with a very much-branched tree-like trunk.
Leaves lanceolate, glaucous ... ...   (59) **dichotoma.**
Leaves ensiform, green ... ...   (60) **Bainesii.**
Species of *Eualoe;* habit not known ... ...   (61) **falcata.**

Subgenus II. PACHIDENDRON. Perianth slightly recurved; pedicels always very short and genitalia much exserted; leaves multifarious, forming a dense rosette.
Leaves lanceolate, very prickly on the upper surface   (62) **ferox.**
Leaves ensiform, not prickly, except a little towards
  the tip beneath:
    Leaves very glaucous, scarcely recurved ...   (63) **africana.**
    Leaves slightly glaucous, scarcely recurved ...   (64) **supralævis.**

Leaves slightly glaucous, much recurved:
    Perianth ¾ in. long   ...   .   (65) **rupestris.**
    Perianth an inch long   ...   .   (66) **Thraskii.**

Subgenus III. GONIALOE. Perianth cylindrical; stamens not exserted; leaves short, imbricated spirally bifarious.

The only species...   ...   .   (67) **variegata.**

Subgenus IV. KUMARA (Rhipidodendron, Willd.). Arborescent, with a much-branched trunk and distichous leaves; perianth straight; inner segments free.

The only species   ...   .   (68) **plicatilis.**

Garden hybrids ...   . $\begin{cases} (69) \textbf{ desmetiana.} \\ (70) \textbf{ insignis.} \end{cases}$

1. A. minima (Baker in Hook. Ic. t. 2423); leaves about a dozen in a sessile rosette, suberect, narrowly linear, channelled deeply down the face, 5–6 in. long, spotted with white; margin armed with abundant, spreading, minute, white teeth; peduncle simple, 6–9 in. long, with several empty bracts in the upper part; flowers many in a dense subcapitate raceme, the lower cernuous; pedicels ascending, ¼–½ in. long; bracts ovate-acuminate, nearly as long as the pedicels; perianth pale-red, cylindrical, 4–5 lines long; tube very short; stamens and style not exserted.

EASTERN REGION: Natal; solitary amongst grass on the South Downs, *Evans,* 409!

2. A. kniphofioides (Baker in Hook. Ic. t. 1939); acaulescent; leaves linear, 1–1½ ft. long, ⅛–¼ in. broad, hardly at all fleshy; margin with minute deltoid spines; inflorescence a lax simple raceme, nearly a foot long; bracts scariose, ovate-acuminate, the lower ½–⅝ in. long; pedicels ascending, the lower as long as the bracts; perianth cylindrical, pale-red, 1¼–1½ in. long; segments linear-oblong, ⅓ in. long, tipped with green; stamens and style shorter than the perianth.

EASTERN REGION: Pondoland; damp grassy places on Mount Enkansweni, near the high road between the rivers Umtamvuna and Emagusheni, 4000 ft.? *Tyson,* 2829!

3. A. Cooperi (Baker in Gard. Chron. 1874, i. 628); acaulescent; produced leaves 8–10, distichous, linear, much dilated at the base, 1½–2 ft. long, an inch broad low down, tapering gradually to the point, thin in texture, dull green, deeply channelled, spotted with white, obscurely striped; margin narrow, white, horny, with minute deltoid teeth; peduncle simple, 1–2 ft. long, with several ovate-cuspidate empty bracts; flowers 20–30 in a dense corymbose raceme; lower pedicels 1–2 in. long; bracts ovate-lanceolate, ¾–1 in. long; perianth cylindrical, 1¼–1½ in. long, reddish-yellow, tipped with green; tube very short; stamens and style included. *Bot. Mag. t.* 6377; *Journ. Linn. Soc.* xviii. 155. *A. schmidtiana, Regel, Gartenfl.* 1879, 97, *t.* 970.

COAST REGION: Port Elizabeth Div.; between Drostdy Farm and the Leadmine, *Burchell,* 4482! and at Cragga Kamma, *Burchell,* 4564!
EASTERN REGION: Natal; Inanda, *Wood,* 843! near Ladysmith, 3000–4000 ft., *Wood,* 5610! and without precise locality, *Cooper,* 1193! 3623!

4. A. microcantha (Haw. Suppl. 105); shortly caulescent; leaves
15–20, dense, multifarious, linear-lanceolate, thin in texture, tapering
gradually to an acute point, 1–1½ ft. long, ¾–1 in. broad low down,
green, channelled down the face, copiously spotted with white;
margin minutely denticulate; peduncle simple, about a foot long,
with copious ovate empty bracts; raceme dense, simple, corymbose; pedicels ascending, 1–1½ in. long, bracts ovate, shorter than
the pedicels; perianth straight, cylindrical, 1¼–1½ in. long, red,
tipped with green; tube very short; stamens and style as long as
the perianth. *Sims in Bot. Mag. t.* 2272; *Link and Otto, Ic.* 87, *t.*
40; *Roem. et Schultes, Syst. Veg.* vii. 697; *Salm-Dyck, Aloe, sect.* xxi.
fig. 1; *Baker in Journ. Linn. Soc.* xviii. 159.

SOUTH AFRICA: without locality.

Introduced into cultivation by Bowie in 1819.

5. A. Kraussii (Baker in Journ. Linn. Soc. xviii. 159); acaulescent; leaves 6–8, subdistichous, linear, 2–3 ft. long, ½–1. in.
broad low down, thin in texture, much spotted in the lower part;
margin distantly and obscurely denticulate; peduncle simple, above
a foot long, with many lanceolate-deltoid empty bracts; flowers
12–20 in a dense corymb; pedicels ½–¾ in. long; bracts lanceolate-
deltoid, much shorter than the pedicels; perianth straight, pale
yellow, ½–⅝ in. long; tube very short; stamens included; style
finally slightly exserted.

VAR. **minor** (Baker); whole plant only 4–5 in. high, with leaves 2–3 in.
long.

KALAHARI REGION: Var. β, Transvaal; stony ridges on mountain tops of the
Saddleback Range, near Barberton, 4800 ft. *Galpin*, 873!

EASTERN REGION: Pondoland, *Bachmann*, 263! Natal; Inanda, *Wood*,
1217! Camperdown, 2000 ft., *Wood*, 1959! and grassy places near the Bay of
Natal, *Krauss*, 275! Swazieland; Piggs Peak, 4500 ft., *Galpin*, 1254!

6. A. myriacantha (Roem. et Schultes, Syst. Veg. vii. 704);
acaulescent; produced leaves 10–12, multifarious, linear, ½ ft. long,
¼–⅓ in. broad, channelled all down the face, dull green, thin in
texture, mottled with small white spots; margin minutely denticulate; peduncle simple, slender, a foot long, with many lanceolate
empty bracts; flowers 10–20 in a dense simple corymb; pedicels ½–¾
in. long; bracts lanceolate-deltoid, ⅓–½ in. long; perianth cylindrical,
rather recurved, ⅝ in. long; tube very short; stamens as long as the
perianth; style finally slightly exserted. *Kunth, Enum.* iv. 515;
Baker in Journ. Linn. Soc. xviii. 156. *Bowiea myriacantha, Haw.
in Phil. Mag.* 1827, 122.

COAST REGION: Albany Div.; grassy mountains near Grahamstown 2000 ft.,
MacOwan, Herb. Aust. Afr., 1554! and Howisons Poort, *Hutton!*

There is an unpublished drawing at Kew, made from the specimens sent home
by Bowie in 1823.

7. A. aristata (Haw. in Phil. Mag. 1825, 280); acaulescent;
rosette 3–4 in. diam.; leaves 50–60, very dense, multifarious, all

ascending, lanceolate, 3–4 in. long, $\frac{1}{2}$ in. broad, $\frac{1}{8}$ in. thick in
the middle, green, narrowed gradually into a long pellucid awn;
teeth deltoid, horny, white, $\frac{1}{16}$ in. long on the 'margin and also
on the keel of the upper half of the back; peduncle simple, 6–9
in. long; raceme simple, lax, 4–6 in. long; pedicels 1–1$\frac{1}{2}$ in. long;
bracts lanceolate-acuminate, much shorter than the pedicels; perianth
straight, cylindrical, 1–1$\frac{1}{4}$ in. long, bright red; segments keeled
with green, longer than the cylindrical tube; stamens and style
included. *Baker in Journ. Linn. Soc.* xviii. 156. *A. longiaristata,*
Roem. et Schultes, Syst. Veg. vi. 684; *Salm-Dyck, Aloe, sect.* xv. *fig.*
7; *Kunth, Enum.* iv. 518.

VAR. *β*, **leiophylla** (Baker in Journ. Linn. Soc. xviii. 156); leaves thinner,
smaller, with smaller teeth, which are almost confined to the margin; perianth
also more slender.

VAR. *γ*, **parvifolia** (Baker); leaves greyish-green, 1$\frac{1}{4}$–2 in. long, $\frac{1}{2}$ in. broad,
as thin as in *leiophylla*, with a shorter pellucid awn, the small deltoid teeth both
on the margin and also forming two rows on the back.

SOUTH AFRICA: without locality, *Zeyher*, 4186! Vars. *β* and *γ*, living plants
introduced and cultivated by *Cooper!*
CENTRAL REGION: Graaff Reinet Div.; stony ridges of the Sneeuw Berg
Range, 2900 ft., *MacOwan*, 1944!

Introduced into cultivation by Bowie in 1824.

8. A. Boylei (Baker in Kew Bullet. 1892, 84); acaulescent; leaves
thin, lanceolate, a foot or more long, an inch broad low down,
tapering gradually to a long point; marginal teeth close, spreading,
small, deltoid, whitish; peduncle robust, about as long as the leaves;
flowers many, aggregated in a dense globose head 5 in. diam.;
pedicels 1$\frac{1}{2}$–2 in. long; bracts ovate-acuminate, the lower an inch or
more long; perianth pale, 1–1$\frac{1}{4}$ in. long; tube scarcely any; segments lanceolate; stamens not longer than the perianth-segments.

KALAHARI REGION: Transvaal; Upper Moodies and Lomatie Valley, near
Barberton, 4000–4500 ft., *Galpin*, 1207!
EASTERN REGION: Griqualand East; near Kokstad, *Haygarth in Herb.*
Wood, 4232! Natal; valley of the Tugela River, *Allison!*

9. A. humilis (Miller, Abr. Gard. Dict. edit. vi. No. 10); acaulescent; leaves 30–40, densely rosulate, multifarious, all ascend-
ing, ovate-lanceolate, very acuminate, 3–4 in. long, $\frac{1}{2}$–$\frac{3}{4}$ in. broad
low down, glaucous green, obscurely lineate, unspotted, $\frac{1}{4}$ in.
thick, slightly tubercled on the face; margin with copious deltoid-
cuspidate white prickles $\frac{1}{12}$ in. long; back with copious, irregularly seriate, white horny tubercles and a few prickles; peduncle
stout, simple, about a foot long, furnished with numerous ovate-
lanceolate empty bracts; raceme lax, simple, about $\frac{1}{2}$ ft. long; pedicels
ascending, $\frac{3}{4}$–1 in. long; bracts large, ovate-lanceolate; perianth
straight, cylindrical, bright red, 1$\frac{1}{4}$–1$\frac{1}{2}$ in. long; tube $\frac{1}{4}$–$\frac{1}{3}$ in. long;
stamens and style finally just exserted. *Thunb. Diss. No.* 6; *Prodr.*
61; *Fl. Cap. edit. Schult.* 311; *Jacq. Hort. Schoenbr.* iv. 10, *t.* 420;
Baker in Journ. Linn. Soc. xviii. 157. *A. perfoliata var. humilis,*

Linn. Sp. Plant. 458. *A. tuberculata, Haw. in Trans. Linn. Soc.* vii. 16 ; *Syn.* 84 (*dwarf form*).

VAR. β, Candollei (Baker in Journ. Linn. Soc. xviii. 157); leaves greener, more distinctly lineate, less tubercled on the back ; prickles fewer; peduncle less bracteate. *A. humilis, DC. Plantes Grasses, t.* 39 ; *Salm-Dyck, Aloe, sect.* xv. *fig.* 1.

VAR. γ, A. incurva (Haw. 'Syn. 85); larger, the leaves broader low down, glaucous green, with an acuminate incurved tip. *Salm Dyck, Aloe, sect.* xv. *fig.* 3. *A. humilis var. incurva, Ker in Bot. Mag. t.* 828.

VAR. δ, A. acuminata (Haw. Syn. 84); leaves ovate-lanceolate, 4–5 in. long, 1¼–1½ in. broad low down, glaucous ; marginal prickles larger, ⅛–¼ in. long. *Kunth, Enum.* iv. 517. *A. humilis, Ker in Bot. Mag. t.* 757.

VAR. ε, A. subtuberculata (Haw. in Phil. Mag. 1825, 280); smaller than the last, with smaller, closer prickles.

VAR. ζ, A. suberecta (Haw. Syn. 84); largest form of all, with leaves 6-7 in. long.

VAR. η, macilenta (Baker) ; leaves thinner, tinged purple, deeply channelled down the face.

VAR. θ, A. echinata (Willd. Enum. 385); leaves smaller than in type, with distinct prickles, not tubercles, on the face. *Salm-Dyck, Aloe, sect.* xv. *fig.* 2 ; *Kunth, Enum.* iv. 516.

SOUTH AFRICA : without locality, living cultivated plants !

COAST REGION : Uitenhage, *Burke !* Albany Div. ; near Grahamstown, *Burke !*

Introduced into cultivation at the beginning of the eighteenth century.

10. A. pratensis (Baker in Journ. Linn. Soc. xviii. 156) ;

acaulescent ; leaves 60–80, ovate-lanceolate, densely rosulate, 4–6 in. long, 1¼–1½ in. broad below the middle, acuminate, glaucous when young only, ⅛ in. thick in the centre, obscurely lineate on both sides, unspotted; margin with copious, lanceolate-deltoid, brown, horny teeth ⅛–¼ in. long ; back smooth or slightly prickly upwards ; peduncle very stout, under a foot long, furnished all the way up with copious, large, ovate-acuminate empty bracts ; raceme dense, simple, ½–1 ft. long ; pedicels ascending, 1–1½ in. long ; bracts ovate with a long cusp, as long as the pedicels ; perianth subcylindrical, straight, pale or bright red tipped with green, 1¼–1½ in. long ; segments much longer than the tube ; stamens and style included or finally just exserted. *Bot. Mag t.* 6705.

SOUTH AFRICA: without locality, *Ecklon and Zeyher,* Hemerocallis No. 31 ! and living cultivated plants !

CENTRAL REGION : Somerset Div. ; summit of Bosch Berg, 4500 ft., *MacOwan,* 1896 !

KALAHARI REGION : Basutoland, living plant in *Hort. Hanbur. !*

11. A. virens (Haw. in Trans. Linn. Soc. vii. 17) ; acaulescent or

obscurely caulescent ; leaves 30–40, densely rosulate, lanceolate, very acuminate, ½ ft. long, 1 in. broad low down, light green, obscurely lineate, tapering gradually from the base to the apex, ¼ in. thick ; margin with pale, horny, deltoid prickles ⅛ in. long ; back minutely tubercled ; peduncle simple, 6–9 in. long, with a few empty bracts ; raceme lax, simple, ½ ft. long ; pedicels ascending, 1–1½ in. long ; bracts ovate-lanceolate, ¼–½ in. long ; perianth straight, cylindrical, bright red, 1½–2 in. long ; segments longer than the tube ; stamens and style finally just exserted. *Haw. Syn.* 83 ; *Ker in Bot. Mag. t.* 1355 ; *Salm-Dyck, Aloe, sect.* xv. *fig.* 8. *Kunth, Enum.* iv. 518; *Baker in Journ. Linn. Soc.* xviii. 157.

VAR. β, **macilenta** (Baker, loc. cit.); leaves smaller, thinner, tinged with purple, concave on the face.

SOUTH AFRICA : without locality.

Introduced into cultivation before 1790.

12. A. Bowiea (Schult. fil. Syst. Veg. vii. 704); acaulescent; leaves 30–40 in a dense multifarious rosette, all ascending or outer reflexing, linear, 4–6 in. long, ½ in. broad low down, tapering gradually to the apex, glaucous green, nearly flat, rounded on the back, copiously dotted with white, often striped with brown towards the base; margin with minute, whitish, deltoid prickles ; peduncle slender, simple, about a foot long, with many small, lanceolate-deltoid, empty bracts ; raceme lax, about ½ ft. long; lower pedicels $\frac{1}{12}$–$\frac{1}{8}$ in. long; bracts small, lanceolate; perianth clavate, greenish, ½ in. long ; tube short, campanulate ; stamens and style exserted ¼–⅓ in. *Salm-Dyck, Aloe, sect.* xiv. *fig.* 1 ; *Kunth, Enum.* iv. 515 ; *Baker in Journ. Linn. Soc.* xviii. 158. *Bowiea africana, Haw. in Phil. Mag.* 1824, 299 & 1827, 123 ; *A. DC. Pl. Rar. Jard. Genev. fasc.* vi. 21, *t.* 2.

SOUTH AFRICA : without locality.

Introduced into cultivation by Bowie in 1822.

13. A. longistyla (Baker in Journ. Linn. Soc. xviii. 158) ; acaulescent; leaves about 30 in a dense rosette, lanceolate, ascending, green, unspotted, 4–6 in. long, an inch broad at the base, tapering gradually to the pungent horny apex ; margin with pale brown, horny, deltoid prickles ⅛ in. long, many of which extend to the back and a few also to the face; peduncle stout, simple, about as long as the leaves ; raceme dense, simple, 3–5 in. long; lower pedicels $\frac{1}{8}$–$\frac{1}{6}$ in. long; bracts lanceolate-deltoid, ¾–1 in. long; perianth slightly curved, 1¾–2 in. long; tube less than half as long as the segments ; stamens slightly exserted ; style finally exserted ½ inch.

CENTRAL REGION: Karoo near Graaff Reinet, 2700 ft., *Bolus,* 689! Jansenville Div.; Zwart Ruggens, 2500–3000 ft., *Drège,* 8640!

Nearly allied to *A. humilis* and *A. virens.*

14. A. Ecklonis (Salm-Dyck, Aloe, sect. xxi. fig. 2); subacaulescent; leaves densely rosulate, ensiform, a foot long, 1¼–1½ in. broad low down, narrowed gradually to the apex, glaucous-green, unspotted ; face flat in the lower half, channelled down the upper half ; margin with close, small, whitish, horny teeth ; peduncle simple, about a foot long, with numerous empty bracts ; raceme dense, capitate, about 3 in. diam.; lower pedicels 1¼–1½ in. long; bracts lanceolate-deltoid, ½ in. long; perianth orange-red, $\frac{3}{4}$–$\frac{7}{8}$ in. long; tube short ; stamens and style distinctly exserted. *Baker in Journ. Linn. Soc.* xviii. 158. *A. claviflora, Burchell, Travels,* i. 272 ?

SOUTH AFRICA: without locality.

Introduced into cultivation by Ecklon about 1836. Burchell's *A. claviflora* is only described very briefly, and no specimens seem to have been kept; it was collected among rocks at Riffle River in Fraserburg Division.

15. **A. brevifolia** (Miller, Gard. Dict. edit. viii. No. 8); old plants with a stem a few inches long, $\frac{1}{2}$ in. diam. below the rosette of leaves; leaves 30–40 in a dense rosette, lanceolate-deltoid, 3–4 in. long, an inch broad at the base, narrowed gradually to the apex, $\frac{1}{4}-\frac{1}{3}$ in. thick in the middle, very glaucous, without either spots or distinct lines; margin with white, lanceolate-deltoid, horny prickles $\frac{1}{12}-\frac{1}{8}$ in. long, which extend to the keel of the upper part of the back; peduncle stout, simple, about a foot long, with many deltoid empty bracts; raceme dense, simple, 6–9 in. long; pedicels $\frac{1}{2}-1$ in. long; bracts deltoid, rather shorter than the pedicels; perianth cylindrical, pale red, $1\frac{1}{4}-1\frac{1}{2}$ in. long; tube cylindrical, shorter than the segments; stamens equalling the perianth; style considerably exserted. *DC. Plantes Grasses, t.* 81; *Haw. Syn.* 80; *Lindl. in Bot. Reg. t.* 996; *Salm-Dyck, Aloe, sect.* xvi. *fig.* 1; *Baker in Journ. Linn. Soc.* xviii. 160, *non Haworth. A. prolifera, Haw. in Trans. Linn. Soc.* vii. 16; *Kunth, Enum.* iv. 519.

VAR. β, **A. postgenita** (Roem. et Schultes, Syst. Veg. vii. 1714); more robust, with leaves 4–5 in. long, $1\frac{1}{4}-1\frac{1}{2}$ in. broad low down. *Kunth, Enum.* iv. 519. *A. prolifera var. major, Salm-Dyck, Cat.* 23; *Haw. Suppl.* 44.

VAR. γ, **A. depressa** (Haw. in Trans. Linn. Soc. vii. 16; Syn. 80, excl. syn.); more robust, with leaves $\frac{1}{3}$ ft. long, 2 in. broad low down, rather less glaucous than in the type, with sometimes a few prickles extending to the upper surface. *Salm-Dyck, Aloe, sect.* xvi. *fig.* 3; *Kunth, Enum.* iv. 519.

SOUTH AFRICA: without locality, living cultivated plants!

Introduced into cultivation at the beginning of the eighteenth century. Two of the varieties are figured by Commelyn in his *Præludia, tt.* 21 *and* 22.

16. **A. Serra** (DC. Plantes Grasses, t. 80); old plants with a short simple stem below the rosette of leaves; leaves 30–40 in a dense rosette, lanceolate-deltoid, 4–5 in. long, $1-1\frac{1}{2}$ in. broad low down, narrowed gradually to the tip, glaucous, unspotted, not distinctly lineate, $\frac{1}{4}-\frac{1}{3}$ in. thick in the middle, with a few white tubercles on the back and sometimes on the face; margin with deltoid, white, horny prickles $\frac{1}{12}-\frac{1}{8}$ in. long, confluent into a horny line; peduncle simple, about a foot long; raceme dense, simple, about $\frac{1}{2}$ ft. long; pedicels $\frac{1}{2}-1$ in. long; bracts about as long as the pedicels; perianth cylindrical, bright red, $1\frac{1}{2}$ in. long; tube cylindrical, shorter than the segments; stamens as long as the perianth; style exserted. *Jacq. Fragm. t.* 61; *Haw. Suppl.* 44; *Roem. et Schultes, Syst. Veg.* vii. 687; *Kunth, Enum.* iv. 519.

SOUTH AFRICA: without locality.

Introduced into cultivation about 1818. Scarcely more than a variety of *A. brevifolia.*

17. **A. lineata** (Haw. in Trans. Linn. Soc. vii. 18; Syn. 79); stem finally $\frac{1}{2}-1$ ft. long below the rosette of leaves, simple, 2 in. diam.; leaves 30–40 in a dense rosette, lanceolate, about a foot long,

2 in. broad low down, narrowed gradually to the apex, $\frac{1}{4}$–$\frac{1}{3}$ in. thick, glaucous-green, unspotted, flat in the lower part of the face, marked with very distinct green vertical lines, rounded and smooth on the back ; margin with close, reddish-brown, deltoid, horny prickles $\frac{1}{12}$–$\frac{1}{8}$ in. long ; peduncle simple, about a foot long ; raceme dense, $\frac{1}{2}$ ft. long ; lower pedicels 1–1$\frac{1}{2}$ in. long ; bracts lanceolate-deltoid, much shorter than the pedicels ; perianth cylindrical, bright red, tipped with green ; tube cylindrical about $\frac{1}{2}$ in. long ; stamens and style scarcely exserted. *Roem. et Schultes, Syst. Veg.* vii. 689 ; *Salm-Dyck, Aloe, sect.* xvii. *fig.* 1 ; *Kunth, Enum.* iv. 520 ; *Baker in Journ. Linn. Soc.* xviii. 159. *A. perfoliata var. lineata, Ait. Hort. Kew.* i. 467.

SOUTH AFRICA : without locality.

Introduced into cultivation about 1789.

18. Aloe heteracantha (Baker in Journ. Linn. Soc. xviii. 161); stem attaining a foot or more in height ; leaves 15–30 in a dense rosette, lanceolate, 6–12 in. long, 1$\frac{1}{2}$–2$\frac{1}{2}$ in. broad at the base, narrowed gradually to the concave apex, green or tinted with red-brown, obscurely lineate towards the base, often with a few irregular small whitish spots, and often marked down the face with one or two slightly-raised longitudinal ribs ; margins usually armed with a few, short, irregular, spreading, deltoid prickles, but some-times quite smooth and unarmed, and without prickles on the upper or under surface ; peduncle simple or branched, 1–3 ft. long, including the 6–10 in. long dense raceme ; bracts oblong-ovate, acute or acuminate, 4–6 lines long ; pedicels $\frac{1}{2}$–1 in. long, ascending ; perianth drooping, cylindrical, 1$\frac{1}{4}$–1$\frac{1}{2}$ in. long, tube very short, outer segments bright red, inner segments nearly white with a red keel ; stamens included, 14–15 lines long. *Bot. Mag. t.* 6863 ; *Gard. Chron.* 1894, xv. 620.

SOUTH AFRICA : without locality.

Described from living cultivated plants. Possibly this may not be a South African species.

19. A. striata (Haw. in Trans. Linn. Soc. vii. 18 ; Syn. 81); old plants with a stem 1–2 ft. long below the dense rosette of leaves ; leaves 12–20, oblong-lanceolate, 1$\frac{1}{2}$–2 ft. long, 4–8 in. broad, glaucous, obscurely spotted and lineate, $\frac{1}{4}$–$\frac{1}{3}$ in. thick in the middle ; margin entire, cartilaginous, white or reddish, a line broad ; peduncle very stout, much branched, bearing 20 or more dense capitate racemes ; lower pedicels $\frac{1}{2}$–$\frac{3}{4}$ in. long ; bracts minute, lanceolate ; perianth bright red, $\frac{7}{8}$–1 in. long ; tube cylindrical, con-stricted above the ovary ; stamens as long as the perianth ; style shortly exserted. *Baker in Journ. Linn. Soc.* xviii. 162. *A. paniculata, Jacq. Fragm.* 48, *t.* 62 ; *Roem. et Schultes, Syst. Veg.* vii. 691 ; *Kunth, Enum.* iv. 522. *A. albo-cincta, Haw. Suppl.* 43 ; *Kunth, Enum.* iv. 525. *A. hanburiana, Naud. in Rev. Hort.* 1875, 165.

VAR. β, **A. rhodocincta** (Hort.); horny border of the leaf pinkish.

VAR. γ, **oligospila** (Baker in Gard. Chron. 1894, xv. 588); stem taller and firmer than in the type; leaves narrower and not so thick, very glaucous, with a few obscure linear blotches on the back, the stripes quite lost when the leaves are mature.

COAST REGION: Albany Div.; precipitous places near Grahamstown, 2200 ft., *MacOwan*, 1144! King Williamstown Div.; near the Chalumna River, living specimens of type and var. γ, introduced and grown by *T. Cooper!*

Introduced into cultivation about 1795. This and some of the other species, Mr. MacOwan says, are fertilized by the *Nectarineæ*, and if the birds are kept off by wire-netting, few or no capsules are produced. This is the only *Aloe* without any teeth to the leaves.

A. Lynchii (Baker in Gard. Chron. 1881, xv. 266) is a hybrid between this very distinct species and *Gasteria verrucosa*, with flowers like those of the present plant but with dull green leaves half an inch thick, narrowed gradually from the base to the tip, and dotted all over with minute whitish spots. Mr. Lynch also crossed it with *A. grandidentata*, and the hybrid flowered at Kew in 1884.

20. A. serrulata (Haw. in Trans. Linn. Soc. vii. 18); mature plants with a stem 1–2 ft. long, 1½–2 in. diam. below the dense rosette of leaves; leaves 12–20, oblong-lanceolate, 6–9 in. long, 2 in. broad low down, narrowed gradually from below the middle to the point, ¼–⅓ in. thick in the centre, pale green with indistinct lines and copious, obscurely seriate, oblong, whitish spots; margin with minute, deltoid, confluent, white, horny prickles; peduncle simple, about a foot long; raceme oblong, half a foot long; lower pedicels ½–¾ in. long; bracts deltoid-cuspidate, as long as the pedicels; perianth bright red, 1¼–1½ in. long, scarcely constricted above the ovary; segments oblong, much shorter than the tube; stamens and style not exserted. *Sims in Bot. Mag. t.* 1415; *Roem. et Schultes, Syst. Veg.* vii. 697; *Salm-Dyck, Aloe, sect.* xx. *fig.* 1; *Kunth, Enum.* iv. 522; *Baker in Journ. Linn. Soc.* xviii. 163. *A. perfoliata var. serrulata, Ait. Hort. Kew.* i. 467.

SOUTH AFRICA: without locality, living cultivated plants!

Introduced into cultivation about 1789. *A. pallescens, Haw. Revis.* 41, is probably a variety with the lines and spots nearly obsolete.

21. A. saponaria (Haw. Syn. 83); mature plants with a stem ½–1 ft. long, 2–3 in. diam. below the dense rosette of leaves; leaves 12–20, lanceolate or oblong-lanceolate, ½–1 ft. long, 2 in. broad low down, ¼ in. thick in the middle, green, or tinged with purple, indistinctly lineate with copious oblong whitish spots; margin with deltoid, cuspidate, horny, reddish-brown prickles ⅛–¼ in. long; peduncle simple or forked, about a foot along; raceme dense, capitate, 3–4 in. long and broad; lower pedicels 1½–2 in. long; bracts deltoid-cuspidate, much shorter than the pedicels; perianth bright red-yellow, 1½–1¾ in. long; tube cylindrical, constricted above the ovary; segments shorter than the tube; stamens not exserted. *Roem. et Schultes, Syst. Veg.* vii. 699; *Kunth, Enum.* iv. 526; *Baker in Journ. Linn. Soc.* xviii. 164. *A. disticha, Miller, Gard. Dict. edit.* viii. *No.* 5. *A. maculosa, Lam. Encyc.* i. 87, *in part. A. umbellata var. minor, DC. Plantes Grasses, t.* 98. *A.*

umbellata, Salm-Dyck, Aloe, sect. xxiii. *fig.* 1. *A. perfoliata var.
saponaria, Ait. Hort. Kew.* i. 467, *excl. syn. A. saponaria var.
minor, Haw. in Trans. Linn. Soc.* vii. 17; *Sims in Bot. Mag.
t.* 1460.

VAR. β, **brachyphylla** (Baker in Journ. Linn. Soc. xviii. 164); leaves 3-4 in.
long, ovate acuminate; prickles smaller than in the type.

COAST REGION : Paarl Mountains, 1000–2000 ft., *Drège,* 8635!   Caledon Div.;
mountains near Genadendal, 1400 ft., *Bolus,* 7430!
EASTERN REGION : Pondoland, *Bachmann,* 264!

Introduced into cultivation early in the eighteenth century, and figured by
Dillenius in the *Hortus Elthamensis, t.* 14, *fig.* 15.

22. A. latifolia (Haw. Syn. 82); stem simple, reaching 1–2 ft. in
length, 2–3 in. diam. below the dense rosette of leaves; leaves
12–20, oblong-lanceolate, ½–1 ft. long, 2–3½ in. broad low down,
¼–⅓ in. thick in the middle, glossy green, obscurely lineate, with
irregular bands of linear-oblong or oblong, whitish blotches, ½–¾ in.
long; margin with deltoid-cuspidate, brown, horny prickles ⅛–⅙ in.
long; peduncle robust, simple or forked, 1–2 ft. long; racemes
4–5 in. long and broad; lower pedicels 1½–2 in. long; bracts very
acuminate, an inch or more long; perianth bright red-yellow,
1¼–1½ in. long; tube longer than the segments, constricted above the
ovary; stamens not exserted. *Roem. et Schultes, Syst. Veg.* vii.
700; *Kunth, Enum.* iv. 526; *Salm-Dyck, Aloe, sect.* xxiii. *fig.* 3;
Baker in Journ. Linn. Soc. xviii. 164. *A. saponaria var. latifolia,
Haw. in Trans. Linn. Soc.* vii. 18; *Sims in Bot. Mag. t.* 1346.

COAST REGION: Cathcart Div.; Tushintush, near Windvogel Mountain,
3500 ft., *Baur,* 1121!
CENTRAL REGION: Karoo near Graaff Reinet, 2700 ft., *Bolus,* 751!

Introduced into cultivation about 1795, probably by Masson.

23. A. leptophylla (N. E. Brown ex Baker in Journ. Linn.
Soc. xviii. 165); stem short, simple, 1½–2 in. diam. below the
rosette of leaves; leaves 12–20, lanceolate, 9–12 in. long, 2–3 in.
broad low down, tapering from the middle to the tip, ⅙ in. thick in
the middle, green or purple tinted, distinctly lineate, with copious,
linear-oblong, whitish blotches; marginal prickles deltoid, ⅛ in. long;
peduncle simple, 1½ ft. long; raceme dense, capitate, about 3 in.
long and broad; pedicels 1–1½ in. long; bracts small, lanceolate-
deltoid, acuminate; perianth 1¼ in. long; segments much shorter
than the cylindrical tube; stamens as long as the perianth. *Baker
in Journ. Bot.* 1889, 44. *A. Cooperi, Hort. de Smet. non Baker.*

VAR. β, **stenophylla** (Baker); leaf as long, but not more than 1¼ in. broad.

SOUTH AFRICA : without locality, *Cooper !*

Described from living plants introduced by Cooper about 1860. It differs from
all its allies by its thinner, very flexible leaves. May be the imperfectly described
A. tenuifolia, Lam. Encyc. i. 87.

24. A. macracantha (Baker in Journ. Linn. Soc. xviii. 167);
stem reaching a length of 2–3 feet and a diameter of 2–3 in. below
the dense rosette of leaves; leaves 20–30, oblong-lanceolate, 1–1½
ft. long, 3–5 in. broad in the lower half, narrowed gradually from
the middle to the tip, ¼ in. thick in the centre, nearly flat on the
face, bright green, with copious, distinct, darker green, vertical lines
and oblong whitish blotches, much paler on the convex back;
margin with horny, brown-tipped, deltoid-cuspidate prickles ¼ in.
long, the outer leaves much recurved; peduncle stout, flattened
downwards, simple or forked, above a foot long; raceme dense,
capitate, about 4 in. long and broad; pedicels 1–1½ in. long; bracts
lanceolate-deltoid, acuminate, shorter than the pedicels; perianth
bright yellow with a tinge of red, 1½–1¾ in. long; tube shorter than
the segments, constricted above the ovary; stamens as long as the
perianth. *Bot. Mag. t.* 6580.

SOUTH AFRICA: without locality, *Cooper!*

Described from living plants introduced by Cooper about 1860.

25. A. obscura (Miller, Gard. Dict. edit. viii. No. 6); stem ½–1 ft.
long below the rosette of leaves, about 2 in. diam.; leaves 15–30,
lanceolate, about a foot long, 2–3 in. broad low down, narrowed
gradually from below the middle to the tip, ¼ in. thick in the middle,
green or more or less glaucous, especially when young, not distinctly
lineate, marked with copious, distinct, oblong, whitish spots; teeth
of the margin deltoid-cuspidate, ¹⁄₁₂–⅛ in. long; peduncle simple or
forked, 1½–2 ft. long; raceme dense, cylindrical, ½–1 ft. long; lower
pedicels 1–1½ in. long; bracts deltoid-cuspidate, much shorter than
the pedicels; perianth bright red, 1¼–1½ in. long; tube slightly constricted above the ovary; segments shorter than the tube; stamens
and style slightly exserted. *Haw. Syn.* 82; *Roem. et Schultes, Syst.
Veg.* vii. 700; *Kunth, Enum.* iv. 526; *Baker in Journ. Linn. Soc.*
xviii. 165. *A. perfoliata var. obscura, Ait. Hort. Kew.* i. 467.
A. maculosa, Lam. Encyc. i. 87, *in part. A. picta, Thunb. Diss.
No.* 4, *in part. DC. Plantes Grasses, t.* 97; *Sims in Bot. Mag.
t.* 1323; *Salm-Dyck, Aloe, sect.* xxiii. *fig.* 2.

CENTRAL REGION: near Graaff Reinet, *Bolus*, 598!

Introduced into cultivation at the beginning of the eighteenth century, and
figured by Dillenius in *Hortus Elthamensis, t.* 15, *fig.* 16.

26. A. grandidentata (Salm-Dyck, Hort. 329; Aloe, sect. xxiii.
fig. 4); stem ½–1 ft. long, simple, 2–2½ in. diam. below the dense
rosette of leaves; leaves 12–20, lanceolate, 1–1½ ft. long, 2½–3 in.
broad in the lower half, narrowed gradually from the middle to the
apex, ¼ in. thick in the middle; face bright green with wavy transverse bands of oblong whitish spots; marginal teeth crowded, deltoid-cuspidate, brown-tipped, ¼ in. long; peduncle robust, 1½–2 ft.
long; racemes 3–7, oblong, the end one 6–8 in. long; lower pedicels
½ in. long; bracts as long as the lower pedicels; perianth pale red or

yellowish-red, 1–1⅛ in. long; tube constricted above the ovary; segments as long as the tube; stamens about as long as the perianth. *Roem. et Schultes, Syst. Veg.* vii. 699; *Kunth, Enum.* iv. 525; *Baker in Journ. Linn. Soc.* xviii. 166.

South Africa: without locality.

Introduced into cultivation about 1820. Both this and *A. latifolia* have been hybridized by Mr. Lynch with *A. striata*.

27. A. Greenii (Baker in Journ. Linn. Soc. xviii. 165); well-grown plants with a short simple stem 2 in. diam. below the dense rosette of leaves; leaves 12–15 in a rosette, lanceolate, 15–18 in. long, 2½–3 in. broad, narrowed gradually from below the middle to an acuminate apex, bright green, with obscure vertical lines and broad irregular wavy bands of confluent oblong whitish spots, ⅓–¾ in. thick in the centre; margin with brown, horny, deltoid-cuspidate prickles ⅛–⅙ in. long; peduncles stiffly erect, about a foot long; panicle about as long as the peduncle, with 5–7 branches; racemes oblong or oblong-cylindrical, the end one 6–9 in. long; lower pedicels ½ in. long; bracts small, lanceolate; perianth pale red, 1¼ in. long; tube much constricted above the ovary; segments shorter than the tube; stamens and style not distinctly exserted. *Bot. Mag. t.* 6520.

South Africa: without locality, *Cooper!*

Described from living plants introduced by Cooper about 1860.

28. A. gasterioides (Baker in Journ. Linn. Soc. xviii. 166); stem short, simple below the dense rosette of leaves; leaves lanceolate, 4–5 in. long, 1½–2 in. broad, green, strongly lineate and marked with copious, oblong, whitish blotches on the face; margin with deltoid, horny prickles ⅛ in. long; peduncle simple, about a foot long; raceme lax, oblong; lower pedicels ½–¾ in. long, much exceeding the small acuminate bracts; perianth bright red, an inch long; tube much constricted in the middle; segments shorter than the tube; stamens not exserted.

South Africa: without locality.

Described from a plant that flowered in the Kew collection in May, 1875. A near ally of *A. obscura* and *A. grandidentata*.

29. A. tricolor (Baker in Bot. Mag. t. 6324); mature plants with a very short simple stem below the dense rosette of leaves; leaves 12–16 in a rosette, lanceolate, 5–6 in. long, 1½–2 in. broad low down, narrowed gradually from below the middle to an acuminate point, ⅓–½ in. thick in the centre, slightly turgid on the face, dead green with irregular transverse bands of small oblong whitish spots; margin with close, deltoid-cuspidate, horny, red-brown prickles $\frac{1}{12}$ in. long, the outer leaves of the rosette much recurved; peduncle purplish, glaucous, 1½ ft. long; racemes 3–5 in a lax deltoid panicle, 3–4 in. long; lower pedicels ¼–⅓ in. long; bracts small, deltoid-cuspidate; perianth bright red, an inch long; tube slender, cylindrical,

longer than the segments ; stamens and style as long as the perianth.
Baker in Journ. Linn. Soc. xviii. 166.

SOUTH AFRICA: without locality.

Described from a living plant that flowered at Kew in April, 1877.

30 **A. microstigma** (Salm-Dyck, Aloe, sect. xxvi. fig. 4) ; stem elongated below the rosette of leaves, slender, simple; leaves 20–30 in a
dense rosette, ensiform, $\frac{1}{2}$–1 ft. long, 1$\frac{1}{2}$ in. broad low down, $\frac{1}{4}$–$\frac{1}{3}$ in.
thick in the centre, glaucous green, with a flat face and numerous
small whitish oblong spots ; margin with small, brown, horny, deltoid-
cuspidate prickles ; peduncle simple, about a foot long, with many
broad empty bracts ; raceme dense, oblong-cylindrical, 6–9 in. long;
pedicels ascending, $\frac{1}{2}$–$\frac{3}{4}$ in. long; bracts oblong, half as long as the
pedicels ; perianth greenish-yellow, $\frac{5}{8}$ in. long ; segments much
longer than the tube ; stamens slightly exserted.   *Baker in Journ.
Linn. Soc.* xviii. 167.   *A. arabica, Salm-Dyck, Cat.* 27 ; *Roem. et
Schultes, Syst. Veg.* vii. *698, excl. syn.*

SOUTH AFRICA: without locality, living cultivated plant !

Introduced into cultivation about 1816.

31. **A. Monteiroi** (Baker in Gard. Chron. 1889, vi. 523) ; stem
very short ; leaves about a dozen in a dense rosette, ensiform, a foot
long, an inch broad low down, tapering gradually to the point,
channelled down the face in the upper half, copiously but obscurely
spotted on both surfaces, armed with middle-sized, not very close,
deltoid-cuspidate teeth ; peduncle simple, terete, 1$\frac{1}{2}$ ft. long ; raceme
moderately dense, 4–6 in. long, 2$\frac{1}{2}$ in. diam. ; pedicels ascending,
$\frac{1}{4}$–$\frac{1}{3}$ in. long; bracts linear ; perianth cylindrical, an inch long, dull
red ; tube constricted above the ovary ; segments oblong, $\frac{1}{3}$ in. long ;
stamens and style considerably exserted beyond the tip of the
perianth-segments.

EASTERN REGION : Delagoa Bay, *Mrs. Monteiro !*

Described from a living plant which flowered at Kew, October, 1889.

32. **A. consobrina** (Salm-Dyck, Aloe, sect. xviii. fig. 3) ; stem
slender, elongated, simple, below the rosette of leaves ; leaves laxly
disposed or more dense, ensiform, 6–8 in. long, an inch broad,
$\frac{1}{4}$ in. thick in the centre, green, with copious, distinct, whitish, round
or oblong spots ; margin with close, red-brown, deltoid-cuspidate
prickles $\frac{1}{12}$–$\frac{1}{8}$ in. long; peduncle above a foot long, slender, simple
or forked ; racemes oblong-cylindrical, 3–4 in. long; pedicels $\frac{1}{4}$–$\frac{1}{3}$ in.
long ; bracts small, lanceolate, acute ; perianth reddish-yellow, 1–1$\frac{1}{4}$
in. long ; tube short, campanulate ; stamens about as long as the
perianth.   *Baker in Journ. Linn. Soc.* xviii. 168.

SOUTH AFRICA : without locality.

Introduced into cultivation about 1845.   A near ally of *A. spicata.*

33. **A. spicata** (Linn. fil. Suppl. 205); stem elongated below the
leaves, simple ; leaves 12–20, laxly disposed, patent or the lower

recurved, ensiform, 1–1$\frac{1}{2}$ ft. long, 1$\frac{1}{2}$–2 in. broad low down, $\frac{1}{3}$ in. thick in the centre, tapering gradually from the middle to a long point, green with a slight glaucous tinge, obscurely lineate towards the base, mottled with copious, irregular, whitish, oblong spots; margin with spreading, horny, deltoid-cuspidate prickles $\frac{1}{8}$–$\frac{1}{6}$ in. long; peduncle slender, simple or forked, above a foot long; raceme oblong, $\frac{1}{2}$ ft. long; pedicels ascending, $\frac{1}{2}$–$\frac{3}{4}$ in. long; bracts small, lanceolate; perianth bright yellow, tinted with red when young, 1$\frac{1}{4}$ in. long; tube short; stamens slightly exserted. *Thunb. Fl. Cap. edit. Schult.* 309; *Haw. Syn.* 76; *Kunth, Enum.* iv. 530; *Bentl. and Trimen, Medic. Plants, t.* 284.

SOUTH AFRICA: without locality.

Interior of the colony, according to Thunberg. Rare in cultivation; introduced in 1795. My description of the flower was taken from a plant in Mr. Peacock's collection in 1879, the same, I believe, from which Bentley and Trimen's figure was drawn.

34. A. tenuior (Haw. in Phil. Mag. 1825, 281); stem many yards long when fully developed, sarmentose; branches terete, $\frac{1}{4}$–$\frac{1}{3}$ in. diam.; internodes $\frac{1}{2}$–1 in. long, obscurely striped with green; leaves linear, 6–8 in. long, about $\frac{1}{2}$ in. broad low down, $\frac{1}{12}$ in. thick in the centre, plain green, not auricled at the base; all the marginal teeth very minute; peduncles slender, simple, 6–9 in. long; racemes moderately dense, 4–6 in. long; lower pedicels $\frac{1}{4}$–$\frac{1}{3}$ in. long; bracts very minute; perianth cylindrical, pale yellow, $\frac{1}{3}$–$\frac{1}{2}$ in. long; segments short, ovate; stamens and style both distinctly exserted. *Roem. et Schultes, Syst. Veg.* vii. 704; *Salm-Dyck, Aloe, sect.* xxv. *fig.* 3; *Kunth, Enum.* iv. 529; *Baker in Journ. Linn. Soc.* xviii. 169.

COAST REGION: Albany Div.; Karoo near Fort Brown, 1500 ft., *MacOwan*, 1140! Bedford Div.; near the Fish River, *Burke!* Bathurst Div.; near the Fish River, *Drège*, 3525! Fort Beaufort Div.; banks of the Kat River, *Zeyher*, 1052!

Introduced into cultivation by Bowie in 1820.

35. A. ciliaris (Haw. in Phil. Mag. 1825, 281); stem many yards long when fully developed, sarmentose; branches terete, $\frac{1}{4}$–$\frac{1}{3}$ in. diam.; leaves laxly disposed over a foot or more, with internodes $\frac{1}{2}$–$\frac{3}{4}$ in. long, obscurely striped with green; blade linear, amplexicaul, 4–6 in. long, $\frac{1}{2}$–$\frac{3}{4}$ in. broad, acuminate, green, flat on the face in the lower half, $\frac{1}{12}$ in. thick in the middle, neither spotted nor striped; teeth of the margin minute, deltoid, white, growing gradually larger towards the base of the leaf; peduncle slender, simple, lateral, 3–9 in. long; raceme lax, oblong, 2–4 in. long; pedicels $\frac{1}{6}$–$\frac{1}{4}$ in. long; bracts minute, lanceolate; perianth cylindrical, bright red, 1–1$\frac{1}{4}$ in. long; tube cylindrical; segments oblong; stamens not exserted; style finally exserted. *Roem. et Schultes, Syst. Veg.* vii. 703; *Salm-Dyck, Aloe, sect.* xxv. *fig.* 1; *Kunth, Enum.* iv. 529; *Baker in Journ. Linn. Soc.* xviii. 169.

COAST REGION: Bathurst Div.; between Riet Fontein and Kowie River,

Burchell, 3993 ! Albany Div.; hills near Grahamstown, 2000–2200 ft., *Mac-Owan*, 1146 ! *Bolus*, 2689 ! and *MacOwan and Bolus, Herb. Norm.*, 298 !

Introduced into cultivation by Bowie about 1820.

36. A. striatula (Haw. in Phil. Mag. 1825, 281); stems many yards long when fully developed, sarmentose; branches $\frac{1}{4}$–$\frac{1}{2}$ in. diam.; internodes $\frac{1}{2}$–1 in. long, conspicuously striped with green; leaves spreading, linear, 6–9 in. long, $\frac{1}{2}$–$\frac{5}{8}$ in. broad, acuminate, green, $\frac{1}{12}$ in. thick in the centre, not auricled at the base; all the marginal prickles minute and deltoid; peduncle lateral, slender, simple, 6–9 in. long; raceme oblong, 3–6 in. long; pedicels $\frac{1}{8}$–$\frac{1}{4}$ in. long; bracts minute; perianth cylindrical, yellow, tinged with green, 1–1$\frac{1}{4}$ in. long; tube cylindrical; segments short, ovate; stamens and style distinctly exserted. *Roem. et Schultes, Syst. Veg.* vii. 703; *Kunth, Enum.* iv. 529 ; *Baker in Journ. Linn. Soc.* xviii. 169.

CENTRAL REGION : Somerset Div.; in a woody ravine near Bruintjes Hoogte, *Burchell*, 3115 !

Introduced into cultivation by Bowie in 1823.

37. A. gracilis (Haw. in Phil. Mag. 1825, 280); stems elongated below the leaves; leaves laxly disposed, lanceolate, spreading, 6–10 in. long, an inch broad at the base, unspotted, $\frac{1}{12}$ in. thick in the centre; margin with minute, deltoid prickles; peduncle simple, $\frac{1}{2}$–1 ft. long; raceme dense, simple, 2–3 in. long; lower pedicels $\frac{1}{4}$ in. long; bracts lanceolate, as long as the pedicels; perianth cylindrical, yellow, 1$\frac{1}{4}$–1$\frac{1}{2}$ in. long; tube short, campanulate; segments three or four times as long as the tube ; stamens as long as the perianth. *Roem. et Schultes, Syst. Veg.* vii. 706 ; *Kunth, Enum.* iv. 531; *Baker in Journ. Linn. Soc.* xviii. 170.

COAST REGION : Simons Bay, *C. Wright!*

Introduced into cultivation by Bowie in 1823, but soon lost.

38. A. Macowani (Baker in Journ. Linn. Soc. xviii. 170); stems elongated, sarmentose; leaves laxly disposed, lanceolate acuminate, a foot or more long, 1$\frac{1}{2}$–2 in. broad below the middle, plain green, distinctly lineate only at the very base; margin with very minute deltoid prickles; peduncle simple, moderately stout, about a foot long; racemes moderately dense, $\frac{1}{2}$ ft. long; flowers very deflexed; pedicels very short; bracts minute; perianth cylindrical, 1–1$\frac{1}{4}$ in. long; segments about as long as the cylindrical tube; stamens slightly exserted ; style finally exserted $\frac{1}{3}$–$\frac{1}{4}$ in.

CENTRAL REGION : Somerset Div.; in woods on the sides of Bosch Berg, 3500 ft., *MacOwan*, 1915 !

39. A. aurantiaca (Baker in Gard. Chron. 1892, xi. 780); stems suberect, several feet long, $\frac{3}{4}$ in. diam.; leaves approximate ; internodes white, with faint stripes of green ; blade lanceolate, 8–9 in. long, 1$\frac{1}{2}$ in. broad low down, plain green; margin with small, pale, deltoid teeth ; peduncle as long as the leaves; raceme very dense, 5–6 in. long, 2 in. diam.; flowers all deflexed, bright yellow, just tinged with red when young; pedicels very short; bracts minute; perianth 1½ in. long; tube short, campanulate; stamens and style much exserted.

SOUTH AFRICA : without locality.

Described from a plant flowered by Marchese Hanbury at La Mortola, June, 1892.

40. A. distans (Haw. Syn. 78) ; stem simple, elongated, suberect ; leaves laxly disposed over the top one or two feet of the stem, ovate-lanceolate, all ascending, 3–4 in. long, 1½ 2 in. broad at the base, narrowed gradually to the tip, dull green, with a slight glaucous tint, concave on the face, ¼ in. thick in the middle, neither spotted nor lineate, rounded on the back; margin with deltoid, white, horny prickles 1/12–⅛ in. long, which extend to the keel towards the top of the back ; peduncle above a foot long, simple ; raceme dense, capitate, 3–4 in. diam. ; lower pedicels 1–1¼ in. long; bracts small, lanceolate-deltoid ; perianth reddish-yellow, 1¼–1½ in. long; tube short ; segments elongated ; stamens and style about as long as the perianth. *Roem. et Schultes, Syst. Veg.* vii. 714; *Salm-Dyck, Aloe, sect.* xxiv. *fig.* 1 ; *Kunth, Enum.* iv. 528; *Baker in Journ. Linn. Soc.* xviii. 171. *A. mitræformis var. angustior, Lam. Encyc.* i. 87. *A. mitræformis var. brevifolia, Sims in Bot. Mag. t.* 1362. *A. brevifolia, Haw. in Trans. Linn. Soc.* vii. 23, *non Syn.* 80.

SOUTH AFRICA: without locality, living cultivated plants !
Introduced into cultivation about 1735.

41. A. mitriformis (Miller, Gard. Dict. edit. viii. No. 1); stem 3 or 4 feet long when developed, simple, 1–2 in. diam. ; leaves laxly disposed, all ascending, ovate-lanceolate, ½ ft. long, 2–3 in. broad low down, ¼–⅓ in. thick in the middle, firm in texture, dull green, with a slight glaucous tinge, neither spotted nor distinctly striped, pungent at the tip ; margin with pale, deltoid, horny prickles 1/12–⅛ in. long ; peduncle stout, above a foot long, often forked ; raceme dense, corymbose, half a foot long and broad ; lower pedicels 1¼–1½ in. long ; bracts small, lanceolate-deltoid ; perianth cylindrical, bright red, 1½–1¾ in. long; tube short ; segments lanceolate, tipped with green; stamens about as long as the perianth. *Lam. Encyc.* i. 87 ; *DC. Plantes Grasses, t.* 99 ; *Sims in Bot. Mag. t.* 1270 ; *Haw. Syn.* 77 ; *Baker in Journ. Linn. Soc.* xviii. 171. *A. xanthacantha, Salm-Dyck, Aloe, sect.* xxiv. *fig.* 3.

VAR. β, **A. Commelyni** (Willd. in Ges. Naturf. Fr. Berl. Mag. v. 282) ; smaller, with more glaucous concave leaves, 4–5 in. long, with a few prickles on the back; marginal prickles white, deltoid. *Salm-Dyck, Aloe, sect.* xxiv. *fig.* 5. *A. mitræformis var. humilior, Haw. in Trans. Linn. Soc.* vii. 23.

VAR. γ, **A. spinulosa** (Salm-Dyck, Obs. 4; Aloe, sect. xxiv. fig. 6); leaves copiously prickly all over the back, sometimes recurved.

VAR. δ, **A. flavispina** (Haw. in Trans. Linn. Soc. vii. 22 ; Syn. 77); leaves narrower, more lanceolate, scarcely above an inch broad ; prickles straw-coloured. *Salm-Dyck, Aloe, sect.* xxiv. *fig.* 2 ; *Kunth, Enum.* iv. 527.

VAR. ε, **A. xanthacantha** (Willd. in Ges. Naturf. Fr. Berl. Mag. v. 282) ; leaves thicker and more spreading; prickles small, deltoid, straw-yellow. *A. mitræ-formis, Salm-Dyck, Aloe, sect.* xxiv. *fig.* 4.

VAR. ζ, **pachyphylla** (Baker in Journ. Linn. Soc. xviii. 172); leaves thick, spreading, dull purplish on both sides, nearly flat on the face, with only one or two reduced prickles on the back near the top; marginal prickles small, deltoid, the same colour as the leaf.

SOUTH AFRICA: without locality, living cultivated plants!
COAST REGION: New Kloof, near Tulbagh, *MacOwan*, 2985!

Introduced into cultivation at the beginning of the eighteenth century. The type is figured by Dillenius in *Hortus Elthamensis, t.* 17, *fig.* 19.

42. A. albispina (Haw. in Trans. Linn. Soc. vii. 22; Syn. 78); stem simple, elongated, suberect, 1–1½ in. diam. ; leaves laxly disposed, ovate-lanceolate, all ascending, 6–8 in. long, 2 in. broad low down, plain green; back slightly prickly; margin with close, deltoid-cuspidate, whitish, horny prickles ⅛–⅙ in. long; peduncle simple, 1½ ft. long; raceme dense, ½ ft. long, 4 in. diam. ; lower pedicels 1¼–1½ in. long ; bracts lanceolate-deltoid, half as long as the pedicels; perianth cylindrical, bright red, 1½ in. long; tube ⅓ in. long; stamens about as long as the perianth. *Roem. et Schultes, Syst. Veg.* vii. 712 ; *Kunth, Enum.* iv. 527 ; *Baker in Journ. Linn. Soc.* xviii. 172.

SOUTH AFRICA: without locality.

Introduced into cultivation by Masson in 1796. Scarcely more than a variety of *A. mitriformis.*

43. A. nobilis (Haw. Syn. 78) ; stem less elongated and stouter than in *A. mitriformis,* rarely forked; leaves rather lax, ascending, lanceolate, about 1 ft. long, 2½–3 in. broad low down, ¼–⅓ in. thick in the centre, green, without spots or stripes, flat in the lower half; margin with brown, deltoid, horny prickles ⅛–⅙ in. long, a few of which extend to the back of the leaf; peduncle simple, 1½ ft. long ; raceme dense, 6–8 in. long; lower pedicels 1½–2 in. long; bracts oblong, much shorter than the pedicels; perianth cylindrical, bright red, 1¼–1½ in. long; tube ¼–⅓ in. long; segments lanceolate, tipped with green: stamens as long as the perianth; style exserted. *Roem. et Schultes, Syst. Veg.* vii. 713 ; *Kunth, Enum.* iv. 528 ; *Salm-Dyck, Aloe, sect.* xxiv. *fig.* 7. *A. mitræformis var. spinosior, Haw. in Trans. Linn. Soc.* vii. 23.

SOUTH AFRICA: without locality.
Introduced into cultivation about 1800.

44. A. Brownii (Baker in Journ. Bot. 1889, 44); stem tall, simple, below the rather lax rosette of leaves, 2–3 in. thick; leaves lanceolate, about a foot and a half long, 3–4 in. broad low down, plain green, neither spotted nor striped, ¼ in. thick in the middle, flat on the face in the lower third, narrowed gradually from the middle to the pungent tip; margin with close, deltoid-cuspidate, brown-tipped prickles ⅙–¼ in. long; peduncle stout, simple, above a foot long, with many ovate empty bracts ; raceme dense, simple,

4–8 in. long; pedicels $\frac{1}{2}$–$\frac{3}{4}$ in. long; bracts ovate-oblong, nearly as long as the pedicels; perianth bright red-yellow, cylindrical, 1–1$\frac{1}{4}$ in. long, cut down very nearly to the base; stamens slightly exserted; style exserted $\frac{1}{2}$ in. *A. nobilis var. densifolia, Baker in Journ. Linn. Soc.* xviii. 172.

SOUTH AFRICA: without locality.

Has been in cultivation at Kew for many years. My description is taken from a plant that flowered in 1885. I have named it after Mr. N. E. Brown, who was the first to claim for it specific distinctness.

45. A. sigmoidea (Baker in Journ. Linn. Soc. xviii. 177); stem simple, elongated below the dense rosette of leaves; leaves lanceolate, 1–1$\frac{1}{2}$ ft. long, 3 in. broad, $\frac{1}{4}$ in. thick in the centre, dull green, slightly glaucous, unspotted, unarmed on the face, slightly prickly and tubercled towards the apex on the back, flat at the base, recurved at the apex; margin with close, spreading, deltoid-cuspidate prickles $\frac{1}{8}$–$\frac{1}{6}$ in. long; flowers unknown.

EASTERN REGION: Amatonga country, *Cooper !*

Described from a living plant in Mr. Cooper's collection in 1879. Most resembles *A. salmdyckiana* in habit and leaf.

46. A. cæsia (Salm-Dyck, Cat. 29; Aloe, sect. xvii. fig. 3); stem reaching a length of 10–12 ft. below the dense rosette of leaves; leaves 30–40, lanceolate, acuminate, 1–1$\frac{1}{2}$ ft. long, 2–3 in. broad low down, very glaucous, unspotted, not lineate, $\frac{1}{4}$–$\frac{1}{3}$ in. thick in the centre, margined with red when young; marginal prickles reddish, deltoid, $\frac{1}{12}$–$\frac{1}{8}$ in. long; peduncle simple, about a foot long, with many deltoid empty bracts; raceme dense, $\frac{1}{2}$ ft. long; pedicels 1–1$\frac{1}{4}$ in.; bracts deltoid, half as long as the pedicels; perianth cylindrical, bright red, 1$\frac{1}{4}$ in. long; segments about as long as the tube; stamens slightly exserted. *Roem. et Schultes, Syst. Veg.* vii. 706; *Kunth, Enum.* iv. 531; *Baker in Journ. Linn. Soc.* xviii. 172.

SOUTH AFRICA: without locality, living cultivated plants!

Introduced into cultivation about 1815.

47. A. succotrina (Lam. Encyc. i. 85); stem when mature 3–4 ft. long below the dense rosette of leaves, simple or forked; leaves 30–40, ensiform, 1$\frac{1}{2}$–2 ft. long, 1$\frac{1}{2}$–2 in. broad low down, $\frac{1}{4}$ in. thick in the centre, tapering gradually from below the middle to the apex, slightly glaucous, not prickly on the back, sometimes slightly spotted; margin with close, spreading, white, deltoid teeth $\frac{1}{12}$ in. long, turning purple when old; peduncle simple, above a foot long, with several deltoid empty bracts; raceme dense, a foot long; pedicels ascending, $\frac{3}{4}$–1 in. long; bracts oblong, reddish, half as long as the pedicels; perianth cylindrical, reddish, 1$\frac{1}{4}$ in. long; tube very short; stamens scarcely exserted. *Baker in Journ. Linn. Soc.* xviii. 173. *A. soccotrina, DC. Plantes Grasses, t.* 85; *Haw. Syn.* 75; *Salm-Dyck, Aloe, sect.* xxii. *fig.* 1. *A. soccotorina, Kunth, Enum.*

iv. 524.   A. perfoliata var. succotrina, Curt. in Bot. Mag. t. 472.
A. vera, Miller, Gard. Dict. edit. viii. No. 15, non Linn.

SOUTH AFRICA : without locality, living cultivated plants !

Introduced into cultivation at the beginning of the eighteenth century.

48. A. purpurascens (Haw in Trans. Linn. Soc. vii. 20); stem
2–3 ft. long below the dense rosette of leaves, simple or forked;
leaves 40–60 in a dense rosette, ensiform, $1\frac{1}{2}$ ft. long, 2 in. broad at
the base, $\frac{1}{4}$ in. thick in the middle, very glaucous, often with a few
scattered white spots, not striped, tapering gradually from the base to
the apex, turning dark purple when old ; margin with deltoid white
horny prickles $\frac{1}{12}$–$\frac{1}{8}$ in. long ; peduncle stout, simple, $1\frac{1}{2}$ ft. long;
raceme dense, simple, a foot long ; pedicels erecto-patent, $\frac{3}{4}$ in. long;
bracts oblong, purplish, shorter than the pedicels ; perianth straight,
cylindrical, reddish, 1–$1\frac{1}{4}$ in. long; tube very short ; stamens
scarcely exserted.   Haw. Syn. 75 ; Salm-Dyck, Aloe, sect. xxii.
fig. 2 ; Kunth, Enum. iv. 524.   A. sinuata, Thunb. Diss. No. 5, ex
parte.   A. perfoliata var. purpurascens, Ait. Hort. Kew. i. 466.   A.
socotrina var. purpurascens, Gawl. in Bot. Mag. t. 1474.   A. ramosa,
Haw. in Trans. Linn. Soc. vii. 26.

SOUTH AFRICA : without locality, living cultivated plants !

Introduced into cultivation about 1780.   Scarcely more than a variety of
A. succotrina.

49. A. arborescens (Miller, Gard. Dict. edit. viii. No. 3); stem
slender, reaching a length of many feet below the dense rosette of
leaves ; leaves ensiform, very glaucous, $1\frac{1}{2}$–2 ft. long, 2 in. broad,
acuminate, neither spotted nor striped, $\frac{1}{4}$–$\frac{1}{3}$ in. thick in the centre,
margined with close deltoid-cuspidate prickles $\frac{1}{8}$–$\frac{1}{6}$ in. long; peduncle
stout, simple or forked, above a foot long ; racemes dense, $\frac{1}{2}$–1 ft.
long; pedicels about an inch long ; bracts ovate or oblong, shorter
than the pedicels ; perianth bright red, cylindrical, $1\frac{1}{4}$–$1\frac{1}{2}$ in. long;
tube very short ; stamens about as long as the perianth.   DC. Plantes
Grasses, t. 38 ; Haw. Syn. 76 ; Sims in Bot. Mag. t. 1306 ; Andr.
Bot. Rep t. 468 ; Roem. et Schultes, Syst. Veg. vii. 708 ; Salm-
Dyck, Aloe, sect. xxvi. fig. 3 ; Kunth, Enum. iv. 529 ; Baker in
Journ. Linn. Soc. xviii. 175.   A. perfoliata var. arborescens, Ait.
Hort. Kew. i. 466.   A. fruticosa, Lam. Encyc. i. 87.   A. arborea,
Medic. Beobacht. 305.   Catevala arborescens, Medic. Theod. 67.

VAR. β, A. frutescens (Salm-Dyck, Cat. 30) ; stems more slender; leaves
smaller (10–15 in. long, 1–$1\frac{1}{2}$ in. broad), more glaucous and not so dense. Haw.
Suppl. 46 ; Kunth, Enum. iv. 530.

SOUTH AFRICA : without locality, living cultivated plants !

Introduced into cultivation about the year 1700.

50. A. pluridens (Haw. in Phil. Mag. 1824, 299) ; stem simple,
reaching a length of 10–12 ft. below the dense rosette of leaves and
a diameter in cultivation of 4–6 inches ; leaves 30–40 in a rosette,
ensiform, very acuminate, $1\frac{1}{2}$–2 ft. long, $1\frac{1}{2}$–2 in. broad, $\frac{1}{8}$–$\frac{1}{4}$ in.

thick in the middle, bright green, without spots or lines; margin
with crowded deltoid prickles $\frac{1}{12}$–$\frac{1}{8}$ in. long, those of the upper half
conspicuous, ascending; peduncle simple, a foot long, with many
deltoid empty bracts; raceme dense, simple, $\frac{1}{2}$ ft. long; lower pedicels
1–1$\frac{1}{4}$ in. long; bracts deltoid, half as long as the pedicels; perianth
cylindrical, reddish-yellow, 1–1$\frac{1}{4}$ in. long; tube very short; stamens
finally slightly exserted. *Roem. et Schultes, Syst. Veg.* vii. 709 ;
Kunth, Enum. iv. 530; *Baker in Journ. Linn. Soc.* xviii. 176.
A. Atherstonei, Baker in Journ. Linn. Soc. xviii. 170.

COAST REGION: Uitenhage Div.; Stink Poort on stony hills, 1000 ft., *Bolus*,
2688 !

CENTRAL REGION: Somerset Div.; in woods at the foot of Bosch Berg, 3000
ft., *MacOwan*, 1825 !

Introduced into cultivation by Bowie about 1820.

51. A. speciosa (Baker in Journ. Linn. Soc. xviii. 178); habit of
A. africana, reaching a height of 20–25 feet ; leaves in a dense rosette,
15–18 in. long, 2 in. broad low down, ensiform, glaucous, with
smaller and weaker prickles than in *A. africana;* peduncle short,
stout, simple ; raceme very dense, simple, a foot long, 4 in. diam.
when expanded, all except the upper flowers deflexed; pedicels $\frac{1}{6}$–$\frac{1}{4}$
in. long; bracts suborbicular, obtuse, $\frac{1}{2}$ in. long and broad ; perianth
cylindrical, 1$\frac{1}{4}$ in. long, rose-tinted in an early stage only, greyish-
white when mature ; tube scarcely any; stamens and style bright
red, conspicuously exserted.

CENTRAL REGION : Somerset Div. ; precipices near Somerset East and at Hell
Poort, near the Great Fish river, 2500 ft., *MacOwan*, 1922 !

52. A. longiflora (Baker in Gard. Chron. 1888, iv. 756) ; stem
simple, 1$\frac{1}{2}$ ft. long below the rosette of leaves ; rosette spread laxly
over the upper foot; leaves spreading, ensiform, 15–18 in. long,
1$\frac{1}{2}$ in. broad low down, tapering gradually to a long point, unspotted,
glossy green, flat on the face in the lower half ; marginal teeth
crowded, spreading, green, deltoid, $\frac{1}{12}$–$\frac{1}{8}$ in. long; inflorescence
a dense simple raceme 6–8 in. long, longer than its peduncle ; bracts
ovate, scariose, $\frac{1}{4}$ in. long; pedicels erecto-patent, not longer than
the bract; perianth cylindrical, primrose-yellow, 1$\frac{1}{2}$ in. long ; tube
oblong, $\frac{1}{3}$ in. long; segments ligulate, connivent, tipped with green ;
style and stamens exserted $\frac{1}{2}$ inch.

SOUTH AFRICA ? without locality.

Described from a plant that flowered in the Botanic Garden at Glasnevin in
December, 1888.

53. A. chloroleuca (Baker in Gard. Chron. 1877, viii. 38) ; stem
reaching a length of several feet in cultivation below the dense
rosette of leaves, simple, 3–4 in. diam.; leaves ensiform, 1$\frac{1}{2}$–2 ft.
long, 1$\frac{1}{2}$–2 in. broad at the base, $\frac{1}{4}$ in. thick in the middle, tapering
gradually from the base to the recurved apex, slightly glaucous,
unspotted ; margin with deltoid-cuspidate brown horny prickles

⅛ in. long; peduncle stout, branched; racemes dense, about a foot
long; pedicels ascending, ¼–⅓ in. long; bracts ovate, acute, about as
long as the pedicels; perianth cylindrical, yellowish-white, about
an inch long; tube very short; stamens and style conspicuously
exserted. *Baker in Journ. Linn. Soc.* xviii. 177.

SOUTH AFRICA? without locality.

Described from a living plant that flowered in the Kew collection in 1877.

54. A. platylepis (Baker in Gard. Chron. 1877, viii. 38); stem
reaching a length of 10–12 ft. in cultivation below the dense rosette
of leaves, 3–4 in. diam., simple or forked; leaves ensiform,
1½ ft. long, 3 in. broad at the base, ¼–⅓ in. thick in the centre,
tapering gradually from the base to the recurved apex, very glaucous,
unspotted, not prickly either on the back or face; margin with
deltoid-cuspidate brown horny prickles ⅛–⅙ in. long; peduncle
stout, sometimes with 3–6 branches; racemes dense, ½–1 ft. long;
pedicels ascending, ¼–⅓ in. long; bracts ovate-oblong, as long as the
pedicels; perianth cylindrical, 1–1¼ in. long, pale-red or yellow;
tube short, campanulate; stamens and style distinctly exserted.
Baker in Journ. Linn. Soc. xviii. 177.

SOUTH AFRICA? without locality.

Described from living plants in the Kew collection in 1877.

55. A. fulgens (Todaro, Hort. Bot. Panorm. ii. 40, t. 33); stem
simple, elongated; leaves lanceolate, densely rosulate, spreading,
glaucous, about a foot long, 3 in. broad low down, unspotted;
margin with close, brown, deltoid-cuspidate teeth; peduncle about
2 ft. long, bearing about three dense long racemes; pedicels
erecto-patent, ⅓–½ in. long; bracts ovate, about as long as the
pedicels; perianth cylindrical, deep red, about an inch long; tube
short, campanulate; stamens much exserted.

SOUTH AFRICA? without locality.

Cultivated in the Botanic Garden at Palermo.

56. A. drepanophylla (Baker in Gard. Chron. 1875, iii. 814);
stem slender, simple, reaching in cultivation a length of 7–8 ft.
below the dense rosette of leaves; leaves ensiform, slightly falcate,
1½–2 ft. long, 2–2½ in. broad low down, ¼ in. thick in the middle,
tapering gradually from the base to the apex, very glaucous,
without either spots or lines, not at all prickly on the
surfaces; margin with small red-brown horny prickles; peduncle
short, simple; raceme dense, a foot long; pedicels ascending, ½ in.
long; bracts oblong, as long as the pedicels; perianth cylindrical,
an inch long, red-tinted only at first, whitish when mature; tube
scarcely any; stamens much exserted. *Baker in Journ. Linn. Soc.*
xviii. 176.

COAST REGION: Alexandria Div.; Zuurberg Range, *Cooper!*

CENTRAL REGION: Somerset Div., *Cooper!*

Described from living plants, introduced into cultivation about 1860 by
T. Cooper.

57. A. salmdyckiana (Schultes fil. in Roem. et Schultes, Syst. Veg. vii. 710); stem reaching a length of 10–12 ft. below the dense rosette of leaves; leaves 30–40 in a rosette, ensiform, $1\frac{1}{2}$–2 ft. long, 3 in. broad low down, $\frac{1}{4}$–$\frac{1}{3}$ in. thick in the centre, narrowed gradually from the base to the apex, dull green, hardly at all glaucous when mature, neither spotted nor striped, slightly prickly on the keel towards the apex, not prickly on the face; margin with deltoid-cuspidate, brown, horny prickles $\frac{1}{8}$–$\frac{1}{6}$ in. long; peduncle simple, above a foot long; raceme dense, simple, a foot long; pedicels 1–$1\frac{1}{2}$ in. long, cernuous; bracts oblong, shorter than the pedicels; perianth bright red, cylindrical, $1\frac{1}{4}$–$1\frac{1}{2}$ in. long; tube short, campanulate; stamens exserted $\frac{1}{4}$–$\frac{1}{2}$ in. *Salm-Dyck, Aloe, sect.* xxvii. *fig.* 1; *Kunth, Enum.* iv. 530; *Baker in Journ. Linn. Soc.* xviii. 177. *A. africana, Salm-Dyck, Cat.* 31, *non Miller. Pachidendron Principis, Haw. Revis.* 37.

SOUTH AFRICA: without locality.

Originally described from the Vienna collections in 1815.

58. A. nitens (Baker in Journ. Linn. Soc. xviii. 170); stem simple, elongated below the dense rosette of leaves; leaves ensiform, 18–20 in. long, 2 in. broad low down, bright green, neither spotted nor striped, $\frac{1}{4}$–$\frac{1}{3}$ in. thick in the middle; margin with brown, horny, deltoid-cuspidate prickles $\frac{1}{8}$ in. long; flowers unknown.

SOUTH AFRICA: without locality.

Sent alive to the Kew collection by Sir H. Barkly in 1878.

59. A. dichotoma (Linn. fil. Suppl. 206); arborescent, with a very much branched trunk 20–30 ft. long, 3–4 ft. diam.; leaves in a dense rosette at the tip of the branches, lanceolate, 6–8 in. long, 1–$1\frac{1}{4}$ in. broad low down, $\frac{1}{4}$–$\frac{1}{3}$ in. thick in the centre, very glaucous, unspotted, nearly flat on the face, rounded on back, sometimes smooth, sometimes studded over with copious minute prickles; margin with small, white, deltoid teeth; peduncle short, stout, branched; racemes lax, 3–6 in. long, with a stout, sulcate rachis; pedicels ascending, $\frac{1}{4}$–$\frac{1}{3}$ in. long; bracts minute, lanceolate; perianth oblong, bright yellow, about an inch long; tube much shorter than the segments, the three outer of which are valvate in bud; style and stamens distinctly exserted. *Thunb. Diss. No.* 1; *Fl. Cap. edit.* ii. 309; *Paterson, Travels, tt.* 2–5; *Haw. Syn.* 75; *Kunth, Enum.* iv. 534; *J. C. Brown in Gard. Chron.* 1873, 712, *fig.* 137; *Dyer in Gard. Chron.* 1874, i. 567, *figs.* 118, 121; *Baker in Journ. Linn. Soc.* xviii. 178. *Rhipidodendron dichotomum, Willd. in Ges. Naturf. Fr. Berl. Mag.* v. 166.

WESTERN REGION: Great Namaqualand, *Schinz,* 1! Little Namaqualand; near Ookiep, *MacOwan,* 2257! and *MacOwan and Bolus, Herb. Norm.,* 800! Modder Fontein, *Whitehead!* Bockland, *Thunberg!* and without precise locality, *Barkly!*

Re-introduced into English gardens by Sir H. Barkly in 1877. Flowered in Cape botanic garden in 1878.

60. A. Bainesii (Dyer in Gard. Chron. 1874, i. 568, figs. 119–120);
arborescent, with a trunk reaching a height of 50–60 ft. and a
diameter of 4–5 ft.; diameter of leafy head 20 ft. or more; leaves
ensiform, 2–3 ft. long in the young state, but much shorter in the
mature plant, $1\frac{1}{2}$–2 in. broad low down, tapering gradually to the
apex, green, unspotted, not prickly on the keel towards the tip;
margin with distant, deltoid, horny prickles $\frac{1}{12}$–$\frac{1}{8}$ in. long; peduncle
and rachis very stout and woody; racemes dense, oblong, 3–4 in.
long; pedicels stout, erecto-patent, $\frac{1}{8}$–$\frac{1}{6}$ in. long; bracts minute,
lanceolate; perianth oblong, $1\frac{1}{2}$ in. long, $\frac{1}{2}$ in. diam., salmon-pink
fading into whitish, tipped with green; tube campanulate; outer
segments valvate; stamens and style much exserted. *Baker in
Journ. Linn. Soc.* xviii. 178; *Bot. Mag. t.* 6848.

VAR. β, **A. Barberæ** (Dyer, loc. cit. fig. 122); leaf broader, with smaller and
more distant prickles than in the type. *A. Zeyheri, Hort. non Salm-Dyck.*

EASTERN REGION: Natal, *Baines!* Var. β, Griqualand East; Shawbury,
Mrs. Barber! MacOwan, 2528! *Baur!*

Flowered in the Cape botanic garden in June, 1878, and again in 1884.

61. A. falcata (Baker in Journ. Linn. Soc. xviii. 181); habit
of growth not known; leaves lanceolate, much curved, 8–12 in. long,
$1\frac{1}{4}$–$1\frac{1}{2}$ in. broad low down, tapering gradually to a brown, horny apex,
unspotted; margin with deltoid-cuspidate, brown, horny prickles
$\frac{1}{8}$–$\frac{1}{6}$ in. long; peduncle short, stout, bearing 3–6 branches; racemes
dense, 4–8 in. long; pedicels ascending, $\frac{1}{3}$–$\frac{1}{2}$ in. long; bracts ovate-
cuspidate, rather longer than the pedicels; perianth cylindrical, an
inch long; tube cylindrical; segments oblong, $\frac{1}{4}$ in. long; stamens as
long as the perianth; style finally exserted $\frac{1}{4}$ in.

CENTRAL REGION: Calvinia Div.; near Kamos, Komseep, and Springbok
Kuil, 2000–3000 ft., *Zeyher,* 1678!

Known to me from dried specimens only.

62. A. ferox (Miller, Gard. Dict. edit. viii. No. 22); stem simple,
reaching in cultivation a length of 10–15 ft. and a diameter of 4–6
in. below the dense rosette of leaves; leaves 30–50, densely aggre-
gated, lanceolate, $1\frac{1}{2}$–2 ft. long, 4–6 in. broad, $\frac{1}{3}$–$\frac{1}{2}$ in. thick in the
middle, very rigid in texture, dull in colour and very glaucous,
pungent at the apex, furnished with copious prickles on both back
and face; margin with brown-tipped, deltoid-cuspidate prickles
$\frac{1}{8}$–$\frac{1}{6}$ in. long; peduncle stout, much branched; racemes very dense,
a foot or more long; pedicels very short; bracts ovate, acute, $\frac{1}{4}$ in.
long; perianth red, clavate, $1\frac{1}{4}$ in. long; tube short; stamens and
style bright red, much exserted. *Lam. Encyc.* i. 87; *DC. Plantes
Grasses, t.* 32; *Haw. in Trans. Linn. Soc.* vii. 21; *Syn.* 76; *Sims in
Bot. Mag. t.* 1975; *Salm-Dyck, Aloe, sect.* xxvii. *fig.* 5; *Gard.
Chron.* 1875, iii. 243, *fig.* 44. *A. perfoliata, var. ferox, Ait. Hort.
Kew.* i. 467. *A. muricata* and *A. horrida, Haw. in Trans. Linn. Soc.*
vii. 25 & 27. *Pachidendron ferox, Haw. Revis.* 38.

VAR. β, **A. subferox** (Spreng. Syst. Veg. ii. 73); leaves less prickly on the face.

A. pseudo-ferox, Salm-Dyck, Cat. 31.   *Pachidendron pseudo-ferox, Haw. Revis.* 38.

VAR. γ, **incurvata** (Baker in Journ. Linn. Soc. xviii. 180); leaves thinner, remarkably incurved, not prickly on the face.

SOUTH AFRICA: without locality, living cultivated plants!

Introduced into cultivation at the beginning of the eighteenth century. *Var. incurvata,* at present known only from young plants in Mr. Cooper's collection, will probably prove a distinct species.

63. A. africana (Miller, Gard. Dict. edit. viii. No. 4); stem reaching a length of 20 ft. below the dense rosette of leaves; leaves ensiform, $1\frac{1}{2}$–2 ft. long, $2\frac{1}{2}$–3 in. broad low down, $\frac{1}{3}$–$\frac{1}{2}$ in. thick in the centre, tapering gradually from the base to the apex, very glaucous, scarcely at all recurved, without prickles on the face and with only a few towards the tip on the back; margin with copious, deltoid-cuspidate, brown, horny prickles $\frac{1}{6}$ in. long; peduncle stout, simple, a foot long, with copious, ovate, empty bracts; raceme dense, a foot long; pedicels very short; bracts ovate, $\frac{1}{4}$–$\frac{1}{3}$ in. long; perianth bright yellow tipped with green, much recurved, $1\frac{1}{4}$–$1\frac{1}{2}$ in. long; tube cylindrical; segments as long as the tube; stamens and style conspicuously exserted. *Haw. in Trans. Linn. Soc.* vii. 21; *Syn.* 76; *Roem. et Schultes, Syst. Veg.* vii. 709; *Salm-Dyck, Aloe, sect.* xxvii. *fig.* 2; *Kunth, Enum.* iv. 532; *Baker in Journ. Linn. Soc.* xviii. 180. *A. africana var. angustior, Sims in Bot. Mag.* t. 2517. *A. Bolusii, Baker, l.c.* 179. *Pachidendron africanum, Haw. Revis.* 36.

SOUTH AFRICA: without locality, living cultivated plants!

Introduced into cultivation early in the eighteenth century.

64. A. supralævis (Haw. in Trans. Linn. Soc. vii. 22; Syn. 77); stem 5–6 ft. long in cultivation when mature below the dense rosette of leaves; leaves ensiform, $1\frac{1}{2}$–2 ft. long, $2\frac{1}{2}$–3 in. broad low down, $\frac{1}{3}$–$\frac{1}{2}$ in. thick in the centre, tapering gradually from below the middle to the pungent apex, unspotted, glaucous, with a few horny prickles on the face and upper part of the keel beneath; margin with close, deltoid-cuspidate, horny prickles $\frac{1}{8}$–$\frac{1}{6}$ in. long; peduncle stout, branched; racemes dense, $\frac{1}{2}$–1 ft. long; pedicels very short; bracts ovate, acute, $\frac{1}{4}$–$\frac{1}{3}$ in. long; perianth reddish, cylindrical, $\frac{7}{8}$–1 in. long; tube as long as the segments; stamens and style much exserted. *Roem. and Schultes, Syst. Veg.* vii. 711; *Salm-Dyck, Aloe, sect.* xxvii. *fig.* 6; *Kunth, Enum.* iv. 533; *Baker in Journ. Linn. Soc.* xviii. 180. *Pachidendron supralæve, Haw. Revis.* 39.

VAR. β, **Hanburii** (Baker); leaves without any prickles on the upper surface; pedicels nearly obsolete; flowers yellow.

SOUTH AFRICA: without locality. Var. β, *Hort. Hanbury!*

Introduced into cultivation at the beginning of the eighteenth century.

65. A. rupestris (Baker); leaves ensiform, recurved, sickle-shaped, glaucous, $1\frac{1}{2}$ ft. long, 2 in. broad low down, narrowed gradually to the point, not prickly on either surface; marginal teeth small, deltoid, with brown horny tip; inflorescence much branched,

candelabriform; racemes dense, 6–8 in. long, above 2 in. diam.;
pedicels very short; bracts very small; perianth reddish-yellow,
slightly recurved, $\frac{3}{4}$ in. long; tube scarcely any; segments linear-
oblong, with a distinct green keel; stamens nearly twice as long as
the perianth. *A. pycnantha, MacOwan herb.*

WESTERN REGION: Little Namaqualand, in rocky ground between Port
Nolloth and Spektakel, *MacOwan, Herb. Austr. Afr.*, 1556.

66. **A. Thraskii** (Baker in Journ. Linn. Soc. xviii. 180); stem
simple in the cultivated plant below the dense rosette of leaves,
3–4 in. diam., very tuberous at the base; leaves ensiform, 2 ft. long,
$2\frac{1}{2}$–3 in. broad low down, $\frac{1}{4}$–$\frac{1}{3}$ in. thick in the centre, slightly
glaucous, unspotted, much recurved, channelled down the face all
the way to the base; margin with small brown-tipped, deltoid-
cuspidate prickles; peduncle stout, simple, about a foot long; raceme
dense, 4–6 in. long; pedicels very short; bracts ovate, acute, $\frac{1}{4}$–$\frac{1}{3}$ in.
long; perianth yellow, slightly recurved, an inch long; tube very
short; segments obtuse; stamens and style much exserted.

KALAHARI REGION: Orange Free State, *Cooper!*

Described from a living plant introduced into cultivation by Cooper about 1860.
It is mentioned by name in Gard. Chron. 1876, v. 400, and is the central smooth-
leaved Aloe in a figure of a group of species made from Mr. J. Peacock's
collection (t. 75).

67. **A. variegata** (Linn. Sp. Plant. 321); rosette of leaves sessile, spirally trifarious, 6·9 in. long, 4–5 in. diam.; leaves 6–8 in a
row, dense, erecto-patent, lanceolate-navicular, 4–5 in. long, an inch
broad, concave on the face, acutely keeled, bright green, with irregular cross bands of confluent oblong whitish spots; margins whitish,
horny and denticulate; peduncle simple, terete, 6–8 in. long, with
2–3 empty bracts; raceme lax, simple, oblong, 3–4 in. long; pedicels
erecto-patent, reddish, $\frac{1}{4}$–$\frac{1}{3}$ in. long; bracts lanceolate-deltoid, as long
as the pedicels; perianth cylindrical, reddish, $1\frac{1}{4}$ in. long; tube
elongated; segments oblong, $\frac{1}{4}$–$\frac{1}{3}$ in. long; stamens just exserted.
Miller, Gard. Dict. edit. viii. *No.* 9; *Thunb Diss. No.* 12; *Bot. Mag.
t.* 513; *DC. Plantes Grasses, t.* 21; *Haw. Syn.* 81; *Salm-Dyck,
Aloe,* sect. xx. *fig.* 2; *Kunth, Enum.* iv. 523; *Baker in Journ. Linn.
Soc.* xviii. 179. *A. punctata, Haw. in Trans. Linn. Soc.* vii. 26.

SOUTH AFRICA: without locality. A note in Burchell's MS. catalogue at
Kew, under No. 2239/2, states that this species grows plentifully on the Pellat
Plains, near Kuruman, in Bechuanaland, but he did not dry any specimens.

Introduced into cultivation at the beginning of the eighteenth century.
Figured by Commelyn in his *Præludia, t.* 28.

68. **A. plicatilis** (Miller, Gard. Dict. edit. viii. No. 7); stem
shrubby, very much branched, reaching even in cultivation a height
of 10–12 ft. and a thickness of a foot; leaves 10–30, distichous,
ligulate, obtuse, 6–9 in. long, $1\frac{1}{4}$–$1\frac{1}{2}$ in. broad, spreading, very
glaucous, unspotted, with a narrow whitish margin denticulate
towards the tip; peduncle simple, about a foot long; raceme
20–30-flowered, $\frac{1}{2}$ ft. long; pedicels $\frac{1}{2}$–$\frac{3}{4}$ in. long; bracts small, deltoid;

perianth cylindrical, reddish-yellow, 1½–1¾ in. long; inner segments
nearly free; outer united to the middle; stamens as long as the
perianth. *Curt. Bot. Mag. t.* 457; *DC. Plantes Grasses, t.* 75;
Jacq. Hort. Schoenbr. t. 423; *Haw. Syn.* 74; *Salm-Dyck, Aloe,
sect.* xxviii. *fig.* 1. *A. disticha, var. plicatilis, Linn. Sp. Plant. edit.*
1, i. 321. *Kumara disticha, Medic. Theod.* 70, *t.* 4. *A. tripetala,
Medic. Beobacht.* 55. *Rhipidodendron distichum, Willd. in Ges.
Naturf. Fr. Berl. Mag.* v. 165. *R. plicatile, Haw Revis.* 45.

VAR. β, **major** (Salm-Dyck); more robust, with leaves a foot or more long,
above 2 in. broad, and stouter branches and peduncles.

COAST REGION : On rocks behind Tulbagh Waterfall, *MacOwan*, 2259! and
on mountains of New Kloof, near Tulbagh, 850 ft., *MacOwan, Herb. Aust.
Afr.*, 1555!

Introduced into cultivation at the beginning of the eighteenth century.

69. A. desmetiana (Hort.); leaves about 40 in a dense rosette,
lanceolate, 4–5 in. broad, 1¼ in. broad low down, narrowed gradually
to the point, rounded on the back and furnished with copious oblong
greenish-white tubercles, which are less numerous on the face, which
is flat in the lower half; marginal teeth close, white, very minute;
inflorescence a simple lax raceme a foot long; pedicels erecto-patent,
¾–1 in. long; bracts lanceolate-deltoid, as long as the pedicels;
perianth cylindrical, 1¼ in. long; tube scarcely any; stamens not
exserted.

A garden hybrid between *A. variegata* and *A. echinata minor.* Described from
a plant flowered by Mr. J. Corderoy in April, 1891.

70. A. insignis (N. E. Brown in Gard. Chron. 1885, xxiv. 40,
fig. 8); shortly caulescent; leaves 30–40 in a dense rosette, ascending,
lanceolate, 7–11 in. long, 1–1½ in. broad low down, glaucous, flat on
the face with a few raised whitish tubercles, which are more numerous
on the rounded back; marginal teeth deltoid, minute; peduncle
simple, longer than the leaves; raceme moderately dense, a foot
long; pedicels ¾–1 in. long; bracts ovate, as long as the pedicels;
perianth cylindrical, 1¼ in. long, bright red in an early stage; tube
scarcely any; stamens exserted.

A garden hybrid between *A. echinata* and *A. drepanophylla*, raised by Mr. T.
Cooper in 1875. It flowered for the first time in 1885.

IX. APICRA, Willd.

Perianth oblong-cylindrical, with a straight tube and 6 short
oblong subequal segments, spreading at the tip, with 3 green stripes
down the keel. *Stamens* 6, hypogynous, shorter than the perianth;
filaments filiform; anthers oblong, small, versatile, dehiscing introrsely. *Ovary* sessile, oblong-trigonous; ovules numerous, superposed; style short, subulate; stigma capitate. *Capsule* oblong-trigonous, coriaceous, loculicidally 3-valved. *Seeds* compressed; testa
brown; albumen fleshy.

Leafy stem always elongated; leaves short, thick, fleshy, multifarious or quinquefarious; flowers small, whitish, arranged in simple or compound lax sub-spicate racemes; pedicels short, ascending; bracts small, ovate.

DISTRIB. Endemic.

Leaves arranged in five straight or spirally twisted rows:
  Leaves lanceolate-deltoid ... ... (1) **pentagona.**
  Leaves deltoid:
    Upper leaves flat on the face ... ... (2) **turgida.**
    Upper leaves concave on the face ... ... (3) **deltoidea.**
  Leaves multifarious, the spirals quite obliterated:
    Perianth smooth ... ... (4) **spiralis.**
    Perianth rugose:
      Leaves smooth on both back and face:
        Leaves deltoid ... . (5) **foliolosa.**
        Leaves lanceolate-deltoid ... ... (6) **congesta.**
      Leaves rugose with tubercles on the back:
        Leaves lanceolate-deltoid, ½ in. thick... ... (7) **bicarinata.**
        Leaves deltoid, ¼–½ in. thick ... ... (8) **aspera.**

1. **A. pentagona** (Willd. in Ges. Naturf. Fr. Berl. Mag. v. 273); leafy stem ½–1 ft. long, 2½–3 in. diam., leaves included; leaves lanceolate-deltoid, arranged in five regular rows, 1¼–1½ in. long, ½–¾ in. broad, ¼–⅓ in. thick, bright green, flat on the face, rounded on the back, scabrous on the margins, with two obscure keels and a few scattered whitish tubercles on the back, the lower spreading, the upper ones ascending; peduncle 1½ ft. long including inflorescence, simple or forked; racemes 6–9 in. long; pedicels erecto-patent, ⅙–¼ in. long; bracts small, ovate; perianth ½–⅝ in. long, smooth on the outside. *Haw. Suppl.* 62; *Baker in Journ. Linn. Soc.* xviii. 217. *Aloe pentagona, Haw. in Trans. Linn. Soc.* vii. 7; *Ker in Bot. Mag. t.* 1338; *Jacq. Fragm. t.* 111; *Salm-Dyck, Aloe, sect.* i. *fig.* 4. *Haworthia pentagona, Haw. Syn.* 97.

VAR. β, **spirella** (Baker, loc. cit.); leaves smaller, more deltoid, about an inch long, irregularly quinquefarious. *Haworthia spirella, Haw. Syn.* 97. *Aloe spirella, Salm-Dyck, Aloe, sect.* i. *fig.* 3.

VAR. γ, **A. bullulata** (Willd. in Ges. Naturf. Fr. Berl. Mag. v. 273); leaves irregularly quinquefarious, with more numerous tubercles on the back. *Haw. Suppl.* 62. *Aloe bullulata, Jacq. Fragm. t.* 109.

VAR. δ, **Willdenowii** (Baker, loc. cit.); more robust; leaves about 2 in. long, arranged in five spirally twisted rows. *Apicra spiralis, Willd. in Ges. Naturf. Fr. Berl. Mag.* v. 273. *Haworthia spiralis, Haw. Syn.* 97. *Aloe spiralis, Haw. in Trans. Linn. Soc.* vii. 7; *Salm-Dyck, Aloe, sect.* i. *fig.* 5, *non Linn.*

SOUTH AFRICA: without locality, living cultivated plants!

Introduced into cultivation about the beginning of the present century.

2. **A. turgida** (Baker in Journ. Bot. 1889, 44); leafy stem 6–9 in. long, 2–2½ in. diam., leaves included; leaves arranged in five spirally twisted rows, deltoid, an inch long, ¾ in. broad, smooth in the face, scabrous on the margin, quite free from spots or tubercles, the lower spreading, dull green, tinged on the face, rounded on the back, ¼–⅓ in. thick in the middle, the upper pale green with several indistinct vertical ribs of darker green, flat on the face; flowers not seen.

COAST REGION: Albany Div., *Hutton!*

Introduced into cultivation in 1872. Described from living plants in the Kew collection.

A near ally·of *A. deltoidea.*

3. **A. deltoidea** (Baker in Journ. Linn. Soc. xviii. 217); leafy stem $\frac{1}{8}$–1 ft. long, 2 in. diam., leaves included : leaves arranged in five regular rows, all except the uppermost spreading, deltoid, $\frac{3}{4}$–1 in. long, $\frac{1}{6}$–$\frac{1}{4}$ in. thick in the middle, bright shining green, slightly concave on the face, except in the oldest leaves, rounded on the back, distinctly keeled towards the apex, serrulate on the keel and edges, entirely without spots or tubercles; peduncle $1\frac{1}{2}$ ft. long including inflorescence ; racemes 1–4 ; pedicels very short ; bracts lanceolate-deltoid ; perianth $\frac{1}{2}$ in. long, smooth. *Aloe (Apicra) deltoidea, Hook. fil. in Bot. Mag. t.* 6071.

COAST REGION : Alexandria Div. ; Zuurberg Range, stony places at Hell Poort, 2000 ft., *Bolus,* 2687 !

Described from living cultivated plants introduced by Cooper about 1860.

4. **A. spiralis** (Baker in Journ. Linn. Soc. xviii. 217) ; leafy stem $\frac{1}{2}$–1 ft. long, $1\frac{1}{2}$–$1\frac{3}{4}$ in. diam., leaves included ; leaves multifarious, lanceolate-deltoid, 1–$1\frac{1}{4}$ in. long, $\frac{1}{2}$–$\frac{3}{4}$ in. broad, bright shining green, smooth, flat on the face, rounded on the back, obscurely crenulate on the margin, all except the lowest ascending ;· peduncle $1\frac{1}{2}$ ft. long including inflorescence, simple or branched ; racemes $\frac{1}{2}$ ft. long ; pedicels very short ; bracts small, ovate-lanceolate ; perianth rugose, $\frac{1}{2}$ in. long. *Aloe spiralis, Linn. Sp. Plant.* 322 ; *DC. Plantes Grasses, t.* 56 ; *Jacq. Fragm. t.* 110 ; *Ker in Bot. Mag. t.* 1455. *Aloe imbricata, Haw. in Trans. Linn. Soc.* vii. 7 ; *Salm-Dyck, Aloe, sect.* i. *fig.* 1. *Haworthia imbricata, Haw. Syn.* 98. *Apicra imbricata, Willd. in Ges. Naturf. Fr. Berl. Mag.* v. 273.

SOUTH AFRICA : without locality, living cultivated plants !

Introduced into cultivation at the beginning of the eighteenth century. It is figured both by Commelyn, *Præludia, t.* 32, and Dillenius, *Hortus Elthamensis, t.* 13, *fig.* 14.

5. **A. foliolosa** (Willd. in Ges. Naturf. Fr. Berl. Mag. v. 274) ; leafy stem $\frac{1}{2}$–1 ft. long, 1–$1\frac{1}{4}$ in. diam. including leaves ; leaves crowded, multifarious, deltoid, squarrose, $\frac{1}{2}$–$\frac{3}{4}$ in. long and broad, $\frac{1}{8}$–$\frac{1}{6}$ in. thick in the centre, dull green, flat on the face, rugose on the margin, obliquely keeled on the back, without spots or tubercles ; peduncle slender, simple, above a foot long ; raceme lax, 6–9 in. long ; lower pedicels $\frac{1}{6}$–$\frac{1}{4}$ in. long ; bracts small, ovate-lanceolate ; perianth smooth, $\frac{1}{2}$ in. long. *Haw. Suppl.* 64 ; *Baker in Journ. Linn. Soc.* xviii. 218. *Haworthia foliolosa, Haw. Syn.* 99. *Aloe foliolosa, Haw. in Trans. Linn. Soc.* vii. 7 ; *Ker in Bot. Mag. t.* 1352 ; *Salm-Dyck, Aloe, sect.* ii. *fig.* 4 ; *Kunth, Enum.* iv. 495.

SOUTH AFRICA : without locality.

Introduced into cultivation by Masson about 1795.

6. **A. congesta** (Baker in Journ. Linn. Soc. xviii. 218) ; leafy stem
½–1 ft. long, 3 in. diam. including leaves ; leaves multifarious, lanceolate deltoid, 1½–1¾ in. long, an inch broad, flat on the face, rounded
on the back, ¼–⅓ in. thick in the middle, irregularly keeled, scabrous
on the edges, with spots or tubercles ; peduncle simple, 1½ ft. long
including the lax raceme ; pedicels very short ; bracts deltoid-cuśpi-
date ; perianth smooth, ½–⅝ in. long. *Aloe congesta, Salm-Dyck,
Aloe, sect.* ii. *fig.* 1.

SOUTH AFRICA : without locality, living cultivated plants !
First described from living plants at Berlin in 1843.

7. **A. bicarinata** (Haw. Suppl. 63) ; leafy stem ½–1 ft. long,
1½–1¾ in. diam. including leaves ; leaves multifarious, lanceolate-
deltoid, ¾–1 in. long, ½ in. broad, dull green, concave on the smooth
face, ⅙ in. thick in the centre, rough with large tubercles with whitish
tips arranged in indistinct, vertical, and transverse rows, and furnished with two indistinct keels towards the apex on the back ;
flowers not seen. *Baker in Journ. Linn. Soc.* xviii. 219. *Aloe
bicarinata, Roem. et Schultes, Syst. Veg.* vii. 652 ; *Kunth, Enum.*
iv. 496.

SOUTH AFRICA : without locality.
Described from a drawing of Haworth's type made at Kew in the year 1818.

8. **A. aspera** (Willd. in Ges. Naturf. Fr. Berl. Mag. v. 274) ;
leafy stem 4–6 in. long, about 1 in. diam. including leaves ; leaves
multifarious, deltoid, ½–⅝ in. long and broad, flat and smooth on the
face, ¼–⅓ in. thick, rounded on the back and rough with concolorous tubercles ; margins crenulate ; peduncle slender, simple, about
a foot long including the raceme ; lower pedicels ¼–⅓ in. long ;
bracts small, lancèolate-deltoid ; perianth ½ in. long, smooth. *Haw.
Suppl.* 63 ; *Baker in Journ. Linn. Soc.* xviii. 218. *Haworthia
aspera, Haw. Syn.* 90. *Aloe aspera, Haw. in Trans. Linn. Soc.* vii.
6 ; *Salm-Dyck, Aloe, sect.* ii. *fig.* 2 ; *Kunth, Enum.* iv. 496.

VAR. β, **major** (Haw. Suppl. 63) ; more robust ; leaves ¾–1 in. long, not quite
so thick, the face flat in the old ones, rather concave in the young ones ; tubercles
either concolorous or tipped with white. *Baker, loc. cit.* 219.

SOUTH AFRICA : without locality, living cultivated plants !
Introduced into cultivation by Masson about 1795.

X. **HAWORTHIA**, Duval.

Perianth oblong-cylindrical, with a straight tube and a bilabiate
limb with six subequal oblong segments, the three lower reflexing
more than in the three upper. *Stamens* 6, hypogynous, shorter than
the perianth ; filaments filiform ; anthers small, oblong, versatile,
dehiscing introrsely. *Ovary* sessile, oblong-trigonous ; ovules
numerous, superposed ; style subulate ; stigma capitate. *Capsule*

oblong-trigonous, chartaceous, loculicidally 3-valved. *Seeds* compressed, acutely angled ; testa brown, membranous ; albumen fleshy ; embryo axile.

Leafy stem short or elongated; leaves short, thick, fleshy, generally multifarious, often tubercled, sometimes toothed or ciliated on the edges and keel ; flowers small, whitish, ribbed with green or reddish-brown, arranged in simple or panicled racemes ; pedicels short, ascending ; bracts small.

DISTRIB. A single species in Angola.

SERIES 1. Leafy stem elongated.

1. TRIQUETRÆ. Leaves regularly trifarious, unspotted.

Leaves ¼ in. thick, the old ones flat on the face :
    Leaves 1¼–1¾ in. long ... ... (1) **cordifolia.**
    Leaves under an inch long ... ... (2) **asperiuscula.**
Leaves ⅛ in. thick, all concave on the face ... ... (3) **viscosa.**

2. TORTUOSÆ. Leaves spirally trifarious, rough, with raised tubercles.

Leaves rough, with small concolorous tubercles ... (4) **tortuosa.**
Leaves very rough, with larger white-tipped tubercles .. (5) **subrigida.**

3. PAPILLOSÆ. Leaves multifarious, with raised or immersed spots.

Tubercles white, prominently raised :
    Leaves 3–4 in. long ... ... (6) **papillosa.**
    Leaves 1–1½ in. long ... ... (7) **Reinwardtii.**
Tubercles greenish-white, prominently raised ... ... (8) **Cassytha.**
Tubercles whitish, slightly raised :
    Leaves 2–2½ in. long ... ... (9) **coarctata.**
    Leaves 1¼–1½ in. long ... ... (10) **Greenii.**
Tubercles greenish, immersed ... ... (11) **Peacockii.**

4. HYBRIDÆ. Leaves multifarious, unspotted.

Leaves rough, with raised papillæ :
    Papillæ larger, tipped white ... . (12) **hybrida**
    Papillæ smaller, concolorous :
      Old leaves red-brown ... ... (13) **rigida.**
      Old leaves blackish ... ... (14) **nigra.**
Leaves smooth, pale green, obscurely lineate on the
  back ... ... (15) **glauca.**

SERIES 2. Leafy stem not elongated; leaves multifarious, not ciliated on the margin.

5. MARGARITIFERÆ. Leaves entire, furnished with raised white tubercles.

Tubercles of the leaf confluent into horizontal bands :
  Bands contiguous and regular :
    Leaf very acuminate, 2¼–3 in. long ... ... (16) **attenuata.**
    Leaf less acuminate, 1¼–1½ in. long ... ... (17) **fasciata.**
  Bands more distinct and less regular ... ... (18) **subfasciata.**
Tubercles not confluent into regular bands :
  Tubercles large and middle-sized, not very close :
    Tubercles abundant on both sides of the leaf... (19) **margaritifera.**
    Tubercles mainly confined to the back of the
    leaf :
      Tubercles contiguous :
        Leaf 3–3¼ in. long ... ... (20) **semiglabrata.**
        Leaf 2–3 in. long ... ... (21) **subattenuata.**
      Tubercles fewer and more distant ... (22) **glabrata.**
    Tubercles small, crowded on both sides of the
    leaf :
      Tubercles very minute ... ... (23) **Radula.**

Tubercles not so minute :
 Tubercles of upper surface green ... (24) **subulata.**
 Tubercles of both surfaces tipped
  with white ... ... (25) **rugosa.**

6. VIRESCENTES. Leaves firm in texture, neither lineate, nor distinctly tubercled, nor conspicuously recurved.

Leaves smooth on both surfaces ... ... (26) **albicans.**
Leaves scabrous on both surfaces :
 Leaves 10–12 to a rosette :
  Leaves 1½–2 in. long .. .. (27) **scabra.**
  Leaves 3–4 in. long ... ... (28) **sordida.**
 Leaves about 20 to a rosette ... ... (29) **icosiphylla.**
 Leaves 30–40 to a rosette ... ... (30) **Tisleyi.**

7. RECURVÆ. Leaves short, very thick, much recurved.

Leaves rigid, scabrous on the back ... ... (31) **recurva.**
Leaves softer in texture, not scabrous on the back :
 Leaves ⅛–½ in. thick in the centre :
  Face of leaf rough ... ... (32) **asperula.**
  Face of leaf smooth ... ... (33) **retusa.**
 Leaves about ¼ in. thick in the centre :
  Leaf oblong-lanceolate ... ... (34) **turgida.**
  Leaf obovate-cuneate ... ... (35) **cuspidata.**

8. MUCRONATÆ. Leaves smooth, pale green, not recurved, limpid and lineate towards the tip.

Leaves under ½ in. broad ... ... (36) **reticulata.**
Leaves ¾ in. broad :
 Leaves oblanceolate-oblong, distinctly awned ... (37) **altilinea.**
 Leaves obovate, indistinctly awned ... ... (38) **cymbiformis.**

SERIES 3. Leafy stem short ; leaves multifarious, margined with distinct teeth or bristles.

9. CHLORACANTHÆ. Leaves firm in texture, not lineate, denticulate on the margin.

Leaves lanceolate ... ... (39) **angustifolia.**
Leaves ovate-lanceolate ... ... (40) **chloracantha.**

10. TESSELLATÆ. Leaves very thick, recurved, firm in texture, lineate, denticulate on the margin.

Leaves deltoid ... ... (41) **tessellata.**
Leaves lanceolate-deltoid ... ... (42) **venosa.**

11. DENTICULATÆ. Leaves lineate on the face in the upper half, denticulate on the margins.

Leaves deltoid, much recurved, ⅜–¼ in. thick ... ... (43) **mirabilis.**
Leaves ¼ in. thick, flat on the face or little recurved :
 Green lines of the tip 1–2 :
  Leaves about an inch long ... ... (44) **affinis.**
  Leaves 1½–2 in. long ... ... (45) **bilineata.**
 Green lines of the tip several ... ... (46) **columnaris.**
Leaves ⅛–¼ in. thick, not recurved :
 Leaves ovate-lanceolate ... ... (47) **subregularis.**
 Leaves oblanceolate-oblong, ⅓–½ in. broad :
  Leaves ½–¾ in. long ... ... (48) **atrovirens.**
  Leaves 1–1½ in. long :
   Leaves 20–30 to a rosette ... ... (49) **lætevirens.**
   Leaves 40–50 to a rosette ... ... (50) **polyphylla.**
 Leaves oblong-lanceolate, ⅓–¾ in. broad :
  Apical awn short ... ... (51) **denticulata.**
  Apical awn long ... ... (52) **vittata.**

12. PALLIDÆ. Leaves pale green, furnished with a long pellucid awn and margined with pellucid deltoid-cuspidate teeth about half a line long.

Leaves not pellucid and lineate at the tip ... ... (53) **minima.**
Leaves pellucid and lineate at the tip, not recurved :
  Leaves $\frac{1}{8}$ in. thick ... ... (54) **translucens.**
  Leaves $\frac{1}{4}$ in. thick :
    Leaves $\frac{1}{3}$–$\frac{1}{4}$ as broad as long ... ... (55) **pallida.**
    Leaves $\frac{1}{5}$ as broad as long :
      Leaves an inch long ... ... (56) **pilifera.**
      Leaves $1\frac{1}{2}$–$1\frac{3}{4}$ in. long ... ... (57) **Cooperi.**
Leaves pellucid and lineate at the tip, recurved ... (58) **sessiliflora.**

13. ARACHNOIDEÆ. Leaves pale green, not recurved, furnished with a long pellucid awn, and margined with pellucid teeth $\frac{1}{12}$–$\frac{1}{8}$ in. long.

Leaves not pellucid and lineate at the tip :
  Leaves 1–$1\frac{1}{4}$ in. long without the awn ... ... (59) **setata.**
  Leaves 2 in. long without the awn ... ... (60) **xiphiophylla.**
Leaves pellucid and lineate at the tip :
  Teeth lanceolate-cuspidate ... ... (61) **arachnoides.**
  Teeth lanceolate-subulate ... ... (62) **Bolusii.**

14. LINEARIFOLIÆ. Leaves linear.

Perianth-segments about as long as the tube ... ... (63) **stenophylla.**
Perianth-segments $\frac{1}{3}$–$\frac{1}{4}$ as long as the tube ... ... (64) **tenuifolia.**

1. **H. cordifolia** (Haw. Suppl. 60); leafy stem 6–8 in. long, 2–$2\frac{1}{2}$ in. diam. including leaves; leaves regularly trifarious, all ascending, ovate, acute, $1\frac{1}{2}$–$1\frac{3}{4}$ in. long, an inch broad, dark green, without spots or tubercles, $\frac{1}{3}$ in. thick, the young ones concave, the old nearly flat on the face, scabrous, rounded on the back and obscurely keeled towards the top; peduncle slender, simple, $\frac{1}{2}$ ft. long; raceme very lax, few-flowered, about $\frac{1}{2}$ ft. long; pedicels erecto-patent, $\frac{1}{3}$–$\frac{1}{2}$ in. long; bracts small, ovate; perianth $\frac{3}{4}$ in. long; segments half as long as the keel, tinged with red. *Baker in Journ. Linn. Soc.* xviii. 200. *Aloe cordifolia, Roem. et Schultes, Syst. Veg.* vii. 653; *Salm-Dyck, Aloe, sect.* iii. *fig.* 1; *Kunth, Enum.* iv. 496.

SOUTH AFRICA: without locality.

First described in 1819 from living plants in the Kew collection sent from the Cape in 1818 by Dr. Mackrell, of Norwich.

2. **H. asperiuscula** (Haw. Suppl. 60); leafy stem 4–6 in. long, 1–$1\frac{1}{4}$ in. diam. including leaves; leaves ovate, all ascending, regularly trifarious, ovate, under an inch long, $\frac{1}{2}$–$\frac{5}{8}$ in. broad, $\frac{1}{3}$ in. thick, dark green, without any spots or tubercles, the young ones rather concave, the old nearly flat on the face, scabrous on both sides, semicircular on the back with an obscure keel towards the tip; peduncle slender, simple, $\frac{1}{2}$ ft. long; raceme very lax, few-flowered; pedicels erecto-patent, $\frac{1}{3}$–$\frac{1}{2}$ in. long; bracts ovate, minute; perianth $\frac{3}{4}$ in. long; limb half as long as the tube. *Baker in Journ. Linn. Soc.* xviii. 200. *Aloe asperiuscula, Roem. et Schultes, Syst. Veg.* vii. 653; *Salm-Dyck, Aloe, sect.* iii. *fig.* 2; *Kunth, Enum.* iv. 496.

SOUTH AFRICA: without locality.

Like the last, first described in 1819 from living plants in the Kew collection sent from the Cape by Dr. Mackrell in 1818.

3. H. viscosa (Haw. Syn. 90) ; leafy stem $\frac{1}{2}$–1 ft. long, $1\frac{1}{4}$–$1\frac{1}{2}$ in. diam., leaves included ; leaves regularly trifarious, all ascending, ovate, 1–$1\frac{1}{4}$ in. long, $\frac{1}{2}$ in. broad, $\frac{1}{6}$ in. thick, dull green, all deeply concave on the face, without spots or tubercles, rugose, sometimes but not always viscous, rounded on the back, acutely keeled towards the apex ; peduncle slender, simple, $\frac{1}{2}$ ft. long ; raceme lax, few-flowered, 3–6 in. long; pedicels erecto-patent, $\frac{1}{6}$–$\frac{1}{4}$ in. long ; bracts minute, ovate ; perianth $\frac{3}{4}$ in. long ; limb one-third as long as the tube. *Baker in Journ. Linn. Soc.* xviii. 200. *Aloe viscosa, Linn. Sp. Plant.* 460; *Miller, Dict. edit.* viii. *No.* 11; *Thunb. Diss. No.* 2; *DC. Plantes Grasses, t.* 16 ; *Bot. Mag. t.* 814; *Salm-Dyck, Aloe, sect.* iii. *fig.* 3. *Apicra viscosa, Willd. in Ges. Naturf. Fr. Berl. Mag.* v. 274. *Aloe triangularis, Lam. Encyc.* i. 89.

VAR. β, **H. indurata** (Haw. Revis. 49) ; leaves fewer, thicker, longer, less concave on the face. *Aloe viscosa indurata, Salm-Dyck, Aloe, sect.* iii. *fig.* 3 β.

VAR. γ, **H. concinna** (Haw. Suppl. 59) ; leaves longer and more recurved than in the type. *Aloe concinna, Roem. et Schultes, Syst. Veg.* vii. 653; *Salm-Dyck, Aloe, sect.* iii. *fig.* 4.

VAR. δ, **H. pseudo-tortuosa** (Haw. Suppl. 59); rows of leaves more or less twisted spirally. *Aloe subtortuosa, Salm-Dyck, Aloe, sect.* iii. *fig.* 5. *A. pseudo-tortuosa, Salm-Dyck, Cat.* 8. *Apicra tortuosa, Willd. in Ges. Naturf. Fr. Berl. Mag.* v. 274.

VAR. ε, **H. torquata** (Haw. in. Phil. Mag. 1827, 123) ; leaves ovate-lanceolate, $1\frac{1}{2}$–2 in. long, spirally trifarious. *Aloe torquata, Salm-Dyck, Aloe, sect.* iii. *fig.* 6.

SOUTH AFRICA : without locality, living cultivated plants !

CENTRAL REGION : stony mountain sides near Graaff Reinet, 2700 ft., *Bolus,* 559 !

Introduced into cultivation at the beginning of the eighteenth century. Figured both by Dillenius and Commelyn.

4. H. tortuosa (Haw. Syn. 90); leafy stem $\frac{1}{2}$ ft. long, 2–3 in. diam. including leaves ; leaves all ascending, arranged in three spirally twisted rows, ovate-lanceolate, $1\frac{1}{2}$–2 in. long, $\frac{3}{4}$ in. broad, $\frac{1}{4}$–$\frac{1}{3}$ in. thick, dark green, unspotted, concave on the face, rounded and acutely keeled on the back, very rough on both sides with minute concolorous tubercles ; peduncle slender, simple, or forked ; racemes very lax, subsecund, 6–9 in. long; pedicels $\frac{1}{8}$–$\frac{1}{4}$ in. long, erecto-patent ; bracts minute, ovate ; perianth $\frac{5}{8}$ in. long ; segments one-third as long as the tube, tinged with red. *Baker in Journ. Linn. Soc.* xviii. 201. *Aloe tortuosa, Haw. in Trans. Linn. Soc.* vii. 7 ; *Roem. et Schultes, Syst. Veg.* vii. 655 ; *Salm-Dyck, Aloe, sect.* iv. *fig.* 2. *A. rigida, Bot. Mag. t.* 1337, *non DC.*

VAR. β, **major** (Salm-Dyck) ; more robust ; leaves 2–$2\frac{1}{2}$ in. long, an inch broad. *Aloe tortuosa major, Salm-Dyck, Aloe, sect.* iv. *fig.* 2 β.

VAR. γ, **H. curta** (Haw. Suppl. 60) ; a dwarf form, with greenish-black leaves about an inch long.

VAR. δ, **H. tortella** (Haw. Suppl. 61) ; leafy stem very twisted, branched at the base ; leaves blackish-green.

SOUTH AFRICA : without locality, living cultivated plants !

Introduced into cultivation by Masson in 1797.

5. H. subrigida (Baker in Journ. Linn. Soc. xviii. 201) ; leafy stem 4–6 in. long, 2½–3 in. broad including leaves; leaves spirally trifarious, ovate-lanceolate, squarrose, 1½–2 in. long, ½–¾ in. broad, dark green, concave on the face, rounded and keeled towards the apex on the back, very rough on both sides with crowded white-tipped tubercles; peduncle simple or branched; racemes lax, ½ ft. long; pedicels short; bracts minute, ovate; perianth ¾ in. long; limb one-third as long as the tube. *Aloe subrigida, Salm-Dyck, Aloe, sect. iv. fig.* 1. *A. pseudo-rigida, Salm-Dyck, Cat.* 9 *and* 41. *Apicra pseudo-rigida, Haw. Suppl.* 62. *A. rigida, Jacq. Fragm. t.* 108, *non DC.*

SOUTH AFRICA: without locality, living cultivated plants!

Introduced into cultivation early in the present century. First sent to Kew by Prince Salm-Dyck in 1822. Intermediate between *H. tortuosa* and *H. rigida.*

6. **H. papillosa** (Haw. Suppl. 58); leafy stem ½–1 ft. long, 5–6 in. diam., leaves included; leaves multifarious, ascending, ovate-lanceolate, acuminate, 3–4 in. long, 1¼–1½ in. broad, ⅓–½ in. thick, glaucous green, rather convex on the face, rounded on the back, obscurely keeled towards the apex, furnished on both sides with scattered raised white tubercles a line in diameter; peduncle branched, a foot or more long; racemes 6–9 in. long; pedicels short, erecto-patent; bracts minute, deltoid; perianth ⅝ in. long; limb half as long as the tube. *Baker in Journ. Linn. Soc.* xviii. 201. *Aloe papillosa, Salm-Dyck, Cat.* 7; *Aloe, sect.* vi. *fig.* 4; *Kunth, Enum.* iv. 501. *Apicra margaritifera maxima, Willd. in Ges. Naturf. Fr. Berl. Mag.* v. 269.

VAR. β, **semipapillosa** (Haw. Revis. 55); smaller, with leaves scarcely at all tubercled on the face.

SOUTH AFRICA: without locality, living cultivated plants!

Introduced into cultivation about 1750.

7. **H. Reinwardti** (Haw. Revis. 53); leafy stem 4–6 in. long, 1½–2 in. diam., leaves included, copiously stoloniferous; leaves all ascending, multifarious, ovate-lanceolate, acuminate, 1–1½ in. long, ½ in. broad, ¼ in. thick, pale green when young, reddish-brown when old, turgid on the face, rounded on the back, obscurely keeled, with 9–11 vertical rows of raised white tubercles ⅓ lin. diam.; peduncle slender, simple, ½ ft. long; raceme lax, about ½ ft. long; pedicels very short; bracts minute, deltoid; perianth ⅝ in. long; limb half as long as the tube. *Baker in Journ. Linn. Soc.* xviii. 202. *Aloe Reinwardti, Salm-Dyck, Obs.* 37; *Aloe, sect.* vi. *fig.* 16; *Kunth, Enum.* iv. 506.

VAR. β, **minor** (Hort.); smaller, with leaves ¾–1 in. long; tubercles like those of the type.

VAR. γ, **major** (Hort.); tufts taller than in the type; tubercles larger and more raised.

SOUTH AFRICA: without locality, living cultivated plants!

Introduced into cultivation by Herr Reinwardt in 1818.

8. **H. Cassytha** (Baker); leafy stem 4 in. long, 2 in. diam., leaves

included, stoloniferous at the base; leaves ovate, multifarious, all ascending, dull green, 1–1¼ in. long, ¾ in. broad, concave and slightly tubercled on the face, ¼ in. thick in the centre, very rough and whitish on the margin, rounded on the back with copious raised tubercles with a whitish tip, faintly doubly keeled towards the apex; flowers not seen.

SOUTH AFRICA : without locality.

Described from a living plant in the Kew collection, received from Pfersdorff in 1875. It may be an *Apicra* near *aspera* and *bicarinata*.

9. **H. coarctata** (Haw. in Phil. Mag. 1824, 301); leafy stem 4–8 in. long, 2–3 in. diam., leaves included, copiously stoloniferous; leaves multifarious, ascending, ovate-lanceolate, acuminate, 2–2½ in. long, ¾ in. broad, ¼–⅓ in. thick, dull green, purplish when old, flattish on the face, rounded on the back, keeled in the upper half, with about seven vertical rows of slightly raised greenish-white tubercles; peduncle slender, simple, ½ ft. long; raceme lax, few-flowered, 4–6 in. long; pedicels ⅙–¼ in. long, erecto-patent; bracts small, deltoid; perianth ¾ in. long; limb half as long as the tube. *Baker in Journ. Linn. Soc.* xviii. 202. *Aloe coarctata, Roem. et Schultes, Syst. Veg.* vii. 647; *Salm-Dyck, Aloe, sect.* vi. *fig.* 17; *Kunth, Enum.* iv. 506.

SOUTH AFRICA : without locality, living cultivated plants !

Introduced into cultivation by Bowie in 1823.

10. **H. Greenii** (Baker in Journ. Linn. Soc. xviii. 202); leafy stem 6–8 in. long, 2 in. diam., including leaves; leaves ascending, multifarious, lanceolate-deltoid, 1¼–1½ in. long, ¾ in. broad, ¼–⅓ in. thick in the centre, dull green, with a smooth flat face and a rounded back with a faint central keel, and seven indistinct vertical ribs with as many irregular rows of immersed whitish tubercles; flowers not seen.

SOUTH AFRICA: without locality, *Cooper!*

Described from living plants in the Kew and Peacock collections. Introduced by Mr. T. Cooper about 1860.

11. **H. Peacockii** (Baker in Journ. Linn. Soc. xviii. 202); leafy stem 6–9 in. long, 2–2½ in. diam., leaves included; leaves multifarious, deltoid, all ascending, about an inch long and broad, ⅙–¼ in. thick, green, concave on the face, rounded on the back with a slightly eccentric keel in the upper half, slightly scabrous on the margin, covered on both sides with small, round, immersed, greenish-white spots; flowers unknown.

SOUTH AFRICA : without locality.

Described from living plants in the Kew and Peacock collections in 1879.

12. **H. hybrida** (Haw. Revis. 51); fleshy stem 3–4 in. long, 2–3 in. diam., leaves included; leaves crowded, multifarious, ovate-lanceolate, 1–1½ in. long, ½–⅝ in. broad, ¼ in. thick, a very dull

green, nearly flat on the face, rounded on the back, with a faint
eccentric keel, scabrous on the margin, rough all over with large irregular raised papillæ tinged white at the tip ; peduncle branched, $1\frac{1}{2}$–2 ft.
long including the inflorescence ; racemes very lax, $\frac{1}{2}$ ft. long ; pedicels
$\frac{1}{8}$–$\frac{1}{6}$ in. long; bracts small, ovate; perianth $\frac{3}{4}$ in. long; limb $\frac{1}{3}$ as long
as the tube. *Baker in Journ. Linn. Soc.* xviii. 203. *Aloe hybrida,*
Salm-Dyck, Cat. 7 ; *Aloe, sect.* iv. *fig.* 4 ; *Kunth, Enum.* iv. 499.

SOUTH AFRICA : without locality, living cultivated plants !

First described from the collection of the Emperor Francis I., at Vienna, in
1814. It was received at Kew from Prince Salm-Dyck and drawn in 1822. May
be a garden hybrid between *H. rigida* and *H. Radula.*

13. **H. rigida** (Haw. Revis. 49) ; leafy stem 3–4 in. long, 2–3 in.
diam., leaves included; leaves dense, multifarious, ovate-lanceolate, re-
curved, $1\frac{1}{2}$–2 in. long, $\frac{5}{8}$–$\frac{3}{4}$ in. broad, $\frac{1}{4}$ in. thick, flat or slightly concave
on the face, rounded on the back, with 1–2 faint eccentric keels, very
dull green, red-brown when old, scabrous all over with minute concolorous papillæ ; peduncle $\frac{1}{2}$–1 ft. long, simple or branched ; racemes
lax, $\frac{1}{2}$ ft. long; pedicels $\frac{1}{8}$–$\frac{1}{6}$ in. long ; bracts minute, deltoid; perianth
$\frac{5}{8}$ in. long; limb half as long as the tube. *Baker in Journ. Linn.*
Soc. xviii. 203. *Aloe rigida, DC. Plantes Grasses, t.* 62 ; *Roem. et*
Schultes, Syst. Veg. vii. 655 ; *Salm-Dyck, Aloe, sect.* iv. *fig.* 3 ; *Kunth,*
Enum. iv. 499. *A. cylindracea, var. rigida, Lam. Encyc.* i. 89.

VAR. β, **H. expansa** (Haw. Syn. 91) ; leaves less scabrous. *Aloe expansa, Haw*
in Trans. Linn. Soc. vii. 8; *Lodd. Bot. Cab. t.* 1430. *Aloe rigida expansa,*
Salm-Dyck, Aloe, sect. iv. *fig.* 3 β. *Apicra patula, Willd. in Ges. Naturf. Fr.*
Berl. Mag. v. 272, *excl. syn.*

SOUTH AFRICA : without locality, living cultivated plants !

Introduced into cultivation about 1795.

14. **H. nigra** (Baker in Journ. Linn. Soc. xviii. 203) ; leafy stem
about $\frac{1}{2}$ ft. long, 3 in. diam., leaves included ; leaves multifarious,
ovate-lanceolate, acuminate, recurved, $1\frac{1}{2}$–2 in. long, $\frac{3}{4}$–1 in. broad,
$\frac{1}{6}$–$\frac{1}{4}$ in. thick, concave on the face, rounded and obliquely keeled on
the back, blackish-green, unspotted, rough, with small crowded concolorous tubercles; peduncle simple, $\frac{1}{2}$ ft. long ; raceme lax, simple,
$\frac{1}{2}$ ft. long; pedicels $\frac{1}{12}$–$\frac{1}{8}$ in. long ; bracts minute, deltoid ; perianth
$\frac{3}{4}$ in. long; limb half as long as the tube. *Apicra nigra, Haw. in Phil.*
Mag. 1824, 302. *Aloe nigra, Roem. et Schultes, Syst. Veg.* vii.
657 ; *Kunth, Enum.* iv. 495.

SOUTH AFRICA : without locality, living cultivated plants !

Introduced into cultivation by Bowie about 1822.

15. **H. glauca** (Baker in Journ. Linn. Soc. xviii. 203) ; leafy
stem 2–3 in. long, 1–$1\frac{1}{2}$ in. diam., leaves included ; leaves multifarious, crowded, all ascending, oblong-lanceolate, $\frac{1}{2}$–$\frac{3}{4}$ in. long,
$\frac{1}{4}$–$\frac{1}{3}$ in. broad, $\frac{1}{6}$ in. thick, pale green when young, dull purple when
old, unspotted, smooth, flat on the face, and concolorous, rounded on
the back, with a central keel down the upper half and 5–7 indistinct vertical ribs of darker green ; peduncle simple, $\frac{1}{2}$ ft. long ;

raceme simple, ½ ft. long ; pedicels short; bracts minute, ovate ; perianth ¾ in. long ; limb half as long as the tube.

KALAHARI REGION : Orange Free State, *Cooper !*
Described from a living cultivated plant introduced by Cooper in 1862.

16. **H. attenuata** (Haw. Syn. 92) ; leafy stem 2½–3 in. long, the rosette of leaves 4–5 in. broad ; leaves 30–40, multifarious, lanceolate-deltoid, acuminate, the outer recurved, 2½–3 in. long, ¾ in. broad, ⅓ in. thick, bright dark green, flat on the face and scabrous with minute whitish tubercles, rounded on the back, keeled towards the apex, with middle-sized white tubercles confluent in a series of regular transverse bands ; peduncle ½–1 ft. long, simple or forked ; racemes lax, about ½ ft. long ; pedicels short ; bracts minute, deltoid ; perianth ¾ in. long; limb half as long as the tube. *Baker in Journ. Linn. Soc.* xviii. 203. *Aloe attenuata, Haw. in Trans. Linn. Soc.* vii. 11; *Salm-Dyck, Aloe, sect.* vi. *fig.* 12; *Kunth, Enum.* iv. 505. *Aloe Radula, Ker in Bot. Mag. t.* 1345, *non Jacq. Apicra attenuata, Willd. in Ges. Naturf. Fr. Berl. Mag.* v. 270.

VAR. β, **H. clariperla** (Haw. in Phil. Mag. 1826, 186); pearly tubercles rather larger and less regularly seriate. *Aloe attenuata clariperla, Salm-Dyck, Aloe, sect.* vi. *fig.* 12 β.

SOUTH AFRICA: without locality, living cultivated plants !
Introduced into cultivation about 1790.

17. **H. fasciata** (Haw. Revis. 54) ; leafy stem short; rosette of leaves about 3 in. diam. ; leaves 40–60, multifarious, ascending, lanceolate-deltoid, 1¼–1½ in. long, ½ in. broad, ⅛ in. thick, glaucous-green, flat on the face or rather concave, without any tubercles, rounded on the back and keeled upwards, with about 20 transverse bands of white middle-sized tubercles ; peduncle simple or forked, ½ ft. long ; racemes lax, few-flowered ; pedicels ⅙–¼ in. long ; bracts minute, deltoid ; perianth ⅝–¾ in. long; limb half as long as the tube. *Baker in Journ. Linn. Soc.* xviii. 204. *Aloe fasciata, Salm-Dyck, Hort.* 326 ; *Aloe, sect.* vi. *fig.* 15 ; *Kunth, Enum.* iv. 506. *Apicra fasciata, Willd. in Ges. Naturf. Fr. Berl. Mag.* v. 272.

VAR. β, **major** (Salm-Dyck, Aloe, sect. vi. fig. 15 β); more robust, with a longer leafy stem 3–4 in. long, and larger, thicker leaves.

SOUTH AFRICA: without locality, living cultivated plants!
Introduced into cultivation at the beginning of the present century.

18. **H. subfasciata** (Baker in Journ. Linn. Soc. xviii. 204); leafy stem short ; rosette of leaves 5–6 in. diam. ; leaves 30–40, crowded, multifarious, lanceolate-deltoid, very acuminate, 3–4 in. long, an inch broad, ¼–⅓ in. thick, bright green, flat and unspotted on the face, rounded on the back and.keeled towards the apex, with the middle-sized white tubercles aggregated into irregular transverse bands ; peduncle branched, about a foot long ; racemes lax, 6–9 in. long; pedicels ⅙–¼ in. long; bracts small, deltoid ; perianth ¾ in. long; limb half as long as the tube. *Aloe subfasciata, Salm-Dyck, Hort.*

325; *Aloe, sect.* vi. *fig.* 14 ; *Kunth, Enum.* iv. 505. *H. fasciata var. major, Haw. Revis.* 54.

SOUTH AFRICA : without locality, living cultivated plants !
First seen in the Vienna collections in 1814. *H. argyrostigma, Hort. Paris.* has narrower leaves than in the type, and smaller, more crowded tubercles.

19. **H. margaritifera** (Haw. Suppl. 55) ; leafy stem short ; rosette of leaves 3–4 in. diam. ; leaves 30–40 in a dense multifarious rosette, lanceolate-deltoid, 2½–3 in. long, 1–1¼ in. broad, ¼–⅓ in. thick, green, flat on the face, rounded on the back and keeled towards the tip, furnished on both sides with copious scattered large white tubercles ; peduncle branched, a foot or more long ; racemes ½ ft. long, denser than in most of the other species ; pedicels very short ; bracts small, ovate ; perianth ⅝ in. long ; limb half as long as the tube. *Baker in Journ. Linn. Soc.* xviii. 204. *Aloe pumila var. margaritifera, Linn. Sp. Plant. edit.* i. 322. *A. margaritifera, Miller, Gard. Dict. edit.* viii. *No.* 14 ; *Ait. Hort. Kew.* i. 468 ; '*Salm-Dyck, Aloe, sect.* vi. *fig.* 5 ; *Kunth, Enum.* iv. 502. *H. major, Duval, Pl. Succ. Hort. Alenç.* 7 ; *Haw. Syn.* 92.

VAR. β, **H. erecta** (Haw. Revis. 55) ; leaves rather smaller ; tubercles more crowded and rather smaller. *Aloe africana margaritifera minor, Dill. Hort. Elth. t.* 16, *fig.* 17. *Aloe erecta, Salm-Dyck, Aloe, sect.* vi. *fig.* 7. *H. minor, Duval, Pl. Succ. Hort. Alenç.* 7.

VAR. γ, **H. granata** (Haw. Suppl. 57) ; much smaller ; leaves more deltoid, 1½–2 in. long, with tubercles much smaller and more crowded. *Aloe africana margaritifera minima, Dill. Hort. Elth. t.* 16, *fig.* 18. *A. granata, Salm-Dyck, Aloe, sect.* vi. *fig.* 6. *Apicra granata, Willd. in Ges. Naturf. Fr. Berl. Mag.* v. 269. *Aloe magaritifera var. minima, Ker in Bot. Mag. t.* 1360. *H. brevis, Haw. Suppl.* 57.

VAR. δ, **H. semimargaritifera** (Haw. Suppl. 53) ; leaves the same size and shape as in the type, strongly keeled on the back and faintly on the face, the tubercles on the back fewer than in the type, and on the face almost restricted to the keel. *Aloe subalbicans, Salm-Dyck, Aloe, sect.* vi. *fig.* 1.

VAR. ε, **H. corallina** (Hort. Peacock.) ; leaves 2½ in. long, scabrous and sparsely tubercled on the face, those of the back middle-sized and aggregated into irregular transverse bands.

SOUTH AFRICA : without locality, living cultivated plants !
COAST REGION: Worcester Div. ; Stony Karoo around Ashton, 800 ft., *Mac-Owan, Herb. Aust. Afr.*, 1557 !

Introduced into cultivation early in the eighteenth century. Mr. Thos. Cooper, in his garden at Reigate, has raised from the seeds of *H. erecta* forms resembling closely *H. papillosa, fasciata, subulata,* and *rugosa.* These are all figured in a paper by Mr. N. E. Brown, in Gard. Chron. 1878, ix. p. 820, figs. 140 to 145, and a good series of dried specimens of them is preserved in the Kew Herbarium.

20. **H. semiglabrata** (Haw. Suppl. 55) ; leafy stem short ; rosette of leaves 3–4 in. diam. ; leaves 30–40, crowded, multifarious, ascending, lanceolate-deltoid, acuminate, 3–3½ in. long, ¾–1 in. broad, ¼–⅓ in. thick, bright green, flat and almost without tubercles on the face, rounded on the back and keeled towards the apex, with copious middle-sized white tubercles in very irregular cross rows, specially developed in the central half of the leaf ; peduncle branched, 1½–2 ft. long including inflorescence ; racemes moderately dense, about ½ ft.

long; pedicels very short; bracts deltoid, as long as the lower
pedicels; perianth ¾ in. long; limb half as long as the tube. *Baker
in Journ. Linn. Soc.* xviii. 205. *Aloe semiglabrata, Salm-Dyck, Hort.*
321; *Aloe, sect.* vi. *fig.* 2; *Kunth, Enum.* iv. 500.

SOUTH AFRICA: without locality.

Introduced into cultivation about 1810. Scarcely more than a variety of
H. margaritifera.

21. H. subattenuata (Baker in Journ. Linn. Soc. xviii. 205);
leafy stem short; rosette of leaves 4–5 in. diam.; leaves 40–50,
crowded, multifarious, lanceolate-deltoid, acuminate, 2–3 in. long,
¾–1 in. broad, ·¼ in. thick, green when young, dull purple when old,
flattish on the face and almost destitute of tubercles, rounded on the
back, keeled towards the apex, with copious middle-sized scattered
white tubercles; peduncle branched, 1½ ft. long including inflorescence; racemes moderately dense, about ½ ft. long; pedicels very
short; bracts small, deltoid; perianth ¾ in. long; limb half as long
as the tube. *Aloe subattenuata, Salm-Dyck, Hort.* 324; *Aloe, sect.* vi.
fig. 11; *Kunth, Enum.* iv. 504. *H. Radula var. magniperlata,
Haw. Revis.* 54.

SOUTH AFRICA: without locality, living cultivated plants!

First described from the collection of the Emperor Francis, at Vienna, in 1814.
Closely allied to *H. margaritifera.*

22. H. glabrata (Baker in Journ. Linn. Soc. xviii. 206); leafy
stem short; rosette of leaves 5–6 in. diam.; leaves 30–40 in a dense
multifarious rosette, lanceolate-deltoid, acuminate, 4–5 in. long, an
inch broad, ¼–⅓ in. thick, glaucous green, nearly flat and without
tubercles on the face, rounded on the back and keeled towards the
apex, with distant scattered middle-sized whitish tubercles, the upper
part of the margin a continuous white horny line; peduncle branched,
2–3 ft. long including inflorescence; racemes about ½ ft. long;
pedicels ⅛–⅙ in. long; bracts deltoid, as long as the pedicels; perianth
⅝ in. long; limb half as long as the tube. *Aloe glabrata, Salm-Dyck,
Hort.* 325; *Aloe, sect.* vi. *fig.* 13; *Kunth, Enum.* iv. 505.

VAR. β, **perviridis** (Salm-Dyck, Aloe, sect. vi. fig. 13 β); tubercles much more
crowded, greenish-white.

VAR. γ, **concolor** (Salm-Dyck, Aloe, sect. vi. fig. 13 γ); leaves scabrous on the
back with minute concolorous tubercles.

SOUTH AFRICA: without locality, living cultivated plants!

First seen in the Berlin collections about 1830. Var. *concolor* is probably a
garden hybrid, perhaps with *H. scabra.*

23. H. Radula (Haw. Syn. 93); leafy stem short; rosette 5–6 in.
diam.; leaves 30–40, crowded in a multifarious rosette, lanceolate-
deltoid, very acuminate, recurved, 2½–3 in. long, ¾ in. broad, ⅙–¼ in.
thick, green, flat on the face, rounded on the back and keeled upwards, scabrous all over both surfaces with minute white tubercles;
peduncle simple or branched, ½–1 ft. long; racemes very lax;
pedicels short; bracts small, deltoid; perianth ¾ in. long; limb half
as long as the tube. *Baker in Journ. Linn. Soc.* xviii. 206. *Aloe*

Radula, Jacq. Hort. Schoenbr. iv. 11, *t.* 422; *Salm-Dyck, Aloe,*
sect. vi. *fig.* 8; *Kunth, Enum.* iv. 504. *Apicra Radula, Willd. in*
Ges. Naturf. Fr. Berl. Mag. v. 270.

SOUTH AFRICA : without locality, living cultivated plants !
Introduced into cultivation at the beginning of the present century.

24. H. subulata (Baker in Journ. Linn. Soc. xviii. 206); leafy
stem short ; rosette of leaves 5–6 in. diam. ; leaves 30–40 in a dense
multifarious rosette, lanceolate-deltoid, very acuminate, 3–4 in. long,
$\frac{3}{4}$–1 in. broad, $\frac{1}{4}$ in. thick, green when young, red-brown when old,
flat and scabrous on the face, with concolorous tubercles rounded on
the back and keeled towards the apex, rugose all over the back with
crowded, small white-tipped tubercles ; peduncle branched, 2–3 ft.
long including the inflorescence ; racemes lax, about $\frac{1}{2}$ ft. long ;
pedicels $\frac{1}{6}$–$\frac{1}{4}$ in. long ; bracts small, deltoid ; perianth $\frac{3}{4}$ in. long ; limb
half as long as the tube. *Aloe subulata, Salm-Dyck, Hort.* 324; *Aloe,*
sect. vi. *fig.* 10; *Kunth, Enum.* iv. 504. *H. Radula var. lævior, Haw.*
Revis. 54.

SOUTH AFRICA : without locality.
First described from the Vienna collections in 1814.

25. H. rugosa (Baker in Journ. Linn. Soc. xviii. 206) ; leafy
stem short; rosette of leaves 5–6 in. diam.; leaves 30–50, crowded,
multifarious, very acuminate, lanceolate-deltoid, 3–4 in. long, $\frac{3}{4}$–1 in.
broad, $\frac{1}{4}$ in. thick, green, flat on the face, rounded on the back and
keeled upwards, scabrous all over both faces with whitish tubercles,
larger and less crowded than those of *H. Radula;* peduncle branched,
2–3 ft. long including inflorescence ; racemes lax, about $\frac{1}{2}$ ft. long;
pedicels $\frac{1}{8}$–$\frac{1}{6}$ in. long ; bracts small, deltoid ; perianth $\frac{5}{8}$ in. long ; limb
half as long as the tube. *Aloe rugosa, Salm-Dyck, Hort.* 323 ;
Aloe, sect. vi. *fig.* 9 ; *Kunth, Enum.* iv. 504. *H. Radula var.*
asperior, Haw. Revis. 54.

SOUTH AFRICA : without locality.
First described from the collection of the Emperor Francis, at Vienna, in 1814.

26. H. albicans (Haw. Syn. 91); leafy stem short ; rosette of
leaves 5–6 in. diam. ; leaves about 30 in a dense multifarious
rosette, ovate-lanceolate, $2\frac{1}{2}$–3 in. long, $1\frac{1}{4}$–$1\frac{1}{2}$ in. broad, $\frac{1}{4}$–$\frac{1}{3}$ in.
thick, rather recurved, whitish-green, quite smooth on both surfaces,
rigid in texture, concave on the face, rounded on the back, keeled
towards the apex ; margins smooth, whitish, horny ; peduncle
simple or branched ; racemes $\frac{1}{2}$ ft. long ; pedicels $\frac{1}{6}$–$\frac{1}{4}$ in. long ; bracts
small, ovate ; perianth $\frac{3}{4}$ in. long; limb half as long as the tube.
Baker in Journ. Linn. Soc. xviii. 207. *Aloe albicans, Haw. in*
Trans. Linn. Soc. vii. 8 ; *Ker in Bot. Mag. t.* 1452; *Salm-Dyck,*
Aloe, sect. v. *fig.* 1 ; *Kunth, Enum.* iv. 500. *Apicra albicans, Willd.*
in Ges. Naturf. Fr. Berl. Mag. v. 271. *H. lævis and H. ramifera,*
Haw. Revis. 52. *Aloe marginata, Lam. Encyc.* i. 89. *A. lævigata,*
Roem. et Schultes, Syst. Veg. vii. 636.

VAR. β, **H. virescens** (Haw. Revis. 52); leaves darker green, with a few white tubercles beneath in the line of the keel. *Aloe virescens, Roem. et Schultes, Syst. Veg.* vii. 637; *Kunth, Enum.* iv. 500.

SOUTH AFRICA: without locality.

Introduced into cultivation early in the eighteenth century. Figured twice by Commelyn.

27. H. scabra (Haw. Suppl. 58; Revis. 51); leafy stem very short; rosette of leaves about 2 in. diam.; leaves about 10, multifarious, ovate, ascending, $1\frac{1}{2}$–2 in. long, $\frac{3}{4}$–1 in. broad, $\frac{1}{4}$–$\frac{1}{3}$ in. thick, dull green, flat or slightly concave on the face, rounded on the back and obliquely keeled towards the tip, rough all over with small crowded concolorous tubercles; peduncle simple, slender, under a foot long; raceme lax, few-flowered; pedicels $\frac{1}{8}$–$\frac{1}{6}$ in. long; bracts very small, ovate; perianth $\frac{5}{8}$ in. long; limb half as long as the tube. *Baker in Journ. Linn. Soc.* xviii. 207. *Aloe scabra, Roem. et Schultes, Syst. Veg.* vii. 644; *Salm-Dyck, Aloe, sect.* vii. *fig.* 1; *Kunth, Enum.* iv. 507.

SOUTH AFRICA: without locality.

Introduced into cultivation in England about 1818, probably by Bowie.

28. H. sordida (Haw. Revis. 51); leafy stem short; rosette about 4 in. diam.; leaves 10–12, multifarious, ascending, lanceolate-deltoid, very acuminate, 3–4 in. long, 1–$1\frac{1}{4}$ in. broad, $\frac{1}{4}$ in. thick, concave on the face, rounded on the back, dull green, slightly scabrous, especially beneath; peduncle slender, simple; raceme lax, about $\frac{1}{2}$ ft. long; pedicels short; bracts minute; perianth $\frac{3}{4}$ in. long; limb half as long as the tube. *Baker in Journ. Linn. Soc.* xviii. 207. *Aloe sordida, Roem. et Schultes, Syst. Veg.* vii. 644; *Salm-Dyck, Aloe, sect.* vii. *fig.* 2; *Kunth, Enum.* iv. 507.

SOUTH AFRICA: without locality.

Introduced into cultivation in England about 1820.

29. H. icosiphylla (Baker in Journ. Linn. Soc. xviii. 207); leafy stem short; rosette 3 in. diam.; leaves about 20 in a dense multifarious rosette, lanceolate-deltoid, $1\frac{1}{4}$–$1\frac{1}{2}$ in. long, $\frac{3}{4}$ in. broad, $\frac{1}{8}$ in. thick in the centre, slightly recurved, purplish-green, concave on the face, rounded on the back and keeled upwards, scabrous all over both surfaces with minute concolorous papillæ; peduncle simple, $\frac{1}{2}$ ft. long; raceme drooping, lax, $\frac{1}{2}$ ft. long; lower pedicels $\frac{1}{8}$ in. long; bracts deltoid, as long as the pedicels; perianth $\frac{5}{8}$–$\frac{3}{4}$ in. long; limb a third as long as the tube.

SOUTH AFRICA: without locality.

Described from living plants cultivated at Kew in 1872. Most resembles *H. glabrata var. concolor.*

30. H. Tisleyi (Baker in Journ. Linn. Soc. xviii. 208); leafy stem short; rosette of leaves $2\frac{1}{2}$–3 in. diam.; leaves 30–40 in a dense multifarious rosette, lanceolate-deltoid, 1–$1\frac{1}{4}$ in. long, $\frac{5}{8}$ in.

broad, ¼ in. thick, dull green, fading to red-brown, flat on the
face, rounded and obscurely and obliquely keeled upwards on the
back, scabrous all over with raised concolorous papillæ; flowers not
seen.

SOUTH AFRICA: without locality.

Described from living plants cultivated at Kew in 1879, received from Mr.
Tisley. Like the last, a near ally of *H. glabrata var. concolor.*

31. **H. recurva** (Haw. Syn. 94; Revis. 51); leafy stem very
short; rosette 2–3 in. diam.; leaves 12–15, multifarious, deltoid,
much recurved, rigid in texture, 1¼–1½ in. long, ¾ in. broad, ¼–⅓ in.
thick in the centre, nearly flat on the face, pale green, with vertical
lines of deeper green, rounded on the back, keeled upwards, dark
green, scabrous with small crowded concolorous papillæ; peduncle
simple, very slender; raceme lax, few-flowered; pedicels ⅛–⅙ in.
long; bracts minute; perianth ¾ in. long; limb one-third the length
of the tube. *Baker in Journ. Linn. Soc.* xviii. 208. *Aloe recurva,
Haw. in Trans. Linn. Soc.* vii. 10; *Bot. Mag. t.* 1353; *Salm-Dyck,
Aloe, sect.* vii. *fig.* 3; *Kunth, Enum.* iv. 507. *Apicra recurva, Willd.
in Ges. Naturf. Fr. Berl. Mag.* v. 270.

SOUTH AFRICA: without locality, a living cultivated plant!

Introduced into cultivation by Masson in 1795. The only living plant we
have had at Kew of late years is one sent home by Sir H. Barkly in 1876.

32. **H. asperula** (Haw. in Phil. Mag. 1824, 300); leafy stem very
short; rosette 2½–3 in. diam.; leaves 10–12, multifarious, deltoid,
very recurved, 1–1½ in. long, ¾ in. broad, ⅓–½ in. thick in the centre,
pale green on both sides, scabrous on the face, with minute concolorous papillæ and marked in the upper half with 7–9 vertical
pale green lines, rounded and smooth on the back and keeled in the
upper half; peduncle slender, simple; raceme lax, few-flowered;
pedicels very short; bracts minute, deltoid; perianth ⅝ in. long;
limb ⅓–½ as long as the tube. *Baker in Journ. Linn. Soc.* xviii.
208. *Aloe asperula, Roem. et Schultes, Syst. Veg.* vii. 635; *Salm-
Dyck, Aloe, sect.* ix. *fig.* 2; *Kunth, Enum.* iv. 508.

SOUTH AFRICA: without locality.

Introduced into cultivation in England in 1823, probably by Bowie.

33. **H. retusa** (Haw. Syn. 95); leafy stem short; rosette 2–3 in.
diam.; leaves 10–15, multifarious, deltoid, very recurved, 1–1½ in.
long, ¾ in. broad, ⅓–½ in. thick in the centre, pale green, smooth on
both surfaces; marked on the upper half of the face with pale
vertical lines, cuspidate at the tip, rounded on the back and keeled
in the upper half; peduncle simple; raceme lax, ½ ft. long; pedicels
very short; bracts small, deltoid; perianth ¾ in. long; limb half as
long as the tube. *Baker in Journ. Linn. Soc.* xviii. 208. *Aloe
retusa, Linn. Sp. Plant.* 322; *Thunb. Diss. No.* 15; *DC. Plantes
Grasses, t.* 45; *Curt. in Bot. Mag. t.* 455; *Salm-Dyck, Aloe, sect.*

ix. *fig.* 3; *Kunth, Enum.* iv. 509. *Apicra retusa, Willd. in Ges. Naturf. Fr. Berl. Mag.* v. 271. *Catevala retusa, Medic. Theod.* 68.

VAR. β, H. mutica (Haw. Revis. 55); leaves not cuspidate at the apex.

SOUTH AFRICA : without locality, living cultivated plants!
Introduced into cultivation at the beginning of the eighteenth century.

34. H. turgida (Haw. Suppl. 52); leafy stem very short; rosettes cæspitose, 2–2½ in. diam. ; leaves 20–30 in a dense multifarious rosette, oblong-lanceolate, much recurved, smooth, pale green, ¾–1 in. long, ¼–⅓ in. broad at the middle, ¼ in. thick, acute, marked down the upper half of the face with 5–7 vertical lines of paler green, rounded on the back and keeled in the upper half; peduncle slender, simple ; raceme lax, few-flowered ; pedicels very short ; bracts small, ovate ; perianth ⅝–¾ in. long ; limb half as long as the tube. *Baker in Journ. Linn. Soc.* xviii. 209. *Aloe turgida, Roem. et Schultes, Syst. Veg.* vii. 635 ; *Salm-Dyck, Aloe, sect.* ix. *fig.* 5 ; *Kunth, Enum.* iv. 509.

SOUTH AFRICA : without locality, living cultivated plants !
Introduced into cultivation by Bowie in 1818.

35. H. cuspidata (Haw. Suppl. 51 ; Revis. 58); leafy stem very short ; rosette 2–2½ in. diam.; leaves 20–30, dense, multifarious, obovate-cuneate, much recurved, about an inch long, ¾ in. broad, ¼–⅓ in. thick in the centre, pale green, smooth, colourless at the apex, with reticulated green vertical ribs, distinctly aristate, rounded on the back, keeled in the upper half ; peduncle simple, 8–10 in. long ; raceme lax, ½ ft. long ; lower pedicels 1/12–⅛ in. long ; bracts deltoid-cuspidate ; perianth ½–⅝ in. long; limb half as long as the tube. *Baker in Journ. Linn. Soc.* xviii. 209. *Aloe cuspidata, Roem. et Schultes, Syst. Veg.* vii. 639; *Kunth, Enum.* iv. 510.

SOUTH AFRICA : without locality, living cultivated plants!
Introduced by Bowie in 1818.

36. H. reticulata (Haw. Syn. 94 ; Revis. 57) ; leafy stem short ; rosettes cæspitose, 2–2½ in. diam. ; leaves 30–40 in a dense multi-farious rosette, oblong-lanceolate, 1–1¼ in. long, under ½ in. broad at the middle, ⅛ in. thick, pale glaucous green on both sides, flattish on the face, minutely scabrous on the margin, rounded and keeled on the back, lineolate, especially on the back in the upper half, with anastomosing ribs of darker green, not distinctly aristate ; peduncle simple, slender, ½ ft. long ; raceme lax, few-flowered, 4–6 in. long ; lower pedicels ⅛–⅙ in. long ; bracts small, deltoid ; perianth ⅝–¾ in. long ; limb half as long as the tube. *Baker in Journ. Linn. Soc.* xviii. 210. *Aloe reticulata, Haw. in Trans. Linn. Soc.* vii. 9 ; *Lodd. Bot. Cab. t.* 1354 ; *Salm-Dyck, Aloe, sect.* x. *fig.* 1 ; *Kunth, Enum.* iv. 510. *Apicra reticulata, Willd. in Ges. Naturf. Fr. Berl. Mag.* v. 272. *Aloe Pumilio, Jacq. Hort. Schoenbr.* iv. 11, *t.* 421. *A. herbacea, DC. Hort. Monspel.* 76. *A. arachnoides var. reticulata, Bot. Mag. t.* 1314.

SOUTH AFRICA : without locality, living cultivated plants !
Introduced into cultivation about 1794.

37. **H. altilinea** (Haw. in Phil. Mag. 1824, 301); leafy stem short; rosettes cæspitose, 3–4 in. diam.; leaves about 30, multifarious, ascending, oblanceolate-oblong, acute, distinctly aristate, $1\frac{1}{2}$–2 in. long, $\frac{3}{4}$ in. broad above the middle, $\frac{1}{6}$ in. thick, flat on the face, rounded on the back, keeled in the upper half, pale green, smooth on both sides, pellucid on the upper, with 5–7 vertical green ribs; peduncle simple, slender, 6–9 in. long; raceme lax, few-flowered; pedicels very short; bracts small, ovate; perianth $\frac{5}{8}$ in. long; limb about half as long as the tube. *Baker in Journ. Linn. Soc.* xviii. 209. *Aloe altilinea, Roem. et Schultes, Syst. Veg.* vii. 638; *Salm-Dyck, Aloe, sect.* xi. *fig.* 3; *Kunth, Enum.* iv. 512. *H. mucronata, H. limpida and H. aristata, Haw. Suppl.* 50–51.

SOUTH AFRICA: without locality, living cultivated plants!
COAST REGION: Uitenhage Div.; near the Zwartkops River, *Zeyher*, 1053! Stockenstrom Div.; stony ridges near Seymour, *Scully*, 98!

Introduced into cultivation about 1820.

38. **H. cymbiformis** (Haw. Syn. 93); rosettes shortly caulescent, dense, 3–4 in. diam.; leaves 20–25; multifarious, obovate, acute, 1–$1\frac{1}{2}$ in. long, $\frac{3}{4}$ in. broad, $\frac{1}{6}$–$\frac{1}{5}$ in. thick, pale green, slightly concave, rounded on the back, keeled upwards, not distinctly aristate, pale glaucous green, marked especially in the upper half with indistinct anastomosing vertical lines of darker green; peduncle simple, $\frac{1}{2}$ ft. long; raceme lax, $\frac{1}{2}$ ft. long; lower pedicels $\frac{1}{6}$–$\frac{1}{4}$ in. long; bracts small, ovate; perianth $\frac{5}{8}$ in. long; segments nearly as long as the tube. *Baker in Journ. Linn. Soc.* xviii. 209. *H. concava, Haw. Revis.* 58. *Aloe cymbiformis, Haw. in Trans. Linn. Soc.* vii. 8; *Bot. Mag. t.* 802. *A. cymbæfolia, Schrad. Neues Journ.* ii. 17, *t.* 2; *Jacq. Fragm. t.* 112, *fig.* 1; *Salm-Dyck, Aloe, sect.* xi. *fig.* 1; *Kunth, Enum.* iv. 511. *Apicra cymbæfolia, Willd. in Ges. Naturf. Fr. Berl. Mag.* v. 271.

VAR. β, **H. obtusa** (Haw. in Phil. Mag. 1825, 232); smaller, with leaves not more than an inch long, darker in colour, and more distinctly striped. *Aloe hebes, Roem. et Schultes, Syst. Veg.* vii. 637; *Kunth, Enum.* iv. 511.

VAR. γ, **H. planifolia** (Haw. in Phil. Mag. 1825, 282); leaves rather thicker, flat on the face. *Aloe planifolia, Roem. et Schultes, Syst. Veg.* vii. 638; *Salm-Dyck, Aloe, sect.* xi. *fig.* 2.

SOUTH AFRICA: without locality, living cultivated plants!
CENTRAL REGION: Somerset Div., *Bowker!*

Introduced into cultivation by Masson in 1795.

39. **H. angustifolia** (Haw. in Phil. Mag. 1825, 283); leafy stem short; rosette about 3 in. diam.; leaves about 20, multifarious, ascending, lanceolate, acuminate, $1\frac{1}{2}$–2 in. long, $\frac{1}{4}$–$\frac{1}{3}$ in. broad, $\frac{1}{8}$ in. thick, moderately firm in texture, a uniform pale green without lines, flat on the face, convex on the back, with 1–3 faint keels upwards, minutely ciliated on the margin; peduncle slender, simple, about $\frac{1}{2}$ ft. long; raceme lax, simple, few-flowered; pedicels very short; bracts small, deltoid; perianth $\frac{5}{8}$–$\frac{3}{4}$ in. long; limb nearly as long as the tube. *Baker in Journ. Linn. Soc.* xviii. 210. *Aloe stenophylla*,

Roem. et Schultes, Syst. Veg. vii. 641 ; *Salm-Dyck, Aloe, sect.* xiii.
fig. 2 ; *Kunth, Enum.* iv. 514.

SOUTH AFRICA : without locality.
Introduced into cultivation by Bowie in 1824.

40. H. chlorocantha (Haw. Revis. 57) ; leafy stem short ; rosettes
about 2 in. diam. ; leaves 20–30, ovate-lanceolate, ascending, multifarious, 1–1½ in. long, ½–¾ in. broad, ⅛ in. thick, a uniform dead
green on both sides, firm in texture, flat on the face, rounded on
the back and keeled in the upper half ; margin ciliated with
minute deflexed teeth ; peduncle slender, simple, about ½ ft. long ;
raceme lax, simple, few-flowered, under ½ ft. long ; pedicels very
short ; bracts small, deltoid ; perianth ⅝ in. long ; limb half as long
as the tube. *Baker in Journ. Linn. Soc.* xviii. 211. *Aloe chlorocantha, Roem. et Schultes, Syst. Veg.* vii. 641 ; *Salm-Dyck, Aloe,
sect.* xiii. *fig.* 1 ; *Kunth, Enum.* iv. 514.

SOUTH AFRICA : without locality, living cultivated plants !
Introduced into cultivation by Bowie in 1819.

41. H. tessellata (Haw. in Phil. Mag. 1824, 300) ; rosette of leaves
sessile, about 2 in. long, 3–4 in. diam. ; leaves 12–15, multifarious,
deltoid, recurved, 1¼–1½ in. long, ¾–1 in. broad, ¼–⅓ in. thick, firm
in texture, flattish on the face, marked with six pale green anastomosing vertical lines on a darker shining green ground, dull green on
the rounded back and rough all over with small raised coriaceous
tubercles, cuspidate at the apex ; margin ciliated with minute reflexed teeth ; peduncle slender, simple, about ½ ft. long ; raceme lax,
simple, secund, few-flowered ; pedicels 1/12–⅓ in. long ; bracts minute,
deltoid ; perianth ¾ in. long ; limb half as long as the tube. *Baker
in Journ. Linn. Soc.* xviii. 211. *Aloe tessellata, Roem. et Schultes,
Syst. Veg.* vii. 635 ; *Salm-Dyck, Aloe, sect.* viii. *fig.* 1 ; *Kunth, Enum.*
iv. 508.

VAR. β, **H. parva** (Haw. in Phil. Mag. 1824, 301) ; leaves shorter, 1–1¼ in.
long and broad ; rosette 2 in. diam. *Aloe parva, Roem. et Schultes, Syst. Veg.*
vii. 653 ; *Salm-Dyck, Aloe, sect.* viii. *fig.* 2.

VAR. **inflexa** (Baker in Journ. Linn. Soc. xviii. 211) ; rosette 2–3 in. long,
2 in. diam. ; leaves dull purplish-green, with a concave face and inflexed
margins.

SOUTH AFRICA : without locality, living cultivated plants !
Introduced into cultivation by Bowie about 1822.

42. H. venosa (Haw. Revis. 51) ; rosette of leaves 1½ in. long,
3½–4½ in. diam. ; leaves 12–15, multifarious, lanceolate-deltoid,
rather recurved, 2½–3 in. long, ⅝ in. broad, ¼ in. thick, firm in
texture, flat on the face, dull green with a purplish tinge, with
about five pale green vertical anastomosing lines, rounded on the
back, keeled and rugose in the upper half, cuspidate at the apex ;
margin minutely denticulate ; peduncle slender, simple, ½ ft. long ;
raceme lax, simple, 6–9 in. long ; pedicels 1/12–⅛ in. long ; bracts

minute, deltoid; perianth $\frac{5}{8}$ in. long; limb half as long as the tube. *Baker in Journ. Linn. Soc.* xviii. 211. *Aloe venosa, Lam. Encyc.* i. 89; *Kunth, Enum.* iv. 514. *A. tricolor, Haw. in Trans. Linn. Soc.* vii. 25. *Apicra tricolor, Willd. in Ges. Naturf. Fr. Berl. Mag.* v. 271. *H. distincta,* N. E. *Brown in Gard. Chron.* 1876, vi. 130, *fig.* 30. (*Aloe africana humilis, etc., Commel. Prælud. t.* 29.)

CENTRAL REGION: Graaff Reinet Div., *Bolus !*

First introduced into cultivation at the beginning of the eighteenth century. Re-introduced by Bolus in 1875.

43. H. mirabilis (Haw. Syn. 95); leafy stem short; rosette 2 in. diam.; leaves about 20, multifarious, deltoid, much recurved, $1–1\frac{1}{4}$ in. long, $\frac{3}{4}$ in. broad, $\frac{3}{8}–\frac{1}{2}$ in. thick in the centre, firm in texture, flat on the much-recurved face, smooth, marked in the upper half with 3–5 pale vertical lines, rounded and scabrous with tubercles on the back, keeled towards the apex, turning red-brown when old, denticulate on the edges and keel; peduncle slender, simple, about $\frac{1}{2}$ ft. long; raceme lax, few-flowered; lower pedicels $\frac{1}{6}–\frac{1}{4}$ in. long; bracts small, lanceolate-deltoid; perianth $\frac{5}{8}$ in. long; limb half as long as the tube. *Baker in Journ. Linn. Soc.* xviii. 212. *Aloe mirabilis, Haw. in Trans. Linn. Soc.* vii. 9; *Bot. Mag. t.* 1354; *Salm-Dyck, Aloe, sect.* ix. *fig.* 1; *Kunth, Enum.* iv. 508. *Apicra mirabilis, Willd. in Ges. Naturf. Fr. Berl. Mag.* v. 269. *Haworthia multifaria, Haw. in Phil. Mag.* 1824, 300.

SOUTH AFRICA: without locality, living cultivated plants!

Introduced into cultivation by Masson in 1795.

44. H. affinis (Baker in Journ. Linn. Soc. xviii. 213); rosette about an inch long, $2–2\frac{1}{2}$ in. diam.; leaves about 20, multifarious, oblong-lanceolate, acute, an inch long, half an inch broad, $\frac{1}{3}$ in. thick, dull green, flat on the face, rather recurved upwards, pellucid towards the tip with 1–2 short green vertical lines, not awned, rounded on the back and keeled upwards, denticulate on the edges; peduncle simple, 6–9 in. long; raceme laxly 4–6-flowered; pedicels $\frac{1}{12}–\frac{1}{8}$ in. long; bracts lanceolate; perianth $\frac{5}{8}$ in. long; limb nearly as long as the tube.

SOUTH AFRICA: without locality.

Described from a living plant sent to Kew in 1875, by Mr. McGibbon, of the Cape Botanic Garden.

45. H. bilineata (Baker in Journ. Linn. Soc. xviii. 213); rosette of leaves about $1\frac{1}{2}$ in. long, 3 in. diam.; leaves about 15, multifarious, oblong-lanceolate, $1\frac{1}{2}–2$ in. long, $\frac{1}{2}$ in. broad, $\frac{1}{3}$ in. thick, dull green, flat on the face, pellucid in the upper third with 1–2 short vertical green lines, not awned, rounded on the back, keeled upwards, denticulate on the edges and keel; peduncle simple, 8–9 in. long; raceme laxly 5–6-flowered; pedicels $\frac{1}{12}–\frac{1}{8}$ in. long; bracts lanceolate; perianth $\frac{5}{8}$ in. long; segments keeled with red, nearly as long as the tube.

SOUTH AFRICA : without locality.

Described from a living plant sent to Kew in 1875, by Mr. McGibbon, of the Cape Botanic Garden.

46. H. columnaris (Baker in Journ. Bot. 1889, 45); leafy stem short ; rosette 3 in. diam.; leaves about 30, multifarious, obovate, cuneate, all ascending, not recurved, $\frac{3}{4}$–1 in. long, about $\frac{1}{2}$ in. broad, $\frac{1}{3}$ in. thick, minutely cuspidate, dull green, pellucid towards the apex for $\frac{1}{4}$ inch with greenish-brown vertical lines ; margin furnished with minute lanceolate or lanceolate-deltoid, deflexed or spreading pellucid teeth ; peduncle simple, about $\frac{1}{2}$ ft. long ; raceme simple, nearly a foot long ; lower pedicels short ; bracts lanceolate-deltoid, $\frac{1}{4}$–$\frac{1}{3}$ in. long; perianth $\frac{5}{8}$ in. long ; limb half as long as the tube.

SOUTH AFRICA : without locality.

Described from a living plant in the Kew collection in 1884, received from Messrs. Veitch, of Exeter.

47. H. subregularis (Baker in Saund. Ref. Bot. t. 232) ; rosette of leaves 1 in. long, $2\frac{1}{2}$–3 in. diam.; leaves 20–30, multifarious, ovate-lanceolate, ascending, $1\frac{1}{4}$–$1\frac{1}{2}$ in. long, $\frac{5}{8}$ in. broad, $\frac{1}{4}$ in. thick, glaucous green, flat on the face, lineate in the upper half and furnished with a few small whitish tubercles, rounded and distinctly keeled on the back ; margin denticulate; peduncle simple, $\frac{1}{2}$ ft. long ; raceme lax, simple, $\frac{1}{2}$ ft. long ; lower pedicels $\frac{1}{4}$–$\frac{1}{3}$ in. long ; bracts as long as the pedicels; perianth $\frac{3}{4}$ in. long ; limb nearly regular, half as long as the tube. *Baker in Journ. Linn. Soc.* xviii. 212.

SOUTH AFRICA : without locality, *Cooper !*

Described from living cultivated plants introduced by Cooper about 1860.

48. H. atrovirens (Haw. Revis. 57) ; rosettes copiously stoloni-ferous, about an inch long, 2 in. diam.; leaves 30–40, dense, multi-farious, oblong-lanceolate, $\frac{1}{2}$–$\frac{3}{4}$ in. long, $\frac{1}{4}$–$\frac{1}{3}$ in. broad, $\frac{1}{8}$ in. thick, dull green, reddish-brown when old, firm in texture, turgid on the face, pellucid towards the tip, with 3–5 vertical anastomosing green lines, rounded and keeled on the back and scabrous with tubercles, dentate on the margins and keel ; peduncle slender, simple $\frac{1}{2}$ ft. long ; raceme lax, simple, few-flowered, secund ; pedicels very short ; bracts minute, deltoid ; perianth $\frac{5}{8}$ in. long ; limb nearly as long as the tube. *Baker in Journ. Linn. Soc.* xviii. 212. *H. pumila, Haw. Syn.* 95. *Aloe atrovirens, DC. Plantes Grasses, t.* 51 ; *Salm-Dyck, Aloe, sect.* x. *fig.* 2. *Aloe pumila* ∈, *Linn. Sp. Plant.* 323. *A. herbacea, Miller, Gard. Dict. edit.* viii. *No.* 18. *A. arachnoides, var. pumila, Willd. Sp. Plant.* ii. 188 ; *Bot. Mag. t.* 1361. *Apicra atrovirens, Willd. in Ges. Naturf. Fr. Berl. Mag.* v. 168. *Catevala atroviridis, Medic. Theod.* 69.

SOUTH AFRICA: without locality, living cultivated plants !

Introduced into cultivation early in the eighteenth century.

49. H. lætevirens (Haw. Suppl. 53); rosette about an inch long, 2–3 in. diam.; leaves 20–30, dense, multifarious, ascending, oblanceolate-oblong, 1–1½ in. long, ⅜ in. broad, ⅛ in. thick, pale green, turning reddish when old, flat on the face, with 3–5 indistinct lines towards the apex, rounded and keeled on the back, with a distinct pellucid awn at the tip, denticulate on the margins and keel; peduncle simple, slender, ½ ft. long; raceme lax, simple, few-flowered; pedicels very short; bracts minute, deltoid; perianth ¾ in. long, limb half as long as the tube. *Baker in Journ. Linn. Soc.* xviii. 212. *Aloe lætevirens, Link, Enum.* i. 335; *Salm-Dyck, Aloe, sect.* x. *fig.* 3; *Kunth, Enum.* iv. 511.

SOUTH AFRICA: without locality, living cultivated plants!
CENTRAL REGION : Graaff Reinet Div., *Bolus!*

Introduced into cultivation by Bowie in 1818.

50. H. polyphylla (Baker in Journ. Linn. Soc. xviii. 213); leafy stem short; rosette 3–3½ in. diam.; leaves 40–50, dense, multifarious, oblong-lanceolate, 1½ in. long, ½ in. broad, ⅙–⅕ in. thick, green, flat on the face, not pellucid but paler towards the tip, with 7–9 close faint vertical lines, apex very acuminate and aristate; back rounded with 1–3 keels in the upper third; margins distinctly dentate; peduncle simple, a foot long; raceme simple, 12–15 in. long; lower pedicels ⅙–⅛ in. long; bracts lanceolate-deltoid; perianth ⅝–¾ in. long; limb shorter than the tube.

SOUTH AFRICA : without locality, *Cooper!*
Described from living cultivated plants introduced by Cooper about 1860.

Mr. N. E. Brown thinks this is a mere variety of *H. altilinea, Haw.*

51. H. denticulata (Haw. Revis. 58); leafy stem short; rosettes 3–4 in. diam.; leaves 20–30, dense, multifarious, oblong-lanceolate, cuneate in the lower half, 1¼–1½ in. long, ½–¾ in. broad at the middle, ⅙–¼ in. thick, pale green, colourless towards the tip with 5–7 green vertical lines, shortly awned, rounded on the back and keeled towards the apex; margins minutely denticulate; peduncle simple, about ½ ft. long; raceme lax, simple, about ½ ft. long; pedicels very short; bracts lanceolate-deltoid; perianth ⅝ in. long; limb nearly as long as the tube. *Baker in Journ. Linn. Soc.* xviii. 213. *Aloe denticulata, Roem. et Schultes, Syst. Veg.* vii. 639. *A. altilinea var. denticulata, Salm-Dyck, Aloe, sect.* xi. *fig.* 3 β; *Kunth, Enum.* iv. 512.

SOUTH AFRICA : without locality.

Introduced into cultivation by Bowie about 1818.

52. H. vittata (Baker in Saund. Ref. Bot. t. 263); leafy stem short; rosette 3–3½ in. diam.; leaves 20–30, dense, multifarious, oblong-lanceolate, very acuminate, 1½–1¾ in. long, ½–⅝ in. broad at the middle, ¼ in. thick, pale green, nearly flat on the face, limpid towards the apex on both sides with about five short vertical stripes, narrowed gradually into a long pellucid awn, rounded on the back and keeled towards the apex; margins denticulate; peduncle simple, stiff, about $\frac{1}{2}$ ft. long; raceme simple, dense upwards; pedicels very short; bracts lanceolate-deltoid; perianth $\frac{5}{8}$–$\frac{3}{4}$ in. long; limb half as long as the tube. *Baker in Journ. Linn. Soc.* xviii. 214.

SOUTH AFRICA : without locality, *Cooper!*

Described from living cultivated plants introduced by Cooper about 1860.

53. H. minima (Baker in Journ. Linn. Soc. xviii. 215); leafy stem very short; rosettes cæspitose, $1\frac{1}{4}$–$1\frac{1}{2}$ in. diam.; leaves 40–50, multifarious, oblong-lanceolate, $\frac{1}{2}$–$\frac{3}{4}$ in. long, $\frac{1}{6}$–$\frac{1}{4}$ in. broad, $\frac{1}{12}$ in. thick, pale glaucous-green, turgid on the face, not pellucid and lineate at the apex, rounded on the back and keeled upwards, tipped with a large pellucid awn; margins and keel ciliated with pellucid lanceolate teeth $\frac{1}{2}$ line long; peduncle simple, 3–4 in. long; raceme 10–12-flowered; lower pedicels $\frac{1}{6}$–$\frac{1}{3}$ in. long; bracts lanceolate-deltoid, as long as the pedicels; perianth $\frac{1}{2}$ in. long; limb half as long as the tube.

SOUTH AFRICA : without locality, *Tuck!*

Described from a living cultivated plant introduced by Tuck in 1866.

54. H. translucens (Haw. Suppl. 52); leafy stem short; rosette $2\frac{1}{2}$–3 in. diam.; leaves 30–40, dense, multifarious, ascending, lanceolate, $1\frac{1}{2}$–2 in. long, $\frac{1}{2}$ in. broad, $\frac{1}{3}$ in. thick, very acuminate, pale glaucous-green, often tinged with purple, turgid on the face, rounded and keeled upwards on the back, pellucid in the upper third with 5–7 green vertical lines, tipped with a pellucid awn, margined with pellucid deltoid cuspidate teeth $\frac{1}{2}$ line long; peduncle simple, $\frac{1}{2}$ ft. long; raceme simple, lax, few-flowered; lower pedicels $\frac{1}{8}$–$\frac{1}{6}$ in. long; bracts lanceolate-deltoid; perianth $\frac{5}{8}$ in. long; limb nearly as long as the tube. *Baker in Journ. Linn. Soc.* xviii. 214. *Aloe translucens, Ait. Hort. Kew. edit.* 2, ii. 300; *Salm-Dyck, Aloe, sect.* xii. *fig.* 1; *Haw. in Trans. Linn. Soc.* vii. 10. *A. arachnoides var. translucens, Ker in Bot. Mag. t.* 1417. *Apicra translucens, Willd. in Ges. Naturf. Fr. Berl. Mag.* v. 168. *H. pellucens, Haw. Syn.* 96.

COAST REGION : Algoa Bay, *Cooper!*

Introduced into cultivation by Masson in 1795.

55. H. pallida (Haw. Revis. 56); rosette 1 in. long, 2–3 in. diam.; leaves about 30, multifarious, ascending, lanceolate, 1–$1\frac{1}{4}$ in. long, $\frac{1}{3}$ in. broad, $\frac{1}{4}$ in. thick, pale green, turgid on the face, the upper third with vertical lines of darker green, the tip furnished with a large pellucid awn, the back rounded and doubly keeled towards the apex, the edges ciliated with deltoid-cuspidate pellucid teeth $\frac{1}{2}$ line long; peduncle simple, about a foot long; raceme lax, few-flowered, 4–6 in. long; pedicels short; bracts as long as the lower pedicels; perianth $\frac{5}{8}$ in. long; segments nearly regular, about as long as the tube. *Baker in Journ. Linn. Soc.* xviii. 214. *Aloe* *pallida*, *Roem. et Schultes*, *Syst. Veg.* vii. 641; *Kunth*, *Enum.* iv. 514.

SOUTH AFRICA: without locality.

Introduced into cultivation by Bowie about 1820.

56. H. pilifera (Baker in Saund. Ref. Bot. t. 234); rosette of leaves an inch long, 2 in. diam.; leaves 20–30, multifarious, ovate-oblong, an inch long, $\frac{1}{2}$ in. broad, $\frac{1}{4}$ in. thick, pale green, pellucid and lineate, with green vertical lines in the upper third, rounded on the back, furnished with a large pellucid awn; margins ciliated with pellucid deltoid-cuspidate teeth $\frac{1}{2}$ line long; peduncle simple, $\frac{1}{2}$ ft. long; raceme lax, simple, $\frac{1}{2}$ ft. long; pedicels very short; bracts lanceolate, $\frac{1}{4}$ in. long; perianth $\frac{3}{4}$ in. long; limb half as long as the tube. *Baker in Journ. Linn. Soc.* xviii. 214.

SOUTH AFRICA: without locality, *Cooper!*

COAST REGION: received at Kew from King Williamstown in 1883.

Introduced into cultivation by Cooper about 1860.

57. H. Cooperi (Baker in Saund. Ref. Bot. t. 233); rosette of leaves $1\frac{1}{2}$ in. long, $2\frac{1}{2}$–3 in. diam.; leaves 30–40, dense, multifarious, oblong-lanceolate, $1\frac{1}{2}$–$1\frac{3}{4}$ in. long, $\frac{5}{8}$ in. broad, $\frac{1}{4}$ in. thick, pale green, pellucid in the upper half with green vertical lines, flat on the face, convex on the back and keeled upwards, furnished with a large pellucid awn; margins and keel ciliated with pellucid deltoid-cuspidate teeth $\frac{1}{2}$ line long; peduncle simple, about a foot long; raceme simple, $\frac{1}{2}$ ft. long; pedicels very short; bracts deltoid, $\frac{1}{6}$–$\frac{1}{4}$ in. long; perianth $\frac{3}{4}$ in. long; limb half as long as the tube. *Baker in Journ. Linn. Soc.* xviii. 215.

SOUTH AFRICA: without locality, *Cooper!*

Introduced into cultivation by Cooper in 1860.

58. H. sessiliflora (Baker); rosette sessile, $2\frac{1}{2}$–3 in. diam.; leaves about 20, ovate, 1–$1\frac{1}{4}$ in. long, $\frac{5}{8}$–$\frac{3}{4}$ in. broad, much recurved, $\frac{1}{4}$ in. thick in the centre, bright green, pellucid in the upper third with 7 green vertical lines, tipped with a short pellucid awn; margins and keel closely ciliated with deflexed lanceolate pale teeth about $\frac{1}{2}$ line long; peduncle slender, simple, under a foot long; flowers arranged in a lax spike half a foot long, all perfectly sessile; bracts small, deltoid; perianth $\frac{3}{4}$ in. long; limb as long as the tube.

SOUTH AFRICA: without locality, *Cooper!*

Described from a living plant at Kew. Differs from all the other *Pallidæ* by its recurved leaves.

59. H. setata (Haw. Suppl. 52; Revis. 56); leafy stem short; rosette $2\frac{1}{2}$–3 in. diam.; leaves 30–40, dense, multifarious, all ascending, oblong-lanceolate, 1–$1\frac{1}{4}$ in. long, $\frac{1}{2}$–$\frac{5}{8}$ in. broad, and $\frac{1}{6}$–$\frac{1}{5}$ in. thick at the middle, pale glaucous green, turgid on the face, not pellucid at the apex, rounded on the back and tipped with a long pellucid awn;

margin ciliated with spreading lanceolate-cuspidate teeth $\frac{1}{12}$ in.
long, not so numerous as in *arachnoides* and *Bolusii*; peduncle
simple, $\frac{1}{2}$ ft. long ; raceme lax, simple, $\frac{1}{2}$ ft. long; pedicels very short ;
bracts small, deltoid ; perianth $\frac{1}{2}$–$\frac{5}{8}$ in. long; limb half as long as the
tube. *Baker in Journ. Linn. Soc.* xviii. 216. *Aloe setosa, Roem.
et Schultes, Syst. Veg.* vii. 641 ; *Salm-Dyck, Aloe, sect.* xii. *fig.* 3 ;
Kunth, Enum. iv. 513.

VAR. β, **nigricans** (Haw. Revis. 56) ; leaves darker green ; teeth more slender
and more numerous.

SOUTH AFRICA : without locality.

Introduced into cultivation by Mackrell in 1818.

60. H. xiphiophylla (Baker) ; leaves 50 or more in a dense
sessile rosette, lanceolate, 2 in. long, $\frac{1}{3}$ in. broad low down, pale
green, not lineate, flat on the face, acutely keeled on the back,
tapering to a pellucid awn $\frac{1}{4}$–$\frac{1}{3}$ in. long; margin with copious
pellucid prickles $\frac{1}{12}$–$\frac{1}{8}$ in. long, which extend to the upper part
of the keel; peduncle $\frac{1}{2}$ ft. long ; raceme lax, cernuous, 3–4 in.
long; pedicels very short ; bracts ovate, rather longer than the
pedicels; perianth-tube oblong, $\frac{1}{3}$ in. long ; segments of the limb
linear, obtuse, rather shorter than the tube ; stamens and style not
exserted from the perianth-tube.

COAST REGION : Uitenhage Div., *Howlett!*

Described from a living plant cultivated at Kew, introduced by Mr. Howlett in
1895.

61. H. arachnoides (Haw. Syn. 96) ; leafy stem about 1$\frac{1}{2}$ in.
long ; rosettes 3–4 in. diam. ; leaves 30–40, dense, multifarious,
oblong-lanceolate, 1$\frac{1}{2}$–2 in. long, $\frac{1}{2}$–$\frac{5}{8}$ in. broad at the middle, $\frac{1}{8}$ in.
thick, pale glaucous green, pellucid and lineate on both sides in the
upper third, flat on the face, rounded and keeled upwards on the
back, tipped with a large pellucid awn ; margin and keel ciliated
with pellucid, lanceolate acuminate, spreading teeth $\frac{1}{12}$–$\frac{1}{8}$ in. long;
peduncle simple, $\frac{1}{2}$–1 ft. long ; raceme lax, many-flowered, $\frac{1}{2}$ ft. long;
pedicels very short ; bracts deltoid, exceeding the pedicels. *Baker
in Journ. Linn. Soc.* xviii. 215. *Aloe arachnoides, Miller, Gard.
Dict. edit.* viii. *No.* 17 ; *Thunb. Fl. Cap. edit. Schult.* 311 ; *Haw. in
Trans. Linn. Soc.* vii. 10 ; *DC. Plantes Grasses, t.* 50 ; *Bot. Mag. t.*
756 ; *Salm-Dyck, Aloe, sect.* xii. fig. 2 ; *Kunth, Enum.* iv. 513.
Apicra arachnoides, Willd. in Ges. Naturf. Fr. Berl. Mag. v. 168.
Catevala arachnoidea, Medic. Theod. 68. *Aloe pumila, var. arach-
noidea, Linn. Sp. Plant,* 322. (*A. africana humilis arachnoidea,
Commel. Prælud. t.* 27.)

COAST REGION : Uitenhage Div.; Karoo near Zwartkops Zoutpan, *Thunberg!*
Introduced into cultivation at the beginning of the eighteenth century.

62. H. Bolusii (Baker in Journ. Linn. Soc. xviii. 215) ; leafy
stem about 1$\frac{1}{2}$ in. long; rosettes 2$\frac{1}{2}$–3 in. diam. ; leaves 30–40,

dense, multifarious, oblong-lanceolate, 1–1¼ in. long, ½ in. broad at
the middle, ⅛ in. thick, pale glaucous green, pellucid and lineate on
both sides in the upper third, flat on the face, rounded on the back
and keeled in the upper half, tipped with a large pellucid awn;
margin and keel ciliated with spreading pellucid lanceolate-subulate
teeth ⅛–⅙ in. long; peduncle simple, ½ ft. long; raceme simple,
many-flowered, ½ ft. long; lower pedicels ¹⁄₁₂–⅛ in. long; bracts
small, deltoid; perianth ⅝ in. long; limb half as long as the tube.

CENTRAL REGION : common about Graaff Reinet, *Bolus*, 158 !
Introduced into cultivation by Bowie in 1823.

63. H. stenophylla (Baker in Hook. Ic. Pl. t. 1974); bulb ovoid;
tunics few, thin, ovate; leaves 4, stiffly erect, narrow linear, 6–8 in.
long, with a prominent midrib and revolute entire margins;
peduncle slender, a foot long; raceme lax, simple, 3–4 in. long;
pedicels short, erecto-patent, articulated at the middle; lower bracts
lanceolate, ¼ in. long; perianth dull pinkish, the tube cylindrical,
¼ in. long; segments linear, subequally falcate, rather shorter than
tube; stamens included in the perianth-tube. *H. Saundersiæ, Baker
in Hook. Ic. Pl. sub t. 1974, name only.*

KALAHARI REGION: Transvaal; grassy slopes of the Saddleback Range,
near Barberton, 3600 ft., *Galpin*, 858! near Steyn, *Mrs. K. Saunders*, 9 !

64. H. tenuifolia (Engl. in Jahrb. x. 2, t. 1) ; rootstock bulbous;
leaves linear from a dilated scariose base, ½ ft. or more long, ⅛–¼ in.
broad, minutely setoso-dentate; peduncle above a foot long; raceme
dense upwards; pedicels very short; bracts ovate or oblong, acute,
¼–⅓ in. long; perianth nearly an inch long; segments ⅓–¼ as long as
the tube.

KALAHARI REGION　Bechuanaland ; near Kuruman, 3600 ft., *Marloth*, 1049 !

XI. **BULBINELLA**, Kunth.

Perianth campanulate, marcescent, persistent; segments distinct,
oblong, subequal, 1-nerved. *Stamens* 6, hypogynous or attached to
the base of the perianth-segments; filaments filiform or slightly
flattened ; anthers oblong, small, dorsifixed, versatile, dehiscing
introrsely. *Ovary* sessile, subglobose, 3-celled ; ovules 2, collateral,
erect ; style cylindrical; stigma capitate. *Capsule* globose, dehiscing
loculicidally. *Seeds* often solitary ; testa black; albumen fleshy.

Root of fleshy fascicled fibres; leaves all radical, terete or flat; scape leafless,
usually simple; flowers small, whitish or yellow, racemose; bracts solitary;
pedicels articulated at the apex.

DISTRIB. Two other species, one in New Zealand, the other in the Campbell
and Auckland islands.

A a 2

Racemes simple :
  Leaves very slender, subterete, 1-3-nerved :
    Lower pedicels ½–⅓ in. long :
      Root crowned with bristles   ...   . (1) **triquetra.**
      Root crowned with a membranous sheath   ... (2) **peronata.**
      Lower pedicels 1–2 in. long   ...   . (3) **filiformis.**
  Leaves linear, 6–12-nerved :
      Flowers whitish ...   . (4) **caudata.**
      Flowers bright yellow ...   . (5) **gracilis.**
  Leaves linear or lanceolate, many-nerved :
      Flowers whitish ; stamens half as long as the
        perianth...   ...   . (6) **carnosa.**
      Flowers bright yellow; stamens as long as the
        perianth...   ...   . (7) **robusta.**
  Racemes copiously panicled   ...   . (8) **Burkei.**

1. B. triquetra (Kunth, Enum. iv. 573) ; root of numerous fleshy fibres, its apex crowned with a dense mass of bristles ; leaves numerous, slender, terete, firm, glabrous, persistent, 4–6 in. long ; peduncle slender, terete, simple, ½–1 ft. long ; raceme dense, oblong or cylindrical, 1–3 in. long, ½ in. diam. ; pedicels ascending, lower ¼–⅓ in. long ; bracts minute, ovate or ovate-lanceolate ; perianth whitish, rarely bright yellow, ⅛ in. long ; stamens and pistil nearly as long as the perianth ; capsule globose, as long as the perianth. *Anthericum triquetrum, Linn. fil. Suppl.* 202 ; *Thunb. Prodr.* 62 ; *Fl. Cap. edit. Schult.* 317 ; *Baker in Journ. Linn. Soc.* xv. 293. *Bulbine triquetra, Roem. et Schultes, Syst. Veg.* vii. 451. *Bulbinella setifolia, Kunth, Enum.* iv. 569. *B. capillaris, Kunth, Enum.* iv. 572. *Phalangium capillare, Poir. Encyc.* vii. 247. *Anthericum capillare, Roem. et Schultes, Syst. Veg.* vii. 457.

VAR. β, **trinervis** (Baker) ; taller, more robust, with obscurely 3-nerved leaves. *Anthericum triquetrum var. trinervis, Baker in Journ. Linn. Soc.* xv. 294.

SOUTH AFRICA : without locality, *Masson! Oldenburg! Harvey,* 109 !
  COAST REGION : Malmesbury Div., *Zeyher,* 1690! near Groene Kloof, *Mac-Owan,* 2483! between Groene Kloof and Saldanha Bay, below 500 ft., *Drège,* 1503! Malmesbury, *Bachmann,* 1821! 2073! near Cape Town, *Bolus,* 3763! Simons Bay, *Wright!* Var. β, Caledon Div. ; Nieuw Kloof, Houw Hoek Mts., *Burchell,* 8043!
  CENTRAL REGION : Var. β, Tulbagh Div. ; Cold Bokkeveld, *Thunberg!*
  WESTERN REGION : Little Namaqualand, *Zeyher,* 1691! *Whitehead!*

2. B. peronata (Kunth, Enum. iv. 570) ; leaves numerous, terete, very slender, 3–4 in. long, surrounded at the base by a reticulated membranous sheath ; peduncle simple, a foot long ; raceme dense, oblong-cylindrical ; lower pedicels ⅙ in. long ; bracts ovate-cuspidate ; perianth yellow, ⅙ in. long ; segments oblong, obtuse ; stamens about as long as the perianth. *Anthericum peronatum, Baker in Journ. Bot.* 1872, 137 ; *Journ. Linn. Soc.* xv. 294.

COAST REGION : Tulbagh Div. ; Roodesand, near Tulbagh, below 1000 ft., *Drège,* 955.
  Known to me only by description.

3. B. filiformis (Kunth, Enum. iv. 572) ; leaves subterete, glabrous, 15–18 in. long, obscurely 3-nerved ; peduncle simple,

slender, flexuose, about a foot long ; raceme very lax, few-flowered ;
pedicels erect, 1½–2 in. long; bracts minute, lanceolate ; perianth
bright yellow, ¼ in. long ; segments oblong-lanceolate ; keel greenish ;
stamens nearly as long as the perianth. *Anthericum filiforme, Ait.
Hort. Kew.* i. 451 ; *edit.* 2, ii. 267 ; *Willd. Sp. Plant.* ii. 135 ;
Roem. et Schultes, Syst. Veg. vii. 456. *Phalangium filiforme, Poir.
Encyc.* v. 247. *Anthericum Aitoni, Baker in Journ. Linn. Soc.* xv
294.

SOUTH AFRICA : without locality.

Described from a type specimen of Aiton's plant in the herbarium of the British
Museum, dried from Kew Gardens in the year 1778. Introduced by Masson.

4. B. caudata (Kunth, Enum. iv. 572) ; root a dense mass of
fleshy fibres, crowned by a ring of copious bristles ; leaves linear,
moderately firm in texture, 1–1½ ft. long, ⅛–⅙ in. broad at the middle,
tapering gradually to the point, 8–12-nerved, ½ in. broad at the
clasping base ; peduncle slender, stramineous, 1–2 ft. long; raceme
dense, oblong, 2–4 in. long, about an inch in diameter ; lower pedicels
about ¼ in. long; bracts minute, ovate or ovate-lanceolate ; perianth
whitish, ⅙ in. long; segments oblong, obtuse ; keel ultimately brownish ;
stamens as long as the perianth; capsule ⅙ in. long. *Anthericum
Cauda-felis, Linn. fil. Suppl.* 202 ; *Willd. Sp. Plant.* ii. 146 ; *Baker
in Journ. Linn. Soc.* xv. 295. *Bulbine Cauda-felis, Roem. et
Schultes, Syst. Veg.* vii. 450. *Anthericum caudatum, Thunb. Prodr.*
63 ; *Fl. Cap. edit. Schult.* 321.

VAR. β, **B. ciliolata** (Kunth, Enum. iv. 570) ; leaves narrower and firmer in
texture, with more incurved margins ; raceme about ½ in. diam. *Anthericum
ciliolatum, Baker in Journ. Bot.* 1872, 137 ; *Journ. Linn. Soc.* xv. 294.

COAST REGION: Malmesbury Div.; Zwartland, *Thunberg !* between Groene
Kloof and Saldanha Bay, below 500 ft., *Drège !* Cape Div.; Koeberg and Tyger-
berg, *MacOwan*, 2500 ! Tulbagh Div.; Winter Hoek Mountain, 5000 ft., *Bolus*
5269 ! Var. β, Tulbagh Div.; Warm Bokkeveld near Ceres, 1800 ft., *Bolus*
2620 ! Uniondale Div.; Brak Kloof near Uniondale, *Bolus*, 2494 !
WESTERN REGION: Little Namaqualand, *Morris in Herb. Bolus*, 5808
Scully, 14! Var. β, Little Namaqualand, between Uitkomst and Geelbeks
Kraal, 2000–3000 ft., *Drège*, 2668 !

5. B. gracilis (Kunth, Enum. iv. 571) ; root of cylindrical fibres,
without any bristles at the crown; leaves weak, linear, ½ ft. long,
½–1 line broad, faintly 6–8-nerved ; peduncle simple, slender, terete,
1–1½ ft. long ; raceme oblong, 1–1½ in. long, 1 in. diam. ; pedicels
erecto-patent, lower ¼–⅓ in. long ; bracts minute, ovate ; perianth
bright yellow, ⅙ in long; segments oblong, obtuse ; stamens nearly
as long as the perianth. *Anthericum gracile, Baker in Journ. Bot.*
1872, 137 ; *Journ. Linn. Soc.* xv. 295. *A. nutans, Thunb. Prodr.*
63 ; *Fl. Cap. edit. Schult.* 319 ; *Baker in Journ. Linn. Soc.* xv. 294,
non Jacq.

SOUTH AFRICA : without locality, *Thunberg !*
WESTERN REGION: Little Namaqualand ; Kamies Bergen, 3000–4000 ft.,
Drège, 2670a ! Near Mieren Kasteel, *Drège*, 2670b !

6. B. carnosa (Baker) ; root of fascicled fleshy fibres, not bristly
at the crown; leaves 6–8, linear, glabrous, a foot long, $\frac{1}{2}$ in. broad
at the base, acuminate, channelled down the face ; peduncle simple,
slender, fragile, 15–18 in. long ; raceme lax, 6–8 in. long when fully
expanded, 1–1$\frac{1}{4}$ in. diam.; lower pedicels $\frac{1}{2}$ in. long ; bracts minute,
deltoid; perianth whitish, $\frac{1}{4}$ in. long; segments oblong, obtuse,
keeled with brown ; stamens not more than half as long as the
perianth ; filaments flattened ; ovules globose ; style short. *Anthericum carnosum, Baker in Journ. Linn. Soc.* xv. 296.

EASTERN REGION : Natal; on damp rocks between the rivers Umzinto and
Iffafa, *Gerrard and McKen,* 1890 !

7. B. robusta (Kunth, Enum. iv. 571) ; root of numerous cylindrical fibres, crowned by a dense mass of bristles ; produced leaves
6–10, linear, moderately firm in texture, acuminate, $\frac{1}{2}$–1$\frac{1}{2}$ ft. long,
$\frac{1}{4}$–$\frac{1}{3}$ in. broad at the middle, many-nerved, $\frac{2}{3}$ in. broad at the clasping
dilated base ; peduncle simple, terete, $\frac{1}{2}$–1$\frac{1}{2}$ ft. long ; raceme dense,
2–4 in. long, above an inch in diameter ; lower pedicels $\frac{1}{2}$–$\frac{3}{4}$ in. long ;
bracts minute, lanceolate ; perianth bright yellow, $\frac{1}{6}$ in. long ; segments oblong, obtuse ; stamens as long as the perianth ; capsule as
long as the perianth. *Anthericum setosum, Willd. ex Roem. et
Schultes, Syst. Veg.* vii. 473; *Baker in Journ. Bot.* 1872, 137 ;
Journ. Linn. Soc. xv. 295.

VAR. β, **B. latifolia** (Kunth, Enum. iv. 572); taller (sometimes 3 ft. high),
more robust, with larger flowers and lanceolate-acuminate leaves an inch broad
low down. *Anthericum floribundum, Ait. Hort. Kew.* i. 447; *edit.* 2, ii. 267;
Willd. Sp. Plant. ii. 137 ; *Baker in Journ. Linn. Soc.* xv. 296. *Trachyandra (?)
floribunda, Kunth, Enum.* iv. 583.

COAST REGION : Malmesbury Div.; Dassenberg, below 500 ft., *Drège,* 486 !
Table Mountain near Cape Town, *Harvey,* 108 ! Worcester Div.; Drakenstein
Mts , 2000–3000 ft., *Drège,* 8763 ! Caledon Div.; Zwartberg, 1000–2000 ft.,
Zeyher, 4211 ! Var. β, Malmesbury, *Bachmann,* 1978 ! Swampy places on Table
Mountain, 2400–2500 ft., *Bolus,* 4588 ! and *MacOwan, Herb. Aust. Afr.,* 1558 !
Summit of Kasteelberg, *Zeyher,* 5050 ! Caledon Div., *Ecklon and Zeyher,* 132 !
CENTRAL REGION : Somerset Div. ; Bruintjes Hoogte, lower part, *Burchell,*
3021 !
WESTERN REGION : Var. β, Little Namaqualand, between Uitkomst and
Geelbeks Kraal, 2000–3000 ft., *Drège,* 2667a !

8. B. Burkei (Benth. in Gen. Plant. iii. 784); root (not seen)
crowned with intermatted bristles 3–4 in. long; produced leaves
6–8, linear, rigid, 1–1$\frac{1}{2}$ ft. long, channelled down the face, scabrous
on the margin ; peduncle 3 ft. long including the inflorescence ;
racemes numerous, erect, very lax, $\frac{1}{2}$–1 ft. long; pedicels erecto-
patent, $\frac{1}{2}$–$\frac{3}{4}$ in. long ; bracts very minute; perianth white, turbinate,
$\frac{1}{12}$ in. long ; segments oblong, obtuse, with a distinct brown keel ;
stamens not exserted; capsule globose, stipitate, $\frac{1}{12}$ in. diam. *Anthericum Burkei, Baker in Journ. Bot.* 1872, 140 ; *Journ. Linn.
Soc.* xv. 298.

KALAHARI REGION : Transvaal ; Apies River, *Burke !*

XII. **BULBINE**, Linn.

Perianth membranous, marcescent, persistent; segments distinct, subequal, 1-nerved, spreading or reflexing when fully expanded. *Stamens* hypogynous or attached to the base of the segments, much shorter than the perianth; filaments filiform, densely bearded; anthers oblong, dorsifixed, versatile, dehiscing introrsely. *Ovary* sessile, globose, 3-celled; ovules superposed, horizontal; style short, filiform; stigma capitate. *Capsule* globose or turbinate, membranous, loculicidally 3-valved. *Seeds* angled by pressure; testa black, opaque; albumen fleshy.

Acaulescent or caulescent herbs, furnished often with a tuberous rootstock; leaves fleshy, subterete, linear or lanceolate; flowers simply racemose, usually bright yellow; pedicels solitary, articulated at the apex; bracts membranous, persistent.

DISTRIB. Two species Australian and two of the Cape species extend to the mountains of Equatorial Africa.

Leaves subterete :
  Perennial, caulescent :
    Stems elongated; leaves green   ...   (1) **caulescens.**
    Stems short; leaves glaucous   ...   (2) **rostrata.**
  Perennial, acaulescent :
    Leaves very slender, not fleshy :
      Whole plant a few inches high ...   (3) **minima.**
      Whole plant 1–1½ ft. high   ...   (4) **filifolia.**
    Leaves fleshy, $\frac{1}{12}$–$\frac{1}{4}$ in. diam. :
      Perianth dull pale yellow   ...   (5) **pallida.**
      Perianth whitish, keeled with
        brown...   ...   .   (6) **parviflora.**
      Perianth bright yellow :
        Raceme lax, about an inch in
          diameter :
          Rootstock fusiform, with
            slender root-fibres   ...   (7) **cæspitosa.**
          Rootstock globose, with
            stout, fleshy root-
            fibres...   ...   ...   (8) **favosa.**
        Raceme dense upwards, 1½–2
          in. diam. ...   .   (9) **asphodeloides.**
    Leaves fleshy, ¼ in. diam. low down   (10) **longiscapa.**
    Leaves fleshy, very stout :
      Leaves semicircular on the back :
        Rootstock oblong, præmorse   (11) **præmorsa.**
        Rootstock globose, not præ-
          morse   ...   .   (12) **pugioniformis.**
      Leaves sulcate on both sides   ...   (13) **bisulcata.**
  Annual, acaulescent   ...   .   (14) **annua.**
Leaves thin, linear or linear-lorate :
  Bracts minute, deltoid :
    Flowers whitish   ...   .   (15) **urgineoides.**
    Flowers greenish-yellow   ...   ...   (16) **laxiflora.**
    Flowers bright yellow :
      Rachis of raceme straight   ...   (17) **nutans.**
      Rachis of raceme zigzag   ...   (18) **flexicaulis.**
  Bracts large, much exceeding the buds :
    Pedicels ½ in. long   ...   .   (19) **narcissifolia.**
    Pedicels 1 in. long   ...   .   (20) **densiflora.**

Leaves thick, very small	(21) **mesembryanthemoides.**
Leaves thick, lanceolate or oblong-lanceolate :	
Raceme lax	(22) **alooides.**
Raceme dense ... ˙	(23) **latifolia.**
Leaves thin, oblong-lanceolate	(24) **natalensis.**
Leaves thin, lorate-oblong, densely ciliated ...	(25) **brunsvigiæfolia.**

1. B. caulescens (Linn. Hort. Cliff. 122) ; perennial, caulescent, without any tuberous rootstock ; stems sometimes a foot long, branched or simple, $\frac{1}{4}$–$\frac{1}{3}$ in. diam., with the lower leaves, $\frac{1}{2}$–1 in. apart, the upper crowded ; leaves 8–12, distichous, subterete, bright green, 6–9 in. long in the cultivated plant, $\frac{1}{4}$ in. diam., flat in the lower part of the face, turgid in the upper part, semicircular on the back ; peduncle simple, about a foot long ; raceme $\frac{1}{2}$–1 ft. long, an inch in diameter ; lower pedicels $\frac{1}{2}$ in. long, bracts small, deltoid-cuspidate, fimbriate at the enlarged base ; perianth bright yellow, $\frac{1}{3}$ in. long ; capsule globose, the size of a pea ; seeds 3–4 in a cell. *Baker in Journ. Linn. Soc.* xv. 343. *B. frutescens, Willd. Enum.* 372 ; *Roem. et Schultes, Syst. Veg.* vii. 442 ; *Kunth, Enum.* iv. 564. *Anthericum frutescens, Linn. Sp. Plant.* 310 ; *DC. Plantes Grasses, t.* 14 ; *Red. Lil. t.* 284. (*Dill. Hort. Elth. t.* 231, *fig.* 298.)

VAR. **B. incurva** (Roem. et Schultes, Syst. Veg. vii. 446) ; habit much more slender, with leaves 1$\frac{1}{2}$–2 in. long and a lax raceme. *Kunth, Enum.* iv. 566. *Anthericum incurvum, Thunb. Prodr.* 62 ; *Fl. Cap. edit. Schult.* 319.

SOUTH AFRICA : without locality, Var. β, *Thunberg !*
COAST REGION : Uitenhage Div. ; between Zwartkops River and Sunday River, *Zeyher*, 1057 ! near the Zwartkops River, *Zeyher*, 1059 ! *Ecklon and Zeyher*, 138 ! Port Elizabeth, sand-hills along the coast, *E.S.C.A. Herb.*, 271 ! King Williamstown Div. ; Keiskamma Hoek, *Cooper*, 3306 ! Var. β, Uitenhage Div. ; near the Sunday River, *Zeyher*, 1058 ! Albany Div. ; between Grahamstown and Blue Krantz, *Burchell*, 3618 !
CENTRAL REGION : Prince Albert Div. ; Zwartbulletje 2500–3000 ft., *Drège*, 8762c ! Somerset Div. ; banks of the Little Fish River, 2500 ft., *MacOwan*, 1893 !

B. incurva is perhaps rather the wild original of the species, grown in arid soil, than a real variety.

2. B. rostrata (Willd. Enum. 372) ; perennial, shortly caulescent ; stem not more than a few inches long ; leaves about 8, subterete, glaucous, $\frac{1}{2}$–1 ft. long, $\frac{1}{4}$ in. diam. low down, tapering to the point ; peduncle slender, $\frac{1}{2}$–1 ft. long ; raceme $\frac{1}{2}$–1 ft. long, an inch in diameter ; lower pedicels $\frac{1}{2}$ in. long ; bracts minute, deltoid-cuspidate ; perianth bright yellow, $\frac{1}{4}$ in. long ; capsule the size of a pea. *Roem. et Schultes, Syst. Veg.* vii. 442 ; *Kunth, Enum.* iv. 564 ; *Baker in Journ. Linn. Soc.* xv. 343. *Anthericum rostratum, Jacq. Coll. Suppl.* 82 ; *Ic.* ii. 17, *t.* 403 ; *Willd. Sp. Plant.* ii. 143.

COAST REGION : Albert Div. ; near Bushman River, below 1000 ft., *Drège*, 8742 !

Described from a living plant at Kew. From dried specimens I cannot distinguish it from *B. caulescens.*

3. B. minima (Baker in Journ. Linn. Soc. xv. 344); acaulescent, perennial, with a globose tuber $\frac{1}{2}$–$\frac{3}{4}$ in. diam.; produced leaves about 6, very slender, subulate, 1–1$\frac{1}{2}$ in. long; peduncle very slender, 2–3 in. long; racemes lax, few-flowered, 1–3 in. long; lower pedicels $\frac{1}{4}$–$\frac{1}{3}$ in. long; bracts minute, deltoid; perianth bright yellow, $\frac{1}{6}$ in. long; segments oblanceolate; capsule not seen.

COAST REGION: Tulbagh Div.; near Tulbagh, under 1000 ft., *Drège*, 953!

4. B. filifolia (Baker in Journ. Linn. Soc. xv. 344); acaulescent, perennial, with a globose tuber $\frac{3}{4}$–1 in. diam., crowned by a dense ring of erect bristles 1–1$\frac{1}{2}$ in. long; leaves 6–8, subterete, firm in texture, erect, $\frac{1}{2}$–1 ft. long, $\frac{1}{4}$ lin. diam., tapering to the point, flat on the face downwards; peduncle very slender, stiffly erect, 1–1$\frac{1}{2}$ ft. long; raceme lax, 1$\frac{1}{2}$–2 in. long; lower pedicels erecto-patent, $\frac{1}{6}$–$\frac{1}{4}$ in. long; bracts deltoid-cuspidate, minute; perianth pale yellow, under $\frac{1}{4}$ in. long; segments lanceolate, spreading when fully expanded.

CENTRAL REGION: Graaff Reinet Div.; on the summit of the Tandjes Berg, 4900 ft., *Bolus*, 762.

5. B. pallida (Baker in Journ. Bot. 1876, 184); perennial, acaulescent, with an oblong tuberous rootstock of firm texture; leaves 6–10, subterete from a clasping membranous deltoid base, 3–4 in. long, fleshy in texture, 1 lin. diam. at the middle; peduncle slender, fragile, erect, $\frac{1}{2}$ ft. long; raceme lax, 2–3 in. long, under an inch in diameter; lower pedicels $\frac{1}{2}$–$\frac{3}{4}$ in. long; bracts deltoid, with a long cusp, deeply fimbriated at the white membranous base; perianth pale yellow, $\frac{1}{4}$ in. long; segments oblanceolate; capsule not seen.

EASTERN REGION: Tembuland; Tabase near Bazeia, 2500 ft., *Baur*, 417!

6. B. parviflora (Baker); acaulescent, perennial; base of the stem surrounded by several large membranous clasping sheath-leaves; produced leaf linear-subulate, not fleshy, 4 in. long, a line broad, flat below the apex; peduncle above a foot long, stiffly erect, terete; raceme cylindrical, dense upwards, $\frac{1}{2}$ ft. long, $\frac{1}{2}$ in. diam.; pedicels erecto-patent, articulated at the apex, $\frac{1}{6}$–$\frac{1}{4}$ in. long; bracts lanceolate from a deltoid base, nearly or quite as long as the perianth; perianth $\frac{1}{6}$ in. long; segments whitish, with a distinct brown keel; stamens nearly as long as the perianth; anthers small, cordate-ovate.

COAST REGION: Worcester Div.; Drakenstein Mountains near Bains Kloof, 1600 ft., *Bolus*, 4070!

7. B. cæspitosa (Baker); acaulescent, perennial, cæspitose, with a short fusiform rootstock, sending out slender cylindrical root-fibres; leaves densely tufted, subulate from a thin deltoid clasping base, $\frac{1}{2}$ ft. long, 1 lin. diam. at the middle; peduncle $\frac{1}{2}$–1 ft. long; raceme lax, finally $\frac{1}{2}$ ft. long, 1 in. diam.; lower pedicels $\frac{1}{2}$ in. long; bracts deltoid-cuspidate, with a dentate white base; perianth bright yellow, ¼ in. long; capsule turbinate, ⅙ in. diam.

KALAHARI REGION: Bechuanaland; at the source of the Moshowa River, near Takun, *Burchell*, 2279! Orange Free State; Caledon River, *Burke!*

8. **B. favosa** (Roem. et Schultes, Syst. Veg. vii. 444); perennial, acaulescent, with a large tuberous rootstock; leaves rosulate, sub-terete, slender, ½ ft. long, ¹⁄₁₂–⅛ in. diam. low down; peduncle slender, fragile, 1–1½ ft. long; raceme lax, 4–6 in. long, an inch in diameter; lower pedicels ¼ in. long; bracts minute, deltoid-cuspidate; perianth bright yellow, ¼ in. long; capsule the size of a pea. *Kunth, Enum.* iv. 564; *Baker in Journ. Linn. Soc.* xv. 343. *Anthericum favosum, Thunb. Prodr.* 63; *Fl. Cap. edit. Schult.* 321.

COAST REGION: Cape Flats, *Thunberg! MacOwan and Bolus, Herb. Norm.,* 1386! near Kenilworth, *Bolus*, 7055! Table Mountain, 800–1500 ft., *Bolus,* 4663!

9. **B. asphodeloides** (Roem. et Schultes, Syst. Veg. vii. 444); acaulescent, perennial, with a short, thick, tuberous rootstock; leaves 10–20, subterete from a dilated clasping base ½ in. broad, ½–1 ft. long, ¹⁄₁₂–⅛ in. thick at the middle, glaucous, tapering to the point, nearly flat on the face; peduncle ½–1 ft. long, ancipitous towards the base; raceme dense, sometimes ½ ft. long, 1½–2 in. diam.; lower pedicels erecto-patent, ½–¾ in. long; bracts lanceolate-deltoid with a long cusp; perianth bright yellow, ¼–⅓ in. long; capsule turbinate, ⅙ in. diam.; seeds 3–4 in a cell. *Kunth, Enum.* iv. 564; *Baker in Journ. Linn. Soc.* xv. 345. *Anthericum asphodeloides, Linn. Sp. Plant.* 311; *Miller, Gard. Dict. edit.* viii. *No.* 6; *Jacq. Hort. Vind.* ii. 85, *t.* 181. *B. graminea, Haw. Revis.* 33? *B. abyssinica, A. Rich. Fl. Abyss.* ii. 334, *t.* 97.

COAST REGION: Cape Flats and neighbourhood of Cape Town, *Zeyher*, 1692! *Ecklon and Zeyher*, 136! *Pappe!* Caledon Div.; Hartebeest River, *Zeyher,* 4227! near the Zonder Einde River, *Zeyher*, 4228! Riversdale Div.; between Zoetmelks River and Little Vet River, *Burchell*, 6288!
CENTRAL REGION: Somerset Div.; near Little Fish River, *MacOwan*, 1834! Colesberg Div., *Shaw!* Hanover Div.; near Hanover, *Shaw*, 53!
KALAHARI REGION: Griqualand West; between Spuigslang and the Vaal River, *Burchell*, 1711! near the Asbestos Mountains, *Burchell*, 2002! 2089 Orange Free State; near Nelsons Kop, *Cooper*, 878! Bechuanaland; on the rocky ridge at Takun, *Burchell*, 2258! between Kuruman and the Vaal River, *Cruickshank in Herb. Bolus*, 2551! Transvaal; near Pretoria, *Rehmann*, 4300! (a form with very slender leaves).
EASTERN REGION: Tembuland; near Bazeia, 2000 ft., *Baur*, 99! Natal; Inanda, *Wood*, 1008! without precise locality, *Gerrard*, 357!

Also Abyssinia, Angola, and highlands of the Zambesi.

10. **B. longiscapa** (Willd. Enum. 372); acaulescent, perennial, with a small tuberous rootstock; leaves 12–20, subterete, fleshy, very glaucous, 8–12 in. long, ¼ in. diam. low down, flattish on the face, rounded on the back; peduncle a foot or more long; raceme dense upwards, a foot or more long, 1½ in. diam.; central pedicel

½–¾ in. long; bracts deltoid-cuspidate, ⅙–¼ in. long, with a dentate white base ; perianth bright yellow, ½ in. long ; segments oblanceolate, reflexing when fully expanded; capsule the size of a large pea. *Roem. et Schultes, Syst. Veg.* vii. 443 ; *Kunth, Enum.* iv. 564 ; *Baker in Journ. Linn. Soc.* xv. 344. *Anthericum longiscapum, Jacq. Coll. Suppl.* 84 ; *Ic.* ii. 17, *t.* 404 ; *Red. Lil. t.* 423 ; *Gawl. in Bot. Mag. t.* 1339. *A. asphodeloides, Ait. Hort. Kew.* i. 450, *non Linn. A. altissimum, Miller, Gard. Dict. edit.* viii. *No.* 8 ; *Ic.* 26, *t.* 39. *A. Lagopus, Thunb. Prodr.* 63.

SOUTH AFRICA : without locality, *Thunberg ! Hort. Jacquin !*
CENTRAL REGION : Somerset Div., *Bowker !*

11. B. præmorsa (Roem. et Schultes, Syst. Veg. vii. 446); acaulescent, perennial, with an oblong præmorse rootstock and thick fleshy root-fibres ; leaves 8–12, subterete, distichous, 1–1½ ft. long, ¼–⅓ in. diam. at the middle, deeply channelled down the face ; peduncle subterete, 1–1½ ft. long; raceme dense upwards, a foot long, 1½ in. diam.; lower pedicels ½ in. long; bracts minute, lanceolate-deltoid ; perianth bright yellow, ⅓ in. long. *Kunth, Enum.* iv. 567 ; *Baker in Journ. Linn. Soc.* xv. 346. *Anthericum præmorsum, Jacq. Coll. Suppl.* 85 ; *Ic.* ii. 17, *t.* 406.

SOUTH AFRICA : without locality, *Thunberg !*

12. B. pugioniformis (Link, Enum. i. 329) ; acaulescent, perennial, with a large globose tuberous rootstock and thick fleshy root-fibres; leaves subterete, glaucous, 1–1½ ft. long, ⅓–½ in. thick in the middle, deeply channelled down the face, rounded on the back ; peduncle stout, subterete, 1–1½ ft. long; raceme dense upwards, ½–1 ft. long, 1–1¼ in. diam.; lower pedicels ½ in. long ; bracts small, lanceolate-cuspidate ; perianth bright yellow, ¼–⅓ in. long; segments oblong, obtuse. *Roem. et Schultes, Syst. Veg.* vii. 444 ; *Kunth, Enum.* iv. 564 ; *Baker in Journ. Linn. Soc.* xv. 346. *Anthericum pugioniforme, Jacq. Coll. Suppl.* 83 ; *Ic.* ii. 17, *t.* 405 ; *Andr. Bot. Rep. t.* 386 ; *Bot. Mag. t.* 1454.

COAST REGION : Riversdale Div.; near Zoetmelks River, *Burchell*, 6711 !

13. B. bisulcata (Haw. in Phil. Mag. 1827, 121); acaulescent, perennial, with a large tuberous rootstock ; leaves green, subulate, fleshy, a foot long, channelled down both sides, but more deeply down the face ; peduncle shorter than the leaves; inflorescence and flowers of *B. pugioniformis. Roem. et Schultes, Syst. Veg.* vii. 445 ; *Kunth, Enum.* iv. 565 ; *Baker in Journ. Linn. Soc.* xv. 346.

SOUTH AFRICA : without locality.

Described from a plant in the garden of Mr. Sweet. Unknown to me. There are no specimens preserved in the London herbaria.

14. B. annua (Willd. Enum. 372) ; annual, acaulescent, without any rootstock ; root-fibres slender ; leaves 12–20, erect, weak, subulate from a dilated clasping base, ½–1 ft. long, 1/12–⅛ in. thick in the middle, green, slightly channelled in the lower part of the face; flowering stems often 3–4 from one root; peduncle subterete, 6–9 in. long; raceme very lax, 10–15-flowered, finally 4–6 in. long, 1½–2 in. diam.; pedicels ascending; lower ¾–1 in. long; bracts small, lanceolate-deltoid, cuspidate; perianth bright yellow, ¼ in. long; capsule subglobose, ⅙ in. diam. ; seeds 3–4 in a cell. *Roem. et Schultes, Syst. Veg.* vii. 445; *Kunth, Enum.* iv. 566; *Baker in Journ. Linn. Soc.* xv. 346. *Anthericum annuum, Linn. Sp. Plant.* 311; *DC. Plantes Grasses, t.* 8; *Red. Lil. t.* 397; *Bot. Mag. t.* 1451.

COAST REGION: Cape Div.; Camps Bay, *Burchell*, 351! Kalk Bay, *Pappe!*

15. B. urgineoides (Baker in Journ. Linn. Soc. xv. 348); acaulescent, perennial, with a tuberous rootstock an inch in diameter, with many fleshy root-fibres from its base; leaves 5–6, fleshy, green, lanceolate, a foot long, 1¼–1½ in. broad low down, narrowed gradually to the point; peduncle stout, terete, a foot long; raceme dense upwards, ½ ft. long, 1½ in. diam.; pedicels ascending, lower ½–⅝ in. long; bracts deltoid, minute; perianth whitish, ¼ in. long; segments oblong, obtuse, with a brown keel; capsule not seen.

WESTERN REGION: Little Namaqualand, *Whitehead!* in Dublin herbarium.

May be the imperfectly known *B. dubia, Roem. et Schultes, Syst. Veg.* vii. 450, described from an incomplete specimen in the herbarium of Zuccarini.

16. B. laxiflora (Baker in Journ. Linn. Soc. xv. 347); acaulescent, perennial; leaves linear, acuminate, fleshy, concave within, glaucous, 6–8 in. long, ¼–⅓ in. broad at the middle, tapering gradually to the point; peduncle subterete, nearly a foot long, glaucous; racemes very lax, ½–1 ft. long, 2 in. diam.; pedicels ½–¾ in. long, decurved after flowering; bracts deltoid, very minute, white; perianth yellow, with green midribs to the segments, ½ in. long; the three interior segments obovate, spreading, thrice as broad as the narrow reflexed outer segments; stamens half as long as the perianth-segments.

SOUTH AFRICA : without locality, *Burchell*, bulb No. 198!

Described from a specimen dried by Burchell, from his garden at Fulham, in July, 1817, with particulars of colour, &c., taken from Burchell's MS. description of the living plant.

17. B. nutans (Roem. et Schultes, Syst. Veg. vii. 447); acaulescent, perennial, with a small rootstock and numerous thick basal root-fibres; leaves 6–8, erect, linear from a broad clasping base, 6–9 in. long, ¼–⅓ in. broad at the middle, deeply channelled down the face; peduncle ½–1 ft. long; raceme dense upwards, ½ ft. or more long, 1–1¼ in. diam.; pedicels ascending, ⅓–½ in. long; bracts lanceolate-deltoid, minute, entire; perianth bright yellow, ¼ in. long; capsule ovoid, membranous, greenish, ⅓ in. long, with several seeds in a cell. *Kunth, Enum.* iv. 567; *Baker in Journ. Linn. Soc.* xv. 348. *Anthericum nutans, Jacq. Coll. Suppl.* 86; *Ic.* ii. 17, *t.* 407.

SOUTH AFRICA : without locality, *Ecklon and Zeyher*, 153!
COAST REGION : hills by the sea-shore at Sea Point, near Cape Town, 100 ft. *Bolus*, 3709!

18. B. flexicaulis (Baker) ; acaulescent, perennial, with a small forked rootstock; leaves 4–6, narrow linear, thin, fleshy, erect, 3–4 in. long, about a line broad at the middle; peduncle slender, fragile, terete, ½ ft. long; raceme very lax, with a zigzag rachis, 3–6 in. long, under an inch in diameter; pedicels ascending, ¼–⅓ in. long; bracts minute, deltoid; perianth bright yellow, under ¼ in. long; capsule globose, membranous, ⅙ in. diam.

COAST REGION : Uitenhage Div. ; among shrubs, on hills by the Zwartkops River, *Pappe*, 80!

19. B. narcissifolia (Salm-Dyck, Hort. Dyck. 334); acaulescent, perennial, with a dense tuft of cylindrical root-fibres ; leaves 6–9, linear-lorate from a dilated base, thin, fleshy, glaucescent, 6–12 in. long, ¼–½ in. broad at the middle; peduncle subterete, 1–1½ ft. long ; raceme short, dense, 1–1½ in. diam.; pedicels ascending, lower ½–¾ in. long; bracts lanceolate-deltoid, membranous, whitish, ¼–⅓ in. long, much exceeding the buds; perianth bright yellow, ⅓ in. long ; segments oblong; capsule not seen. *Kunth, Enum.* iv. 567 ; *Baker in Journ. Linn. Soc.* xv. 347.

SOUTH AFRICA : without locality, *Ecklon and Zeyher*, 133 !
COAST REGION : Stockenstrom Div. ; Seymour, *Scully*, 71 !
CENTRAL REGION : Somerset Div. ; Bosch Berg, 3000 ft., *MacOwan*, 1583 !
KALAHARI REGION : Transvaal ; Matebe Valley, *Holub*, 1587 ! 1589 !
EASTERN REGION : Natal ; Colenso, on a grassy hill, 3000 ft., *Wood*, 4026 !

20. B. densiflora (Baker in Journ. Linn. Soc. xv. 347) ; acaulescent, perennial, with a thick tuberous rootstock of firm texture ; leaves numerous, thin, linear, suberect, 6–9 in. long, ¼–⅓ in. broad at the middle ; peduncle subterete, 1–1½ ft. long ; raceme short, very dense, 1½ in. diam.; pedicels ascending, sometimes an inch long ; bracts lanceolate-deltoid, cuspidate, whitish, membranous, ⅓–½ in. long, much exceeding the buds ; perianth bright yellow, ¼ in. long ; capsule not seen.

CENTRAL REGION : Somerset Div., *Bowker !*   Bedford Div., *Burke !*

21. B. mesembryanthemoides (Haw. in Phil. Mag. 1825, 31) ; acaulescent, perennial, with a tuberous rootstock ¼ in. diam., forking into fibres at the base ; leaves few, very fleshy, ⅓–½ in. long, oblong or obovate-spathulate, flattish on the face, hemispherical on the back ; peduncle very slender, stiffly erect, 2–3 in. long ; raceme very lax, 3–6-flowered ; pedicels erecto-patent, ¼–⅓ in. long ; bracts deltoid, very minute ; perianth bright yellow, ⅙ in. long ; segments oblanceolate, reflexing. *Roem. et Schultes, Syst. Veg.* vii. 448 ; *Kunth, Enum.* iv. 568 ; *Baker in Journ. Linn. Soc.* xv. 348.

COAST REGION : Uitenhage Div. ; among shrubs near the Zwartkops River, *Zeyher*, 1068 !   Albany Div., *Bowker !*

There is a drawing at Kew of a plant from Bowie, grown at Kew in the year 1825.

22. B. alooides (Willd. Enum. 372); acaulescent, perennial, with a thick tuberous rootstock ; leaves 6–12, dense, rosulate, thick,

fleshy, bright green, lanceolate, 6–9 in. long, an inch broad low down,
tapering gradually to a long point; peduncle slender, terete, a foot long;
raceme lax, ½–1 ft. long, 1–1½ in. diam.; pedicels very slender,
ascending, lower ½ ft. long; bracts minute, deltoid-cuspidate;
perianth bright yellow, ¼ in. long; segments oblong, obtuse, reflexing;
capsule not seen.   *Roem. et Schultes, Syst. Veg.* vii. 448; *Kunth,
Enum.* iv. 567; *Baker in Journ. Linn. Soc.* xv. 348.   *Anthericum
alooides, Linn. Sp. Plant,* 311; *Miller, Dict. edit.* 6, *No.* 5; *Thunb.
Prodr.* 62; *Fl. Cap. edit. Schult.* 319; *Red. Lil. t.* 283; *Bot. Mag.
t.* 1317; *Lodd. Bot. Cab. t.* 996.   *Bulbine acaulis, Linn. Hort.
Cliff.* 123.   *B. Zeyheri, Baker in Journ. Linn. Soc.* xv. 347.   (*Dill.
Hort. Elth. t.* 232, *fig.* 299.)

SOUTH AFRICA: without locality, *Zeyher,* 4219! *Pappe!* and cultivated
specimens!
COAST REGION: Clanwilliam Div.; between Oliphants River and Kanagas
Berg, *Zeyher,* 4222! Mountains near Cape Town, *Thunberg!* Swellendam Div.;
Hessequas Kloof, *Zeyher,* 4224!

23. B. latifolia (Roem. et Schultes, Syst. Veg. vii. 447); acaulescent, perennial, cæspitose, with a stout rootstock of firm texture;
leaves 12–20 in a dense rosette, oblong-lanceolate, thick, fleshy,
glaucescent on the rounded back, a foot long, 2–3 in. broad low
down, tapering gradually to a point, the outer flat, the inner concave
on the face; peduncle above a foot long, ancipitous towards the
base; raceme dense upwards, a foot long, 1½ in. diam.; pedicels
⅓–½ in. long; bracts small, deltoid-cuspidate; perianth bright yellow,
⅓ in. long; stamens more than half as long as the perianth-segments.
Kunth, Enum. iv. 567; *Baker in Journ. Linn. Soc.* xv. 349.
Anthericum latifolium, Linn. fil. Suppl. 202; *Thunb. Prodr.* 63;
Fl. Cap. edit. Schult. 323; *Jacq. Coll.* iii. 249; *Ic.* ii. 17, *t.* 408.
B. macrophylla, Salm-Dyck, Hort. Dyck. 333; *Kunth, Enum.* iv. 568.
B. Mettinghii, Tenore, Syll. 563?

CENTRAL REGION: Somerset Div.; Bosch Berg, 3000 ft., *MacOwan,* 1894!

24. B. natalensis (Baker); acaulescent, perennial, with a tuberous
rootstock of firm texture, nearly an inch in diameter; leaves numerous,
thin, ascending, oblong-lanceolate, acuminate, 8–9 in. long, 1¼–1½ in.
broad low down; peduncle ½–1 ft. long; raceme moderately dense,
finally ½ ft. long; pedicels ascending, lower ½–⅝ in. long; bracts
lanceolate from a deltoid base, ⅙ in. long, protruded beyond the buds;
perianth ⅙ in. long; segments bright yellow with a broad green
keel; stamens much shorter than the perianth-segments; capsule
subglobose, ⅛ in. diam.

EASTERN REGION: Natal; Inanda, *Wood,* 553!

25. B. (?) brunsvigiæfolia (Baker); leaf thin, lorate-oblong, 6–8 in.
long, 2 in. broad, subobtuse, minutely cuspidate, glabrous on the
surface, densely persistent, shortly ciliated on the edge; peduncle
stout, a foot long; flowers many in a lax raceme 3–4 in. long;

pedicels ascending or spreading, $\frac{1}{2}$–$\frac{3}{4}$ in. long; bracts membranous, ovate-lanceolate, $\frac{1}{3}$–$\frac{1}{2}$ in. long; perianth $\frac{1}{2}$ in. long; segments oblanceolate, obtuse, distinctly keeled with green; stamens $\frac{1}{8}$ in. long; filaments densely stuppose in the upper half; anthers small, oblong; style filiform, $\frac{1}{6}$ in. long; stigma capitate.

WESTERN REGION: Little Namaqualand; Modder Fontein, 1500–2000 ft., *Drège*, 2674!

XIII. BOWIEA, Harv.

Flowers polygamous. *Perianth* campanulate, marcescent; segments distinct, lanceolate, reflexing. *Stamens* perigynous, much shorter than the perianth; filaments slightly flattened; anthers oblong, versatile, dehiscing introrsely. *Ovary* sessile, ovoid, 3-celled; ovules numerous, horizontal, superposed; style short, cylindrical; stigma obscurely 3-lobed. *Capsule* membranous, dehiscing loculicidally. *Seeds* oblong, compressed; testa lax, black, shining; albumen fleshy.

DISTRIB. Endemic.

1. **B. volubilis** (Harv. in Bot. Mag. t. 5619); rootstock a green, globose, tuber-like bulb 4–6 inches in diameter, with a few thick distichous tunics; produced leaves 1–2, small, subterete, erect, vanishing very early; stem sarmentose, reaching a length of 6–8 ft., producing copious, spreading, pinnate branches, with slender, terete, arcuate-ascending branchlets 1–1$\frac{1}{2}$ in. long, with a minute, lanceolate bract at the base; flowers greenish, produced from the main stem on long arcuate pedicels, some imperfect, $\frac{1}{3}$–$\frac{1}{2}$ in. diam.; capsule ovoid, brownish, $\frac{1}{3}$–$\frac{1}{2}$ in. long. *Baker in Journ. Linn. Soc.* xiii. 291; *Irmisch in Abhandl. Nat. Ver. Bremen*, vi. 433, t. 5.

COAST REGION: Stockenstrom Div.; Katberg, *Hutton!*
KALAHARI REGION: Orange Free State, *Cooper*, 3609! Transvaal, *Nelson*, 89!
EASTERN REGION: Transkei, 3000 ft., *Bowker!* Griqualand East; near Kokstad, 5300 ft., *Tyson! MacOwan and Bolus, Herb. Norm. Aust. Afr.*, 482! Natal, *McKen*, 32! *Cooper*, 3263!

XIV. SCHIZOBASIS, Baker.

Perianth campanulate, marcescent, persistent; segments distinct, subequal, 1-nerved. *Stamens* attached to the base of the perianth-segments; filaments slightly flattened; anthers oblong, dorsifixed, versatile, dehiscing introrsely. *Ovary* sessile, 3-celled; ovules few in a cell, superposed; style cylindrical; stigma 3-lobed. *Capsule* membranous, loculicidally 3-valved. *Seeds* 1–3 in a cell, turgid or angled by pressure; testa black, lax; albumen fleshy.

Rootstock a globose bulb with membranous tunics; leaves radical, evanescent;

stems climbing or erect; inflorescence various; bracts minute; pedicels solitary, articulated at the apex.

DISTRIB. A single species in Angola.

Stems climbing ...   .   (1) **cuscutoides.**
Stems erect:
   Inflorescence a simple raceme   ...   .   (2) **flagelliformis.**
   Inflorescence a panicle with racemose branches...   (3) **Macowani.**
   Inflorescence a lax intricate corymbose panicle...   (4) **intricata.**

1. S. cuscutoides (Benth. Gen. Plant, iii. 786); bulb ovoid-sub globose, tunicated, with pale red flesh and abundance of viscous juice; leaves unknown; stems annual, filiform, climbing, forming a lax intricately branched mass; flowers solitary, lateral or terminal, on deflexed or erecto-patent slender pedicels $\frac{1}{6}$–$\frac{1}{4}$ in. long; perianth $\frac{1}{8}$ in. long; segments oblong, obtuse, white, with a distinct brown keel; stamens nearly as long as the perianth; capsule globose, $\frac{1}{6}$ in diam. *Asparagus cuscutoides, Burchell ex Baker in Journ. Linn. Soc.* xiv. 606.

CENTRAL REGION: Hopetown Div.; by the Orange River, near Vissers Drift, *Burchell*, 2673!
KALAHARI REGION: Orange Free State; near Boshof, *Mrs. Barber!*

2. S. flagelliformis (Baker in Journ. Linn. Soc. xv. 261); bulb and leaves unknown; stem erect, 1–1$\frac{1}{2}$ ft. long including the inflorescence; raceme sometimes a foot long; rachis slightly curved; pedicels ascending, lower 1–6 in. long; bracts minute, deltoid; perianth $\frac{1}{8}$ in. long; segments lanceolate, with a distinct greenish-brown keel; stamens much shorter than the perianth-segments; capsule unknown. *Anthericum flagelliforme, Baker in Journ. Bot.* 1872, 140.

KALAHARI REGION: Transvaal; Apies River, *Burke!*

3. S. Macowani (Baker in Journ. Bot. 1873, 105); bulb globose, 1–1$\frac{1}{2}$ in. diam., with brown membranous tunics produced an inch above its apex; leaves unknown; peduncle slender, curved, 3–4 in. long; panicle deltoid, 3–4 in. long and broad, formed of 4–10 lax ascending racemes 1–2 in. long, $\frac{1}{2}$ in. diam.; pedicels erecto-patent, $\frac{1}{8}$–$\frac{1}{4}$ in. long; bracts minute, deltoid; perianth campanulate, $\frac{1}{8}$ in. long; segments oblong, obtuse; stamens a little shorter than the perianth; capsule greenish, $\frac{1}{6}$ in. diam. *Baker in Journ. Linn. Soc.* xv. 261.

CENTRAL REGION: Somerset Div.; rocky places near the Little Fish River, in the vicinity of Somerset East, 3000 ft., very rare, *MacOwan*, 1847!

4. S. intricata (Baker in Journ. Bot. 1874, 368); mature bulb globose, 1–1$\frac{1}{2}$ in. diam.; leaves 4–10, terete, erect, 3–4 in. long, vanishing before the stems appear; stems erect, 3–12 in. long including the inflorescence; peduncle as long as the panicle, sometimes spirally twisted; panicle 3–6 in. long and broad, very lax, corymbose, with zigzag rachises and slender, spreading, or ascending pedicels $\frac{1}{2}$–1 in. long, articulated at the apex; perianth campanulate, $\frac{1}{12}$ in. long; segments oblong, obtuse, white, with a distinct brown keel; stamens nearly as long as the perianth; ovules about 6 in a cell; capsule not seen. *Baker in Journ. Linn. Soc.* xv. 261. *Anthericum intricatum, Baker in Journ. Bot.* 1872, 140. *Asparagus micranthus, Thunb. herb.*

SOUTH AFRICA: without locality, *Zeyher*, 4284!

CENTRAL REGION: Somerset Div.; stony places on flat-topped rocks near Little Fish River, 3000 ft., *MacOwan*, 2131! Colesberg Div., *Shaw!* Albert Div.; by the Orange River near Sand Drift, *Burke*, 370!

KALAHARI REGION: Transvaal; Queens River Valley near Barberton, 2300 ft., *Galpin*, 525!

XV. ERIOSPERMUM, Jacq.

Perianth campanulate, marcescent, persistent; segments distinct, subequal, 1-nerved. *Stamens* 6, attached to the base of the perianth-segments; filaments lanceolate or filiform; anthers ovoid, basifixed, dehiscing introrsely. *Ovary* subsessile, globose, 3-celled; ovules few in a cell, superposed; style subulate; stigma capitate, entire. *Capsule* coriaceous, loculicidally 3-valved down to the base, the valves persistent. *Seeds* few, densely and persistently pilose; embryo long, cylindrical, often projecting beyond the fleshy albumen.

Rootstock large and tuberous; leaves usually solitary, produced after the flowers; racemes simple; pedicels solitary; bracts minute; flowers whitish or tinged with green, yellow or claret-purple.

DISTRIB. Several species in Tropical Africa.

Leaf solitary, without facial processes:
  Leaf oblong or lanceolate:
    Pedicels obsolete ... ... (1) **cernuum.**
    Lower pedicels $\frac{1}{8}-\frac{1}{4}$ in. long:
      Leaf glabrous, lanceolate ... ... (2) **calcaratum.**
      Leaf glabrous, oblong ... ... (3) **Haygarthii.**
      Leaf ligulate, densely hairy ... ... (4) **villosum.**
    Lower pedicels 1–1½ in. long:
      Perianth $\frac{1}{6}$ in. long, tuber small, globose:
        Peduncle solitary ... ... (5) **parvifolium.**
        Peduncles 2–3-nate... ... ... (6) **porphyrovalve.**
      Perianth $\frac{1}{4}$ in. long, tuber large ... (7) **lanceæfolium.**
    Lower pedicels 2–3 in. long, perianth 4–6
      lines long ... ... (8) **luteo-rubrum.**
    Lower pedicels 4–6 in. long, perianth 3–4
      lines long ... ... (9) **Burchellii.**
  Leaf cordate-ovate, hairy:
    Pedicels $\frac{1}{8}-\frac{1}{4}$ in. long:
      Leaf $\frac{1}{2}$ in. long ... ... (10) **microphyllum.**
      Leaf 3–4 in. long ... ... (11) **brevipes.**
    Pedicels $\frac{1}{2}$–1 in. long:
      Inner and outer segments similar ... (12) **pubescens.**
      Inner and outer segments different ... (13) **natalense.**
    Pedicels 1½–2 in. long ... ... (14) **lanuginosum.**
  Leaf cordate-ovate, glabrous:
    Pedicels short:
      Raceme short, corymbose ... ... (15) **corymbosum.**
      Raceme elongated:
        Bracts small... ... ... (16) **Bellendeni**
        Bracts larger ... ... (17) **Cooperi.**

Pedicels elongated:
Raceme 2–3 in. long:
Raceme deltoid ... ... (18) **thyrsoideum.**
Raceme not deltoid ... ... (19) **albucoides.**
Raceme $\frac{1}{2}$–1 ft. long ... ... (20) **latifolium.**
Leaf solitary, with a single compound process from its base ... ... (21) **paradoxum.**
Leaf solitary, its face covered with processes:
Processes simple, glabrous:
Raceme subspicate ... ... (22) **bowieanum.**
Raceme lax ... ... (23) **proliferum.**
Processes simple, strigose ... ... (24) **folioliferum.**
Processes pinnatifid ... ... (25) **alcicorne.**
Leaf unknown:
Perianth $\frac{1}{6}$ in. long:
Outer segments purplish ... ... (26) **spirale.**
Outer segments green ... ... (27) **confertum.**
Perianth $\frac{1}{3}$ in. long ... ... (28) **ornithogaloides.**
Leaves two or more to a flower stem:
Leaves small, petioled ... ... (29) **tenellum.**
Leaves large, sessile:
Flowers yellow ... ... (30) **Mackenii.**
Flowers white ... ... (31) **Galpini.**

1. **E. cernuum** (Baker); tuber small, subglobose; leaf not produced with the flowers, oblong-lanceolate, glabrous, firm in texture, 1–2 in. long, $\frac{1}{4}$–$\frac{1}{3}$ in. broad; petiole as long as the blade; peduncle very slender, glabrous, 6–8 in. long, straight or flexuose; flowers few in a lax spike, cernuous, seen only in an advanced state; perianth-segments oblong, $\frac{1}{6}$ in. long; lobes of the capsule oblong, as long as the perianth-segments.

Coast Region: Cape Peninsula; hill-sides and sand-dunes near Hout Bay, 400–500 ft., *Bolus*, 7238! *Schlechter*, 427! Kenilworth Flats, *Schlechter*, 456!

2. **E. calcaratum** (Baker in Journ. Linn. Soc. xv. 264); tuber large, oblong, with many rounded lobes; leaf solitary, produced after the flowers, sessile, lanceolate, glabrous, 2 in. long, $\frac{1}{2}$ in. broad, obsoletely papillose, with longitudinal depressed lines; peduncle slender, $\frac{1}{2}$ ft. long; raceme lax, cylindrical, 3–4 in. long; pedicels $\frac{1}{6}$–$\frac{1}{4}$ in. long; bracts minute, deltoid, the lower distinctly spurred, like those of an *Urginea;* perianth campanulate, $\frac{1}{6}$ in. long; segments oblanceolate, obtuse, white, with a broad green keel; stamens nearly as long as the perianth; filaments lanceolate. *Gard. Chron.* 1875, iii. 716.

Central Region: Graaff Reinet, frequent under bushes to the south of the town, 2600–2700 ft., *Bolus!*

Described from a plant cultivated at Kew in 1874.

3. **E. Haygarthii** (Baker); tuber small, oblong; leaves hyster anthous, small, oblong, distinctly petioled, glabrous, cuneate at the base, under an inch long; peduncle slender, 3–4 in. long; raceme moderately dense, 6–8-flowered, an inch long; pedicels ascending, the lower $\frac{1}{4}$–$\frac{1}{3}$ in. long; bracts small, ovate, not spurred; perianth

$\frac{1}{6}$ in. long; segments oblong, obtuse, white, with a dark brown keel; stamens half as long as the perianth; style short.

EASTERN REGION: Griqualand East; Vaal Bank, *Haygarth in Herb. Wood*, 4221!

4. E. villosum (Baker); petiole invested by the thick outer tunics of the corm to a length of 4 inches; blade ligulate, obtuse, 4–5 in. long, $\frac{1}{4}$–$\frac{1}{3}$ in. broad, narrowed to the base, moderately firm, densely hairy on both surfaces; flowers not seen.

WESTERN REGION: Little Namaqualand; Kasteel Poort near Klip Fontein, 3000 ft., *Bolus*, 6609!

5. E. parvifolium (Jacq. Collect. Suppl. 74; Ic. t. 422); tuber small, depresso-globose, with brown skin and white flesh; leaf solitary, produced after the flowers, oblong-lanceolate, 2 in. long, $\frac{1}{2}$ in. broad, shortly petioled; peduncle slender, $\frac{1}{2}$–1 ft. long; raceme lax, reaching $\frac{1}{2}$ ft. long, 1$\frac{1}{2}$–2 in. broad; pedicels ascending or spreading; lower $\frac{1}{2}$–1 in. long; bracts minute, deltoid; perianth campanulate, $\frac{1}{6}$ in. long; segments oblong, obtuse, white, with a green or brown keel; stamens as long as the perianth; filaments scarcely flattened; capsule turbinate; hair of seeds brownish-white. *Roem. et Schultes, Syst. Veg.* vii. 504; *Kunth, Enum.* iv. 650; *Baker in Journ. Linn. Soc.* xv. 264. *E. dregeanum, Presl. Bemerk.* 113.

COAST REGION: Paarl Div.; Klein Drakenstein Mts., below 1000 ft., *Drège!* Stellenbosch Div.; between Stellenbosch and Bottelary Hill, *Burchell*, 8343! Uniondale Div.; Lange Kloof, between Wagenbooms River and Apies River, *Burchell*, 4943!

6. E. porphyrovalve (Baker in Journ. Bot. 1891, 71); tuber globose, under $\frac{1}{2}$ in. diam., crowned with fine brown fibres; leaf solitary, small, lanceolate, rigidly coriaceous, crisped, glabrous, not seen fully developed; peduncles very slender, 2–3-nate, flexuose, 2–3 in. long; racemes lax, 1–2 in. long; pedicels ascending, lower 1–1$\frac{1}{2}$ in., upper $\frac{1}{4}$–$\frac{1}{3}$ in. long; bracts minute, ovate; perianth $\frac{1}{6}$ in. long; segments oblanceolate, obtuse, white with a red-brown keel; stamens shorter than the perianth; filaments lanceolate; capsule obovoid-cuneate, $\frac{1}{4}$ in. long; valves dark purple.

KALAHARI REGION: Transvaal; Houtbosch, *Rehmann*, 5765! Lake Chrissie, *Elliot*, 1602!

7. E. lanceæfolium (Jacq. Collect. Suppl. 72; Ic. t. 421); tuber large, with a brown skin, red flesh, and copious wiry root-fibres; leaf solitary, produced after the flowers, oblong-lanceolate, 5–6 in. long, 1–1$\frac{1}{2}$ in. broad, shortly petioled, glabrous; peduncle terete, above a foot long; raceme lax, $\frac{1}{2}$ ft. long, 1–1$\frac{1}{2}$ in. diam.; pedicels sometimes an inch long; bracts deltoid, minute; perianth campanulate, $\frac{1}{4}$ in. long; segments oblong, acute, white with a claret-brown keel; stamens shorter than the perianth; filaments lanceolate; capsule turbinate; hair of seeds brownish. *Red. Lil. t.* 394; *Ait.*

Hort. Kew. edit. 2, ii. 256 ; *Roem. et Schultes, Syst. Veg.* vii. 503 ; *Kunth, Enum.* iv. 651 ; *Baker in Journ. Linn. Soc.* xv. 264.

SOUTH AFRICA : without locality.

A specimen (*Bolus*, 6610 !) from a tuber collected in Little Namaqualand, and cultivated by Mr. Bolus in his garden near Cape Town in 1885, appears to be this species, but the blade of the leaf is only 2¼ inches long and 14 lines broad, and the perianth-segments ⅓ inch long.

8. E. luteo-rubrum (Baker) ; rootstock not seen ; leaf solitary, lanceolate, rigidly coriaceous, glabrous, shortly petioled, 3–4 in. long, ⅓ in. broad below the middle, narrowed gradually to the base and apex ; peduncle flexuose, 4–5 in. long ; raceme lax, subcorymbose, 4–5 in. long ; pedicels ascending ; lower 2–3 in. long, upper ½ in. ; bracts minute, deltoid ; perianth bright yellow, tinged with red, ⅓–½ in. long ; segments oblong, obtuse, yellow keeled with reddish-brown ; stamens half as long as the perianth ; filaments slightly flattened.

KALAHARI REGION : Transvaal ; summit of Saddleback Range, near Barberton, on stony ground, 4500–5000 ft., *Galpin*, 528 !
EASTERN REGION : Natal ; Inanda, *Wood*, 1346 !

9. E. Burchellii (Baker) ; tuber oblong ; leaf solitary, produced after the flowers, petiolate, 5–6 in. long, 8–10 lines broad, lanceolate, tapering to an acute apex and cuneately narrowed below the middle to an acute base, rigidly coriaceous, many nerved, glabrous ; peduncle 3 in. long ; raceme lax, corymbose, 5–6 in. long, 4 in. broad ; lower pedicels arcuate, ascending, 4–5 in. long ; bracts deltoid, minute ; perianth campanulate, ¼–⅓ in. long ; segments oblanceolate, obtuse, white keeled with brown ; stamens shorter than the perianth ; filaments lanceolate ; capsule turbinate, ⅓ in. long ; hair of the seeds soft, white.

KALAHARI REGION : Griqualand West ; near the Asbestos Mountains, between Witte Water and Riet Fontein, *Burchell*, 2008 ! Transvaal ; Mooi River Burke ! Magalies Berg, *Burke* ! near Johannesburg, *Adlam*, 4 !
EASTERN REGION : Natal, near Ladysmith, 4000 ft., *Wood*, 4239 !

10. E. microphyllum (Baker) ; tuber small, oblong ; leaves hysteranthous, long-petioled, cordate-ovate, ½ in. long, minutely ciliated ; peduncle slender, flexuose, glabrous, 3–4 in. long ; raceme lax, 1–2 in. long ; pedicels ascending, the lower ¼ in. long ; bracts minute, ovate, dark brown ; perianth ⅙ in. long ; segments oblong, white, with a dark brown keel ; stamens much shorter than the perianth ; style short.

EASTERN REGION : Natal ; Weenen County, South Downs, 5000–6000 ft., *Wood*, 4394 !

11. E. brevipes (Baker in Journ. Linn. Soc. xv. 263) ; tuber as large as a potato, with a whitish epidermis ; leaves solitary, produced after the flowers, cordate-ovate, bright green, 3–4 in. long, glabrous on the face when mature, hairy on the back ; petiole longer than the blade; peduncle slender, 1–2 ft. long; raceme dense, cylindrical, 6–9 in. long, under an inch in diameter; pedicels $\frac{1}{8}$–$\frac{1}{4}$ in. long, erecto-patent; bracts lanceolate-deltoid; perianth campanulate, $\frac{1}{6}$ in. long; segments oblong, white keeled with green; stamens shorter than the perianth; filaments white, lanceolate; capsule globose, $\frac{1}{6}$ in. diam.; hair of seeds brownish white. *Gard. Chron.* 1880, xiv. 231.

COAST REGION: Humansdorp Div.; between Galgebosch and Melk River, *Burchell,* 4779! Algoa Bay, *Cooper!* British Kaffraria; sand-flats near the Bushman River, 500 ft., *Baur,* 1029! and without precise locality, *Mrs. Hutton.*
CENTRAL REGION: Somerset Div., *Bowker!*

12. **E. pubescens** (Jacq. Hort. Schoenbr. iii. 8, t. 265); tuber large, oblong, with a brown skin, many rounded lobes and copious wiry root-fibres; leaf solitary, produced after the flowers, cordate-ovate, hairy, 3–4 in. long, with a channelled petiole as long as the blade; peduncle slender, stiffly erect, $\frac{1}{2}$–1 ft. long; raceme lax, finally 4–6 in. long; arcuate pedicels $\frac{1}{2}$–1 in. long; bracts minute, deltoid; perianth campanulate, $\frac{1}{6}$–$\frac{1}{4}$ in. long; segments oblong, obtuse, white, keeled with green; stamens shorter than the perianth; filaments lanceolate; seeds densely clothed with white hairs. *Lindl. Bot. Reg. t.* 578; *Roem. et Schultes, Syst. Veg.* vii. 504; *Kunth, Enum.* iv. 653; *Baker in Journ. Linn. Soc.* xv. 263.

SOUTH AFRICA: without locality, *Zeyher,* 4279!
COAST REGION: Cape Flats, below 100 ft., *Bolus,* 3916!
CENTRAL REGION: Somerset Div.; summit of Bosch Berg, 4500 ft., *MacOwan,* 1822!
EASTERN REGION: Tembuland; Bazeia Mountains, 3500–4000 ft., *Baur,* 496!

13. **E. natalense** (Baker); rootstock a large tuber; leaf solitary, produced after the flowers; blade cordate-ovate, membranous, hairy, 3–4 in. long and broad; petiole rather longer than the blade, densely pilose; peduncle a foot or more long; raceme lax, $\frac{1}{2}$ ft. long in flower, a foot in fruit, 2 in. diam.; pedicels erecto-patent, an inch long; bracts minute, deltoid; perianth broadly campanulate, $\frac{1}{6}$ in. long; segments broad, oblong, obtuse; outer all brown; inner white with a brown keel; stamens shorter than the perianth; filaments lanceolate; capsule globose, $\frac{1}{4}$ in. diam.; seeds densely coated with whitish-brown hairs.

EASTERN REGION: Pondoland; near Fort Donald, 3500 ft., *Tyson,* 1667! Griqualand East; near Clydesdale, *MacOwan and Bolus, Herb. Norm. Aust. Afr.,* 1205! Natal; around Durban, 120 ft., *Wood,* 107! *MacOwan, Herb. Aust. Afr.,* 1559! Umgeni, *Rehmann,* 8562! Inanda, *Wood,* 256!
Closely allied to *E. lanuginosum, Jacq.*

14. **E. lanuginosum** (Jacq. Hort. Schoenbr. iii. 7, t. 264); tuber the size of a large potato, grey on the outside, purplish within, with copious wiry root-fibres; leaf solitary, produced after the flower, cordate-ovate, hairy, 3–4 in. long, with a short channelled petiole;

peduncle slender, terete, a foot or more long ; raceme lax, a foot long,
3–4 in. diam. ; pedicels spreading or ascending, $1\frac{1}{2}$–2 in. long; bracts
deltoid, minute ; perianth dull pale yellow, $\frac{1}{4}$ in. long; segments
oblong, acute, keeled with green ; stamens a little shorter than the
perianth ; filaments lanceolate ; seeds brown, clothed with white
hairs. *Roem. et Schultes, Syst. Veg.* vii. 504; *Kunth, Ennm.* iv.
653 ; *Baker in Journ. Linn. Soc.* xv. 263.

SOUTH AFRICA : without locality.

Known to me only from Jacquin's figure.

15. E. corymbosum (Baker in Journ. Linn. Soc. xv. 266); tuber
small, globose, crowned with a dense ring of flexuose fine brown
fibres 1–$1\frac{1}{2}$ in. long; leaf solitary, cordate-ovate, $\frac{1}{2}$ in. long, glabrous,
with a petiole clasping the stem 1–$1\frac{1}{2}$ in. long; peduncle 1–2 in.
long ; raceme dense, corymbose, 6–10-flowered ; pedicels $\frac{1}{6}$–$\frac{1}{4}$ in.
long; bracts very minute, deltoid; perianth campanulate, $\frac{1}{6}$ in. long;
segments oblong, obtuse, white with a greenish-brown keel; stamens
shorter than the perianth ; filaments lanceolate ; capsule and seeds
unknown.

KALAHARI REGION: Griqualand West; Dutoits Pan, *Tuck!* *MacOwan*,
1969! Bechuanaland ; Batlapin Territory, *Holub!*

16. E. Bellendeni (Sweet, Hort. Brit. edit. 2, 529, name only);
tuber large, globose, with many rounded lobes, a brown skin and
copious wiry root-fibres ; leaf solitary, produced after the flowers,
cordate-ovate, glabrous, shortly petioled, 3–4 in. long ; peduncle
stiffly erect, a foot long; raceme dense, cylindrical, 3–4 in. long in
flower, under an inch in diameter ; pedicels ascending, $\frac{1}{8}$–$\frac{1}{4}$ in. long ;
bracts very minute ; perianth campanulate, $\frac{1}{6}$ in. long; segments
oblong, obtuse, with a broad green or brown keel; stamens shorter
than the perianth ; filaments lanceolate ; capsule turbinate, $\frac{1}{4}$ in.
diam. ; hair of seeds soft, brownish. *Baker in Journ. Linn. Soc.*
xv. 265. *E. latifolium, Ker in Bot. Mag. t.* 1382, *non Jacq.*

SOUTH AFRICA : without locality, *Masson!*
COAST REGION : Uitenhage, *Ecklon and Zeyher*, 739 ! near Port Elizabeth,
Zeyher, 4281 !
CENTRAL REGION : Cave Mountain, near Graaff Reinet, 4400 ft., *Bolus*, 517 !
near Somerset East, 2500 ft., *MacOwan*, 1888 !
KALAHARI REGION : Griqualand West; between Griqua Town and Witte
Water, *Burchell*, 1995 ! Transvaal ; Lomatie Valley, near Barberton, 4000 ft.,
Galpin, 1052 !
EASTERN REGION : Pondoland, *Bachmann*, 278 !

17. E. Cooperi (Baker in Journ. Linn. Soc. xv. 265); tuber unknown ; leaf developed after the flowers, solitary, cordate-ovate, with
a long petiole ; peduncle above a foot long ; raceme short and very
dense in the flowering stage; lower pedicels $\frac{1}{8}$–$\frac{1}{6}$ in. long ; bracts
lanceolate-deltoid, nearly as long as the pedicels ; perianth campanulate, $\frac{1}{6}$ in. long ; segments oblong, obtuse ; outer all brown ; inner
white with a brown keel; stamens shorter than the perianth ;

filaments lanceolate; capsule obconic, $\frac{1}{4}$ in. long; hair of the seeds brownish white.

KALAHARI REGION: Basutoland, *Cooper*, 3307! 3310!
EASTERN REGION : Natal; in a valley near Van Reenens Pass, 5000–6000 ft., *Wood*, 4519!

18. **E. thyrsoideum** (Baker); tuber globose, 2–3 in. diam., very firm in texture; leaf produced after the flowers, solitary, cordate-ovate, glabrous, 3–4 in. long, shortly petioled; peduncle slender, flexuose, 5–6 in. long; raceme lax, thyrsoid, 2–3 in. long; bracts minute, deltoid, white with a brown keel; pedicels arcuate, ascending; lower above an inch long, upper $\frac{1}{4}$ in.; perianth campanulate, $\frac{1}{6}$ in. long; segments oblong, obtuse, brownish; stamens shorter than the perianth; filaments lanceolate.

CENTRAL REGION: Stony ground on Bruintjes Hoogte, 3500 ft., *MacOwan*, 1863! 2198!

19. **E. albucoides** (Baker in Journ. Linn. Soc. xv. 265); tuber oblong, an inch in diameter, with a grey skin; leaf solitary, developed after the flowers, cordate-ovate, glabrous, 2 in. long and broad, shortly petioled; peduncle very slender, 6–9 in. long; raceme lax, 2–3 in. long, 1–1$\frac{1}{4}$ in. diam.; pedicels ascending, arcuate, $\frac{3}{4}$–1 in. long; bracts deltoid, very minute; perianth campanulate, $\frac{1}{6}$ in. long; segments oblong, obtuse, yellowish-white with a green keel; inner obovate, cucullate at the apex; stamens much shorter than the perianth; filaments lanceolate; style as long as the globose ovary.

SOUTH AFRICA : without locality, *Cooper !*
Described from a plant that flowered at Kew in 1873, received from Mr. Cooper.

20. **E. latifolium** (Jacq. Collect. Suppl. 73; Ic. ii. 18, t. 420); tuber large, subglobose, with brown skin, reddish flesh and copious wiry root-fibres; leaf solitary, produced after the flowers, cordate-ovate, glabrous, 3–4 in. long, shortly petioled; peduncle slender, terete, a foot or more long; raceme lax, $\frac{1}{2}$–1 ft. long; lower pedicels arcuate, 1$\frac{1}{2}$–2 in. long; bracts very minute, deltoid; perianth campanulate, $\frac{1}{6}$ in. long; segments oblong, acute, white keeled with claret-brown; stamens $\frac{2}{3}$ the length of the perianth; filaments lanceolate; capsule globose, $\frac{1}{4}$ in. diam.; hair of seeds brownish-white. *Thunb. Fl. Cap. edit. Schult.* 317; *Ait. Hort. Kew. edit.* 2, ii. 256; *Roem. et Schultes, Syst. Veg.* vii. 502; *Kunth, Enum.* iv. 652; *Baker in Journ. Linn. Soc.* xv. 265. *Ornithogalum capense, Linn. Sp. Plant.* 308; *Thunb. Prodr.* 62. (*Commel. Hort. Amstel.* ii. t. 88; *Breyn. Cent. t.* 41).

COAST REGION : Stellenbosch Div.; between Jonkers Valley and Stellenbosch, *Burchell*, 8335! Uniondale Div.; Lange Kloof, *Burchell*, 4941! 5017! Ultenhage, *Ecklon and Zeyher*, 537!

21. **E. paradoxum** (Gawl. in Bot. Mag. sub t. 1382); tuber large, irregular in shape, with a brown skin; leaf solitary, produced after

the flowers, cordate-ovate, shortly petioled, coriaceous, pilose, with a branched process 3–4 in. long issuing from its base, bearing numerous erecto-patent, compound, or simple linear pinnæ; peduncle short, pilose; raceme 6–7-flowered; pedicels very short; bracts minute, ovate; perianth $\frac{1}{3}$ in. long; segments elliptic, obtuse, white keeled with green; stamens about half as long as the perianth; filaments linear. *Kunth, Enum.* iv. 654; *Baker in Journ. Linn. Soc.* xv. 267. *Ornithogalum paradoxum, Jacq.· Collect. Suppl.* 81, *t.* 1. *Thaumaza paradoxa, Salisb. Gen.* 15.

SOUTH AFRICA : without locality.

Known to me only from Jacquin's figure.'

22. E. bowieanum (Baker in Journ. Linn. Soc. **xv.** 267); tuber large, oblong; leaf solitary, shortly petioled, cordate-ovate, glabrous, $\frac{1}{2}$ in. long, the face covered with clavate, erect, glabrous processes $\frac{1}{4}$–$\frac{1}{2}$ in. long; peduncle 1$\frac{1}{2}$–2 in. long; raceme dense, subspicate, 5–6-flowered; perianth $\frac{1}{6}$–$\frac{1}{5}$ in. long; segments oblong, white keeled with purple; stamens shorter than the perianth; filaments lanceolate.

SOUTH AFRICA : without locality.

Described from a drawing of a plant received from Bowie, cultivated at Kew in 1822.

23. E. proliferum (Baker in Journ. Linn. Soc. xv. 267); tuber large, depresso-globose, with a brown skin; leaf produced after the flowers, solitary, obcordate, glabrous, $\frac{1}{2}$ in. long, furnished on the face with numerous, slender, cylindrical, glabrous processes two or three times as long as the blade; peduncle 3–4 in. long; raceme lax, 2–3 in. long, 9–12-flowered; pedicels ascending, $\frac{1}{4}$–$\frac{3}{4}$ in. long; bracts minute, deltoid; perianth campanulate, $\frac{1}{6}$ in. long; segments oblong, obtuse, white keeled with green; stamens much shorter than the perianth; filaments lanceolate. *E. folioliferum, Gawl. in Bot. Reg. t.* 795; *Kunth, Enum.* iv. 654, *non Andrews.*

SOUTH AFRICA : without locality.

There is a drawing at Kew of a plant gathered by Bowie in 1821.

24. E. folioliferum (Andr. Bot. Rep. t. 521); tuber large, oblong; leaf contemporary with the flowers, solitary, ovate, acute, rounded at the base, 2–2$\frac{1}{2}$ in. long, purple on the back, furnished on the face with numerous, erect, simple, cylindrical, strigose processes $\frac{1}{4}$–$\frac{3}{4}$ in. long; peduncle 6–15 in. long; raceme lax, 6–9-flowered; pedicels erecto-patent, $\frac{1}{4}$–$\frac{3}{4}$ in. long; bracts minute, deltoid; perianth campanulate, $\frac{1}{6}$ in. long; segments oblong, yellowish-white keeled with green; stamens shorter than the perianth; filaments linear. *Phylloglottis foliolifera, Salisb. Gen.* 15.

WESTERN REGION : Little Namaqualand ; Kasteel Poort, near Klip Fontein, 3000 ft., *Bolus,* 6608! and *MacOwan and Bolus, Herb. Norm. Aust. Afr.,* 1389!

25. **E. alcicorne** (Baker); tuber and flowers unknown; leaf with a petiole 2–3 in. long and an orbicular, coriaceous, glabrous blade $\frac{1}{3}$ in. long and broad, from the face of which issues a dense mass of linear processes about $\frac{1}{2}$ in. long, with 1–2 ascending linear pinnæ from the middle of the margin.

CENTRAL REGION: Graaff Reinet Div.; near Zuurpoort, on the Sneeuw Berg Range, 5000 ft., *Bolus*, 838!

26. **E. spirale** (Berg. in Roem. et Schultes, Syst. Veg. vii. 1696); tuber globose, the size of a large pea, with grey skin and filiform root-fibres; leaf linear, known only in a rudimentary state; peduncle flexuose, 2–3 in. long; raceme corymbose, 2–8-flowered; lower pedicels above $\frac{1}{2}$ in. long; bracts minute, ovate; perianth campanulate, $\frac{1}{6}$ in. long; segments oblong, obtuse, outer purplish; inner white with a purplish keel; stamens included; filaments lanceolate; capsule globose; hair of seeds brownish. *Kunth, Enum.* iv. 654. *Bolus in Hook. Ic. t.* 2260. *Anthericum spirale, Linn. Mant. alt.* 224; *Roem. et Schultes, Syst. Veg.* vii. 481.

SOUTH AFRICA: without locality, *Bergius*!

COAST REGION: Cape Flats; Kenilworth Racecourse, *Schlechter*, 600! near Wynberg, *MacOwan and Bolus, Herb. Norm. Austr. Afr.*, 1388!

27. **E. confertum** (Baker in Engl. Jahrb. xv., Beibl. 35, 5); tuber globose, $\frac{1}{4}$–$\frac{3}{4}$ in. diam.; leaves hysteranthous, unknown; peduncle very slender, more or less spirally twisted, 2–3 in. long; flowers 2–8, corymbose; pedicels ascending, the lower an inch or more long; bracts ovate, minute; perianth campanulate, $\frac{1}{6}$ in. long; segments oblong, the outer all green, the inner white, with a distinct green keel; stamens nearly or quite as long as the perianth; filaments linear; style short.

COAST REGION: Malmesbury Div.; near Hopefield, *Bachmann*, 1817! 1818!

28. **E. ornithogaloides** (Baker in Journ. Linn. Soc. xv. 266); tuber small, irregular in form, horizontal, $\frac{1}{3}$ in. thick, with a brown skin; leaf unknown; peduncle very slender, stramineous, 4–5 in. long; raceme dense, corymbose, 3–4-flowered; pedicels $\frac{1}{8}$–$\frac{1}{4}$ in. long; bracts small, ovate, very convex on the back; perianth campanulate, $\frac{1}{3}$ in. long; segments oblong, obtuse, white with a very distinct brown keel; stamens half as long as the perianth; filaments lanceolate; capsule and seeds not known.

EASTERN REGION: Natal; Fields Hill, Pinetown, 1500 ft., *Sanderson*, 905!

29. **E. tenellum** (Baker); tuber globose, under $\frac{1}{2}$ in. diam., densely coated with fine matted brown fibres; leaves 2–3 to a stem, contemporary with the flowers, lanceolate, rigidly coriaceous, glabrous, about an inch long, $\frac{1}{8}$–$\frac{1}{6}$ in. broad, with a petiole nearly as long as the blade; peduncle very slender, 3–4 in. long; racemes very lax, 1–2 in. long, 4–8-flowered; pedicels ascending; lower

$\frac{3}{4}$–$1\frac{1}{4}$ in. long; bracts ovate, minute; perianth $\frac{1}{6}$ in. long; segments oblanceolate, obtuse, white, with a red-brown keel; stamens much shorter than the perianth; filaments lanceolate; capsule unknown.

KALAHARI REGION: Transvaal; Pretoria Div., Wonderboompoort, *Rehmann*, 4468! Bechuanaland; Bakwena Territory, 3500 ft., *Holub!*

30. **E. Mackenii** (Baker in Journ. Linn. Soc. xv. 266); tuber globose, firm in texture, $1\frac{1}{2}$ in. diam., with a brown skin and copious, long, slender root-fibres; leaves 2–3 to a stem, developed with the flowers, oblong-spathulate, sessile, glabrous, reaching 6 in. long, $1\frac{1}{2}$ in. broad at the middle; peduncle a foot or more long; raceme moderately dense, 2–4 in. long, above an inch in diameter; pedicels erecto-patent; lower $\frac{1}{2}$–1 in. long; bracts small, ovate; perianth campanulate, $\frac{1}{3}$ in. long; segments oblanceolate-oblong, pale yellow, with a broad greenish keel; stamens shorter than the perianth; filaments slightly flattened. *Bulbine Mackenii, Hook. fil. in Bot. Mag. t.* 5955.

EASTERN REGION: Transkei; on flats by the Bashee River, *Barber*, 18! Tembuland; Tabase, near Bazeia, 2000–2500 ft., *Baur*, 377! Natal; Inanda, *Wood*, 259! and without precise locality, *Mrs. K. Saunders!*

31. **E. Galpini** (Schinz. in Bull. Herb. Boiss. iv. 416); rootstock globose, 1 in. diam., crowned with wiry fibres; leaves 2, contemporary with the flowers, oblong, firm, glabrous, $1\frac{1}{2}$–2 in. long; petiole sheathing, nearly as long as the blade; peduncle straight, slender, 6–8 in. long; flowering raceme lax, 4–5 in. long; pedicels erecto-patent, $\frac{1}{4}$–$\frac{1}{3}$ in. long; bracts ovate, acute, minute, persistent; perianth $\frac{1}{6}$–$\frac{1}{4}$ in. long; segments linear-oblong, yellow; stamens $\frac{2}{3}$ the length of the perianth; filaments rather flattened; anthers oblong, small; capsule obovoid, as long as the perianth.

KALAHARI REGION: Transvaal; banks of the Queens River, near Barberton 2300 ft., *Galpin*, 1135!

XVI. ANTHERICUM, Linn.

Perianth marcescent, not twisted after flowering; segments distinct, patent, subequal, oblong or oblanceolate, closely 3–5-nerved on the keel. *Stamens* 6, hypogynous or adnate to the very base of the segments; filaments filiform or slightly compressed; anthers linear or oblong, dorsifixed, versatile, dehiscing introrsely. *Ovary* sessile, globose; ovules several in a cell, superposed; style subulate, rather declinate; stigma capitate. *Capsule* coriaceous, loculicidally 3-valved. *Seeds* triquetrous; testa black, opaque; embryo cylindrical; albumen firm in texture.

Rootstock obscure; root-fibres wiry or cylindrical; radical leaves usually linear or subterete; racemes simple or panicled; pedicels articulated at the middle or apex, often 2–3-nate; bracts small, scariose; flowers small, white, the segments keeled with green or brown.

DISTRIB. Species about 60. Also Tropical Africa, Europe, and a few in America.

Subgenus PHALANGIUM. Pedicels articulated at or below the middle, lower usually 2–3-nate; filaments and style smooth.

- Filaments longer than the anthers:
  - Pedicels all solitary ... ... (1) undulatum.
  - Lower pedicels 2–3-nate:
    - Leaves $\frac{1}{8}$ in. broad ... ... (2) rigidum.
    - Leaves $\frac{1}{2}$ in. broad ... ... (3) pachyphyllum.
- Filaments as long as the anthers ... ... (4) longistylum.
- Filaments shorter than the large anthers:
  - Bracts small ... ... (5) angulicaule.
  - Bracts large, scariose:
    - Keel of segments 3-nerved ... ... (6) Cooperi.
    - Keel of segments 5-nerved ... ... (7) anceps.

Subgenus DILANTHES. Pedicels articulated at or below the middle; lower usually 2–3-nate; filaments and style scabrous, with raised points.

- Perianth $\frac{1}{2}$ in. long:
  - Leaves $\frac{1}{2}$–$\frac{3}{4}$ in. broad, rigidly coriaceous ... (8) trichophlebium.
  - Leaves linear, $\frac{1}{4}$ in. broad:
    - Leaf-sheaths short:
      - Racemes not capitate:
        - Pedicels very short ... ... (9) polyphyllum.
        - Pedicels $\frac{1}{4}$–$\frac{1}{2}$ in. long ... ... (10) triflorum.
      - Racemes capitate .. .. (11) capitatum.
    - Leaf-sheaths as long as the blade ... (12) Saundersiæ.
  - Leaves short, lorate ... ... (13) crassinerve.
  - Leaves narrow linear, $\frac{1}{8}$–$\frac{1}{6}$ in. broad:
    - Leaves pubescent... ... ... (14) transvaalense.
    - Leaves glabrous ... ... (15) nudicaule.
  - Leaves subterete:
    - Lower pedicels geminate ... ... (16) Bolusii.
    - Lower pedicels 3–4-nate ... ... (17) fasciculatum.
- Perianth $\frac{1}{4}$–$\frac{1}{3}$ in. long:
  - Pedicels and rachis smooth:
    - Leaves narrow linear, complicate ... (18) Galpini.
    - Leaves firm, narrow linear, flat:
      - Racemes simple or forked at the base:
        - Lower pedicels geminate ... (19) Schultesii.
        - Lower pedicels solitary ... (20) multisetosum.
      - Racemes many ... ... (21) patulum.
    - Leaves firm, linear, $\frac{1}{2}$–$\frac{3}{4}$ in. broad ... (22) robustum.
    - Leaves moderately firm, linear ... ... (23) pulchellum.
  - Pedicels and rachis scabrous ... ... (24) viscosum.

Subgenus TRACHYANDRA. Pedicels solitary, articulated at the apex; filaments and style scabrous.

- Leaves subterete, $\frac{1}{24}$–$\frac{1}{8}$ in. diam.:
  - Leaves spirally contorted:
    - Dwarf:
      - Raceme simple ... ... (25) serpentinum.
      - Racemes panicled ... . (26) flexifolium.
    - Tall ... ... (27) Pappei.

Leaves straight :
 Stem and pedicels nearly or quite smooth :
  Bracts small ; root not crowned with
   the bristly relics of old leaves :
    Dwarf :
     Lower pedicels ⅓–¾ in. long (28) **pudicum.**
     Lower pedicels ¼–⅓ in. long (29) **micranthum.**
    Tall ; raceme simple :
     Leaves short ... ... (30) **brevifolium.**
     Leaves a foot long ... (31) **chlamydophyllum.**
    Tall ; racemes 1–3 ... ... (32) **brachypodum.**
    Tall ; racemes numerous, pani-
     cled ... ... (33) **elongatum.**
  Bracts small ; old leaves splitting up
   into bristles :
    Leaves rigid, slender ... ... (34) **Macowani.**
    Leaves rigid, narrow linear,
     rather twisted ... ... (35) **subcontortum.**
    Leaves not rigid, 1/12–⅛ in. diam. (36) **pubescens.**
  Bracts large :
    Dwarf ... ... (37) **longepedunculatum.**
    Tall ; leaves and flowers gla-
     brous :
     Pedicels short ... ... (38) **tabulare.**
     Pedicels long ... ... (39) **brevicaule.**
    Tall ; leaves and flowers very
     hairy... ... ... (40) **canaliculatum**
 Stem and pedicels persistently scabrous :
  Dwarf ; racemes short :
   Ovules 2 in a cell ... ... (41) **Kunthii.**
   Ovules 4 in a cell ... ... (42) **asperatum.**
  Tall ; racemes long... ... ... (43) **scabrum.**
Leaf flat, linear or ensiform :
 Dwarf :
  Glabrous ... ... (44) **involucratum.**
  Hairy ; peduncle short ; raceme simple,
   dense ... ... (45) **hispidum.**
  Hairy ; peduncle longer than panicle .. (46) **thyrsoideum.**
 Tall ; bracts small :
  Peduncle smooth :
   Leaves glabrous :
    Leaves ⅓–½ in. broad ... ... (47) **revolutum.**
    Leaves ¼–⅓ in. broad ... ... (48) **falcatum.**
   Leaves glabrous on the face, muri-
    cated on the edge... ... ... (49) **longifolium.**
   Leaves very hairy on the face :
    Racemes 1–3 ... ... (50) **hirsutum.**
    Racemes 6–8 ... ... (51) **pilosum.**
  Peduncle scabrous :
   Capsule smooth ... ... (52) **muricatum.**
   Capsule muricated ... ... (53) **Gerrardi.**
 Tall ; bracts large ... ... (54) **ciliatum.**
Leaves urceolate at the tip... ... ... (55) **paradoxum.**

1. A. undulatum (Jacq. Collect. Suppl. 87 ; Ic. ii. 18, t. 411) ;
root-fibres slender ; crown of the root not setose ; radical leaves
linear, glabrous, acuminate, a foot long, ⅙–¼ in. broad ; peduncle
terete, above a foot long, with a single much-reduced leaf ; raceme

simple or forked, lax, $\frac{1}{2}$ ft. long; pedicels all solitary, articulated at
the middle, the lower spreading, $\frac{1}{2}-\frac{3}{4}$ in. long; lower bracts lanceolate,
as long as the pedicels; perianth white, $\frac{1}{2}$ in.
long; segments obtuse,
with a closely 3-nerved greenish keel; stamens half as long as the
perianth; filaments exceeding the linear-oblong anthers; style $\frac{1}{2}$ in.
long. *Baker in Journ. Bot.* 1872, 138; *Journ. Linn. Soc.* xv. 304.
Phalangium undulatum, Poir. Encyc. v. 242. *Anthericum graminifolium, Willd. Sp. Plant.* ii. 139; *Roem. et Schultes, Syst. Veg.* vii.
463. *Chlorophytum* (?) *graminifolium, Kunth, Enum.* iv. 606.

SOUTH AFRICA: without locality.
Known to me only from Jacquin's figure.

2. A. rigidum (Baker in Journ. Bot. 1872, 141; Journ. Linn.
Soc. xv. 303); radical leaves linear, rigid in texture, glabrous, a
foot long, $\frac{1}{8}$ in. broad, with 6–8 raised veins on each side of the
prominent midrib; peduncle 1$\frac{1}{2}$–2 ft. long, ancipitous in the lower
half, with 1–2 much-reduced leaves; racemes lax, sparingly panicled,
the central one 2–3 in. long; bracts minute, deltoid, red-brown;
pedicels $\frac{1}{12}-\frac{1}{8}$ in. long, articulated at the middle, the lower 2–3-nate;
perianth $\frac{1}{4}$ in. long; segments oblanceolate, with a 3-nerved brown
keel; stamens rather shorter than the perianth; filaments twice as
long as the small oblong anthers; ovules 6 in a cell. *Chlorophytum* (?)
rigidum, Kunth, Enum. iv. 604.

COAST REGION: Tulbagh Div., Great Winter Hoek Mountain, 2000–3000 ft.,
Drège, 8738!

3. A. pachyphyllum (Baker in Journ. Linn. Soc. xv. 304);
radical fibres long, wiry, slender; root crowned with a ring of
bristles; radical leaves about 8, clasping the base of the stem, linear,
firm in texture, the central ones 6–8 in. long, $\frac{1}{2}$ in. broad, with about
20 scabrous nerves on each side of the prominent midrib; peduncle
simple, ancipitous, leafless, 1–1$\frac{1}{2}$ ft. long; raceme lax, simple,
3–6 in. long; bracts small, deltoid, scariose; pedicels $\frac{1}{8}-\frac{1}{4}$ in. long,
articulated below the middle, the lower 3–4-nate; perianth $\frac{1}{2}$ in. long;
segments oblanceolate, with a greenish 3-nerved keel; stamens
shorter than the perianth; filaments twice as long as the linear-
oblong anthers; capsule ovoid-triquetrous, obtuse, $\frac{1}{6}$ in. long.

COAST REGION: Albany Div.; Grahamstown! (Collector not stated on
label).

4. A. longistylum (Baker in Journ. Linn. Soc. xv. 305); radi-
cal leaves densely tufted, linear-subulate, glabrous, a foot long, $\frac{1}{8}-\frac{1}{6}$ in.
broad above the clasping base, with 6–8 prominent nerves on each
side of the distinct midrib; peduncle leafless, $\frac{1}{2}$ ft. long below the
inflorescence; racemes copiously panicled; central 6–9 in. long;
bracts small, deltoid, whitish, scariose; pedicels $\frac{1}{8}-\frac{1}{4}$ in. long, articulated below the middle, the lower 2–3-nate; perianth $\frac{1}{2}$ in. long;
segments oblanceolate, with a 3-nerved green keel; stamens shorter
than the perianth; filaments as long as the linear-oblong anthers;

style declinate, $\frac{1}{2}$ in. long; capsule ovoid-triquetrous, rigidly coriaceous, $\frac{1}{6}$–$\frac{1}{4}$ in. long.

KALAHARI REGION: Transvaal, *Baines!*

5. **A. angulicaule** (Baker in Journ. Linn. Soc. xv. 305); root-fibres slender and wiry; radical leaves 6–8, linear, glabrous, firm in texture, 1–1½ ft. long, $\frac{1}{8}$–$\frac{1}{2}$ in. broad, strongly closely ribbed; peduncle strongly angled, leafless, 1–2 ft. long; racemes 1–3, lax, 2–4 in. long; bracts small, deltoid-cuspidate, scariose, dark brown; pedicels $\frac{1}{8}$–$\frac{1}{4}$ in. long, articulated below the middle, the lower 2–3-nate; perianth white, $\frac{1}{3}$ in. long; segments oblong, with a closely 3-nerved green keel; stamens shorter than the perianth; filaments shorter than the linear-oblong anthers, which are $\frac{1}{6}$ in. long; style declinate, $\frac{1}{3}$ in. long.

COAST REGION: Uitenhage Div.; sandy hills by the Zwartkops River, *Zeyher!* King Williamstown Div.; Keiskamma, *Mrs. Hutton!* KALAHARI REGION: Transvaal; near Lydenberg, *Atherstone!* Yster Spruit, *Nelson, 324!* EASTERN REGION: Natal, *Sanderson,* 261! *Mrs. K. Saunders!*

6. **A. Cooperi** (Baker in Journ. Linn. Soc. xv. 304); root-fibres very slender; root crowned with a ring of bristles; root-leaves about 6, linear, glabrous, firm in texture, 6–9 in. long, $\frac{1}{8}$–$\frac{1}{6}$ in. broad, with 12–15 nerves without a distinct midrib; peduncle slender, stiffly erect, simple, leafless, 1–1½ ft. long; raceme dense, simple, few-flowered; bracts scariose, whitish, lanceolate-deltoid or deltoid, persistent, the lower $\frac{1}{2}$ in. long; pedicels $\frac{1}{4}$–$\frac{1}{3}$ in. long, articulated at the middle, the lower 3–4-nate; perianth $\frac{1}{2}$–$\frac{1}{2}$ in. long; segments with a closely 3-nerved brown keel; stamens rather shorter than the perianth; anthers linear-oblong, $\frac{1}{3}$ in. long; filaments very short; capsule oblong-triquetrous, rigidly coriaceous, $\frac{1}{4}$ in. long.

KALAHARI REGION: Basutoland, *Cooper,* 3302! EASTERN REGION: Natal, *Cooper,* 1004!

7. **A. anceps** (Baker in Journ. Linn. Soc. xv. 305); radical leaves 5–6, linear, glabrous, moderately firm in texture, 6–9 in. long, $\frac{1}{6}$ in. broad, with about 30 conspicuous close ribs; peduncle 6–9 in. long, simple, strongly angled; raceme dense, simple, 3 in. long; bracts large, white, scariose, acuminate, $\frac{1}{2}$ in. long; pedicels short, articulated at the middle, the lower 2–3-nate; perianth $\frac{1}{3}$ in. long; segments oblong, tipped with red-brown, with a 5-nerved green keel; filaments shorter than the linear-oblong anthers, which are $\frac{1}{6}$ in. long; style declinate, $\frac{1}{4}$ in. long.

KALAHARI REGION: Transvaal, *Baines!*

8. **A. trichophlebium** (Baker); rhizome stout, creeping along the surface of the ground, with abundant wiry root-fibres, the old leaves remaining as bristles; radical leaves about 6, oblong-lanceolate, rigidly coriaceous, densely pubescent, 3–4 in. long, $\frac{1}{2}$–$\frac{3}{4}$ in.

broad, with 30–40 prominently-raised hairy ribs and a thickened hairy margin; peduncle simple, leafless, densely pilose, rather shorter than the leaves; raceme dense, simple, oblong, 1½–2 in. long; bracts lanceolate, densely pilose, like the leaves in texture; pedicels ⅛–¼ in. long, articulated at the middle; perianth ½ in. long; segments oblanceolate, with a closely 3-nerved keel; stamens rather shorter than the perianth; filaments scabrous, exceeding the anthers, which are ⅜ in. long; style ½ in. long.

KALAHARI REGION: Transvaal; hills above Aapies River near Pretoria, *Rehmann,* 4314!

9. **A. polyphyllum** (Baker); leaves many to each stem, linear, firm, glabrous, strongly ribbed, a foot long, ¼ in. broad; whole plant 2 ft. high; peduncle stout, terete; panicle of about 3 racemes, the end one 3–4 in. long, dense upwards; pedicels short, articulated near the base; bracts small; perianth ½ in. long; segments oblong, white, with a distinct brown keel; stamens nearly as long as the perianth; anthers large; style much exserted.

KALAHARI REGION: Transvaal; on stony hills, De Kaap Valley, near Barberton, 2300 ft., *Galpin,* 1149!

10. **A. triflorum** (Ait. Hort. Kew. i. 448); root-fibres dense, fleshy, cylindrical or nodose; root crowned with fine flexuose bristles; root-leaves linear, moderately firm in texture, 6–9 in. long, ¼–⅓ in. broad at the middle, minutely ciliated on the edges, with 8–10 distinct nerves on each side of the midrib; peduncle ½–1 ft. long, angled downwards, with 2–3 rudimentary leaves; raceme simple, lax, 2–3 in. long; bracts small, deltoid-cuspidate; pedicels ¼–½ in. long, articulated at the middle, the lower 2–3-nate; perianth white, ½ in. long; segments oblanceolate, with a closely 3-nerved keel; stamens rather shorter than the perianth; filaments scabrous, much longer than the oblong anthers; capsule ovoid-triquetrous, ⅓ in. long. *Willd. Sp. Plant.* ii. 140; *Roem. et Schultes, Syst. Veg.* vii. 466. *Chlorophytum* (?) *triflorum, Kunth, Enum.* iv. 606. *Anthericum bipedunculatum, Jacq. Collect. Suppl.* 88; *Ic.* ii. 18, *t.* 410. *Phalangium pedunculatum, Baker in Journ. Linn. Soc.* xv. 315. *Phalangium bipedunculatum, Poir. Encyc.* v. 244. *P. triflorum, Pers. Syn.* i. 368. *Anthericum pauciflorum, Thunb. Prodr.* 63; *Fl. Cap. edit. Schult.* 320. *Trachyandra pauciflora, Kunth, Enum.* iv. 584. *T.* (?) *brehmeana, Kunth, Enum.* iv. 586. *Chlorophytum brehmeanum, Roem. et Schultes, Syst. Veg.* vii. 454.

VAR. β, minor (Baker, loc. cit.); a dwarf form, 4–6 in. high, with narrower canaliculate leaves and lax 2–4-flowered raceme with all the pedicels solitary.

COAST REGION: Clanwilliam Div., *Zeyher!* Malmesbury Div.; Malmesbury, *Bachmann,* 805! 806! near Groene Kloof, below 1000 ft., *Drège,* 8722a! *Bolus,* 4348! Zwartland, *Thunberg!* Cape Div.; Lion Mountain, *Thunberg!* Yzerplaat, near Zout River, *Zeyher,* 4659! Near Tulbagh, *Pappe!* Worcester, *Zeyher!* Var. β, Paarl Div.; between Paarl and Lady Grey Railway Bridge, *Drège,* 8723a! Uitenhage Div.; by the Zwartkops River, *Pappe!*

KALAHARI REGION: Var. β, Transvaal; Saddleback Range, near Barberton, 4500 ft., *Galpin,* 1025!

11. A. capitatum (Baker) ; densely tufted ; old leaves splitting into fibres ; produced leaves 3–4 to a stem, linear, firm, strongly ribbed, glabrous, $\frac{1}{2}$ ft. long, $\frac{1}{4}$–$\frac{1}{3}$ in. broad ; peduncle naked, slender, stiffly erect, terete, 6–9 in. long ; flowers many, crowded into a single globose head ; pedicels erecto-patent, articulated at the middle, the lower $\frac{1}{4}$–$\frac{1}{3}$ in. long ; bracts large, ovate, membranous ; perianth $\frac{1}{2}$ in. long ; segments oblong, white, with a keel of 3 distinct brown ribs ; stamens half as long as the perianth ; anthers small, oblong ; style short.

EASTERN REGION : Natal ; Van Reenens Pass, 5000–6000 ft., *Wood,* 4795 !

12. A. Saundersiæ (Baker) ; radical leaves about 6, linear, the central ones with a flat blade half a foot long and sheathing the base of the stem for an equal length, moderately firm in texture, $\frac{1}{2}$ in. broad, with 10–12 pilose raised nerves on each side of the prominent midrib ; peduncle 1$\frac{1}{2}$ ft. long, leafless, strongly angled ; racemes 2, 1–1$\frac{1}{2}$ in. long ; bracts small, deltoid ; pedicels erecto-patent, $\frac{1}{4}$–$\frac{1}{3}$ in. long, articulated at the middle, the lower 2–3-nate ; perianth $\frac{1}{3}$ in. long ; segments oblanceolate, with a closely 3-nerved keel ; stamens much shorter than the perianth ; filaments scabrous, exceeding the oblong anthers ; style declinate, $\frac{1}{4}$ in. long.

EASTERN REGION : Natal, *Mrs. K. Saunders.*
Closely allied to *A. triflorum.*

13. A. crassinerve (Baker in Journ. Bot. 1891, 71) ; root-fibres cylindrical ; old leaves rather fibrous ; leaves lorate, glabrous, very rigid, 3 in. long, $\frac{1}{3}$–$\frac{1}{2}$ in. broad, with thick veins and a very thick margin ; . peduncle simple, $\frac{1}{2}$ ft. long, with 2 empty membranous bracts ; raceme lax, simple, 3–6 in. long ; pedicels erecto-patent, the lower $\frac{1}{2}$–$\frac{3}{4}$ in. long, articulated at the middle ; bracts ovate-acuminate, the lower $\frac{1}{2}$–$\frac{3}{4}$ in. long ; perianth $\frac{1}{3}$–$\frac{1}{2}$ in. long, white tinged with red ; stamens shorter than the perianth ; filaments scabrous, rather exceeding the linear-oblong anthers ; style exserted.

WESTERN REGION : Little Namaqualand ; near Ookiep,. 3000 ft., *Bolus,* 6600 ! *Scully,* 114 !

14. A. transvaalense (Baker) ; root-stock crested with copious fibres ; leaves many, rigid, narrowly linear, $\frac{1}{8}$ in. broad, erect, 2–6 in. long, channelled down the face, clothed with soft spreading hairs ; peduncle simple, flexuose, pubescent, $\frac{1}{2}$–1 ft. long ; raceme lax, simple, 3–6 in. long ; rachis densely pubescent ; pedicels all single, very short, articulated at the middle ; bracts lanceolate from an ovate base, the lower $\frac{1}{4}$–$\frac{1}{3}$ in. long ; perianth $\frac{1}{2}$ in. long ; segments oblong, white, with a distinctly 3-nerved brown keel ; stamens nearly as long as the perianth ; anthers large, oblong.

KALAHARI REGION : Transvaal ; Saddleback Range, near Barberton, 3500 ft., *Galpin,* 1035 !

15. A. nudicaule (Baker) ; root-fibres long and slender, bearing small tubers at the tip ; old leaf-bases breaking up into fibres ; produced leaves firm, linear, glabrous, 6–9 in. long, $\frac{1}{6}$ in. broad; peduncle stiffly erect, simple, ancipitous, without any reduced leaf; raceme dense, short, simple, few-flowered; pedicels $\frac{1}{8}$–$\frac{1}{4}$ in. long, erecto-patent, articulated at the middle; bracts ovate-lanceolate, the lower $\frac{1}{4}$–$\frac{1}{3}$ in. long; perianth $\frac{1}{2}$–$\frac{5}{8}$ in. long; segments oblong, with a closely 3-nerved green or red-brown midrib; stamens half as long as the perianth; anthers oblong, small; capsule oblong, shorter than the perianth, with the valves strongly ribbed transversely.

EASTERN REGION : Griqualand East; on hillsides near Clydesdale, 2500 ft., *Tyson*, 1054 !

16. A. Bolusii (Baker) ; root-fibres short, cylindrical; old leaves not splitting up into fibres ; leaves 8–10, terete, glabrous, $\frac{1}{2}$ ft. long, $\frac{1}{16}$ in. diam.; peduncle glabrous, leafless, 6–9 in. long; raceme lax, simple, 5–6 in. long; flowers all single; pedicels erecto-patent, articulated at the middle, the lower 2-nate, $\frac{1}{2}$–$\frac{3}{4}$ in. long; bracts ovate-acuminate, $\frac{1}{4}$–$\frac{1}{2}$ in. long; perianth $\frac{1}{2}$ in. long; segments linear-oblong, spreading, with a 3-nerved brown midrib; filaments rough, $\frac{1}{4}$ in. long : anthers small, oblong; style $\frac{1}{3}$ in. long.

WESTERN REGION : Little Namaqualand ; Naries, near Spektakel, 3300 ft., *Bolus*, 6601!

17. A. fasciculatum (Baker in Journ. Linn. Soc. xv. 316); leaves numerous, spreading, subterete, $1\frac{1}{2}$–3 in. long, $\frac{1}{12}$–$\frac{1}{8}$ in. broad, channelled down the face, moderately firm in texture, not rigid, closely 10–15-nerved, minutely ciliated on the margin ; peduncle slender, simple, leafless, about 2 in. long ; raceme simple, 1–$1\frac{1}{2}$ in. long ; bracts small, deltoid ; pedicels $\frac{1}{4}$–$\frac{1}{3}$ in. long, articulated at the middle; lower 2–3-nate; perianth $\frac{1}{2}$ in. long; segments oblanceolate, with a closely 3-nerved green keel ; stamens shorter than the perianth ; filaments papillose, 2–3 times the length of the oblong anthers, which are $\frac{1}{8}$ in. long; style declinate, $\frac{1}{3}$ in. long.

KALAHARI REGION : Plains of the Vaal River, *Bowker*, 4!

18. A. Galpini (Baker); whole plant 6–9 in. high; old leaves splitting up into fibres ; produced leaves many, narrowly linear, rigid, complicate, 4–5 in. long, some falcate, ciliated with fine deflexed hairs ; peduncle slender, terete ; panicle of 3–4 sparse lax racemes ; pedicels short, erecto-patent, articulated below the middle, the lower ones 2-nate ; bracts linear, from an ovate base, lower $\frac{1}{4}$ in. long; perianth $\frac{1}{3}$ in. long; segments oblong, white, with a distinct brown keel ; stamens shorter than the perianth ; anthers small.

KALAHARI REGION : Transvaal ; grassy plains around Barberton, 2900 ft., *Galpin*, 1160 !

19. A. Schultesii (Baker in Journ. Bot. 1872, 140); root-fibres cylindrical ; relics of old leaves not breaking up into bristles ; root-leaves 6–10, linear, rigid in texture, 3–9 in. long, $\frac{1}{6}$ in. broad, with 10–12 crowded raised nerves on each side of the midrib, minutely ciliated on the margin ; peduncle $\frac{1}{2}$–1 ft. long, with 2–3 rudimentary leaves; racemes one or few, lax, 1–3 in. long; bracts minute, deltoid; pedicels $\frac{1}{4}$–$\frac{1}{2}$ in. long, articulated at the middle, lower geminate; perianth $\frac{1}{3}$ in. long; segments oblanceolate, with a closely 3-nerved green or brown keel; stamens rather shorter than the perianth; filaments scabrous, much exceeding the oblong anthers; capsule globose-triquetrous. *Baker in Journ. Linn. Soc.* xv. 315. *Trachyandra Schultesii, Kunth, Enum.* iv. 586. *Chlorophytum dubium, Roem. et Schultes, Syst. Veg.* vii. 455.

COAST REGION: Paarl Div.; western slopes of the Drakenstein Mts., near Bains Kloof, 1600 ft., *Bolus*, 4071! Tulbagh, *Pappe!* Stellenbosch Div.; western slopes of Hottentots Holland Mts. near Lowrys Pass, 700 ft., *Bolus*, 5560! Caledon Div.; near Villiersdorp, 1300 ft., *Bolus*, 5268!

20. **A. multisetosum** (Baker); root-fibres cylindrical; rootstock crowned with a dense ring of long bristles; produced root-leaves few, narrowly linear, rigid, glabrous, shorter than the stem, $\frac{1}{8}$ in. broad; peduncle naked, simple, rather compressed, not angled, under a foot long; raceme simple, short, dense, few-flowered; bracts small, brown, ovate-cuspidate; pedicels short, erecto-patent, all solitary; perianth-segments oblanceolate-oblong, obtuse, $\frac{1}{3}$ in. long, white with a 3-nerved green keel; stamens much shorter than the perianth; anthers linear-oblong; filaments scarcely flattened, not scabrous.

EASTERN REGION: Swazieland; Havelock Concession, 4000 ft., *Galpin*, 1013!

21. **A. patulum** (Baker); whole plant 1$\frac{1}{2}$ ft. high; rootstock crested with fibres; leaves many, linear, moderately firm, glabrous, a foot long, $\frac{1}{6}$ in. broad; peduncle slender, distinctly 2-edged; panicle of 4–5 racemes 1–2 in. long, the central branches spreading at a right angle; pedicels short, geminate, articulated near the base; bracts small, ovate; perianth $\frac{1}{4}$ in. long; segments oblong, obtuse, white with a distinct greenish-brown keel; stamens $\frac{2}{3}$ the length of the perianth; anthers large.

KALAHARI REGION: Transvaal; Saddleback Range, near Barberton, 4000 ft., *Galpin*, 1232!

22. **A. robustum** (Baker); leaves linear, firm, glabrous, strongly ribbed, 2–3 ft. long, $\frac{3}{4}$ in. broad at the middle; peduncle 3–4 ft. long, stiffly erect, acutely angled, naked; panicle $\frac{1}{2}$ ft. long; racemes 3–5, moderately dense, the lateral erecto-patent; bracts small, ovate, nearly black: pedicels erecto-patent, articulated at the base, $\frac{1}{8}$–$\frac{1}{4}$ in. long, many lower geminate; perianth $\frac{1}{4}$–$\frac{1}{3}$ in. long; segments linear-oblong, with a 3-nerved green keel; anthers linear, $\frac{1}{4}$ in. long, exceeding the filaments; style long, overtopping the anthers.

EASTERN REGION: Natal; near Stanger, 500 ft., *Wood*, 3972!

23. **A. pulchellum** (Baker in Journ. Bot. 1872, 140, excl. syn.); root-fibres long and slender; relics of old leaves breaking up into fibres; root-leaves linear, moderately firm in texture, not rigid, $\frac{1}{2}$–1 ft. long, $\frac{1}{6}$ in. broad, with 10–12 scabrous veins on each side of

the midrib; peduncle leafless, $\frac{1}{2}$-$1\frac{1}{2}$ ft. long; racemes one or few, lax, 1-3 in. long; bracts minute, deltoid; pedicels $\frac{1}{8}$-$\frac{1}{4}$ in. long, articulated at or below the middle, lower 2-3-nate; perianth $\frac{1}{3}$ in. long; segments oblanceolate, with a closely 3-nerved green or brown keel; stamens rather shorter than the perianth; filaments scabrous, much exceeding the oblong anthers; capsule globose-triquetrous, emarginate, $\frac{1}{6}$ in. diam.

COAST REGION: Queenstown Div.; Shiloh, 3500 ft., *Baur*, 937!

CENTRAL REGION: Albert Div.; Burghers Dorp, *Cooper*, 788!

KALAHARI REGION: Basutoland, *Cooper*, 3200! near the Vaal River, *Nelson*, 185!

EASTERN REGION: Transkei; Tsomo Flats, 1500-2000 ft., *Baur*, 722! Tembuland; Xongora, *Baur*, 152! Tabase, near Bazeia, 2500 ft., *Baur*, 376! Pondoland, *Bachmann*, 275! Natal; Inanda, *Wood*, 118! 264! 366! near Durban, *Wood*, 3197! Durban Bay, *Krauss*, 74! and without precise locality, *Cooper*, 3295! *Plant*, 78! *Gerrard*, 554! Delagoa Bay, *Forbes!*

24. **A. viscosum** (Baker in Journ. Bot. 1872, 141); root-fibres long and slender; relics of old leaves not breaking up into threads; root-leaves linear, rigid in texture, $\frac{1}{2}$ ft. long, $\frac{1}{8}$-$\frac{1}{6}$ in. broad, with 8-10 scabrous raised nerves on each side of the midrib; peduncle scabrous, shorter than the leaves; racemes several, lax, 1-2 in. long; rachises very scabrous; bracts small, white, ovate, scariose; pedicels $\frac{1}{8}$-$\frac{1}{6}$ in. long, articulated at the middle; perianth $\frac{1}{3}$ in. long; segments oblanceolate, with a closely 3-5-nerved keel; stamens much shorter than the perianth; filaments scabrous, exceeding the oblong anthers; capsule globose-triquetrous, $\frac{1}{4}$ in. diam. *Baker in Journ. Linn. Soc.* xv. 316. *Chlorophytum* (?) *viscosum, Kunth, Enum.* iv. 605.

WESTERN REGION: Little Namaqualand, near the mouth of the Orange River, *Drège*, 2673!

25. **A. serpentinum** (Baker); root-fibres thick and fleshy; crown not bristly; leaves about 5, terete, firm in texture, 3-4 in. long, $\frac{1}{2}$ line broad, glabrous, very much twisted spirally; peduncle very slender, stiffly erect, 3-4 in. long; raceme simple, 1-2 in. long, dense upwards; pedicels solitary, articulated at the apex, lower $\frac{1}{4}$ in. long; bracts large, lanceolate-deltoid, cuspidate; perianth $\frac{1}{4}$ in. long; segments very narrow, white, keeled with brown; stamens half as long as the perianth; filaments papillose.

COAST REGION: Riet Valley, near Cape Town, *Ecklon and Zeyher!* Malmesbury, *Bachmann*, 870!

26. **A. flexifolium** (Linn. fil. Suppl. 201); root-fibres thick and fleshy; crown not bristly; radical leaves 6-12, terete, firm in texture, 3-4 in. long, under a line in diameter, very much twisted spirally, glabrous or hispid; peduncle slender, terete, 2-5 in. long; inflorescence a rhomboid panicle 2-3 in. long with a straight or zigzag rachis; racemes very lax, few-flowered; bracts very minute, ovate; pedicels very short; perianth $\frac{1}{4}$ in. long; segments oblanceolate, with a 3-nerved brown keel; stamens shorter than the perianth;

filaments papillose.   *Thunb. Prodr.* 62 ; *Fl. Cap. edit. Schult.* 318 ;
Roem. et Schultes, Syst. Veg. vii. 462 ; *Baker in Journ. Linn. Soc.*
xv. 307.   *Trachyandra* (?) *flexifolia, Kunth, Enum.* iv. 579.

SOUTH AFRICA : without locality, *Thunberg !*

WESTERN REGION : Little Namaqualand ; sandy places near Kook Fontein,
3000 ft., *Bolus*, 6605 ! between Zwart Doorn River and Groen River, *Drège*,
2671a !

27. **A.ʼ Pappei** (Baker) ; root-fibres thick and fleshy ; crown not
bristly ; leaves 1–4 to a root, terete, firm in texture, pubescent, 6–9 in.
long, $\frac{1}{2}$ line broad, straight or slightly contorted spirally ; peduncle
slender, subterete, pubescent, 6–10 in. long ; racemes 1–3, very lax,
the central one 3–4 in. long ; bracts small, deltoid-cuspidate ; pedicels
solitary, $\frac{1}{8}$–$\frac{1}{4}$ in. long ; perianth $\frac{1}{8}$ in. long ; segments very distinctly
keeled with green ; stamens half as long as the perianth ; filaments
papillose.

COAST REGION : Tulbagh, *Pappe !*

28. **A. pudicum** (Baker in Journ. Linn. Soc. xv. 308) ; rootstock
vertical, cylindrical, 3 in. long ; root-fibres slender, wiry ; crown not
bristly ; leaves subterete, deciduously hairy, 3–4 in. long, $\frac{1}{16}$–$\frac{1}{12}$ in.
broad, not rigid, dark green ; peduncle slender, ancipitous, hairy,
1$\frac{1}{2}$–2 in. long ; racemes 1–2, very lax, 2–4 in. long ; bracts minute,
deltoid ; lower pedicels $\frac{1}{2}$–$\frac{3}{4}$ in. long ; perianth $\frac{1}{4}$ in. long ; segments
oblanceolate, with a closely 3-nerved keel ; stamens much shorter
than the perianth ; filaments muricated ; capsule globose, $\frac{1}{6}$ in. diam.

COAST REGION : Uitenhage Div. ; among shrubs near the Zwartkops River,
Zeyher, 1070 !

29. **A. micranthum** (Baker in Journ. Bot. 1891, 71) ; roots not
seen ; leaves subterete, spreading, 3–4 in. long, very slender, glabrous,
obscurely bristle-ciliated towards the base ; stem slender, glabrous,
2–3 in. long ; raceme very lax, simple, 2–3 in. long ; pedicels solitary,
ascending, articulated at the apex, the lower $\frac{1}{4}$–$\frac{1}{3}$ in. long ; bracts
small, ovate-cuspidate ; perianth $\frac{1}{4}$ in. long ; segments linear-oblong,
white, with a distinct brown keel ; stamens $\frac{1}{3}$ shorter than the segments ; filaments very scabrous ; anthers oblong, small ; style short.

KALAHARI REGION : Griqualand West ; Dutoits Pan, near Kimberley, *Elliot*,
1220 !

30. **A. brevifolium** (Thunb. Prodr. 62) ; leaves 5–6 to a root,
linear-subulate, firm in texture, glabrous, 2–3 in. long, $\frac{1}{8}$ in. broad ;
stem slender, terete, leafless, $\frac{1}{2}$ ft. long ; raceme simple, very lax,
$\frac{1}{2}$ ft. long ; bracts deltoid-cuspidate, very minute ; lower pedicels
slender, spreading, $\frac{1}{2}$–$\frac{5}{8}$ in. long ; perianth $\frac{1}{4}$ in. long ; segments
oblanceolate ; stamens shorter than the perianth ; capsule very small,
subglobose.   *Fl. Cap. edit. Schult.* 319 ; *Baker in Journ. Bot.* 1872,
139 ; *Journ. Linn. Soc.* xv. 298.   *Bulbine brevifolia, Roem. et*

Schultes, Syst. Veg. vii. 451.   *Bulbinella brevifolia, Kunth, Enum.* iv. 573.

SOUTH AFRICA : without locality, *Thunberg !*

31. **A. chlamydophyllum** (Baker) ; root-fibres long, rather thickened ; leaves a dozen or more in a tuft, subterete, a foot long, $\frac{1}{12}$ in. diam., moderately firm in texture, glabrous, each with a membranous funnel-shaped sheath at the base ; peduncle simple, glabrous, half as long as the leaves ; raceme simple, 3 in. long, dense upwards, bracts lanceolate, $\frac{1}{6}$ in. long ; pedicels erecto-patent, $\frac{1}{2}$ in. long, articulated at the apex ; perianth white, $\frac{1}{3}$ in. long ; segments oblanceolate, with a 3-nerved brown keel ; stamens nearly as long as the perianth ; filaments slightly scabrous.

COAST REGION : On rocky mountain slopes in Tulbagh Kloof, *MacOwan*, 2603 !

32. **A. brachypodum** (Baker) ; root-fibres slender, wiry ; old leaves not breaking up into bristles ; root-leaves linear, glabrous, thick in texture, with thickened margins, nearly flat, a foot long, $\frac{1}{8}$ in. broad ; peduncle slender, smooth, subterete, 1–1$\frac{1}{2}$ ft. long ; racemes 1–3, very lax, the end one finally 6–9 in. long ; bracts very minute, deltoid-cuspidate ; pedicels erecto-patent, not more than $\frac{1}{12}$–$\frac{1}{6}$ in. long ; perianth $\frac{1}{4}$ in. long ; segments oblanceolate, white, with a distinct brown keel ; stamens much shorter than the perianth ; filaments muricated ; capsule subglobose, $\frac{1}{6}$ in. diam.

COAST REGION : Sand-dunes near Cape Town, under 100 ft., *Bolus*, 3921 ! mountains near Simons Town, 1000–2000 ft., *Ecklon and Zeyher*, 106 !

33. **A. elongatum** (Willd. Sp. Plant. ii. 136) ; root-fibres fleshy, cylindrical ; crown not bristly ; root-leaves subterete, 1–1$\frac{1}{2}$ ft. long, $\frac{1}{12}$–$\frac{1}{8}$ in. broad, glabrous or deciduously pilose ; peduncle stout, ancipitous, 6–9 in. long, naked or slightly muricated ; panicle 1–1$\frac{1}{2}$ ft. long, made up of many lax racemes, the end one 6–8 in. long ; bracts minute, deltoid ; pedicels solitary, $\frac{1}{4}$–$\frac{1}{3}$ in. long ; perianth $\frac{1}{3}$–$\frac{1}{2}$ in. long ; segments oblanceolate, with a closely 3-nerved green keel ; stamens a little shorter than the perianth ; filaments muricated ; anthers minute, oblong ; capsule turbinate, $\frac{1}{6}$ in. diam. *Roem. et Schultes, Syst. Veg.* vii. 456. *A. filiforme, Thunb. Prodr.* 62, *ex parte, non Ait.* *A. jacquinianum, Roem. et. Schultes, Syst. Veg.* vii. 462 ; *Baker in Journ. Linn. Soc.* xv. 308. *Trachyandra Jacquinii, Kunth, Enum.* iv. 578. *T. elongata, Kunth, Enum.* iv. 584.

VAR. β, **holostachyum** (Baker) ; raceme simple. *A. jacquinianum, Schult. fil. var. affinis, Baker in Journ. Linn. Soc.* xv. 308, *partly.*

VAR. γ, **A. flexifolium** (Jacq. Collect. Suppl. 93 ; Ic. ii. t. 412, non Linn. fil.) ; leaves flaccid, flexuose, pubescent.

COAST REGION : Cape Div. ; hills below Table Mountain, *Thunberg !* Cape Flats, *Pappe ! Bolus.* 3750 ! Wynberg, *Burchell*, 887 ! *Bolus*, 2848 !
CENTRAL REGION : Cradock Div., *Cooper*, 1301 ! Colesberg, *Shaw !*
KALAHARI REGION : Carnarvon Div. ; Buffels Bout, *Burchell*, 1602 !
EASTERN REGION : Var. β, Griqualand East ; Clydesdale, 2500 ft., *Tyson*, 2122 ! Natal, *Gerrard*, 552 ! Berea near Durban, 100 ft., *Wood*, 72 !

34. A. Macowani (Baker in Journ. Linn. Soc. xv. 309); root-fibres slender, wiry; old leaves breaking up into a dense mass of rigid bristles; leaves subterete, rigid in texture, $\frac{1}{2}$–1 ft. long, $\frac{1}{2}$ lin. diam., glabrous, sometimes flexuose; peduncle slender, wiry, 3–9 in. long; racemes lax, usually simple, 3–6 in. long; bracts minute, deltoid or the lowest lanceolate; lower pedicels $\frac{1}{2}$–$\frac{3}{4}$ in. long, erecto-patent; perianth $\frac{1}{3}$ in. long; segments oblanceolate, with a closely 3-nerved green keel; stamens rather shorter than the perianth; filaments papillose; capsule globose, $\frac{1}{6}$ in. diam.

Coast Region: Albany Div.; Grahamstown, *MacOwan*, 64!

Central Region: Graaff Reinet Div.; Karoo near Graaff Reinet, 2600 ft., *Bolus*, 740! Koudevelds Berg, 4000 ft., *Bolus*, 2592! Prince Albert Div., *Drège*, 578b!

Kalahari Region: Basutoland, *Cooper*, 747! 3298! Bechuanaland, Chooi Desert, *Burchell*, 2342! Transvaal; Matebe Valley, *Holub!*

Eastern Region: Tembuland; Bazeia, 2000 ft., *Baur*, 753!

35. A. subcontortum (Baker); whole plant under a foot long; root crowned with copious fibres; leaves many, narrowly linear, hairy, rather flexuose, 4–6 in. long, $\frac{1}{15}$ in. diam.; panicle of several lax racemes; pedicels articulated at the apex, lower $\frac{1}{2}$ in. long; bracts minute; perianth $\frac{1}{3}$ in. long; segments linear-oblong, white with a brown keel; stamens shorter than the perianth; anthers small.

Eastern Region: Griqualand East, 4300 ft., *MacOwan and Bolus, Herb. Norm.*, 1206!

36. A. pubescens (Baker in Journ. Linn. Soc. xv. 309); root-fibres slender; relics of old leaves breaking up into bristles; radical leaves subterete, $\frac{1}{2}$–1 ft. long, $\frac{1}{12}$–$\frac{1}{8}$ in. broad, dark green, not rigid in texture, clothed with fine, soft, spreading, deciduous hairs; peduncle 3–6 in. long, ancipitous, softly pilose; racemes 1–2, the end one very lax, 4–6 in. long; bracts minute, deltoid or lanceolate-deltoid; pedicels solitary, lower $\frac{1}{3}$–$\frac{1}{2}$ in. long; perianth $\frac{1}{3}$ in. long; segments oblanceolate, with a closely 3-nerved green keel; stamens nearly as long as the perianth; filaments muricated; capsule globose, $\frac{1}{8}$ in. diam. *A. jacquinianum, Schult. fil. var. affinis, Baker in Journ. Linn. Soc.* xv. 308, *partly. Trachyandra affinis, Kunth, Enum.* iv. 579.

Coast Region: Albany Div.; Fish River Heights, *Hutton!*

Central Region: near Somerset East, 2200 ft., *MacOwan*, 1589!

Kalahari Region: Orange Free State, *Cooper*, 3308!

37. A. longepedunculatum (Steud. in Roem. et Schultes, Syst. Veg. vii. 457, 1692); root-fibres thick and fleshy; crown not bristly; leaves subterete, glabrous, not rigid, 3–9 in. long, $\frac{1}{2}$ in. diam.; peduncle slender, 1–2 in. long; racemes 1–3, very lax, 1–2 in. long; bracts ovate-cuspidate, the lower $\frac{1}{4}$–$\frac{1}{2}$ in. long; pedicels solitary, the lower $\frac{1}{2}$–1 in. long; perianth $\frac{1}{3}$ in. long; segments oblanceolate, with a distinct, closely 3-nerved brown keel; stamens much shorter than the perianth; filaments muricated; capsule globose, $\frac{1}{6}$ in. diam. *Baker in Journ. Bot.* 1872, 138; *Journ. Linn.*

Soc. xv. 308.   *A. filiforme, Thunb. Prodr.* 62, *ex parte. A. vermicularis, Soland. in herb. Banks. Trachyandra longepedunculata, Kunth, Enum.* iv. 584.

SOUTH AFRICA: without locality, *Oldenburg! Ludwig! Bergius!*
COAST REGION : Malmesbury, *Bachmann,* 472 ! Cape Div.; hills below Table Mountain, *Thunberg!* sandy places near Green Point, *Zeyher,* 4245! near Cape Town, *Bolus,* 3754 ! 3764 ! Simons Bay, *Wright,* 229!

38. A. tabulare (Baker); root-fibres long, slender, wiry; old leaves not breaking up into bristles; radical leaves linear-subulate from a clasping deltoid base, subrigid, glabrous, straight, erect, $\frac{1}{2}$–1 ft. long, $\frac{1}{12}$–$\frac{1}{8}$ in. broad ; peduncle 6–9 in. long, leafless, glabrous, smooth ; racemes 1–2, the end much the largest, dense upwards, finally 4–6 in. long; pedicels solitary, erecto-patent, the lowest $\frac{1}{4}$–$\frac{1}{3}$ in. long ; bracts lanceolate from a deltoid base, as long as or longer than the pedicels ; perianth $\frac{1}{3}$ in. long ; segments oblanceolate, white, with a 3-nerved pale brown keel; stamens much shorter than the perianth ; filaments muricated ; capsule globose, $\frac{1}{6}$ in. diam.

COAST REGION : Table Mountain, 2400 ft., *Bolus,* 4726 !
Closely allied to *A. brevicaule* and *A. canaliculatum.*

39. A. brevicaule (Baker in Journ. Linn. Soc. xv. 298); root-fibres slender and wiry ; crown of the root not bristly ; leaves sub-terete, firm in texture, channelled down the face, glabrous, $\frac{1}{2}$ ft. long, $\frac{1}{8}$ in. broad ; stem much shorter than the leaves ; racemes 1–2, the end one very lax, finally a foot long ; bracts lanceolate-deltoid, $\frac{1}{6}$–$\frac{1}{4}$ in. long; pedicels solitary, finally decurved, $\frac{1}{2}$–$\frac{3}{4}$ in. long; perianth $\frac{1}{5}$ in. long ; segments white, keeled with brown ; stamens shorter than the perianth ; capsule globose, $\frac{1}{8}$ in. diam.

SOUTH AFRICA : without locality, *Thunberg!*

40. A. canaliculatum (Ait. Hort. Kew. i. 448); root-fibres long, cylindrical; relics of old leaves not breaking up into bristles ; leaves subterete, a foot or more long, $\frac{1}{16}$–$\frac{1}{12}$ in. broad, channelled down the face, not rigid, scabrous, with sharp, spreading points ; peduncle stout, leafless, $\frac{1}{2}$–1 ft. long, densely pubescent ; raceme simple, dense upwards, finally 4–6 in. long; pedicels solitary, ascending, densely pilose, the lower $\frac{1}{2}$–$\frac{3}{4}$ in. long ; bracts lanceolate, pubescent, nearly or quite as long as the pedicels ; perianth densely pubescent, $\frac{1}{2}$ in. long ; segments oblanceolate, with a closely 3-nerved pale brown keel ; stamens much shorter than the perianth ; filaments muricated ; capsule globose, $\frac{1}{3}$–$\frac{1}{2}$ in. diam., densely pubescent. *Willd. Sp. Plant.* ii. 141; *Roem. et Schultes, Syst. Veg.* vii. 460 ; *Baker in Journ. Linn. Soc.* xv. 309 ; *Bot. Reg. t.* 877 ; *Bot. Mag. t.* 1124. *Bulbine canaliculata, Spreng. Syst.* ii. 86. *Phalangium canaliculatum, Poir. Encyc.* v. 249. *Trachyandra canaliculata, Kunth, Enum.* iv. 578.

COAST REGION : Cape Flats, *Pappe! Bolus,* 3749 ! Simons Day, *Wright,* 214 !

41. A. Kunthii (Baker) ; root-fibres long and slender ; leaves linear-subulate, 4–6 in. long, hispid, dilated to a hyaline base ; stems 3 from a root, 2–3 in. long, scabrous ; racemes 1–2, subflexuose, 3–4 in. long ; pedicels scabrous, finally ½ in. long ; bracts minute, ovate ; perianth ¼ in. long ; segments oblong, with a 3-nerved keel ; stamens shorter than the perianth ; filaments muricated ; capsule subglobose ; cells 2-seeded. *Trachyandra humilis, Kunth, Enum.* iv. 574.

SOUTH AFRICA : without locality, *Drège*, 8734.

Known to me only from Kunth's description. I cannot trace the number in Drège's list of stations. There is already an Abyssinian *Anthericum humile*, *Hochst.*

42. A. asperatum (Baker in Journ. Bot. 1872, 138 ; Journ. Linn. Soc. xv. 310) ; leaves subterete, rigid, light green, 7–8 in. long, scabrous, with minute, spreading, glandulose bristles ; peduncle 4 in. long, subterete ; raceme lax, simple, 4–5 in. long ; bracts minute, deltoid ; pedicels solitary, scabrous, ⅓ in. long ; perianth ⅓ in. long ; segments with a closely 3-nerved brown keel ; stamens rather shorter than the perianth ; filaments muricated ; ovules 4 in a cell. *Trachyandra asperata, Kunth, Enum.* iv. 574.

CENTRAL REGION : Albert Division, 4500–5000 ft., *Drège*, 8735 !

43. A. scabrum (Linn. fil. Suppl. 202) ; leaves subterete, not rigid, 1½ ft. long. $\frac{1}{12}$–⅛ in. broad, channelled down the face, scabrous downwards with raised points ; peduncle compressed, flexuose, ½–1 ft. long, scabrous with raised points ; racemes few, very lax, finally ½–1 ft. long ; rachises persistently scabrous ; bracts minute, lanceolate-deltoid ; pedicels ½–¾ in. long, finally spreading or decurved ; perianth ⅓ in. long : segments with a distinct 3-nerved greenish keel ; stamens much shorter than the perianth ; filaments muricated ; capsule globose, ⅙ in. diam. *Thunb. Prodr.* 63 ; *Fl. Cap. edit. Schult.* 320 ; *Baker in Journ. Linn. Soc.* xv. 310. *Bulbine scabra, Roem. et Schultes, Syst. Veg.* vii. 451. *Trachyandra scabra, Kunth, Enum.* iv. 585.

COAST REGION : Sandy plains between Cape Town and Hottentots Holland, *Thunberg !*

44. A. involucratum (Baker in Journ. Linn. Soc. xv. 311) ; rootstock elongated, vertical ; outer rudimentary leaves ovate, membranous ; radical leaves flat, linear, glabrous, 4–5 in. long, ⅙ in. broad, moderately firm in texture ; peduncle slender, terete, glabrous, 1–2 in. long ; racemes simple, very lax, few-flowered, finally 3–4 in. long ; bracts small, deltoid cuspidate ; pedicels ascending or spreading, the lower ⅓–½ in. long ; perianth ½ in. long ; segments whitish, oblanceolate, with a distinct greenish keel ; stamens half as long as the perianth ; filaments papillose ; capsule not seen.

WESTERN REGION : Little Namaqualand near Mieren Kasteel, *Drège*, 2681 !

45. A. hispidum (Linn. Sp. Plant. edit. 2, 446) ; root-fibres thick and fleshy ; outer rudimentary leaves large and membranous ; root-leaves 2–4, linear, flat, glabrous, margined with purple, 4–8 in. long, $\frac{1}{8}$–$\frac{1}{3}$ in. broad, often spirally twisted towards the top ; peduncle hairy upwards, 1–6 in. long ; raceme dense, simple, 1–2 in. long ; bracts large, lanceolate-deltoid, entirely membranous, whitish ; pedicels slender, hairy, lower 1–1$\frac{1}{2}$ in. long ; perianth hispid, $\frac{1}{3}$ in. long ; segments oblanceolate, white with a distinct brown keel ; stamens shorter than the perianth ; filaments muricated ; capsule globose, smooth. *Thunb. Prodr.* 63 ; *Fl. Cap. edit. Schult.* 321 ; *Willd. Sp. Plant.* ii. 145 ; *Jacq. Collect. Suppl.* 91 ; *Ic.* ii. 17, *t.* 409. *A. squameum, Linn. fil. Suppl.* 202 ; *Roem. et Schultes, Syst. Veg.* vii. 481. *Arthropodium hispidum, Spreng. Syst.* ii. 87. *Trachyandra hispida, Kunth, Enum.* iv. 575. *Phalangium squameum, Poir. Encyc.* v. 246. *Bulbinella* (?) *squamea, Kunth, Enum.* iv. 573. *Anthericum undulatum, Thunb. Prodr.* 63 ; *Fl. Cap. edit. Schult.* 321 ; *Willd. Sp. Plant.* ii. 140 ; *Roem. et Schultes, Syst. Veg.* vii. 470. *Trachyandra undulata, Kunth, Enum.* iv. 583.

Coast Region : Malmesbury, *Bachmann,* 471 ! sandy places near Cape Town, *Thunberg ! Zeyher,* 1693 ! 5012 ! *Bolus,* 3731 ! *Cooper,* 3296 !

46. A. thyrsoideum (Baker in Journ. Bot. 1872, 139); root-fibres thick and fleshy ; outer rudimentary leaves ovate, membranous, obtuse ; radical leaves linear, 5–6 in. long, $\frac{1}{8}$ in. broad, clothed throughout with conspicuous, spreading, pellucid hairs ; peduncle hairy, 2–4 in. long ; inflorescence a deltoid panicle about 2 in. long ; rachis densely pubescent ; racemes short, lax ; bracts deltoid-cuspidate, $\frac{1}{8}$–$\frac{1}{6}$ in. long ; pedicels erecto-patent, lower $\frac{1}{4}$ in. long ; perianth glabrous, $\frac{1}{4}$–$\frac{1}{3}$ in. long ; segments oblanceolate, with a distinct brownish or greenish keel ; stamens rather shorter than the perianth ; filaments muricated ; capsule not seen.

Central Region : Tulbagh Div. ; near Yuk River Hoogte, *Burchell,* 1231 !

47. A. revolutum (Linn. Sp. Plant. 310 ; J. Commel. Hort. i. 67, t. 34) ; root-fibres slender and wiry ; outer rudimentary leaves large and membranous ; root-leaves numerous, linear, glabrous, moderately firm in texture, channelled down the face, 1–2 ft. long, $\frac{1}{6}$–$\frac{1}{4}$ in. broad ; peduncle stout, smooth, $\frac{1}{2}$–1 ft. long ; racemes numerous, lax, 3–6 in. long ; rachis smooth ; bracts minute, deltoid-cuspidate ; pedicels solitary, the lower $\frac{1}{2}$ in. long ; perianth glabrous, $\frac{1}{3}$ in. long ; segments white, with a distinct 3-nerved brown keel ; stamens shorter than the perianth ; filaments muricated ; capsule globose, glabrous, $\frac{1}{6}$ in. diam. *Thunb. Prodr.* 62 ; *Fl. Cap. edit. Schult.* 318 ; *Ait. Hort. Kew.* i. 447 ; *Gawl. in Bot. Mag. t.* 1044. *Phalangium revolutum, Pers. Syn.* i. 368 ; *Haw. Syn.* 63. *Anthericum divaricatum, Jacq. Hort. Schoenbr.* iv. 7, *t.* 414. *Trachyandra revoluta* and *T. divaricata, Kunth, Enum.* iv. 579, 580.

South Africa : without locality, *Oldenburg !*

COAST REGION : Cape Div.; sand-dunes near Green Point, *Thunberg!* between Cape Town and Salt River, *Burchell*, 897 !
EASTERN REGION : sandy places near King Williams Town, 1250 ft., *Tyson*, 3072 !

48. A. falcatum (Linn. fil. Suppl. 202); root-fibres slender, cylindrical; outer rudimentary leaves membranous; radical leaves flat, spreading, linear, glabrous, a foot or more long, $\frac{1}{4}$–$\frac{1}{3}$ in. broad; peduncle $\frac{1}{2}$–$1\frac{1}{2}$ ft. long, glabrous or slightly hairy ; racemes 1–5, lax, the end one finally 1–$1\frac{1}{2}$ ft. long; bracts lanceolate or deltoid-cuspidate, the lower $\frac{1}{4}$–$\frac{1}{3}$ in. long ; pedicels ascending or finally decurved, the lower $\frac{1}{2}$–$\frac{3}{4}$ in. long; perianth $\frac{1}{3}$ in. long, glabrous ; segments oblanceolate, white, with a distinct brown keel; stamens rather shorter than the perianth ; filaments muricated ; capsule turbinate-oblong, glabrous, $\frac{1}{3}$ in. long. *Thunb. Prodr.* 63 ; *Fl. Cap. edit. Schult.* 323 ; *Willd. Sp. Plant.* ii. 138. *A. vespertinum, Jacq. Hort. Schoenbr.* i. 44, *t.* 85 ; *Willd. Sp. Plant.* ii. 139 ; *Gawl. in Bot. Mag. t.* 1040. *Bulbine falcata, Roem. et Schultes, Syst. Veg.* vii. 451. *Trachyandra falcata, Kunth, Enum.* iv. 586. *Phalangium vespertinum, Poir. Encyc.* v. 249. *Trachyandra vespertina, Kunth, Enum.* iv. 581.

COAST REGION: Saldanha Bay, *Thunberg!* Caledon Div.; Zwart River, *Zeyher!* Bathurst Div.; near Theopolis, *Burchell*, 4076!
KALAHARI REGION : Griqualand West; near the Asbestos Mountains, between Witte Water and Riet Fontein, *Burchell*, 2012 !

49. A. longifolium (Jacq. Collect. Suppl. 92 ; Ic. ii. 18, t. 413); root-fibres long and slender ; outer rudimentary leaves membranous ; root-leaves linear, flat, 1–$1\frac{1}{2}$ ft. long, moderately firm in texture, muricated mainly on the margin ; peduncle stout, glabrous or pubescent, $\frac{1}{2}$–1 ft. long ; racemes numerous, lax, 2–4 in. long ; bracts small, lanceolate or deltoid ; pedicels ascending, the lower $\frac{1}{2}$ in. long ; perianth $\frac{1}{3}$ in. long ; segments oblanceolate, white, with a 3-nerved green or brownish keel; stamens much shorter than the perianth ; filaments scabrous ; capsule globose, glabrous, $\frac{1}{6}$ in. diam. *Willd. Sp. Plant.* ii. 139; *Roem. et Schultes, Syst. Veg.* vii. 464. *A. fimbriatum, Thunb. Prodr.* 63 ; *Fl. Cap. edit. Schult.* 322 ; *Roem. et Schultes, Syst. Veg.* vii. 465. *Phalangium longifolium, Poir. Encyc.* v. 243. *Trachyandra longifolia and T. fimbriata, Kunth, Enum.* iv. 582, 583.

VAR. *β*, Burchellii (Baker in Journ. Linn. Soc. xv. 312) ; leaves spreading ; stems decumbent, viviparous, pubescent ; pedicels spreading or cernuous, the lower an inch long.

COAST REGION : Malmesbury, *Bachmann*, 802 ! sandy places near Cape Town, *Thunberg!* Devils Mountain, near Cape Town, 600 ft., *Bolus*, 3809 ! Stellenbosch, *Sanderson*, 987 ! Caledon Div.; Zonder Einde River, *Gill!* Var. *β*, Bathurst Div.; near Port Alfred, *Burchell!*

50. A. hirsutum (Thunb. Prodr. 63) ; root-fibres slender, wiry ; old leaves not breaking up into bristles ; root-leaves flat, linear, $\frac{1}{2}$–$1\frac{1}{2}$ ft. long, $\frac{1}{4}$–$\frac{1}{2}$ in. broad, firm in texture, finely and persistently

hairy all over both surfaces ; peduncle slender, pubescent, $\frac{1}{2}$-1 ft.
long ; racemes 1-3, very lax, the end one finally 6-9 in. long ;
bracts minute, deltoid-cuspidate ; pedicels solitary, ascending or
cernuous, the lower $\frac{1}{3}$-$\frac{1}{2}$ in. long ; perianth glabrous, $\frac{1}{3}$ in. long ;
segments white, with a distinct brown keel ; stamens much shorter
than the perianth ; filaments muricated ; capsule globose, glabrous,
$\frac{1}{6}$-$\frac{1}{4}$ in. diam. *Fl. Cap. edit. Schult.* 322 ; *Willd. Sp. Plant.* ii.
140 ; *Roem. et Schultes, Syst. Veg.* vii. 460. *Baker in Journ. Linn.
Soc.* xv. 313. *Trachyandra hirsuta* and *T. corymbosa, Kunth, Enum.*
iv. 577.

South Africa : without locality, *Thunberg ! Oldenburg !*
Coast Region : Malmesbury Div. ; near Groene Kloof, under 1000 ft.,
Drège, 1494 ! Cape Div. ; rooky places, Kamps Bay, *Pappe !* clayey soil near
Wynberg, *Zeyher,* 4657 ! Devils Mountain, 700 ft., *Bolus,* 3794 ! Riversdale
Div. ; near Zoetmelks River, *Burchell,* 6654 ! 6797 ! Bathurst Div. ; near
Theopolis, *Burchell,* 4118 ! (a form with prostrate viviparous flowering stems).

51. **A. pilosum** (Baker) ; leaves 5-6 to a stem, erect, linear, very
acutely keeled, above a foot long, $\frac{1}{4}$-$\frac{1}{3}$ in. broad, clothed with fine,
soft, spreading hairs ; peduncle 1$\frac{1}{2}$ ft. long, hairy like the leaves ;
panicle of 6-8 lax racemes ; pedicels ascending, articulated at the
apex, the lower $\frac{1}{4}$-$\frac{1}{3}$ in. long ; bracts minute, ovate ; perianth $\frac{1}{2}$ in.
long ; segments thin, oblanceolate, white, with a brown keel ;
stamens rather shorter than the perianth ; anthers small, oblong ;
capsule small, globose.

Eastern Region : Griqualand East ; near Clydesdale, 2700 ft., *Tyson,* 2114 !
MacOwan and Bolus, Herb. Norm. Aust. Afr., 1207 !

52. **A. muricatum** (Linn. fil. Suppl. 202) ; root-fibres very
numerous, slender, cylindrical ; outer rudimentary leaves large,
ovate, whitish, membranous ; root-leaves linear-ensiform, 1-1$\frac{1}{2}$ ft.
long, $\frac{1}{2}$ in. broad, firm in texture, scabrous, closely ribbed ; peduncle
stout, hispid, 1-1$\frac{1}{2}$ ft. long ; racemes numerous, lax, the central ones
6-8 in. long ; bracts deltoid-cuspidate, the lower $\frac{1}{4}$-$\frac{1}{3}$ in. long ;
pedicels solitary, ascending, $\frac{1}{4}$-$\frac{1}{2}$ in. long ; perianth glabrous, $\frac{1}{3}$ in.
long ; segments oblanceolate, white, with a distinct brown keel ;
stamens much shorter than the perianth ; filaments muricated ;
capsule subglobose, glabrous. *Thunb. Prodr.* 63 ; *Fl. Cap. edit.
Schult.* 322 ; *Willd. Sp. Plant.* ii. 145 ; *Roem. et Schultes, Syst.
Veg.* vii. 459. *Arthropodium muricatum, Spreng. Syst.* ii. 87.
Trachyandra muricata, Kunth, Enum. iv. 576.

Coast Region : hills and stony places below Table Mountain, on Devils
Mountain, and on Lion Mountain near Green Point, *Thunberg ! Ecklon and
Zeyher !*

53. **A. Gerrardi** (Baker in Journ. Bot. 1872, 137) ; root-fibres
slender, wiry ; old leaves breaking up into a mass of bristles ; root-
leaves 6-10, linear, hispid, rarely glabrous, $\frac{1}{2}$-1 ft. long, $\frac{1}{6}$ in. broad,
bright green, moderately firm in texture, triquetrous on the back,

deeply channelled down the face ; peduncle ½ ft. long, muricated with copious whitish papillæ ; racemes 4–10, very lax, the end one 4–5 in. long ; rachises very scabrous ; bracts minute, deltoid-cuspidate ; pedicels ¼–½ in. long, finally decurved ; perianth pubescent, 4–5 lines long ; segments white, with a closely 3-nerved brown keel ; stamens much shorter than the perianth ; filaments muricated ; capsule globose, densely and persistently muricated. *Baker in Gard. Chron.* 1876, vi. 100.

SOUTH AFRICA : without locality, *Hutton!*
COAST REGION : Uitenhage Div. ; near the Zwartkops River, *Ecklon and Zeyher*, 110 ! Albany Div. ; near Grahamstown, 2000 ft., *MacOwan*, 1454 !
KALAHARI REGION : Transvaal ; Saddleback Range near Barberton, 3500 ft., *Galpin*, 1036 !
EASTERN REGION : Tembuland ; hills near Bazeia, 2000 ft., *Baur*, 310 ! Griqualand East ; near Clydesdale, 2500 ft., *Tyson*, 2123 ! Natal ; Inanda, *Wood*, 432 ! Tintern, 5000–6000 ft., *Evans*, 364 ! Zululand, *Gerrard*, 1527 !

54. A. ciliatum (Linn. fil. Suppl. 202) ; root-fibres some fleshy, others slender ; outer rudimentary leaves large, membranous ; leaves weak, linear, glabrous, finely ciliated, 1–2 ft. long, ⅓–½ in. broad ; peduncle ½ ft. long, glabrous or pubescent ; racemes simple or forked, lax, the central a foot or more long ; bracts ovate-cuspidate, greenish, ½ in. long ; pedicels ½–1 in. long, at first ascending, finally deflexed ; perianth ⅓ in. long ; segments lanceolate, with a closely 3-nerved green keel ; stamens much shorter than the perianth ; filaments papillose ; capsule glabrous, globose, ⅓ in. diam. *Thunb. Prodr.* 63 ; *Fl. Cap. edit. Schult.* 324 ; *Willd. Sp. Plant.* ii. 146. *A. blepharophoron, Roem. et Schultes, Syst. Veg.* vii. 461. *Bulbine ciliata, Link, Enum.* i. 329 ; *Roem. et Schultes, Syst. Veg.* vii. 450. *Trachyandra ciliata, T. blepharophora and T. bracteosa, Kunth, Enum.* iv. 578, 582, 585.

COAST REGION : Malmesbury Div. ; between Groene Kloof and Dassenberg, below 500 ft., *Drège*, 1493 ! Cape Div. ; Kalk Bay. *Pappe!* Simons Bay, *Wright*, 225 ! Lion Mountain, *Thunberg!* Knysna Div. ; near Goukamma River, *Burchell*, 5601 !
WESTERN REGION : Little Namaqualand, *Whitehead!*

55. A. paradoxum (Roem. et Schultes, Syst. Veg. vii. 459) ; rootstock oblique ; root-fibres long and slender ; outer rudimentary leaves membranous ; root-leaves 2, linear, very hairy, 3½–4½ in. long, forming a sort of cup at the tip ⅓ in. diam., of which the upper edge is the longest, revolute and tipped with a setaceous awn ; peduncle hispid, 2 in. long ; raceme dense, simple, an inch long ; pedicels reflexed after flowering, finally ¾ in. long ; bracts as long as the pedicels ; perianth hairy, ½ in. long ; stamens much shorter than the perianth ; filaments muricated. *Baker in Journ. Linn. Soc.* xv. 314. *Trachyandra paradoxa, Kunth, Enum.* iv. 576.

SOUTH AFRICA : without locality.

Described by the younger Schultes from a specimen in the herbarium of Zuccarini. Can it be an abnormal form of *A. hispidum ?*

XVII. CHLOROPHYTUM, Ker.

Perianth marcescent, persistent, not twisted after flowering; segments distinct, subequal, patent, 3–7-nerved. *Stamens* hypogynous, or adnate to the base of the perianth-segments; filaments filiform, often dilated above the middle; anthers linear-oblong, dorsifixed, dehiscing introrsely. *Ovary* sessile, ovoid, 3-celled; ovules several in a cell, superposed; style filiform; stigma capitate. *Capsule* deeply 3-lobed, acutely angled, loculicidally 3-valved. *Seeds* discoid, thin; testa black; embryo cylindrical; albumen firm in texture.

Rootstock very small; root-fibres wiry or fleshy; radical leaves linear and sessile, or broader and petioled; racemes simple or panicled; pedicels articulated; flowers small, whitish.

DISTRIB. Many species in Tropical Africa and Tropical Asia. A few in Australia and South America.

Racemes simple, not crested :
    Dwarf ; leaves lanceolate...   ...   .   (1) **modestum.**
    Tall ; leaves linear...   ...   .   (2) **vaginatum.**
    Tall ; leaves lanceolate, straight   ...   ...   (3) **Bowkeri.**
    Tall ; leaves lanceolate, curved ...   ...   (4) **drepanophyllum.**
Racemes panicled (rarely simple), not crested :
    Dwarf ; leaves lanceolate, ciliated   ...   ...   (5) **crispum.**
    Tall ; leaves not ciliated :
      Leaves linear :
        Leaves rigid, very narrow   ...   ...   (6) **pulchellum.**
        Leaves thin, $\frac{1}{4}$–$\frac{1}{2}$ in. broad   ...   ...   (7) **delagoense.**
      Leaves thin, lanceolate :
        Raceme forked at the base or simple ...   (8) **inornatum.**
        Racemes copiously panicled :
          Pedicels short   ...   .   (9) **elatum.**
          Pedicels long   ...   .   (10) **macrosporum.**
Racemes crowned with a rosette of reduced leaves ...   (11) **comosum.**

1. C. modestum (Baker in Journ. Linn. Soc. xv. 329); root-fibres many, fleshy; radical leaves 12–20, lanceolate, membranous, glabrous, 6–9 in. long, $\frac{1}{3}$–$\frac{1}{2}$ in. broad, narrowed gradually from the middle to the base, with about a dozen distinct ribs; peduncle slender, simple, arcuate, 3–4 in. long; raceme lax, simple, 2–3 in. long; bracts ovate, $\frac{1}{6}$–$\frac{1}{4}$ in. long; pedicels ascending, articulated at the middle, $\frac{1}{8}$–$\frac{1}{6}$ in. long, lower geminate; perianth whitish, $\frac{1}{4}$ in. long; segments oblanceolate ; stamens nearly as long as the perianth ; capsule obovoid-triquetrous ; seeds 3–4 in a cell.

EASTERN REGION: Natal, *Krauss*, 177 !

2. C. vaginatum (Baker); leaves 3–4 to a stem, long-sheathing, the free blade linear, glabrous, $\frac{1}{2}$–1 ft. long, $\frac{1}{4}$–$\frac{1}{2}$ in. broad ; peduncle 2–2$\frac{1}{2}$ ft. long ; raceme dense, simple, 3–6 in. long ; pedicels erecto-patent, the lower $\frac{1}{8}$–$\frac{1}{4}$ in. long ; bracts large, lanceolate from an ovate base, protruding beyond the buds; perianth $\frac{1}{4}$–$\frac{1}{3}$ in. long : segments oblong, white, with an obscure brown keel ; stamens $\frac{2}{3}$ the length of the perianth ; anthers small ; filaments lanceolate.

EASTERN REGION : Natal; Weenen County, 4000 ft., *Wood*, 4425 !

3. C. Bowkeri (Baker in Journ. Linn. Soc. xv. 332); root-fibres
fleshy, cylindrical; leaves lanceolate, 2–3 ft. long, 1–1½ in.
broad at the middle, narrowed gradually to the apex and clasping base,
moderately firm in texture, with 30–40 close distinct ribs; inflorescence 4–5 ft. long including the peduncle; raceme simple, 1–1½ ft.
long, stiffly erect; lower internodes 1–1½ in. long; bracts deltoid or
lanceolate, membranous, ¼–1 in. long; pedicels ⅛–¼ in. long, articulated at the middle, lower 3–4-nate; perianth white, ⅓–½ in. long;
segments 3–5-nerved; stamens nearly as long as the perianth;
anthers linear-oblong, shorter than the filaments; capsule obovoid,
emarginate, ½ in. long. *Ref. Bot. t.* 352.

SOUTH AFRICA: without locality, *Bowker!*
EASTERN REGION: Natal; Inanda, *Wood*, 1228!

4. C. drepanophyllum (Baker); root-fibres fleshy; basal scales
ovate, membranous; leaves lanceolate, glabrous, sickle-shaped,
moderately firm in texture, 9–12 in. long, an inch broad; peduncle
stout, leafless, ½ ft. long; raceme simple, 6–8 in. long, with a stout
pilose rachis: bracts scariose, ovate, acute, ⅓ in. long; pedicels
shorter than the bracts; perianth ½ in. long; segments linear-oblong,
rather pilose; stamens ⅓ shorter than the perianth; anthers large,
lanceolate.

WESTERN REGION: Little Namaqualand; near Nababeep, 3000 ft., *Bolus*,
6584!

5. C. crispum (Baker in Journ. Linn. Soc. xv. 331); root-fibres 3–4
in. long, cylindrical from a slender base; leaves lanceolate, densely
rosulate, 2–3 in. long, ⅙–¼ in. broad, sessile, crisped, ciliated on the
margin, distinctly 9-nerved; inflorescence ½–1 ft. long including the
stem, copiously panicled; racemes very lax; bracts minute, deltoid;
pedicels ⅛–¼ in. long, articulated at the middle, lower 2–3-nate;
perianth whitish, ¼–⅓ in. long; segments oblanceolate, 3-nerved;
stamens much shorter than the perianth; anthers oblong, small; capsule globose, ¼ in. diam. *Anthericum crispum, Thunb. Prodr.* 63;
Fl. Cap., edit. Schult. 324. *Bulbine crispa, Roem. et Schultes, Syst.*
Veg. vii. 448; *Kunth, Enum.* iv. 568.

SOUTH AFRICA: without locality, *Thunberg!*
COAST REGION: Uitenhage Div.; sand-hills near the Zwartkops River,
Zeyher, 1069!
CENTRAL REGION(?): Somerset Div. (?) *Bowker!*

6. C. pulchellum (Kunth, Enum. iv. 605); root-fibres slender;
leaves 5–6, rigid, erect, linear, glabrous, 9–12 in. long, ⅙ in.
broad, with a thickened edge and 8–12 close distinct ribs; peduncle
slender, 1½ ft. long, with a small reduced leaf from the middle;
racemes few, short, very lax; lower pedicels 2–3-nate, erecto-patent,
⅛–⅙ in. long; bracts minute, ovate, imbricated; perianth ⅙ in. long;
segments linear-oblong, whitish, with a broad red-brown keel;
stamens nearly as long as the perianth; anthers large, lanceolate;

filaments papillose; capsule oblong, emarginate, acutely angled, $\frac{1}{6}$ in. long, with the small valves prominently ribbed transversely.

SOUTH AFRICA : without locality, *Lalande!*

Described from Kunth's type specimen in the Berlin herbarium.

7. C. delagoense (Baker) ; leaves linear, membranous, glabrous, tapering to the apex, 8–9 in. long, $\frac{1}{4}$–$\frac{1}{3}$ in. broad; racemes lax, 4–6 in. long, probably panicled ; lower flowers geminate ; pedicels articulated at the middle, the lower $\frac{1}{6}$–$\frac{1}{4}$ in. long ; lower bracts linear, $\frac{1}{2}$–1 in. long, upper small, ovate ; perianth white, $\frac{1}{4}$ in. long ; segments oblong-lanceolate, laxly 3-nerved on the keel ; stamens as long as the perianth ; anthers oblong, much shorter than the filaments.

EASTERN REGION: Delagoa Bay, *Mrs. Monteiro!*

8. C. inornatum (Gawl. in Bot. Mag. t. 1071) ; root-fibres cylindrical, densely fascicled.; leaves lanceolate, membranous, glabrous, obscurely petioled, a foot long, an inch broad at the middle ; peduncle $1\frac{1}{2}$–2 ft. long, with several reduced leaves ; raceme lax, simple or forked at the base, a foot long ; pedicels erecto-patent, $\frac{1}{4}$–$\frac{1}{3}$ in. long, the lower 3–4-nate ; bracts small, ovate ; perianth $\frac{1}{3}$–$\frac{1}{2}$ in. long ; segments oblong, with a broad 3-nerved green keel : filaments $\frac{1}{6}$ in. long, as long as the anthers. *Roem. et Schultes, Syst. Veg.* vii. 453 ; *Kunth, Enum.* iv. 603 ; *Baker in Journ. Linn. Soc.* xv. 324.

WESTERN REGION : Little Namaqualand ; near Modderfontein, 3000 ft., *Bolus,* 6585 !

Also Sierra Leone and Guinea.

9. C. elatum (R. Br. Prodr. 277) ; rootstock short, woody ; root-fibres cylindrical ; leaves 12–20, lanceolate, firm in texture, $1\frac{1}{2}$–2 ft. long, an inch broad, narrowed to the apex and dilated clasping base, with about 30 distinct ribs ; stem 3–4 ft. long including the inflorescence, which is copiously panicled ; racemes lax, ascending, $\frac{1}{2}$–1 ft. long ; bracts small ; pedicels slender, 3–6-nate, $\frac{1}{8}$–$\frac{1}{4}$ in. long, articulated at the middle ; perianth whitish, $\frac{1}{3}$ in. long ; segments oblanceolate, 3–5-nerved ; stamens rather shorter than the perianth ; anthers linear-oblong ; capsule $\frac{1}{4}$ in. diam. ; seeds 5–6 in a cell. *Spreng. Syst.* ii. 88 ; *Roem. et Schultes, Syst. Veg.* vii. 454 ; *Kunth, Enum.* iv. 604 ; *Baker in Journ. Linn. Soc.* xv. 330 ; *Ref. Bot.* t. 216. *Anthericum elatum,* Ait. Hort. Kew, i. 448. *Phalangium elatum,* Red. Lil. t. 191. *P. fastigiatum, Poir. Encyc.* v. 246. *Asphodelus capensis, Linn. Syst. Veg. edit.* 10, 982. (*Asphodelus foliis planis, &c., Miller, Ic. t.* 56.)

VAR. β, C. Burchellii (Baker in Journ: Linn. Soc. xv. 330) ; a shade variety with thin leaves and a less compound inflorescence.

COAST REGION : Alexandria Div. ; Addo, 1000–2000 ft., *Drège,* 8719 ! Var. β. Albany Div. ; Blue Krantz, *Burchell,* 3650 ! 3650/1 !

CENTRAL REGION : Somerset Div. ; woods at the foot of the Bosch Berg, 2800 ft., *MacOwan,* 1967 !

KALAHARI REGION : Orange Free State, *Cooper,* 3592 !

EASTERN REGION : Pondoland ; woods at Emagushen, 3000 ft., *Tyson,* 3154 !

Var. β, Tembuland; Bazeia, 2500 ft., *Baur*, 131! Natal; Inanda, *Wood*, 842 partly!

Anthericum variegatum, Floral Mag., 1875, *t.* 152, is a form with variegated leaves.

10. C. macrosporum (Baker in Journ. Linn. Soc. xv. 330); leaves lanceolate, sessile, moderately firm in texture, glabrous, 1½–2 ft. long, under an inch broad, with about a dozen distinct ribs; stem 3 ft. long below the panicle, stout, stiffly erect; panicle deltoid, 1½–2 ft. long, with a few spreading branches; bracts small; pedicels 3–6-nate, articulated at the middle, some finally an inch long; perianth whitish, ⅓ in. long; segments oblong, with a broad green keel; stamens nearly as long as the perianth; anthers small, linear-oblong; capsule ½ in. long, oblong, very acutely angled; seeds orbicular, 5–6 in a cell, ⅙ in. diam.

KALAHARI REGION: South African Gold Fields, *Baines*. (Perhaps intertropical).

11. C. comosum (Baker in Journ. Linn. Soc. **xv.** 329); root-fibres cylindrical; radical leaves 10–12, linear, glabrous, moderately firm in texture, 1–1½ ft. long, ¼–½ in. broad at the middle, narrowed gradually to the apex and clasping base, distinctly 12–15-nerved; stem 1½–2 ft. long including the inflorescence, simple or branched; racemes very lax, reaching a foot in length, with a tuft of linear leaves 1–3 in. long from the apex and sometimes also from the upper nodes; bracts small, deltoid-cuspidate; pedicels 2–4-nate, ⅙–¼ in. long, articulated at the middle; perianth ⅓ in. long, whitish; outer segments oblanceolate, inner oblong, 3-nerved in the centre; stamens nearly as long as the perianth; anthers small, oblong; capsule ⅓ in. diam., deeply 3-lobed; seeds 3–5 in a cell. *Anthericum comosum, Thunb. Prodr.* 63; *Fl. Cap. edit. Schult.* 323. *Phalangium comosum, Poir. Encyc.* v. 252. *Cæsia comosa, Spreng. Syst.* ii. 88; *Kunth, Enum.* iv. 610. *Hartwegia comosa, Nees in Nova Acta* xv. 2, 373; *Kunth, Enum.* iv. 607. *Anthericum sternbergianum, Roem. et Schultes, Syst. Veg.* vii. 1693. *Phalangium viviparum, Hort.*

SOUTH AFRICA: without locality, *Ecklon and Zeyher*, 112!
COAST REGION: Uniondale Div.; Lange Kloof, *Thunberg!* Stockenstrom Div.; summit of Katberg, *Scully*, 137!
CENTRAL REGION: Somerset, *Elliot*, 328!
EASTERN REGION: Natal; Inanda, *Wood*, 842 partly! and without precise locality, *Cooper*, 3313!

XVIII. CÆSIA, R. Br.

Perianth marcescent, persistent, spirally twisted when faded; segments oblanceolate, subequal, 3-nerved, connate at the very base, *Stamens* 6, attached to the base of the perianth-segments; filaments filiform; anthers small, oblong, dorsifixed, dehiscing introrsely.

Ovary sessile, globose, 3-celled ; ovules 2 in a cell, collateral ; style filiform ; stigma capitate. *Capsule* globose, loculicidally 3-valved. *Seeds* triquetrous, solitary in the cells; testa black, crustaceous, papillose ; albumen fleshy.

Rootstock obscure; leaves linear, graminoid ; racemes usually panicled, very lax; pedicels filiform, articulated at the apex, often 2-3-nate ; flowers small, tender, lilac-blue.

DISTRIB. Also Australian, and one species in Madagascar.

Raceme simple ...   .   (1) eckloniana.
Racemes copiously panicled :
    Stems with single leaves from 1-2 nodes   ...   ...   (2) dregeana.
    Stems with tufts of leaves from many nodes   ...   ...   (3) Thunbergii.

1. C. eckloniana (Roem. et Schultes, Syst. Veg. vii. 1691) ; densely cæspitose, with scarcely any rootstock ; root-fibres slender, wiry ; leaves subterete, densely tufted, erect, 6-9 in. long, under a line in diameter, channelled down the lower part of the face, rounded and strongly 5-ribbed on the back ; peduncle simple, very slender, 1-3 in. long; raceme very lax, simple, 6-8 in. long ; lower internodes $1\frac{1}{2}$-2 in. long ; bracts minute, deltoid; pedicels decurved after the flower falls, lower 1-2-nate, $\frac{1}{4}$ in. long ; perianth bright lilac, $\frac{1}{4}$ in. long ; stamens half as long as the perianth. *Kunth, Enum.* iv. 609. *Anthericum scilliflorum, Ecklon, Herb. Cap. No.* 35 ; *Baker in Journ. Linn. Soc.* xv. 298. *A. Zeyheri, Baker, loc. cit.*

SOUTH AFRICA : without locality, *Zeyher*, 4234 ! *Ecklon*, 35 !
COAST REGION : Cape Peninsula, near Simons Town, 800 ft., *Bolus*, 4692 !

2. C. dregeana (Kunth, Enum. iv. 611) ; densely cæspitose, with scarcely any rootstock ; root-fibres tufted, slender, wiry ; root-leaves linear, graminoid, erect, flat, firm in texture, a foot long, $\frac{1}{8}$ in. broad, with numerous fine ribs; inflorescence 1-2 ft. long including the short stem, copiously panicled, with reduced leaves from the lower nodes; racemes very lax, 3-6 in. long; bracts minute, deltoid ; pedicels $\frac{1}{4}$-$\frac{1}{2}$ in. long, finally recurved, lower geminate ; perianth bright lilac, $\frac{1}{4}$-$\frac{1}{3}$ in. long ; segments oblanceolate ; stamens half as long as the perianth. *Anthericum dregeanum, Baker in Journ. Bot.* 1872, 139 ; *Journ. Linn. Soc.* xv. 299.

COAST REGION : Clanwilliam Div. ; near Groen River and Watervals River, and at Uien Vallei, 2000-3000 ft., *Drège*, 8768 ! Ezels Bank, 3000-4000 ft., *Drège*, 8767 ! Paarl Div.; Great Drakenstein Mountains, under 1000 ft., *Drège*, 8769 ! Mountains above Worcester, *Rehmann*, 2544 !

3. C. Thunbergii (Kunth, Enum. iv. 610) ; densely cæspitose, with scarcely any rootstock ; root-fibres wiry ; leaves linear, graminoid, persistent, glabrous, 1-2 ft. long, $\frac{1}{4}$ in. broad, finely many-nerved ; inflorescence 3-4 ft. long including the stem, copiously panicled, with tufts of leaves from many of the lower nodes ; racemes $\frac{1}{2}$-1 ft. long; lower internodes $1\frac{1}{2}$-2 in. long ; bracts small, lanceolate-deltoid ; pedicels $\frac{1}{2}$-$\frac{3}{4}$ in. long, finally decurved, lower 2-3-nate ; perianth purplish, $\frac{1}{4}$-$\frac{1}{3}$ in. long; segments oblanceolate ; stamens half as long as the perianth ; capsule globose,

$\frac{1}{6}$ in. diam. *Anthericum contortum, Linn. fil. Suppl.* 202; *Thunb. Prodr.* 63; *Fl. Cap. edit. Schult.* 319; *Roem. et Schultes, Syst. Veg.* vii. 480, 1695; *Baker in Journ. Linn. Soc.* xv. 299.

COAST REGION : Clanwilliam Div.; Blue Berg, 2000–3000 ft., *Drège,* 1504b! Cape Div.; flats near Rondebosch, *Burchell,* 726! vineyards and fields below Table Mountain, *Thunberg!* Paarl Mountains, 900 ft., *Bolus,* 5565! Worcester Div. ; Dutoits Kloof, 2000–3000 ft., *Drège,* 1504a! George Div.; near Touw River, *Burchell,* 5748! on the Post Berg near George, *Burchell,* 5951!

XIX. NANOLIRION, Benth.

Perianth marcescent, spirally twisted after flowering; segments distinct, subequal, oblanceolate, obscurely 3-nerved. *Stamens* 6, hypogynous or the inner attached to the base of the perianth-segments; filaments flattened; anthers ovoid-oblong, dorsifixed, with introrse dehiscence. *Ovary* sessile, 3-celled; ovules 2 in a cell, collateral, erect; style filiform; stigma capitate. *Capsule* subglobose, loculicidally 3-valved. *Seeds* not seen.

DISTRIB. Endemic, closely allied to the Australian and New Zealand mono-typic genus *Herpolirion, Hook. fil.*

1. **N. capense** (Benth. in Gen. Plant. iii. 793); whole plant an inch high, consisting of dense tufts crowded on a slender creeping rootstock, which sends out wiry root-fibres; leaves 20 or more to a rosette, subterete, firm in texture, strongly ribbed, persistent, rounded on the back, channelled down the face; flowers 1–3 on a very short peduncle from the centre of the rosette of leaves; pedicels $\frac{1}{4}$–$\frac{1}{3}$ in. long; perianth dull blue, $\frac{1}{4}$–$\frac{1}{3}$ in. long; stamens half as long as the perianth; capsule $\frac{1}{5}$ in. long and broad. *Oliv. in Hook. Ic. t.* 1726. *Herpolirion capense, Bolus in Journ. Linn. Soc.* xviii. 395.

COAST REGION : Tulbagh Div.; on the rocky summit of Winterhoek Mountain, near Tulbagh, 6500 ft. *Bolus,* 5170 !

XX. AGAPANTHUS, L'Herit.

Perianth gamophyllous, infundibuliform; segments oblong, sub-equal, much longer than the tube. *Stamens* 6, inserted at the throat of the perianth-tube; filaments filiform, as long as the segments; anthers small, oblong, dorsifixed, dehiscing introrsely. *Ovary* narrow, sessile, 3-celled; ovules numerous, superposed; style filiform; stigma minute. *Capsule* coriaceous, elongate-oblong, loculicidally 3-valved. *Seeds* flat; testa lax, blackish, produced upwards into an oblong wing.

DISTRIB. Endemic.

1. **A. umbellatus** (L'Herit. Sert. Angl. 17); rootstock short, tuberous; root-fibres fleshy, cylindrical; radical leaves 6–8, lorate, green, fleshy, 1–2 ft. long, 1–1$\frac{1}{2}$ in. broad; peduncle leafless, stout,

terete, 3–4 ft. long; flowers 30–50 in an umbel, bright blue, rarely white; spathe-valves 2, broad, membranous, falling early; pedicels 1½–3 in. long, articulated at the apex; perianth 1½–2 in. long; segments ⅓–½ in. broad, at least twice as long as the tube, distinctly keeled; capsule 1½ in. long; seeds ½ in. long. *Ait. Hort. Kew.* i. 414; *Bot. Mag. t.* 500; *Red. Lil. t.* 4; *Kunth, Enum.* iv. 479; *Baker in Journ. Linn. Soc.* xi. 369. *Crinum africanum, Linn. Sp. Plant.* 292. *Mauhlia africana, Dahl, Obs.* 26. *M. linearis, Thunb. Prodr.* 60; *Fl. Cap. edit. Schult.* 308; *Diss. Nov. Gen.* 113. (*C. Commel. Hort. Amstel.* ii. *t.* 67. *Tulbaghia, Heister, Brunsvigia* 10 *in nota; Fabric. Helmst.* 4).

VAR. β, **A. multiflorus** (Willd. Enum. 353 in nota); a more robust variety, with more numerous flowers (60–80) and broader leaves. *Kunth, Enum.* iv. 480.

VAR. γ, **A. præcox** (Willd. Enum. 353); an early flowering form (at the end of June in English gardens) with narrower leaves, a shorter peduncle, and narrower perianth-segments. *Kunth, Enum.* iv. 480. *A. umbellatus var. minimus, Lindl. Bot. Reg. t.* 699.

VAR. δ, **A. minor** (Lodd. Bot. Cab. t. 42); a dwarf form, with leaves not above ½ in. broad, flowers ¾–1 in. long, pedicels 1–1½ in. long. *A. umbellatus var. minor, Red. Lil. t.* 403.

VAR. ε, **Leichtlinii** (Baker in Gard. Chron. 1878, x. 428); differs from *minor* by its shorter, broader leaves, larger flowers, denser umbel; pedicels not more than ½ in. long.

The above varieties are connected by gradual intermediates. Besides these, *A. giganteus* (*Hort. Weiner Illust. Gart.-Zeit.* 1880, 119) is a very robust form, with 150–200 flowers and leaves 2 in. broad, and *A. mooreanus, Hort.*, a form with the narrow leaves of *minor*, but flowers as large as in the type.

COAST REGION: Uitenhage Div.; Van Stadens Berg, *MacOwan*, 1914! Port Elizabeth, *Drège*, 4510! Var. γ, Cape Div.; Table Mountain, *Ecklon*, 7b! Var δ, Cape Peninsula, plains above Simons Bay, *Milne*, 167! Riversdale Div.; Lange Bergen, *Burchell*, 7138! George Div.; Honing Klip, *Drège*, 8594!

CENTRAL REGION: Somerset Div.; Bosch Berg, 4000 ft., *MacOwan*, 1914!

KALAHARI REGION: Transvaal; Houtbosch, *Rehmann*, 5798! Saddleback Range, near Barberton, 4000–4600 ft., *Galpin*, 1224! Var. δ, Orange Free State, *Cooper*, 3270! Transvaal; near Lydenburg, *Atherstone!*

EASTERN REGION: Natal; Tugela River, *Allison!* Var. δ, Natal; Ingote, *Sutherland!* Var. δ, Natal; Enon, *Mrs. K. Saunders in Herb. Wood*, 3153! and without precise locality, *Cooper*, 3272!

XXI. TULBAGHIA, Linn.

Perianth gamophyllous, hypocrateriform; tube oblong or cylindrical; segments oblong or lanceolate, subequal, patent, generally shorter than the tube, which has a corona at the throat which is either entire and annular or composed of three distinct processes placed opposite the inner segments. *Stamens* 6, biseriate, inserted on the tube of the perianth; filaments short; anthers oblong, dorsifixed, dehiscing introrsely. *Ovary* sessile, 3-celled; ovules many in a cell, superposed; style short, columnar; stigma capitate. *Capsule* oblong, chartaceous, loculicidally 3-valved. *Seeds* oblong, compressed; testa loose, blackish; albumen fleshy.

Rootstock tuberous; leaves radical, linear or lorate, fleshy; peduncle slender, naked; flowers small, arranged in an umbel with two membranous spathe-valves.

DISTRIB. : Two or three species occur on the mountains of Tropical Africa.

Corona cup-shaped, with an entire or crenate edge :
　Corona $\frac{1}{12}$-$\frac{1}{8}$ in. long :
　　Leaves very narrow :
　　　Perianth greenish　...　...　...　...　(1) **acutiloba.**
　　　Perianth white　...　...　...　...　(2) **leucantha.**
　　Leaves linear or lorate　..　...　...　...　(3) **alliacea.**
　Corona very short :
　　Perianth-segments a little longer than the corona　(4) **dregeana.**
　　Perianth-segments twice as long as the corona　...　(5) **natalensis.**
　　Perianth-segments at least four times the length
　　　of the corona :
　　　　Leaves linear, a foot long　...　...　...　(6) **hypoxidea.**
　　　　Leaves very narrow, 2–3 in. long　...　...　(7) **pauciflora.**
　Coronal-lobes bifid, united at the base　...　...　...　(8) **capensis.**
　Coronal-lobes distinct, entire or emarginate :
　　Perianth $\frac{1}{2}$ in. long　...　...　...　...　...　(9) **cepacea.**
　　Perianth $\frac{3}{4}$ in. long　...　...　...　...　...　(10) **violacea.**

1. T. acutiloba (Harv. Thes. Cap. t. 180); root of many fleshy fibres ; corm ampullæform, with brown, membranous tunics ; leaves 4–6, linear, moderately firm in texture, 4–6 in. long, $\frac{1}{2}$–1 line broad ; peduncle slender, terete, $\frac{1}{2}$–1 ft. long ; umbel 2–6-flowered ; spathe-valves green, lanceolate ; pedicels $\frac{1}{4}$–1 in. long ; perianth $\frac{1}{3}$ in. long ; tube oblong, greenish ; segments lanceolate, nearly as long as the tube ; corona annular, dark purple, $\frac{1}{8}$ in. long, crenate at the margin ; upper stamens reaching to the throat of the corona. *Baker in Journ. Linn. Soc.* xi. 371.

VAR. β, **curta** (Baker, loc. cit.) ; perianth-segments much shorter, not above $\frac{1}{12}$ in. long.

VAR. γ, **major** (Baker, loc. cit.) ; more robust than the type, with peduncles 1$\frac{1}{2}$ ft. long, leaves nearly a foot long, and more numerous, rather larger flowers.

SOUTH AFRICA : without locality, *Sieber*, 262 ! *Pappe !* Var. γ, *Zeyher !*
COAST REGION : Table Mountain, *Ecklon*, 94 ! Stockenstrom Div., Katberg, 3000–4000 ft., *Drège!* Queenstown Div., *Cooper*, 463 ! Var. β, Stellenbosch Div. ; Somerset West, *Drège*, 1516 !
CENTRAL REGION : near Somerset East, 3000 ft., *MacOwan*, 1582 !
KALAHARI REGION : Var. γ, Griqualand West ; Upper Campbell, *Burchell*, 1829 !
EASTERN REGION : Tembuland ; near Gatberg, 4000 ft., *Baur*, 736 ! Natal ; near Durban, *Wood*, 43 ! Inanda, *Wood*, 173 ! 201 ! and without precise locality, *Buchanan !*

2. T. leucantha (Baker) ; root with many long fibres ; leaves 5–6, linear, glabrous, 5–6 in. long, $\frac{1}{12}$–$\frac{1}{8}$ in. broad ; peduncle slender, rather longer than the leaves ; umbel 4–6-flowered ; pedicels $\frac{1}{2}$–$\frac{3}{4}$ in. long ; perianth whitish, $\frac{1}{3}$ in. long ; segments linear, as long as the oblong tube ; corona nearly as long as the perianth lobes, crenate.

KALAHARI REGION : Transvaal ; Bosch Veldt, between Kleinsmit and Kamel Poort, *Rehmann*, 4842 !
EASTERN REGION : Griqualand East ; Zuurberg Range, 3500 ft., *MacOwan and Bolus, Herb. Norm. Aust. Afr.*, 1208 ! *Tyson.* Natal ; Umzinyati Falls, *Wood*, 1200 ! near Tugela River, *Wood*, 4408 !

3. T. alliacea (Linn. fil. Suppl. 193); corm globose; root-fibres
fleshy, cylindrical ; outer rudimentary leaves brown, membranous;
produced leaves 6–8, erect, fleshy, $\frac{1}{2}$–1$\frac{1}{2}$ ft. long, $\frac{1}{6}$–$\frac{1}{3}$ in. broad;
peduncle fragile, terete, 1–2 ft. long ; umbel 6–10-flowered ; pedicels
1–2 in. long ; spathe-valves ovate or ovate-lanceolate, membranous ;
perianth greenish, $\frac{1}{3}$ in. long; segments lanceolate, about half as long
as the tube ; corona annular, purplish-brown or yellow, shorter than
the perianth-segments, crenate on the margin; upper stamens
reaching up into the corona ; capsule ovoid, splitting the marcescent
perianth-tube, finally $\frac{1}{2}$ in. long. *Thunb. Prodr.* 60 ; *Fl. Cap. edit.
Schult.* 306 ; *Baker in Journ. Linn. Soc.* xi. 371, *non Kunth.
T. brachystemma, Kunth, Enum.* iv. 483. *T. affinis, Link, Enum.*
i. 310 ; *Roem. et Schultes, Syst. Veg.* vii. 994.

Var. β, **T. ludwigiana** (Harv. in Bot. Mag. t. 3547); leaves lorate, $\frac{1}{2}$ ft.
long, $\frac{1}{3}$–$\frac{3}{4}$ in. broad; corona bright yellow, urceolate, with inflexed edges.

Coast Region: Malmesbury Div.; Groene Kloof, 300 ft., *Bolus,* 4347 !
Cape Div. ; Muizenberg, 1100 ft., *Bolus,* 4649 ! Albany Div.. *Cooper,* 3279 !
Stockenstrom Div. ; Katberg, 3000–4000 ft., *Drège !* King Williamstown Div. ;
Keiskamma, *Mrs. Hutton !* Var. β, Keiskamma, *Mrs. Hutton !*
Kalahari Region : Transvaal ; hills near Pretoria, 4000 ft.; *McLea in Herb.
Bolus,* 3091 !
Eastern Region : Natal, *Sanderson,* 429 ! Inanda, *Wood,* 257 !

The leaves have usually a strong alliaceous scent, but in a plant grown by
Mr. Elwes this was entirely wanting.

4. T. dregeana (Kunth, Enum. iv. 483); outer rudimentary
leaves membranous, brown, 1–1$\frac{1}{2}$ in. long; produced leaves linear,
moderately firm in texture, 3–6 in. long, $\frac{1}{12}$ in. broad; peduncle
slender, under a foot long; umbel 4–8-flowered; spathe-valves
lanceolate ; pedicels $\frac{1}{4}$–1 in. long ; perianth greenish, $\frac{1}{4}$–$\frac{1}{3}$ in. long ;
segments lanceolate, $\frac{1}{2}$–1 line long ; corona annular, crenate, brownish,
rather shorter than the perianth-segments ; upper stamens reaching
to the throat of the corona. *Baker in Journ. Linn. Soc.* xi. 371.

Western Region: Little Namaqualand ; near Lily Fontein and Ezels
Fontein, *Drège,* 2658 !

Perhaps only a montane variety of *T. acutiloba,* from which it differs by its
shorter perianth-segments and corona.

5. T. natalensis (Baker in Gard. Chron. 1891, ix. 668) ; leaves
6–8 to a tuft, linear, bright green, $\frac{1}{2}$–1 ft. long at the flowering time,
channelled down the face, with an alliaceous scent when broken ;
scape terete, above a foot long ; umbel 6–10-flowered ; pedicels
$\frac{1}{4}$–$\frac{1}{2}$ in. long; spathe-valves 2, lanceolate, an inch long; perianth
white, tube campanulate, $\frac{1}{6}$ in. long ; segments obovate, half as long
again as the tube ; corona greenish-white, half as long as the
perianth-segments, deeply lobed ; anthers reaching halfway up the
corona.

Eastern Region : Natal, described from a living plant in May, 1891, sent to
Kew by Mr. J. M. Wood.

6. T. hypoxidea (Smith in Rees Cyclop.); leaves linear, about a foot long, $\frac{1}{6}-\frac{1}{4}$ in. broad; peduncle about as long as the leaves; umbel 6–8-flowered; pedicels $\frac{1}{2}$–1 in. long; perianth $\frac{1}{2}$ in. long; segments linear, about as long as the tube; corona annular, sub-entire, not above a quarter as long as the perianth-segments. *Kunth, Enum.* iv. 482 ; *Baker in Journ. Linn. Soc.* xi. 372.

SOUTH AFRICA: without locality.

Described from the type specimen in the Smithian herbarium.

7. T. pauciflora (Baker in Engl. Jahrb. xv. Heft 3, 6); corm oblong, $\frac{1}{4}$ in. diam.; leaves 5–6, filiform, very slender, glabrous, 2–3 in. long; peduncles 1–2 from a corm, about as long as the leaves; flowers 1–3 to an umbel; spathe-valves lanceolate, $\frac{1}{4}-\frac{1}{3}$ in. long; pedicels $\frac{1}{4}-\frac{1}{2}$ in. long; perianth-tube oblong, $\frac{1}{6}$ in. long; segments linear, as long as the tube, white, with a brown keel; corona entire, very short; style very short; capsule globose, $\frac{1}{6}$ in. diam.

SOUTH AFRICA : without locality ; Hemerocall. 6, *Ecklon and Zeyher !*

8. T. capensis (Jacq. Hort. Vind. ii. t. 115); corm globose; root-fibres fleshy, cylindrical ; outer rudimentary leaves short, mem-branous ; produced leaves 8–12, linear, fleshy, a foot or more long, $\frac{1}{3}-\frac{1}{2}$ in. broad ; peduncle $1\frac{1}{2}$–2 ft. long; umbel 6–8-flowered ; spathe-valves small, membranous, lanceolate ; pedicels $\frac{3}{4}$–1 in. long; perianth purplish-green, $\frac{1}{3}$ in. long; segments lanceolate, about half as long as the tube ; coronal-lobes purplish-brown, deeply bifid, as long as the perianth-segments; upper stamens placed opposite the top of the staminodia. *Linn. Mant. alt.* 223 ; *Baker in Journ. Linn. Soc.* xi. 370. *T. alliacea, Gawl. in Bot. Mag. t.* 806 ; *Kunth, Enum.* iv. 481 (*excl. syn.*), *non Linn. fil.*

VAR. β, **gracilis** (Baker, loc. cit.) ; much less robust, with leaves not more than 3–4 in. long, $\frac{1}{8}-\frac{1}{6}$ in. broad.

SOUTH AFRICA : without locality.

WESTERN REGION : Var. β, Little Namaqualand ; Hardeveld, 2000–3000 ft., *Zeyher*, 4268 !

9. T. cepacea (Linn. fil. Suppl. 194); rudimentary basal leaves brown, membranous ; produced leaves 4–6, narrow linear, 4–8 in. long, $\frac{1}{12}$ in. broad ; peduncle 1–$1\frac{1}{2}$ ft. long; flowers 6–12 in an umbel ; spathe-valves lanceolate, lilac-tinted ; pedicels $\frac{1}{4}-\frac{1}{2}$ in. long ; perianth bright lilac, $\frac{1}{2}$ in. long; segments oblong-lanceolate, half as long as the cylindrical tube, which is only $\frac{1}{2}$ lin. diam. ; coronal-lobes distinct to the base; upper stamens placed in the tube below the throat. *Thunb. Prodr.* 60 ; *Fl. Cap. edit. Schult.* 306 ; *Willd. Sp. Plant.* ii. 34 ; *Kunth, Enum.* iv. 484 (*excl. var. β*) ; *Baker in Journ. Linn. Soc.* xi. 372. *Omentaria cepacea, Salisb. Gen.* 88.

SOUTH AFRICA : without locality, *Masson !*

COAST REGION : Uitenhage Div.; Van Stadens Berg, *Burchell*, 4741 ! *Ecklon and Zeyher*, 645 ! *Zeyher !*

CENTRAL REGION : arid places in Kannaland and Hantam, *Thunberg !*

10. T. violacea (Harv. in Bot. Mag. t. 3555); basal rudimentary leaves brown, membranous; leaves 4–6, linear, flaccid, 8–12 in. long, $\frac{1}{8}$–$\frac{1}{4}$ in. broad; peduncle 1–2 ft. long; umbel 10–20-flowered; spathe-valves ovate-lanceolate, $\frac{1}{2}$–$\frac{3}{4}$ in. long, tinged with lilac; pedicels $\frac{1}{2}$–1$\frac{1}{2}$ in. long; perianth bright lilac, $\frac{3}{4}$ in. long; segments oblong-lanceolate, half as long as the cylindrical tube; coronal-lobes $\frac{1}{12}$–$\frac{1}{8}$ in. long, truncate or emarginate; upper stamens not reaching to the throat of the perianth-tube. *Kunth, Enum.* iv. 485; *Baker in Journ. Linn. Soc.* xi. 372.

VAR. β, **minor** (Baker, loc. cit.); peduncle 4–5 in. long; umbel 3–6-flowered; pedicels very short; perianth-segments lanceolate, more than half as long as the tube.

VAR. γ, **obtusa** (Baker, loc. cit.); segments of the perianth-limb oblong, obtuse, $\frac{1}{8}$–$\frac{1}{8}$ in. broad. *T. cepacea var. robustior, Kunth, Enum.* iv. 484.

COAST REGION: Port Elizabeth, *E.S.C.A. Herb.*, 262! Albany Div.; Bothas Hill, *MacOwan*, 914! King Williamstown Div.; Keiskamma Hoek, *Cooper*, 544!

CENTRAL REGION: Albert Div.; by the Orange River, *Burke!* Var. γ, Somerset Div.; between the Zuurberg and Klein Bruntjes Hoogte, 2000–2500 ft., *Drège!*

EASTERN REGION: Kaffrarian Mountains, *Mrs. Barber*, 41!

XXII. ALLIUM, Linn.

Perianth polyphyllous, marcescent; segments subequal, patent or permanently connivent. *Stamens* 6, attached to the base of the perianth-segments; stamens filiform or flattened, the three inner tricuspidate in the section *Porum;* anthers oblong, dorsifixed, dehiscing introrsely. *Ovary* sessile, or shortly stipitate; ovules usually 2 in a cell; style filiform, usually inserted into a central hollow of the ovary; stigma capitate or obscurely tricuspidate. *Capsule* globose, small, membranous, dehiscing loculicidally. *Seeds* angled; testa black, membranous; albumen fleshy.

Rootstock a tunicated bulb or short rhizome; leaves all radical, but often sheathing the base of the peduncle for some distance; flowers umbellate; spathe-valves usually two, membranous; odour alliaceous.

DISTRIB. Species 250–300, concentrated in the North temperate zone of both hemispheres.

1. A. dregeanum (Kunth, Enum. iv. 382); bulb ovoid, under an inch in diameter; tunics pale, membranous; peduncle terete, 1$\frac{1}{2}$–2 ft. long; leaves 3–5, linear, 8–12 in. long, $\frac{1}{6}$ in. broad, with long basal sheaths; umbel globose, many-flowered, usually but not always with a number of large brown ovoid bulbillæ; pedicels generally about $\frac{1}{2}$ in. long, sometimes 1–2 in.; perianth permanently campanulate, $\frac{1}{6}$ in. long, white tinged with pink; segments much imbricated, ovate, acute, 3-nerved on the keel; stamens shorter than the segments, 3 inner tricuspidate, 3 outer ovate in the lower half; ovary ovoid-trigonous; ovules 2 in a cell; style shorter than the ovary.

COAST REGION: Malmesbury Div.; near Hopefield, *Bachmann*, 863! Rivers-

dale Div. ; near Zoetmelks River, *Burchell*, 6626! Queenstown Div. ; Storm Berg Range, 5000–6000 ft., *Drège*, 8660a!
CENTRAL REGION : mountains near Graaff Reinet, 3200 ft., *Bolus*, 648!
WESTERN REGION : Little Namaqualand, *Scully*, 215!
KALAHARI REGION : Orange Free State, Caledon River, *Burke!*

This seems to me quite distinct from the European *A. Scorodoprasum, Linn.*, with which Regel (*Monogr. All.* 42) unites it. Drège's 8661, from the Zwartkops River, referred by Kunth to *A. Scorodoprasum*, I have not seen. Burchell's 1547 from Carnarvon has the robust habit, large bulb, and broad leaves of *A. Ampeloprasum*, but the specimens are in young bud only.

XXIII. MASSONIA, Thunb.

Perianth gamophyllous ; tube cylindrical ; segments subequal, linear or lanceolate, spreading or reflexed. *Stamens 6*, inserted in a single row at the throat of the perianth-tube ; filaments united in a cup at the base ; anthers oblong, dorsifixed, dehiscing introrsely. *Ovary* sessile, 3-celled ; ovules many in a cell, superposed ; style cylindrical ; stigma capitate. *Capsule* obovate-triquetrous, membranous, loculicidally 3-valved. *Seeds* globose ; testa black, membranous ; albumen moderately firm in texture.

Rootstock a tunicated bulb ; leaves two, broad, opposite ; flowers in a sessile or nearly sessile globose capitulum, surrounded by a number of ovate or oblong, imbricated, membranous bracts ; pedicels short, not articulated, subtended by smaller bracts ; perianth usually white.

DISTRIB. Endemic.

Leaves bristly on the surface :
  Filaments very short ... ... (1) **tenella.**
  Filaments as long as or longer than the
    perianth-segments :
      Leaves lanceolate ... ... (2) **setulosa.**
      Leaves oblong ... ... (3) **echinata.**
      Leaves round :
        Perianth ½ in. long ... ... (4) **hirsuta.**
        Perianth an inch long ... ... (5) **muricata.**
Leaves rough with pustules over the face :
  Leaves lanceolate... ... ... (6) **pauciflora.**
  Leaves small, obovate :
    Perianth-segments as long as the tube ... (7) **Schlechtendahlii.**
    Perianth-segments half as long as the tube (8) **latebrosa.**
  Leaves large, round-oblong :
    Perianth an inch long ... ... (9) **pustulata.**
    Perianth ½ in. long ... ... (10) **longipes.**
Leaves glabrous on the face :
  Filaments not above $\frac{1}{12}$–$\frac{1}{8}$ in. long :
    Anthers small :
      Filaments free to the base ... ... (11) **jasminiflora.**
      Filaments joined in a cup at the base (12) **Bowkeri.**
    Anthers large ... ... (13) **brachypus.**
  Filaments about ¼ in. long :
    Perianth-segments linear :
      Perianth-segments not as long as the
        tube ... ... (14) **Dregei.**
      Perianth-segments as long as the tube (15) **Huttoni.**

Perianth-segments lanceolate, shorter than
　the tube :
　　Perianth-tube ¼ in. long :
　　　Leaves as long as broad ...　　...　(16) **versicolor.**
　　　Leaves half as long as broad　...　(17) **amygdalina.**
　　Perianth-tube above ½ in. long　...　(18) **Greenii.**
　Perianth-segments lanceolate, as long as
　　the tube :
　　　Pedicels very short :
　　　　Leaves suborbicular　　...　　...　(19) **calvata.**
　　　　Leaves lanceolate　　　...　　...　(20) parvifolia.
　　　Pedicels ¼–½ in. long ...　　...　　...　(21) **orientalis.**
　Perianth-segments lanceolate, longer than
　　the tube ...　　...　...　　...　　...　(22) **concinna.**
Filaments ½–¾ in. long :
　Perianth-segments much shorter than the
　　tube :
　　Leaves oblong-lanceolate :
　　　Leaves an inch broad　　...　　...　(23) **pedunculata.**
　　　Leaves 3–4 in. broad　　...　　...　(24) **longifolia.**
　　Leaves broad-ovate or suborbicular :
　　　Filaments white :
　　　　Segments half as long as the
　　　　　tube　　...　　...　　...　(25) **candida.**
　　　　Segments ⅓–¼ as long as the
　　　　　tube　　...　　...　　...　(26) **læta.**
　　　Filaments reddish-yellow　　...　(27) **cordata.**
　　　Filaments bright-red　　...　　...　(28) **sanguinea.**
　Perianth-segments about as long as the
　　tube :
　　Filaments red :
　　　Leaves lanceolate-oblong　　...　(29) **lanceæfolia.**
　　　Leaves suborbicular :
　　　　Leaves 2 in. long and broad　(30) **namaquensis.**
　　　　Leaves 6–10 in. long and
　　　　　broad　　...　　...　　...　(31) **latifolia.**
　　　Filaments bright yellow　　...　(32) **nervosa.**
　　　Filaments greenish-white　　...　(33) **obovata.**

1. M. tenella (Soland. ex Baker in Journ Linn. Soc. xi. 389);
bulb ovoid, ½ in. diam. ; leaves lanceolate, thick in texture, suberect,
an inch long, bristly on the upper part of the face, narrowed at the
base into a sheath clasping the peduncle which is longer than the
blade; capitulum small, few-flowered ; outer bracts ovate, acute, ⅓ in.
long ; pedicels very short; perianth whitish, ⅓ in. long ; segments
lanceolate, ascending, not more than half as long as the tube ; filaments under a line long; anthers oblong, nearly as long as the
filaments.

CENTRAL REGION: Aliwal North Div. ; Witbergen, 7000–8000 ft., *Drège*,
3509 !

Figured in Masson's drawings at the British Museum from a plant obtained in
Bokkeland, that flowered in England in July, 1794.

2. M. setulosa (Baker in Journ. Linn. Soc. xi. 389); bulb ovoid,
½ in. diam. ; leaves thick in texture, lanceolate, acute, 1–1¼ in. long,
4½–5 lin. broad, gradually narrowed to the base, furnished with a
few short whitish bristles on the upper surface ; capitulum small,

sessile, 10–12-flowered ; outer bracts oblong, acute, ½ in. long ; outer
pedicels ⅛–¼ in. long ; perianth whitish, ½ in. long ; segments
lanceolate, reflexing, rather shorter than the slender tube ; filaments
whitish, ¼–⅓ in. long ; anthers oblong, minute.

SOUTH AFRICA : without locality, *Ecklon and Zeyher !*
No specimen at Kew.

3. **M. echinata** (Linn. fil. Suppl. 193) ; bulb ovoid, under an
inch in diameter ; tunics dull brown ; leaves thin, oblong-spathulate,
obtuse, 3–4 in. long, 1½–2 in. broad at the middle, clothed over both
surfaces with conspicuous whitish bristles ; capitulum an inch in
diameter, shortly peduncled ; outer bracts green, oblong, acute, ½ in.
long ; outer pedicels ½–¾ in. long ; perianth whitish, ⅔ in. long ;
segments linear, reflexing, rather shorter than the cylindrical tube ;
filaments slender, ½ in. long ; anthers oblong, minute.   *Thunb.
Prodr.* 60 ; *Fl. Cap. edit. Schult.* 308 ; *Ait. Hort. Kew. edit.* 2, ii.
210 : *Kunth, Enum.* iv. 296 ; *Baker in Journ. Linn. Soc.* xi. 389.

SOUTH AFRICA : without locality, *Masson,* sketched in the year 1794 !
CENTRAL REGION : Bokkeland Berg, *Thunberg !*
WESTERN REGION : Little Namaqualand ; Hardeveld, 2000–3000 ft., *Zeyher,*
1717 !

4. **M. hirsuta** (Link et Otto, Abbild. i. t. 1) ; bulb ovoid, ½ in.
diam. ; leaves round, obtuse, thick in texture, 1–2 in. long and broad,
clothed over the face on both sides with whitish bristles ; capitulum
sessile, ¾–1 in. diam. ; outer bracts oblong, pubescent, much imbricated, ½ in. long ; outer pedicels ⅛–¼ in. long ; perianth whitish,
½ in. long ; segments lanceolate, reflexing, rather shorter than the
slender tube ; stamens as long as the perianth-segments ; anthers
oblong, minute.   *Roem. et Schultes, Syst. Veg.* iv. 296 ; *Kunth,
Enum.* iv. 296 ; *Baker in Journ. Linn. Soc.* xi. 388.

SOUTH AFRICA : without locality, *Masson !*
COAST REGION : Uitenhage Div. ; Zwartkops River, *Zeyher,* 4273 ! *Ecklon
and Zeyher,* 130 !

5. **M. muricata** (Gawl. in Bot. Mag. t. 559) ; bulb ovoid, 1 in.
diam. ; leaves thin, round-cordate, 3–4 in. long and broad, densely
setose on the upper surface towards the margin ; capitulum dense,
globose, subsessile ; outer bracts oblong-lanceolate, an inch long ;
outer pedicels ¼–½ in. long ; perianth whitish, an inch long ; segments
lanceolate, reflexing, half as long as the tube ; filaments as long as
the perianth-tube, united into a distinct cup at the base ; anthers
small, oblong.   *Ait. Hort. Kew. edit.* 2, ii 210 ; *Kunth, Enum.* iv.
296 ; *Baker in Journ. Linn. Soc.* xi. 389.

SOUTH AFRICA : without locality, *Masson.*

Introduced by Masson into cultivation in 1790. It has long been lost, and I
am not aware that there is any specimen in existence.

6. **M. pauciflora** (Ait. Hort. Kew. edit. 2, ii. 210) ; leaves lanceolate or elliptical, tubercled ; tubercles not setose ; segments of the

perianth-limb ovate. *Roem. et Schultes, Syst. Veg.* vii. 987 ; *Kunth, Enum.* iv. 296.

SOUTH AFRICA : without locality, *Masson.*

Introduced into cultivation by Masson in 1790. Known to me only from Aiton's brief description.

7. **M. Schlechtendahlii** (Baker in Journ. Bot. 1874, 5) ; bulb ovoid, $\frac{3}{4}$–1 in. diam ; leaves obovate-spathulate, cuspidate, glabrous, 2–3 in. long, 1–1$\frac{1}{4}$ in. broad, rough with pustules over the face ; capitulum sessile, 10–12-flowered ; outer bracts oblong spathulate, acute, $\frac{1}{2}$–$\frac{3}{4}$ in. long ; outer pedicels $\frac{1}{6}$–$\frac{1}{4}$ in. long ; perianth white, $\frac{5}{8}$ in. long ; segments linear-lanceolate, as long as the tube ; filaments longer than the segments, subulate down to the base ; anthers small, oblong.

SOUTH AFRICA : without locality.

Described from a specimen sent by Schlechtendahl to the herbarium of De Candolle.

8. **M. latebrosa** (Masson ex Baker in Journ. Bot. 1886, 336) ; bulb not seen ; leaves oblong or obovate, erecto-patent, 2–2$\frac{1}{2}$ in. long, 1–1$\frac{1}{4}$ in. broad, acute, scabrous and streaked vertically with purple ; capitulum sessile, under an inch in diam. ; flowers white ; bracts lanceolate ; tube cylindrical, $\frac{1}{3}$–$\frac{1}{2}$ in. long, $\frac{1}{8}$ in. diam. at the throat ; segments lanceolate, reflexing, $\frac{1}{6}$ in. long ; stamens erect, $\frac{1}{4}$ in. long, connate in a ring at the base.

CENTRAL REGION : Bokkeveld, Aug., 1792, *Masson.*

Described from a drawing at the British Museum.

9. **M. pustulata** (Jacq. Coll. iv. 177 ; Hort. Schoenbr. iv. 27, t. 454) ; bulb ovoid, 1 in. diam. ; leaves broad oblong, 5–6 in. long, 3–4 in. broad, acute, narrowed to the base, distantly ribbed, rough with tubercles on the face ; capitulum shortly peduncled, under 2 in. diam. ; outer bracts ovate-lanceolate, an inch long ; outer pedicels $\frac{1}{4}$–$\frac{1}{8}$ in. long ; perianth white, an inch long ; segments linear-lanceolate, reflexing, shorter than the tube ; filaments whitish, about $\frac{1}{2}$ in. long ; anthers small, oblong. *Red. Lil. t.* 183 *(excl. syn.)* ; *Gawl. in Bot. Mag. t.* 642 ; *Roem. et. Schultes, Syst. Veg.* vii. 987 ; *Kunth, Enum.* iv. 296. *M. scabra, Andr. Bot. Rep. t.* 220.

SOUTH AFRICA : without locality, *Masson.*

Introduced into cultivation by Masson 1790. There is a sketch made from a garden plant in 1792 amongst the Masson drawings at the British Museum.

10. **M. longipes** (Baker) ; bulb globose, 1 in. diam., with a neck 2 in. long ; leaves 2, broad ovate or oblong, about 3 in. long, 2 in. broad, minutely pustulate or smooth on the face ; capitulum dense, globose, 2 in. diam. ; outer bracts oblong, acute, membranous, greenish, an inch long ; pedicels $\frac{1}{2}$ in. long ; perianth white, with a cylindrical tube $\frac{1}{3}$ in. long and lanceolate reflexing segments a little shorter than the tube ; stamens twice as long as the segments, connate at the base.

COAST REGION: Caledon Div.; near Danger Point. Brought to the Cape Town garden by a visitor in 1884, *Bolus*, 5973!

11. **M. jasminiflora** (Burchell ex Baker in Journ. Linn. Soc. xi. 390); bulb small, globose, white; leaves 2, ovate, subacute, thin glabrous, not spotted, 2–3 in. long, 1–1¾ in. broad; capitulum sessile, an inch or more in diameter; outer bracts ovate, acute, much imbricated, ½ in. long; outer pedicels ¼–⅓ in. long; perianth white, ½ in. long; segments lanceolate, reflexing, half as long as the slender cylindrical tube; filaments ligulate, ⅛ in. long; anthers small, oblong, blackish-purple. *Podocallis nivea, Salisb. Gen.* 17?

COAST REGION: Queenstown Div.; Bowkers Kop, 4000 ft., *Galpin*, 1817!

KALAHARI REGION: Bechuanaland; Pellat Plains, at Jabiru Fontein near Takun, *Burchell*, Bulb No. 7!

Described from a specimen cultivated by Burchell in his garden at Fulham, dried Nov. 1818. Flowered at Kew, Oct. 1894; presented by Rev. J. Miles, who received it from the Orange Free State. Flowers scented like a ripe pear, according to Burchell.

12. **M. Bowkeri** (Baker in Journ. Linn. Soc. xi. 390); bulb ovoid, 1 in. diam; leaves round-oblong, thin, glabrous, obtuse, 1–1¼ in. long, ¾–1 in. broad; capitulum globose, sessile, 15–20-flowered; outer bracts lanceolate, acuminate, ½–⅝ in. long; outer pedicels very short; perianth white, ½ in. long; segments lanceolate, reflexing, half as long as the stout cylindrical tube, which is a line in diameter; filaments fleshy, ligulate, ⅛ in. long, connate into a distinct cup; anthers small, oblong.

KALAHARI REGION: Griqualand West; Klip Drift, *Mrs. Barber*, 30! Orange Free State, *Bowker!*

The flowers are sweetly scented, according to Mrs. Barber.

13. **M. brachypus** (Baker in Journ. Bot. 1874, 368); bulb ovoid, 1½ in. diam.; leaves oblong, obtuse, thin, glabrous, ½ ft. long, 3–3½ in. broad; capitulum shortly peduncled; outer bracts ovate or oblong, acuminate, above an inch long; pedicels very short; perianth white, ½ in. long; segments lanceolate, longer than the tube, which is infundibuliform in the upper, cylindrical in the lower half; filaments very short; anthers linear-oblong, ⅙ in. long.

SOUTH AFRICA: without locality.

Described from a plant cultivated at Kew in Feb. 1874.

14. **M. Dregei** (Baker); bulb globose, ¾ in. diam.; leaves round-oblong, glabrous, 2 in. long, exclusive of the sheathing base, which is half as long as the blade; capitulum globose, under an inch in diameter; outer bracts oblong, subacute, ½ in. long; pedicels very short; perianth white; tube slender, ⅓ in. long; segments linear, shorter than the tube; filaments ¼ in. long; anthers minute, oblong.

COAST REGION: Clanwilliam Div.; Lange Vallei, under 1000 ft., *Drège*, 2688!

15. M. Huttoni (Baker in Journ. Linn. Soc. xi. 390); bulb ovoid, 1 in. diam.; leaves round-ovate, thin in texture, glabrous, $1\frac{3}{4}$–2 in. long, 1–$1\frac{1}{4}$ in. broad, subobtuse, narrowed into a petiole, which sheaths the peduncle, which is $\frac{1}{2}$–$\frac{3}{4}$ in. long; capitulum 15–20-flowered; outer bracts broad ovate, acute, $\frac{1}{2}$–$\frac{5}{8}$ in. long; outer pedicels $\frac{1}{6}$–$\frac{1}{4}$ in. long; perianth whitish, $\frac{1}{2}$ in. long; segments linear, reflexing, as long as the tube; filaments filiform, $\frac{1}{4}$ in. long, free down to the base; anthers small.

COAST REGION: Albany, *Hutton!*

No specimen at Kew.

16. M. versicolor (Baker in Journ. Bot. 1876, 184); bulb ovoid, under an inch in diameter; leaves ovate or roundish, acute, glabrous, $1\frac{1}{2}$–2 in. long and broad; capitulum dense, sessile, globose, 1 in. diam.; outer bracts ovate, acute, much imbricated, $\frac{1}{2}$ in. long; outer pedicels very short; perianth white; tube cylindrical, $\frac{1}{4}$ in. long; segments lanceolate, reflexing, shorter than the tube; filaments cylindrical, $\frac{1}{6}$ in. long; anthers oblong, minute.

CENTRAL REGION: Somerset Div.; banks of the Little Fish River, 2500 ft., *MacOwan*, 2178!

17. M. amygdalina (Baker in Gard. Chron. 1889, vi. 715); habit of *M. versicolor;* leaves 2, ovate, acute, glabrous, decumbent, about 2 in. long by an inch broad; flowers white, smelling strongly of almonds, forming a dense, globose, sessile, central head 1 in. diam.; pedicels short; perianth-tube cylindrical, above $\frac{1}{4}$ in. long; segments lanceolate, reflexing, two-thirds the length of the tube; filaments erect, slightly connate at the base, $\frac{1}{4}$ in. long.

SOUTH AFRICA: without locality.

Imported by Mr. Jas. O'Brien; flowered by Sir C. W. Strickland in Dec. 1889.

18. M. Greenii (Baker); bulb ovoid, $\frac{1}{3}$ in. diam., with a long neck; leaves 2, ovate, glabrous, 3–4 in. long, 2 in. broad; capitulum dense, globose; outer bracts ovate, $\frac{1}{2}$ in. long; pedicels short; perianth tube above $\frac{1}{2}$ in. long; segments lanceolate, half as long as the tube; stamens as long as the segments.

KALAHARI REGION: Griqualand West; stony places near Kimberley, *MacOwan*, 2842!

19. M. calvata (Baker in Journ. Bot. 1878, 321); bulb ovoid or globose, under an inch in diameter; leaves ovate or orbicular, thin, obtuse, glabrous when mature, slightly tuberculato-hispid when young, 2–3 in. long and broad; capitulum dense, sessile, 1 in. diam.; outer bracts ovate, acute, much imbricated, under $\frac{1}{2}$ in. long; pedicels very short; perianth white, $\frac{1}{2}$ in. long; segments lanceolate, as long as the tube; filaments as long as the perianth-segments; anthers oblong, minute.

CENTRAL REGION: mountains near Graaff Reinet, 4400–5500 ft., *Bolus*, 749!

20. M. parvifolia (Baker in Engl. Jahrb. xv. Heft 3, 8); bulb globose, ½ in. diam.; leaves lanceolate, acute, firm, petioled, glabrous, an inch long, ¼ in. broad at the middle; capitulum dense, sessile, ¾–1 in. diam.; pedicels short; outer bracts large, oblong, green, membranous; perianth white; tube oblong, ¼ in. long; segments lanceolate, as long as the tube; filaments thick, as long as the perianth-segments.

SOUTH AFRICA: without locality; *Ecklon and Zeyher, Asphod.* 25!
No specimen at Kew.

21. M. orientalis (Baker in Journ. Bot. 1878, 321); bulb ovoid, 1 in. diam.; leaves orbicular or ovate, thin, glabrous, obtuse or subacute, 2–3 in. long and broad; capitulum dense, globose, sub-sessile, 1–1½ in. diam.; outer bracts oblong, acute, ciliated at the apex, ¾ in. long; outer pedicels ¼–½ in. long; perianth white, ½ in. long; segments lanceolate, reflexing, as long as the tube; filaments ¼–⅓ in. long; anthers oblong, minute.

COAST REGION: Sand-dunes at Port Elizabeth, *Bolus,* 2239! *Holub!*

22. M. concinna (Baker); bulb globose, ½ in. diam.; leaves 2, suborbicular, naked, 1–1¼ in. long and broad; capitulum dense, globose, 1–1½ in. diam.; bracts oblong, acute, ½ in. long; perianth ⅓ in. long; segments lanceolate, exceeding the subcampanulate tube; filaments ¼ in. long.

COAST REGION: Stockenstrom Div.; stony ground above the waterfall, Elands River, *Scully,* 54!

23. M. pedunculata (Baker in Engl. Jahrb. xv. Heft 3, 8); bulb globose, 1 in. diam.; leaves 2, oblong-lanceolate, membranous, glabrous, 3–4 in. long, an inch broad at the middle; peduncle 3–4 in. long, nearly hidden by the sheathing bases of the leaves; capi-tulum globose, 1 in. diam., not surrounded by large bracts; pedicels short; perianth white; tube slender, cylindrical, 8–9 lines long; segments linear, half as long as the tube; filaments twice as long as the perianth-segments; anthers small, oblong, bluish; style overtop-ping the anthers.

COAST REGION: Malmesbury Div.; near Hopefield, *Bachmann,* 2043!

24. M. longifolia (Jacq. Hort. Schoenbr. iv. 29, t. 457); bulb ovoid, 1 in. diam.; leaves oblanceolate-oblong, cuspidate, thin, glabrous, finally 8–9 in. long, 3–4 in. broad; peduncle about an inch long, hidden by the sheathing base of the leaves; capitulum globose, 2 in. diam.; outer bracts ovate, acute, an inch long; outer pedicels ½–¾ in. long; perianth white, ¾–⅞ in. long; segments lanceolate, reflexing, half as long as the tube; filaments whitish, much longer than the perianth-segments; anthers oblong, minute. *Roem. et Schultes, Syst. Veg.* vii. 990; *Kunth, Enum.* iv. 297; *Baker in Journ. Linn. Soc.* xi. 391.

SOUTH AFRICA: without locality.
Known to me only from Jacquin's figure.

25. M. candida (Burch. ex Kunth, Enum. iv. 297); bulb ovoid, 1 in. diam. ; leaves broad, ovate, obtuse, thin, glabrous, finally 3–5 in. long and broad ; capitulum dense, sessile, 2–3 in. diam. ; outer bracts ovate, acute, much imbricated, an inch long; pedicels finally $\frac{3}{4}$–1 in. long; perianth white, $\frac{3}{4}$–$\frac{7}{8}$ in. long; segments lanceolate, reflexing half as long as the tube ; filaments slender, whitish, under $\frac{1}{2}$ in. long; anthers oblong, minute; capsule turbinate, acutely triquetrous, $\frac{3}{4}$ in. long. *Baker in Journ. Linn. Soc.* xi. 392. *M. longifolia var. candida, Burch. in Bot. Reg. t.* 694.

COAST REGION : Mossel Bay Div. ; little Brak River, *Burchell*, 6197/5 !

26. M. læta (Masson ex Baker in Journ. Bot. 1886, 336); bulb not seen ; leaves ovate, subobtuse, glabrous, spreading, 3–4 in. long, 2 in. broad, distinctly and distantly ribbed vertically; capitulum dense, sessile, about an inch in diameter ; flowers white ; bracts ovate ; tube infundibuliform, $\frac{1}{2}$ in. long ; segments very short, lanceolate, reflexing ; filaments stout, erect, $\frac{1}{3}$ in. long.

WESTERN REGION : Little Namaqualand ; summit of Kamies Bergen, *Masson.*

Described from a sketch, now at the British Museum, made in the year 1794.

27. M. cordata (Jacq. Hort. Schoenbr. iv. 30, t. 459); bulb ovoid, 1 in. diam. ; leaves cordate, suborbicular, cuspidate, thin, glabrous, 6–8 in. long, 4–5 in. broad ; capitulum dense, globose, 2–3 in. diam. ; bracts ovate, acute, 1 in. long; outer pedicels $\frac{1}{2}$ in. long ; perianth white, $\frac{3}{4}$–$\frac{7}{8}$ in. long; segments lanceolate, reflexing, half as long as the tube ; filaments reddish-yellow, $\frac{1}{2}$ in. long; anthers oblong, minute. *Roem. et Schultes, Syst. Veg.* vii. 989 ; *Kunth, Enum.* iv. 297 ; *Baker in Journ. Linn. Soc.* xi. 391.

SOUTH AFRICA : without locality.

Known to me only from Jacquin's figure.

28. M. sanguinea (Jacq. Hort. Schoenbr. iv. 31, t. 461); bulb ovoid, 1 in. diam. ; leaves round-cordate, thin, glabrous, cuspidate, conspicuously lineate, 4–6 in. long, 3–4 in. broad ; capitulum dense, sessile, 2 in. diam. ; outer bracts ovate-lanceolate, an inch long ; outer pedicels $\frac{1}{2}$ in. long; perianth white, $\frac{3}{4}$–$\frac{7}{8}$ in. long; segments lanceolate, reflexing, about half as long as the tube ; filaments bright red, $\frac{1}{2}$ in. long; anthers oblong, minute ; capsule turbinate, acutely triquetrous, an inch long. *Roem. et Schultes, Syst. Veg.* vii. 989 ; *Kunth, Enum.* iv. 297 ; *Baker in Journ. Linn. Soc.* xi. 391. *M. latifolia, Gawl. in Bot. Mag. t.* 848, *non Linn. fil.*

VAR. β, M. coronata (Jacq. Hort. Schoenbr. iv. 30, t. 460); leaves more obtuse, not lineate ; filaments claret-red. *Kunth, loc. cit.*

SOUTH AFRICA : without locality.

29. M. lanceæfolia (Jacq. Hort. Schoenbr. iv. 29, t. 436) ; bulb ovoid, 1$\frac{1}{2}$ in. diam. ; leaves oblanceolate-oblong, thin, glabrous, acute, finally nearly a foot long, 4–5 in. broad ; peduncle 1$\frac{1}{2}$–2 in. long, hidden by the sheathing bases of the leaves ; capitulum dense, 1$\frac{1}{2}$ in.

diam.; outer bracts ovate, acute, an inch long; outer pedicels $\frac{1}{2}$ in. long; perianth white, under an inch long; segments lanceolate, reflexing, as long as the tube; filaments bright red, $\frac{1}{2}$–$\frac{5}{8}$ in. long; anthers oblong, minute. *Roem. et Schultes, Syst. Veg.* vii. 991; *Kunth, Enum.* iv. 297; *Baker in Journ. Linn. Soc.* xi. 391.

SOUTH AFRICA: without locality.
Known to me only from Jacquin's figure.

30. M. namaquensis (Baker); bulb globose, with a neck an inch long; leaves 2, suborbicular, glabrous, 2 in. long and broad; capitulum dense, globose, 1 in. diam.; outer bracts ovate-cuspidate, 1–1$\frac{1}{4}$ in. long; pedicels short; perianth white, an inch long; tube cylindrical; segments linear, nearly as long as the tube; stamens as long as the segments; filaments reddish; capsule obovoid, acutely triquetrous, $\frac{1}{3}$ in. long.

WESTERN REGION: Little Namaqualand; near Kook Fontein, 3000 ft., *Bolus*, 6596! *Scully*, 27!

31. M. latifolia (Linn. fil. Suppl. 193); bulb ovoid, 1 in. diam.; leaves ovate, subacute, thin, glabrous, finally 6–10 in. long; capitulum dense, sessile, 2 in. diam.; outer bracts ovate, acute, 1–1$\frac{1}{2}$ in. long; outer pedicels $\frac{1}{2}$ in. long; perianth whitish, an inch long; segments linear-lanceolate, reflexing, about as long as the tube; filaments bright red, longer than the perianth-segments; anthers small, oblong, yellow; capsule turbinate, membranous, acutely trique rous. *Thunb. Prodr.* 60; *Fl. Cap edit. Schult.* 307; *Jacq. Hort. Schoenbr.* iv. 28, *t.* 455; *Willd. Sp. Plant.* ii. 28; *Ait. Hort. Kew. edit.* 2, ii. 209 (*excl. syn.*); *Roem. et Schultes, Syst. Veg.* vii. 988.; *Kunth, Enum.* iv. 296. *M. grandifolia, Gawl. in Bot. Mag. sub t.* 991. *M. depressa, Houtt. Handl.* xii. 424, *t.* 85, *fig.* 1.

COAST REGION: Caledon Div., *Zeyher*!
CENTRAL REGION: Worcester Div.; Roggeveld, *Thunberg*. Somerset Div., *Bowker*! near Graaff Reinet, 3200 ft., *Bolus*, 802! Colesberg Div.; near Schuil Hoek Berg, *Burchell*, Bulb 72!
WESTERN REGION: Little Namaqualand; between Pedros Kloof and Lily Fontein, 3000–4000 ft., *Drège*, 2683c!

32. M. nervosa (Hornem. Hort. Hafn. Suppl. 39); leaves ovate-lanceolate, smooth, mucronate, with about 15 distinct ribs; perianth-tube pale violet; segments pale stramineous, not reflexing; stamens orange-yellow, twice as long as the perianth-segments, united in a violet-blue cup at the base. *Kunth, Enum.* iv. 298.

SOUTH AFRICA: without locality.
Known to me only from Horneman's description.

33. M. obovata (Jacq. Hort. Schoenbr. iv. 29, t. 458); bulb ovoid, under 1 in. diam.; leaves obovate, thin, green, glabrous, cuspidate, 6–8 in. long, 3–4 in. broad; capitulum subsessile, 1$\frac{1}{2}$ in. diam.; outer bracts ovate, acute, an inch long; outer pedicels $\frac{1}{2}$ in. long; perianth white, an inch long; segments linear-lanceolate, reflexing, as long as the tube; filaments greenish-white, $\frac{1}{2}$ in. long; anthers small, oblong. *Roem. et Schultes, Syst. Veg.* vii. 991; *Kunth,*

Enum. iv. 297; *Baker in Journ. Linn. Soc.* xi. 391. *M. grandi-flora, Lindl. Bot. Reg. t.* 958.

SOUTH AFRICA : without locality.

Known to me only from the figures cited.

XXIV. DAUBENYA, Lindl.

Perianth gamophyllous; tube cylindrical; limb oblique, the outer half elongated, especially in the outer flowers of the capitulum; segments more or less unequal, some oblong-unguiculate, others lanceolate or deltoid. *Stamens* 6, inserted at different heights near the base of the perianth-segments; filaments short, filiform; anthers oblong, dorsifixed, dehiscing introrsely. *Ovary* sessile, 3-celled; ovules several in each cell, superposed; style cylindrical; stigma capitate. *Capsule* and *seeds* unknown.

Habit of *Massonia*, from which it differs by its irregular flowers.

DISTRIB. Endemic.

Capitulum subsessile :
Flowers yellow ...	...	...	...	...	...	(1) **aurea.**
Flowers bright red	...	...	...	...	...	(2) **coccinea.**
Peduncle 2–4 in. long	...	...	...	...	...	(3) **fulva.**

1. **D. aurea** (Lindl. Bot. Reg. t. 1813); leaves thin, rather fleshy oblong, glabrous, 4 in. long, 2 in. broad, many-ribbed; peduncle very short; capitulum globose, 3 in. diam.; outer bracts several, oblong, membranous; flowers nearly sessile, bright yellow, outer very irregular, with three subequal oblong-spathulate segments longer than the tube and three small lanceolate inner segments; stamens half as long as the outer segments; inner flowers with all the segments linear and shorter than the stamens. *Kunth, Enum.* iv. 301; *Baker in Journ. Linn. Soc.* xi. 394.

SOUTH AFRICA : without locality.

Figured from a plant cultivated by Messrs. Young, of Epsom, in June, 1835.

2. **D. coccinea** (Harv. ex Baker in Journ. Linn. Soc. xi. 395); leaves oblong, glabrous, firm in texture, 4–5 in. long, 1½–1¾ in. broad, conspicuously ribbed vertically, crenulate on the margin; capitulum subsessile ; outer bracts obovate, membranous, an inch long; flowers bright scarlet, the outer ones 3 in. long, with the oblong-spathulate segments of the lower lip 1¼–1½ in. long, the other three linear and very small; stamens only $\frac{1}{12}-\frac{1}{8}$ in. long; central flowers with 6 ascending, unequal, linear segments.

SOUTH AFRICA : without locality, *Harvey!*

Described from a specimen gathered by Dr. Harvey, now in the Herbarium of Trinity College, Dublin.

3. **D. fulva** (Lindl. Bot. Reg. 1839, t. 53); leaves oblong, thin but fleshy, glabrous, many-ribbed, 4–5 in. long, with inflexed edges towards the base ; peduncle 2–4 in. long; capitulum globose, 2½–3

in. diam.; outer bracts oblong, membranous, greenish, $\frac{3}{4}$ in. long, much shorter than the flowers; pedicels very short; flowers bright red, outer 2–2$\frac{1}{2}$ in. long, with the oblong-spathulate outer segments of the lower lip $\frac{1}{2}$–$\frac{3}{4}$ in. long, and the three inner segments minute linear uniseriate or biseriate; filaments $\frac{1}{12}$–$\frac{1}{4}$ in. long; inner flowers with all the segments linear, $\frac{1}{8}$–$\frac{1}{6}$ in. long. *Kunth, Enum.* iv. 300; *Baker in Journ. Linn Soc.* xi. 395.

SOUTH AFRICA : without locality.

Figured from a specimen in the garden of Mr. R. Barchard at Wandsworth in 1839. I have seen it only in the herbarium of Mr. Wilson Saunders.

XXV. WHITEHEADIA, Harv.

Perianth persistent, with a very short tube and 6 subequal, spreading, 3–5-nerved segments. *Stamens* 6, inserted at the throat of the perianth-tube; filaments erect, connate at the base in a nectariferous cup; anthers linear-oblong, dorsifixed, versatile, dehiscing introrsely. *Ovary* sessile, globose, trilocular; ovules many, superposed; style subulate; stigma capitate. *Capsule* subglobose, acutely triquetrous, membranous, loculicidally 3-valved. *Seeds* lagenæform; testa black, shining; albumen cartilaginous.

DISTRIB. Endemic.

1. **W. latifolia** (Harvey, Cape Gen. edit. 2, 396); bulb globose, 1$\frac{1}{2}$–2 in. diam.; tunics brown, membranous; leaves 2, basal, spreading, round-oblong, subacute or emarginate, 6–8 in. long; peduncle subclavate, $\frac{1}{2}$ ft. long; raceme dense, subspicate, 3–6 in. long; bracts round-cuspidate, 1–1$\frac{1}{4}$ in. long, the upper empty; perianth greenish, $\frac{3}{8}$ in. long; capsule $\frac{3}{4}$ in. long. *W. bifolia, Baker in Journ. Linn. Soc.* xiii. 226. *Eucomis bifolia, Jacq. Collect.* iv. 215; *Ic.* ii. 21, *t.* 449; *Willd. Sp. Plant.* ii. 92; *Bot. Mag. t.* 840; *Roem. et Schultes, Syst. Veg.* vii. 624; *Kunth, Enum.* iv. 303. *Basilæa bifolia, Poir. Encyc. Suppl.* i. 591. *Melanthium massoniæfolium, Andr. Bot. Rep. t.* 368.

WESTERN REGION : Little Namaqualand, *Whitehead!* among rocks near Klip Fontein, 3000 ft., *Bolus,* 6565 !

XXVI. POLYXENA, Kunth.

Perianth gamophyllous, persistent; tube cylindrical or infundi-buliform; segments equal, lanceolate, shorter than the tube. *Stamens* 6, biseriate or uniseriate at the throat of the tube; filaments filiform, distinct or connate at the base; anthers small, oblong, dorsifixed, versatile, dehiscing introrsely. *Ovary* sessile, ampullæform, trilocular; ovules many, superposed; style subulate; stigma capitate. *Capsule* membranous, loculicidally 3-valved. *Seeds* small, globose; testa black, shining; albumen firm in texture.

Rootstock a tunicated bulb; leaves 2, usually erect; inflorescence corymbose; pedicels short; bracts not involucrant.

DISTRIB. Endemic.

Here, as throughout, I have followed Mr. Bentham in classification, but my own view would be to unite, at any rate, the subgenus *Astemma* with *Massonia*.

Subgenus ASTEMMA. Filaments uniserial at the throat of the perianth-tube.

Filaments scarcely longer than the perianth-segments ... (1) **comata.**
Filaments much longer than the perianth-segments:
　Leaves much overtopping the corymb, thin:
　　Filaments white ...　...　...　...　... (2) **angustifolia.**
　　Filaments bright red ...　...　...　... (3) **Burchellii.**
　Leaves but little overtopping the corymb:
　　Segments half as long as the tube ...　... (4) **rugulosa.**
　　Segments much shorter than the tube ...　... (5) **marginata.**
　Leaves lying flat on the ground ...　...　... (6) **hæmanthoides.**

Subgenus EUPOLYXENA. Filaments biserial.

Filaments slightly biserial:
　Perianth-segments much shorter than the tube:
　　Flowers white ...　...　...　...　... (7) **odorata.**
　　Flowers lilac ...　...　...　...　... (8) **pygmæa.**
　Perianth-segments nearly as long as the tube ... (9) **Bakeri.**
Filaments conspicuously biserial ...　...　...　... (10) **uniflora.**

1. **P. comata** (Baker); bulb globose, above an inch in diam.; root-fibres numerous; leaves 2, oblong, oblong-lanceolate or lorate, $\frac{1}{4}$–1 ft. long, thin, glabrous; peduncle 1–1$\frac{1}{2}$ in. long, hidden by the clasping bases of the leaves; corymb dense, many-flowered, 1$\frac{1}{2}$–2 in. diam; pedicels very short; bracts small, lanceolate, membranous; perianth white, fragrant; tube very slender, 1–1$\frac{1}{2}$ in. long; segments lanceolate, reflexing, $\frac{1}{6}$ in. long; filaments a little longer than the perianth-segments, uniseriate at the throat of the tube; anthers small, oblong; capsule globose, $\frac{1}{4}$ in. diam. *Massonia comata, Burch. ex Baker in Journ. Linn. Soc.* xi. 392.

CENTRAL REGION: Colesberg Div.; near Ruigte Fontein, *Burchell*, Bulb 46! Carolus Poort, *Burchell*, 2751!
KALAHARI REGION: Orange Free State; in damp localities near running water, *Mrs. Barber!*

2. **P. angustifolia** (Baker); bulb globose, 1 in. diam.; leaves 2, erect, thin, glabrous, oblong-spathulate, $\frac{1}{2}$ ft. long, 1–1$\frac{1}{2}$ in. broad above the middle, narrowed to a clasping petiole; peduncle 1–2 in. long; corymb many-flowered, 1–2 in. diam.; lower pedicels $\frac{1}{4}$–$\frac{1}{3}$ in. long; bracts greenish, lanceolate; flowers white, fragrant; perianth-tube cylindrical, $\frac{3}{4}$ in. long; segments lanceolate, reflexing, $\frac{1}{4}$–$\frac{1}{3}$ in. long; filaments white, uniserial, arcuate, $\frac{1}{2}$ in. long; anthers small, oblong. *Massonia angustifolia, Linn. fil. Suppl.* 193; *Ait. Hort. Kew.* i. 405, *t.* 4; *Ker in Bot. Mag. t.* 736; *Red. Lil. t.* 392; *Kunth, Enum.* iv. 298. *M. lanceolata, Thunb. Prodr.* 60; *Fl. Cap. edit. Schult.* 308; *Diss. Nov. Gen.* ii. 40. *M. Zeyheri, Kunth, Enum.* iv. 298.

SOUTH AFRICA: without locality, *Zeyher!*
CENTRAL REGION: Fraserberg Div.; summit of a mountain, Ouderste Rogge Veld, *Thunberg.*

Introduced into cultivation by Masson in 1775.

3. P. Burchellii (Baker); bulb globose, above an inch in diameter; root-fibres very numerous; leaves 2, thin, erect, glabrous, oblong-lanceolate, acute, $\frac{1}{2}$ ft. long, $1\frac{1}{2}$–2 in. broad above the middle, narrowed suddenly to a short sheathing base; peduncle an inch long; corymb dense, few-flowered; pedicels very short; outer bracts lanceolate, $\frac{1}{2}$ in. long; perianth-tube whitish, cylindrical, $\frac{1}{2}$ in. long; segments lanceolate, nearly as long as the tube; filaments uniserial, bright red, twice as long as the perianth-segments; anthers bright yellow, oblong. *Massonia Burchellii, Baker in Journ. Linn. Soc.* xi. 393.

South Africa: without locality, *Burchell!*

4. P. rugulosa (Baker); bulb oblong; leaves 2, oblong, obtuse, narrowed to a clasping base, scarcely overtopping the flowers; peduncle short; flowers densely corymbose; pedicels very short, bracteate; perianth-tube cylindrical, $\frac{1}{2}$ in. long; segments lanceolate, half as long as the tube; filaments rose-red, uniseriate, distinct, longer than the perianth-segments; anthers oblong, yellow. *Massonia rugulosa, Lichtenst. ex Kunth, Enum.* iv. 299.

South Africa: without locality, *Lichtenstein*, in Berlin Herbarium.

5. P. marginata (Baker); bulb ovoid-subglobose; leaves oblong, lacunoso-rugose beneath, rather longer than the peduncle, undulated on the margin; peduncle short; corymb dense; pedicels very short; bracts lanceolate, membranous; perianth whitish; tube cylindrical; segments lanceolate, rather shorter than the tube; filaments uniseriate, pale red, twice as long as the perianth-segments; anthers yellowish. *Massonia undulata, Willd. ex Kunth, Enum.* iv. 299.

South Africa: without locality, *Willdenow Herb.*, 6373.

6. P. hæmanthoides (Baker in Hook. Ic. t. 1727); bulb ovoid, above 1 in. diam.; leaves 2, humifuse, oblong, acute, smooth, 2–4 in. long, with a scabrous horny edge; capitulum dense, globose; pedicels very short; bracts small, white, membranous, ovate-lanceolate; perianth pale; tube subcylindrical, $\frac{1}{2}$ in. long; segments oblong-lanceolate, much shorter than the tube; filaments bright red, $\frac{1}{2}$ in. long; anthers small, yellow, oblong, versatile.

Central Region: Fraserburg Div.; on the Nieuwveld near Fraserburg, 4200 ft., *Bolus*, 5493!

7. P. odorata (Baker); bulb ovoid, $\frac{1}{2}$ in. diam.; root-fibres few; leaves 2, lanceolate, erect, glabrous, 3–5 in. long, $\frac{1}{3}$–$\frac{2}{3}$ in. broad; peduncle about an inch long; flowers white, fragrant, 6–10 in a dense corymb; pedicels short; bracts small, lanceolate, membranous, perianth-tube narrowly infundibuliform, $\frac{2}{3}$ in. long; segments lanceolate, falcate, $\frac{1}{5}$ in. long.; stamens slightly biseriate, inserted near the throat of the tube; filaments distinct; anthers small, oblong. *Massonia odorata, Hook. fil. in Bot. Mag. t.* 5891; *Baker in Journ. Bot.* 1874, 5.

Central Region: Colesberg, *Arnot !*
Described from a plant that flowered at Kew in October, 1870.

8. P. pygmæa (Kunth, Enum. iv. 294) ; bulb ovoid, ½ in. diam. ;
leaves 2, oblong-lanceolate, erect, glabrous, narrowed gradually into a
clasping base; peduncle about an inch long; corymb few-flowered;
pedicels ¼–⅓ in. long; bracts small, lanceolate, membranous;
perianth lilac; tube cylindrical, ½ in. long, dilated at the throat;
segments oblong-lanceolate, falcate, ⅙ in. long; stamens slightly
biseriate, distinct; filaments filiform, as long as the segments;
anthers minute, oblong. *Polyanthes pygmæa, Jacq. Colleot. Suppl.*
56; Ic. ii. 15, *l.* 380; *Willd. Sp. Plant.* ii. 165. *Mauhlia ensifolia,*
Thunb. Prodr. 60, *t.* 1; *Fl. Cap. edit. Schult.* 308. *Agapanthus*
ensifolius, Willd. Sp. Plant. ii. 48. *Massonia violacea, Andr. Bot.*
Rep. t. 46; *Red. Lil. t.* 386. *M. ensifolia, Gawl. in Bot. Mag.*
t. 554; *Ait. Hort. Kew. edit.* 2, ii. 211 ; *Roem. et Schultes, Syst. Veg.*
vii. 992. *Hyacinthus bifolius, Bout. in Cav. Anal.* v. 14, *t.* 41, *fig.* 1.
Massonia ovata, E. Meyer in herb. Drège.

COAST REGION : between the Sunday River and Fish River, *Thunberg.*
Uitenhage Div., *Ecklon and Zeyher,* 757 !
CENTRAL REGION : Worcester Div.; Constable, 3000–3500 ft., *Drège,* 2187 !
WESTERN REGION : Little Namaqualand; Hardeveld, 3000–4000 ft., *Zeyher,*
1716 !
KALAHARI REGION : Griqualand West ; Klip Drift, in rocky localities by the
Vaal River, *Mrs. Barber !*

9. P. Bakeri (Durand and Schinz, Consp. Fl. Afric. 366); bulb
ovoid-subglobose; leaves 2, elliptical, obtuse, glabrous, coriaceous but
somewhat fleshy, a little overtopping the corymb; peduncle short,
slender; corymb dense, many-flowered; pedicels as long as the flowers;
bracts long, linear, membranous, dilated at the tip; perianth-tube
slender, elongated; segments linear, but little shorter than the tube;
stamens slightly biseriate ; filaments distinct, the three upper slightly
exceeding the perianth-segments. *Massonia pygmæa, Schlecht. ex*
Kunth, Enum. iv. 298 ; *Baker in Journ. Linn. Soc.* xi. 393.

SOUTH AFRICA : without locality, *Mund and Maire,* in Berlin Herbarium.

10. P. uniflora (Baker); bulb small, ovoid; leaves 2, linear,
erect, glabrous, 3–4 in. long, scarcely a line broad; peduncle
1-flowered, an inch long; perianth pale lilac; tube very slender,
1¼ in. long; segments lanceolate, reflexing, a quarter as long as the
tube; stamens conspicuously biseriate, 3 ·inserted low down in the
tube, 3 at its throat; filaments ⅛ in. long. *Massonia uniflora,*
Soland. ex Ker in Bot. Reg. sub t. 694 (*name only*); *Baker in*
Journ. Linn. Soc. xi. 393.

SOUTH AFRICA : without locality, *Masson !*

XXVII. LACHENALIA, Jacq.

Perianth gamophyllous, persistent ; tube campanulate ; three outer
segments oblong, slightly gibbous near the apex; three inner usually
longer, obtuse, spathulate, spreading upwards. *Stamens* 6, inserted
in the perianth-tube ; filaments filiform ; anthers small, oblong, ver-

satile, dehiscing introrsely. *Ovary* ovoid, 3-celled; ovules many, superposed; style long, slender, stigma capitate. *Capsule* obovoid-triquetrous, membranous or subcoriaceous, loculicidally 3-valved. *Seeds* turgid; funiculus often long; testa black, crustaceous; albumen moderately firm.

Rootstock a tunicated bulb; leaves usually two, lorate-lanceolate, suberect, clasping the base of the peduncle, often spotted, rarely one or several; peduncle leafless; inflorescence a simple raceme or spike; bracts solitary, persistent, not spurred; flowers very various in size and colour, upper minute, abortive.

DISTRIB. Endemic.

Subgenus EULACHENALIA. Perianth cylindrical, equally rounded at the base; segments dimorphic.

Inner segments scarcely longer than the outer	...	(1) **pendula.**
Inner segments a little longer than the outer	...	(2) **rubida.**
Inner segments much longer than the outer	...	(3) **tricolor.**

Subgenus COELANTHUS. Perianth ventricose, oblique at the base; segments dimorphic.

The only species ... ... (4) **reflexa.**

Subgenus ORCHIOPS. Perianth oblong or oblong-cylindrical, equally rounded at the base.

Inflorescence subspicate :
  Flowers all erecto-patent:
    Leaves subterete ... ... (5) **orthopetala.**
    Leaves lorate-lanceolate :
      Leaves smooth on the face :
        Perianth ¼ in. long ... ... (6) **orchioides.**
        Perianth ⅓–½ in. long ... ... (7) **glaucina.**
      Leaves pustulate on the face ... (8) **liliflora.**
    Leaves oblong, obtuse, crisped ... ... (9) **undulata.**
    Leaves ovate, densely bristly ... ... (10) **trichophylla.**
  Flowers cernuous ... ... (11) **Bowkeri.**
Inflorescence laxly racemose ... ... (12) **patula.**

Subgenus CHLORIZA. Perianth campanulate or oblong-campanulate, equally rounded at the base; segments dimorphic.

Inflorescence spicate or subspicate :
  Leaves many, subterete ... ... (13) **contaminata.**
  Leaves 2, linear-complicate ... ... (14) **Bachmanni.**
  Leaves 2, flat :
    Peduncle short ... ... (15) **carnosa.**
    Peduncle elongated :
      Leaves smooth :
        Flowers white ... ... (16) **fistulosa.**
        Flowers bright lilac ... ... (17) **lilacina.**
      Leaves pustulate ... ... (18) **pustulata.**
Inflorescence racemose :
  Leaf usually single :
    Leaf smooth :
      Leaf subulate :
        Lower flowers cernuous ... ... (19) **rhodantha.**
        Lower flowers suberect or
          spreading ... ... (20) **Zeyheri.**
      Leaf linear, or linear-subulate :
        Perianth ¼ in. long ... ... (21) **convallarioides.**
        Perianth ⅓–½ in. long ... ... (22) **unifolia.**
      Leaf lanceolate ... ... (23) **anguinea.**
    Leaf hispid, linear ... ... (24) **hirta.**
    Leaf hispid, orbicular ... ... (25) **Massoni.**

Leaves usually 2; stamens short:
  Leaves subterete ... ... (26) **campanulata.**
  Leaves lanceolate or lorate, smooth:
    Flowers all ascending:
      Raceme dense ... ... (27) **isopetala.**
      Raceme lax:
        Flowers whitish ... ... (28) **mediana.**
        Flowers bright red... ... (29) **rosea.**
    Lower flowers spreading:
      Perianth ¼ in. long ... ... (30) **Youngii.**
      Perianth ½ in. long ... ... (31) **succulenta.**
  Leaves lanceolate or lorate, pustulate:
    Pedicels short ... ... (32) **pallida.**
    Pedicels long ... ... (33) **Cooperi.**
Leaves usually 2; stamens much exserted:
  Leaves subterete ... ... (34) **juncifolia.**
  Leaves lorate or lanceolate:
    All the flowers ascending ... ... (35) **purpureo-cærulea.**
    Lower flowers cernuous:
      Pedicels short ... ... (36) **unicolor.**
      Pedicels long:
        Flowers violet ... ... (37) **violacea.**
  Leaves spreading, oblong:
    Perianth ⅓ in. long ... ... (38) **latifolia.**
    Perianth ¼–½ in. long:
      Leaf 2–2½ in. long ... ... (39) **bowieana.**
      Leaf 4–5 in. long ... ... (40) **nervosa.**
Leaves numerous ... ... (41) **polyphylla.**

Subgenus BRACHYSCYPHA. Perianth cylindrical; segments all nearly uniform; inflorescence capitate.

The only species ... ... (42) **pusilla.**

1. **L. pendula** (Ait. Hort. Kew. i. 461; edit. 2, ii. 288); bulb globose, 1 in. diam.; peduncle, including inflorescence, ½–1 ft. long, more robust than in the two next species; leaves 2, rarely 1, lorate-lanceolate, thin, ½–1 ft. long, 1½–2 in. broad at the middle; raceme few- or many-flowered, 2–6 in. long, all except the upper flowers more or less cernuous; pedicels very short; bracts small, deltoid; perianth cylindrical, 1¼–1½ in. long, ¼–⅓ in. diam.; outer segments linear-oblong, yellow passing upwards into red, not spotted; inner but little longer, bright red-purple at the truncate tip, ¼ in. broad; stamens reaching the tip of the perianth; style finally just exserted. *Thunb. Prodr.* 64; *Fl. Cap. edit. Schult.* 328; *Willd. Sp. Plant.* ii. 180; *Andr. Bot. Rep. t.* 41; *Sims in Bot. Mag. t.* 590; *Roem. et Schultes, Syst. Veg.* vii. 614; *Tratt. Archiv. t.* 151-152; *Kunth, Enum.* iv. 291; *Baker in Journ. Linn. Soc.* xi. 403.

COAST REGION: Cape Div.; near Hout Bay, 1100 ft., *MacOwan and Bolus, Herb. Norm. Aust. Afr.*, 795! Cape Flats, *Burchell*, 8573! *Rogers!* Stellenbosch, Div., *Zeyher!* Hottentots Holland, *Ecklon and Zeyher*, 32!

2. **L. rubida** (Jacq. Collect. Suppl. 60; Ic. ii. 17, t. 398); bulb globose, about ½ in. diam.; peduncle, including inflorescence, 6–9 in. long; leaves usually 2, lanceolate, finally 5–6 in. long, about an inch broad at the middle, spotted, much narrowed to a long clasping

base; raceme 6–20-flowered, all the flowers except the uppermost more or less drooping; pedicels very short; bracts minute, deltoid; perianth cylindrical, about an inch long, $\frac{1}{6}$ in. diam.; outer segments bright red, tipped with green; inner $\frac{1}{6}$–$\frac{1}{4}$ in. longer, yellow below the exposed tip, where they are $\frac{1}{6}$ in. broad; stamens reaching to the tip of the outer perianth-segments; style finally just exserted. *Willd. Sp. Plant.* ii. 179; *Ait. Hort. Kew, edit.* 2, ii. 288; *Roem. et Schultes, Syst. Veg.* vii. 615; *Tratt. Archiv. t.* 145; *Kunth, Enum.* iv. 291; *Baker in Journ. Linn. Soc.* xi. 404.

VAR. β, **L. tigrina** (Jacq. Collect. Suppl. 67; Ic. ii. 17, t. 399); outer perianth-segments with dense minute spots of bright red on a yellowish ground. *Roem et Schultes, Syst. Veg.* vii. 615; *Tratt. Archiv. t.* 146; *Kunth, Enum.* iv. 291.

VAR. γ, **L. punctata** (Jacq. Collect. ii. 323; Ic. ii. 16, t. 397); outer segments with spots of dark red on a paler red ground; inner pale yellow, spotted towards the tip with red. *Willd. Sp. Plant.* ii. 180; *Roem. et Schultes, Syst. Veg.* vii. 616; *Tratt. Archiv. t.* 147; *Kunth, Enum.* iv. 291.

COAST REGION: Malmesbury, *Bachmann*, 1877! Cape Div.; Muizenberg, 100–200 ft., *Bolus*, 4889! *MacOwan*, 2551! *Pappe! MacOwan and Bolus, Herb. Norm. Austr. Afr.*, 796! Flats between Tygerberg and Blueberg, under 500 ft., *Drège!* Var. γ, Clanwilliam, *Mader*, 2161!

CENTRAL REGION: Var. γ, Calvinia Div.; Bitter Fontein, *Zeyher*, 1697!

WESTERN REGION: Little Namaqualand, *Whitehead!*

3. **L. tricolor** (Thunb. Prodr. 64; Fl. Cap. edit. Schult. 327); bulb globose, $\frac{3}{4}$–1 in. diam.; peduncle, including inflorescence, $\frac{1}{2}$–1 ft. long; leaves usually 2, lanceolate-lorate, 6–9 in. long, about 1 in. broad, thin, often spotted, with purple or darker green; raceme few- or many-flowered, usually 3–4, rarely 6–9, in. long; lower flowers cernuous; lower pedicels $\frac{1}{6}$–$\frac{1}{4}$ in. long; bracts small, deltoid; perianth cylindrical, $\frac{3}{4}$–1 in. long, $\frac{1}{6}$ in. diam.; outer segments oblong, $\frac{1}{3}$ in. long, yellow, tipped with green; inner $\frac{1}{4}$–$\frac{1}{3}$ in. longer, purplish-red at the spreading truncate tip, which is $\frac{1}{4}$ in. broad; stamens reaching to the tip of the inner segments; style finally a little exserted. *Curt. in Bot. Mag. t.* 82; *Red. Lil. t.* 2; *Lodd. Bot. Cab. t.* 767; *Kerner, Hort. t.* 176; *Tratt. Archiv. t.* 148; *Roem. et Schultes, Syst. Veg.* vii. 612; *Kunth, Enum.* iv. 290; *Baker in Journ. Linn. Soc.* xi. 404. *Phormium aloides,, Linn. fil. Suppl.* 205, *ex parte.*

VAR. β, **L. quadricolor** (Jacq. Collect. Suppl. 62; Ic. ii. 16, t. 396); perianth with a red base and greenish-yellow middle; outer segments tipped with green; inner with red-purple. *Andr. Bot. Rep. t.* 148; *Lodd. Bot. Cab. t.* 746; *Reich. Flor. Exot. t.* 119; *Tratt. Archiv. t.* 153; *Kunth, Enum.* iv. 291. *L. superba, Hort.*

VAR. γ, **L. luteola** (Jacq. Collect. iv. 148; Ic. ii. 16, t. 395); perianth lemon-yellow, tinged with green towards the tip. *Red. Lil. t.* 297; *Kunth, Enum.* iv. 290; *Lodd. Bot. Cab. t.* 734; *Tratt. Archiv. tt.* 149, 150; *Fl. des Serres, t.* 1873. *L. quadricolor var. luteola, Sims in Bot. Mag. t.* 1704. *L. macrophylla, Lemaire, Ill. Hort.* ii. *Misc.* 99. *L. tricolor var. luteola, Ait. Hort. Kew. edit.* 2, ii. 288; *Bot. Mag. t.* 1020.

VAR. δ, **L. Nelsoni** (Hort. in Floral Mag. n.s. t. 452); perianth bright yellow, both rows of segments faintly tinged with green. *Berl. Gartenzeit.* 1882, 421, *with figure.*

VAR. ε, **L. aurea** (Lindl. in Gard. Chron. 1856, 404, f. 176; 1872, 291, f. 109);

perianth bright orange-yellow. *Florist,* 1871, 265, *L. tricolor var. aurea, Hook. fil. in Bot. Mag. t.* 5992.

COAST REGION: Malmesbury Div.; Zwartland, *Thunberg!* Cape Div.; near Salt River, *Thunberg!* near Cape Town, *Cooper,* 3303! 3546! Var. β, Malmesbury Div.; Zwartland, *Zeyher* 1696! Groene Kloof, *Drège,* 8624! Saldanha Bay, *Ecklon and Zeyher!* Kulebas Kraal, *Pappe!* Var. γ, Cape Flats, *Pappe!* Var. ε, Paarl Mountains, 1750 ft., *MacOwan and Bolus,* Herb. Norm. Austr. *Afr.,* 504! Port Elizabeth, at Hill Park, *Herb. E.S.C.A.* 491!

The varieties are connected by intermediate stages. Several hybrids between *L. pendula* and the varieties of *L. tricolor* are in cultivation, the finest of which is *L. Cammi, Hort.,* which combines the bright yellow flowers of *L. aurea* with the habit of *L. pendula. L. Comessi, Hort.,* differs from *L. Nelsoni* by the outer segments being rather longer in proportion to the inner.

4. L. reflexa (Thunb. Prodr. 64; Fl. Cap. edit. Schult. 327); bulb globose, $\frac{1}{2}$–$\frac{3}{4}$ in. diam.; leaves 2, falcate, thin, lanceolate, 4–6 in. long, clasping the base of the stem for 1–2 inches; peduncle 2–6 in. long including the spike; spike usually few-, rarely many-flowered; rachis flexuose; flowers all erecto-patent; bracts small, deltoid-cuspidate; perianth yellowish, about an inch long, finally $\frac{1}{4}$ in. diam.; tube ventricose, oblique at the base, as long as the outer segments, becoming scariose; inner segments $\frac{1}{12}$–$\frac{1}{8}$ in. longer than the outer, spreading at the very tip, where they are $\frac{1}{8}$ in. broad; stamens included; style finally exserted. *Roem. et Schultes, Syst. Veg.* vii. 617; *Baker in Journ. Linn. Soc.* xi. 404. *Cœlanthus complicatus, Willd. ex Roem. et Schultes, Syst. Veg.* vii. xlvi. *in note; Kunth, Enum.* iv. 282.

COAST REGION: Cape Div.; sandy places near Green Point, *Pappe!* Flats between Tygerberg and Blueberg, *Drège!* Cape Flats, *Bolus,* 2838! Swellendam Div.; on mountain ridges by the lower part of Zonder Einde River, *Zeyher,* 4292! Alexandria Div.; Zuurberg Range, *Bolus,* 2635!

Mr. F. W. Moore, of Glasnevin, has sent to Kew a fine garden hybrid between this and *L. tricolor var. aurea.* We have also received the same hybrid from Messrs. Dammann, of Naples, under the name of *L. regeliana.*

5. L. orthopetala (Jacq. Collect. iii. 240; Ic. ii. 15, t. 383); bulb small, globose; leaves 4–5, erect, linear, subulate, glabrous, 4–6 in. long, channelled down the face; peduncle flexuose, 3–6 in. long; flowers generally many, in a dense subspicate raceme 1–4 in. long, all permanently ascending; bracts deltoid-amplexicaul; perianth oblong, $\frac{1}{3}$–$\frac{1}{2}$ in. long, whitish or tinged with red on the upper part of the back of the segments; inner segments linear-oblong, a little longer than the outer, scarcely at all spreading; style exserted. *Roem. et Schultes, Syst. Veg.* vii. 601; *Kunth, Enum.* iv. 286; *Tratt. Archiv. t.* 164; *Baker in Journ. Linn. Soc.* xi. 405. *L. angustifolia, Herb. Drège. Scillopsis orthopetala, Lemaire in Ill. Hort.* iii. *Misc.* 34.

COAST REGION: Malmesbury Div.; between Groene Kloof and Saldanha Bay, below 500 ft., *Drège!* Cape Div.; Koeberg, 500 ft., *MacOwan and Bolus,* Herb. Norm. Austr. Afr., 503! Tygerberg, *MacOwan,* 2522!

A form sent in 1891 by Mr. F. W. Moore from the Glasnevin Garden has pedicels $\frac{1}{8}$–$\frac{1}{4}$ in. long.

6. L. orchioides (Ait. Hort. Kew. i. 460; edit. 2, ii. 284); bulb globose, ¾–1 in. diam.; leaves usually 2, lorate, suberect, smooth, often spotted, an inch broad, clasping the base of the stem; peduncle, including the inflorescence, ½–1 ft. long, often spotted; spike 1–6 in. long, about an inch in diam.; flowers all permanently ascending; bracts small, deltoid-amplexicaul; perianth oblong, about ⅓ in. long, ⅙ in. diam., white, yellow, red, or blue; tube campanulate; outer segments oblong; inner 1/12–⅛ in. longer, spreading at the tip; stamens and style included. *Jacq. Collect.* iii. 241; *Ic.* ii. 16, *t.* 390; *Thunb. Prodr.* 64; *Fl. Cap. edit. Schult.* 327; *Willd. Sp. Plant.* ii. 172; *Gawl. in Bot. Mag. tt.* 854 *and* 1269; *Tratt. Archiv. t.* 154; *Roem. et Schultes, Syst. Veg.* vii. 603; *Kunth, Enum.* iv. 284; *Baker in Journ. Linn. Soc.* xi. 405; *Saund. Ref. Bot. t.* 171. *L. pulchella, Kunth, Enum.* iv. 284. *L. mutabilis, Sweet, Brit. Flow. Gard. ser,* 2, *t.* 129; *Lodd. Bot. Cab. t.* 1076; *Kunth, Enum.* 285. *Hyacinthus orchioides, Linn. Sp. Plant.* 318. *Phormium hyacinthoides, Linn. fil. Suppl.* 204. *Muscari orchioides, Miller, Dict. edit.* viii. *No.* 5. *Orchiastrum Aitoni, virentiflavum, pulchellum, and mutabile, Lemaire in Ill. Hort.* ii. *Misc.* 100.

SOUTH AFRICA : without locality, *Masson! Forster! Zeyher,* 1695! 4287!
COAST REGION : Malmesbury, *Bachmann,* 485! 2062! Cape Div.; near Cape Town, *Bolus,* 2839! Table Mountain, *MacGillivray,* 475! Simons Bay, *Wright!* Tygerberg, *Pappe!* Brak Fontein, near Matroos Bay, *Ecklon and Zeyher,* 41! Tulbagh, *Pappe!* Caledon Div.; between Zwartberg and Ganse Kraal, 1000–2000 ft., *Zeyher,* 4286! Swellendam Div.; on dry hills by Breede River, *Burchell,* 7478! mountain ridges by the lower part of Zonder Einde River, *Zeyher,* 4289! Albany Div.; near Grahamstown, *MacOwan,* 1337! *Galpin,* 228!
WESTERN REGION : Little Namaqualand ; near Ookiep, *Morris in Herb. Bolus,* 5803! *Scully,* 86!

The most striking named colour-forms are *atroviolacea* (hyacinth-blue), *virenti-flava* (greenish-yellow), and *mutabilis* (inner segments dull yellow, tipped with red-brown).

7. L. glaucina (Jacq. Collect. Suppl. 59; Ic. ii 16, t. 391); bulb globose, about 1 in. diam.; leaves usually two, lorate, smooth, often spotted, about an inch broad, clasping the base of the stem; stem ½–1 ft. long including the inflorescence; spike moderately dense, many-flowered, 1–4 in. long; flowers all permanently erecto-patent; bracts minute, deltoid; perianth oblong-cylindrical, ½–¾ in. long, ⅙ in. diam., white, red, yellow, or tinged with blue; inner segments ⅙–¼ in. longer than the outer, spreading, especially the lowest; stamens and style reaching the tip of the inner segments. *Willd. Sp. Plant.* ii. 171; *Tratt. Archiv. t.* 155; *Roem. et Schultes, Syst. Veg.* vii. 603; *Hook. in Bot. Mag. t.* 3552; *Kunth, Enum.* iv. 284; *Baker in Journ. Linn. Soc.* xi. 405. *L. sessiliflora, Andr. Bot. Rep. t.* 460. *L. pallida, Lindl. in Bot. Reg. tt.* 1350 *and* 1945; *Kunth, Enum.* iv. 284; *Baker in Saund. Ref. Bot. t.* 170, *and Journ. Linn. Soc.* xi. 405, *non Ait. Orchiastrum glaucinum, Lemaire in Ill. Hort.* ii. *Misc.* 100.

COAST REGION : Table Mountain, *Ecklon,* 448! near Cape Town, *Bolus,* 4600! Stellenbosch, *Sanderson,* 950! by the Berg River, near Paarl, *Drège,* 1492b!

Swellendam Div. ; on mountain ridges by the lower part of Zonder Einde River, *Zeyher,* 4288 !
WESTERN REGION : Little Namaqualand ; near Mieren Kasteel, *Drège !*

8. L. liliflora (Jacq. Collect. Suppl. 66 ; Ic. ii. 16, t. 387) ; bulb globose ; leaves 2, lanceolate, falcate, 6–9 in. long, about an inch broad, densely pustulate on the face ; peduncle, including the inflorescence, about a foot long ; flowers 12–20 in a dense subspicate raceme ; pedicels very short ; bracts minuto, amplexicaul ; perianth white, ¾ in. long ; all the segments spreading, the outer ⅛ in. longer than the inner ; stamens as long as the inner segments. *Willd. Sp. Plant.* ii. 176 ; *Roem. et Schultes, Syst. Veg.* vii. 601 ; *Tratt. Archiv.* t. 137 ; *Kunth, Enum.* iv. 286 ; *Baker in Journ. Linn. Soc.* xi. 406. *Scillopsis liliiflora, Lemaire in Ill. Hort.* iii. *Misc.* 35.

COAST REGION : Paarl Div. ; between Paarl and Lady Grey Railway Bridge, *Drège,* 1492a !

9. L. undulata (Masson ex Baker in Journ. Bot. 1886, 336) ; bulb not seen ; leaves 2, oblong, obtuse, suberect, much crisped towards the margin, glabrous, 3–4 in. long, an inch broad ; peduncle green, 4 in. long ; flowers in a lax spike 3 in. long ; outer segments tinged with green, ¼ in. long ; inner white, tinged with claret-purple, about ½ in. long ; stamens as long as the outer segments.

SOUTH AFRICA : without locality, *Masson !*
Described from a drawing at the British Museum.

10. L. trichophylla (Baker in Journ. Bot. 1874, 368) ; bulb small, globose ; leaf solitary, ovate, 1½ in. long, clasping the base of the stem, brownish, densely clothed with spreading bristles on the face and margin ; peduncle slender, reddish, ½ ft. long ; flowers about 20 in a moderately dense spike 2–3 in. long, all erecto-patent, many upper minute and sterile ; bracts broad, amplexicaul ; perianth cylindrical, ¾ in. long, ⅙ in. diam., white or tinged with bright red ; inner segments ⅛–⅙ in. longer than the outer, spreading at the tip ; stamens as long as the inner segments.

COAST REGION : Clanwilliam, *Mader in herb. MacOwan,* 2167 !

11. L. Bowkeri (Baker) ; bulb not seen ; leaf solitary, thin, lanceolate, falcate, glabrous, 4–6 in. long ; peduncle, including the inflorescence, 3–6 in. long ; flowers 4–12 in a moderately dense spike, all quite sessile and more or less cernuous ; bracts minute, deltoid ; perianth cylindrical, ⅓–½ in. long, ⅙ in. diam. ; tube campanulate ; outer segments oblong, whitish tinged with red ; inner ⅛ in. long, white, tinged with red on the keel of the truncate tip ; stamens included ; style finally exserted.

CENTRAL REGION ? Somerset Div. ? *Bowker !*

12. L. patula (Jacq. Collect. iv. 149 ; Ic. ii. 15, t. 384) ; bulb small, globose, leaves 2, lanceolate, smooth, unspotted, falcate, 4–6 in. long ; peduncle slender, spotted ; flowers 15–20 in a raceme

2–3 in. long ; lower pedicels ¼ in. long ; bracts deltoid, minute ; perianth oblong-cylindrical, ⅔ in. long, ⅙ in. diam., white tinged with pink ; inner segments much longer than the outer, broad at the tip and very spreading ; stamens included. *Willd. Sp. Plant.* ii. 175 ; *Tratt. Archiv. t.* 156 ; *Roem. et Schultes, Syst. Veg.* vii. 602 ; *Kunth, Enum.* iv. 288 ; *Baker in Journ. Linn. Soc.* xi. 406. *Scillopsis patula, Lemaire in Ill. Hort.* iii. *Misc.* 35.

SOUTH AFRICA : without locality.

Known to me only from Jacquin's figure.

13. L. contaminata (Ait. Hort. Kew. i. 460 ; edit. 2, ii. 285) ; bulb globose, 1 in. diam. ; leaves 6–10, erect, semiterete, glabrous, 6–9 in. long, tapering gradually to the point, channelled down the face, plain or spotted ; peduncle 3–6 in. long ; flowers in a dense subspicate raceme 1–3 in. long, about ⅖ in. diam., all erecto-patent ; bracts minute, adnate ; perianth campanulate, ⅙ in. long and broad, white or tinged with bright red, especially on the back of the outer segments ; inner segments a little longer than the outer, spreading at the tip ; stamens as long as the perianth or slightly exserted. *Thunb. Prodr.* 64 ; *Willd. Sp. Plant.* ii. 174 ; *Bot. Mag. t.* 1401 ; *Baker in Journ. Linn. Soc.* xi. 406. *L. hyacinthoides, Jacq. Collect. Suppl.* 58 ; *Ic.* ii. 15, *t.* 382 ; *Willd. Sp. Plant.* ii. 173 ; *Roem. et Schultes, Syst. Veg.* vii. 599 ; *Tratt. Archiv. t.* 163 ; *Kunth, Enum.* iv. 285. *L. angustifolia, Jacq. Collect. Suppl.* 57 ; *Ic.* ii. 15, *t.* 381 ; *Willd Sp. Plant.* ii. 173 ; *Bot. Mag. t.* 735 ; *Red. Lil. t.* 162 ; *Tratt. Archiv. t.* 161 ; *Kunth, Enum. loc. cit. L. albida, Tratt. Archiv. t.* 162. *Himas angustifolia, and hyacinthoides, Salisb. Gen.* 21. *Scillopsis angustifolia, contaminata, and hyacinthoides, Lemaire in Ill. Hort.* iii. *Misc.* 33–35.

SOUTH AFRICA : without locality, *Zeyher,* 4296 !

COAST REGION : Malmesbury Div.; Klip Fontein, *Zeyher,* 4295 ! Groene Kloof, *Zeyher ! Ecklon and Zeyher,* 53 ! Malmesbury, *Bachmann,* 1231 ! Cape Flats, near Kuyls River, *Pappe!* near Tulbagh, *MacOwan,* 2898 ! *MacOwan and Bolus, Herb. Norm. Aust. Afr.,* 945 ! Worcester, *Zeyher !* Caledon Div.; near the Hot Spring at Caledon, *Zeyher ! Ecklon and Zeyher,* 52 !

14. L. Bachmanni (Baker in Engl. Jahrb. xv. Heft 3, 8) ; bulb not seen ; leaves 2, linear, glabrous, conduplicate, 4–8 in. long above the clasping base ; peduncle 4–8 in. long ; raceme dense, subspicate, 1½–2 in. long ; pedicels scarcely any ; bracts oblong, minute ; perianth campanulate, ¼ in. long ; segments oblong, subequal, white, with a bright red keel ; stamens as long as the perianth.

COAST REGION : Malmesbury Div.; near Hopefield, *Bachmann,* 1232 !

Not in Kew Herbarium.

15. L. carnosa (Baker in Journ. Linn. Soc. xi. 407) ; bulb globose, 1 in. diam. ; leaves 2, subcoriaceous, lorate or ovate, clasping the short peduncle and entirely hiding it, then spreading, 4 in. long, above an inch broad ; peduncle an inch long ; flowers in a dense spike 3 in. long, 1 in. diam. ; bracts lanceolate from a deltoid base, tinged red, ¼ in. long, those of the upper half of the spike protruded beyond the buds and squarrose ; perianth dull-coloured, oblong, ¼ in. long; inner segments scarcely longer than the outer, not spreading at the tip ; stamens included.

WESTERN REGION : Little Namaqualand, between Uitkomst and Geelbeks Kraal, 2000–3000 ft., *Drège*, 2689a !

16. **L. fistulosa** (Baker in Gard. Chron. 1884, xxi. 668) ; bulb ½ in. diam. ; leaves 2, lanceolate, smooth, bright green, erect, ½–1 in. broad; peduncle unspotted, 5–6 in. long ; flowers in a lax spike 2–3 in. long, ¾ in. diam., all quite sessile, erecto-patent ; bracts minute, deltoid ; perianth oblong-campanulate, pale lilac, ¼ in. long, ⅙ in. diam. ; inner segments a little longer than the outer, spreading at the top ; stems included.

SOUTH AFRICA : without locality, cultivated specimen !

Described from a living plant flowered by Mr. T. S. Ware, at Tottenham, April, 1884 ; and another that flowered at Kew in July, 1886.

17. **L. lilacina** (Baker in Gard. Chron. 1884, xxi. 668); bulb small ; leaves 2, lanceolate, smooth, as long as the peduncle ; peduncle about ½ ft. long, finely spotted with red ; flowers in a moderately dense spike 2–3 in. long, ¾–⅞ in. broad, horizontal ; bracts minute, deltoid ; perianth campanulate, lilac, turning to bright claret-red when dried, ⅓ in. long, ⅙ in. diam. ; inner segments ⅛ in. longer than the outer, very spreading ; stamens as long as the outer segments.

SOUTH AFRICA : without locality, cultivated specimen !

Described from a living plant flowered by Mr. T. S. Ware, at Tottenham, in April, 1884.

18. **L. pustulata** (Jacq. Collect. iii. 244 ; iv. 220, t. 2, fig. 5 ; Ic. ii. 15, t. 386) ; bulb ½ in. diam. ; leaves 2, fleshy, lanceolate, falcate, 6–9 in. long, ½–1 in. broad, rough, with pustules on the face ; peduncle 3–6 in. long ; flowers in a dense subspicate raceme 2–3 in. long, nearly an inch in diameter, lower patent, upper erecto-patent ; pedicels very short, erecto-patent ; bracts minute ; perianth oblong, ⅓ in. long, ⅙ in. diam., white or faintly tinged with red ; outer segments a little longer than the inner, spreading at the tip ; stamens slightly exserted. *Bot. Mag. t.* 817 ; *Willd. Sp. Plant.* ii. 176 ; *Roem. et Schultes, Syst. Veg.* vii. 609 ; *Andr. Bot. Rep. t.* 350 ; *Tratt. Archiv tt.* 135–136 ; *Kunth, Enum.* iv. 287 ; *Baker in Journ. Linn. Soc.* xi. 407. *L. reclinata, Dietr. Lexic. Nachtr.* iv. 292. *L. pyramidalis, Dehnh. Revist. Nap.* i. 3, 162 ; *Walp. Ann.* i. 853. *Chloriza pustulata, Salisb. Gen.* 21. *Scillopsis pustulata, Lemaire in Ill. Hort.* iii. *Misc.* 34.

COAST REGION : Clanwilliam Div. ; near the Oliphants River, *Ecklon and Zeyher,* 40 ! 46 ! Malmesbury Div. ; Zwartland, *Ecklon and Zeyher,* 39 ! Cape Div. ; Mountains near Cape Town, *Ecklon and Zeyher,* 37 ! sea-shore near Cape Town, *Bolus,* 4740 ! Paarl Div. ; Great Drakenstein Mountains, and foot of Paarl Mountains, *Drège,* 8628 ! Tulbagh Div. ; Witsenberg Range, 2000 ft.,

Bolus, 5389! Caledon Div.; by the Zonder Einde River, near Appels Kraal, *Zeyher*, 4289! Uitenhage Div.; near the Zwartkops River, *Ecklon and Zeyher*, 38! CENTRAL REGION: Carnarvon Div.; Klip Fontein, *Burchell*, 1534!

This is perhaps synonymous with *L. pallida*, *Ait.*

19. L. **rhodantha** (Baker); bulb ovoid, $\frac{1}{2}$ in. diam.; leaf solitary, smooth, erect, 3–5 in. long, $\frac{1}{2}$ lin. broad, not clasping the base of the peduncle; peduncle slender, bright red, 3–4 in. long; raceme dense, about an inch long, $\frac{5}{8}$–$\frac{3}{4}$ in. diam.; lower flowers cernuous; pedicels erecto-patent, $\frac{1}{8}$–$\frac{1}{6}$ in. long; bracts minute, spreading, ovate; perianth campanulate, bright red, $\frac{1}{6}$ in. long and broad; segments about equal in length; stamens bright red, distinctly exserted.

CENTRAL REGION: Graaff Reinet Div.; grassy slopes of the Oude Berg, 4300 ft., *Bolus*, 719!

20. L. **Zeyheri** (Baker in Journ. Linn. Soc. xi. 407); bulb small, globose; leaf usually solitary, $\frac{1}{2}$–1 ft. long, subterete, tapering gradually to the point, channelled down the face; peduncle slender, 4–9 in. long; flowers in a dense narrow raceme 1–1$\frac{1}{2}$ in. long, $\frac{1}{2}$–$\frac{5}{8}$ in. diam., all erecto-patent or the lowest spreading; pedicels erecto-patent; bracts minute; perianth campanulate, white or tinged with bright red towards the tip of the segments, $\frac{1}{6}$–$\frac{1}{4}$ in. long, $\frac{1}{6}$ in. diam.; outer segments ovate, inner not longer; stamens included.

COAST REGION: Tulbagh Div.; between the Witsenberg and Schurfdeberg Ranges, *Zeyher*, 1694! *Pappe!*

21. L. **convallarioides** (Baker in Journ. Linn. Soc. xi. 407); bulb globose, $\frac{1}{3}$–$\frac{1}{2}$ in. diam.; leaf solitary, linear, erect, $\frac{1}{12}$–$\frac{1}{8}$ in. broad, channelled down the face; peduncle slender, 4–6 in. long; flowers in a lax raceme an inch long, $\frac{1}{2}$ in. diam., upper horizontal, lower cernuous; pedicels $\frac{1}{12}$–$\frac{1}{8}$ in. long; bracts minute; perianth campanulate, $\frac{1}{8}$ in. long, white tinged with red; inner segments not longer than the outer, not spreading at the tip; stamens included.

VAR. β, **robusta** (Baker, loc. cit.); more robust; leaves solitary, rarely 2, 4–7 in. long, $\frac{1}{4}$–$\frac{1}{3}$ in. broad; flowers 20–30.

COAST REGION: Var. β, Albany, *Williamson!* Not at Kew.
EASTERN REGION: Transkei, *Bowker*, 444!

22. L. **unifolia** (Jacq. Hort. Schoenbr. i. 43, t. 83); bulb small, globose; leaf long, linear-subulate, erect, clasping the base of the stem for 2–3 inches, conspicuously branded with brown towards the base; peduncle, including the inflorescence, $\frac{1}{2}$–1 ft. long; flowers 6–20, arranged in a lax raceme; pedicels ascending, lower $\frac{1}{6}$–$\frac{1}{4}$ in. long; bracts minute, deltoid; perianth oblong, about $\frac{1}{2}$ in. long, $\frac{1}{6}$ in. diam., white, or more or less tinged with red or blue; tube campanulate; outer segments oblong, inner protruding $\frac{1}{12}$–$\frac{1}{8}$ in., broad and spreading at the tip; stamens as long as the inner segments. *Bot. Mag. t.* 766; *Willd. Sp. Plant.* ii. 178; *Tratt. Archiv. t.* 158; *Roem. et Schultes, Syst. Veg.* vii. 611; *Kunth, Enum.* iv. 289; *Baker in Journ. Linn. Soc.* xi. 406. *Monoestes unifolia, Salisb. Gen.* 21. *Scillopsis unifolia, Lemaire in Ill. Hort.* iii. *Misc.* 34.

VAR. β, L. **Wrightii** (Baker in Journ. Bot. 1878, 322); perianth shorter and more campanulate, about $\frac{1}{3}$ in. long, $\frac{1}{6}$ in. diam.; pedicels as long as the flowers.

VAR. γ, **Rogersii** (Baker); perianth as in the last variety, but pedicels shorter and free portion of the blade of the leaf broader.

VAR. δ, **Pappei** (Baker); like the last variety in flower and pedicels, but with two leaves.

SOUTH AFRICA: without locality, *Masson! Niven! Thom!* Var. β, *Grey* Var. γ, *Rogers!*

COAST REGION: Clanwilliam Div., *Zeyher!* Malmesbury Div.; Groene Kloof, *MacOwan*, 2501! *Bolus*, 4353! Cape Div.; near Cape Town, *Bolus* 4353! *MacOwan and Bolus, Herb. Norm. Aust. Afr.*, 370! *Pappe!* Paarl Div.; near Paarl Mts., below 1000 ft., *Drège*, 477! Var. β, Cape Div.; Simons Bay, *Wright*, 219! Camps Bay, *Ecklon and Zeyher*, 130! Caledon Div.; on the Zwartberg, near the Hot Spring, *Zeyher!* Var. γ, Tulbagh, *Pappe!* Var. δ, Cape Flats, *Pappe!*

WESTERN REGION: Little Namaqualand; near Klip Fontein, 3000 ft., *Bolus*, 6592!

23. **L. anguinea** (Sweet, Brit. Flow. Gard. t. 179); bulb $\frac{1}{2}$ in. diam.; leaf single, smooth, lanceolate, 6–9 in. long, about an inch broad, spotted with cross-bars of darker green on the back; peduncle 4–6 in. long, spotted with purple; raceme $2\frac{1}{2}$–3 in. long; pedicels cernuous at the tip, $\frac{1}{3}$–$\frac{1}{2}$ in. long; perianth campanulate, whitish tipped with green, $\frac{1}{3}$ in. long; inner segments scarcely longer than the outer; stamens much exserted, rather declinate. *Kunth, Enum.* iv. 289; *Baker in Journ. Linn. Soc.* xi. 408. *Scillopsis anguinea, Lemaire in Ill. Hort.* iii. *Misc.* 34.

SOUTH AFRICA: without locality.

Known to me only from the figure cited, which was drawn from a plant in Mr. Colvill's garden in 1825, gathered by Synnot.

24. **L. hirta** (Thunb. Prodr. 64); bulb $\frac{1}{3}$–$\frac{1}{2}$ in. diam.; leaf solitary, linear, 5–8 in. long, clasping the base of the stem, clothed with spreading stramineous bristles; peduncle slender, spotted, 6–10 in. long; flowers 6–20 in a lax raceme 1–3 in. long, 1 in. diam., upper ascending, lower horizontal; lower pedicels $\frac{1}{4}$–$\frac{1}{3}$ in. long; bracts minute; perianth oblong-campanulate, $\frac{1}{3}$ in. long, $\frac{1}{6}$ in. diam., white or tinged with red; inner segments a little longer than the outer; stamens included. *Fl. Cap. edit. Schult.* 327; *Willd. Sp. Plant.* ii. 178; *Roem. et Schultes, Syst. Veg.* vii. 611; *Kunth, Enum.* iv. 289; *Baker in Journ. Linn. Soc.* xi. 408.

SOUTH AFRICA: without locality, *Masson! Thunberg!*

COAST REGION: Malmesbury Div.; Groene Kloof, *Ecklon and Zeyher*, 47! *MacOwan and Bolus, Herb. Norm. Aust. Afr.*, 797!

WESTERN REGION: Little Namaqualand; *Pappe*, 4291! *Zeyher!*

25. **L. Massoni** (Baker in Journ. Bot. 1886, 336); bulb ovoid, $\frac{3}{4}$ in. diam.; leaf single, suberect, clasping the stem at the base, flat above, orbicular, $1\frac{1}{2}$ in. long and broad, covered over the surface with white bristly hairs; peduncle purple, 4 in. long; raceme lax, 2 in. long; lower pedicels $\frac{1}{8}$–$\frac{1}{6}$ in. long; bracts minute, ovate; perianth

with purple-tinted outer segments $\frac{1}{2}$ in. long, inner white, $\frac{1}{6}$ in. longer; stamens as long as the outer segments.

WESTERN REGION. Little Namaqualand, *Masson!*

Described from a drawing at the British Museum, made in August, 1793.

26. L. campanulata (Baker in Journ. Bot. 1874, 6); bulb globose, $\frac{1}{2}$ in. diam.; leaves usually 2, rarely 1, subterete, tapering to the point, equalling or overtopping the peduncle; peduncle slender, 3–9 in. long; flowers in a dense raceme 1–1$\frac{1}{2}$ in. long, $\frac{1}{2}$–$\frac{5}{8}$ in. diam., upper erecto-patent, lower horizontal; lower pedicels $\frac{1}{12}$ in. long; bracts minute; perianth campanulate, $\frac{1}{8}$–$\frac{1}{6}$ in. long, white tinged with red; outer segments ovate; inner not longer; stamens included.

CENTRAL REGION: Somerset Div.; rocky places on the summit of Bosch Berg, 4800 ft., *MacOwan*, 1836!

May be a variety of *L. Zeyheri.*

27. L. isopetala (Jacq. Collect. Suppl. 68; Ic. ii. 17, t. 401); bulb globose, $\frac{3}{4}$–1 in. diam.; leaves 2, lanceolate, smooth, glabrous, suberect, 6–9 in. long, about an inch broad; peduncle, including the inflorescence, about a foot long; raceme dense, 2–6 in. long; flowers all erecto-patent; pedicels short; bracts deltoid, minute; perianth oblong, $\frac{1}{4}$–$\frac{1}{3}$ in. long, $\frac{1}{6}$ in. diam., white or tinged with red; inner segments very little longer than the outer, scarcely at all spreading; stamens as long as the inner segments. *Willd. Sp. Plant.* ii. 179; *Roem. et Schultes, Syst. Veg.* vii. 610; *Tratt. Archiv. t.* 165; *Kunth, Enum.* iv. 286; *Baker in Journ. Linn. Soc.* xi. 406. *Scillopsis isopetala, Lemaire in Ill. Hort.* iii. *Misc.* 35.

SOUTH AFRICA: without locality, *Thom*, 736! 979! *Grey!*
COAST REGION: Cape Div.? low ground near North Hoek Forest, *Milne*, 200
Mossel Bay Div.; Little Brak River, *Burchell*, 6188!

28. L. mediana (Jacq. Collect. iii. 242; Ic. ii. 16, t. 392); bulb small, globose; leaves 2, smooth, lorate, above a foot long, an inch broad; peduncle flexuose, unspotted, a foot long; raceme lax, 2–3 in. long, 1 in. diam.; all the flowers erecto-patent; pedicels short; bracts ovate; perianth oblong-campanulate, $\frac{1}{3}$ in. long, white, with the base tinted blue and the tip of all the segments green; inner segments a little longer than the outer; stamens just exserted, *Roem. et Schultes, Syst. Veg.* vii. 604; *Tratt. Archiv. t.* 159; *Kunth, Enum.* iv. 287; *Baker in Journ. Linn. Soc.* xi. 408. *Scillopsis mediana, Lemaire in Ill. Hort.* iii. *Misc.* 35.

COAST REGION: Cape Div.; Rondebosch, 50 ft., *Bolus*, 4829! Yzer Plaat, near Salt River, *Zeyher*, 4660! Caledon Div.; among shrubs near the Hot Spring at Caledon (a form with the lower flowers tipped with red), *MacOwan, Herb. Aust. Afr.*, 1560!

29. L. rosea (Andr. Bot. Rep. t. 296); bulb $\frac{1}{2}$ in. diam.; leaves 1–2, smooth, lanceolate, 6–9 in. long, $\frac{1}{2}$–$\frac{3}{4}$ in. broad at the middle; peduncle as long as the leaves, not spotted; raceme lax, 2–3 in. long, 1 in. diam.; flowers all ascending; pedicels erecto-patent, $\frac{1}{12}$–$\frac{1}{3}$ in.

long ; bracts minute ; perianth oblong-campanulate, $\frac{1}{4}$ in. long, bright
red ; inner segments slightly longer than the outer ; stamens not
exserted.   *Tratt. Archiv, t.* 157 ; *Roem. et Schultes, Syst. Veg.* vii.
602 ; *Kunth, Enum.* iv. 286.   *L. bifolia, Gawl. in Bot. Mag. t.* 1611 ;
Lodd. Bot. Cab. t. 920 ; *Roem. et Schultes, Syst. Veg.* vii. 602 ; *Kunth,
Enum.* iv. 286.   *Scillopsis rosea* and *S. bifolia, Lemaire in Ill. Hort.*
iii. *Misc.* 35.

CoAST REGION : Caledon Div. ; among shrubs on the Zwartberg, *Templeman
in Herb. MacOwan,* 2081 !

30. L. Youngii (Baker); bulb globose ; leaves 2, lanceolate,
ascending, glabrous, $\frac{1}{2}$ ft. long; peduncle 2–4 in. long ; raceme dense,
2–3 in. long, 1 in. diam.; central and lower flowers horizontal ; pedicels erecto-patent, $\frac{1}{12}$–$\frac{1}{8}$ in. long ; bracts minute, ovate ; perianth
campanulate, bright red-purple, $\frac{1}{4}$ in. long ; outer and inner segments
nearly equal; stamens as long as the perianth or slightly exserted.

CoAST REGION: George Div. ; Montagu Pass, 1200 ft., *Young in Herb. Bolus,*
5545 !

31. L. succulenta (Masson ex Baker in Journ. Bot. 1886, 336) ;
bulb $\frac{1}{2}$ in. diam. ; leaves lanceolate, purple-tinted, 4–5 in. long, $\frac{1}{3}$ in.
broad ; peduncle slender, purple, as long as the leaves ; raceme lax,
2 in. long; lower flowers spreading ; pedicels erecto-patent, lower
$\frac{1}{6}$ in. long ; bracts small, deltoid ; perianth white tinged with claret,
$\frac{1}{2}$ in. long ; outer segments $\frac{1}{8}$–$\frac{1}{6}$ in. longer than the inner ; stamens
included.

CoAST REGION: Clanwilliam Div.; near Oliphants River, *Masson !*
Described from a drawing at the British Museum, made in the year 1793.

32. L. pallida (Ait. Hort. Kew. i. 460 ; edit. 2, ii. 285) ; bulb
globose, $\frac{1}{2}$ in. diam. ; leaves 2, lorate, pustulate on the face, 6–9 in.
long, $\frac{1}{2}$–$1\frac{1}{4}$ in. broad ; peduncle $\frac{1}{2}$ ft. long, not spotted ; raceme
moderately dense, 2–3 in. long, 1 in. diam.; lower flowers horizontal ;
pedicels short, erecto-patent ; bracts deltoid, minute ; perianth white,
campanulate, $\frac{1}{4}$ in. long ; the three outer segments with a green
gibbosity at the tip ; inner segments a little longer than the outer ;
stamens as long as the inner segments.   *Red. Lil. t.* 22 ? *Bot. Reg.
t.* 287 ? *Tratt. Archiv. t.* 160 ? *non Kunth.   L. lucida, Gawl. in Bot.
Mag. t.* 1372 ; *Roem. et Schultes, Syst. Veg.* vii. 605 ; *Kunth, Enum.*
iv. 287 ; *Baker in Journ. Linn. Soc.* xi. 408.   *L. racemosa, Gawl.
in Bot. Mag. t.* 1517 ; *Roem. et Schultes, Syst. Veg.* vii. 608 ; *Kunth,
Enum.* iv. 287 ; *Baker in Journ. Linn. Soc.* xi. 408.   *L. odoratissima, Baker in Gard. Chron.* 1884, xxi. 668.   *Platyestes racemosa,
Salisb. Gen.* 21.   *Scillopsis racemosa, Lemaire in Ill. Hort.* iii.
Misc. 34.   *Chloriza lucida, Salisb. Gen.* 21.   *Scillopsis lucida,
Lemaire in Ill. Hort.* iii. *Misc.* 34.

SOUTH AFRICA : without locality, cultivated specimens !

The identity of *L. racemosa* and *L. pallida* is rendered certain by a specimen
of *L. pallida* obtained from the Goodenough herbarium, which is identical with
the figure of *L. racemosa* in the Bot. Mag., and with my *L. odoratissima,* the

leaf showing traces of having been pustulate. The plant figured by Redouté, Trattinnick, and in the Botanical Register, is doubtfully the same as *L. pallida* of Aiton, having no pustules on the leaves.

33. L. Cooperi (Baker in Journ. Linn. Soc. xi. 409); bulb $\frac{1}{2}$ in. diam.; leaves 2, lanceolate, $\frac{1}{2}$ ft. long, under an inch broad, pustulate on the upper part of the face; peduncle 3–4 in. long; raceme 20–30-flowered, 3–4 in. long, $1\frac{1}{4}$ in. diam.; lower pedicels $\frac{1}{4}$–$\frac{1}{3}$ in. long; bracts minute, deltoid; perianth oblong-campanulate, reddish-white, $\frac{1}{3}$ in. long, $\frac{1}{6}$–$\frac{1}{4}$ in. diam.; inner segments distinctly longer than the outer; stamens as long as the inner segments.

SOUTH AFRICA: without locality, *Cooper!* No specimen at Kew.

Described from a living plant in the garden of Mr. Wilson Saunders, at Reigate, about 1870, introduced by Mr. T. Cooper.

34. L. juncifolia (Baker in Journ. Linn. Soc. xi. 409); bulb small; leaves 2, subterete, smooth, 4–8 in. long, $\frac{1}{8}$–$\frac{1}{4}$ in. diam., channelled down the face; peduncle 3–4 in. long; raceme 6–12-flowered, $\frac{3}{4}$–$1\frac{1}{2}$ in. long, $\frac{3}{4}$ in. diam.; pedicels $\frac{1}{8}$–$\frac{1}{6}$ in. long; bracts deltoid, minute; perianth oblong-campanulate, white tinged with red, $\frac{1}{5}$ in. long, $\frac{1}{8}$ in. diam.; segments equal in length; stamens much exserted.

SOUTH AFRICA: without locality, *Ecklon and Zeyher*, 51! Not in Kew Herbarium.

35. L. purpureo-cærulea (Jacq. Collect. Suppl. 63; Ic. ii. 16, t. 388); bulb globose, $\frac{1}{2}$ in. diam.; leaves 2, lorate, pustulate on the face, 4–6 in. long, $\frac{1}{2}$–1 in. broad; peduncle robust, $\frac{1}{2}$ ft. long; raceme moderately dense, 4–5 in. long, 1 in. diam.; all the flowers erecto-patent; pedicels short; bracts deltoid, minute; perianth campanulate, blue-purple, $\frac{1}{4}$ in. long; inner segments a little longer than the outer, spreading widely at the tip; stamens much exserted. *Willd. Sp. Plant.* ii. 176; *Gawl. in Bot. Mag. t.* 745; *Andr. Bot. Rep. t.* 251; *Roem. et Schultes, Syst. Veg.* vii. 606; *Kunth, Enum.* iv. 288; *Baker in Journ. Linn. Soc.* xi. 409. ' *L. botryoides, Tratt. Archiv. t.* 140. *Platyestes purpureo-cærulea, Salisb. Gen.* 21. *Scillopsis purpureo-cærulea, Lemaire in Ill. Hort.* iii. *Misc.* 35.

COAST REGION: Malmesbury Div.; near Kalebas Kraal, *MacOwan*, 2604!

36. L. unicolor (Jacq. Collect. Suppl. 61; Ic. ii. 16, t. 389); bulb $\frac{1}{2}$–1 in. diam.; leaves 2, lorate, smooth or pustulate on the upper part of the face, 4–6 in. long, $\frac{1}{2}$–1 in. broad; peduncle robust, stiffly erect, 3–6 in. long; raceme dense, 2–5 in. long, $\frac{3}{4}$–1 in. broad; lower flowers horizontal or cernuous; pedicels $\frac{1}{4}$–$\frac{1}{6}$ in. long; bracts deltoid, minute; perianth campanulate, bright red, $\frac{1}{4}$ in. long; inner segments slightly longer than the outer; stamens much exserted. *Gawl. in Bot. Mag. t.* 1373; *Tratt. Archiv. t.* 138; *Roem. et Schultes, Syst. Veg.* vii. 607; *Kunth, Enum.* iv. 288. *L. versicolor, var. unicolor, Baker in Journ. Linn. Soc.* xi. 409. *Scillopsis unicolor, Lemaire in Ill. Hort.* iii. *Misc.* 34.

VAR. β, **L. purpurea** (Jacq. Collect. Suppl. 65; Ic. ii. 16, t. 393); outer segments white tipped with green, inner purple. *Tratt. Archiv. t.* 143; *Willd. Sp. Plant.* ii. 177; *Roem. et Schultes, Syst. Veg.* vii. 609; *Kunth, Enum.* iv. 289. *Scillopsis purpurea, Lemaire in Ill. Hort.* iii. *Misc.* 35.

VAR. γ, **L. fragrans** (Jacq. Hort. Schoenbr. i. 43, t. 82); habit more slender; perianth white, slightly tinged with red. *Willd. Sp. Plant.* ii. 176; *Lodd. Bot. Cab. t.* 1140; *Tratt. Archiv. t.* 141; *Kunth, Enum.* iv. 287. *Scillopsis fragrans, Lemaire in Ill. Hort.* iii. *Misc.* 35.

SOUTH AFRICA: without locality, *MacOwan,* 2283!

COAST REGION: Clanwilliam Div.; near the Oliphants River, *Ecklon and Zeyher,* 43! 44! *Zeyher!* Malmesbury Div.; Groene Kloof, *Ecklon and Zeyher,* 54! *Bolus,* 4352! *Bachmann,* 1234! Tulbagh Div.; Winterhoek, *Pappe!* Steendaal, *Pappe!* Mossel Bay Div.; Between Zout River and Duyker River. *Burchell,* 6346! George Div.; Montagu Pass, 1200 ft., *Young in Herb. Bolus,* 5545!

WESTERN REGION: Little Namaqualand; near Ookiep, *Morris in Herb. Bolus,* 5804! near Modder Fontein, 3000 ft., *Bolus,* 6591!

37. L. violacea (Jacq. Collect. iv. 147; Ic. ii. 16, t. 394); bulb $\frac{1}{2}$ in. diam.; leaves 2, lorate, smooth, spotted, $\frac{1}{2}$ ft. long, an inch broad; peduncle curved, $\frac{1}{2}$–1 ft. long; raceme lax, 4–5 in. long, $1\frac{1}{4}$–$1\frac{1}{2}$ in. diam.; flowers many, cernuous; pedicels $\frac{1}{4}$–$\frac{1}{3}$ in. long; bracts minute; perianth campanulate, $\frac{1}{4}$ in. long; outer segments greenish, inner rather longer, violet, spreading at the tip; stamens much exserted. *Willd. Sp. Plant.* ii. 177; *Tratt. Archiv. t.* 166; *Roem. et Schultes, Syst. Veg.* vii. 610; *Kunth, Enum.* iv. 289; *Baker in Journ. Linn. Soc.* xi. 410. *L. bicolor, Lodd. Bot. Cab. t.* 1129; *Kunth, loc. cit. Scillopsis violacea, Lemaire in Ill. Hort.* iii. *Misc.* 35.

SOUTH AFRICA: without locality, cultivated specimen!

38. L. latifolia (Tratt. Archiv. t. 143); bulb large, globose; leaves 2, thin, smooth, oblong, 3–4 in. long, 1–$1\frac{1}{2}$ in. broad; peduncle unspotted, erect, 4–6 in. long; raceme moderately dense, 3–4 in. long, under an inch in diameter; lower flowers horizontal or cernuous; pedicels short, erecto-patent; perianth bright red, campanulate, $\frac{1}{6}$ in. long; segments subequal; stamens bright red, much exserted. *L. fragrans, Andr. Bot. Rep. t.* 302, *non Jacq.*

COAST REGION: Swellendam Div.; among shrubs by the Buffeljagts River, *Zeyher,* 4290!

39. L. bowieana (Baker in Journ. Linn. Soc. xi. 410); bulb large, globose; leaves 2, ovate-oblong, spreading, 2–$2\frac{1}{2}$ in. long, 1–$1\frac{1}{4}$ in. broad, smooth, spotted with red-brown; peduncle 5–6 in. long; raceme 20–30-flowered, 3–4 in. long, $1\frac{1}{4}$–$1\frac{1}{2}$ in. diam.; pedicels $\frac{1}{8}$ in. long; bracts deltoid, minute; perianth oblong-campanulate, $\frac{1}{4}$–$\frac{1}{3}$ in. long, white, tinged with red; inner segments a little longer than the outer; stamens much exserted.

SOUTH AFRICA: without locality, *Bowie!*

Described from a drawing of a plant grown at Kew about 1820, introduced by Bowie.

40. L. nervosa (Gawl. in Bot. Mag. t. 1497); bulb large, globose; leaves 2, oblong, thin, smooth, spreading, 4–5 in. long, $1\frac{1}{2}$–2 in. broad at the middle, conspicuously vertically veined; peduncle stout,

stiffly erect, 3–6 in. long; raceme dense, 4–6 in. long, 1–1½ in. diam.;
pedicels short, erecto-patent; bracts deltoid; perianth campanulate,
¼ in. long, bright red; inner segments scarcely longer than the outer,
spreading only at the tip; stamens much exserted. *Roem. et
Schultes, Syst. Veg.* vii. 607; *Kunth, Enum.* iv. 288; *Baker in
Journ. Linn. Soc.* xi. 410. *Platyestes nervosa, Salisb. Gen.* 21.
Scillopsis nervosa, Lemaire in Ill. Hort. iii. *Misc.* 34.

South Africa: without locality, *Bowie* (a drawing in the Kew Herbarium)!
Coast Region: Mossel Bay Div.; dry hills on the East side of Gauritz River,
Burchell, 6438!

41. L. polyphylla (Baker in Engl. Jahrb. xv. Heft 3, 7); bulb
small, ovoid; leaves 6–8, slender, subterete, erect, smooth, shorter
than the peduncle; peduncle slender, unspotted, 4–8 in. long;
flowers 8–15 in a moderately dense raceme ½–1 in. long, ⅝–¾ in.
diam., lower cernuous; pedicels erecto-patent, ⅛–⅙ in. long; bracts
minute; perianth campanulate, ⅙ in. long and broad, white, tinged
with red; inner segments slightly longer than the outer; stamens
distinctly exserted.

Coast Region: Tulbagh, *Ecklon and Zeyher,* 50!

42. L. pusilla (Jacq. Collect. Suppl. 71; Ic. ii. 15, t. 385); bulb
globose, ½ in. diam.; leaves about 4, lanceolate, coriaceous, glabrous,
with a thickened margin, 2–3 in. long, narrowed gradually from the
middle to a short clasping petiole, spotted with brown or becoming
red-brown when exposed; peduncle ½–1½ in. long; flowers 6–12 in
a dense corymb, permanently erect, pale lilac; pedicels 1/12–⅛ in. long;
bracts small, deltoid; perianth cylindrical, ¼–⅓ in. long; tube very
short; inner segments obscurely spathulate, slightly longer than the
outer; stamens and style much exserted. *Willd. Sp. Plant.* ii. 175;
Tratt. Archiv. t. 144; *Roem. et Schultes, Syst. Veg.* vii. 616; *Kunth,
Enum.* iv. 292. *Massonia undulata, Thunb. Prodr.* 60; *Fl. Cap.
edit. Schult.* 308; *Diss. Nov. Gen.* ii. 41; *Willd. Sp. Plant.* ii. 28;
Ait. Hort. Kew. edit. 2, ii. 211; *Roem. et Schultes, Syst. Veg.* vii.
992; *Kunth, Enum.* iv. 300. *Brachyscypha undulata, Baker in
Journ. Linn. Soc.* xi. 394.

South Africa: without locality, *Drège,* 2687! *Zeyher,* 4275! *Ecklon and
Zeyher,* 23! and a specimen cultivated by *Jacquin!*
Coast Region: Malmesbury Div.; near Hopefield, *Bachmann,* 1791!

XXVIII. DRIMIA, Jacq.

Perianth gamophyllous, deciduous, cut away round the base; tube
campanulate; segments subequal, linear-oblong, spreading or reflexing, generally cucullate at the apex. *Stamens* 6, inserted at the
throat of the perianth-tube, shorter than the segments, sometimes
slightly declinate; filaments filiform; anthers small, oblong, versatile, dehiscing introrsely. *Ovary* sessile, ovoid, 3-celled; ovules

many, superposed; style filiform; stigma capitate, faintly 3-lobed.
Capsule ovoid-triquetrous, membranous, loculicidally 3-valved.
Seeds discoid, sometimes winged; testa black, lax; albumen fleshy;
embryo cylindrical.

Rootstock a tunicated bulb; leaves broad and rather fleshy or narrow and
rigid, often produced at a different season to the flowers; inflorescence a simple
raceme; bracts membranous, persistent; flowers whitish or reddish-white, often
tinged with green.

DISTRIB. Seven species in Tropical Africa.

Subgenus EUDRIMIA. Perianth-segments channelled down the face, cucullate
at the tip.
    Leaves contemporary with the flowers:
        Leaves narrow linear, stiffly erect:
            Bracts much shorter than the pedicels   ...   (1) media.
            Bracts as long as the pedicels   ...   ...   (2) rigidifolia.
        Leaves lorate-lanceolate:
            Upper bracts about ½ in. long   ...   ...   (3) robusta.
            Upper bracts about an inch long   ...   ...   (4) altissima.
    Leaves produced after the flowers:
        Raceme oblong or cylindrical:
            Leaves pilose on both surface and margin   ...   (5) villosa.
            Leaves persistently ciliated:
                Dwarf   ...   .   (6) pusilla.
                Tall...   ...   .   (7) ciliaris.
            Leaves glabrous:
                Leaves narrow linear, ⅛ in. broad   ...   (8) purpurascens.
                Leaves linear, ¼-⅓ in. broad   ...   ...   (9) angustifolia.
                Leaves lanceolate:
                    Racemes lax, few-flowered ...   ...   (10) haworthioides.
                    Racemes dense, many-flowered:
                        Pedicels much shorter than
                            flowers   ...   .   (11) Burchellii.
                        Pedicels about as long as flowers   (12) elata.
                    Leaves unknown ...   .   (13) pauciflora.
        Racemes congested, globose, dense:
            Pedicels ⅛-⅓ in. long:
                Head of flowers 1 in. diam.   ...   ...   (14) sphærocephala.
                Head of flowers 1½-2 in. diam. ...   ...   (15) neriniformis.
            Pedicels ½-¾ in. long ...   .   (16) capitata.
Subgenus LEDEBOURIOPSIS. Perianth-segments nearly flat.
    Flowers reddish; pedicels long:
        Perianth ¼ in. long ...   .   (17) hyacinthoides.
        Perianth 1-1¼ in. long   ...   .   (18) macrantha.
    Flowers greenish-yellow; pedicels short:
        Leaves contemporary with the flowers:
            Leaf terete, usually solitary   ...   .   (19) anomala.
            Leaves 2, linear   ...   .   (20) Cooperi.
            Leaves 3-4, lanceolate...   ...   .   (21) chlorantha.
        Leaves produced after the flowers ...   .   (22) Bolusii.

1. D. media (Jacq. Collect. Suppl. 40; Ic. ii. 15, t. 375); bulb
subglobose, 1½-2 in. diam.; leaves many, contemporary with the
flowers, narrow linear, subtriquetrous, stiffly erect, glabrous, rather
scabrous on the margin, 1-1½ ft. long, $\frac{1}{12}-\frac{1}{8}$ in. broad low down;
peduncle slender, fragile, 1-2 ft. long; raceme lax, oblong, 2-5 in.

long, 1½ in. diam.; lower pedicels ⅓–½ in. long; bracts small, lanceolate from a deltoid base; perianth about ½ in. long, reddish-white;
segments 3–4 times as long as the tube; stamens much shorter than
the segments; style just reaching to the tip of the perianth-segments.
Willd. Sp. Plant. ii. 166; *Roem. et Schultes, Syst. Veg.* vii. 596;
Baker in Journ. Linn. Soc. xi. 420. *Hyacinthus medius, Poir.*
Encyc. Suppl. iii. 120. *Idothea media, Kunth, Enum.* iv. 342.

SOUTH AFRICA: without locality, *Zeyher*, 4250!
COAST REGION: near Cape Town, *Bolus*, 2835! Tulbagh, *Elliot*, 223!
Worcester Div.; Dutoits Kloof, 2000–3000 ft., *Drège*, 1496! Stellenbosch Div.;
between Lowrys Pass and Jonkers Hoek, *Burchell*, 8322! Port Elizabeth Div.;
near Cragga Kamma, *Burchell*, 4593!
EASTERN REGION: Swaziland; Havelock Concession, *Saltmarshe in Herb.*
Galpin, 1056!

2. D. rigidifolia (Baker in Journ. Linn. Soc. xi. 420); bulb ovoid,
1½–2 in. diam.; leaves many, contemporary with the flowers, narrow
linear, glabrous; stiffly erect, 1½ ft. long, ⅛–⅙ in. broad low down;
peduncle as long as the leaves; raceme oblong, moderately dense,
3–6 in. long, 1½–2 in. diam.; pedicels ¼–½ in. long; bracts lanceolate, about as long as the pedicels; perianth ⅝ in. long, reddish-
white; segments convolute, 3–4 times the length of the tube;
stamens rather shorter than the perianth-segments; style not protruded beyond the tip of the perianth-segments.

CENTRAL REGION? Somerset Div.? *Bowker!*

Originally described from a specimen dried by Dr. Harvey from the garden of
Baron Ludwig.

3. D. robusta (Baker in Saund. Ref. Bot. t. 190); bulb ovoid,
3–4 in. diam.; leaves 6–8, contemporary with the flowers, lorate-
lanceolate, rather fleshy, flaccid, green, 1½–2 ft. long, 1½–2 in. broad
at the middle; peduncle 3–5 ft. long, very robust, ½–¾ in. diam.
low down; raceme lax, cylindrical, finally 1–2 ft. long, about 2 in.
diam.; pedicels spreading, about ½ in. long; bracts lanceolate, about
as long as the pedicels; perianth dull reddish-white, ½–⅝ in. long;
tube short, campanulate; segments convolute; stamens nearly as
long as the perianth-segments; style finally ½ in. long, thickened
upwards, protruded beyond the tip of the perianth-segments;
capsule oblong, ⅓ in. long. *Baker in Journ. Linn. Soc.* xi. 421.

KALAHARI REGION: Orange Free State, *Cooper*, 3288!
EASTERN REGION: Griqualand East; mountain sides near Clydesdale, 3000–
3500 ft., *Tyson*, 1108! *MacOwan and Bolus, Herb. Norm. Austr. Afr.*, 1209!

4. D. altissima (Hook. in Bot. Mag. t. 5522, non Gawl.); bulb
ovoid, 3–4 in. diam.; leaves 8–10, contemporary with the flowers,
lorate-lanceolate, rather fleshy, slightly glaucous, 1½–2 ft. long,
1½–2 in. broad; peduncle very robust, 3–4 ft. long; raceme cylindrical, dense upwards, lax downwards, finally a foot or more long;
lower pedicels patent, ¾–1 in. long; bracts lanceolate, very acuminate,
as long as the pedicels, protruding far beyond the upper buds;
perianth dull reddish-white; tube ⅙ in. long; stamens rather de-

clinate, much shorter than the perianth-segments; style finally $\frac{1}{2}$ in. long. *Baker in Journ. Linn. Soc.* xi. 421.

EASTERN REGION: Natal; Inanda, *Wood*, 199!

Originally described and figured from a plant sent to Kew by Mr. Sanderson.

5. **D. villosa** (Lindl. in Bot. Reg. t. 1346); bulb ovoid, above an inch in diameter; leaves 3–4, produced after the flowers, lorate, glaucous, undulated, about a foot long, $\frac{1}{2}$–$\frac{3}{4}$ in. broad, more or less pilose on the surface and finely ciliated on the margin; peduncle slender, 1$\frac{1}{2}$–2 ft. long; raceme oblong, lax in the lower half, 4–5 in. long, 1$\frac{1}{2}$–2 in. diam.; lower pedicels $\frac{1}{4}$–$\frac{1}{2}$ in. long; bracts lanceolate, $\frac{1}{2}$ in. long, conspicuously protruded beyond the buds; perianth $\frac{1}{2}$–$\frac{5}{8}$ in. long, greenish-purple; tube campanulate, $\frac{1}{8}$ in. long; segments canaliculate, reflexing; stamens shorter than the perianth-segments; style protruded beyond the tip of the perianth-segments. *Baker in Journ. Linn. Soc.* xi. 422. *Idothea villosa, Kunth, Enum.* iv. 343.

EASTERN REGION: Griqualand East; mountain slopes near Clydesdale, 2700–5000 ft., *Tyson*, 1671! 2168!

6. **D. pusilla** (Jacq. Collect. Suppl. 42; Ic. ii. 15, t. 374); bulb ovoid, 1 in. diam.; leaves 4–6, produced after the flowers, lanceolate, ciliated on the margin, 3–4 in. long, $\frac{1}{4}$–$\frac{1}{3}$ in. broad, narrowed gradually to the base; peduncle 2–6 in. long; raceme lax, cylindrical, 1$\frac{1}{2}$–3 in. long, 1$\frac{1}{4}$ in. broad; pedicels $\frac{1}{8}$–$\frac{1}{6}$ in. long; bracts deltoid, minute; perianth greenish, $\frac{1}{2}$–$\frac{5}{8}$ in. long; segments cucullate, 3 times the length of the campanulate tube; stamens and style reaching to the tip of the perianth-segments. *Willd. Sp. Plant.* ii. 165; *Roem. et Schultes, Syst. Veg.* vii. 598; *Baker in Journ. Linn. Soc.* xi. 421. *Hyacinthus pusillus, Poir. Encyc. Suppl.* iii. 120. *D. humilis, Berg. ex Eckl. Top. Verz.* 2. *D. eckloniana, Roem. et Schultes, Syst. Veg.* vii. 1710. *Idothea pusilla and I. humilis, Kunth, Enum.* iv. 344.

VAR. β, **setosa** (Baker in Journ. Linn. Soc. xi. 422); peduncle densely beset with whitish bristles.

SOUTH AFRICA: without locality, *Banks!* *Ecklon*, 89! *Rogers!* Var. β, *Ecklon and Zeyher*, 28!

COAST REGION: Cape Flats near Rosebank, under 100 ft., *Bolus*, 3923!

7. **D. ciliaris** (Jacq. Collect. Suppl. 41; Ic. ii. 15, t. 377); bulb globose, 2–3 in. diam.; leaves 4–5, produced after the flowers, linear, thin, flaccid, ciliated, 6–9 in. long, $\frac{1}{4}$–$\frac{1}{3}$ in. broad; peduncle slender, 1–1$\frac{1}{2}$ ft. long; raceme lax, cylindrical, 6–8 in. long, 1$\frac{1}{4}$–1$\frac{1}{2}$ in. diam.; lower pedicels $\frac{1}{4}$ in. long; bracts minute, deltoid, spurred near the base; perianth greenish, $\frac{1}{2}$–$\frac{5}{8}$ in. long; segments cucullate, 2–3 times as long as the tube, reflexing; stamens nearly as long as the segments; style overtopping the anthers. *Willd. Sp. Plant.* ii. 165; *Gawl. in Bot. Mag. t.* 1444; *Roem. et Schultes, Syst. Veg.* vii. 597; *Baker in Journ. Linn. Soc.* xi. 421. *Hyacinthus ciliaris,*

Poir. Encyc. Suppl. iii. 120. *Idothea ciliaris, Kunth, Enum.* iv. 343.

CENTRAL REGION : stony hills near Graaff Reinet, 2500 ft., *Bolus*, 781 !

8. **D. purpurascens** (Jacq. fil. Eclog. 48, t. 30); bulb globose; leaves 9–10, produced after the flowers, linear, glabrous, ½ ft. long. ⅙ in. broad; peduncle a foot long, terete, purplish, glaucous; raceme oblong; flowers about 30; pedicels as long as the flowers; bracts linear-lanceolate, half as long as the pedicels, gibbous near the base; perianth purplish-green, ½ in. long; tube short; segments canaliculate; stamens nearly as long as the perianth-segments; style exserted. *Roem. et Schultes, Syst. Veg.* vii. 598; *Baker in Journ. Linn. Soc.* xi. 421. *Idothea purpurascens, Kunth, Enum.* iv. 342.

SOUTH AFRICA: without locality.

Known only from Jacquin's description and figure.

9. **D. angustifolia** (Baker); bulb globose, 3 in. diam.; leaves about 6, produced after the flowers, linear, thin, flaccid, glabrous, 2–2½ ft. long, ¼–⅓ in. broad; peduncle 2–3 ft. long; raceme very lax, a foot or more long; lower pedicels curved, 1–1½ in. long; bracts lanceolate, deciduous, ¼–⅓ in. long, protruded a little beyond the buds; perianth greenish, ½–⅝ in. long; tube campanulate, ⅛ in. long; segments channelled; stamens shorter than the perianth-segments; style reaching to the tip of the segments. *D. elata β Cooperi, Baker in Journ. Linn. Soc.* xi. 422.

CENTRAL REGION; Somerset Div.; at the foot of Bosch Berg, *Tuck in Herb. MacOwan*, 2211! Tarkastad Div.; near Tarka, *Cooper*, 387!
EASTERN REGION: Natal; valley near Colenso, 3000 ft., *Wood*, 4027!

10. **D. haworthioides** (Baker in Gard. Chron. 1875, iii. 366); rootstock epigæous, consisting of a rosette of oblong-spathulate fleshy scale-leaves; proper leaves 3–4, produced in February, lanceolate, thin, glabrous, 2–3 in. long, ¼–⅓ in. broad; peduncle produced in December, slender, terete, a foot long; raceme very lax, oblong, 2–3 in. long, an inch broad; pedicels ¼ in. long, erecto-patent; bracts minute, spurred at the base; perianth greenish-white, ⅓ in. long; segments cucullate, 3 times the length of the campanulate tube; stamens shorter than the perianth-segments.

CENTRAL REGION : stony places, Graaff Reinet, 2500–3200 ft., *Bolus*, 40!

Described from a plant that flowered at Kew, Dec., 1874.

11. **D. Burchellii** (Baker in Saund. Ref. Bot. iii. App. 2); bulb ovoid, 2–3 in. diam.; leaves 5–6, produced in April and May, lorate, glabrous, thin, flaccid, 6–10 in. long, about an inch broad; peduncle 1–1½ ft. long; raceme moderately dense, cylindrical, finally a foot long, 1–1¼ in. diam.; pedicels very short, at most $\frac{1}{12}$–⅛ in. long; bracts small, lanceolate; perianth ⅓–½ in. long, dull-red tinged with green; segments channelled, twice as long as the tube; stamens shorter than the perianth-segments; style ⅛ in. long. *Journ. Linn. Soc.* xi. 420.

COAST REGION: Uitenhage Div.; between Galgebosch and Melk River, *Burchell*, 4769! and without precise locality, *Ecklon and Zeyher*, 954!

CENTRAL REGION: Somerset Div.; stony places on the sides of Bruintjes Hoogte, 3500 ft., *MacOwan*, 1835a! 1854! Richmond Div.; northern slopes of the Sneeuw Berg Range, near Nieuw Berg, 4800 ft., *Bolus*, 2573!

12. **D. elata** (Jacq. Collect. Suppl. 38; Ic. ii. 15, t. 373); bulb globose, about 3 in. diam.; tunics brown, imbricated; leaves 5–6, produced after the flowers, lanceolate, glaucescent, glabrous, sub-erect, a foot or more long, $\frac{1}{2}$–$\frac{3}{4}$ in. broad; peduncle stiffly erect, $1\frac{1}{2}$–2 ft. long; raceme moderately dense, cylindrical, $\frac{1}{2}$–$1\frac{1}{2}$ ft. long, $1\frac{1}{2}$–2 in. diam.; lower pedicels about $\frac{1}{2}$ in. long; bracts lanceolate, $\frac{1}{3}$–$\frac{1}{2}$ in. long; perianth $\frac{1}{2}$–$\frac{5}{8}$ in. long, greenish-purple; tube campanulate, $\frac{1}{6}$ in. long; segments canaliculate, reflexing; stamens a little shorter than the perianth-segments; style reaching the tip of the perianth-segments; capsule subglobose, deeply lobed laterally, $\frac{1}{2}$–$\frac{5}{8}$ in. long. *Willd. Sp. Plant.* ii. 165; *Red. Lil. t.* 430; *Bot. Mag. t.* 822; *Roem. et Schultes, Syst. Veg.* vii. 597; *Baker in Journ. Linn. Soc.* xi. 422. *Hyacinthus elatus, Poir. Encyc. Suppl.* iii. 120. *Idothea elata, Kunth, Enum.* iv. 343. *Drimia concolor, Baker, loc. cit.*

SOUTH AFRICA: without locality, *Zeyher*, 1713!

COAST REGION: Malmesbury Div., *Bachmann*, 1033! Cape Flats, near Rosebank, below 100 ft., *Bolus*, 3922! Stockenstrom Div.; summit of Chumie Peak, *Scully*, 166!

KALAHARI REGION: Orange Free State; Caledon River, *Burke!*

EASTERN REGION: Tembuland; hillsides, Bazeia, 2000 ft., *Baur*, 418! Natal; Umsondus River, *Rehmann*, 7632! and without precise locality, *Sanderson*, 464! *Buchanan! Mrs. K. Saunders!*

13. **D. pauciflora** (Baker in Engl. Jahrb. xv. Heft 3, 6); bulb and leaves unknown; peduncle long, stiff; raceme very lax, $\frac{1}{2}$ ft. long, 8–10-flowered; lower pedicels ascending, $1\frac{1}{2}$ in. long in flower, 2–3 in. in fruit; upper ascending, much shorter; bracts very small, deciduous; perianth $\frac{5}{8}$–$\frac{3}{4}$ in. long; tube short, campanulate; lobes linear-conduplicate, with a white edge and brown keel; ovary linear-oblong, narrowed to a style $\frac{1}{4}$ in. long; capsule oblong, deeply lobed, 1–$1\frac{1}{4}$ in. long; seeds many, thin, flat, black, $\frac{1}{8}$–$\frac{1}{4}$ in. diam.

SOUTH AFRICA: without locality, *Ecklon and Zeyher*, 102!

14. **D. sphærocephala** (Baker); bulb globose, 1 in. diam.; leaves hysteranthous, not seen; peduncle terete, a foot long; flowers very numerous, aggregated into a globose umbel 1 in. diam.; pedicels $\frac{1}{8}$–$\frac{1}{4}$ in. long; bracts lanceolate, longer than the pedicels; perianth white, tinged with red, $\frac{1}{4}$ in. long; tube short, campanulate; segments deeply channelled, 2–3 times the length of the tube; stamens as long as the perianth; anthers small, oblong; ovary globose; style straight, subulate.

KALAHARI REGION: Transvaal; rocky mountain sides at the back of Concession Creek, near Barberton, 4500 ft., *Galpin*, 1020!

15. D. neriniformis (Baker); bulb ovoid, 1 in. diam.; scales large, ovate; leaves 3, not produced with the flowers, thin, ciliated with fine long spreading hairs; peduncle slender, glabrous, 12–18 in. long; flowers in a dense, globose, congested, capitate raceme 1½–2 in. diam.; pedicels ¼–⅓ in. long; bracts linear; perianth white, ⅓ in. long; tube very short; segments linear, cucullate at the tip; stamens as long as the perianth; anthers small; style ¼ in. long.

EASTERN REGION: Natal; Van Reenens Pass, in marshy ground, 5000–6000 ft., *Wood*, 4794!

16. D. capitata (Baker); bulb ovoid, under an inch in diameter; leaves linear, villose, produced after the flowers; peduncle slender, 1–1½ ft. long; flowers in a very dense, globose, congested, capitate raceme 2 in. diam.; pedicels slender, ½–¾ in. long; bracts lanceolate, with a terete basal spur, sometimes ¼ in. long; perianth reddish-green, ⅓–½ in. long; tube ⅛–⅙ in. long; segments very canaliculate and cucullate, reflexing, 2–3 times as long as the tube; stamens much shorter than the perianth-segments; style overtopping the stamens.

KALAHARI REGION: Orange Free State, *Cooper*, 885!
EASTERN REGION: Tembuland; Bazeia, 2000–2500 ft., *Baur*, 1160!

17. D.? hyacinthoides (Baker in Journ. Bot. 1874, 6); bulb and leaves not seen; peduncle 1–1½ ft. long; raceme oblong, moderately dense, 3–6 in. long, 1½–2 in. diam.; lower pedicels ½–1¼ in. long; bracts small, deltoid, furnished with a terete basal spur, which in the lowest is ⅙ in. long; perianth campanulate, ⅓ in. long, reddish-purple; segments oblong, not channelled down the face nor cucullate at the apex, twice as long as the tube; filaments lanceolate, shorter than the oblong anthers; style as long as the ovary, reaching to the tip of the perianth-segments.

COAST REGION: Albany Div.; shady valleys near Grahamstown, 2000 ft., *MacOwan*, 1465!

18. D.? macrantha (Baker in Engl. Jahrb. xv. Heft 3, 7); bulb large, scaly; leaf linear, glabrous, hysteranthous, 1½–2 ft. long; peduncle with inflorescence 4–6 ft. long; raceme lax, ½ ft. or more long; pedicels ascending or spreading, cernuous at the apex, lower 1½–2 in. long; bracts small, deciduous; perianth reddish, 1–1¼ in. long; segments oblanceolate, ⅙ in. broad, with a distinct closely 3-nerved keel; stamens nearly an inch long; filaments slightly flattened; style slender, above ½ in. long; stigma capitate. *Ornithogalum? macranthum, Baker in Journ. Linn. Soc.* xiii. 280.

COAST REGION: Uitenhage Div.; Van Stadens River, below 200 ft., *Drège*, 2204! Queenstown Div.; flats by the Zwart Kei River, 4000 ft., *Drège*, 3531! Komgha Div.; damp valleys between Komgha and the mouth of the Kei River, 1500 ft., *Flanagan*, 468!
EASTERN REGION: Griqualand East; river banks near Kokstad, *Haygarth in Herb. Wood*, 4211!

19. D.? anomala (Benth. in Gen. Plant. iii. 808); bulb globose, green, 2 in. diam.; leaves solitary, rarely 2, contemporary with the

flowers, perfectly terete, glaucescent, $1\frac{1}{2}$–2 ft long, $\frac{1}{6}$ in. diam. low
down, tapering gradually to the point; peduncle slender, $1\frac{1}{2}$ ft. long;
raceme moderately dense, cylindrical, 6–12 in. long, under an inch
in diameter; pedicels $\frac{1}{6}$–$\frac{1}{4}$ in. long, articulated at the apex; bracts
minute, ovate, lowest minutely spurred, deciduous; perianth cam-
panulate, greenish yellow, $\frac{1}{6}$–$\frac{1}{4}$ in. long; segments oblong, 3–4 times
the length of the tube; stamens half as long as the segments;
filaments linear; style as long as the ovary. *Ornithogalum? anoma-*
lum, Baker in Saund. Ref. Bot. t. 178; *Journ. Linn. Soc.* xiii. 284.
Urginea eriospermoides, Baker in Gard. Chron. 1887, ii. 126.

SOUTH AFRICA: without locality, cultivated specimens!
CENTRAL REGION: Somerset Div.; sandy places near Somerset East, 2800 ft.,
MacOwan, 1853! Graaff Reinet Div.; sides of the Oude Berg, 3750 ft., *Bolus*,
270! Murraysburg Div.; stony places near Murraysburg, 4000 ft., *Tyson*,
165!

20. **D.? Cooperi** (Benth. in Gen. Plant. iii. 808); bulb globose;
leaves 2, contemporary with the flowers, linear, glabrous, 6–8 in. long,
$\frac{1}{4}$–$\frac{1}{3}$ in. broad; peduncle 1–$1\frac{1}{2}$ ft. long; raceme lax, cylindrical,
$\frac{1}{2}$–1 ft. long, under an inch in diameter; pedicels cernuous at the
apex, $\frac{1}{6}$–$\frac{1}{4}$ in. long; bracts small, ovate, concave or obscurely spurred;
perianth campanulate, greenish-yellow, $\frac{1}{4}$ in. long; tube short, green;
segments oblong; stamens shorter than the segments; filaments
linear; style as long as the ovary. *Ornithogalum? Cooperi, Baker*
in Journ. Linn. Soc. xiii. 284.

SOUTH AFRICA: without locality, *Bowker! Mrs. Barber!* and a cultivated
specimen introduced by *Cooper!*

21. **D.? chlorantha** (Baker); bulb not seen; leaves 3–4, contemporary with the flowers, lanceolate, $1\frac{1}{2}$–2 ft. long, an inch broad,
narrowed to the acute tip and long sheath, glabrous; peduncle $2\frac{1}{2}$ ft.
long; raceme long, very lax; pedicels erecto-patent, lower $\frac{1}{2}$ in. long;
bracts long, linear from an ovate base, caducous; perianth green, $\frac{1}{4}$ in.
long; tube short; segments oblong, nearly flat; stamens nearly as
long as the perianth; anthers large.

KALAHARI REGION: Transvaal; hill-side among rocks, in damp, shady
places, near Sheba Battery, Avoca, 2000 ft., *Galpin*, 1191!

22. **D.? Bolusii** (Baker); bulb exactly like that of *D. hawor-*
thioides; leaves 5–6, produced after the flowers, lorate, obtuse, thin,
fleshy; peduncle slender, a foot long; raceme very lax, slender, $\frac{1}{2}$ ft.
long, under an inch in diameter; pedicels erecto-patent, $\frac{1}{4}$–$\frac{1}{3}$ in. long;
bracts minute, lanceolate-deltoid, spurred above the base; perianth
campanulate, $\frac{1}{6}$ in. long; tube short: segments oblanceolate-spathulate,
whitish, with a broad green keel; stamens shorter than the segments.
Ornithogalum? haworthioides, Baker in Journ. Bot. 1878, 322.

CENTRAL REGION: Graaff Reinet Div.; Cave Mountains, 2900 ft., *Bolus*,
814! Not in Kew Herbarium.

XXIX. RHADAMANTHUS, Salisb.

Perianth gamophyllous, campanulate; segments ovate, obtuse, equal, erect, about as long as the tube. *Stamens* 6, inserted below the throat of the perianth-tube, subuniseriate; filaments short, erect, flattened downwards; anthers oblong, connivent, dehiscing introrsely. *Ovary* sessile, ovoid, 3-celled; ovules many, superposed; style short, columnar; stigma capitate. *Capsule* membranous, ovoid, loculicidally 3-valved. *Seeds* discoid, winged, 6–8 in a cell; testa black; albumen firm in texture.

DISTRIB. Endemic.

Peduncle 3–6 in. long ; flowers unilateral   ...   ... (1) **convallarioides.**
Peduncle 16–18 in. long ; flowers not unilateral...   ... (2) **cyanelloides.**

1. R. convallarioides (Salisb. Gen. 37); rootstock a tunicated bulb ; leaves 9–10, produced after the flowers, linear-subulate, 2½–3 in. long; peduncle stiff, slender, naked, 3–6 in. long; raceme lax, 6–20-flowered ; pedicels ¼–⅓ in. long; bracts minute, deltoid ; perianth whitish, ¼ in. long, with 6 green-purple stripes. *Baker in Journ. Linn. Soc.* xi. 434. *Hyacinthus convallarioides, Linn. fil. Suppl.* 204; *Thunb. Prodr.* 64 ; *Fl. Cap. edit. Schult.* 326 ; *Jacq. Hort. Schoenbr.* i. 42, *t.* 81 ; *Willd. Sp. Plant.* ii. 168 ; *Roem. et Schultes, Syst. Veg.* vii. 584 ; *Kunth, Enum.* iv. 305.

SOUTH AFRICA: without locality, *Masson!*
COAST REGION : Robertson Div. ; Karoo near Montagu, *Bolus,* 7567 !
CENTRAL REGION : Karoo below the Rogge Veld, *Thunberg.*

2. R. ? cyanelloides (Baker) ; rootstock, leaf, fruit and seeds unknown ; inflorescence and perianth as in *Eriospermum ;* peduncle slender, naked, above a foot long; raceme lax, oblong, above 2 in. long; pedicels erecto-patent, ¼–⅓ in. long; bracts membranous, pale, tricuspidate, persistent, the central cusp large, linear ; perianth campanulate, ¼ in. long, cut down to the base; segments oblong-lanceolate, white, with a distinct 1-nerved brown keel ; stamens like those of *Cyanella,* as long as the perianth, with very short filaments and six equal large cylindrical anthers permanently connivent in a cone and dehiscing by apical pores ; style not protruded beyond the tips of the anthers.

COAST REGION : Komgha Div.; grassy valleys near Prospect Farm, Komgha, 2100 feet, *Flanagan,* 573 !

Most likely a new genus.

XXX. LITANTHUS, Harv.

Perianth gamophyllous, deciduous ; tube oblong; segments short, ovate, obtuse, ascending. *Stamens* 6, uniseriate, inserted below the throat of the perianth-tube; filaments very short; anthers with a produced emarginate connective, dehiscing introrsely. *Ovary* ovoid, sessile, 3-celled; ovules many, superposed ; style cylindrical; stigma capitate. *Capsule* ovoid, membranous, loculicidally 3-valved. *Seeds* compressed ; testa black, opaque.

DISTRIB. Endemic ; monotypic.

1. L. pusillus (Harv. in Hook. Journ. 1844, 315, t. 9); bulb
white, globose, $\frac{1}{4}$–$\frac{1}{2}$ in. diam.; leaves usually 2, produced after the
flowers, setaceous, finally 4–6 in. long; peduncle filiform, 1–3 in.
long, 1–2-flowered, with a pair of small connate spurred bracteoles
at the fork or just below the cernuous flower; perianth $\frac{1}{4}$ in. long,
white with a tinge of pink; segments keeled with green; capsule
erect, $\frac{1}{6}$ in. long. *Hook. fil. in Bot. Mag. t.* 5995; *Baker in Journ.
Linn. Soc.* xi. 419.

SOUTH AFRICA: without locality, *Drège*, 8514 !
COAST REGION: Uitenhage Div.; hills near the Zwartkops River, *Zeyher*,
1067 ! 4253. *Ecklon and Zeyher*, 11! British Kaffraria, *Bowker and Mrs. Barber*,
767!
EASTERN REGION : Swaziland ; rocky ridge above Komassan River, Havelock
Concession, 3600 ft., *Saltmarshe in Herb. Galpin*, 1042 !

XXXI. DIPCADI, Medic.

Perianth gamophyllous, marcescent, deciduous; tube oblong-
cylindrical; segments dimorphic; outer reflexing, convolute, often
longer than the inner; inner connivent, many-nerved on the back,
spreading only at the tip. *Stamens* 6, uniseriate, inserted at or
below the throat of the tube; filaments short, filiform; anthers
linear, dorsifixed, dehiscing introrsely. *Ovary* sessile or substipitate,
ovoid, trilocular; ovules several in a cell, superposed; style cylindrical; stigma capitate. *Capsule* subquadrate, membranous, loculicidally 3-valved. *Seeds* discoid; testa black; albumen firm in
texture; embryo cylindrical.

Rootstock a tunicated bulb; leaves basal, usually linear or subterete; peduncle
naked; flowers usually green, arranged in a lax raceme; pedicels articulated at
the apex; bracts membranous.

DISTRIB. Also Tropical Africa, Madagascar, the Mediterranean region, and
India. Species about 30.

Subgenus TRICHARIS. Inner and outer segments of the perianth about equal
in length.

Leaves glabrous :
　　Leaves two or three, linear　　...　　...　　...　　...　(1) hyacinthoides.
　　Leaves numerous, filiform :
　　　　Leaves straight :
　　　　　　Perianth $\frac{1}{4}$ in. long　...　　...　　...　　...　(2) gracillimum.
　　　　　　Perianth $\frac{3}{4}$–$\frac{7}{8}$ in. long :
　　　　　　　　Leaves glabrous ...　　...　　...　　...　(3) polyphyllum.
　　　　　　　　Leaves shortly pilose on the edge
　　　　　　　　　　and keel　　...　　...　　...　　...　(4) Marlothii.
　　　　Leaves spirally twisted　　...　　...　　...　　...　(5) spirale.
　　Leaves densely bristly on the back :
　　　　Leaves straight :
　　　　　　Bracts shorter than the lower pedicels　...　　...　(6) setosum.
　　　　　　Bracts longer than the lower pedicels　...　　...　(7) Readei.
　　　　Leaves usually spirally twisted...　　...　　..　　...　(8) ciliare.

Subgenus UROPETALUM. Outer falcate segments distinctly longer than the connivent inner ones.

Leaves linear, $\frac{1}{4}$–$\frac{1}{2}$ in. broad :
  Leaves not crisped :
    Outer perianth-segments $\frac{1}{2}$–$\frac{3}{4}$ in. long :
      Leaves rigid in texture   ...   .   (9) **rigidifolium.**
      Leaves moderately firm, spirally twisted   ...   (10) **volutum.**
      Leaves fleshy, flaccid   ...   .   (11) **umbonatum.**
    Outer perianth-segments $\frac{3}{4}$–1 in. long ...   ...   (12) **viride.**
    Outer perianth-segments 1–1$\frac{1}{2}$ in. long ...   ...   (13) **elatum.**
  Leaves crisped   ...   .   (14) **crispum.**
  Leaves lanceolate or lorate :
    Pedicels short, ascending   ...   .   (15) **bakerianum.**
    Pedicels long, spreading   ...   .   (16) **glaucum.**

1. D. hyacinthoides (Baker in Journ. Linn. Soc. xi. 398); bulb ovoid, $\frac{1}{2}$–$\frac{3}{4}$ in. diam; leaves 2–3, linear-subulate, glabrous, $\frac{1}{2}$–1 ft. long; peduncle slender, terete, $\frac{1}{2}$–1 ft. long ; raceme lax, subsecund, 4–10-flowered; lower pedicels $\frac{1}{6}$–$\frac{1}{3}$ in. long; bracts lanceolate-deltoid, acuminate, $\frac{1}{4}$–$\frac{1}{3}$ in. long; perianth green, $\frac{1}{2}$–$\frac{3}{4}$ in. long; outer segments ligulate-convolute, falcate from the base, about as long as the oblong-cylindrical tube ; inner lanceolate, nearly as long as the outer; stamens inserted below the throat of the tube; filaments flattened ; anthers linear, $\frac{1}{6}$ in. long, not reaching to the tip of the inner segments. *Uropetalum hyacinthoides, Spreng. Syst.* iv. *Cur. Post.* 135 ; *Roem. et Schultes, Syst. Veg.* vii. 618; *Kunth, Enum.* iv. 378. *Polemannia hyacinthiflora, Berg. in Linnæa,* i. 250. *Hyacinthus brevifolius, Thunb. Prodr.* 63 ; *Fl. Cap. edit. Schult.* 325. *Scilla brevifolia, Roem. et Schultes, Syst. Veg.* vii. 574. *Peribæa* (?) *brevifolia, Kunth, Enum.* iv. 294.

SOUTH AFRICA : without locality, *Masson ! Drège,* 1517! *Zeyher,* 1700 !
COAST REGION : British Kaffraria, *Cooper,* 3291 ! Mossel Bay Div. ; Little Brak River, *Burchell,* Bulb 104 !
CENTRAL REGION : Kannaland, Hantam, and Rogge Veld, *Thunberg !*
KALAHARI REGION : Transvaal ; grassy slopes of the Saddleback Range, near Barberton, 4500 ft., *Galpin,* 583 !
EASTERN REGION : Griqualand East ; meadows around Clydesdale, 2500 ft., *Tyson,* 2124 ! 3609 ! Natal ; Inanda, *Wood,* 283 !

2. D. gracillimum (Baker) ; bulb ovoid, $\frac{1}{2}$ in. diam. ; leaves 5–6, very slender, filiform, glabrous, much shorter than the peduncle; peduncle very slender, 4–6 in. long; raceme 2–5-flowered, lax, secund ; lower pedicels $\frac{1}{8}$–$\frac{1}{6}$ in. long; bracts lanceolate, acuminate, a little longer than the pedicels ; perianth green, $\frac{1}{3}$ in. long; outer segments as long as the tube, falcate from the base ; inner oblong, connivent, nearly as long as the outer ; stamens inserted at the throat of the tube, not reaching nearly to the tip of the segments.

CENTRAL REGION : Colesberg, *Shaw !*

3. D. polyphyllum (Baker); bulb globose, $\frac{3}{4}$–1 in. diam; leaves numerous, filiform, flaccid, glabrous, 6–15 in. long; peduncle slender, fragile, 1$\frac{1}{2}$–2 ft. long ; flowers 8–9 in a very lax subsecund raceme ; lower pedicels $\frac{1}{4}$–$\frac{1}{3}$ in. long ; bracts lanceolate, acuminate, deciduous, as long as the lower pedicels ; perianth bright green, $\frac{3}{4}$–$\frac{7}{8}$ in. long ;

outer segments linear-subterete, as long as the oblong-cylindrical tube; inner lanceolate, nearly as long as the outer; stamens inserted at the throat of the tube, not reaching nearly to the tip of the segments.

SOUTH AFRICA : without locality, cultivated specimen !
EASTERN REGION : Natal ; Groenberg, *Wood,* 1166 !

4. **D. Marlothii** (Engl. Jahrb. x. 3) ; bulb ovoid ; leaves linear-complicate, 8–10 in. long, shortly pilose on the edge and keel ; peduncle twice as long as the leaves ; raceme 4 in. long, 6–7-flowered ; bracts scariose, ovate-lanceolate, $\frac{1}{3}$ in. long ; pedicels spreading, as long as the bracts ; perianth green, $\frac{2}{3}$ in. long ; segments subequal, cohering in the lower third ; capsule sessile, oblong, $\frac{1}{3}$ in. diam.

KALAHARI REGION : Bechuanaland, 4000 ft., *Marloth,* 1041. Not in Kew Herbarium.

Very near D. polyphyllum.

5. **D. spirale** (Baker in Eng. Jahrb. xy. Heft 3, 7) ; bulb ovoid, $\frac{1}{2}$–$\frac{3}{4}$ in. diam. ; leaves 3–4, firm, subulate, glabrous, 2–3 in. long, spirally twisted in the upper half ; peduncle glabrous, $1\frac{1}{2}$–2 in. long ; raceme lax, an inch long ; pedicels erecto-patent, $\frac{1}{8}$–$\frac{1}{6}$ in. long ; bracts ovate-lanceolate, as long as the pedicels ; perianth greenish, $\frac{1}{2}$ in. long ; tube cylindrical, much longer than the lobes ; outer lobes linear-oblong, spreading, but little longer than the inner ; capsule globose, deeply 3-lobed.

CENTRAL REGION : Calvinia Div. ; Hantam Mountains, *Meyer !* Not in Kew Herbarium.

6. **D. setosum** (Baker in Journ. Linn. Soc. xi. 398) ; bulb $\frac{1}{2}$–$\frac{3}{4}$ in. diam. ; leaves 4–5, linear, firm in texture, $\frac{1}{6}$–$\frac{1}{4}$ in. broad, clothed mainly on the back with yellowish setæ ; peduncle 4–6 in. long ; raceme laxly 5–8-flowered ; lower pedicels $\frac{1}{3}$–$\frac{2}{3}$ in. long ; bracts deltoid, acuminate, $\frac{1}{6}$–$\frac{1}{4}$ in. long ; perianth greenish-yellow, $\frac{5}{8}$–$\frac{3}{4}$ in. long ; outer segments linear-convolute, as long as the tube ; inner as long as the outer ; anthers subsessile at the throat of the perianth-tube.

WESTERN REGION : Little Namaqualand ; Modderfontein, *Whitehead !* Not in Kew Herbarium.

7. **D. Readii** (Baker) ; allied to *D. setosum ;* habit more slender ; leaves linear, 3–4 in. long, $\frac{1}{12}$ in. broad, setose on the back ; raceme lax ; lower pedicels $\frac{1}{8}$–$\frac{1}{6}$ in. long ; bracts lanceolate, acuminate, longer than the pedicels ; perianth $\frac{3}{4}$–$\frac{7}{8}$ in. long ; outer segments subulate, falcate. *D. setosum var. Readii, Baker in Journ. Linn. Soc.* xi. 398.

EASTERN REGION : Natal ; Fuller's Farm, *Reade,* 94 ! Not in Kew Herbarium.

8. **D. ciliare** (Baker in Journ. Linn. Soc. xi. 398) ; bulb ovoid, $\frac{1}{2}$–$\frac{3}{4}$ in. diam. ; leaves 4–8, linear, usually conspicuously twisted spirally, 3–4 in. long, ciliated on the back and edge, especially

towards the base; peduncle slender, fragile, 4–8 in. long; flowers
4–8 in a lax secund raceme; lower pedicels $\frac{1}{8}$–$\frac{1}{4}$ in. long; bracts
lanceolate-deltoid, cuspidate, $\frac{1}{4}$–$\frac{1}{3}$ in. long; perianth green, $\frac{1}{2}$–1 in.
long; outer segments linear-subterete, about as long as the tube;
inner lanceolate, nearly as long as the outer; stamens inserted below
the throat of the tube; filaments and linear anthers each $\frac{1}{8}$ in. long;
capsule $\frac{1}{2}$ in. diam.; seeds about 12 in a cell. *Uropetalum ciliare,*
Ecklon and Zeyher ex Harv. Thes. Cap. ii. 45, *t. 170.*

COAST REGION: Lion Mountain, near Cape Town, 250 ft., *Bolus*, 519!
Uitenhage Div. ; *Ecklon and Zeyher*, 48!
CENTRAL REGION: near Somerset East, 2000–3000 ft., *MacOwan*, 1666b!
Bowker !   Cradock Div., *Cooper*, 493!
KALAHARI REGION : Transvaal; Apies River, *Burke !*
EASTERN REGION : Tembuland ; Cenduli, near Bazeia, 2000 ft., *Baur*, 805 !

9. D. rigidifolium (Baker in Journ. Linn. Soc. xi. 399); bulb
ovoid, $\frac{1}{2}$ in. diam.; leaves 2–3, linear, erect, glabrous, firm in texture,
persistent, 6–8 in. long, $\frac{1}{3}$–$\frac{1}{2}$ in. broad low down; ribs close, distinct;
margin thickened; peduncle less than twice as long as the leaves;
raceme laxly 5–6-flowered; lower pedicels $\frac{1}{8}$–$\frac{1}{4}$ in. long; bracts
lanceolate, acuminate, $\frac{1}{3}$–$\frac{1}{2}$ in. long; perianth green, $\frac{5}{8}$ in. long; outer
segments linear-convolute, as long as the tube; inner lingulate,
connivent, $\frac{1}{12}$ in. shorter than the outer; stamens inserted at the
throat of the tube.

KALAHARI REGION : Transvaal; Apies River, *Burke !*

10. D. volutum (Baker) ; bulb ovoid-conical, 1 in. diam.; leaves
3, linear, 3–6 in. long, moderately firm in texture, pilose, spirally
twisted in the upper half; peduncle 6–8 in. long; flowers 8–12 in a
lax raceme 3–4 in. long; pedicels arcuate, $\frac{1}{8}$–$\frac{1}{4}$ in. long; bracts ovate-
cuspidate, lower $\frac{1}{6}$ in. long; perianth greenish, $\frac{3}{4}$ in. long; outer
segments $\frac{1}{4}$ in. long, longer than the inner, linear, reflexing.

WESTERN REGION: Little Namaqualand, *Scully*, 214!

11. D. umbonatum (Baker in Journ. Linn. Soc. xi. 400); bulb
globose, $\frac{3}{4}$ in. diam.; leaves 2–3, fleshy, erect, flaccid, linear, glabrous,
$\frac{1}{2}$–1 ft. long, $\frac{1}{8}$–$\frac{1}{6}$ in. broad low down; peduncle slender, 1–1$\frac{1}{2}$ ft.
long; raceme very lax, 6–12-flowered; lower pedicels $\frac{1}{4}$–$\frac{1}{3}$ in. long;
bracts small, lower ovate-lanceolate, $\frac{1}{4}$ in. long; perianth green,
$\frac{1}{2}$–$\frac{3}{4}$ in. long; outer segments linear-convolute, as long as the cylindrical tube, $\frac{1}{8}$–$\frac{1}{6}$ in. longer than the short, oblong, connivent, green-
white inner ones; stamens inserted at the throat of the tube; fila-
ments very short; capsule subquadrate, obscurely stipitate; seeds
12–15 in a cell. *Uropetalum umbonatum, Baker in Saund. Ref. Bot.*
t. 17.

KALAHARI REGION : Transvaal; lower hill slopes, 2900 ft., Barberton, *Galpin,*
763! hills near Pretoria, *Elliot*, 1285! 1689!
EASTERN REGION : Natal; at the foot of Table Mountain, 1000–1500 ft.,
Krauss, 437! Clairmont, *Wood*, 1763! and without precise locality, *Sanderson,*
509!

12. D. viride (Moench, Meth. Suppl. 267); bulb globose or ovoid, $\frac{3}{4}$–1 in. diam.; leaves 3–6, linear, glabrous, fleshy, flaccid, $\frac{1}{2}$–1 ft. long, $\frac{1}{8}$–$\frac{1}{4}$ in. broad; peduncle stiff, slender, $\frac{1}{2}$–1$\frac{1}{2}$ ft. long; raceme very lax, 6–15-flowered; lower pedicels $\frac{1}{4}$–$\frac{1}{3}$ in. long; bracts large, ovate-lanceolate, with a long cusp; perianth bright green, 1–1$\frac{1}{2}$ in. long; inner segments and tube together $\frac{1}{2}$–$\frac{5}{8}$ in. long; outer segments filiform, reflexing, $\frac{1}{2}$–1 in. long; stamens inserted at the throat of the tube; filaments shorter than the anthers; capsule obscurely stipitate, $\frac{1}{2}$–$\frac{5}{8}$ in. long and broad; seeds about 15 in a cell. *Baker in Journ. Linn. Soc.* xi. 401. *Hyacinthus viridis, Linn. Sp. Plant. edit.* 2, 454; *Jacq Ic. t.* 66; *Red. Lil. t.* 203. *Lachenalia viridis, Thunb. Prodr.* 64; *Ait. Hort. Kew.* i. 462; *edit.* 2, ii. 285; *Willd. Sp. Plant.* ii. 174. *Zuccagnia viridis, Thunb. Fl. Cap. edit. Schult.* 328. *Phormium viride, Thunb. Nov. Gen.* 127 *in syn. Uropetalum viride, Gawl. in Bot. Reg. sub t.* 156; *Roem. et Schultes, Syst. Veg.* vii. 620; *Kunth, Enum.* iv. 379. *Dipcadi filamentosum, Medic. in Ust. Ann.* ii. 13.

COAST REGION: Riversdale Div.; near Kafferkuils River, *Thunberg.* Uitenhage Div., *Zeyher! Ecklon and Zeyher,* 49!

CENTRAL REGION: Somerset Div.; at the foot of the Bosch Berg, 2500 ft., *MacOwan,* 1924! Graaff Reinet Div.; Karoo near Graaff Reinet, 2600 ft., *Bolus,* 739! Albert Div., *Cooper,* 3289!

KALAHARI REGION: Griqualand West; Asbestos Mountains, *Burchell,* 2048! Orange Free State; Caledon River, *Burke!* and without precise locality, *Cooper,* 888! Transvaal; damp grassy hollows near Barberton, 2800 ft., *Galpin,* 1154!

EASTERN REGION: Pondoland, *Bachmann,* 282! 283! Natal; Inanda, *Wood,* 266! 384! 1423! Mooi River district, 4000–5000 ft, *Wood,* 5623! Nottingham, *Buchanan,* 144!

13. D. elatum (Baker); bulb not seen; leaves 4, firm, erect, linear, glabrous, a foot long, $\frac{1}{6}$ in. broad; peduncle slender, terete, 2 ft. long; raceme lax, a foot long; pedicels erecto-patent, lower $\frac{1}{2}$ in. long; bracts lanceolate-acuminate, lower $\frac{1}{2}$ in. long; perianth greenish; tube subcampanulate, $\frac{1}{5}$ in. long; inner segments oblong, as long as the tube; outer setaceous, 1–1$\frac{1}{2}$ in. long, longer in the upper flowers; stamens rather shorter than the inner segments; capsule $\frac{1}{2}$ in. long and broad.

EASTERN REGION: Griqualand East; moist places around Clydesdale, 2500 ft., *Tyson,* 2107!

14. D. crispum (Baker in Journ. Linn. Soc. xi. 399); bulb globose, 1 in. diam.; leaves 3–5, linear, membranous, flaccid, glabrous, 4–5 in. long and crisped in the wild plant, above a foot long, $\frac{1}{2}$ in. broad and plane when cultivated; peduncle slender, $\frac{1}{2}$–1 ft. long; raceme very lax, subsecund, 10–15-flowered; lower pedicels $\frac{1}{4}$–$\frac{1}{3}$ in. long; lower bracts lanceolate, shorter than the pedicels; upper small, deltoid; perianth green, $\frac{3}{4}$–$\frac{7}{8}$ in. long; tube constricted above the ovary; outer segments linear-convolute, $\frac{1}{4}$–$\frac{1}{3}$ in. long; inner connivent, $\frac{1}{6}$ in. shorter; stamens inserted at the throat of the tube; filaments very short. *Uropetalum crispum, Burch. ex Gawl. in Bot. Reg. sub t.* 156 (*name only*).

CENTRAL REGION: Hopetown Div.; near the Orange River between Puff-adder Halt and Bare Station, *Burchell*, 2682! and by the Orange River near Vissers Drift, *Burchell*, Bulb 34!

15. D. bakerianum (Bolus in Journ. Linn. Soc. xviii. 394); bulb globose, 1 in. diam.; leaves 4–6, lanceolate, acute, thin, glabrous, 6–9 in. long, about an inch broad low down; peduncle 2–3 in. long; raceme very lax, 4–6 in. long; pedicels very short, ascending, lower $\frac{1}{8}$–$\frac{1}{6}$ in. long; bracts ovate-lanceolate, cuspidate, much longer than the pedicels; perianth green, $\frac{3}{4}$ in. long; outer segments ligulate-convolute, as long as the tube; inner oblong, connivent, $\frac{1}{12}$–$\frac{1}{8}$ in. shorter; stamens inserted at the throat of the tube; filaments a little shorter than the anthers; capsule sessile, $\frac{5}{8}$–$\frac{3}{4}$ in. long and broad; seeds 15–20 in a cell.

CENTRAL REGION: Murraysburg Div.; Karoo-like plains near Murraysburg, 3500 ft., *Bolus*, 2059!

16. D. glaucum (Baker in Journ. Linn. Soc. xi. 401); bulb large, globose; leaves 5–6, thin, glabrous, lorate-lanceolate, glaucous, a foot long, $2\frac{1}{2}$–3 in. broad; peduncle stout, terete, 2 ft. long; raceme lax, 15–18 in. long, 4 in. broad when expanded, 30–40-flowered; pedicels patent, $1\frac{1}{2}$–2 in. long; bracts small, lanceolate, membranous; perianth bright green, $1\frac{1}{4}$ in. long; tube cylindrical, $\frac{1}{2}$ in. long; outer segments linear, longer than the tube; inner connivent, oblong-lanceolate, shorter than the tube; stamens inserted near the throat of the tube; filaments about as long as the anthers; capsule obscurely stipitate, $\frac{3}{4}$ in. long and broad; seeds 15–20 in a cell. *Uropetalum glaucum*, *Burch. in Bot. Reg. t.* 156; *Roem. et Schultes, Syst. Veg.* vii. 619; *Kunth, Enum.* iv. 379.

KALAHARI REGION: Griqualand West; between the Asbestos Mountains and Witte Water, *Burchell*, 2066!

XXXII. GALTONIA, Dcne.

Perianth gamophyllous, marcescent; tube oblong or clavate; outer segments oblong; inner obovate. *Stamens* 6, inserted below the throat of the perianth-tube; filaments filiform; anthers linear-oblong, dorsifixed, versatile, dehiscing introrsely. *Ovary* sessile, oblong-trigonous; ovules many, superposed; style cylindrical; stigma capitate. *Capsule* oblong, membranous, loculicidally 3-valved. *Seeds* very numerous, angled by pressure; testa black, membranous; albumen fleshy; embryo cylindrical, as long as the albumen.

Rootstock a tunicated bulb; leaves large, lanceolate, fleshy; peduncle naked; raceme lax; pedicels articulated at the apex; bracts large, membranous; flowers large, white, or tinged with green.

DISTRIB. Endemic.

Perianth-tube oblong:
    Stamens inserted high up in the tube ... ... (1) **candicans**.
    Stamens inserted low down in the tube ... ... (2) **princeps**.
Perianth-tube clavate ... ... (3) **clavata**.

1. G. candicans (Dcne. in Flore des Serres, xxiii. 33); bulb large, globose; leaves 4–6, lanceolate, fleshy, flaccid, glaucescent, 2–3 ft. long, 1–2 in. broad; peduncle stout, terete, rather longer than the leaves; raceme lax, $\frac{1}{2}$–1 ft. long; flower-pedicels cernuous at the apex, 1–2 in. long; bracts large, lanceolate; perianth pure white, 1–1$\frac{1}{2}$ in. long; tube oblong; segments longer than the tube; stamens about $\frac{1}{2}$ in. long; capsule as long as the perianth, erect. *Rev. Hort.* 1882, 32. *Hyacinthus candicans, Baker in Saund. Ref. Bot. t.* 174; *Journ. Linn. Soc.* xi. 425; *Flore des Serres, t.* 2172-3; *Gard. Chron.* 1871, 380; 1872, 1099; *Berl. Monatschr.* 1878, *t.* 2; *Rev. Hort.* 1880, 469, *fig.* 97.

CENTRAL REGION: Aliwal North Div.; on the Witte Bergen, 7000–8000 ft., *Drège,* 3529!

KALAHARI REGION: Orange Free State; near Nelsons Kop, *Cooper,* 3285!

EASTERN REGION: Natal; on the Drakensberg, 6000–7000 ft., *Evans,* 360! sources of the Tugela River, *M'Ken,* 5! *Allison!* Olivers Hoek Pass, *Wood,* 3498!

2. G. princeps (Dcne. loc. cit.); habit and leaves of *G. candicans;* raceme shorter and flowers fewer; perianth 1–1$\frac{1}{4}$ in. long, tinged with green on the outside; segments a little shorter than, or at most as long as, the oblong perianth-tube; stamens inserted below the middle of the perianth-tube; filaments flattened and dilated at the base; anthers smaller than in *G. candicans. Hyacinthus princeps, Baker in Saund. Ref. Bot. t.* 175; *Journ. Linn. Soc.* xi. 426.

EASTERN REGION: Transkei: Tsomo River, *Mrs. Barber,* 884! Tembuland; Bazeia Mountain, 2000–3000 ft., *Baur,* 23! Pondoland; near Fort Donald, 3500 ft., *Tyson,* 1665! Natal; near Pietermaritzburg, 2400–2800 ft., *Adlam,* 134! near Howick, *Wood,* 3520! Zululand, *Gerrard,* 2149!

3. G. clavata (Baker in Bot. Mag. t. 6885); bulb ovoid, 3–4 in. diam.; tunics brown, splitting into fibres at the top; leaves 6–8, lanceolate, acute, flaccid, glaucous green, 2–2$\frac{1}{2}$ ft. long, 1–2 in. broad, with a distinct whitish margin; peduncle stiff, terete, about 2 ft. long; raceme lax, $\frac{1}{2}$ ft. long; flower-pedicels 1–2 in. long, cernuous at the apex; bracts lanceolate; perianth-tube clavate, an inch long; segments oblong, obtuse, half as long as the tube, greenish in the central third outside; stamens inserted at the throat of the perianth-tube; filaments flattened, half as long as the segments; anthers linear-oblong.

SOUTH AFRICA: without locality.

Described from a cultivated plant in 1881, received from Mr. C. Ayres of Cape Town.

XXXIII. ALBUCA, Linn.

Perianth polyphyllous, persistent; 3 outer segments oblong, more or less spreading; 3 inner shorter, permanently connivent, cucullate, and furnished with a large gland at the apex. *Stamens* 6, hypogynous, erect, all fertile, or the anthers of the 3 outer, small and imperfect or entirely absent; filaments often winged and dilated at the base; anthers oblong, versatile, dehiscing introrsely. *Ovary* sessile, oblong; ovules many, superposed; style usually obconico-prismatic, rarely cylindrical; stigma 3-lobed. *Capsule* ovoid-triquetrous, membranous, loculicidally 3-valved. *Seeds* compressed, sometimes almost winged; testa black, membranous; albumen fleshy.

Rootstock a tunicated bulb; leaves all radical, flat, tapering to a long point or terete; inflorescence a simple raceme; bracts acuminate, persistent, not spurred; flowers large, yellow or white, broadly keeled with green or reddish-brown, rarely entirely green.

HAB. Also Tropical Africa and Arabia.

Subgenus EUALBUCA. Style prismatic, narrowed to the base; three outer stamens without anthers, or with the anthers rudimentary or barren.

Flowers white, banded with green :
  Tall, glabrous   ...   .   (1) altissima.
  Dwarf, glandular-pubescent   ...   .   (2) glandulosa.
Flowers pale yellow, banded with green :
  Leaves linear :
    Tunics of the bulb not splitting into fibres
      at the tip :
        Peduncle 2–3 ft. long   ...   ...   (3) major.
        Peduncle about a foot long :
          Bracts 1–1½ in. long ...   .   (4) flaccida.
          Bracts ¼–½ in. long   ...   .   (5) minor.
    Tunics of the bulb splitting into long fibres
      at the tip   ...   .   (6) Cooperi.
  Leaves subterete :
    Perianth ½ in. long :
      Leaves glabrous   ...   .   (7) Massoni.
      Leaves pubescent :
        Leaves 7–8   ...   .   (8) trichophylla.
        Leaves 4   ...   .   (9) minima.
    Perianth ¾ in. long :
      Tunics of the bulb not splitting into fibres :
        Mature leaves glabrous   ...   ...   (10) juncifolia.
        Mature leaves pubescent   ...   ...   (11) namaquensis.
      Tunics of the bulb splitting into fibres :
        Pedicels short, cernuous   ...   ...   (12) fibrosa.
        Pedicels long, erect :
          Lower pedicels 1–1½ in. long   ...   (13) exuviata.
          Lower pedicels 3–4 in. long   ...   (14) corymbosa.
  Flowers almost entirely green   ...   .   (15) viridiflora.

Subgenus FALCONERA. Style prismatic, narrowed to the base; stamens all bearing anthers.

Flowers white, banded with green :
  Leaves lanceolate or lorate :
    Perianth 1½ in. long   ...   .   (16) Nelsoni.
    Perianth an inch long   ...   .   (17) crinifolia.
  Leaves linear, acuminate :
    Pedicels permanently erect :
      Leaves ½–¾ in. broad   ...   .   (18) fastigiata.
      Leaves ¼–½ in. broad   ...   .   (19) caudata.
      Leaves 1/12–⅛ in. broad   ...   .   (20) pachychlamys.
    Pedicels cernuous at the tip   ...   .   (21) Baurii.

Leaves subterete :
  Leaves glabrous :
    Leaves usually 3 ... ... (22) **humilis.**
    Leaves 12–15 ... ... (23) **polyphylla.**
  Leaves viscous ... ... (24) **viscosa.**
Flowers yellow, banded with green :
  Leaves linear or lanceolate :
    Pedicels short, cernuous at the tip ... ... (25) **fragrans.**
    Pedicels erect:
      Tunics not split into fibres at the tip ꞉
        Leaves straight ; pedicels long ... (26) **aurea.**
        Leaves twisted ; pedicels short ... (27) **tortuosa.**
      Tunics split into fibres at the tip .. ... (28) **setosa.**
  Leaves subterete :
    Leaves circinate, densely glandular-pubescent.... (29) **spiralis.**
    Leaves straight, glabrous when mature:
      Leaves very slender... ... ... (30) **tenuifolia.**
      Leaves more robust :
        Outer tunics truncate at the apex ... (31) **Macowani.**
        Outer tunics produced and barred across (32) **collina.**
Subgenus LEPTOSTYLA. Style cylindrical.
The only species ... ... (33) **Shawii.**
Imperfectly known ... ... (34) **parviflora.**

1. A. altissima (Dryand. in Vet. Acad. Nya Handl. Stockh. 1784, 292); bulb depresso-globose, $1\frac{1}{2}$–2 in. diam. ; tunics membranous ; leaves 5–6, lanceolate, $1\frac{1}{2}$–2 ft. long, $1\frac{1}{2}$–2 in. broad low down, glaucous, tapering gradually to a long point, with incurved edges ; peduncle robust, 2–3 ft. long ; raceme lax, 1–$1\frac{1}{2}$ ft. long; pedicels ascending, cernuous at the tip when in flower, lower 2–3 in. long; bracts lanceolate, lower an inch long; flowers inodorous ; perianth $\frac{3}{4}$–1 in. long, white, broadly keeled with green ; outer stamens sterile ; style prismatic, as long as or shorter than the ovary; stigma tricuspidate. *Jacq. Collect.* ii. 264 ; *Ic.* i. 7, *t.* 63 ꞉ *Willd. Sp. Plant.* ii. 98 ; *Roem. et Schultes, Syst. Veg.* vii. 494 ; *Kunth. Enum.* iv. 373 ; *Baker in Journ. Linn. Soc.* xiii. 286. *A. alba, Lam. En-yc.* i. 76. *A. cornuta, Red. Lil. t.* 70 ; *Kunth, Enum. loc. cit.*

SOUTH AFRICA: without locality, cultivated specimens !
COAST REGION : Mossell Bay Div. : Little Brak River, *Burchell*, 6197/4 !
CENTRAL REGION : Somerset Div. ; Brak River, between Somerset East and Bruintjes Hoogte, *MacOwan*, 2130 !

2. A. glandulosa (Baker in Gard. Chron. 1875, iii. 814); bulb globose, 1 in. diam. ; tunics membranous, not splitting into fibres at the top ; leaves 2–3, $\frac{1}{2}$ ft. long, linear at the base, $\frac{1}{4}$–$\frac{1}{3}$ in. broad, tapering to a semiterete point, rounded on the back, bright green, finely glandular ; peduncle a foot long, terete, densely glandular ; raceme corymbose, 3-flowered in the only specimen seen ; pedicels erect, $\frac{1}{2}$–$\frac{3}{4}$ in. long ; bracts nearly or quite as long as the pedicels ; flowers very fragrant, permanently erect ; perianth $\frac{3}{4}$ in. long, white, with broad glandular bands of green ; outer stamens sterile ; style prismatic, as long as the ampullæform ovary.

SOUTH AFRICA : without locality. No specimen in Kew Herbarium.

Described from a plant sent by Prof. MacOwan that flowered at Kew in April, 1875,

3. A. major (Linn. Sp. Plant. edit. 2, 438); bulb globose, 1½–2 in.
diam., copiously soboliferous; tunics membranous; leaves 6–10,
linear, glabrous, 1–1½ ft. long, ¾–1 in. broad near the base, tapering
gradually to a long point; peduncle 2–3 ft. long, moderately stout;
raceme very lax, ½–1 ft. long; pedicels ascending, cernuous at the tip
in the flowering stage, lower 2–3 in. long; bracts lanceolate-acuminate, 1–1½ in. long; flowers inodorous; perianth ¾–1 in. long, pale
yellow, broadly banded with green; outer stamens sterile; style
prismatic, ⅓ in. long; apex conic, subentire. *Dryand. in Vet. Acad.*
Nya Handl. Stockh. 1784, 293; *Jacq. Collect. Suppl.* 96; *Ic.* ii. 21,
t. 443; *Thunb. Prodr.* 65; *Fl. Cap. edit. Schult.* 330; *Willd. Sp.*
Plant. ii. 99; *Red. Lil. t.* 69; *Bot. Mag. t.* 804; *Lodd. Bot. Cab.*
t. 1191; *Roem. et Schultes, Syst. Veg.* vii. 495, 1696; *Kunth, Enum.*
iv. 374; *Baker in Journ. Linn. Soc.* xiii. 286. *A. lutea, Lam.*
Encyc. i. 76. *Ornithogalum canadense, Linn. Sp. Plant. edit.* i. 308
(*Moris. Hist. Univ. Sect.* iv. *t.* 24, *fig.* 7).

SOUTH AFRICA: without locality, *Zeyher,* 4190! and cultivated specimens!
COAST REGION: Cape Div.; Kalk Bay, *Pappe!* at the foot of Table Mountain, *MacOwan, Herb. Aust. Afr.,* 1561!

4. A. flaccida (Jacq. Collect. iv. 201; Ic. ii. 21, t. 444); bulb
small, ovoid; tunics membranous, not splitting into fibres at the tip;
leaves about 4, linear, glabrous, erect, flaccid, 1–1½ ft. long, ⅓–½ in.
broad low down; peduncle slender, 1–1½ ft. long; raceme lax, many-
flowered, ½ ft. long; pedicels spreading, cernuous at the tip; lower
1½–2 in. long; bracts lanceolate-acuminate, 1–1½ in. long; flowers
inodorous; perianth ¾–1 in. long, pale yellow, broadly banded with
green; outer stamens sterile; style prismatic, longer than the ovary;
stigma entire. *Willd. Sp. Plant.* ii. 99; *Roem. et Schultes, Syst.*
Veg. vii. 496; *Kunth, Enum.* iv. 374; *Baker in Journ. Linn.*
Soc. xiii. 287, *ex parte, non Ref. Bot. t.* 334.

SOUTH AFRICA: without locality.

Known to me only from Jacquin's figure and description.

5. A. minor (Linn. Sp. Plant. edit. 2, 438); bulb ovoid, 1 in.
diam.; tunics membranous; leaves 5–6, linear, glabrous, about a foot
long, ½ in. broad low down, tapering gradually into a long subterete
point; peduncle slender, 1–1½ ft. long; raceme very lax, 6–12-
flowered; pedicels ascending, cernuous at the tip in the flowering
stage, lower sometimes 2–3 in. long; bracts ovate-lanceolate,
acuminate, ⅓–½ in. long; flowers inodorous; perianth ¾ in. long,
pale yellow, broadly banded with green; outer stamens sterile; style
prismatic, as long as the ovary, not tricuspidate at the apex. *Dryand.*
in Vet. Acad. Nya Handl. Stockh. 1784, 294; *Thunb. Prodr.* 65;
Fl. Cap. edit. Schult. 330; *Willd. Sp. Plant.* ii. 100; *Gawl. in*
Bot. Mag. t. 720; *Red. Lil. t.* 21; *Roem. et Schultes, Syst. Veg.* vii.
495; *Baker in Saund. Ref. Bot. t.* 239; *Journ. Linn. Soc.* xiii.
287. *A. lutea* β, *Lam. Encyc.* i. 76. *A. coarctata, Dryand. in*
Vet. Acad. Nya Handl. Stockh. 1784, 295; *Kunth, Enum.* iv. 375.

Ornithogalum canadense β, *Linn. Sp. Plant. edit.* i. 308 (*Herm. Parad. p.* 209).

COAST REGION: Cape Div.; Table Mountain, *MacGillivray*, 477!
CENTRAL REGION: Somerset Div.; near Somerset East, *MacOwan*, 1851!

6. A. Cooperi (Baker in Journ. Bot. 1874, 366); bulb ovoid,

¾–1 in. diam.; tunics splitting into long fine fibres at the tip; leaves about 4 to a bulb, linear, acuminate, glabrous, ½–1 ft. long, ¼ in. broad low down; peduncle slender, ½–1 ft. long; flowers inodorous, 4–8 in a lax raceme; flower-pedicels cernuous at the apex, lower 1–1½ in. long; bracts ovate or ovate-lanceolate, ½–¾ in. long; perianth ¾–1 in. long, pale yellow, broadly banded with green turning to red-brown; outer stamens sterile; style prismatic, rather longer than the ovary. *A. flaccida, Baker in Ref. Bot. t.* 334, *non Jacq.*

SOUTH AFRICA: without locality, *Forster! Harvey*, 812! *Zeyher*, 1714!
COAST REGION: Cape Div.; sides of Table Mountain, *Pappe!*

7. A. Massoni (Baker in Journ. Bot. 1886, 336); bulb small;

leaves 6–7, terete, glabrous, ½ ft. long, under a line in diameter; peduncle slender, terete, about as long as the leaves; raceme rhomboid, 10–12-flowered, about 2 in. long and broad; pedicels erecto-patent, lower an inch long; bracts lanceolate, lower ½ in. long; perianth greenish, ½ in. long; inner segments rather shorter than the outer.

COAST REGION: Clanwilliam Div.; near Oliphants River, *Masson!*

Described from a drawing at the British Museum. It has no dissections, so that the character of the stamens is not shown.

8. A. trichophylla (Baker in Gard. Chron. 1889, vi. 94); bulb

ovoid, ½–⅝ in. diam.; outer tunics membranous, not splitting into fibres at the tip; leaves 7–8, erect, subterete, papillate, deeply channelled down the face, 9–10 in. long, $\frac{1}{16}$ in. diam.; peduncle slender, terete, flexuose, as long as the leaves, papillate; flowers few, arranged in a lax raceme ½ ft. long; pedicels ascending, papillate, the lower an inch long; bracts small, lanceolate from a broad base; perianth bright yellow, ½ in. long; outer segments oblong, faintly green on the outside; stamens nearly as long as the segments; outer anthers rudimentary; style prismatic, as long as the ovary. *A. Elliotii, Baker in Journ. of Bot.* 1891, 71.

KALAHARI REGION: Orange Free State, *Cooper*, 887! Transvaal; crevices of sandstone rocks near Lake Chrissie, *Elliot*, 1597!
EASTERN REGION: Natal; Drakensberg Range, *Wood*, 3445! *Evans*, 375!

Described from a cultivated plant sent from Natal by Mr. Adlam to Cambridge Botanic Garden.

9. A. minima (Baker); bulb subglobose, ⅓ in. diam.; outer tunics

membranous, brown; old leaves not splitting into fibres; produced leaves 4, erect, terete, very slender, finely pubescent, 3–6 in. long; peduncle very slender, pubescent, 4 in. long; flowers 4 in a lax raceme; pedicels and rachis pubescent; lower pedicel spreading, an

inch long; perianth bright yellow, $\frac{1}{2}$ in. long; segments oblong, keeled
with green; stamens rather shorter than the perianth; alternate
anthers rudimentary; style $\frac{1}{4}$ in. long.

KALAHARI REGION: Orange Free State; stony, grassy summit of Quaqua
Mountain, Witzies Hoek, 7000 ft., *Thode*, 61!

Not in Kew Herbarium.

10. A. juncifolia (Baker in Gard. Chron. 1876, v. 534); bulb
globose, an inch in diameter; tunics membranous, not splitting into
fibres at the top; leaves 6–10, terete, erect, bright green, 6–15 in.
long, deeply channelled down the face, glabrous when mature,
obscurely pubescent in an early stage; peduncle slender, terete,
glaucous, about a foot long; raceme lax, few-flowered; pedicels
cernuous at the tip, lower 1–1$\frac{1}{2}$ in. long; bracts ovate-lanceolate,
$\frac{1}{3}$–$\frac{1}{2}$ in. long; flowers inodorous; perianth $\frac{3}{4}$ in. long, pale yellow,
banded with green; outer stamens sterile; style prismatic, as long
as the ovary. *Bot. Mag. t.* 6395.

COAST REGION: Swellendam Div.; on mountain ridges along the lower part
of the Zonder Einde River, *Zeyher*, 4196!

Described from a living plant from Mr. Hutton that flowered at Kew in March,
1876.

11. A. namaquensis (Baker); bulb small; rudimentary leaves
many, membranous; leaves 10–12, subterete, 4–6 in. long, densely
pubescent, spirally twisted at the tip; peduncle 6–9 in. long; raceme
lax, 4–8 in. long; pedicels erecto-patent, lower 2 in. long; bracts
oblong-cuspidate, lower an inch long; perianth $\frac{5}{8}$–$\frac{7}{8}$ in. long, probably
yellow; outer segments brown in the central half; outer anthers
rudimentary; filaments flattened; style prismatic, as long as the ovary.

WESTERN REGION: Little Namaqualand, *Scully*, 109!

12. A. fibrosa (Baker in Gard. Chron. 1874, ii. 386); bulb small,
ovoid; outer tunics crowned with dense persistent bristles 1$\frac{1}{2}$–2 in.
long; leaves 2, terete, bright green, glabrous, a foot or more long,
$\frac{1}{8}$ in. thick at the base, deeply channelled down the face; peduncle
slender, flexuose, terete, glaucous, a foot long; raceme laxly 3–4-
flowered; pedicels cernuous, $\frac{1}{8}$–$\frac{1}{3}$ in. long; bracts lanceolate, $\frac{1}{4}$–$\frac{1}{3}$ in.
long; perianth under an inch long, pale yellow, broadly keeled with
green; outer stamens sterile; style prismatic, rather longer than the
oblong ovary.

SOUTH AFRICA: without locality. No specimen in Kew Herbarium.

Described from a plant that flowered at Kew in August, 1874, sent by Mr.
Bennett.

13. A. exuviata (Baker); bulb ovoid, with a long neck, firm in
texture, $\frac{1}{2}$–$\frac{3}{4}$ in. diam.; outer tunics splitting into long, fine, persistent
fibres connected by transverse bars; leaves 2–4, subterete, slender,
glabrous, $\frac{1}{2}$ ft. long; peduncle slender, terete, 4–6 in. long; racemes
3–4-flowered, corymbose; pedicels erecto-patent, lower 1–1$\frac{1}{2}$ in.
long; bracts lanceolate, $\frac{1}{2}$–$\frac{3}{4}$ in. long; perianth $\frac{3}{4}$ in. long, yellow,

broadly banded with green ; outer stamens sterile ; style prismatic, as long as the ovary.

CENTRAL REGION: stony places near Somerset East, 2500 ft., *MacOwan*, 1830 !

14. **A. corymbosa** (Baker in Gard. Chron. 1886, xxvi. 38) ; bulb globose, above 1 in. diam., neck very short ; outer tunics splitting into fibres ; leaves 6–8, glabrous, terete, convex on both back and face, 1–2 ft. long, ⅛ in. broad low down, tapering to the apex ; peduncle terete, ½ ft. long ; flowers 5–10 in a lax corymb, scented, erect ; lower pedicels 3–4 in. long ; bracts ovate-lanceolate or ovate ; perianth an inch long, yellow, banded with green ; inner stamens sterile ; style prismatic, as long as the ovary.

COAST REGION : Port Elizabeth, *Wilson!*

Described from a living plant sent by Dr. John Wilson, of St. Andrew's, in July, 1886. See Dr. Wilson's paper on the fertilization of this plant in *Trans. Bot. Soc. Edinb.* xvi. 365.

15. **A. viridiflora** (Jacq. Collect. Suppl. 98; Ic. ii. 21, t. 446) ; bulb globose, 1–1½ in. diam. ; tunics not splitting into fibres at the top ; leaves 6–9, linear-subulate, channelled down the face, a foot long, densely glandular-pubescent ; peduncle a foot long, glandular towards the base ; raceme laxly 6–10-flowered ; pedicels cernuous at the apex, the lower 1½–2 in. long ; bracts lanceolate, ¾–1 in. long ; flowers inodorous ; perianth ¾–1 in. long ; the segments green, with a pale yellow margin ; outer stamens sterile ; style prismatic, as long as the ovary. *Willd. Sp. Plant.* ii. 100 ; *Bot. Mag. t.* 1656 ; *Roem. et Schultes, Syst. Veg.* vii. 496 ; *Kunth, Enum.* iv. 374; *Baker in Journ. Linn. Soc.* xiii. 287.

SOUTH AFRICA : without locality.

Known to me only from Jacquin's figure and description.

16. **A. Nelsoni** (N. E. Brown in Gard. Chron. 1880, xiv. 198, t. 41) ; bulb globose, as large as an apple ; tunics thick, green, not split into fibres at the top ; leaves 4–6, lanceolate, acuminate, bright green, glabrous, 3–4 ft. long, 2 in. broad low down ; peduncle stout, stiffly erect, 2–3 ft. long ; raceme lax, a foot or more long ; pedicels ascending, lower 2–3 in. long ; bracts lanceolate, the lower 2½–3 in. long ; flowers erect ; perianth 1½ in. long, white, broadly banded with greenish- or reddish-brown ; stamens all fertile ; style prismatic, longer than the ovary. *Hook. fil. in Bot. Mag. t.* 6649.

EASTERN REGION : Natal ; in an open glade near the Umlazi River, *Nelson!*

Described from a plant cultivated at Kew.

17. **A. crinifolia** (Baker) ; bulb not seen ; leaves lorate, thin, membranous, distinctly veined, resembling those of *Crinum zeylanicum*, 2–3 ft. long, 3–3½ in. broad at the middle, narrowed gradually to the base, and furnished with a distinct midrib throughout ; peduncle 2 ft. long, moderately robust ; raceme lax, ½ ft. long ; pedicels ascending, lower 2 in. long ; bracts lanceolate-acuminate, longer than the pedicels ; perianth 1 in. long, white, banded with

reddish-brown ; stamens all fertile ; style $\frac{1}{2}$ in. long, nearly twice as long as the ovary.

EASTERN REGION : Natal ; Inanda, *Wood*, 750 !

18. **A. fastigiata** (Dryand. in Vet. Acad. Nya Handl. Stockh. 1784, 296) ; bulb globose, 2–3 in. diam. ; tunics not splitting into fibres at the top ; leaves 5–6, linear, glabrous, comparatively firm in texture, 1$\frac{1}{2}$–2 ft. long, $\frac{1}{2}$–$\frac{3}{4}$ in. broad low down, acuminate ; peduncle much shorter than the leaves ; raceme few-flowered, corymbose ; pedicels ascending or spreading, lower 4–5 in. long ; bracts ovate-lanceolate, acuminate, $\frac{3}{4}$–1 in. long ; flowers inodorous, permanently erect ; perianth $\frac{3}{4}$–1 in. long, white, broadly banded with green, fading to reddish-brown ; stamens all fertile ; style prismatic, as long as the ovary ; stamens conic, subentire. *Thunb. Prodr.* 65 ; *Fl. Cap. edit. Schult.* 331 ; *Willd. Sp. Plant.* ii. 101 ; *Andr. Bot. Rep. t.* 450 ; *Red. Lil. t.* 474 ; *Bot. Reg. t.* 277 ; *Roem. et Schultes, Syst. Veg.* vii. 498 ; *Kunth, Enum.* iv. 375 ; *Baker in Ref. Bot. t.* 44 ; *Journ. Linn. Soc.* xiii. 287. *Falconera fastigiata, Salisb. Gen.* 36.

VAR. β, floribunda (Baker); raceme longer, less corymbose, with more numerous flowers ; lower pedicels not so long.

SOUTH AFRICA : without locality, cultivated specimens collected by *Cooper !* and from the *Goodenough Herbarium !*
CENTRAL REGION : Var. β, Somerset Div. ; sides of the Bosch Berg, 3000 ft., *MacOwan*, 1832 !

19. **A. caudata** (Jacq. Collect. iv. 203 ; Ic. ii. 20, t. 442) ; bulb globose, 2–3 in. diam. ; tunics not splitting into fibres at the top ; leaves 5–6, linear, glabrous, acuminate, comparatively firm in texture, 1–1$\frac{1}{2}$ ft. long, $\frac{1}{2}$ in. broad low down ; peduncle moderately robust, 1–1$\frac{1}{2}$ ft. long ; raceme lax, corymbose, 6–9 in. long ; pedicels ascending, lower 3–4 in. long ; bracts lanceolate, acuminate, 1–1$\frac{1}{2}$ in. long ; flowers inodorous, permanently erect ; perianth $\frac{3}{4}$–1 in. long, white, broadly banded with green ; stamens all fertile ; style prismatic, $\frac{1}{4}$–$\frac{1}{3}$ in. long ; stigmatose apex conical, entire. *Willd. Sp. Plant.* ii. 102 ; *Roem. et Schultes, Syst. Veg.* vii. 497 ; *Kunth, Enum.* iv. 375 ; *Baker in Ref. Bot. t.* 45 ; *Journ. Linn. Soc.* xiii. 288.

COAST REGION : Albany Div. ; near Grahamstown, *Burke !*
CENTRAL REGION : Somerset Div., at the foot of Bruintjes Hoogte Mountain, 3500 ft , *MacOwan*, 1962 !

20. **A. pachychlamys** (Baker); bulb large, with many thick tunics, crowned with wiry fibres ; leaves few, short, narrow linear, $\frac{1}{12}$–$\frac{1}{6}$ in. broad, glabrous, channelled down the face ; peduncle 3–6 in. long ; raceme 6–8 in. long ; lower pedicels 1$\frac{1}{2}$–2 in. long ; bracts ovate or lanceolate ; perianth $\frac{5}{8}$ in. long ; segments white, with a broad brown keel ; stamens all antheriferous ; style prismatic

KALAHARI REGION : Transvaal ; mountain sides, Highland Creek, near Barberton, 4500 ft., *Galpin*, 630 !
EASTERN REGION : Swaziland ; Havelock Concession, on hills 3500–4000 ft. *Saltmarshe in Herb. Galpin*, 991 !

21. A. Baurii (Baker); bulb ovoid, 1–1½ in. diam.; tunics splitting into copious persistent fibres at the top; leaves 5–6, linear-acuminate, glabrous, 1–1½ ft. long, ⅓–½ in. broad low down; peduncle moderately stout, above a foot long; raceme short, few-flowered; pedicels ascending, cernuous at the apex, lower 1½ in. long; bracts lanceolate-acuminate, lower ¾–1 in. long; perianth ¾ in. long, white, broadly banded with green; stamens all fertile; style comparatively slender, rather longer than the ovary.

EASTERN REGION: Tembuland: Bazeia, along river-banks, 2000–3000 ft., *Baur*, 414!

22. A. humilis (Baker in Kew Bullet. 1895, 153); bulb ovoid, ⅓–½ in. diam.; leaves usually 3, narrow linear, erect, glabrous, 3–6 in. long, ⅓–½ lin. broad; peduncle slender, 1½–2 in. long; flowers 2–3, corymbose; pedicels erecto-patent, lower an inch long; bracts small, lanceolate or ovate-acuminate; perianth ½ in. long; segments oblong, white, with a broad, red-brown, many-nerved keel; stamens all antheriferous, rather shorter than the perianth; filaments dilated at the base; style prismatic, as long as the ovary.

EASTERN REGION: Natal; wet rocks at the top of Tabamhlopi Mountain, 6600 ft., *Evans*, 361!

23. A. polyphylla (Baker in Gard. Chron. 1874, i. 471); bulb ovoid, 1½–2 in. diam.; tunics not splitting into fibres at the apex; leaves 12–15, subulate, bright green, glabrous, 2–3 in. long, not more than $\frac{1}{24}$ in. diam.; peduncle slender, terete, glaucous, ½ ft. long; raceme few-flowered, corymbose; pedicels erecto-patent, the lower above an inch long; bracts ovate-lanceolate, ⅓–½ in. long; perianth permanently erect, ¾ in. long, white, banded with green, fading to reddish-brown; stamens all fertile; style prismatic, as long as the ovary.

CENTRAL REGION: Somerset Div; on the summit of Bosch Berg, 4500 ft., *MacOwan*, 2071!

Described from a plant sent by Professor MacOwan, which flowered at Kew, March, 1874.

24. A. viscosa (Linn. fil. Suppl. 196); bulb ovoid, 1–1½ in. diam.; tunics not splitting into fibres at the apex; leaves numerous, nearly subulate, deeply channelled down the face, ½–1 ft. long, $\frac{1}{12}$–⅙ in. broad, clothed with copious, persistent, sessile glands; peduncle about a foot long, slender, viscose; raceme laxly 6–12-flowered; pedicels ascending, cernuous at the apex; lower 1–2 in. long; bracts small, lanceolate; flowers inodorous; perianth ¾ in. long, white, banded with green; stamens all fertile; style prismatic, as long as the ovary; stigmas conical, entire. *Dryand. in Vet. Acad. Nya Handl. Stockh.* 1784, 297; *Thunb. Prodr.* 65; *Fl. Cap. edit. Schult.* 331; *Jacq. Coll Suppl.* 99; *Ic.* ii. 21, *t.* 445; *Willd. Sp. Plant.* ii. 103; *Kunth, Enum.* iv. 377; *Baker in Journ. Linn. Soc.* xiii. 289. *Falconera viscosa, Salisb. Gen.* 36.

COAST REGION: Mossel Bay Div.; between Hartenbosch and Mossel Bay, *Burchell*, 6222!

25. A. fragrans (Jacq. Hort. Schoenbr. i. 44, t. 84); bulb globose, proliferous, 2 in. diam.; tunic not splitting into fibres at the top; leaves 6–8, linear, flaccid, glabrous, $1\frac{1}{2}$–2 ft. long, $\frac{1}{2}$–$\frac{3}{4}$ in. broad low down; peduncle moderately stout, 1–$1\frac{1}{2}$ ft. long; raceme lax, finally a foot long; pedicels short, cernuous at the tip; bracts lanceolate, $\frac{1}{4}$–$\frac{1}{3}$ in. long; flowers very fragrant; perianth $\frac{3}{4}$ in. long, yellow, banded with green; stamens all fertile; style prismatic, as long as the ovary. *Willd. Sp. Plant.* ii. 103; *Roem. et Schultes, Syst. Veg.* vii. 501; *Kunth, Enum.* iv. 376; *Baker in Journ. Linn. Soc.* xiii. 289.

SOUTH AFRICA: without locality.
Known to me only from Jacquin's figure.

26. A. aurea (Jacq. Coll. iv. 202; Ic. ii. 20, t. 441); bulb globose, $1\frac{1}{2}$–2 in. diam.; tunics not splitting into fibres at the top; leaves 6–9, linear, flaccid, glabrous, $1\frac{1}{2}$–2 ft. long, $\frac{1}{2}$–1 in. broad low down, nearly flat; peduncle moderately stout, 1–$1\frac{1}{2}$ ft. long; raceme very lax, sometimes a foot long; pedicels ascending, permanently erect, lower 4–5 in. long; bracts lanceolate, acuminate, 1–$1\frac{1}{2}$ in. long; flowers inodorous; perianth an inch long, yellow, banded with green, which turns to reddish-brown; stamens all fertile; style prismatic, as long as the ovary. *Willd. Sp. Plant.* ii. 102; *Roem. et Schultes, Syst. Veg.* vii. 499; *Kunth, Enum.* iv. 376; *Baker in Journ. Linn. Soc.* xiii. 288.

CENTRAL REGION: collected between Plettenbergs Beacon and Graaf Reinet, *Burchell,* Bulb 45! and Bulb 201!

27. A. tortuosa (Baker); bulb globose, proliferous, $1\frac{1}{2}$–2 in. diam.; leaves 8–11, linear, glabrous, flaccid, conspicuously spirally twisted, a foot or more long, $\frac{1}{2}$ in. broad low down; peduncle moderately stout, about a foot long; raceme moderately dense, $\frac{1}{2}$ ft. long; pedicels ascending, permanently erect, lower 1–$1\frac{1}{2}$ in. long; bracts ovate-lanceolate, nearly as long as the pedicels; perianth $\frac{3}{4}$–$\frac{7}{8}$ in. long, yellow, broadly striped with green; stamens all fertile; style prismatic, as long as the ovary.

CENTRAL REGION: Jansenville Div.; dry hills by the Sunday River, near Noorsdoorn Plaats, 1500 ft., *Bolus,* 2629!

28. A. setosa (Jacq. Collect. Suppl. 100; Ic. ii. 20, t. 440); bulb globose, 2–3 in. diam.; tunics splitting at the top into copious persistent fibres; leaves 6–9, lanceolate, acuminate, glabrous, 1–$1\frac{1}{2}$ ft. long, $\frac{3}{4}$–1 in. broad low down; peduncle moderately stout, 1–$1\frac{1}{2}$ ft. long; raceme lax, corymbose, 4–6 in. long; pedicels permanently erect, lower 3–4 in. long; bracts small, ovate-lanceolate; flowers faintly scented; perianth $\frac{3}{4}$–1 in. long, yellow, banded with green; stamens all fertile; style prismatic, as long as the ovary. *Willd. Sp. Plant.* ii. 102; *Bot. Mag. t.* 1481; *Roem. et Schultes, Syst. Veg.* vii. 498; *Kunth, Enum.* iv. 375; *Baker in Journ. Linn. Soc.* xiii. 288. *Branciona setosa, Salisb. Gen.* 36.

CENTRAL REGION: Somerset Div.; among shrubs near Somerset East, 2300 ft. *MacOwan,* 1833!

29.. A. spiralis (Linn. fil. Suppl. 196) ; bulb small, ovoid ; tunics not splitting into fibres at the top ; leaves 10–12, subterete, 6–9 in. long, channelled down the face, circinate at the apex, densely glandular-pubescent ; peduncle slender, 6–8 in. long ; raceme laxly 4–6-flowered ; pedicels cernuous at the apex, lower $\frac{3}{4}$–1 in. long ; bracts lanceolate, $\frac{1}{2}$–$\frac{3}{4}$ in. long ; flowers inodorous ; perianth $\frac{5}{8}$–$\frac{3}{4}$ in. long, yellow, banded with green ; stamens all fertile ; style prismatic, as long as the ovary ; stigma tricuspidate. *Thunb. in Vet. Acad. Nya Handl. Stockh.* 1780, 59, *t.* 2, *fig.* 1 ; *Prodr.* 65 ; *Fl. Cap. edit. Schult.* 331 ; *Jacq. Collect. Suppl.* 100 ; *Ic.* ii. 20, *t.* 439 ; *Willd. Sp. Plant.* ii. 104 ; *Roem. et Schultes, Syst. Veg.* vii. 503 ; *Kunth, Enum.* iv. 377 ; *Baker in Journ. Linn. Soc.* xiii. 289. *Falconera spiralis, Salisb. Gen.* 36.

SOUTH AFRICA : without locality, *Harvey !*
COAST REGION : Malmesbury Div. ; summit of Riebeks Castle, *Thunberg.*
Not in Kew Herbarium.

30. A. tenuifolia (Baker in Saund. Ref. Bot. t. 335); bulb ovoid, $\frac{1}{2}$–$\frac{3}{4}$ in. diam.; tunics membranous, not splitting into fibres at the top; leaves 6–9, terete, very slender, 6–9 in. long, $\frac{1}{2}$ lin. diam., faintly channelled down the face, obscurely glandular; peduncle slender, 4–6 in. long; flowers few, laxly corymbose; pedicels erecto-patent, lower 2–3 in. long; lower bracts $\frac{3}{4}$–1 in. long; perianth erect, $\frac{3}{4}$ in. long, pale yellow, banded with green, turning to reddish-brown; stamens all fertile; style triquetrous in the upper half, longer than the ovary. *Baker in Journ. Linn. Soc.* xiii. 288.

SOUTH AFRICA : without locality, *MacOwan.* Not in Kew Herbarium.

Described from a cultivated plant. A specimen collected near Graaff Reinet (*Bolus*, 823) is either this or nearly allied, but the leaves are not glandular.

31. A. Macowani (Baker); bulb globose, $1\frac{1}{2}$ in. diam., producing abundant offsets, forming a large cushion of compacted bulbs ; tunics truncate, not produced above its neck, and not splitting into fibres at the top; leaves 6–8, linear-subulate, bright green, glabrous, 1–$1\frac{1}{2}$ ft. long, $\frac{1}{6}$ in. broad, channelled down the face ; peduncle terete, $\frac{1}{2}$–$1\frac{1}{2}$ ft. long ; raceme lax, few-flowered, subcorymbose, 3–6 in. long ; pedicels erect, lower 2–3 in. long ; bracts lanceolate, acuminate, lower $\frac{3}{4}$–$1\frac{1}{4}$ in. long; perianth $\frac{3}{4}$ in. long, dull yellow, broadly banded with green; stamens all fertile ; style prismatic, as long as the 'ovary. *A. polyphylla, Baker in Journ. Bot.* 1874, 367, *but not of Gard. Chron.* 1874, i. 471.

CENTRAL REGION : Somerset Div. ; near Somerset East, 2200 ft., *MacOwan,* 1840! 1849!

32. A. collina (Baker); outer tunics of the bulb produced much above its neck, barred across as in *Sypharissa*, and splitting into fibres at the tip ; leaves about 4, subterete, glabrous, 1–$1\frac{1}{2}$ ft. long, $\frac{1}{12}$ in. diam. low down ; peduncle curved, slender, 4–6 in. long ; raceme few-flowered, lax, subcorymbose ; pedicels permanently erect,

1-1½ in. long; bracts ovate-lanceolate, acuminate, ½-¾ in. long; perianth ¾-1 in. long, yellow, broadly barred with green; style comparatively slender, as long as the ovary; stigma tricuspidate.

CENTRAL REGION: Graaff Reinet, on stony hills, 2500 ft., *Bolus*, 786!

33. A. Shawii (Baker in Journ. Bot. 1874, 367); bulb ovoid, ¾-1 in. diam.; tunics whitish, membranous, not splitting into fibres at the top; leaves 6-12, filiform, very slender, glabrous, 4-5 in. long; peduncle slender, 6-10 in. long; raceme very lax, 3-9-flowered; pedicels cernuous at the apex, lower 1-1½ in. long; bracts ovate-lanceolate, acuminate, ¼-⅓ in, long; perianth yellow, keeled with green, ½-⅝ in. long; anthers of the outer stamens very small and sterile; style cylindrical, longer than the ovary; stigma distinctly 3-lobed.

COAST REGION: Stutterheim Div.; Kabousie, 3500 ft., *Murray*, 54!
CENTRAL REGION: Colesberg, *Shaw!*
KALAHARI REGION: Vaal River, *Shaw!*

34. A. parviflora (Donn, Hort. Cantab. edit. xi. 123); bulb subglobose; peduncle erect, 6-10 in. long; flowers small, yellowish-green, subsessile; bracts small, narrow, acuminate. *Dietr. Gartenlex. edit.* 2, i. 237; *Roem. et Schultes, Syst. Veg.* vii. 502; *Kunth, Enum.* iv. 377.

SOUTH AFRICA: without locality.

Introduced into cultivation in 1791. Known to me only from the above insufficient description. Probably an *Ornithogalum* or *Urginea.*

XXXIV. URGINEA, Steinh.

Perianth polyphyllous, deciduous; segments subequal, oblong, 1-nerved on the keel in the Cape species. *Stamens* 6, inserted on the base of the perianth-segments; filaments filiform or flattened at the base or middle; anthers linear or linear-oblong, versatile, dehiscing introrsely. *Ovary* sessile, ovoid, 3-celled; ovules superposed, few or many in a cell; style filiform; stigma capitate. *Capsule* globoso-triquetrous, membranous, loculicidally 3-valved. *Seeds* discoid, often winged; testa black, membranous; albumen fleshy.

Rootstock a tunicated bulb; leaves filiform or lanceolate, often not produced till after the flowers; peduncle naked; raceme simple; bracts often spurred either at the base or middle; flowers whitish, with green or purple-brown keels.

HAB. Species about 40. Extends to Tropical Africa, the Mediterranean region, and India.

Flowers very small (⅛-⅙ in. long); bracts minute, deltoid,
    the lower obscurely spurred:
    Leaves terete:
       Tall, raceme 5-6 in. long ... ... (1) **rigidifolia.**
       Dwarf, raceme 2-3 in. long:
          Pedicels ⅛-⅙ in. long ... ... (2) **nematodes.**
          Lower pedicels ½ in. long ... ... (3) **tenella.**

Leaves lanceolate ... ... (4) **pusilla.**
Leaves oblong-spathulate, very short, rigid :
　　Leaves obscurely ciliated ... ... (5) **marginata.**
　　Leaves densely ciliated with black bristles... ... (6) **ciliata.**
Flowers larger ($\frac{1}{3}-\frac{1}{2}$ in. long) ; lower bracts usually
　　conspicuously spurred :
Raceme globose, capitate :
　　Bulb small ($\frac{1}{2}$ in. diam.) ... ... (7) **Eckloni.**
　　Bulb large ($1\frac{1}{2}-2$ in. diam.) ... ... (8) **capitata.**
Raceme oblong, lax, few-flowered :
　　Leaves many ... ... (9) **fragrans.**
　　Leaves few :
　　　　Produced sheaths short ... ... (10) **filifolia.**
　　　　Produced sheaths long ... ... (11) **exuviata.**
Raceme oblong, dense :
　　Leaf terete ... ... (12) **macrocentra.**
　　Leaf linear ... ... (13) **Dregei.**
Raceme lax, cylindrical :
　　Perianth $\frac{1}{6}$ in. long :
　　　　Racemes lax :
　　　　　　Leaves synanthous :
　　　　　　　　Leaves 1-2 ... ... (14) **riparia.**
　　　　　　　　Leaves 4-6 :
　　　　　　　　　　Perianth greenish-white ... (15) **delagoensis.**
　　　　　　　　　　Perianth reddish-white ... (16) **rubella.**
　　　　　　Leaves hysteranthous :
　　　　　　　　Bulb $\frac{1}{2}$ in. diam. ... ... (17) **modesta.**
　　　　　　　　Bulb 1-2 in. diam :
　　　　　　　　　　Bulb crowned with fibres ... (18) **multisetosa.**
　　　　　　　　　　Bulb not crowned with fibres ... (19) **natalensis.**
Raceme dense, cylindrical :
　　Basal spur of lower bracts small ... ... (20) **echinostachya.**
　　Basal spur of lower bracts large, subulate ... (21) **kniphofioides.**
　　　　Perianth $\frac{1}{4}-\frac{1}{2}$ in. long :
　　　　　　Pedicels very short ... ... (22) **Forsteri.**
　　　　　　Pedicels longer, ascending :
　　　　　　　　Flowers lilac ... ... (23) **lilacina.**
　　　　　　　　Flower white, keeled with brown :
　　　　　　　　　　Peduncle with raceme about a
　　　　　　　　　　　　foot long ... ... (24) **Burkei.**
　　　　　　　　　　Peduncle with raceme 4-5 ft.
　　　　　　　　　　　　long ... ... (25) **maritima.**
　　　　　　Pedicels longer, patent :
　　　　　　　　Peduncle with raceme about a foot
　　　　　　　　　　long ... ... (26) **physodes.**
　　　　　　　　Peduncle with raceme 4-6 ft. long... (27) **altissima.**

1. U. rigidifolia (Baker in Journ. Bot. 1878, 323) ; bulb not
seen ; leaves produced after the flowers, terete, glabrous, 8–12 in.
long, $\frac{1}{16}$ in. diam. ; peduncle slender, terete, stiffly erect, under a
foot long ; raceme lax, oblong-cylindrical, 4–6 in. long, 1–1$\frac{1}{4}$ in.
diam. ; pedicels patent, very slender ; bracts small, ovate, lower
with a spur as long as the blade ; perianth whitish, $\frac{1}{6}$ in. long ;
segments oblong, with a broad green keel ; filaments very short,
lanceolate ; anthers oblong.

CENTRAL REGION : on plains of the Karroo, near Graaff Reinet, 2500 ft.,
Bolus, 783 !

2. U. nematodes (Baker in Journ. Linn. Soc. xiii. 218); bulb small, ovoid; leaves 10–12, synanthous, ascending, filiform, glabrous, 1–2 in. long; peduncle slender, terete, stiffly erect, 2–3 in. long; raceme lax, 20–40-flowered, 2–3 in. long, $\frac{1}{2}$ in. diam.; pedicels ascending, $\frac{1}{8}$–$\frac{1}{4}$ in. long; bracts deltoid, very minute; perianth $\frac{1}{4}$ in. long; segments oblong, white, keeled with purple; filaments lanceolate, $\frac{1}{12}$–$\frac{1}{8}$ in. long; capsule unknown. *Anthericum filifolium, Thunb. Prodr.* 62; *Fl. Cap. edit. Schult.* 317, *non Jacq. A. nematodes, Roem. et Schultes, Syst. Veg.* vii. 472. *Ornithogalum Thunbergii, Kunth, Enum.* iv. 369.

SOUTH AFRICA : without locality, *Thunberg !*

3. U. tenella (Baker); bulb globose, $\frac{1}{2}$–$\frac{3}{4}$ in. diam.; leaves synanthous, slender, filiform, glabrous; peduncle slender, 2–3 in. long; raceme lax, 1–2 in. long; pedicels ascending, lower $\frac{1}{2}$ in. long; lower bracts shortly spurred; perianth campanulate, $\frac{1}{6}$–$\frac{1}{4}$ in. long; segments oblong, white with a distinct red-brown keel; stamens shorter than the perianth; anthers small, globose; style short.

EASTERN REGION : Natal; Van Reenens Pass, in crevices of rocks, 5000 ft., *Wood,* 4562 !

4. U. pusilla (Baker in Journ. Linn. Soc. xiii. 217); bulb globose, white, 1–1$\frac{1}{4}$ in. diam.; leaves produced after the scape, about a dozen, lanceolate, spreading, 3–4 in. long, $\frac{1}{4}$–$\frac{1}{2}$ in. broad; peduncle slender, 2–4 in. long, pubescent and purplish downwards; raceme dense, oblong, 1–2 in. long; pedicels slender, patent or deflexed, $\frac{1}{2}$–$\frac{3}{4}$ in. long; bracts minute, deltoid; perianth $\frac{1}{6}$ in. long; segments oblong, keeled with green; filaments lanceolate, shorter than the perianth; capsule globose, $\frac{1}{6}$ in. long; seeds 2–3 in a cell. *Anthericum pusillum, Jacq. Collect. Suppl.* 95; *Ic.* ii. 18, *t.* 417; *Willd. Sp. Plant.* ii. 147; *Roem. et Schultes, Syst. Veg.* vii. 474. *Cæsia pusilla, Spreng. Syst.* ii. 88. *Phalangium pauciflorum, Poir. Encyc.* v. 251. *Idothea* (?) *drimioides, Kunth, Enum.* iv. 345.

COAST REGION : Clanwilliam Div.; Bokkeveld Flats, near "Addies," *Zeyher,* 4247 !

KALAHARI REGION : near Hopetown, *Muskett in Herb. Bolus,* 2570 !

5. U. marginata (Baker in Journ. Linn. Soc. xiii. 218); bulb not seen; leaves about two, oblong-spathulate, obtuse, rigidly coriaceous, glabrous, about 1$\frac{1}{2}$ in. long, $\frac{1}{2}$ in. broad, with a thickened obscurely ciliated margin; peduncle slender, terete, $\frac{1}{2}$ ft. long; raceme many-flowered, subcapitate; pedicels slender, $\frac{1}{2}$–$\frac{3}{4}$ in. long; bracts deltoid, very minute; perianth $\frac{1}{8}$–$\frac{1}{4}$ in. long; segments oblong, keeled with purple; filaments filiform, shorter than the perianth; capsule unknown. *Anthericum marginatum, Thunb. Prodr.* 63; *Fl. Cap. edit. Schult.* 324; *Roem. et Schultes, Syst. Veg.* vii. 474; *Idothea* (?) *marginata, Kunth, Enum.* iv. 346.

CENTRAL REGION : Calvinia Div.; Hantam, *Thunberg !*

6. U. ciliata (Baker in Journ. Linn. Soc. xiii. 218); bulb not seen; leaves two, oblong-spathulate, subacute, rigidly coriaceous, an

inch long, $\frac{1}{2}$ in. broad, with a thickened margin ciliated with dense,
persistent, short, spreading black bristles; peduncle slender, terete,
5–8 in. long; raceme oblong, lax, 2–3 in. long, under an inch in diam.;
pedicels ascending, $\frac{1}{4}$–$\frac{1}{3}$ in. long, cernuous at the apex; bracts deltoid,
very minute; perianth $\frac{1}{8}$ in. long; segments oblong, keeled with
purple; filaments short, filiform; capsule unknown. *Ornithogalum
ciliatum, Linn. fil. Suppl.* 199; *Thunb. Prodr.* 62; *Fl. Cap. edit.
Schult.* 316; *Willd. Sp. Plant.* ii. 117; *Roem. et Schultes, Syst. Veg.*
vii. 528; *Kunth, Enum.* iv. 359.

SOUTH AFRICA: without locality, *Thunberg!*

7. U. Eckloni (Baker in Engl. Jahrb. xv. Beibl. 35, 6); bulb
globose, $\frac{1}{2}$ in. diam. ; leaves hysteranthous, not seen; peduncle very
slender, wiry, 4–8 in. long; raceme dense, capitate, 1 in. diam.;
pedicels erecto-patent, $\frac{1}{4}$–$\frac{1}{3}$ in. long; bracts minute, deltoid, persistent; perianth campanulate, $\frac{1}{8}$ in. long; segments oblong, white
with a brown keel, splitting off from the receptacle at the base;
stamens shorter than the perianth; ovary globose; style short.

SOUTH AFRICA: without locality, *Ecklon and Zeyher*, 128! Not in Kew
Herbarium.

Nearly allied to *U. capitata.*

8. U. capitata (Baker); bulb globose, $1\frac{1}{2}$–2 in. diam.; leaves
6–8, not fully developed till after the flowers fade, linear, a foot
long, $\frac{1}{3}$–$\frac{1}{2}$ in. broad; peduncle lateral, terete, $\frac{1}{2}$–1 ft. long; raceme
many-flowered, capitate, globose, $1\frac{1}{4}$ in. diam.; pedicels at first
$\frac{1}{3}$–$\frac{1}{2}$ in. long, finally $1\frac{1}{2}$ in. long; bracts minute, ovate, deeply
saccate in the middle; perianth $\frac{1}{6}$–$\frac{1}{5}$ in. long, white inside, bright
claret-purple outside; filaments clavate, much shorter than the
perianth; style tricuspidate, as long as the ovoid ovary; capsule
$\frac{1}{3}$ in. long. *Ornithogalum capitatum, Hook. in Bot. Mag. t.* 5388;
Baker in Journ. Linn. Soc. xiii. 284.

COAST REGION: British Kaffraria, *Cooper*, 208! 3275!
KALAHARI REGION: Transvaal; Saddleback Range, near Barberton, 5000 ft.,
Galpin, 987!
EASTERN REGION: Griqualand East; hills around Kokstad, *Tyson*, 1842!
Natal; Inanda, *Wood*, 178!

9. U. fragrans (Steinh. in Ann. Sc. Nat. ser. 2, i. 328); bulb
globose, $1\frac{1}{2}$–2 in. diam., squamose upwards, the tunics scarcely produced above its neck; leaves 12–20, contemporary with the flowers,
subterete, glabrous, firm in texture, 6–8 in. long, $\frac{1}{2}$ lin. diam.;
peduncle slender, terete, glaucous, a foot long; raceme moderately
dense, 4–6 in. long; pedicels $\frac{1}{3}$–$\frac{1}{2}$ in. long; bracts lanceolate, spurred
at the base; perianth white, fragrant, $\frac{1}{2}$ in. long; segments oblong,
with a purplish keel; filaments filiform, half as long as the perianth;
style slightly declinate, $\frac{1}{4}$ in. long; capsule not seen. *Baker in
Journ. Linn. Soc.* xiii. 219. *Anthericum fragrans, Jacq. Hort.
Schoenbr.* i. 45, *t.* 86; *Willd. Sp. Plant.* ii. 135; *Roem. et Schultes,
Syst. Veg.* vii. 470. *Albuca fugax, Gawl. in Bot. Reg. t.* 311.

Ornithogalum fragrans, Kunth, Enum. iv. 366. *Phalangium fragrans, Poir. Encyc.* v. 247. *Sypharissa fragrans, Salisb. Gen.* 38.

SOUTH AFRICA: without locality, *Mass; n!*
COAST REGION: Malmesbury, *Bachmann,* 696! Not in Kew Herbarium.

10. U. filifolia (Steinh. in Ann. Sc. Nat. ser. 2, i. 329); bulb globose, ½ in. diam. ; tunics dark brown ; inner produced an inch or two, with a few cross bands towards the top; leaves 4–6, terete, slender, wiry, glabrous, 1–2 ft. long; peduncle terete, slender, ½–1½ ft. long; raceme lax or moderately dense, 1–3 in. long; pedicels ascending, lower ¼–½ in. long; bracts small, ovate, lower with a cylindrical spur from the base ⅛ in. long; perianth inodorous, ⅓ in. long; segments oblong, white, with a purple-brown keel; filaments filiform, much shorter than the perianth; anthers linear-oblong, ⅛ in. long; style short, declinate; capsule not seen. *Baker in Journ. Linn. Soc.* xiii. 219. *Anthericum filifolium, Jacq. Collect. Suppl.* 93 ; *Ic.* ii. 18, *t.* 414 ; *Willd. Sp. Plant.* ii. 135; *Roem. et Schultes, Syst. Veg.* vii. 471. *Phalangium filifolium, Poir. Encyc.* v. 242. *Albuca filifolia, Gawl. in Bot. Reg. t.* 557. *Ornithogalum filifolium, Kunth, Enum.* iv. 369. *Anthericum spiratum, Thunb. Prodr.* 62 ; *Fl. Cap. edit. Schult.* 317; *Kunth, Enum.* iv. 656.

COAST REGION: Malmesbury Div. ; Klip Fontein near Darling, *Zeyher,* 4249 ! Cape Div. ; hills near Cape Town, *Thunberg!* Hout Bay, 250 ft., *MacOwan, Herb. Aust. Afr.,* 1562! Cape Flats, *Zeyher,* 4250! Worcester Div. ; Dutoits Kloof, 3000–4000 ft., *Drège,* 8744aa ! Riversdale Div. ; between little Vet River and Kampsche Berg, *Burchell.* 6907!

WESTERN REGION : Little Namaqualand; between Buffels River and Pedros Kloof, 2000–3000 ft., *Drège,* 2677b !

11. U. exuviata (Steinh. in Ann. Sc. Nat. ser. 2, i. 330) ; bulb globose, 1–1½ in. diam.; root-fibres numerous, fleshy, cylindrical; inner tunics produced 3–4 inches above the bulb, furnished with numerous strongly-raised cross-bars ; leaves 2–4, terete, glabrous, firm in texture, slender, 1–1½ ft. long; peduncle slender, terete, 1–1½ ft. long; raceme lax, 2–3 in. long; pedicels erecto-patent, lower ⅓–½ in. long ; bracts small, ovate, membranous, lower with a terete basal spur ¼ in. long; perianth ⅓–½ in. long; sepals oblong, white, with a red-brown keel; stamens half as long as the perianth ; filaments filiform ; anthers linear-oblong, ⅛ in. long; style declinate, ¼ in. long; capsule not seen. *Baker in Journ. Linn. Soc.* xiii. 219. *Anthericum exuviatum, Jacq. Collect. Suppl.* 89, *t.* 14, *fig.* 2 ; *Ic.* ii. 18, *t.* 415 ; *Willd. Sp. Plant.* ii. 136 ; *Roem. et Schultes, Syst. Veg.* vii. 471. *Albuca exuviata, Gawl. in Bot. Mag. t.* 871. *Phalangium exuviatum, Poir Encyc.* v. 243. *Ornithogalum* (?) *exuviatum, Kunth, Enum.* iv. 369.

SOUTH AFRICA : without locality, *Zeyher,* 1056! *Bergius!*
COAST REGION : Malmesbury Div., *Bachmann,* 229 ! Cape Div. ; Simons Bay, *Wright.* 221! Worcester Div. ; Dutoits Kloof, 1000–2000 ft., *Drège,* 8744a ! Uitenhage Div.; near the Zwartkops River, *Zeyher,* 4248 ! Port Elizabeth, *E.S.C.A. Herb.,* 39 !

12. U. macrocentra (Baker in Gard. Chron. 1887, i. 702) ; bulb

large, globose; leaf single, terete, 1$\frac{1}{2}$ ft. long, $\frac{1}{6}$ in. diam.; peduncle
stout, terete, rather glaucous, 2$\frac{1}{2}$–3 ft. long; raceme dense,
5–6 in. long; pedicels ascending, the lower $\frac{1}{2}$ in. long; lower bracts
with a convolute spur $\frac{3}{4}$–1 in. long; perianth $\frac{1}{6}$ in. long; segments
white with a brown keel, oblanceolate-oblong; stamens shorter than
the perianth; anthers brown, oblong; style short.

EASTERN REGION: Transkei, *Mrs. Barber*, 895!

13. U. Dregei (Baker); bulb not seen; leaf single, linear, glabrous,
shorter than the peduncle; peduncle slender, fragile, 1–1$\frac{1}{2}$ ft. long;
flowers 12–20 in a dense oblong raceme about an inch long; pedicels
erecto-patent, lower $\frac{1}{8}$–$\frac{1}{6}$ in. long; bracts ovate, minute, lower with
a cylindrical spur as long as the blade; perianth $\frac{1}{6}$ in. long; segments
linear-oblong, white, with a red-brown keel; stamens one-third
shorter than the perianth; anthers small, oblong. *Hyacinthus
convallarioides, E. Meyer in Drège Docum.* 82, *name only.*

COAST REGION: Cape Flats, *Zeyher*, 4251! Worcester Div.; Dutoits Kloof,
3000–4000 ft., *Drège*, 1501! Caledon Div.; mountain sides near Geuadendal,
3600 ft., *Bolus*, 7428!

14. U. riparia (Baker); bulb narrow, ampullæform, 2 in. long;
leaves 2, subterete, glabrous, contemporary with the flowers, shorter
than the peduncle; peduncle slender, stiffly erect, above a foot long;
raceme cylindrical, very lax in the lower half, $\frac{1}{2}$ ft. long, $\frac{1}{2}$–$\frac{5}{8}$ in.
diam.; pedicels slender, ascending, as long as or longer than the
flowers; bracts small, ovate, membranous, lower with a small ovate
obtuse basal spur; perianth campanulate, $\frac{1}{6}$ in. long; segments
oblong, obtuse, with a conspicuous 1-nerved reddish brown keel;
stamens nearly as long as the perianth; filaments linear, white;
anthers oblong, $\frac{1}{12}$ in. long; style very short; capsule globose, $\frac{1}{6}$ in.
long.

EASTERN REGION: Natal, Umzinyati Falls, *Wood*, 1052!

15. U. delagoensis (Baker); bulb oblong-cylindrical, $\frac{3}{4}$ in. diam.;
leaves 5–6, synanthous, linear, glabrous, $\frac{1}{6}$ in. broad; peduncle
moderately stout, stiffly erect, a foot long; raceme lax, cylindrical, a
foot long; pedicels erecto-patent, $\frac{1}{8}$–$\frac{1}{6}$ in. long; bracts ovate, minute,
deciduous, the lower not seen; perianth oblong, greenish-white, $\frac{1}{6}$ in.
long; segments oblong; capsule oblong-clavate, $\frac{1}{3}$ in. long; seeds
flat, oblong, $\frac{1}{6}$ in. long.

EASTERN REGION: between Delagoa Bay and Lobombo Mountains, *Bolus*,
7627!

16. U. rubella (Baker); bulb ovoid, 1 in. diam.; tunics brown,
membranous; leaves synanthous, very slender, subterete, glabrous,
fragile, 4–6 to a bulb, 3–6 in. long; peduncles slender, fragile, 6–9
in. long, 1–4 to a bulb; raceme lax, cylindrical, 3–6 in. long; bracts
small, ovate, membranous, the lower with a distinct subulate spur;
pedicels erecto-patent, articulated at the apex, the lower $\frac{1}{4}$–$\frac{1}{3}$ in.
long; perianth $\frac{1}{6}$ in. long; segments linear-oblong, bright red with

a white edge ; stamens distinctly shorter than the perianth ; anthers small, oblong.

EASTERN REGION: Natal; near the Mooi River, 4000–5000 ft., *Wood*, 5723!

17. **U. modesta** (Baker in Engl. Jahrb. xv. Beibl. 35, 6); bulb globose, $\frac{1}{2}$ in. diam.; leaves hysteranthous, very narrow, glabrous; peduncle very slender, 6–9 in. long; raceme moderately dense, $1\frac{1}{2}$–2 in. long, $\frac{1}{3}$ in. diam.; pedicels erecto-patent, $\frac{1}{8}$–$\frac{1}{6}$ in. long; lower bracts small, ovate, with a cylindrical spur $\frac{1}{4}$ in. long; upper without any spur; perianth campanulate, $\frac{1}{6}$ in. long; segments oblong, white, with a red-brown keel; stamens shorter than the perianth; anthers small, oblong; style as long as the ovary.

EASTERN REGION: Pondoland, *Bachmann*, 273!

18. **U. multisetosa** (Baker); bulb ovoid, tunicated, $1\frac{1}{2}$–2 in. diam., crowned with a dense ring of long bristles; leaves hysteranthous, unknown; peduncle slender, fragile, a foot or more long; raceme lax, cylindrical, 6–8 in. long; bracts small, oblong, membranous, the lower with a distinct oblong spur adpressed to the rachis; pedicels ascending or spreading, $\frac{1}{8}$–$\frac{1}{4}$ in. long, articulated at the apex; perianth campanulate, $\frac{1}{6}$ in. long; segments oblong, white, with a greenish-brown keel; stamens rather shorter than the perianth; filaments linear; anthers small, oblong; capsule $\frac{1}{4}$ in. long.

KALAHARI REGION: Transvaal; mountain tops, Upper Moodies, near Barberton, 4500 ft., *Galpin*, 584!

EASTERN REGION: Natal, near the Mooi River, 4000–5000 ft., *Wood*, 5724!

19. **U. natalensis** (Baker); bulb globose, 1 in. diam., not crowned with fibres; leaves subterete, hysteranthous, slender, rigid; peduncle slender, $\frac{1}{2}$–1 ft. long; raceme lax, cylindrical, 3–4 in. long, about $\frac{1}{2}$ in., diam.; pedicels ascending, lower a little longer than the flowers; bracts minute, ovate, lower with a slender terete basal spur $\frac{1}{4}$–$\frac{1}{3}$ in. long; perianth $\frac{1}{6}$ in. long, splitting off from the base in a cap; segments narrow, membranous, oblanceolate, with a conspicuous brown keel; stamens a little shorter than the perianth; anthers very small; capsule oblong, $\frac{1}{6}$ in. long.

EASTERN REGION: Natal; Inanda, *Wood*, 277! Umzinyati Falls, *Wood*, 1036!

20. **U. echinostachya** (Baker); bulb and hysteranthous leaves not seen; peduncle stout, $1\frac{1}{2}$–2 ft. long; raceme cylindrical, $\frac{1}{2}$–1 ft. long, $\frac{5}{8}$–$\frac{3}{4}$ in. diam., dense in the upper half, lax in the lower half; pedicels very short, at most $\frac{1}{8}$ in. long; bracts linear or lanceolate, acuminate, protruded beyond the buds; lowest with a small obscure obtuse basal spur adpressed to the rachis; perianth campanulate, $\frac{1}{6}$ in. long; segments oblanceolate-oblong, obtuse, dull white, with a dull brown keel; stamens shorter than the perianth; filaments filiform; anthers oblong, small; style short.

EASTERN REGION: Natal; Inanda, *Wood*, 276!

21. U. kniphofioides (Baker); bulb ovoid, 3 in. diam., with many thick tunics; leaves withered at the flowering time, ensiform, firm, glabrous, 2–3 ft. long; peduncle stout, 3–4 ft. long; raceme dense, cylindrical, nearly a foot long, ¾ in. diam.; pedicels very short; bracts lanceolate, ¼ in. long, the lower with a large subulate basal spur; perianth campanulate, greenish-white, ⅙ in. long; stamens nearly as long as the perianth-segments; anthers oblong, small.

EASTERN REGION: Swaziland; Havelock Concession, 3200–5000 ft., *Salt-marshe in Herb. Galpin*, 1055!

22. U. Forsteri (Baker); bulb like that of *U. maritima*, but larger; leaves hysteranthous, lanceolate; peduncle very stout, stiffly erect, 2–3 ft. long; raceme subspicate, dense, cylindrical, sometimes 1–1½ ft. long, 1–1½ in. diam.; pedicels erecto-patent, very short; bracts lanceolate, ¼ in. long, with a small spur about the middle; perianth ⅓ in. long; segments oblong-lanceolate, obtuse, white, with a brown keel; stamens nearly as long as perianth; filaments rather flattened; anthers small, oblong; ovary globose; style slightly declinate, ¼ in. long.

SOUTH AFRICA: without locality, *Forster! Wallich!*
COAST REGION: Malmesbury Div.; near Hopefield, *Bachmann*, 1708! Worcester Div., *Cooper*, 3275!

23. U. lilacina (Baker); bulb ovoid, 1½ in. diam.; leaves hysteranthous, terete; peduncle reaching a length of 4 feet, stout towards the base, tapering upwards; raceme moderately dense, cylindrical, 6–8 in. long, an inch in diameter; pedicels ascending, ¼–½ in. long; bracts small, oblanceolate, membranous, early deciduous, not spurred; flowers fragrant, lilac, turning to pinkish when dried; perianth campanulate, nearly ¼ in. long; segments oblanceolate-oblong, faintly keeled with dull green; stamens much shorter than the perianth; filaments rather flattened; anthers oblong, very small; ovary ovoid; style short.

EASTERN REGION: Natal; Inanda, in boggy ground, *Wood*, 198! 642!

24. U. Burkei (Baker); bulb and hysteranthous leaves not seen; peduncle under a foot long; raceme lax, cylindrical, ½ ft. long, 1 in. diam.; pedicels ascending, lower ½ in. long; bracts lanceolate, small, membranous, deciduous, spurred at the base; perianth ⅓ in. long; segments lanceolate-oblong, white, with a distinct brown keel; stamens rather shorter than the perianth; anthers small, oblong; style ⅛ in. long; capsule ovoid, obtusely trigonous, ½ in. long.

KALAHARI REGION: Transvaal; Magalies Berg, *Burke!* near the Vaal River, *Nelson!*
EASTERN REGION: Delagoa Bay; between Lorenço Marquez and Puzeen's, 100 ft., *Bolus*, 7628!

25. U. maritima (Baker in Journ. Linn. Soc. xiii. 221); bulb large, globose; leaves hysteranthous, lanceolate, 1–1½ ft. long, 2–3 in. broad; peduncle 2–3 ft. long; raceme cylindrical, moderately dense, reaching 1½–2 ft. in length, 1–1½ in. diam.; pedicels ascending,

lower $\frac{1}{2}-\frac{3}{4}$ in. long; bracts lanceolate, $\frac{1}{4}-\frac{1}{3}$ in. long, protruding
beyond the buds, obscurely spurred below the middle; perianth
campanulate, $\frac{1}{4}-\frac{1}{3}$ in. long; segments oblanceolate-oblong, obtuse,
white, with a brown keel; stamens rather shorter than the perianth;
filaments rather flattened; anthers small, oblong; ovary ovoid; style
short, declinate. *U. Scilla, Steinh. in Ann. Sc. Nat. ser.* 2, i. 330.
Scilla maritima, Linn. Sp. Plant. 308; *Red. Lil.* ii. *t.* 116. *Ornithogalum Squilla, Gawl. in Bot. Mag. t.* 918.

CoAST REGION: Cape Div.; Tyger Berg, under 1000 ft., *Drège*, 3527!

I cannot distinguish this from the well-known officinal *Squill* of the Mediterranean region.

26. U. physodes (Baker in Journ. Linn. Soc. xiii. 217); bulb
large, globose, solid, purplish on the outside; leaves 6–9, produced
long after the flowers, lanceolate, glabrous, 6–9 in. long, 1–1$\frac{1}{2}$ in.
broad; peduncle terete, purplish, $\frac{1}{2}$ ft. long; raceme moderately
dense, many-flowered, 3–5 in. long, 2 in. diam.; pedicels subpatent,
$\frac{1}{2}-\frac{3}{4}$ in. long; bracts small, ovate, concave, not distinctly spurred;
perianth whitish, $\frac{1}{4}$ in. long; segments oblong, keeled with purple;
filaments papillose, shorter than the perianth; capsule ovoid, $\frac{1}{3}-\frac{1}{2}$ in.
long; seeds about 2 in a cell. *Anthericum physodes, Jacq. Collect.
Suppl.* 94; *Ic.* ii. 18, *t.* 418; *Willd. Sp. Plant.* ii. 147; *Roem. et
Schultes, Syst. Veg.* vii. 473. *Albuca physodes, Gawl. in Bot. Mag.
t.* 1046. *Cæsia physodes, Spreng. Syst.* ii. 88.

SOUTH AFRICA: without locality.

Known to me only from the figures cited.

27. U. altissima (Baker in Journ. Linn. Soc. xiii. 221); bulb
very large, globose; leaves hysteranthous, lanceolate, glabrous, 1–1$\frac{1}{2}$ ft.
long, 2–3 in. broad; peduncle stout, sometimes 3–4 ft. long; raceme
cylindrical, moderately dense, sometimes 2–3 ft. long, 1$\frac{1}{4}$–1$\frac{1}{2}$ in.
diam.; lower pedicels patent, $\frac{1}{2}$–1 in. long; bracts small, lanceolate,
bent and obscurely spurred below the middle; perianth campanulate,
$\frac{1}{3}$ in. long; segments oblanceolate-oblong, whitish, with a dull brown
keel; stamens shorter than the perianth; filaments slightly flattened;
anthers oblong, $\frac{1}{12}$ in. long; style $\frac{1}{6}$ in. long; capsule small, globose,
obtusely trigonous. *Ornithogalum altissimum, Linn. fil. Suppl.* 199;
Thunb. Prodr. 62; *Fl. Cap. edit. Schult.* 315; *Willd. Sp. Plant.* ii.
119; *Roem. et Schultes, Syst. Veg.* vii. 521; *Kunth, Enum.* iv. 357.
O. giganteum, Jacq. Hort. Schoenbr. i. 45, *t.* 87. ' *Drimia altissima,
Gawl. in Bot. Mag. t.* 1074.

CoAST REGION: Uitenhage Div.; *Ecklon and Zeyher*, 608! between Bethels-
dorp and Uitenhage, *Burchell*, 4403!
CENTRAL REGION: near Graaff Reinet, 2500 ft., *Bolus*, 649!
Also Tropical Africa.

XXXV. VELTHEIMIA, Gleditsch.

Perianth gamophyllous, marcescent; tube long, cylindrical; segments very short, ovate. *Stamens* 6, uniseriate, inserted at the middle
of the perianth-tube; filaments filiform; anthers small, oblong,

dorsifixed, dehiscing introrsely. *Ovary* sessile, oblong, 3-celled ; ovules 2, collateral, placed in the middle of the cells ; style filiform ; stigma small, capitate. *Capsule* large, membranous, turbinate, acutely triquetrous, loculicidally 3-valved. *Seeds* turbinate ; testa thin, black ; albumen horny.

Rootstock a large bulb with membranous tunics ; leaves all radical, membranous, lorate or lorate-oblong ; peduncle stout, terete, naked ; inflorescence a dense raceme ; pedicels short ; bracts lanceolate-acuminate, membranous.

Bracts as long as the flowers ... ... (1) **bracteata.**
Bracts much shorter than the flowers :
 Leaf green, 2–3 in. broad ... ... (2) **viridifolia.**
 Leaf glaucous, 1¼ in. broad ... ... (3) **glauca.**

1 **V. bracteata** (Harv. ex Baker in Journ. Linn. Soc. xi. 411); leaves thin, lorate-oblanceolate, about a foot long, 2 in. broad, narrowed gradually from the middle to the base ; peduncle about a foot long ; raceme very short and dense ; pedicels nearly or quite obsolete ; bracts lanceolate-acuminate, as long as or a little longer than the flower ; perianth yellow, about an inch long ; segments ⅛–⅙ in. long ; stamens half as long as the perianth.

COAST REGION : British Kaffraria, *Cooper*, 320 !

2. **V. viridifolia** (Jacq. Hort. Schoenbr. i. 41, t. 78); bulb ovoid, 2–3 in. diam. ; leaves oblong-lorate, green, membranous, finally a foot long, 2–3 in. broad at the middle ; peduncle 1–1½ ft. long, terete, green, mottled with purple ; raceme very dense, 3–6 in. long ; pedicels very short ; bracts half as long as the flowers ; perianth 1¼–1½ in. long, yellow or tinged with red ; segments ⅙ in. long ; stamens ½ in. long, not protruded out of the tube ; capsule 1½–1¾ in. long, with valves an inch broad. *Willd. Sp. Plant.* ii. 181; *Lodd. Bot. Cab. t.* 1245; *Kunth, Enum.* iv. 281; *Baker in Journ. Linn. Soc.* xi. 411. *Aletris capensis, Linn. Sp. Plant. ed.* 2,456; *Mant.* 367; *Thunb. Prodr.* 60; *Bot. Mag. t.* 501. *V. undulata, Moench, Meth.* 631. *V. capensis, Thunb. Fl. Cap. edit. Schult.* 309; *Red. Lil.* iv. *t.* 193.

COAST REGION : Sand hills near Port Elizabeth, *Burchell*, Bulb 67 ! Albany Div. ; near Grahamstown, 2000 ft., *MacOwan*, 1831 !

3. **V. glauca** (Jacq. Hort. Schoenbr. i. 40, t. 77); bulb ovoid, 2 in. diam. ; leaves about 6, oblanceolate-lorate, acute, glaucous, finally a foot long, 1¼ in. broad ; peduncle not so stout as in the last; raceme very dense, 3–6 in. long; pedicels very short ; bracts half as long as the flowers ; perianth yellow or bright red, an inch long ; segments ¹⁄₁₂ in. long ; stamens reaching to the throat of the perianth-tube. *Willd. Sp. Plant.* ii. 182 ; *Wendl. Collect.* iii. 11, *t.* 78 ; *Bot. Mag. t.* 1091 ; *Red. Lil. t.* 440 ; *Kunth, Enum.* iv. 282 ; *Baker in Journ. Linn. Soc.* xi. 411. *Aletris glauca, Ait. Hort. Kew.* i. 463.

SOUTH AFRICA : without locality, *Zeyher*, 1719 ! *Pappe !*
COAST REGION : Malmesbury Div. ; Saldanha Bay, *Grey !* near Malmesbury, 250 ft., *MacOwan and Bolus, Herb. Norm. Aust. Afr.*, 944 ! near Hopefield, *Bachmann*, 1943 !
WESTERN REGION : Little Namaqualand, *Whitehead !*

XXXVI. HYACINTHUS, Linn.

Perianth gamophyllous, marcescent, finally deciduous; tube campanulate or oblong; segments subequal, falcate, as long as or longer than the tube. *Stamens* 6, inserted at or below the throat of the perianth-tube; filaments filiform; anthers versatile, dehiscing introrsely. *Ovary* ovoid, sessile; ovules few in a cell, superposed; style cylindrical; stigma capitate. *Capsule* membranous, loculicidally 3-valved. *Seeds* globose; testa thin, black; embryo cylindrical, much shorter than the horny albumen.

Rootstock a bulb with membranous tunics; leaves all radical, fleshy, linear or lorate; inflorescence racemose; bracts small, membranous.

DISTRIB. Species about 30, nearly all plants of the Oriental and Mediterranean regions. One Tropical African.

Inflorescence corymbose ... ... (1) **corymbosus.**
Inflorescence an elongated raceme ... ... (2) **Gawleri.**

1. **H. corymbosus** (Linn. Mant. 223); bulb ovoid, ½ in. diam.; neck ½–1 in. long, sheathed by a loose, truncate, membranous tunic; leaves 3–6, linear, channelled down the face, finally 4–5 in. long; peduncle 1–2 in. long; flowers 4–8, corymbose; lower pedicels erecto-patent, finally ¼–⅓ in. long; bracts minute, ovate; perianth red-lilac, ½–⅝ in. long; tube campanulate; segments oblanceolate-oblong, two or three times as long as the tube; stamens distinctly biseriate; upper inserted at the throat of the tube, nearly as long as the segments; capsule small, globose. *Thunb. Prodr.* 64; *Fl. Cap. edit. Schult.* 325; *Jacq. Collect.* iii. 230, *t.* 19, *fig.* 2; *Andr. Bot. Rep.* v. *t.* 345; *Baker in Journ. Linn. Soc.* xi. 426. *Massonia corymbosa, Gawl. in Bot. Mag. t.* 991. *M. linearis, E. Meyer in Drège Docum.* 119, *name only. Scilla corymbosa, Gawl. in Bot. Mag. sub t.* 1468. *Peribæa corymbosa, Kunth, Enum.* iv. 293.

COAST REGION: Cape Div.; sandy places between Lion Mountain and the sea shore, *Thunberg!* near the Lighthouse, *Ecklon,* 111! Cape Flats, *Zeyher,* 1715! Camps Bay, 200 ft., *MacOwan and Bolus, Herb. Norm. Aust. Afr.,* 943! Tulbagh Div.; Roodesand, near Tulbagh, *Drège!*

2. **H. Gawleri** (Baker); bulb globose; rudimentary leaf oblong; produced leaves 4–5, linear, arcuate, 3–4 in. long, channelled down the face; peduncle shorter or longer than the leaves; flowers 6–8 in a lax raceme, the lowest cernuous; lower pedicels ¼ in. long; bracts minute or obsolete; perianth reddish-lilac, ½ in. long; tube very short; segments oblanceolate-oblong; stamens subuniseriate, a little shorter than the perianth-segments. *Scilla brevifolia, Gawl. in Bot. Mag. t.* 1468, *excl. syn. Thunb. S. brachyphylla, Roem. et Schultes, Syst. Veg.* vii. 573. *Peribæa Gawleri, Kunth, Enum.* iv. 293.

SOUTH AFRICA: without locality.

Known to me only from the *Bot. Mag.* figure, which was drawn from a plant at Messrs. Lee & Kennedy's in June, 1812. *Scilla brevifolia, Thunb.,* is *Dipcadi hyacinthoides.*

XXXVII. DRIMIOPSIS, Lindl.

Perianth polyphyllous, persistent, permanently campanulate; segments oblong, 1-nerved; 3 inner cucullate at the tip. *Stamens* 6, inserted at the base of the perianth-segments; filaments short, dilated at the base; anthers small, versatile, dehiscing introrsely. *Ovary* sessile, globose-trigonous, trilocular; ovules 2 in a cell, collateral from the base; style short, cylindrical; stigma capitate. *Capsule* small, globose, membranous, loculicidally 3-valved. *Seeds* 1–2 in a cell, turgid.

Rootstock a tunicated bulb; leaves usually fleshy and flaccid, petioled or sessile; peduncle slender, naked; flowers small, greenish-white, arranged in a dense subspicate raceme; upper imperfect; bracts obsolete.

DISTRIB. Five other species, all tropical African, 4 Eastern, 1 Western.

Leaves distinctly petioled :
  Blade cordate :
    Perianth ⅙ in. long ... ... (1) **maculata.**
    Perianth ⅛ in. long ... ... (2) **minor.**
  Blade narrowed to the base :
    Raceme ½–1 in. long ... ... (3) **Woodii.**
    Raceme 3–5 in. long ... ... (4) **maxima.**
Leaves sessile :
  Leaves erect :
    Dwarf; small-flowered ... ... (5) **Burkei.**
    Tall; large-flowered ... ... (6) **Saundersiæ.**
  Leaves spreading ... ... (7) **humifusa.**

1. D. maculata (Lindl. in Paxt. Flow. Gard. ii. 73, fig. 172); bulb globose, 1 in. diam.; leaves 3–6; petiole deeply channelled, 1–6 in. long; blade thin, cordate-ovate, 3–4 in. long, green, with darker blotches; peduncle 4–12 in. long; raceme dense, 1–2 in. long; perfect flowers greenish, shortly pedicellate; perianth ⅙ in. long; outer segments oblong, inner ovate; upper flowers smaller, whitish; stamens nearly as long as the segments; anthers ovoid, white, ½ lin. long. *Baker in Saund. Ref. Bot.* iii. *t.* 191, *App.* 17; *Journ. Linn. Soc.* xiii. 227.

EASTERN REGION: Natal; Inanda, *Wood,* 233! *Mrs. K. Saunders!* Nottingham, *Buchanan!* near Durban, 150 ft., *MacOwan and Bolus, Herb. Norm. Austr. Afr.,* 1013!

2. D. minor (Baker in Saund. Ref. Bot. iii. t. 192, App. 17); bulb ovoid, ½–¾ in. diam.; leaves 3–4; petiole 1–2 in. long, channelled; blade cordate-ovate, acute, 1½–2 in. long, thin, green, with darker blotches; peduncle slender, 3–6 in. long; raceme dense, ½–1 in. long; lower flowers patent, very shortly pedicellate; perianth $\frac{1}{12}$–⅛ in. long; segments oblong; stamens nearly as long as the perianth. *Journ. Linn. Soc.* xiii. 227.

EASTERN REGION : Natal. No specimen at Kew.
Described from a plant introduced by Cooper.

3. D. Woodii (Baker); bulb ovoid or globose, ¾–1 in. diam.; leaves about 3; petiole 1½–3 in. long; blade oblong-lanceolate, thin,

3–4 in. long, 1 in. broad below the middle; peduncle very slender, 8–12 in. long; raceme dense, $\frac{1}{2}$–1 in. long; upper flowers white, sessile; lower greenish, cernuous, very shortly pedicellate; perianth campanulate, $\frac{1}{8}$ in. long; segments oblong, inner broader; stamens distinctly shorter than the segments; anthers subglobose, white.

EASTERN REGION: Natal; Inanda, *Wood*, 656! Klip River, 3500–4500 ft., *Sutherland!*

4. **D. maxima** (Baker); leaves about 4, contemporary with the flowers, oblong, membranous, blotched with brown, 6–8 in. long, 1$\frac{1}{2}$–2 in. broad at the middle, narrowed gradually into a long channelled petiole; peduncle 9–12 in. long; raceme narrow, 3–5 in. long; pedicels very short; bracts obsolete; perianth oblong, $\frac{1}{4}$ in. long; segments oblong, with a red-purple keel; stamens half the length of the perianth; anthers small, oblong; style short.

EASTERN REGION: Natal; valley near Bothas, 2000 ft., *Wood*, 4773!

5. **D. Burkei** (Baker in Saund. Ref. Bot. iii. App. 17); bulb ovoid, $\frac{1}{2}$ in. diam.; leaves 2–3, thin, fleshy, sessile, oblong-lanceolate, acute, 2–3 in. long, narrowed from below the middle to a clasping base; peduncle 1–2 in. long; raceme very dense, oblong, $\frac{1}{2}$–1 in. long; flowers all whitish; perianth $\frac{1}{12}$ in. long; segments all oblong; stamens included. *Journ. Linn. Soc.* xiii. 228.

KALAHARI REGION: Transvaal; Apies River, *Burke!*

6. **D. Saundersiæ** (Baker); bulb ovoid, $\frac{3}{4}$–1 in. diam.; leaves 3–4, sessile, oblanceolate, thin, $\frac{1}{2}$–1 ft. long, 1–1$\frac{1}{2}$ in. broad above the middle, narrowed gradually to a clasping base $\frac{1}{2}$ in. broad; peduncle a foot or more long; raceme 2–6 in. long, dense upwards, lax downwards; upper flowers sessile; lower cernuous, very shortly pedicellate; perianth nearly $\frac{1}{4}$ in. long; segments all oblong, greenish or brown in the central third, the rest whitish; stamens $\frac{1}{3}$ shorter than the perianth.

EASTERN REGION: Natal; Itafamasi, *Wood*, 774! 938! and without precise locality, *Mrs. K. Saunders!*

7. **D. humifusa** (Baker); bulb ovoid, 1$\frac{1}{4}$–1$\frac{1}{2}$ in. diam.; leaves 2, spreading, sessile, ovate, 3–4 in. long, 1$\frac{1}{2}$–2 in. broad, membranous, pale green, with a few blotches of darker green; peduncle slender, terete, erect, 3–4 in. long; racemes cylindrical, 3–4 in. long; pedicels very short; lower flowers cernuous; bracts obsolete; perianth greenish, campanulate, $\frac{1}{6}$ in. long; segments oblong; stamens half as long as the perianth; ovary sessile, globose; style as long as the ovary. *Scilla humifusa, Baker in Gard. Chron.* 1881, xv. 626.

EASTERN REGION: Natal; cultivated specimens grown by *Mr. W. Bull* at Chelsea! and in the Durban Botanic Garden, *Wood*, 4059!

XXXVIII. EUCOMIS, L'Herit.

Perianth polyphyllous, persistent; segments subequal, 1-nerved, patent, united in a cup at the very base, oblong or oblanceolate-oblong. *Stamens* 6, attached above the base of the segments; filaments deltoid at the base; anthers small, oblong, versatile, dehiscing introrsely. *Ovary* sessile, subglobose; ovules many in a cell, superposed; style cylindrical; stigma capitate or minutely tricuspidate. *Capsule* membranous, loculicidally 3-valved. *Seeds* obovoid, not compressed; testa black or brown; embryo nearly as long as the horny albumen.

Rootstock a large tunicated bulb; leaves multifarious, lorate or oblong; peduncle simple, cylindrical or clavate; inflorescence a raceme, crowned with a coma of leafy empty bracts; flowers greenish.

DISTRIB. One distinct species in the highlands of Central Africa.

Scape cylindrical:
　Pedicels long, erecto-patent:
　　Leaves membranous:
　　　　Perianth-segments concolorous 　　... 　... 　(1) **punctata.**
　　　　Perianth-segments margined with purple 　... 　(2) **bicolor.**
　　Leaves moderately firm 　... 　... 　... 　... 　(3) **pallidiflora.**
　Pedicels short:
　　Flowers greenish-white tinged with purple 　... 　(4) **humilis.**
　　Flowers green:
　　　　Leaf undulated, 2–3 in. broad 　　... 　... 　(5) **undulata.**
　　　　Leaf flat, 1¼–1½ in. broad 　... 　... 　... 　(6) **amaryllidifolia.**
Scape clavate:
　Flowers shortly pedicellate:
　　Leaves ensiform, 2 in. broad 　... 　... 　... 　(7) **robusta.**
　　Leaves lingulate, 3–4 in. broad 　... 　.. 　... 　(8) **regia.**
　Flowers subsessile 　... 　... 　.. 　... 　... 　(9) **nana.**

1. **E. punctata** (L'Herit. Sert. 18, t. 18); bulb globose, 2–3 in. diam.; leaves 6–9, oblanceolate, finally 1½–2 ft. long, 2–3 in. broad above the middle, thin in texture, acute, not undulated, spotted with purple on the back towards the base; peduncle terete, ½–1 ft. long, green, spotted with purple; racemes lax, cylindrical, finally a foot long, 2½–3 in. diam.; pedicels all erecto-patent, ½–1 in. long; lower bracts ovate; the others longer, lanceolate; leaves of the coma 12–20, oblong, acute; perianth green, ½ in. long; segments oblong; stamens ⅓ in. long; filaments dilated only at the base; capsule as long as the perianth, truncate at the apex. *Ait. Hort. Kew.* i. 433; *edit.* 2, ii. 246; *Thunb. Diss. Nov. Gen.* 129; *Fl. Cap. edit. Schult.* 316; *Bot. Mag. t.* 913; *Red. Lil.* iv. *t.* 208; *Kunth, Enum.* iv. 302; *Flore des Serres, t.* 2307. *Asphodelus comosus, Houtt. Handl.* xii. 336, *t.* 83. *Ornithogalum punctatum, Thunb. Prodr.* 62. *Basilœa punctata, Mirb. Hist. Nat. Pl.* viii. 339; *Zuccagn. in Roem. Coll.* 137; *Lam. Ill.* ii. 381, *t.* 239, *fig.* 2. *Fritillaria punctata, Gmel. Syst. Nat.* ii. 545.

VAR. β, **striata** (Willd. Enum. 364); leaves vertically striped with purple on the back. *Gawl. in Bot. Mag. t.* 1539. *E. striata, Donn. Hort. Cant. edit.* 6, 86.

VAR. γ, **concolor** (Baker); leaves and stems always green.

2. **E. bicolor** (Baker in Gard. Chron. 1878, x. 492); bulb large,
globose; leaves 5–6, lorate, thin, unspotted, undulated, finally
1½–2 ft. long, 3–4 in. broad at the middle; peduncle cylindrical,
1–1½ ft. long; raceme dense upwards, lax downwards, ½–1 ft. long,
2½–3 in. diam.; all the pedicels erecto-patent; central ¾–1 in. long;
bracts lanceolate, often tinged red; leaves of the coma 12–20, oblong,
acute, usually margined with red-purple; perianth ⅖–⅝ in. long;
segments oblong, pale greenish-white, margined with red-purple;
filaments red-purple, ¼ in. long, lanceolate-deltoid.   *Bot. Mag. t.* 6816.

EASTERN REGION: Griqualand East; Mount Currie, 2500 ft., *Tyson*, 1436!
Natal; Inanda, *Wood*, 720 ! Charlestown, *Wood*, 5609 ! Tugela River, *Allison*,
23 ! valleys of the Drakensberg, in swamps, 6000–7000 ft., *Evans*, 396 !

Described from a cultivated-plant sent from Natal by Mr. C. Mudd to J. Veitch
and Sons.

3. **E. pallidiflora** (Baker in Gard. Chron. 1887, ii. 154); leaves
lorate-lanceolate, acute, moderately firm, persistent, above 2 ft. long,
2–3 in. broad at the middle, not undulated; peduncle cylindrical,
½ ft. long, ½ in. diam.; raceme moderately dense, 8–9 in. long, 2 in.
diam.; lower pedicels erecto-patent, about ½ in. long; upper sometimes patent; lower bracts ovate; central longer, lanceolate; leaves
of coma about a dozen, ovate, membranous, acute; perianth green,
⅖–⅝ in. long; segments oblong; stamens much shorter than the
perianth; capsule shorter than the perianth.

KALAHARI REGION: Orange Free State, *Cooper*, 1195 !

Described from a plant cultivated by Mr. W. E. Gumbleton, introduced by
Mr. Nelson from S. E. Africa.

4. **E. humilis** (Baker in Kew Bullet. 1895, 152); leaves oblong,
moderately firm, ½ ft. long, 2–3 in. broad at the middle, obtuse, undulated at the margin, narrowed to the base, tinged with purple
above, much spotted with purple beneath; peduncle very short,
cylindrical; raceme dense, oblong, 2–4 in. long, crowned with a
dense coma of small oblong leaves with purple margins; pedicels
very short; bracts large, lanceolate; perianth ½ in. long, greenish-
white, tinged with purple; segments oblong; stamens shorter than
the perianth-segments; filaments purple; anthers small, oblong;
style as long as the ovary.

EASTERN REGION: Natal; summit of Tabamhlopi Mountain, 6000–7000 ft.,
Evans, 398 !

Intermediate between *E. bicolor* and *E. nana*.

5. **E. undulata** (Ait. Hort. Kew. i. 433, edit. 2, ii. 246); bulb
globose, 2–3 in. diam.; leaves 6–9, lorate-lanceolate, finally 1–1½ ft.
long, 2–3 in. broad at the middle, acute or obtuse, thin in texture,
undulated towards the margin, not spotted on the back; peduncle
cylindrical, ½–1 ft. long; raceme dense, oblong, 3–6 in. long, 2 in.

diam.; upper flowers patent, subsessile; lower patent or cernuous, on very short pedicels; lower bracts ovate; leaves of the coma 12–30, oblong, acute, crisped, membranous, 1½–2 in. long; perianth green, ¼ in. long; segments oblong or oblanceolate-oblong; filaments ¼ in. long, lanceolate-deltoid in the lower half; capsule subglobose, ⅓ in. diam. *Willd. Sp. Plant.* ii. 93; *Thunb. Diss. Nov. Gen.* 129; *Fl. Cap. edit. Schult.* 317; *Bot. Mag. t.* 1083; *Kunth, Enum.* iv. 302. *E. regia, Red. Lil.* iii. *t.* 175. *Fritillaria longifolia, Hill. Hort. Kew.* 354; *and edit.* 2, 354, *t.* 15. *Ornithogalum undulatum, Thunb. Prodr.* 62. *Basilæa coronata, Lam. Encyc.* i. 382. *Fritillaria autumnalis, Miller, Dict. edit.* 8, *No.* 10.

COAST REGION: Uitenhage Div., *Ecklon and Zeyher*, 102!

CENTRAL REGION: Somerset Div.; upper part of the Bosch Berg, 4500 ft., *MacOwan*, 1907! Graaff Reinet Div.; south side of the Sneeuw Berg Range, *Burke!* summit of Cave Mountain, 4200 ft., *Bolus*, 564! Colesberg Div.; near the Hondeblats River, *Burchell*, 2701! near Colesberg, *Shaw!*

KALAHARI REGION: Orange Free State; Mud River Drift, *Rehmann*, 3614! without precise locality, *Cooper*, 3285! Transvaal; Mooi River, *Burke!* summit of Saddleback Range, 5000 ft., *Galpin*, 712!

EASTERN REGION: Transkei Mountains, *Mrs. Barber!* Natal; Inanda, *Wood*, 436! 853!

We have at Kew old garden specimens dried by Gouan, Fothergill (1780), and Bishop Goodenough (1781).

6. E. amaryllidifolia (Baker in Gard. Chron. 1878, x. 492); bulb ovoid, 1½–2 in. diam.; leaves 5–6, lorate-lanceolate, obtuse or acute, unspotted, finally 12–15 in. long, 1¼–1½ in. broad at the middle, not crisped; peduncle under a foot long, unspotted, cylindrical; raceme dense, oblong, 2–3 in. long, 1½ in. diam.; pedicels very short; leaves of the coma 15–20, oblong, acute, membranous, crisped; perianth green, ½ in. long; segments oblong; stamens half as long as the perianth.

KALAHARI REGION: Orange Free State; Caledon River, *Burke!*

Described from a plant sent from Prof. MacOwan, which flowered at Kew in August, 1878.

7. E. robusta (Baker in Gard. Chron. 1894, xvi. 562); bulb large; basal leaves ensiform, acute, 2 ft. long, 2 in. broad at the middle, narrowed gradually to the base and apex, very thick in texture, crisped on the margin; peduncle very short, plain green, an inch thick; raceme dense, oblong, 6–8 in. long, crowned with a coma of 20–30 oblong acute reduced leaves; pedicels stout, ascending, ⅓–½ in. long; bracts ovate, acute, plain green, longer than the pedicels; perianth green, ⅝–¾ in. long; segments oblong-spathulate; filaments lanceolate, not more than half as long as the perianth; ovary globose; style short.

EASTERN REGION: Natal; near Koenigsberg, cultivated specimen!

Introduced by Messrs. Dammann & Co., of Naples, in 1894.

8. E. regia (Ait. Hort. Kew. i. 433; edit. 2, ii. 245); bulb globose, 2–3 in. diam.; leaves 6–8, lingulate, obtuse, firmer in texture

than in *E. undulata*, and not undulated towards the edge, finally
1–1½ ft. long, 3–4 in. broad at the middle, narrowed gradually from
the middle to the base; peduncle clavate, 3–6 in. long; raceme dense,
oblong, 3–6 in. long, 1½–2 in. diam.; leaves of the coma 12–20,
oblong, acute, membranous, crisped, 1½–2 in. long; upper flowers
patent; lower cernuous; pedicels very short; perianth green, ½ in.
long; segments oblong; stamens half as long as the perianth. *Willd.
Sp. Plant.* ii. 93; *Roem. et Schultes, Syst. Veg.* vii. 623; *Kunth,
Enum.* iv. 302; *Baker in Journ. Linn. Soc.* xiii. 225. *E. clavata*,
Baker in Saund. Ref. Bot. t. 238.

SOUTH AFRICA: without locality, *Masson!*
KALAHARI REGION: Orange Free State, *Cooper*, 1194! 1196!

9. **E. nana** (Ait. Hort. Kew. i. 432; edit. 2, ii. 245); bulb
globose, 2 in. diam.; leaves about 8, lingulate, obtuse, firmer in
texture than in *E. punctata* and *E. undulata*, suffused with purple on
the back towards the base, not undulated, finally 1½–2 ft. long,
3–4 in. broad above the middle, narrowed gradually from the middle
to the base; peduncle short, clavate, an inch thick at the top, spotted
with purple; raceme dense, oblong, 3–4 in. long, 2–2½ in. diam.;
pedicels scarcely any; leaves of the coma 12–20, oblong, acute;
perianth green, ½ in. long; segments oblong; filaments ¼ in. long,
lanceolate in the lower half. *Jacq. Collect.* iv. 213; *Hort. Schoenbr.*
i. 47, *t.* 92; *Thunb. Fl. Cap. edit. Schult.* 316; *Gawl. in Bot.
Mag. t.* 1495; *Roem. et Schultes, Syst. Veg.* vii. 623. *Ornithogalum
nanum, Thunb. Prodr.* 62. *Basilœa nana, Zuccagn. in Roem. Coll.*
136; *Poir. Encyc. Suppl.* i. 590.

VAR. **purpureocaulis** (Baker); peduncle coloured purple; leaves of the coma
margined with purple. *E. purpureocaulis, Andr. Bot. Rep. t.* 369; *Tratt. Thes.
t.* 55.

SOUTH AFRICA: without locality, *Masson!*
EASTERN REGION: Natal; South Downs, 5000–6000 ft., *Evans*, 401! Tugela
Valley, *Boyle!*

Mainly described from a plant that flowered at Kew in 1878.

XXXIX. SCILLA, Linn.

Perianth polyphyllous, persistent, campanulate or oblong; segments
subequal, 1-nerved, spreading from the base or a short space above
it. *Stamens* 6, attached to the base of the perianth-segments;
filaments filiform, or slightly flattened; anthers oblong, versatile,
dehiscing introrsely. *Ovary* sessile, or shortly stipitate, 3-celled;
ovules 2, collateral, or several, superposed; style filiform; stigma
capitate. *Capsule* loculicidally 3-valved. *Seeds* globose, or angled
by pressure; testa black; albumen firm.

Rootstock a tunicated bulb; leaves all radical, subterete or flat; inflorescence
a simple raceme; bracts small; flowers generally blue or mauve-purple.

DISTRIB. Widely spread through the Old World, and one or two doubtful species in Chili. Species about 80.

Subgenus EUSCILLA. Perianth-segments spreading from the base.

- Leaves subterete ... ... (1) **flexuosa.**
- Leaves linear... ... ... (2) **firmifolia.**
- Leaves lanceolate or lorate:
  - Flowers whitish:
    - Ovary green ... . (3) **rigidifolia.**
    - Ovary lilac... ... ... (4) **versicolor.**
  - Flowers bright blue-lilac:
    - Pedicels $\frac{1}{3}$–$\frac{1}{2}$ in. long ... ... (5) **Kraussii.**
    - Pedicels 1–1$\frac{1}{2}$ in. long ... ... (6) **natalensis.**
  - Flowers mauve-purple:
    - Leaves linear ... ... (7) **Adlami.**
    - Leaves oblong ... ... (8) **lachenalioides.**

Subgenus LEDEBOURIA. Perianth-segments spreading from above the base.

- Flowers small ($\frac{1}{8}$–$\frac{1}{6}$ in. long); more or less tinged with bright purple:
  - Bulb small ($\frac{1}{4}$–$\frac{1}{3}$ in. diam.):
    - Leaves subterete ... ... (9) **leptophylla.**
    - Leaves linear:
      - Leaves 2–3 in. long ... ... (10) **Barberi.**
      - Leaves 1–1$\frac{1}{2}$ in. long ... . (11) **minima.**
    - Leaves lanceolate, sessile:
      - Raceme $\frac{1}{2}$ in. diam. ... ... (12) **exigua.**
      - Raceme $\frac{3}{4}$–1 in. diam.:
        - Raceme oblong ... ... (13) **inandensis.**
        - Raceme globose ... ... (14) **Tysoni.**
    - Leaves lanceolate, petioled ... ... (15) **saturata.**
    - Leaves oblong:
      - Perianth $\frac{1}{8}$ in. long:
        - Raceme $\frac{1}{3}$ in. long and broad ... ... (16) **sphærocephala.**
        - Raceme $\frac{1}{2}$–$\frac{3}{4}$ in. long and broad ... (17) **Sandersoni.**
      - Perianth $\frac{1}{6}$ in. long:
        - Raceme $\frac{1}{2}$–$\frac{5}{8}$ in. long and broad ... (18) **globosa.**
        - Raceme 1–1$\frac{1}{4}$ in. long and broad ... (19) **Baurii.**
  - Bulbs middle-sized (about 1 in. diam.):
    - Racemes very lax... ... ... (20) **revoluta.**
    - Raceme lax, oblong-cylindrical ... ... (21) **humifusa.**
    - Racemes dense, oblong:
      - Leaves linear... .. .. (22) **Macowani.**
      - Leaves lanceolate:
        - Lower pedicels ascending ... ... (23) **concinna.**
        - Lower pedicels cernuous ... ... (24) **Cooperi.**
      - Leaves oblong-lanceolate ... ... (25) **pusilla.**
    - Racemes dense, globose:
      - Leaves lanceolate ... ... (26) **Rogersii.**
      - Leaves oblong or oblong-lanceolate:
        - Pedicels $\frac{1}{8}$–$\frac{1}{6}$ in. long:
          - Peduncle very short ... ... (27) **Leichtlini.**
          - Peduncle 2–3 in. long ... ... (28) **Galpini.**
        - Pedicels $\frac{1}{4}$–$\frac{1}{3}$ in. long ... ... (29) **costachys.**
  - Bulbs large (2–3 in. diam.):
    - Leaves broad ovate ... ... (30) **ovatifolia.**
    - Leaves lanceolate or oblong-lanceolate ... ... (31) **lanceæfolia.**
    - Leaves lorate-oblong ... ... (32) **polyantha.**
- Flowers small, green, not at all, or but slightly tinged with purple:
  - Bulbs $\frac{3}{4}$–1 in. diam.:
    - Leaves linear ... ... (33) **Nelsoni.**

Leaves lanceolate:
  Leaves 1½–3 in. long :
    Bulb ½ in. diam. ... ... (34) **Eckloni.**
    Bulb ¾–1 in. diam. ... ... (35) **Ludwigii.**
  Leaves 6–8 in. long ... ... (36) **prasina.**
  Leaves oblong-lanceolate ... ... (37) **diphylla.**
  Leaves oblong ... ... (38) **paucifolia.**
Bulb 1½–2 in. diam.:
  Leaves linear ... ... (39) **linearifolia.**
  Leaves lanceolate:
    Pedicels ¼–½ in. long :
      Filaments purple ... ... (40) **livida.**
      Filaments pale... ... ... (41) **concolor.**
    Pedicels ½–¾ in. long ... ... (42) **subsecunda.**
  Leaves lorate-lanceolate... ... ... (43) **megaphylla.**
  Leaves oblong-lanceolate ... ... (44) **laxiflora.**
  Leaves oblong ... ... (45) **socialis.**
Bulb 2–4 in. diam. :
  Leaves lanceolate ... ... (46) **tricolor.**
  Leaves oblong-lanceolate ... ... (47) **microscypha.**
Flowers larger, oblong campanulate, ¼–⅓ in. long :
  Leaves produced after the flowers ... ... (48) **undulata.**
  Leaves produced with the flowers :
    Leaves oblong ... ... (49) **spathulata.**
    Leaves oblong-lanceolate ... ... (50) **zebrina.**
    Leaves lanceolate, 3–4 in. long ... .. (51) **lanceolata.**
    Leaves lanceolate, about a foot long :
      Pedicels short :
        Peduncle short... ... ... (52) **subglauca.**
        Peduncle long... ... ... (53) **lorata.**
      Pedicels long :
        Filaments pale... ... ... (54) **pendula.**
        Filaments dark lilac ... ... (55) **floribunda.**
      Leaves lanceolate, 1½–2 ft. long ... ... (56) **princeps.**

1. S. flexuosa (Baker in Journ. Linn. Soc. xiii. 245); bulb not seen ; leaves about 3, subterete, fleshy, smooth, weak, ½–1 ft. long, $\frac{1}{12}$ in. diam.; peduncle slender, much shorter than the leaves ; raceme lax, oblong, 1–1½ in. long, 1 in. diam. ; pedicels ¼–⅓ in. long ; bracts lanceolate, nearly as long as the pedicels ; perianth campanulate, ¼ in. long ; stamens rather shorter than the perianth. *Hyacinthus flexuosus, Thunb. Prodr.* 64; *Fl. Cap. edit. Schult.* 326; *Willd. Sp. Plant.* ii. 168; *Roem. et Schultes, Syst. Veg.* vii. 585.

SOUTH AFRICA : interior region, *Thunberg!*
Described from Thunberg's original specimen.

2. S. firmifolia (Baker in Saund. Ref. Bot. iii. App. 7); bulb globose, 1–1½ in. diam., not crowned with bristles ; leaves 5–6, linear, glabrous, spreading, firm in texture, strongly ribbed, 5–6 in. long; peduncle slender, 6–8 in. long ; raceme moderately dense, cylindrical, 3–4 in. long, ½–⅝ in. diam.; pedicels articulated at the apex, lower ¼ in. long; bracts small, lanceolate; perianth campanulate, ⅛ in. long, white, tinged with red ; segments oblong; filaments much dilated at the base, shorter than the perianth-segments ; anthers oblong ; ovary substipitate, globose; style rather longer than the ovary. *Baker in Journ. Linn. Soc.* xiii. 237.

COAST REGION: Albany Div. ; rocks near New Year River, 2200 ft., *Mac-Owan*, 461!

EASTERN REGION: Tembuland ; between Morley and Umtata River, 1000–2000 ft., *Drège*, 4492!

3. **S. rigidifolia** (Kunth, Enum. iv. 330); bulb ovoid, 1½–1 in. diam., crowned with numerous persistent bristles; leaves 5–6, lanceolate, rigidly coriaceous, strongly ribbed, glabrous, ½–1 ft. long, ½–1 in. broad ; peduncle about as long as the leaves; raceme dense, 3–6 in. long, 2–3 in. broad; pedicels articulated at the apex, lower spreading, 1–1½ in. long ; bracts linear, ¼ in. long; perianth white, tinged with green, campanulate, ⅛–⅙ in. long; segments oblong; filaments linear, shorter than the perianth-segments ; ovary globose, green, stipitate ; ovules 6–8 in a cell; style rather longer than the ovary. *Baker in Journ. Linn. Soc.* xiii. 242. *S. pallidiflora, Baker in Saund. Ref. Bot. t.* 179.

VAR. β, **nervosa** (Baker in Journ. Linn. Soc. xiii. 242); more robust, with broader leaves and longer pedicels. *Ornithogalum nervosum, Burchell, Trav.* i. 537.

VAR. γ, **Gerrardi** (Baker) ; dwarfer, with fewer flowers and hairy linear leaves. *S. Gerrardi, Baker in Journ. Linn. Soc.* xiii. 237.

COAST REGION: King Williamstown Div. ; Keiskamma, *Mrs. Hutton!* Var. γ, Albany Div.; Grahamstown Flats, *MacOwan*, 214! King Williamstown Div.; Keiskamma, *Mrs. Hutton!*

CENTRAL REGION: Aliwal North Div.; Witte Bergen, 5000–6000 ft., *Drège*, 4506c!

KALAHARI REGION: Transvaal ; Apies River, *Burke!* Barberton, 2600–4000 ft., *Galpin*, 620! and without precise locality, *Sanderson!* Var. β, Hopetown Div.; by the Orange River, *Burchell*, 2663! Griqualand West ; between Griqua Town and Witte Water, *Burchell*, 1968! Orange Free State ; near Nelsons Kop, *Cooper*, 882! Bechuanaland ; Batlapin Territory, *Holub!* Chooi Desert, *Burchell*, 2337/2! Var. γ, Orange Free State, *Cooper*, 875! 2980!

EASTERN REGION: Tembuland ; Bazeia, 2000 ft., *Baur*, 577! Griqualand East ; Mountains near Kokstad, 5000 ft. *Tyson*, 1480! Natal, *Buchanan! Plant!* Var. β, Natal ; Inanda, *Wood*, 458! Clairmont, *Wood*, 1944! Var. γ, Tembuland ; Tabase, near Bazeia, 2500 ft., *Baur*, 378! Natal ; Weenen County, South Downs, 5000 ft., *Wood*, 4393! and without precise locality, *Gerrard*, 1829!

4. **S. versicolor** (Baker in Saund. Ref. Bot. t. 305); bulb globose; leaves 6–8, lanceolate, 6–9 in. long ; peduncle terete, as long as the leaves ; raceme conical, moderately dense, 5–6 in. long, 3½–4 in. diam.; lower pedicels patent, 1½–2 in. long ; bracts small, lanceolate ; perianth campanulate, ⅙ in. long, white, tinged with green ; filaments linear, shorter than the perianth-segments ; ovary globose, bright lilac ; ovules about 6 in a cell ; style rather longer than the ovary. *Baker in Journ. Linn. Soc.* xiii. 243.

SOUTH AFRICA : without locality. No specimen at Kew.

Described and figured from a living plant in the collection of Mr. Wilson Saunders at Reigate in 1871, introduced by Mr. Thos. Cooper.

5. **S. Kraussii** (Baker in Journ. Linn. Soc. xiii. 243); bulb globose, 1½–2 in. diam., not crowned with bristles ; leaves about 4, spreading, lanceolate, firm in texture, strongly ribbed, densely pubescent, 2–3 in. long; peduncle slender, 6–9 in. long; raceme

moderately dense, 3–4 in. long, 1–1½ in. diam.; pedicels erecto-patent, articulated at the tip, ⅓–½ in. long; bracts linear, ⅛ in. long; perianth campanulate, bright blue, ⅙ in. long; segments oblong; filaments shorter than the perianth-segments, deltoid at the base; ovary subsessile,' globose; ovules few, superposed.

EASTERN REGION: Pondoland, *Bachmann*, 286! Natal; Inanda, *Wood*, 357! Noodsberg, *Wood*, 928! stony hill near York, 3000–4000 ft., *Wood*, 4304! and without precise locality, *Gerrard*, 740! *Krauss*, 444!

6. **S. natalensis** (Planch. in Flore des Serres, t. 1043); bulb ovoid, 3–4 in. diam., not bristly at the neck; leaves 6–8, lorate-lanceolate, glabrous or pubescent, 1–1½ ft. long, 3–4 in. broad at the middle, narrowed gradually to the base; peduncle stout, stiffly erect, 1–1¼ ft. long; raceme dense, ½–1 ft. long, 2–3 in. diam.; pedicels articulated at the apex, lower 1–1½ in. long; bracts linear, ¼–⅓ in. long; perianth bright blue, ¼–⅓ in. long; segments oblong-lanceolate; filaments linear, shorter than the perianth; ovary globose, shortly stipitate; ovules 10–12 in a cell; style as long as the ovary. *Bot. Mag. t.* 5379; *Baker in Journ. Linn. Soc.* xiii. 243.

VAR. β, sordida (Baker, loc. cit.); smaller, with leaves tinged with brown, and fewer, smaller flowers.

KALAHARI REGION: Orange Free State, *Cooper*, 3278! Transvaal; Saddle-back Range, near Barberton, 2800–4000 ft., *Galpin*, 619!

EASTERN REGION: Tembuland; mountain-sides near Bazeia, 2300–3000 ft., *Baur*, 413! Natal, *McKen! Mrs. K. Saunders!* Var. β, Natal, cultivated specimen!

Var. β is described from a plant that flowered at Kew in 1864. The little known *S. plumbea, Lindl. in Bot. Reg. t.* 1355, described from a drawing made by Sydenham Edwards in Kew Gardens in 1813, is probably a form of this species, and may be the same as var. *sordida.*

7. **S. Adlami** (Baker in Gard. Chron. 1891, ix. 521); bulb ovoid, ½ in. diam.; leaf single, contemporary with the flowers, linear, rather fleshy, glabrous, 8–9 in. long, ½ in. broad above the middle, narrowed gradually to the base; peduncle slender, 3–4 in. long; raceme short; pedicels ¼–½ in. long; bracts deltoid, minute; perianth mauve-purple, ⅙ in. long; segments oblong, spreading from the base; stamens shorter than the perianth; filaments flattened; anthers minute; ovary green, globose; style short.

EASTERN REGION: Natal, cultivated specimen, *Adlam!*

Sent by Mr. J. W. Adlam to Mr. J. H. Tillett, of Sprowston, near Norwich, with whom it flowered April, 1891.

8. **S. lachenalioides** (Baker); bulb globose, 1 in. diam.; leaves 2, subbasal, ascending, oblong, acute, sessile, 3–4 in. long, 1–1½ in. broad; peduncle 3–4 in. long; raceme dense, oblong, 1–1¼ in. long; pedicels very short; bracts minute; perianth oblong, bright mauve-purple, ⅓–½ in. long; segments linear-oblong, with a greenish keel and minutely cucullate tip; stamens more than half as long as the perianth; anthers small, oblong; style as long as the ovary.

EASTERN REGION: Transkei, *Hallack!* Tembuland; Bazeia Mountain, 2500–3500 ft., *Baur*, 549! Griqualand East; Malowe Mountain, near Clydesdale, 4000 ft., *Tyson*, 2878!

9. S. leptophylla (Baker); bulb small, ovoid; leaves 5–6, firm, subterete, glabrous, arcuate, 2 in. long, $\frac{1}{3}$ lin. diam.; peduncle 1–1$\frac{1}{2}$ in. long; raceme dense, subglobose, $\frac{1}{2}$–$\frac{3}{4}$ in. long and broad; pedicels $\frac{1}{8}$–$\frac{1}{6}$ in. long, the lower cernuous; bracts minute, linear; perianth campanulate, bright mauve-purple, $\frac{1}{8}$ in. long; stamens as long as the perianth; filaments bright purple.

KALAHARI REGION : Transvaal ; Drakensberg Range, near the Devils Kantoor, 5200 ft., *Bolus*, 7623 !

10. S. Barberi (Baker in Journ. Linn. Soc. xiii. 247); bulb small; leaves linear, 2–3 in. long, $\frac{1}{8}$ in. broad; peduncle slender, 2–2$\frac{1}{2}$ in. long; raceme lax, oblong, 1–1$\frac{1}{4}$ in. long, $\frac{5}{8}$ in. diam.; pedicels patent, $\frac{1}{6}$–$\frac{1}{4}$ in. long, lower cernuous; perianth campanulate, bright purple inside, the base green outside; stamens nearly as long as the perianth; filaments filiform, bright purple; ovary stipitate, discoid at the base.

EASTERN REGION : Transkei; near the Tsomo River, *Mrs. Barber*, 805 ! Not in the Kew Herbarium.

11. S. minima (Baker in Saund. Ref. Bot. iii. App. 6); bulb ovoid, $\frac{1}{4}$–$\frac{1}{3}$ in. diam.; leaves 2–3, linear, suberect, 1–1$\frac{1}{2}$ in. long; peduncle very slender, rather longer than the leaves; raceme moderately dense, oblong, $\frac{1}{2}$–1 in. long, $\frac{1}{3}$ in. diam.; pedicels $\frac{1}{8}$–$\frac{1}{6}$ in. long, lower cernuous; perianth campanulate, $\frac{1}{12}$ in. long, green, more or less tinged with purple; stamens as long as the perianth; filaments green, linear; ovary globose, stipitate. *Baker in Journ. Linn. Soc. xiii.* 246.

CENTRAL REGION: Aliwal North; rocky heights near Kraai River, 4500 ft., *Drège*, 3510 !

KALAHARI REGION : Transvaal: Magalies Berg, *Burke !*

12. S. exigua (Baker in Journ. Linn. Soc. xiii. 247); bulb small, ovoid ; leaves lanceolate, 1$\frac{1}{2}$–2 in. long, $\frac{1}{4}$ in. broad, striped and spotted on the back with purple ; peduncle slender, erect, 2–3 in. long ; raceme moderately dense, oblong, an inch long, $\frac{1}{2}$ in. diam.; pedicels $\frac{1}{8}$–$\frac{1}{6}$ in. long, lower cernuous ; perianth campanulate, $\frac{1}{8}$ in. long, bright mauve-purple inside, green at the base outside ; stamens shorter than the perianth ; filaments filiform, bright purple; ovary globose, stipitate, discoid at the base.

EASTERN REGION: Natal; near Camperdown, *Sanderson*, 670 ! Not in the Kew Herbarium.

13. S. inandensis (Baker); bulb ovoid, $\frac{1}{2}$–$\frac{3}{4}$ in. diam.; leaves 2–3, sessile, lanceolate, 3 in. long, $\frac{1}{3}$–$\frac{1}{2}$ in. broad at the middle, narrowed gradually to the base and apex; peduncle slender, curved, 3–6 in. long ; raceme moderately dense, oblong, 1–1$\frac{1}{2}$ in. long, 1 in. diam.; central pedicels $\frac{1}{4}$–$\frac{1}{3}$ in. long, lower cernuous ; perianth campanulate, $\frac{1}{8}$–$\frac{1}{6}$ in. long, greenish outside, mauve-purple inside; filaments filiform, bright purple, nearly as long as the perianth; ovary globose, stipitate, discoid at the base.

EASTERN REGION: Natal; Inanda, *Wood*, 630 ! near Tongaati River, *Mrs. K. Saunders !*

14. S. Tysoni (Baker); bulb $\frac{1}{2}$ in. diam.; leaves 2, sessile, curved, lanceolate, 2–3 in. long, $\frac{1}{4}$–$\frac{1}{2}$ in. broad at the middle; peduncle 1$\frac{1}{2}$–2 in. long; raceme dense, globose, $\frac{3}{4}$–1 in. diam.; pedicels $\frac{1}{4}$ in. long; bracts minute; perianth campanulate, bright mauve-purple, $\frac{1}{6}$ in. long; stamens rather shorter than the perianth; filaments bright mauve-purple.

EASTERN REGION: Griqualand East, *Tyson !*

15. S. saturata (Baker in Journ. Bot. 1874, 365); bulb ovoid, $\frac{1}{2}$–$\frac{3}{4}$ in. diam.; leaves 3–4, lanceolate, acuminate, 4–8 in. long, $\frac{1}{4}$–$\frac{1}{3}$ in. broad at the middle, narrowed gradually to a clasping petiole 1–2 in. long; peduncle slender, 3–6 in. long; raceme dense, oblong, 1–1$\frac{1}{2}$ in. long, $\frac{3}{4}$–1 in. diam.; central pedicels $\frac{1}{6}$–$\frac{1}{3}$ in. long, lower flowers cernuous; perianth campanulate, bright mauve-purple inside and out, $\frac{1}{8}$–$\frac{1}{6}$ in. long; stamens nearly as long as the perianth; filaments linear, bright purple; ovary globose, stipitate.

KALAHARI REGION: Orange Free State, *Cooper*, 993! Transvaal; MacMac, *Mudd !*

EASTERN REGION: Pondoland, *Bachmann*, 291! Natal; Inanda, *Wood*, 1056! Umzinyati Valley, in a swamp, *Wood*, 1356! Swaziland; Havelock Concession, 5000 ft., *Saltmarshe in Herb. Galpin*, 1053 !

16. S. sphærocephala (Baker); bulb $\frac{1}{2}$ in. diam.; leaves 3, sessile, oblong, glabrous, thin, under an inch long; peduncle slender, 1$\frac{1}{2}$ in. long; raceme dense, globose, $\frac{1}{2}$ in. long and broad; pedicels very short; bracts minute; perianth campanulate, bright mauve-purple, $\frac{1}{8}$ in. long; stamens as long as the perianth; filaments bright purple.

WESTERN REGION: Little Namaqualand, *Bolus !* No specimen at Kew.

17. S. Sandersoni (Baker in Saund. Ref. Bot. iii. App. 5); bulb small; leaves oblong, acute, 2–3 in. long, $\frac{3}{4}$–1$\frac{1}{4}$ in. broad at the middle; peduncle slender, as long as the leaves; raceme dense, subglobose, $\frac{1}{2}$–$\frac{3}{4}$ in. long and broad; pedicels $\frac{1}{4}$–$\frac{1}{2}$ in. long, lower cernuous; perianth campanulate, bright mauve-purple inside, $\frac{1}{6}$ in. long, green at the base outside, and the segments keeled with green; stamens as long as the perianth; filaments linear, bright purple, ovary stipitate. *Baker in Journ. Linn. Soc. xiii. 246.*

KALAHARI REGION : Transvaal, *Sanderson !*

EASTERN REGION : Tembuland; Bazeia Mountain, 3500–4000 ft., *Baur*, 550 partly! Natal; summit of a hill above York, 4000–5000 ft., *Wood*, 4312!

18. S. globosa (Baker); bulb $\frac{1}{2}$ in. diam.; leaves 2, sessile, ascending, oblong-lanceolate, 2–2$\frac{1}{2}$ in. long; peduncle 1$\frac{1}{2}$ in. long; raceme dense, globose, $\frac{1}{2}$–$\frac{5}{8}$ in. diam.; pedicels ascending, lower rather longer than the flowers; bracts minute; perianth campanulate, greenish-purple, $\frac{1}{6}$ in. long; stamens included.

EASTERN REGION : Griqualand East; near Kokstadt, *Tyson !* No specimen at Kew.

19. S. Baurii (Baker); bulb globose, $\frac{1}{3}$–$\frac{1}{2}$ in. diam.; leaves 1–2, oblong, 1–2 in. long, $\frac{1}{4}$–1 in. broad at the middle; peduncle slender, as long as the leaves; raceme dense, subglobose, 1–1¼ in. long and broad; central pedicels ¼–⅓ in. long, lower cernuous; perianth campanulate, bright mauve-purple inside and out, ⅙ in. long; stamens rather shorter than the perianth; filaments filiform, very dark purple; ovary stipitate, globose.

EASTERN REGION: Tembuland; Bazeia Mountain, 3500–4000 ft., *Baur*, 550 partly!

20. S. revoluta (Baker in Saund. Ref. Bot. iii. App. 6); bulb globose, about 1 in. diam., abundantly stoloniferous; leaves 6–9, spreading, oblong-spathulate, acute, 2½–3 in. long, ½–¾ in. broad at the middle; peduncle slender, curved, 3–6 in. long; raceme lax, oblong, 1–3 in. long, 1–1½ in. diam.; central pedicels spreading, cernuous at the tip, ½–¾ in. long; perianth campanulate, bright mauve-purple inside, green outside, ⅛–⅙ in. long; stamens nearly as long as the perianth; filaments filiform, purple; ovary globose, stipitate, discoid at the base. *Hyacinthus revolutus, Linn. fil. Suppl.* 204; *Thunb. Prodr.* 64; *Fl. Cap. edit. Schult.* 326; *Roem. et Schultes, Syst. Veg.* vii. 585. *Phalangium revolutum, Pers. Syn.* i. 367. *Xeodolon revolutum, Salisb. Gen.* 18. *Drimia* (?) *revoluta, Kunth, Enum.* iv. 341. *D. lanceæfolia* β, *Gawl. in Bot. Mag. t.* 1380. *D. lanceæfolia, Lodd. Bot. Cab. t.* 278. *D. Gawleri and ovalifolia, Schrad. Blumenb.* 30; *Roem. et Schultes, Syst. Veg.* vii. 595–6; *Kunth, Enum.* iv. 339.

SOUTH AFRICA: without locality, interior region, *Thunberg.*
COAST REGION: Caledon Div.; on Donker Hoek Mountain, *Burchell*, 7982!

21. S. humifusa (Baker in Gard. Chron. 1881, xv. 626); bulb ovoid, 1¼–1½ in. diam.; leaves 2, spreading on the surface of the ground, sessile, cordate-oblong, obtuse, 3–4 in. long, more than half as broad, pale green, with a few blotches of darker green; peduncle erect, 3–4 in. long, purple at the base; raceme lax, oblong-cylindrical, 3–4 in. long; pedicels all very short, the lower cernuous; bracts obsolete; perianth campanulate, purplish-green, ⅙ in. long; stamens nearly as long as the perianth; ovary sessile, oblong.

EASTERN REGION: Natal. No specimen at Kew.

Described from a plant flowered by Mr. W. Bull, May, 1881.

22. S. Macowani (Baker in Gard. Chron. 1875, iii. 748); bulb globose, about 1 in. diam; leaves 3–4, linear, erect, 3–6 in. long, ¼ in. broad, channelled down the face, mottled with purple on the back in the lower half; peduncle very slender, 2–4 in. long; raceme moderately dense, oblong, finally 2–3 in. long, ½–⅝ in. diam.; central pedicels ⅙–¼ in. long, lower cernuous at the tip; perianth campanulate, ⅛ in. long, green, more or less tinged with bright purple; filaments nearly as long as the perianth, bright purple; ovary globose, stipitate.

CENTRAL REGION: Somerset Div.; on the Bosch Berg, 4000 ft., *MacOwan*, 1841

23. S. concinna (Baker in Saund. Ref. Bot. t. 235); bulb ovoid, middle-sized ; leaves 3–4, erect, lanceolate, 8–9 in. long, under an inch broad, barred transversely all over the back with purple blotches; peduncle moderately stout, stiffly erect, $\frac{1}{2}$ ft. long; raceme dense, oblong, 1$\frac{1}{2}$–2 in. long, 1–1$\frac{1}{4}$ in. diam. ; pedicels erecto-patent, $\frac{1}{4}$–$\frac{1}{3}$ in. long ; perianth campanulate, bright mauve-purple inside, $\frac{1}{6}$–$\frac{1}{5}$ in. long; stamens shorter than the perianth ; filaments dark purple ; ovary stipitate, discoid at the base.   *Baker in Journ. Linn. Soc.* xiii. 252.

SOUTH AFRICA : without locality, *Cooper.*   No specimen at Kew.

Described from a living plant in the garden of Mr. Wilson Saunders at Reigate about 1870, introduced by Mr. Thos. Cooper.

24. S. Cooperi (Hook. fil. in Bot. Mag. t. 5580); bulb subglobose, 1 in. diam. ; leaves 4–5, erect, lanceolate, 8–10 in. long, under an inch broad, striated on the back with purple and spotted towards the base ; peduncle curved, 4–6 in. long ; raceme moderately dense, oblong, 2–3 in. long, 1–1$\frac{1}{4}$ in. diam. ; central pedicels $\frac{1}{4}$–$\frac{1}{3}$ in. long, lower cernuous; perianth campanulate, $\frac{1}{2}$ in. long, green outside, bright mauve-purple inside; filaments filiform, dark purple ; ovary stipitate, discoid at the base.   *Baker in Saund. Ref. Bot.* iii. *App.* 7 ; *Journ. Linn. Soc.* xiii. 247.

SOUTH AFRICA : without locality, cultivated specimen, *Cooper !*
KALAHARI REGION : Orange River, cultivated specimen, *Hort. Bull !*

25. S. pusilla (Baker in Journ. Bot. 1876, 183); bulb ovoid or oblong, about 1 in. diam., 1 in. long ; tunics many, brown, firm in texture ; leaves 5–6, sessile, oblong-lanceolate, acuminate, 1$\frac{1}{2}$–2 in. long ; peduncle as long as the leaves ; raceme dense, oblong, 1–1$\frac{1}{4}$ in. long, $\frac{1}{2}$ in. diam. ; central pedicels $\frac{1}{12}$–$\frac{1}{8}$ in. long, lower cernuous; perianth campanulate, $\frac{1}{8}$ in. long, mauve-purple; stamens as long as the perianth ; filaments bright purple ; ovary sessile, globose.

EASTERN REGION : Tembuland; hill sides, Bazeia, 2000 ft., *Baur,* 293!

26. S. Rogersii (Baker); bulb globose, 1 in. diam.; leaves 5–6, sessile, lanceolate, 2$\frac{1}{2}$–3 in. long, $\frac{1}{4}$–$\frac{1}{3}$ in. broad at the middle, narrowed gradually to the base and apex; peduncles 2 to a bulb, slender, curved, as long as the leaves ; raceme dense, subglobose, $\frac{3}{4}$ in. long and broad ; central pedicels $\frac{1}{4}$ in. long, lower cernuous ; perianth campanulate, $\frac{1}{8}$ in. long, bright purple inside and out; filaments dark purple, filiform, nearly as long as the perianth; ovary globose, stipitate, discoid at the base.

SOUTH AFRICA: without locality, *Rogers !*

27. S. Leichtlinii (Baker) ; bulb ovoid, 1 in. diam. ; leaves 2–3, spreading, oblong, acute, 2–3 in. long, 1–1$\frac{1}{4}$ in. broad at the middle, copiously marbled with black protuberant spots on a bright green groundwork ; peduncle slender, under an inch long ; raceme dense, globose, $\frac{5}{8}$–$\frac{3}{4}$ in. long and broad; central pedicels $\frac{1}{4}$ in. long, lower

cernuous; perianth campanulate, $\frac{1}{2}$ in. long, bright mauve-purple both
inside and out, down to the very base; stamens nearly as long as the
perianth; filaments filiform, bright purple; ovary globose, stipitate.

SOUTH AFRICA: without locality, cultivated specimen!

Described from a living plant that flowered with Max Leichtlin, at Baden-
Baden, in June, 1878.

28. S. Galpini (Baker); bulb globose, 1 in. diam.; leaves 3–4,
oblong-spathulate, sessile, 2–3 in. long, under an inch broad; peduncle
slender, curved, 2–3 in. long; raceme dense, subglobose, $\frac{3}{4}$ in. diam.;
pedicels $\frac{1}{8}$–$\frac{1}{6}$ in. long, bracts minute; perianth campanulate, bright
mauve-purple, $\frac{1}{3}$ in. long; stamens nearly as long as the perianth;
filaments bright purple.

KALAHARI REGION: Transvaal; summit of Devil's Kantoor, near Barberton,
5600 ft., *Galpin*, 672!

29. S. oostachys (Baker); bulb 1 in. diam.; leaves 2, oblong or
oblong-lanceolate, 3–4 in. long, an inch broad at the middle; peduncle
slender, 3–4 in. long; raceme dense, globose, an inch long and broad;
lower pedicels cernuous, $\frac{1}{4}$–$\frac{1}{3}$ in. long; bracts minute; perianth bright
mauve-purple, $\frac{1}{8}$–$\frac{1}{6}$ in. long; stamens as long as the perianth; filaments
bright mauve-purple.

EASTERN REGION: Natal; Upper Umkomaas, 5000–6000 ft., *Wood*, 4627!

30. S. ovatifolia (Baker in Saund. Ref. Bot. t. 183); bulb globose,
2 in. diam.; leaves 4, spreading, ovate, acute, 3–4 in. long, 2–3 in.
broad, with a distinct cartilaginous margin and copious, round, green
spots on a glaucous green ground; peduncle stout, very short; raceme
dense, oblong, 2–3 in. long, 1$\frac{1}{4}$–1$\frac{1}{2}$ in. diam.; central pedicels $\frac{1}{3}$–$\frac{1}{2}$ in.
long, lower cernuous; perianth campanulate, $\frac{1}{6}$–$\frac{1}{5}$ in. long, bright
mauve-purple inside, green outside at the base; filaments bright
purple; ovary stipitate, globose, discoid at the base. *S. lanceæfolia
var. ovatifolia, Baker in Journ. Linn. Soc.* xiii. 252.

EASTERN REGION: Griqualand East; rugged mountain slopes around Clydesdale, 5000 ft., *Tyson*, 1123!

Described from a plant cultivated by Mr. W. W. Saunders between 1862
and 1870, introduced by Cooper.

31. S. lanceæfolia (Baker in Saund. Ref. Bot. t. 182); bulb
ovoid, 2–3 in. diam.; leaves 6–8, oblong or oblong-lanceolate, acute,
finally 5–6 in. long, 1$\frac{1}{2}$–2 in. broad at the middle, with blotches of
dark green on a pale green ground; peduncles 2–5 to a bulb, curved,
2–4 in. long; raceme dense, oblong, 2–4 in. long, 1$\frac{1}{4}$–1$\frac{1}{2}$ in. diam.;
central pedicels $\frac{1}{3}$–$\frac{1}{2}$ in. long, lower cernuous; perianth oblong-campanulate, green, more or less tinged with mauve-purple, $\frac{1}{6}$ in. long;
filaments bright purple; ovary globose, stipitate. *Baker in Journ.
Linn. Soc.* xiii. 251. *Lachenalia lanceæfolia, Jacq. Collect. Suppl.* 69;
Ic. ii. 17, *t.* 402; *Willd. Sp. Plant.* ii. 178; *Red. Lil.* i. *t.* 59; *Gawl.
in Bot. Mag. t.* 643. *L. maculata, Tratt. Archiv.* ii. 132, *t.* 168.
Scilla maculata, Schrank. Fl. Monac. t. 100. *Drimia lanceæfolia,
Gawl. in Bot. Mag. sub t.* 1380; *Roem. et Schultes, Syst. Veg.* vii.

594; *Schrader, Blumenb.* 29; *Kunth, Enum.* iv. 339. *Sugillaria lanceæfolia, Salisb. Gen.* 18. *Drimia acuminata, Lodd. Bot. Cab. t.* 1041; *Kunth, Enum.* iv. 339.

COAST REGION : Albany Div. ; Fish River Heights, *Hutton !*

CENTRAL REGION : Somerset Div. ; at the foot of the Bosch Berg, 2300 ft., *MacOwan*, 1840 !

KALAHARI REGION : Orange Free State, *Cooper*, 3286 ! Bechuanaland ; near Takun, *Burchell*, 2260 ! 2305 ! Transvaal ; Apies River, *Burke !* MacMac, *Mudd !* near Pretoria, 4000 ft., *McLea in Herb. Bolus*, 3093 ! Saddleback Range, near Barberton, 4000–5000 ft., *Galpin*, 508 !

EASTERN REGION : Natal ; Inanda, *Wood*, 181 ! 267 ! near Durban, *Wood*, 41 ! *Krauss*, 464 ! Bothas Hill, *Wood*, 4776 ! and without precise locality, *Plant*, 102 ! *Cooper*, 3287 !

32. S. polyantha (Baker in Gard. Chron. 1878, ix. 104) ; bulb globose, $1\frac{1}{2}$–2 in. diam. ; leaves 4, lorate-oblong, suberect, 12–15 in. long, 2–$2\frac{1}{4}$ in. broad at the middle, with dark green blotches on the face on a pale green ground, spotted with purple on the back towards the base ; peduncle slender, 6–9 in. long, spotted with purple ; raceme dense, oblong, 4–5 in. long, $1\frac{1}{2}$–2 in. diam. ; central pedicels $\frac{1}{4}$ in. long, lower cernuous ; perianth campanulate, $\frac{1}{6}$ in. long, green outside, mauve-purple inside ; filaments dark purple ; ovary globose, stipitate.

VAR. β, **angustifolia** (Baker) ; leaves lanceolate, under an inch broad at the middle.

KALAHARI REGION : Var. β, Transvaal ; Saddleback Range, near Barberton, 4000–5000 ft., *Galpin*, 1096 !

EASTERN REGION : Griqualand East ; mountain-sides around Clydesdale, 2500 ft., *Tyson*, 2159 ! Natal ; York, cultivated specimen, *Hort. Bull !*

33. S. Nelsoni (Baker) ; bulb small ; leaves linear, erect, channelled down the face, 3–4 in. long, $\frac{1}{8}$–$\frac{1}{6}$ in. broad ; peduncle very slender, shorter than the leaves ; raceme oblong, 1–2 in. long, $\frac{1}{2}$–$\frac{3}{4}$ in. diam. ; central pedicels $\frac{1}{6}$ in. long, lower cernuous ; perianth campanulate, green, $\frac{1}{8}$ in. long ; filaments bright purple ; ovary stipitate, globose.

SOUTH AFRICA : without locality, *Burke !*

KALAHARI REGION : by the Vaal River, *Nelson*, 167 !

The briefly characterized *S. viridiflora, Baker in Journ. Linn. Soc.* xiii. 255 (*Drimia viridiflora, Ecklon ex Kunze in Linnæa*, xx. 10), gathered (in Natal ?) by Gueinzius, is said to have linear-subulate leaves, very short pedicels, and greenish flowers.

34. S. Eckloni (Baker in Engl. Jahrb. xv. Beibl. 35, 7) ; bulb oblong, $\frac{1}{3}$ in. diam. ; leaves 3, ascending, lanceolate, synanthous, $1\frac{1}{2}$–2 in. long, $\frac{1}{4}$ in. broad ; peduncles 2 from a bulb, slender, $1\frac{1}{2}$ in. long ; raceme dense, few-flowered, $\frac{1}{2}$–$\frac{2}{3}$ in. long ; pedicels very short ; bracts ovate, minute ; perianth campanulate, $\frac{1}{12}$ in. long ; segments oblong, green, with a pale margin ; stamens included ; ovary globose ; style short.

SOUTH AFRICA : without locality, *Ecklon and Zeyher*, 12 ! No specimen at Kew.

35. S. Ludwigii (Baker in Saund. Ref. Bot. iii. App. 9) ; bulb ovoid, $\frac{3}{4}$–1 in. diam. ; leaves 5–6, sessile, lanceolate, 2–3 in. long, $\frac{1}{4}$ in. broad at the base, narrowed gradually to the apex ; peduncle 1–2 in. long; raceme moderately dense, oblong or oblong-cylindrical, finally 2–3 in. long, $\frac{1}{3}$–$\frac{1}{2}$ in. diam.; pedicels $\frac{1}{3}$–$\frac{1}{2}$ in. long, lower cernuous; perianth campanulate, green, $\frac{1}{3}$ in. long; filaments purple towards the apex; ovary globose, sessile. *Baker in Journ. Linn. Soc.* xiii. 248. *Drimia Ludwigii, Miquel in Bull. Scien. Phys. Neerl.* 1839, 39. *D. ensifolia, Eckl. in Linnæa,* xx. 235 (*name only*). *Idothea* (?) *Ludwigii, Kunth, Enum.* iv. 681.

COAST REGION: Uitenhage Div.; near the Zwartkops River, *Ecklon and Zeyher*, 10! *Zeyher*, 4262! Cathcart Div.; between the Katberg and Klipplaat River, *Drège*, 8618a!

Drimia angustifolia, Kunth, Enum. iv. 340 (*Drège*, 8618b), which I have not seen, is perhaps a narrow-leaved variety of this species.

36. S. prasina (Baker in Saund. Ref. Bot. iii. App. 10); bulb ovoid, about 1 in. diam.; leaves 5–6, thin, erect, sessile, lanceolate, 6–8 in. long, 1–1$\frac{1}{2}$ in. broad at the middle, sometimes spotted with purple; peduncles slender, 2–4 in. long; raceme moderately dense, oblong, finally 2–3 in. long, $\frac{5}{8}$–$\frac{3}{4}$ in. broad; central pedicels $\frac{1}{2}$ in. long, lower cernuous; perianth campanulate, green, $\frac{1}{3}$–$\frac{1}{2}$ in. long; filaments purple towards the tip; ovary globose, stipitate. *Journ. Linn. Soc.* xiii. 248.

COAST REGION: Albany Div.; Grahams Town, cultivated specimen! "Kaffirland," *Gill*!

CENTRAL REGION: Graaff Reinet Div.; hills near Graaff Reinet, 2500 ft., *Bolus*, 782! Somerset Div.; at the foot of the Bosch Berg, 2500 ft., *MacOwan*, 1842!

KALAHARI REGION: Bechuanaland: between Hamapery and Kosi Fontein, *Burchell*, 2545!

37. S. diphylla (Baker); bulb globose, $\frac{3}{4}$–1 in. diam.; leaves 2, erect, sessile, oblong-lanceolate, 2–4 in. long, $\frac{5}{8}$–$\frac{3}{4}$ in. broad at the middle; peduncle 2–8 in. long; raceme dense, oblong, 1–1$\frac{1}{2}$ in. long; lower pedicels ascending, $\frac{1}{4}$ in. long; bracts minute; perianth campanulate, green, $\frac{1}{3}$ in. long; stamens shorter than the perianth.

KALAHARI REGION: Transvaal; Saddleback Range, near Barberton, 5000 ft., *Galpin*, 1182!

38. S. paucifolia (Baker in Saund. Ref. Bot. t. 181); bulb ovoid, about an inch in diameter; leaves 2–3, spreading, oblong, acute, 2–3 in. long, under an inch broad, spotted with dark green on a pale green ground; peduncles slender, curved, 3–4 in. long; raceme lax, oblong, finally 3–4 in. long, 1$\frac{1}{4}$–1$\frac{1}{2}$ in. diam.; central pedicels $\frac{1}{3}$–$\frac{1}{2}$ in. long, lower cernuous; perianth green, campanulate, $\frac{1}{6}$–$\frac{1}{3}$ in. long; filaments purple in the upper half; ovary globose, stipitate, discoid at the base. *Journ. Linn. Soc.* xiii. 251.

SOUTH AFRICA: without locality. No specimen at Kew.

Described from a plant flowered by Mr. Wilson Saunders at Reigate about 1870, introduced by Mr. Thos. Cooper.

39. S. linearifolia (Baker in Saund. Ref. Bot. t. 184); bulb ovoid, 1$\frac{1}{2}$ in. diam.; leaves 4–5, linear, erect, finally 1–1$\frac{1}{2}$ ft. long, $\frac{1}{3}$–$\frac{3}{4}$ in. broad, tapering gradually to the apex, spotted with purple on the back towards the base; peduncle much curved, 3–4 in. long; raceme dense, oblong, finally 3–4 in. long, above an inch in diameter; pedicels cernuous, $\frac{1}{3}$–$\frac{1}{2}$ in. long; perianth oblong-campanulate, $\frac{1}{6}$ in. long, green, tinged with dull purple; filaments purple; ovary globose, stipitate, discoid at the base. *Journ. Linn. Soc.* xiii. 252.

EASTERN REGION: Natal; Inanda, *Wood*, 1208! and without precise locality, *Buchanan*, 4!

40. **S. livida** (Baker in Gard. Chron. 1883, xx. 166); bulb globose, 1½ in. diam.; leaves 6–8, lanceolate, 6–8 in. long, 1½ in. broad at the middle, bright green, unspotted; peduncle 4–5 in. long; raceme dense, oblong, 4–6 in. long, 1½–2 in. diam.; central pedicels ½ in. long, lower cernuous; perianth oblong-campanulate, green, $\frac{1}{6}$–$\frac{1}{5}$ in. long; filaments purple in the upper half; ovary stipitate, globose.

SOUTH AFRICA: without locality, cultivated specimen, *Hort. Horsman!*
EASTERN REGION: Natal; cultivated specimen, *Hort. Kew!*

41. **S. concolor** (Baker in Saund. Ref. Bot. iii. App. 13); bulb ovoid, 1½–2 in. diam.; leaves 5–6, spreading, lanceolate, 6–8 in. long, 1–1½ in. broad, not spotted; peduncle curved, 3–4 in. long; raceme oblong-cylindrical, finally 3–4 in. long, 1–1¼ in. diam.; central pedicels $\frac{1}{4}$–$\frac{1}{3}$ in. long, lower cernuous; perianth campanulate, green, $\frac{1}{6}$ in. long; filaments not tinted purple; ovary stipitate, globose. *Journ. Linn. Soc.* xiii. 252. *Drimia Cooperi, Baker in Saund. Ref. Bot. t.* 18. *D. dregeana, Kunth, Enum.* iv. 340.

SOUTH AFRICA: without locality, garden specimens, collected by *Cooper!*
COAST REGION: Port Elizabeth Div.; Cragga Kamma, *Burchell*, 4568! Alexandria Div.; Johanas Kloof, between Enon and the Zuurberg Range, 1000–2000 ft., *Drège*, 8616a!
CENTRAL REGION: Somerset Div.; mountain above Commadagga, *Burchell*, Bulb 59!

42. **S. subsecunda** (Baker in Gard. Chron. 1881, xvi. 38); bulb globose, 2 in. diam.; leaves 6–8, lanceolate, ½–1 ft. long, 1–1½ in. broad at the middle, plain green on the face, tinged with purple on the back, and blotched a little towards the base; peduncle very flexuose, drooping, nearly a foot long; raceme dense, subsecund, ½ ft. long; pedicels spreading, $\frac{1}{2}$–$\frac{3}{4}$ in. long; perianth oblong-campanulate, $\frac{1}{5}$ in. long, green, brownish outside towards the base; filaments bright purple; ovary stipitate, discoid at the base.

SOUTH AFRICA: without locality, cultivated specimen, *Bowker!*
EASTERN REGION: Natal, cultivated specimen, *Hort. Ware!*

Described from a plant received from Mr. Bowker, that flowered at Kew in June, 1881.

43. **S. megaphylla** (Baker); bulb large; leaves sessile, lorate-lanceolate, 1½ ft. long, 3–3½ in. broad at the middle, but little narrowed to the base; peduncle 9–12 in. long; raceme lax, 6–8 in. long, 1¼–1½ in. diam.; pedicels ascending or spreading, the central ones $\frac{1}{3}$–$\frac{1}{2}$ in. long; bracts minute, lanceolate; perianth greenish,

campanulate, $\frac{1}{5}$ in. long; stamens as long as the perianth; filaments green.

KALAHARI REGION: Transvaal; damp, grassy hollows near Barberton, 2000–2800 ft., *Galpin*, 1184!

44. S. laxiflora (Baker in Gard. Chron. 1891, ix. 668); bulb globose, $1\frac{1}{2}$ in. diam.; leaves 3–4, oblong-lanceolate, pale green, mottled with darker green, the largest 3 in. long, under an inch broad, not narrowed to the base; peduncle slender, terete, 3 in. long; raceme very lax, oblong, 3 in. long, $1\frac{1}{2}$ in. diam.; bracts minute, deltoid; central pedicels patent, pale mauve, $\frac{1}{2}$–$\frac{3}{4}$ in. long; perianth green, $\frac{1}{5}$ in. long, sharply reflexing when expanded; filaments bright mauve, nearly as long as the segments.

SOUTH AFRICA: without locality. No specimen at Kew.

Described from a plant cultivated at Kew in 1891.

45. S. socialis (Baker in Saund. Ref. Bot. t. 180); bulbs ovoid, gregarious, about $1\frac{1}{2}$ in. diam.; leaves 3–4, spreading, oblong, acute, 3–4 in. long, an inch broad at the middle, spotted with dark green on a pale green ground; peduncle curved, 2–3 in. long; raceme dense, oblong, $1\frac{1}{2}$–2 in. long, about an inch in diameter; pedicels cernuous, $\frac{1}{4}$–$\frac{1}{3}$ in. long; perianth campanulate, green, $\frac{1}{5}$ in. long; filaments purple in the upper half; ovary globose, stipitate, discoid at the base. *Journ. Linn. Soc.* xiii. 251.

EASTERN REGION: Natal, cultivated specimens, *Cooper*, 3634! 3635!

Described from a plant flowered by Mr. Wilson Saunders at Reigate about 1870, introduced by Mr. Thos. Cooper.

46. S. tricolor (Baker in Gard. Chron. 1880, xiv. 230); bulb globose, 2–3 in. diam.; leaves 6–7, lanceolate, ascending, a foot long, 2 in. broad at the middle, dark green with a few blotches of lighter green on the face, suffused all over the back with claret-brown, with a few irregular streaky blotches of green; peduncle curved, 6–8 in. long, claret-brown throughout; raceme dense, 2–3 in. long, $1\frac{1}{2}$–2 in. diam.; central pedicels patent, $\frac{1}{2}$–1 in. long, lower cernuous; perianth oblong-campanulate, $\frac{1}{8}$–$\frac{1}{2}$ in. long, green inside and out; filaments bright mauve-purple; ovary stipitate, globose, discoid at the base.

COAST REGION: Port Elizabeth? No specimen at Kew.

Described from a plant that flowered at Kew in July, 1880, received from Mr. Elwes.

47. S. microscypha (Baker in Gard. Chron. 1881, xvi. 102); bulb ovoid, 4 in. diam.; leaves 6–8, suberect, oblong-lanceolate, acute, recurving, above a foot long, 4–$4\frac{1}{2}$ in. broad at the middle, pale glaucous green on both sides, the outer copiously barred transversely with purple on the back towards the base, and with green higher up; peduncle a foot long, compressed, green, unspotted; raceme 5–6 in. long, 2 in. diam.; pedicels $\frac{1}{2}$–$\frac{3}{4}$ in. long; perianth campanulate, $\frac{1}{5}$ in.

long, green inside and out; filaments not tinged with purple; ovary stipitate, discoid at the base.

SOUTH AFRICA: without locality, cultivated specimen, *Bowker!*

Described from a plant that flowered at Kew in May, 1881, received from Mr. Bowker.

48. **S. undulata** (Baker in Saund. Ref. Bot. iii. App. 11); bulb globose, 1½–2 in. diam., squamose in the upper half; leaves 5–6, produced after the flowers, lanceolate, firm in texture, glaucous, undulated, 4–5 in. long, ¼–⅓ in. broad; peduncle 1½–2 in. long; raceme oblong, moderately dense, 1½–2 in. long, 1–1¼ in. diam.; pedicels horizontal, ¼–⅓ in. long; perianth oblong-campanulate, ¼–⅓ in. long, green, with a slight tinge of purple; filaments greenish-white; ovary globose, stipitate. *Journ. Linn. Soc.* xiii. 249. *Drimia undulata, Jacq. Collect. Suppl.* 41; *Ic.* ii. 15, *t.* 376; *Willd. Sp. Plant.* ii. 166 (*excl. syn.*); *Roem. et Schultes, Syst. Veg.* vii. 598; *Kunth, Enum.* iv. 340.

SOUTH AFRICA: without locality.

Known to me only from Jacquin's figure.

49. **S. spathulata** (Baker in Saund. Ref. Bot. t. 187); bulb ovoid, 2–3 in. diam.; leaves 5–6, oblong-spathulate, 6–8 in. long, 2 in. broad at the middle, spotted with dark green on a pale green ground, barred with purple on the back towards the base; peduncle curved, 2–4 in. long; raceme oblong, finally 4–6 in. long, 1½–2 in. diam.; pedicels ½–¾ in. long, lower cernuous; perianth oblong-campanulate, ¼ in. long, green, tinged with purple; filaments bright purple; ovary stipitate, discoid at the base. *Journ. Linn. Soc.* xiii. 253.

EASTERN REGION: Natal, cultivated specimens!

50. **S. zebrina** (Baker in Saund. Ref. Bot. t. 185); bulb globose, 1½–2 in. diam.; leaves 5–6, oblong-lanceolate, 8–12 in. long, 1½–2 in. broad, green, with a slight glaucous tinge, plain on the face, copiously barred vertically with purple on the back, with transverse bars towards the base; peduncle 4–6 in. long, curved, spotted towards the base; raceme dense, oblong, 3–4 in. long, 1¼–1½ in. diam.; pedicels ¼–⅓ in. long, lower cernuous; perianth oblong-campanulate, ¼ in. long, green, tinged with purple; filaments pale purple; ovary stipitate, globose, discoid at the base. *Journ. Linn. Soc.* xiii. 253.

SOUTH AFRICA: without locality. No specimen at Kew.

Described from a plant flowered by Mr. Wilson Saunders at Reigate in 1870, introduced by Mr. Thomas Cooper.

51. **S. lanceolata** (Baker in Saund. Ref. Bot. iii. App. 14); bulb subglobose, 1¼–1½ in. diam.; leaves 5–6, lanceolate, acute, plain green, 3–4 in. long, under an inch broad; peduncle slender, flexuose, 4–5 in. long; raceme very laxly 8–12-flowered, oblong, 1½–2 in.

long; pedicels subpatent, lower $\frac{1}{3}$–$\frac{1}{2}$ in. long; perianth oblong-
campanulate, $\frac{1}{3}$ in. long, green, tinged with purple; filaments tinged
with purple; ovary globose, stipitate. *Journ. Linn. Soc.* xiii. 254.
Drimia lanceolata, Schrad. Blumenb. 28; *Roem. et Schultes, Syst.
Veg.* vii. 594; *Kunth, Enum.* iv. 339. *Lachenalia reflexa, Andr.
Bot. Rep. t.* 299; *Tratt. Archiv.* ii. 132, *t.* 169.

SOUTH AFRICA: without locality.

Known to me only from the descriptions and figures.

52. S. subglauca (Baker in Saund. Ref. Bot. t. 186); bulb globose, 1$\frac{1}{2}$–2 in. diam.; leaves 5–6, lanceolate, erect, pale glaucous
green, 9–12 in. long, an inch broad, spotted with purple on the back
towards the base; peduncle 3–4 in. long, spotted with purple;
raceme oblong, 3–4 in. long, 1$\frac{1}{2}$ in. diam.; pedicels cernuous, $\frac{1}{3}$–$\frac{1}{2}$ in.
long; perianth oblong-campanulate, $\frac{1}{4}$ in. long, green, tinged with
purple; filaments bright purple; ovary globose, stipitate, discoid at
the base. *Journ. Linn. Soc.* xiii. 253.

SOUTH AFRICA: without locality. No specimen at Kew.

Described from a plant flowered by Mr. Wilson Saunders at Reigate in 1870,
introduced by Mr. Thomas Cooper.

53. S. lorata (Baker in Saund. Ref. Bot. iii. App. 14); bulb
ovoid, 1$\frac{1}{2}$–2 in. diam.; leaves 5–6, lanceolate, acute, erect, 8–12 in.
long, 1 in. broad, slightly narrowed to the base, spotted with purple
on the back low down; peduncle 8–9 in. long, spotted with purple
towards the base; raceme dense, oblong, 3–4 in. long, 1$\frac{1}{2}$–1$\frac{3}{4}$ in.
broad; pedicels patent, $\frac{1}{4}$–$\frac{1}{3}$ in. long, lower cernuous; perianth
oblong-campanulate, $\frac{1}{4}$ in. long, dull purple; filaments dull purple;
ovary globose, stipitate, discoid at the base. *Journ. Linn. Soc.* xiii.
253. *Drimia apertiflora, Baker in Saund. Ref. Bot. t.* 19.

SOUTH AFRICA: without locality. No specimen at Kew.

Described from a living plant that flowered in 1868 with Mr. Wilson Saunders
at Reigate, introduced by Mr. Thomas Cooper.

54. S. pendula (Baker in Saund. Ref. Bot. iii. App. 14); bulb
3 in. diam.; tunics many, thin, brown; produced leaves about 4,
sessile, suberect, lanceolate, 12–16 in. long, 1$\frac{1}{2}$–2$\frac{1}{2}$ in. broad, plain
green on the face, striped and spotted with purple on the back, some-
times up to the top, but more especially low down; peduncle slender,
curved, plain green, 5–6 in. long; raceme oblong, 3–4 in. long, often
subsecund; pedicels slender, sometimes an inch long; flowers fragrant; perianth oblong-campanulate, $\frac{1}{3}$ in. long, green inside and out;
filaments pale lilac; ovary stipitate, discoid at the base. *Journ.
Linn. Soc.* xiii. 254; *Gard. Chron.* 1878, ix. 756. *Drimia pendula,
Burch. ex Baker in Saund. Ref. Bot.* iii. *App.* 14.

CENTRAL REGION: Somerset Div.; cultivated specimens grown from bulbs
collected at Commadagga, *Burchell,* Bulb 55!

Cultivated by Burchell at Fulham, 1818–1821.

55. S. floribunda (Baker in Saund. Ref. Bot. t. 188); bulb sub-globose, 2–3 in. diam.; leaves 5–6, lanceolate, a foot long, 1½–2 in. broad, spotted with dark green on a pale green ground; peduncle nearly a foot long, unspotted; raceme dense, finally 6–8 in. long, 2 in. diam.; pedicels ½–¾ in. long, lower cernuous; perianth oblong-campanulate, ⅓ in. long, green, tinged inside with purple; filaments pale purple in the upper half; ovary stipitate, much produced at the base. *Journ. Linn. Soc.* xiii. 254.

SOUTH AFRICA : without locality, cultivated specimen, *Cooper !*

Described from a living plant flowered by Mr. Wilson Saunders at Reigate in 1870, introduced by Mr. Thomas Cooper.

56. S. princeps (Baker in Saund. Ref. Bot. t. 189); bulb globose, 2–3 in. diam.; leaves 5–6, lorate-lanceolate, 1½–2 ft. long, 2–2¼ in. broad, spotted with dark green on a pale ground; peduncle rather stout, curved, under a foot long; raceme dense, sometimes a foot long, 3–3½ in. diam.; central pedicels patent, 1–1½ in. long, lower cernuous; perianth oblong-campanulate, ⅓ in. long, green tinged with purple; filaments tinged with lilac in the upper half; ovary stipitate, much produced at the base. *Journ. Linn. Soc.* xiii. 254.

SOUTH AFRICA : without locality, cultivated specimen !

XL. ORNITHOGALUM, Linn.

Perianth persistent, campanulate, polyphyllous; segments sub-equal, patent, oblong or oblong-lanceolate, with or without a distinct few-nerved keel. *Stamens* hypogynous; filaments usually more or less flattened, often unequal; anthers oblong, versatile, dehiscing introrsely. *Ovary* sessile, 3-celled; ovules many in a cell, superposed; style short or elongated; stigma capitate. *Capsule* membranous, loculicidally 3-valved. *Seeds* globose or angled by pressure; testa thin, black; albumen firm; embryo short.

Rootstock a tunicated bulb; leaves various in shape and texture, all radical; peduncle leafless; inflorescence a simple elongated or corymbose raceme; bracts scarious, persistent; flowers usually white or yellow, never blue or mauve-purple.

DISTRIB. Widely spread in the Old World. Species about 100.

Subgenus CARUELIA. Perianth-segments not keeled; style very short and stout.

Few-flowered; peduncle short :
   Outer segments spotted at the tip :
      Spot distinct :
         Perianth ⅓–½ in. long   ...   ...   (1) **thunbergianum.**
         Perianth an inch long   ...   ...   (2) **speciosum.**
      Spot faint ...   .   (3) **maculatum.**
   Outer segments concolorous :
      Flowers white :
         Leaves linear, usually 2   ...   ...   (4) **diphyllum.**

Leaves subterete, 3–6 :
　Peduncle and leaves 1–1½ in.
　　long ... ... (5) **pauciflorum.**
　Peduncle and leaves 2–4 in. long (6) **rupestre.**
　Peduncle long ... ... (7) **inconspicuum.**
　Flowers orange-yellow ... ... (8) **aurantiacum.**
Many-flowered ; peduncle short ... (9) **multifolium.**
Many-flowered ; peduncle long :
　Perianth-segments not reflexing :
　　Perianth ½ in. long ... ... (10) **Saundersiæ.**
　　Perianth ¼–1 in. long ... ... (11) **thyrsoides.**
　Perianth-segments reflexing ... ... (12) **revolutum.**

Subgenus CATHISSA. Perianth-segments not distinctly keeled ; style filiform, at least ⅛ in. long.

Flowers few, small :
　Leaves not clasping the base of the stem :
　　Leaves subterete ... ... (13) **schlechterianum.**
　　Leaves linear ... ... (14) **gracile.**
　　Leaves lanceolate ... ... (15) **Rogersii.**
　Leaves clasping the base of the stem :
　　Bracts cordate-ovate ... ... (16) **deltoideum.**
　　Bracts linear-subulate ... ... (17) **pubescens.**
Flowers many, small :
　Leaves not clasping the base of the stem :
　　Leaves subterete :
　　　Bracts linear ... ... (18) **griseum.**
　　　Bracts lanceolate ... ... (19) **oostachyum.**
　　　Bracts ovate or oblong :
　　　　Leaf single ... ... (20) **monophyllum.**
　　　Leaves usually 3 :
　　　　Peduncle ½ ft. long ... (21) **leptophyllum.**
　　　　Peduncle 1–1½ ft. long ... (22) **gracilentum.**
　　Leaves linear :
　　　Perianth ⅕–⅓ in. long ... ... (23) **Zeyheri.**
　　　Perianth ⅓–½ in. long ... ... (24) **natalense.**
　　　Perianth ¼ in. long ... ... (25) **lineare.**
　Leaves clasping the base of the stem :
　　Pedicels short :
　　　Leaf short ... ... (26) **paludosum.**
　　　Leaf ½ ft. long ... ... (27) **inandense.**
　　Pedicels longer ... ... (28) **pilosum.**
Flowers larger, solitary ... ... (29) **tulbaghense.**
Flowers larger, raceme lax, elongated ... ... (30) **hispidum.**
Flowers larger ; raceme dense, subcorymbose :
　Leaves oblong-lanceolate ... ... (31) **Baurii.**
　Leaves linear :
　　Stamens ½ as long as the perianth ... (32) **coarctatum.**
　　Stamens ⅓ as long as the perianth ... (33) **tenellum.**
　Leaves lanceolate ... ... (34) **lacteum.**

Subgenus OSMYNE. Perianth-segments distinctly keeled ; flowers yellow or yellowish.

Leaves subterete :
　Leaves produced after the flowers ... ... (35) **secundum.**
　Leaves contemporary with the flowers :
　　Leaves vittate with white ... ... (36) **vittatum.**
　　Leaves not vittate :
　　　Leaves 1–2 ... ... (37) **barbatum.**
　　　Leaves many ... ... (38) **tuberosum.**

- Leaves linear:
  - Leaves straight:
    - Flowers small ($\frac{1}{4}$–$\frac{1}{3}$ in. long):
      - Pedicels short ... ... (39) **flavovirens.**
      - Pedicels long... ... ... (40) **polyphlebium.**
    - Flowers $\frac{1}{2}$–1 in. long:
      - Style $\frac{1}{5}$–$\frac{1}{4}$ in. long:
        - Flowers pale yellow ... ... (41) **suaveolens.**
        - Flowers yellowish-green:
          - Leaves linear ... ... (42) **prasinum.**
          - Leaves lorate ... ... (43) **xanthochlorum.**
      - Style $\frac{1}{3}$ in. long:
        - Pedicels short ... ... (44) **elatum.**
        - Pedicels long ... ... (45) **Monteiroi.**
  - Leaves circinate at the apex ... ... (46) **spirale.**
- Leaves small, lanceolate or oblong-lanceolate ... (47) **bolusianum.**

Subgenus Beryllis. Perianth-segments distinctly keeled; flowers white or whitish (in *fuscatum* reddish-brown).

- Leaves subterete:
  - Leaf solitary; bracts spurred ... ... (48) **calcaratum.**
  - Leaves 2 or more; bracts not spurred:
    - Leaves very slender:
      - Peduncle shorter than the leaves ... (49) **niveum.**
      - Peduncle as long as the leaves:
        - Perianth $\frac{1}{6}$ in. long ... ... (50) **bulbinelloides.**
        - Perianth $\frac{1}{4}$ in. long:
          - Raceme few-flowered ... (51) **oliganthum.**
          - Raceme many-flowered ... (52) **setifolium.**
      - Peduncle longer than the leaves:
        - Perianth $\frac{1}{6}$ in. long ... ... (53) **subulatum.**
        - Perianth $\frac{1}{4}$ in. long ... ... (54) **tortuosum.**
    - Leaves not so slender:
      - Pedicels very short ... ... (55) **comptum.**
      - Pedicels $\frac{1}{4}$–$\frac{1}{2}$ in. long:
        - Bracts as long as pedicels:
          - Perianth $\frac{1}{6}$ in. long; style very short ... ... (56) **aciphyllum.**
          - Perianth $\frac{1}{3}$ in. long; style as long as ovary ... (57) **graminifolium.**
        - Bracts shorter than pedicels ... (58) **juncifolium.**
- Leaves short, linear:
  - Flowers white:
    - Style short... ... ... (59) **humifusum.**
    - Style as long as the ovary ... ... (60) **Bergii.**
  - Flowers reddish-brown ... ... (61) **fuscatum.**
- Leaves long, linear:
  - Flowers white:
    - Bracts short ... ... (62) **albovirens.**
    - Bracts longer than the pedicels:
      - Leaves 2, narrow linear ... ... (63) **Saltmarshei.**
      - Leaves 3–6, linear ... ... (64) **Eckloni.**
  - Flowers greenish-white:
    - Filaments linear and lanceolate ... ... (65) **chloranthum.**
    - Filaments lanceolate and quadrate at the base ... ... (66) **virens.**
- Leaves long, lorate-lanceolate:
  - Filaments linear and lanceolate:
    - Bracts $\frac{1}{2}$–$\frac{3}{4}$ in. long ... ... (67) **scilloides.**
    - Bracts $\frac{3}{4}$–$1\frac{1}{4}$ in. long ... ... (68) **longebracteatum.**
  - Filaments lanceolate and quadrate at the base (69) **caudatum.**

Leaves short, oblong or oblong-lanceolate:
Margin of the leaf thickened ... ... (70) **ovatum.**
Margin of the leaf papillose ... ... (71) **crenulatum.**
Margin of the leaf glabrous ... ... (72) **Galpini.**
Leaves short, ovate-lanceolate ... ... (73) **trichophyllum.**

1. O. thunbergianum (Baker in Journ. Linn. Soc. xiii. 269); bulb globose, $\frac{1}{2}$ in. diam.; leaves 3–4, linear or lanceolate, glabrous, much shorter than the stem; peduncle slender, 3–12 in. long; flowers 1–6; pedicels very short; bracts ovate-cuspidate, $\frac{1}{4}$–$\frac{1}{2}$ in. long; perianth campanulate, yellowish, $\frac{1}{3}$–$\frac{1}{2}$ in. long; segments oblong, obtuse, $\frac{1}{4}$ in. broad, the three outer spotted at the tip with black; stamens half as long as the perianth; alternate filaments lanceolate; ovary ovoid, $\frac{1}{8}$ in. long; style very short, cylindrical. *O. maculatum, Thunb. Prodr.* 62; *Fl. Cap. edit. Schult.* 314; *Kunth, Enum.* iv. 352, *non Jacq.*

VAR. *β*, **concolor** (Baker); flower without any spots.

SOUTH AFRICA: without locality; Var. *β*, *Forster!*

COAST REGION: Malmesbury Div.; Saldanha Bay, *Thunberg! Pappe!* near Malmesbury, 500 ft., *Bolus*, 4346! Paarl Div.; Paarl Mountains, 1000–2000 ft., *Drège*, 8668!

2. O. speciosum (Baker in Journ. Bot. 1891, 72); bulb globose, 1 in. diam.; leaves 4, short, thick, linear, glabrous; peduncle 6–12 in. long; flowers 3–5 in a raceme with a flexuose rachis; lower pedicels arcuate, $\frac{1}{2}$ in. long; bracts ovate, amplexicaul, acute, an inch long; perianth campanulate, an inch long; segments oblong, obtuse, white, with a distinct purplish-black spot at the tip; stamens $\frac{1}{3}$ the length of the perianth; filaments lanceolate, much longer than the anthers; style short.

WESTERN REGION: Namaqualand, *Scully*, 175! *MacOwan and Bolus, Herb. Norm. Aust. Afr.*, 1390!

3. O. maculatum (Jacq. Collect. ii. 368, t. 18, fig. 3); bulb globose, $\frac{1}{2}$–$\frac{3}{4}$ in. diam.; leaves 4–5, linear, 3–4 in. long, channelled down the face, glabrous; peduncle 2–3 in. long; flowers 2–4, sub-corymbose; pedicels short, ascending; bracts ovate-lanceolate; perianth yellow, $\frac{1}{3}$–$\frac{1}{2}$ in. long; segments oblong, acute, $\frac{1}{6}$–$\frac{1}{5}$ in. broad, the three outer obscurely spotted with brown at the tip; stamens half as long as the perianth; filaments not dilated; ovary ovoid; style very short. *Baker in Journ. Linn. Soc.* xiii 269. *O. notatum, Roem. et Schultes, Syst. Veg.* vii. 528; *Kunth, Enum.* iv. 352. *Phæocles maculata, Salisb. Gen.* 35.

SOUTH AFRICA: without locality.

Known to me only from Jacquin's figure. Perhaps conspecific with *O. thunbergianum.*

4. O. diphyllum (Baker in Kew Bullet. 1895, 153); bulb globose, $\frac{1}{4}$ in. diam.; outer tunics pale, membranous; leaves 2, rarely 3, linear, erect, glabrous, $1\frac{1}{2}$–2 in. long, 1 lin. broad, narrowed very gradually to the base; scape slender, erect, $1\frac{1}{4}$–$1\frac{1}{2}$ in. long; flowers usually 2, rarely 1, erect; pedicels short; bracts lanceolate, membranous; perianth campanulate, pure white, $\frac{1}{4}$ in. long; segments oblong, obtuse, much imbricated; ovary globose; style very short; stamens half as long as the perianth, linear, uniform; anthers oblong, small, yellow.

EASTERN REGION: Natal; summit of Tambamhlopi Mountain, in swamps, 6000–7000 ft., *Evans*, 374!

5. O. pauciflorum (Baker); bulb globose, $\frac{1}{4}$–$\frac{1}{2}$ in. diam.; leaves few, very slender, filiform, glabrous, 1–1½ in. long; peduncle very slender, 1¼ in. long; perianth campanulate, $\frac{1}{4}$ in. long; segments oblong, pure white, without any green keel; stamens $\frac{1}{3}$ shorter than the perianth; filaments linear, uniform; anthers small, oblong; style shorter than the globose ovary.

SOUTH AFRICA: without locality, *Bergius!* in Berlin Herbarium.

6. O. rupestre (Linn. fil. Suppl. 199); bulb ovoid, $\frac{1}{3}$–$\frac{1}{2}$ in. diam.; leaves 3–6, subterete, clasping the base of the stem, 2–3 in. long; peduncle slender, 2–4 in. long; flowers 1–6, subcorymbose; lower pedicels arcuate, $\frac{1}{2}$–$\frac{3}{4}$ in. long; bracts lanceolate, $\frac{1}{4}$–$\frac{1}{2}$ in. long; perianth white, campanulate, $\frac{1}{4}$–$\frac{1}{3}$ in. long; segments oblong, subobtuse, $\frac{1}{8}$ in. broad; stamens less than half as long as the perianth; filaments lanceolate, alternate broader; ovary ovoid; style stout, very short. *Thunb. Prodr.* 61; *Fl. Cap. edit. Schult.* 313; *Willd. Sp. Plant.* ii. 123; *Roem. et Schultes, Syst. Veg.* vii. 526; *Kunth, Enum.* iv. 367; *Baker in Journ. Linn. Soc.* xiii. 271. *O. virgineum, Soland. ex Baker in Journ. Linn. Soc.* xiii. 271.

COAST REGION: Malmesbury Div.; Witte Klip, *Thunberg!* Groene Kloof, *Thunberg! Pappe!* Klip Berg, near Darling, *Drège*, 1512! Tulbagh Div.; New Kloof, near Tulbagh, *Zeyher*, 4205!
WESTERN REGION: Little Namaqualand, near Kook Fontein, *Drège*, 2663! by the Orange River near Verleptpram, under 1000 ft., *Drège*, 2662! and without precise locality, *Morris in Herb. Bolus*, 5806! *Scully*, 143! *Pappe! MacOwan and Bolus, Herb. Norm. Aust. Afr.*, 1391!

7. O. inconspicuum (Baker); bulb small, globose; tunics membranous; leaves 3, erect, glabrous, subterete, much shorter than the peduncle, under 1 lin. diam.; peduncle slender, terete, fragile, a foot or more long; raceme short, dense, few-flowered; bracts lanceolate, membranous, the lower $\frac{1}{4}$–$\frac{1}{3}$ in. long; pedicels short, erecto-patent; perianth $\frac{1}{4}$ in. long; segments oblong, tinged with brown on the back, not distinctly keeled; stamens shorter than the perianth; anthers small, oblong; ovary globose; style short, subulate.

KALAHARI REGION: Transvaal; Lomati Valley, near Barberton, in swamps, 4000 ft., *Galpin*, 1361!

8. O. aurantiacum (Baker in Gard. Chron. 1878, x. 748); bulb ovoid, $\frac{1}{3}$ in. diam.; tunics thin, pale; leaves 2–3, very slender, terete, glabrous, shorter than the peduncle; peduncle very slender, 3–4 in. long; flowers 1–2, bright yellow; pedicels short; bracts ovate-cuspidate, $\frac{1}{4}$–$\frac{1}{3}$ in. long; perianth $\frac{1}{8}$ in. long; segments oblong, acute, $\frac{1}{8}$–$\frac{1}{6}$ in. broad; stamens more than half as long as the segments; filaments equal, linear-subulate; ovary ovoid; style very short.

COAST REGION: Malmesbury Div.; Groene Kloof, cultivated specimen, *Bolus!*

Described from plant that flowered at Kew in November, 1878, introduced by Mr. Bolus.

9. O. multifolium (Baker in Journ. Linn. Soc. xiii. 271); bulb $\frac{1}{2}$ in. diam.; leaves 10–12, suberect, firm in texture, filiform, 2–3 in. long; peduncle 3–6 in. long; raceme dense, subcorymbose; flowers 12–25; lower pedicels $\frac{3}{4}$–1 in. long; bracts lanceolate, acuminate, $\frac{1}{2}$–$\frac{3}{4}$ in. long; perianth white, $\frac{1}{3}$–$\frac{1}{2}$ in. long; segments ovate-lanceolate, $\frac{1}{8}$–$\frac{1}{6}$ in. broad; filaments lanceolate, subequal, $\frac{1}{6}$ in. long; style stout, very short.

WESTERN REGION: Little Namaqualand, *Whitehead!* No specimen at Kew.

10. O. Saundersiæ (Baker in Gard. Chron. 1891, x. 452); bulb large, globose; leaves many to a bulb, lorate, flaccid, bright green, above a foot long, 2 in. broad at the middle; peduncle stout, green, terete, 2–3 ft. long; flowers many, corymbose; pedicels 1$\frac{1}{2}$–2 in. long; bracts lanceolate, greenish, the outer an inch long; perianth white, $\frac{1}{2}$ in. long; segments orbicular, much imbricated, tinged with green outside; stamens half as long as the perianth; filaments uniform, lanceolate; ovary blackish-green; style very short.

KALAHARI REGION: Transvaal; hillsides, amongst rocks in damp places, near Barberton, 3000–3400 ft., *Galpin,* 1205! cultivated specimen, *Mrs. K. Saunders!*

Described from a plant grown at Kew in 1891, received from Mrs. K. Saunders.

11. O. thyrsoides (Jacq. Hort. Vind. iii. 17, t. 28); bulb globose, green, 1–1$\frac{1}{2}$ in. diam.; leaves 5–6, lanceolate, fleshy, $\frac{1}{2}$–1$\frac{1}{2}$ ft. long, $\frac{1}{2}$–1$\frac{1}{2}$ in. broad, obscurely ciliated; peduncle stiffly erect, $\frac{1}{2}$–1$\frac{1}{2}$ ft. long; raceme dense, subcorymbose; pedicels erecto-patent, lower 1–2 in. long; bracts ovate-acuminate, $\frac{1}{2}$–1 in. long; perianth campanulate, $\frac{3}{4}$–1 in. long; segments oblong, subobtuse, $\frac{1}{3}$–$\frac{1}{2}$ in. broad, pure white with a brownish-green claw; stamens not more than $\frac{1}{3}$ as long as the perianth; filaments all dilated at the base, alternate, quadrate, obscurely tricuspidate; ovary ovoid, $\frac{1}{6}$ in. long; style thick and very short. *Thunb. Prodr.* 62; *Fl. Cap. edit. Schult.* 315; *Ait. Hort. Kew.* i. 442, edit. 2, ii. 261; *Roem. et Schultes, Syst. Veg.* vii. 509; *Kunth, Enum.* iv. 353; *Baker in Saund. Ref. Bot. t.* 20; *Journ. Linn. Soc.* xiii. 269. *O. arabicum, Linn. Sp. Plant.* 307, *ex parte. O. Grimaldiæ, Nocca, Pl. Select. Tic. t.* 4. *Aspasia thyrsoides, Salisb. Gen.* 34.

VAR. β, **flavescens** (Lindl. Bot. Reg. t. 305); flowers some white, some fulvous-yellow. *O. flavescens, Jacq. Collect.* iii. 233; *Ic.* ii. 20, t. 437. *Aspasia flavescens, Salisb. Gen.* 34.

VAR. γ, **flavissimum** (Baker); flowers bright lemon-yellow. *O. flavissimum, Jacq. Collect. Suppl.* 75; *Ic.* ii. 20, t. 436; *Andr. Bot. Rep. t.* 505.

VAR. δ, **aureum** (Baker in Journ. Linn. Soc. xiii. 270); flowers fulvous-yellow. *O. aureum, Curt. in Bot. Mag. t.* 190; *Red. Lil.* viii. *t.* 439; *Kunth, Enum.* iv. 352. *O. dubium, Houtt. Handl.* xii. 309, *t.* 82, *fig.* 3. *O. bicolor, Haw. Misc.* 177; *Kunth, Enum.* iv. 370. *Aspasia aurea, Salisb. Gen.* 34.

VAR. ε, **miniatum** (Baker); flowers bright scarlet; leaves short, with a ciliated hyaline margin. *O. miniatum, Jacq. Collect.* iii. 233; *Ic.* ii. 20, 438. *Roem. et Schultes, Syst. Veg.* vii. 1698; *Kunth, Enum.* iv. 352.

COAST REGION: Malmesbury Div.; Groene Kloof, *Zeyher*, 1680! *Ecklon!* Cape Div.; Table Mountain and Lion Mountain, *Ecklon*, 22! Camps Bay, *Zeyher*, 5047! *MacOwan and Bolus, Herb. Norm. Aust. Afr.*, 940! Paarl Div.; near Paarl, *Drège!* Worcester Div., *Cooper*, 1637! 1690! Mossel Bay Div.; Little Brak River, *Burchell*, 6197/6! and Bulb 111! Between Little Brak River and Hartenbosch, *Burchell*, 6198! Riversdale Div.; by the Zoetmelks River, *Burchell*, 6809! Var. γ, Albany Div.; near Grahamstown, 1700 ft., *Galpin*, 311! Var. δ, near Tulbagh, *Ecklon*, 572! *MacOwan and Bolus, Herb. Norm. Aust. Afr.*, 505! Uitenhage Div.; near the Zwartkops River, *Zeyher*, 4201! Port Elizabeth, *E.S.C.A. Herb.*, 85!

CENTRAL REGION: Somerset Div.; sides of the Bosch Berg, 4000 ft., *MacOwan*, 1819! Var. δ, Somerset Div.; stony slopes at the foot of Bruintjes Hoogte, 3000 ft., *MacOwan*, 1818!

12. **O. revolutum** (Jacq. Hort. Schoenbr. i. 46, t. 89); bulb globose, $\frac{3}{4}$–1 in. diam.; leaves 5–6, lanceolate, 6–9 in. long, $\frac{1}{2}$–$\frac{3}{4}$ in. broad, minutely ciliated on the margin; peduncle terete, stiffly erect, a foot long; raceme dense, subcorymbose; pedicels erecto-patent, lower 1–1$\frac{1}{2}$ in. long; bracts ovate-acuminate, $\frac{1}{2}$–$\frac{3}{4}$ in. long; perianth white, $\frac{3}{4}$ in. long; segments oblong, obtuse, $\frac{1}{4}$ in. broad, reflexing nearly from the base; stamens not more than $\frac{1}{2}$ as long as the perianth; filaments dilated at the base, alternate ovate, sometimes obscurely cuspidate; ovary ovoid, $\frac{1}{8}$ in. long; style stout, very short. *Willd. Sp. Plant.* ii. 118; *Gawl. in Bot. Mag. t.* 653; *Bot. Reg. t.* 315; *Roem. et Schultes, Syst. Veg.* vii. 513; *Kunth, Enum.* iv. 354; *Baker in Journ. Linn. Soc.* xiii. 271. *Aspasia revoluta, Salisb. Gen.* 34.

SOUTH AFRICA: without locality, cultivated specimen!

Described from a plant flowered by Mr. T. S. Ware at Tottenham in July, 1878.

13. **O. schlechterianum** (Schinz in Bull. Herb. Boiss. ii. 223); bulb globose, $\frac{1}{3}$ in. diam.; leaf subterete, glabrous, 2 in. long; peduncle 2–5 in. long; flowers 4–5, corymbose; pedicels ascending, the lower $\frac{1}{2}$ in. long; bracts ovate-cuspidate; perianth white, $\frac{1}{6}$ in. long; segments oblong, obscurely keeled; stamens shorter than the perianth; filaments linear; anthers small, oblong; style short.

COAST REGION: Cape Div.; Table Mountain, above Klassenbosch, 1800–2500 ft., *Bolus*, 7054! *Schlechter*, 138!

14. **O. gracile** (Baker in Journ. Bot. 1874, 366); bulb globose, $\frac{1}{4}$ in. diam; leaves generally 2, linear, glabrous, weak, suberect, 3–6 in. long, $\frac{1}{12}$–$\frac{1}{8}$ in. broad; peduncle very slender, 6–8 in. long; raceme dense, 1–6-flowered; pedicels ascending, lower $\frac{1}{12}$–$\frac{1}{6}$ in. long; bracts deltoid, more or less cuspidate; perianth white, $\frac{1}{4}$–$\frac{1}{3}$ in. long; seg-

ments oblong-lanceolate, under $\frac{1}{12}$ in. broad, obscurely 3-nerved down
the centre; stamens half as long as the perianth; alternate filaments
subulate and slightly flattened; style slender, $\frac{1}{8}$ in. long, as long as
the turbinate ovary.

COAST REGION : Uitenhage Div.; at the sources of the Bulk River, *MacOwan*,
1939 !

15. O. **Rogersii** (Baker), bulb globose, $\frac{1}{3}$ in. diam.; leaves 2,
lanceolate, weak, glabrous, spreading from the apex of the bulb,
1–$1\frac{1}{2}$ in. long, $\frac{1}{12}$ in. broad; peduncle very slender, 4–6 in. long;
flowers 6–8 in a raceme about an inch long; pedicels erecto-patent,
lower $\frac{1}{4}$–$\frac{1}{3}$ in. long; bracts lanceolate-acuminate; perianth white,
$\frac{1}{4}$ in. long; segments $\frac{1}{12}$ in. broad, obscurely 3-nerved down the
centre; stamens half as long as the perianth; alternate filaments
lanceolate; style slender, as long as the ovary.

COAST REGION : George Div.; George Mountain, marshy side, 2000 ft.,
Rogers!

16. O. **deltoideum** (Baker in Journ. Linn. Soc. xiii. 281); bulb
globose, $\frac{1}{3}$ in. diam.; leaves 2–4, clasping the lower part of the stem,
superposed, linear-subulate, squarrose, subcoriaceous, very short,
channelled down the face, hispid on both sides; peduncle very
slender, 3–4 in. long; flowers 3–6 in a lax raceme; pedicels ascend-
ing, lower $\frac{1}{4}$–$\frac{1}{3}$ in. long; bracts small, cordate-ovate, membranous;
perianth white, $\frac{1}{3}$ in. long; segments oblong, not keeled, $\frac{1}{12}$ in.
broad; stamens a little shorter than the segments; filaments lanceolate, subequal; style slender, as long as the ovary.

CENTRAL REGION : Calvinia Div.; Hantam Mountains, *Meyer!*
WESTERN REGION: Little Namaqualand; Silver Fontein, near Ookiep, 2000–
3000 ft., *Drège*, 2664!

17. O. **pubescens** (Baker in Journ. Linn. Soc. xiii. 282); bulb
ovoid, $\frac{1}{4}$–$\frac{1}{3}$ in. diam.; leaves 2–3, superposed, clasping the lower part
of the stem, spreading, lanceolate, $\frac{1}{2}$–1 in. long, densely setose on
both surfaces; peduncle flexuose, slender, 4–5 in. long; raceme
laxly 5–6-flowered, an inch long; pedicels ascending, lower $\frac{1}{6}$–$\frac{1}{4}$ in.
long; bracts linear-subulate, $\frac{1}{8}$–$\frac{1}{4}$ in. long; perianth white, $\frac{1}{8}$–$\frac{1}{4}$ in.
long; segments lanceolate, acute, not keeled; filaments $\frac{1}{12}$ in. long,
subequal, a little flattened; style filiform, as long as the ovary.

COAST REGION : Albany Div.; *Williamson!* In Dublin Herbarium.

18. O. **griseum** (Baker in Journ. Linn. Soc. xiii. 281); bulb
globose, brownish, $1\frac{1}{2}$–2 in. diam.; leaves 5–6, linear-subulate,
ascending, glabrous, 6–8 in. long; peduncle slender, a foot long;
flowers about 20, in a raceme 3–4 in. long; pedicels erecto-patent,
lower $\frac{1}{6}$–$\frac{1}{4}$ in. long; bracts linear, $\frac{1}{8}$–$\frac{1}{5}$ in. long; perianth whitish,
concolorous, $\frac{1}{4}$ in. long; segments obtuse, $\frac{1}{12}$ in. broad; filaments
subequal, slightly flattened, much shorter than the segments; style
filiform, as long as the ovary.

SOUTH AFRICA : without locality.

Described from a drawing of a plant grown at Kew in 1823, introduced by
Bowie.

19. O. oostachyum (Baker); bulb small, ovoid; leaves 3–4, slender, erect, terete, glabrous, 3–4 in. long; peduncle slender, 3–6 in. long; raceme dense, oblong, an inch long; pedicels erecto-patent, lower $\frac{1}{8}$ in. long; bracts small, lanceolate; perianth $\frac{1}{8}$ in. long; segments oblong-lanceolate, white, obscurely keeled; stamens half as long as the perianth; filaments uniform, linear; anthers small; style shorter than the ovary.

EASTERN REGION: Griqualand East; Zuurberg Range, 5500 ft., *Tyson*, 1545!

20. O. monophyllum (Baker); bulb globose, $\frac{3}{4}$–1 in. diam.; outer tunics produced over its neck and banded transversely; leaf 1, very slender, terete, glabrous, a foot long; peduncle slender, 6–8 in. long; raceme dense, subsecund, 2–3 in. long; pedicels short, ascending; bracts small, ovate-cuspidate; perianth $\frac{1}{4}$ in. long; segments oblong, acute, white, with an obscure brownish keel; stamens half as long as the perianth; style longer than the ovary.

KALAHARI REGION: Transvaal; Saddleback Range, near Barberton, 5000 ft., *Galpin*, 1051! Swaziland; Havelock Concession, *Saltmarshe*.

21. O. leptophyllum (Baker); bulb globose, $\frac{1}{2}$ in. diam.; outer tunics produced above its neck; leaves 3, basal, erect, subterete, very slender, 6–9 in. long; peduncle slender, terete, often 2–3-nate, $\frac{1}{2}$ ft. long; raceme lax, many-flowered, 1$\frac{1}{2}$–2 in. long; pedicels erecto-patent, lower $\frac{1}{4}$ in. long; bracts ovate, with a long cusp; perianth $\frac{1}{4}$ in. long; segments white, with an indistinct brownish keel; stamens much shorter than the perianth; anthers small; style short.

EASTERN REGION: Natal; near Bothas, 2000 ft., *Wood*, 4774!

22. O. gracilentum (Baker); bulb ovoid, $\frac{1}{3}$ in. diam.; leaves about 3, subterete, slender, glabrous, $\frac{1}{2}$–1 ft. long; peduncle slender, 1–1$\frac{1}{2}$ ft. long; raceme dense upwards, 1–3 in. long, $\frac{1}{2}$ in. diam.; pedicels erecto-patent, lower $\frac{1}{8}$–$\frac{1}{6}$ in. long; bracts oblong or ovate, as long as the pedicels; perianth white, $\frac{1}{4}$ in. long; segments oblanceolate, under $\frac{1}{12}$ in. broad, obscurely 3-nerved down the centre; stamens a quarter shorter than the perianth; filaments lanceolate; style slender, $\frac{1}{12}$ in. long, shorter than the ovary.

CENTRAL REGION: Somerset Div.; on the Bosch Berg, 4500 ft., *MacOwan*, 2216!

23. O. Zeyheri (Baker in Journ. Linn. Soc. xiii. 281); bulb globose, $\frac{1}{2}$ in. diam.; leaves 2–4, linear, glabrous, weak, channelled down the face, $\frac{1}{2}$–1 ft. long; peduncles 1–3 to a bulb, slender, 2–12 in. long; racemes many-flowered, lax, finally 3–6 in. long; pedicels ascending, lower sometimes above an inch long; bracts small, deltoid-cuspidate; perianth whitish, $\frac{1}{5}$–$\frac{1}{4}$ in. long; segments oblong, not keeled, $\frac{1}{12}$ in. broad; stamens half as long as the perianth; filaments subequal, a little flattened; style slender, as long as the ovary.

COAST REGION : Uitenhage Div. ; near the Zwartkops River, *Zeyher,* 1686 ! Alexandria Div. ; Zuurberg Range, 2000–3000 ft., *Drège,* 878b ! Albany Div.; near Grahamstown, 2200 ft., *MacOwan,* 1437 !

CENTRAL REGION : Graaff Reinet Div.; summit of the Oude Berg, in stagnant places, 5000 ft., *Bolus,* 632 ! Albert Div., *Cooper,* 601 !

KALAHARI REGION : Orange Free State ; Caledon River, *Burke !* Transvaal ; Hooge Veld, *Rehmann,* 6861 ! Bosch Veld, *Rehmann,* 4840 !

EASTERN REGION : Natal; near Reenens Pass, in marshy ground, 5000–6000 ft., *Wood,* 4532 !

24. O. natalense (Baker in Kew Bullet. 1893, 210) ; bulb small ; leaves 3, suberect, thin, linear, 3–4 in. long, $\frac{1}{6}$–$\frac{1}{3}$ in. broad, hairy on the surfaces and ciliated on the margin ; peduncle 1–4 in. long ; raceme lax, 1$\frac{1}{2}$–3 in. long ; pedicels ascending, lower $\frac{1}{4}$ in. long ; bracts as long, lanceolate from a broad clasping base ; perianth $\frac{1}{4}$–$\frac{1}{3}$ in. long ; segments oblong-lanceolate, acute, white, with an obscure, brown, 3-nerved keel ; stamens $\frac{1}{3}$ short of the perianth ; filaments linear, pubescent ; style as long as the globose ovary.

EASTERN REGION : Natal; summit of Amawahqua Mountain, 6800 ft., *Wood,* 4567 !

25. O. lineare (Baker) ; bulb small, globose ; leaves 3–4, erect, glabrous, narrow linear, 4–5 in. long ; peduncle 4–5 in. long ; raceme dense, 2–6-flowered ; pedicels erecto-patent, lower $\frac{1}{4}$ in. long ; bracts lanceolate, from an ovate clasping base, rather longer than the pedicels ; perianth $\frac{1}{4}$ in. long ; segments oblong, white, with an obscure brownish keel ; stamens half as long as the perianth ; filaments uniform, linear ; anthers small, oblong.

EASTERN REGION: Natal; Liddesdale, *Wood,* 4267a !

26. O. paludosum (Baker in Journ. Bot. 1874, 366) ; bulb not seen ; leaf clasping the base of the stem, small, lanceolate ; peduncle 8–12 in. long ; raceme lax, many-flowered, finally 3–4 in. long ; pedicels ascending, lower $\frac{1}{4}$ in. long ; bracts lanceolate, as long as the pedicels ; perianth white, $\frac{1}{4}$–$\frac{1}{3}$ in. long ; segments oblanceolate, obtuse, $\frac{1}{12}$ in. broad, not keeled ; stamens a little shorter than the perianth ; filaments linear, subequal ; style slender, $\frac{1}{12}$ in. long.

CENTRAL REGION : Cradock Div.; on the Elands Berg, *Cooper,* 219 !

27. O. inandense (Baker); bulb globose, $\frac{1}{3}$ in. diam. ; produced leaves about 3, clasping the lower part of the stem, thin, pubescent, the upper much the longest, linear, $\frac{1}{2}$ ft. long, $\frac{1}{8}$–$\frac{1}{6}$ in. broad; peduncle slender, 1$\frac{1}{2}$ ft. long; raceme many-flowered, cylindrical, 3 in. long, $\frac{1}{2}$ in. diam.; pedicels erecto-patent, lower $\frac{1}{4}$ in. long; bracts subulate from a deltoid base, sometimes exceeding the pedicels; perianth dull yellowish-white, $\frac{1}{4}$ in. long; segments $\frac{1}{18}$ in. broad, obscurely and laxly 3-nerved in the central third; stamens one-third shorter than the perianth ; filaments linear-lanceolate ; style filiform, as long as the ovary.

EASTERN REGION : Natal; Inanda, *Wood,* 1168 !

28. O. pilosum (Linn. fil. Suppl. 199); bulb ovoid, $\frac{3}{4}$–1 in. diam.; tunics firm, nearly black; leaves 3–4, clasping the base of the stem, linear, firm in texture, with a thickened margin, ciliated, 2–4 in. long, $\frac{1}{8}$–$\frac{1}{6}$ in. broad; peduncle slender, 3–12 in. long; raceme many-flowered, very lax, 3–4 in. long; pedicels ascending, lower $\frac{1}{2}$–1 in. long; bracts ovate-lanceolate, acuminate, lower $\frac{1}{3}$–$\frac{1}{2}$ in. long; perianth white, $\frac{1}{4}$–$\frac{1}{3}$ in. long; segments oblong-lanceolate, $\frac{1}{12}$ in. broad, not distinctly keeled; stamens $\frac{1}{3}$ shorter than the perianth; filaments lanceolate; style slender, $\frac{1}{12}$ in. long. *Thunb. Prodr.* 61; *Fl. Cap.* 313; *Roem. et Schultes, Syst. Veg.* vii. 523; *Kunth, Enum.* iv. 370; *Baker in Journ. Linn. Soc.* xiii. 282. *O. mundianum, Kunth, Enum.* iv. 351.

COAST REGION: Cape Div.; Devils Mountain, *Pappe!* Simons Town, *Schlechter*, 338! Tulbagh Div.; Winter Hoek Mountain, near Tulbagh, 1000 ft., *Bolus*, 5267! Tulbagh, *Pappe!* Worcester Div.; near the Breede River, *Cooper*, 1670! 1695! Riversdale Div.; hills near Zoetmelks River, *Burchell*, 6802!

29. O. tulbaghense (Baker); bulb and leaves not seen; peduncle very slender, $\frac{1}{2}$ ft. long; flowers solitary, terminal; bract small, ovate; perianth pure white, $\frac{5}{8}$ in. long; segments oblong-lanceolate, not keeled, $\frac{1}{6}$ in. broad; stamens half as long as the perianth; filaments linear-subulate; ovary ovoid; style filiform, as long as the ovary.

COAST REGION: Tulbagh Div.; New Kloof near Tulbagh, *Zeyher*, 4205! No specimen at Kew.

30. O. hispidum (Hornem. Hort. Hafn. 330); bulb globose, whitish, $\frac{1}{2}$–1 in. diam.; leaves 2–4, clasping the lower part of the stem, superposed, spreading, lanceolate, pilose, the longest $\frac{1}{2}$ ft. long, $\frac{1}{3}$–$\frac{1}{2}$ in. broad; peduncle slender, $\frac{1}{2}$–1 ft. long; raceme laxly 6–12-flowered, 3–4 in. long; pedicels ascending, lower $\frac{3}{4}$–1 in. long; bracts lanceolate, $\frac{1}{3}$–$\frac{1}{2}$ in. long; perianth whitish, $\frac{5}{8}$–$\frac{3}{4}$ in. long; segments oblong-lanceolate, not distinctly keeled; stamens half as long as the perianth; filaments ovate at the base, subequal; style filiform, as long as the ovary. *Kunth, Enum.* iv. 350; *Baker in Journ. Linn. Soc.* xiii. 282. *Anthericum pilosum, Jacq. Collect. Suppl.* 87; *Ic.* ii. 18, *t.* 416; *Willd. Sp. Plant.* ii. 140; *Roem. et Schultes, Syst. Veg.* vii. 480. *Phalangium pilosum, Poir. Encyc.* v. 244.

COAST REGION: Cape Div.; Table Mountain, near Kirstenbosch, 1500 ft., *Bolus*, 4515! Muizenberg, 1000 ft., *Bolus*, 3306!

31. O. Baurii (Baker); bulb under $\frac{1}{2}$ in. diam.; leaves 2, oblong-lanceolate, subacute, minutely ciliated, suberect, 2$\frac{1}{2}$–3 in. long, $\frac{1}{2}$–$\frac{5}{8}$ in. broad; peduncles 1–2, lateral, rather longer than the leaves; raceme few-flowered, very dense, corymbose; pedicels erecto-patent, at most $\frac{1}{3}$ in. long; bracts ovate-lanceolate, greenish, $\frac{1}{2}$–$\frac{3}{4}$ in. long; perianth pure white, $\frac{1}{3}$ in. long; segments oblong, subobtuse, not keeled, $\frac{1}{6}$ in. broad ; stamens $\frac{1}{4}$ shorter than the perianth ; filaments lanceolate ; style filiform, $\frac{1}{12}$ in. long.

32. O. coarctatum (Jacq. Collect. Suppl. 77 ; Ic. ii. 20, t. 435) ; bulb globose, 1 in. diam. ; leaves 4–5, linear, glabrous, channelled down the face, shorter than the peduncle ; peduncle slender, 1–1$\frac{1}{4}$ ft. long ; raceme dense, subcorymbose ; pedicels erecto-patent, lower 1–1$\frac{1}{2}$ in. long ; bracts lanceolate-acuminate, $\frac{1}{2}$–$\frac{3}{4}$ in. long ; perianth white, $\frac{1}{2}$–$\frac{3}{4}$ in. long ; segments oblong, subacute, $\frac{1}{6}$ in. broad ; stamens not more than $\frac{1}{3}$ as long as the segments ; filaments all dilated at the base, alternate obscurely tricuspidate ; ovary ovoid ; style filiform, $\frac{1}{8}$ in. long. *Willd. Sp. Plant.* ii. 125 ; *Ræm. et Schultes, Syst. Veg.* vii. 511 ; *Kunth, Enum.* iv. 353 ; *Baker in Journ. Linn. Soc.* xiii. 270.

SOUTH AFRICA : without locality, *Zeyher*, 1679 !
COAST REGION : Cape Div. ; Cape Flats, *Pappe!* Albany Div. ; in marshy ravines around Grahamstown, 1500–2500 ft., *Galpin*, 312 !

33. O. tenellum (Jacq. Collect. ii. 316 ; Ic. ii. 19, t. 427) ; bulb globose, 1 in. diam. ; leaves 4–5, linear, glabrous, $\frac{1}{2}$–1 ft. long ; peduncle slender, stiffly erect, a foot or more long ; raceme dense, subcorymbose ; pedicels erecto-patent, lower 1–2 in. long ; bracts ovate-acuminate, $\frac{1}{2}$–$\frac{3}{4}$ in. long ; perianth pure white, $\frac{1}{2}$–$\frac{3}{4}$ in. long ; segments oblong, $\frac{1}{6}$ in. broad, not keeled ; stamens half as long as the perianth ; alternate filaments dilated at the base ; style slender, $\frac{1}{12}$ in. long. *Willd. Sp. Plant.* ii. 121 ; *Roem. et Schultes, Syst. Veg.* vii. 526 ; *Kunth, Enum.* iv. 358 ; *Baker in Journ. Linn. Soc.* xiii. 283.

COAST REGION : Swellendam Div. ; between Duivenhoeks River and Grootvaders Bosch, *Burchell*, 7216 !

Perhaps only a slender, narrow-leaved variety of *O. lacteum.*

34. O. lacteum (Jacq. Collect. Suppl. 76 ; Ic. ii. 20, t. 434) ; bulb globose, 1–1$\frac{1}{2}$ in. diam. ; leaves 6–12, lanceolate, thin, hairy on the margin, $\frac{3}{4}$–1 in. broad ; peduncle stout, erect, 1–1$\frac{1}{2}$ ft. long ; raceme dense, many-flowered, subcorymbose ; lower pedicels reaching 1$\frac{1}{2}$–2 in. long ; bracts large, ovate-acuminate ; perianth pure white, $\frac{3}{4}$ in. long ; segments oblong, not keeled, $\frac{1}{4}$ in. broad ; stamens $\frac{1}{3}$–$\frac{1}{2}$ as long as the perianth ; filaments alternately lanceolate and lanceolate with an ovate base ; style slender, $\frac{1}{8}$–$\frac{1}{6}$ in. long. *Willd. Sp. Plant.* ii. 117 ; *Andr. Bot. Rep. t.* 274 ; *Bot. Mag. t.* 1134 ; *Red. Lil.* vii. *t.* 418 ; *Lodd. Bot. Cab. t.* 1159 ; *Ræm. et Schultes, Syst. Veg.* vii. 513, 1698 ; *Kunth, Enum.* iv. 354 ; *Baker in Journ. Linn. Soc.* xiii. 283. *Aspasia lactea, Salisb. Gen.* 34.

VAR. β, conicum (Baker in Journ. Linn. Soc. xiii. 284) ; more slender, with a laxer inflorescence and narrower, more acute perianth-segments. *O. conicum, Jacq. Collect.* iii. 232 ; *Ic.* ii. 19, *t.* 428 ; *Roem. et Schultes, Syst. Veg.* vii. 514 ; *Hook. in Bot. Mag. t.* 3538 ; *Kunth, Enum.* iv. 354.

COAST REGION: Cape Div.; near Camps Bay, *MacOwan*, 2652! Table
Mountain, 1800 ft., *Bolus*, 4919! Lion Mountain, *Ecklon*, 569! Paarl Div.;
near Paarl, *Drège*!
WESTERN REGION : Little Namaqualand; near Ookiep, *Bolus*, 5805!
EASTERN REGION : Tembuland; Bazeia, 2000 ft., *Baur*, 508!

35. O. secundum (Jacq. Collect. Suppl. 79; Ic. ii. 20, t. 433);
bulb ovoid $\frac{3}{4}$–1 in. diam.; leaves 5–6, produced after the flowers,
subterete, channelled down the face, scabrous on the margin;
peduncle slender, 3–4 in. long; flowers 1–6, in a very lax, often
secund raceme; pedicels erecto-patent, lower $\frac{3}{4}$–1 in. long; bracts small,
lanceolate; perianth yellow, $\frac{1}{3}$–$\frac{1}{2}$ in. long; segments distinctly
keeled; stamens $\frac{2}{3}$ the length of the perianth; filaments subequal,
slightly flattened; style as long as the ovary. *Willd. Sp. Plant.* ii.
122; *Roem. et Schultes, Syst. Veg.* vii. 527; *Kunth, Enum.* iv. 369.
Urginea (?) *secunda, Baker in Journ. Linn. Soc.* xiii. 222. *Monotassa
secunda, Salisb. Gen.* 37.

COAST REGION : Paarl Div.; Paarl Mountains, 2000–3000 ft, *Drège*, 1513a!

36. O. vittatum (Kunth, Enum. iv. 367); bulb globose, 1 in.
diam.; leaves 6–8, subterete, glaucous green, 6–9 in. long, channelled
down the face, distinctly keeled with white; peduncle slender,
stiffly erect, 6–9 in. long; raceme laxly 4–12-flowered; pedicels
cernuous at the apex, lower $\frac{1}{2}$–$\frac{3}{4}$ in. long; bracts linear from an ovate
base, $\frac{1}{3}$–$\frac{1}{2}$ in. long; perianth yellow, $\frac{1}{3}$–$\frac{1}{2}$ in. long; segments $\frac{1}{6}$–$\frac{1}{8}$ in.
broad, distinctly keeled; stamens more than half as long as the
perianth; alternate filaments lanceolate and tricuspidate; style twice
as long as the ovary, $\frac{1}{4}$–$\frac{1}{3}$ in. long. *Baker in Journ. Linn. Soc.* xiii.
278. *Albuca vittata, Gawl. in Bot. Mag. t.* 1329; *Roem. et Schultes,
Syst. Veg.* vii. 500. *Tæniola vittata, Salisb. Gen.* 35.

SOUTH AFRICA : without locality, cultivated specimens!
Described from a plant that flowered at Kew in June, 1878.

37. O. barbatum (Jacq. Hort. Schoenbr. i. 47, t. 91); bulb
globose, 1 in. diam.; leaves 1–2, subulate, weak, glabrous, 6–9 in.
long, channelled down the face; peduncle slender, a foot long;
raceme laxly 5–6-flowered; pedicels erecto-patent, lower $\frac{1}{4}$ in. long;
bracts lanceolate, $\frac{1}{4}$–$\frac{1}{3}$ in. long; perianth yellow, $\frac{1}{2}$ in. long; segments
oblong, distinctly keeled; stamens $\frac{2}{3}$ the length of the perianth;
filaments subequal, lanceolate; style $\frac{1}{6}$–$\frac{1}{4}$ in. long. *Willd. Sp. Plant.*
ii. 122; *Roem. et Schultes, Syst. Veg.* vii. 524; *Kunth, Enum.* iv.
366; *Baker in Journ. Linn. Soc.* xiii. 278.

COAST REGION : Cape Div.; Cape Flats, *Pappe!*

38. O. tuberosum (Miller, Gard. Dict. ed. 8, No. 10); bulb ovoid,
1–1$\frac{1}{2}$ in. diam.; leaves 6–12, subterete, glabrous, 6–9 in. long,
channelled down the face; peduncle $\frac{1}{2}$–1$\frac{1}{2}$ ft. long; raceme lax, few-
flowered; pedicels ascending, lower 1–1$\frac{1}{4}$ in. long; bracts large,
lanceolate; perianth pale yellow, $\frac{3}{4}$ in. long; segments with a distinct 5-nerved keel; stamens more than half as long as the perianth;

filaments all lanceolate ; style ¼ in. long. *Baker in Journ. Linn. Soc.* xiii. 279. *O. polyphyllum, Jacq. Collect. Suppl.* 79 ; *Ic.* ii. 19, t. 430 ; *Willd. Sp. Plant.* ii. 123 ; *Roem. et Schultes, Syst. Veg.* vii. 524 ; *Kunth, Enum.* iv. 366. *O consanguineum, Kunth, Enum.* iv. 368.

SOUTH AFRICA : without locality, *Zeyher ! Ecklon*, 281, *Mund and Maire.*

39. O. flavovirens (Baker in Journ. Bot. 1874, 366) ; bulb globose, 1 in. diam. ; leaves 5–6, linear, thin, glabrous, 1–1½ ft. long ; peduncle 2 ft. long ; raceme 3–4 in. long, dense upwards, lax down-wards ; pedicels ascending, articulated at the apex, lower ⅙–¼ in. long ; bracts linear, acuminate, projecting far beyond the buds, lower an inch long ; perianth yellowish-green, ¼–⅓ in. long ; segments linear-oblong, ⅛ in. broad, with a closely 5-nerved indistinct green keel ; stamens ⅔ the length of the perianth ; filaments equal, lanceolate ; style as long as the ovary.

CENTRAL REGION : Somerset Div. ; near Somerset East, in swamps, 2800 ft., *MacOwan*, 1852 !
EASTERN REGION : Tembuland ; Bazeia, 2000 ft., *Baur*, 570 !

40. O. polyphlebium (Baker) ; bulb not seen ; leaves linear, glabrous, channelled down the face, about as long as the peduncle ; peduncle slender, 4–6 in. long ; raceme many-flowered, dense, sub-corymbose ; pedicels erecto-patent, lower 1–1¼ in. long ; bracts lanceolate-acuminate, ½–¾ in. long ; perianth pale yellowish-green, ¼–⅓ in. long ; keel green, laxly 7–9-nerved ; stamens ⅔ the length of the perianth ; filaments subequal, all with a dilated base ; style ⅛–⅙ in. long.

KALAHARI REGION : Bechuanaland ; Batlapin territory, *Holub !*
Nearly allied to *O. prasinum, Lindl.*

41. O. suaveolens (Jacq. Collect. ii. 316 ; Ic. ii. 19, t. 431) ; bulb globose, 1–1½ in. diam. ; leaves 3–6, linear, glabrous, a foot long ; peduncle a foot or more long, stiffly erect ; raceme lax, few- or many-flowered, 3–6 in. long ; pedicels erecto-patent, lower 1½–2 in. long ; bracts large, lanceolate ; flowers yellow, sweet-scented, ½–¾ in. long ; segments oblong ; keel distinct, closely 5–7-nerved, soon turning brown ; stamens ⅔ the length of the perianth ; filaments subequal, lanceolate at the base ; style ⅛–¼ in. long. *Willd. Sp. Plant.* ii. 122 ; *Roem. et Schultes, Syst. Veg.* vii. 523 ; *Baker in Journ. Linn. Soc.* xiii. 279. *O. odoratum, Jacq. Collect. Suppl.* 78 ; *Ic.* ii. 20, t. 432 ; *Willd. Sp. Plant.* ii. 121 ; *Andr. Bot. Rep. t.* 260. *O. albucoides, Thunb. Fl. Cap. edit. Schult.* 314 ; *Kunth, Enum.* iv. 367. *Anthericum albucoides, Ait. Hort. Kew.* i. 449 ; edit. 2, ii. 269. *Phalangium albucoides, Poir. Encyc.* v. 249.

COAST REGION : Piquetberg Div. ; near Piquetberg, *Thunberg !* Malmesbury Div. ; Saldanha Bay, *Thunberg !* Paarde Berg, *Thunberg !* between Groene Kloof and Saldanha Bay, *Drège*, 1511 Cape Div. ; Muizenberg, 1000 ft., *Bolus*, 3298 !

42. O. prasinum (Lindl. in Bot. Reg. t. 158); bulb ovoid, 1½–2 in. diam.; leaves 6–9, linear, glabrous, glaucous green, deeply channelled down the face, 1 ft. long, usually twisted; peduncle ½–1½ ft. long; raceme dense, many-flowered, oblong or deltoid; pedicels spreading or ascending, lower 1–2 in. long; bracts linear, acuminate, ½–1 in. long; perianth dull greenish-yellow, ⅝ in. long; segments with a closely 5-nerved green keel; stamens ⅔ the length of the perianth; filaments subequal, lanceolate at the base; style ¼ in. long, exceeding the ovary. *Kunth, Enum.* iv. 360; *Baker in Journ. Linn. Soc.* xiii. 280.

Central Region: Colesberg Div.; near the Hondeblats River, *Burchell*, Bulb 49! near Schuilhoek Berg, *Burchell*, Bulb 71! near the Orange River, *Knobel!*

Central Region: Griqualand West; between Griqua Town and Witte Water, *Burchell*, 1966! Transvaal; Apies River, *Burke!*

43. O. xanthochlorum (Baker); bulb large, globose; leaves thick, fleshy, lorate, glabrous, under a foot long, an inch broad; peduncle stout, a foot long; raceme dense, 6–8 in. long; pedicels stout, erecto-patent, lower 1–1¼ in. long; bracts large, ovate-lanceolate, projecting beyond the buds, lower 1½ in. long; perianth yellowish-green, ½ in. long; stamens ⅓ in. long; filaments linear, uniform; anthers small, oblong; style as long as the ovary, under ¼ in. long.

Western Region: Little Namaqualand; Kaus Mountain, 2500 ft., *Bolus*, 6598!

44. O. elatum (Baker); bulb not seen; leaves linear, glabrous, a foot or more long; peduncle stout, stiffly erect, 3 ft. long; raceme ½ ft. long, dense at the top, lax in the lower part; pedicels ascending, articulated at the apex, ¼–⅓ in. long; bracts lanceolate-acuminate, ¾–1 in. long, projecting far beyond the buds; perianth yellowish, an inch long; segments with a distinct closely 7–9-nerved keel, soon turning brown; stamens nearly as long as the perianth; filaments linear; style above ½ in. long.

Coast Region: Albany Div., *Cooper*, 3280!

45. O. Monteiroi (Baker); bulb not seen; leaves linear, complicate, firm in texture, above a foot long, strongly ribbed, minutely ciliated on the margin; peduncle stiffly erect, above a foot long; raceme lax, many-flowered, a foot long; pedicels ascending, articulated at the apex, lower ¾–1 in. long; bracts linear-lanceolate, as long as the pedicels; perianth yellow, ⅝–¾ in. long; segments with a 5-7-nerved distinct keel; stamens a little shorter than the perianth; filaments all lanceolate at the base; style ⅓–½ in. long.

Eastern Region: Delagoa Bay, *Monteiro!*

Closely allied to *O. suaveolens*, *Jacq.*

46. O. spirale (Schinz in Bull. Herb. Boiss. iv. App. iii. 42); bulb ovoid-oblong, above 1 in. diam.; leaves linear-lanceolate, channelled down the face, 8 in. long, ⅕ in. broad, circinnate at the apex;

peduncle rigid, 4–5 in. long; raceme lax, 20-flowered; pedicels patent, ½ in. long; bracts scariose, acuminate; perianth ⅓–¾ in. long; segments $\frac{1}{12}$–⅛ in. broad, yellow, 3–4-nerved; filaments dilated at the base.

WESTERN REGION: Great Namaqualand; Rehoboth, *Fleck*, 890. Not in Kew Herbarium.

The locality is just north of the Tropic, but I include the species here, as it will doubtless be found to occur within the limits of the South African flora.

47. **O. bolusianum** (Baker in Journ. Linn. Soc. xiii. 279); bulb depresso-globose, an inch in diameter; leaves 2–4, clasping the base of the stem, lanceolate or oblong-lanceolate, spreading, pilose; peduncle slender, 6–8 in. long; raceme lax, few-flowered; pedicels ascending, lower ¼–⅓ in. long; bracts small, deltoid-cuspidate; perianth yellowish green, ¼ in. long; segments oblong, with a laxly 3-nerved green keel; stamens more than half as long as the perianth; filaments lanceolate; style shorter than the ovary.

CENTRAL REGION: hills near Graaff Reinet, 2600 ft., *Bolus*, 96! 778!

There is also a drawing at Kew, made from a plant introduced by Bowie, cultivated at Kew in 1823.

48. **O. calcaratum** (Baker in Gard. Chron. 1874, i. 723); bulb ovoid, ¾ in. diam.; leaf solitary, terete, glabrous, 7–8 in. long, with many fine ribs; peduncle very slender, terete, a foot long; raceme laxly 8–10-flowered, 1–1½ in. long; pedicels erecto-patent, ⅛–⅙ in. long; bracts minute, deltoid, with a large spur at the base; perianth white, ⅙ in. long; segments oblong, ⅙ in. long, distinctly keeled with green; stamens nearly as long as the perianth; filaments not distinctly flattened; style as long as the ovary.

SOUTH AFRICA: without locality, *MacOwan.*

Described from a plant cultivated by Mr. W. Wilson Saunders at Reigate in 1872, received from Prof. MacOwan. No specimen at Kew.

49. **O. niveum** (Ait. Hort. Kew. i. 440; edit. 2, ii. 257); bulb globose, ⅓–½ in. diam.; leaves 5–6, filiform, glabrous, spreading, ½ ft. long; peduncle slender, much shorter than the leaves; raceme lax, few-flowered; pedicels ascending, lower ⅙–¼ in. long; bracts small, ovate-cuspidate; perianth whitish, ¼–⅓ in. long; segments oblong, distinctly keeled with green; stamens ⅔ the length of the perianth; filaments linear and lanceolate; style $\frac{1}{12}$ in. long. *Willd. Sp. Plant.* ii. 115; *Bot. Reg. t.* 235; *Roem. et Schultes, Syst. Veg.* vii. 527; *Kunth, Enum.* iv. 358; *Baker in Journ. Linn. Soc.* xiii. 274. *Urophyllon niveum, Salisb. Gen.* 36.

SOUTH AFRICA: without locality, *Masson!*
CENTRAL REGION: Alexandria Div.; Zwartwarter Poort, *Burchell*, 3367!

50. **O.? bulbinelloides** (Baker); rootstock unknown; leaves subterete, ½ ft. long, channelled down the face, glabrous; peduncle terete, as long as the leaves; raceme lax, 2–3 in. long; pedicels

ascending, $\frac{1}{4}$–$\frac{1}{3}$ in., finally $\frac{1}{2}$ in. long ; bracts ovate-acuminate, half as long as the pedicels ; perianth white, $\frac{1}{6}$ in. long ; segments oblong, outer 5-nerved, inner 3-nerved ; stamens half as long as the perianth ; filaments equal, lanceolate-acuminate ; style as long as the ovary. *Bulbinella* (?) *ornithogaloides, Kunth, Enum.* iv. 693. *Anthericum ornithogaloides, Baker in Journ. Linn. Soc.* xv. 294.

CENTRAL REGION: Aliwal North Div.; banks of the Orange River near Aliwal North, *Drège*, 8695.

No specimen at Kew.

51. O. oliganthum (Baker); bulb globose, $\frac{1}{3}$–$\frac{1}{2}$ in. diam.; outer tunics produced beyond the top and breaking up into fine fibres ; leaves 6–9, filiform, very slender, glabrous, 3–6 in. long ; peduncle very slender, about as long as the leaves ; raceme very lax, few-flowered ; pedicels ascending, $\frac{1}{6}$ in. long ; bracts small, ovate, cuspidate ; perianth $\frac{1}{4}$ in. long ; segments oblanceolate, distinctly keeled ; stamens nearly as long as the perianth ; filaments lanceolate ; style as long as the ovary.

EASTERN REGION: Natal; Fields Hill, 1000 ft., *Wood*, 1973 !

Very near *O. niveum.*

52. O. setifolium (Kunth, Enum. iv. 351) ; leaves about 10, rigid, filiform, subterete, channelled down the face, equalling or over-topping the scape ; peduncle 3–4 in. long ; raceme many-flowered, $1\frac{1}{4}$–$2\frac{1}{4}$ in. long ; pedicels $\frac{1}{12}$–$\frac{1}{8}$ in. long ; bracts ovate, acuminate, longer than the pedicels ; perianth $\frac{1}{4}$ in. long ; segments oblong ; filaments linear and lanceolate ; style shorter than the ovary.

SOUTH AFRICA: without locality, *Drège*, 8674.

No specimen at Kew.

53. O. subulatum (Baker in Gard. Chron. 1874, i. 723) ; bulb depresso-globose, $\frac{1}{2}$ in. diam.; leaves 3–6, erect, filiform, glabrous, 2–6 in. long ; peduncle slender, terete, erect, $\frac{1}{2}$–1 ft. long ; raceme moderately dense, $1\frac{1}{2}$–3 in. long ; pedicels short, ascending ; bracts minute, deltoid-cuspidate ; perianth $\frac{1}{6}$ in. long, whitish ; segments oblong, distinctly keeled with green ; stamens $\frac{3}{4}$ the length of the perianth, alternately linear and lanceolate ; style as long as the ovary.

CENTRAL REGION: Somerset Div.; near Somerset East, 2800 ft., *MacOwan*, 1965! Graaff Reinet Div.; stony hills near Graaff Reinet, 2500–2600 ft., *Bolus*, 795!

EASTERN REGION: Tembuland; Bazeia, on moist flats near rivers, 2000 ft., *Baur*, 392!

Described from a plant flowered by Mr. Wilson Saunders at Reigate in Sept., 1872.

54. O. tortuosum (Baker); bulb ovoid or subglobose, $\frac{1}{3}$–$\frac{1}{2}$ in. diam.; leaves 6–8, filiform, glabrous, $1\frac{1}{2}$–2 in. long, straight or spirally twisted ; peduncle very slender, 4–6 in. long ; raceme lax, 1–2 in. long ; pedicels ascending, lower $\frac{1}{8}$–$\frac{1}{4}$ in. long ; bracts minute, ovate, cuspidate; perianth white, $\frac{1}{4}$ in. long; segments keeled with reddish-brown; stamens nearly as long as the perianth; filaments flattened; style as long as the ovary.

COAST REGION: Tulbagh, *Pappe!*

55. O. comptum (Baker in Journ. Linn. Soc. xiii. 274); bulb sub-globose, $\frac{1}{2}$–1 in. diam.; leaves 6–12, linear-subulate, rigid in texture, glabrous, strongly ribbed, 2–6 in. long, often spirally twisted; peduncle $\frac{1}{2}$–1 ft. long; raceme dense, 1–3 in. long; pedicels very short; bracts small, lanceolate from a deltoid base; perianth brownish-white, $\frac{1}{4}$ in. long; segments distinctly keeled with green, turning to red-brown; stamens nearly as long as the perianth; filaments lanceolate, subequal; style as long as the ovary.

COAST REGION: Knysna Div.; hills near Plettenberg Bay, *Pappe!* Uitenhage Div.; near the Zwartkops River, *Zeyher*, 4208! *Ecklon and Zeyher*, 939! 942! 1066!

CENTRAL REGION: Somerset Div.; on the Bosch Berg, 3000–4500 ft., *MacOwan*, 1843! 2067!

EASTERN REGION: Tembuland; Tabase, near Bazeia, 2500 ft., *Baur*, 393!

56. O. aciphyllum (Baker in Journ. Bot. 1874, 365); bulb not seen; leaves 3, subterete, erect, rigid, glabrous, a foot long, dilated for several inches at the base, and clasping the stem; peduncle longer than the leaves; raceme lax, 3–4 in. long; pedicels ascending, lower $\frac{1}{4}$ in. long; bracts ovate, acuminate, as long as the pedicels; perianth $\frac{1}{6}$ in. long; segments with a distinct 3-nerved red-brown keel; stamens $\frac{2}{3}$ the length of the segments; filaments alternately linear and lanceolate; style very short.

CENTRAL REGION: Colesberg, *Shaw!*

57. O. graminifolium (Thunb. Prodr. 61); bulb globose, $\frac{1}{2}$–1 in. diam.; outer tunics produced far above its neck; leaves 4–6, sub-terete, glabrous, $\frac{1}{2}$–1 ft. long; raceme lax, 2–3 in. long; pedicels ascending, lower $\frac{1}{4}$–$\frac{1}{2}$ in. long; bracts ovate, acuminate, as long as the pedicels; perianth white, $\frac{1}{4}$ in. long; segments with a rather in-distinct 3-nerved green or reddish-brown keel; stamens $\frac{2}{3}$ the length of the perianth; filaments slightly flattened; style as long as the ovary. *Fl. Cap. edit. Schult.* 313; *Roem. et Schultes, Syst. Veg.* vii. 526; *Kunth, Enum.* iv. 358; *Baker in Journ. Linn. Soc.* xiii. 274. *O. Rudolphi, Jacq. Eclog.* i. 31, 151, *t.* 20; *Roem. et Schultes, Syst. Veg.* vii. 525. *O. tenuifolium, Red. Lil.* vi. *t.* 312, *non Gussone. O. rupestre, Rudolphi in Schrad. Journ.* 1799, ii. 281, *non Linn. fil. O. juncifolium, Guwl. in Bot. Mag. t.* 972, *non Jacq. O. dregeanum, Kunth, Enum.* iv. 351. *O. canaliculatum, Lagas. Gen.* 14; *Kunth, Enum.* iv. 370.

SOUTH AFRICA: without locality, *Thunberg!*

COAST REGION: Cape Flats, *Zeyher*, 1681! Paarl Div.; Klein Drakenstein Mountains, under 500 ft., *Drège*, 1508! Tulbagh Div.; New Kloof, near Tulbagh, *Zeyher*, 1682! Winterhoek Mountain, *Pappe!*

CENTRAL REGION: Murraysburg Div.; Koudeveld Mountains, 5000 ft., *Bolus*, 1490!

EASTERN REGION: Natal; Inanda, *Wood*, 658!

58. O. juncifolium (Jacq. Hort. Schoenbr. i. 46, t. 90); bulb large, globose; leaves 12–20, subulate, glabrous, under a foot long; scape flexuose, about a foot long; raceme moderately dense, ½–1 ft. long; pedicels, ascending or spreading, lower ¼ in. long; bracts lanceolate-acuminate, much shorter than the pedicels; perianth white, ¼ in. long; segments distinctly keeled; stamens ⅔ the length of the perianth; filaments alternately linear and lanceolate; style ⅓ in. long. *Willd. Sp. Plant.* ii. 123; *Roem. et Schultes, Syst. Veg.* vii. 524; *Kunth, Enum.* iv. 358; *Baker in Journ. Linn. Soc.* xiii. 274.

SOUTH AFRICA: without locality.

Known to me only from Jacquin's figure. May be only a robust garden-grown form of *O. graminifolium.*

59. O. humifusum (Baker in Gard. Chron. 1874, i. 500); bulb globose, ½–¾ in. diam., copiously bulbilliferous; leaves 1–3, linear or lanceolate, finely pilose or glabrous, varying in length from one to nine inches; peduncle slender, 1–9 in. long; raceme lax, few-flowered; pedicels ascending, lower ¼–½ in. long; bracts ovate cuspidate, as long as or shorter than the pedicels; perianth white, ¼–⅓ in. long; segments distinctly or indistinctly keeled with green; stamens ⅔ the length of the perianth; alternate filaments linear and lanceolate; style shorter than the ovary.

SOUTH AFRICA: without locality, cultivated specimen!
CENTRAL REGION: Somerset Div.; rocky ledges near the Little Fish River, 2500 ft., *MacOwan,* 1838! 1897!

Described from a plant that flowered at Kew in March, 1874, received from Mr. Wilson Saunders.

60. O. Bergii (Schlecht. in Linnæa, 1826, 253); bulb globose, ¼–½ in. diam.; leaves 2–4, clasping the base of the stem for an inch or more, linear, spreading or ascending, 1–3 in. long, densely pilose; peduncle slender, ½–1 ft. long; raceme moderately dense, few- or many-flowered, 1–3 in. long; pedicels ascending, lower ¼–½ in. long; bracts lanceolate from an ovate base, nearly or quite as long as the pedicels; perianth white, ¼–⅓ in. long; segments oblong-lanceolate, distinctly keeled with red-brown; stamens more than half as long as the perianth; inner filaments much dilated at the base, sometimes distinctly tricuspidate; style as long as the ovary. *Roem. et Schultes, Syst. Veg.* vii. 507, 1697; *Kunth, Enum.* iv. 350; *Baker in Journ. Linn Soc.* xiii. 275. *O. miniatum, Schinz in Bull. Herb. Boiss.* ii. 223, *non Jacq. O. ciliatum, Eckl. ex Roem. et Schultes, Syst. Veg.* vii. 1697, *non Linn. fil.*

COAST REGION: Cape Div.; Table Mountain, *Ecklon,* 572! Lion Mountain, *Schlechter,* 133! Cape Flats, *Harvey,* 872! Paarl Div.; Paarl Mountains, *Drège,* 1506a! Drakenstein Mountains, near Wellington, *Rehmann,* 2274! Tulbagh Div.; near Tulbagh Waterfall, 800 ft., *Bolus,* 5265! Winterhoek Mountain, 2500 ft., *Bodkin in Herb. Bolus,* 5428! New Kloof, *Zeyher,* 1682! Swellendam Div.; hill near Swellendam, *Kitching!* Mossel Bay Div.; east side of Gauritz River, *Burchell,* 6447! near Uitenhage, *Burchell,* 4274!

61. O. fuscatum (Jacq. Collect. Suppl. 80 ; Ic. ii. 19, t. 429) ; bulb subglobose, 1 in. diam.; leaves 2–3, linear, glabrous, 3–4 in. long, $\frac{1}{4}-\frac{1}{3}$ in. broad ; peduncle slender, $\frac{1}{2}$–1 ft. long ; raceme 12–20-flowered, 3–5 in. long ; pedicels ascending, $\frac{1}{3}-\frac{1}{2}$ in. long ; bracts lanceolate-acuminate, nearly as long as the pedicels ; perianth reddish-brown, $\frac{1}{3}$ in. long; segments oblong, with a distinct green keel ; stamens much shorter than the perianth; filaments alternately linear and lanceolate ; style as long as the ovary. *Willd. Sp. Plant.* ii. 122; *Roem. et Schultes, Syst. Veg.* vii. 523 ; *Kunth, Enum.* iv. 366 ; *Baker in Journ. Linn. Soc.* xiii. 273. *Ardernia fuscata, Salisb. Gen.* 35.

SOUTH AFRICA : without locality.

Known to me only from Jacquin's figure.

62. O. albovirens (Baker in Gard. Chron. 1878, x. 364); bulb globose, green, 1 in. diam. ; leaves 3–4, linear, acuminate, bright green, glabrous, 1–1½ ft. long, $\frac{1}{2}-\frac{3}{4}$ in. broad low down ; peduncle slender, terete, 1–1½ ft. long ; raceme dense, oblong, about 2 in. long ; pedicels ascending, lower $\frac{1}{2}-\frac{3}{4}$ in. long ; bracts linear, much shorter than the pedicels ; perianth whitish, $\frac{1}{4}-\frac{1}{3}$ in. long ; segments broadly keeled with green ; stamens $\frac{2}{3}$ the length of the perianth ; filaments subequal, linear ; style as long as the ovary.

SOUTH AFRICA : without locality.

Described from a plant that flowered at Kew in 1875, received from Mr. Cordukes, of Natal. No specimen at Kew.

63. O. Saltmarshei (Baker); bulb ovoid, 1 in. diam. ; leaves 2, narrow linear, glabrous, erect, $\frac{1}{2}$ ft. long ; scape a foot long ; raceme oblong, very dense, an inch long ; pedicels very short ; bracts large, lanceolate ; perianth $\frac{1}{2}$ in. long ; segments oblong, obtuse, with a distinct brown keel ; stamens nearly as long as the perianth-segments ; filaments alternately linear and lanceolate ; style longer than the globose ovary.

EASTERN REGION : Swaziland ; Komassan Range, Havelock Concession, 4000 ft., *Saltmarshe in Herb. Galpin,* 1057 !

64. O. Eckloni (Schlecht. in Linnæa, xxv. 177) ; bulb globose, 1–1½ in. diam. ; leaves 5–6, linear, acuminate, glabrous, 1–1½ ft. long, $\frac{1}{2}$–1 in. broad low down ; peduncle slender, 1–1½ ft. long ; raceme dense upwards, finally 4–6 in. long ; pedicels ascending, lower $\frac{1}{4}-\frac{1}{3}$ in. long ; bracts lanceolate-setaceous, $\frac{1}{2}$ in. long, protruding beyond the buds ; perianth pure white, $\frac{1}{3}$ in. long ; segments distinctly keeled with green ; stamens $\frac{2}{3}$ the length of the perianth ; filaments subequal, lanceolate ; style $\frac{1}{12}$ in. long. *Baker in Journ. Linn. Soc.* xiii. 276. *O. acuminatum (canaliculatum on the plate), Baker in Saund. Ref. Bot. t.* 177, *non Schur.*

COAST REGION : British Kaffraria, *Cooper,* 3277 !

CENTRAL REGION : Somerset Div. ; at the foot of the Bosch Berg, 3000 ft., *MacOwan,* 1651 ! rocky banks of Little Fish River, 2300 ft., *MacOwan,* 1839 !

KALAHARI REGION : Griqualand West, by the Vaal River, *Burchell,* 1777 ! Transvaal; hills above Apies River, *Rehmann,* 4308 ! Saddleback Range, near

Barberton, 4500 ft., *Galpin*, 1147! Lomati Valley, near Barberton, in swamps, 4000 ft., *Galpin*, 1054!

EASTERN REGION: Tembuland; Bazeia Mountain, 2500–3000 ft., *Baur*, 571! Griqualand East; near Clydesdale, 2500 ft., *Tyson*, 2119! Natal; Inanda, *Wood*, 245! 1073! near Durban, *Wood*, 139! 4080! *MacOwan and Bolus, Herb. Norm. Austr. Afr.*, 1012!

Also in Tropical Africa.

65. **O. chloranthum** (Baker in Gard. Chron. 1875, iv. 323); bulb ovoid, $1\frac{1}{2}$–2 in. diam.; leaves 3–4, suberect, linear, acuminate, bright green, glabrous, channelled down the face, 12–16 in. long, $\frac{1}{2}$ in. broad; peduncle terete, glabrous, $1\frac{1}{2}$–2 ft. long; flowers about 20 in an oblong raceme 3–4 in. long, $1\frac{1}{2}$ in. diam.; pedicels patent or erecto-patent, lower $\frac{1}{4}$–$\frac{1}{3}$ in. long; bracts lanceolate-acuminate, longer than the pedicels, protruding conspicuously beyond the buds; perianth greenish, $\frac{1}{2}$ in. long; segments oblong, obtuse, with a 3-nerved keel filling the central third; stamens nearly as long as the perianth; inner filaments lanceolate, outer linear; ovary globose; style slender, $\frac{1}{6}$ in. long.

SOUTH AFRICA: without locality.

Described from a living plant sent to Kew by Mr. Wilson Saunders, which flowered in July, 1875.

66. **O. virens** (Lindl. in Bot. Reg. t. 814); bulb globose, 1–$1\frac{1}{2}$ in. diam.; leaves 5–6, linear, acuminate, glabrous, $1\frac{1}{2}$–2 ft. long, $\frac{3}{4}$–1 in. broad low down; peduncle slender, a foot or more long; raceme dense, 3–4 in. long; pedicels ascending, lower $\frac{1}{4}$ in. long; bracts linear-subulate; perianth greenish-white, $\frac{1}{4}$–$\frac{1}{3}$ in. long; segments broadly keeled with green; stamens $\frac{2}{3}$ the length of the perianth; alternate filaments quadrate at the base; style $\frac{1}{12}$ in. long. *Roem. et Schultes, Syst. Veg.* vii. 507; *Kunth, Enum.* iv. 358; *Baker in Journ. Linn. Soc.* xiii. 275.

EASTERN REGION: Delagoa Bay, *Forbes!*

Habit and leaf of *O. Eckloni*, from which it differs by its stamens.

67. **O. scilloides** (Jacq. Hort. Schoenbr. i. 46, t. 88); bulb ovoid, 2–3 in. diam.; leaves 5–6, lanceolate-lorate, acuminate, glabrous, $1\frac{1}{2}$–2 ft. long, 1–$1\frac{1}{2}$ in. broad low down; peduncle robust, terete, $1\frac{1}{2}$–2 ft. long; raceme dense, a foot or more long; pedicels ascending, lower $\frac{1}{2}$ in. long; bracts lanceolate-setaceous, $\frac{1}{2}$–$\frac{3}{4}$ in. long, projecting beyond the buds; perianth white, $\frac{1}{3}$ in. long; segments keeled with green; stamens $\frac{2}{3}$ the length of the perianth; filaments linear and lanceolate; style $\frac{1}{8}$ in. long. *Willd. Sp. Plant.* ii. 119; *Cav. Descr.* 125; *Roem. et Schultes, Syst. Veg.* vii. 522; *Kunth, Enum.* iv. 358; *Baker in Journ. Linn. Soc.* xiii. 276.

SOUTH AFRICA: without locality. No specimen at Kew.

68. **O. longebracteatum** (Jacq. Hort. Vind. iii. t. 29); bulb 2–3 in. diam.; leaves 5–6, lorate-lanceolate, acuminate, glabrous, $1\frac{1}{2}$–2 ft. long, 1–$1\frac{1}{2}$ in. broad low down; peduncle robust, terete,

$1\frac{1}{2}$–2 ft. long; raceme dense upwards, $\frac{1}{2}$ ft. or more long; pedicels ascending or spreading, lower $\frac{3}{4}$–$1\frac{1}{4}$ in. long; bracts lanceolate-setaceous, $\frac{3}{4}$–1 in. long, projecting conspicuously beyond the buds; perianth whitish, $\frac{1}{3}$ in. long; segments broadly keeled with green; stamens $\frac{3}{4}$ the length of the perianth; alternate filaments linear and lanceolate; style $\frac{1}{8}$ in. long. *Willd. Sp. Plant.* ii. 120; *Red. Lil. t.* 120; *Roem. et Schultes, Syst. Veg.* vii. 521; *Kunth, Enum.* iv. 357; *Baker in Journ. Linn. Soc.* xiii. 277. *O. bracteatum, Thunb. Prodr.* 62; *Fl. Cap. edit. Schult.* 314.

SOUTH AFRICA: without locality, *Thunberg!*

COAST REGION: Riversdale Div.; Vet River, *Gill!* Queenstown Div.; Zwart Kei River, 4000 ft., *Drège*, 3532b!

CENTRAL REGION: Somerset Div.; in woods at the foot of the Bosch Berg, 3000 ft., *MacOwan*, 1844!

69. O. caudatum (Ait. Hort. Kew. i. 442; edit. 2, ii. 261); bulb large, ovoid; leaves 5–6, lorate-lanceolate, acuminate, glabrous, $1\frac{1}{2}$–2 ft. long, 1–$1\frac{1}{2}$ in. broad low down; peduncle stout, erect, $1\frac{1}{2}$–3 ft. long; raceme dense, $\frac{1}{2}$–1 ft. long; pedicels ascending, lower $\frac{1}{2}$ in. long; bracts lanceolate-setaceous, $\frac{1}{2}$–$\frac{3}{4}$ in. long, projecting beyond the buds; perianth white, $\frac{1}{3}$ in. long; segments distinctly keeled with green; stamens $\frac{3}{4}$ the length of the perianth; filaments alternately lanceolate and quadrate at the base; style $\frac{1}{12}$–$\frac{1}{8}$ in. long. *Jacq. Collect.* ii. 315; *Ic.* ii. 19, *t.* 423; *Willd. Sp. Plant.* ii. 125; *Bot. Mag. t.* 805; *Roem. et Schultes, Syst. Veg.* vii. 520; *Kunth, Enum.* iv. 357; *Baker in Saund. Ref. Bot. t.* 262; *Journ. Linn. Soc.* xiii. 276. *O. Massoni, Gmel. Syst.* i. 551.

SOUTH AFRICA: without locality, cultivated specimens!

70. O. ovatum (Thunb. Prodr. 62); bulb not seen; leaves 2, oblong, subacute, rigidly coriaceous, glabrous, $1\frac{1}{2}$–2 in. long; margin thickened; peduncle short, slender; raceme dense, 6–20-flowered, 1–$1\frac{1}{4}$ in. long; pedicels ascending, lower $\frac{1}{4}$ in. long; bracts lanceolate, as long as the pedicels; perianth white, $\frac{1}{4}$ in. long; segments linear-oblong, distinctly keeled with green; stamens $\frac{3}{4}$ the length of the perianth; filaments subequal, lanceolate; style $\frac{1}{12}$ in. long. *Fl. Cap. edit. Schult.* 315; *Willd. Sp. Plant.* ii. 117; *Roem. et Schultes, Syst. Veg.* vii. 528; *Kunth, Enum.* iv. 359; *Baker in Journ. Linn. Soc.* xiii. 278.

SOUTH AFRICA: without locality, *Thunberg!*

71. O. crenulatum (Linn. fil. Suppl. 198); bulb oblong, $\frac{1}{2}$ in. diam.; leaves 2, oblong, obtuse, or subacute, subcoriaceous, glabrous, 1–$1\frac{1}{4}$ in. long, rugose on the margins; peduncle stiffly erect, 2–4 in. long; raceme lax, 4–8-flowered; pedicels ascending, lower $\frac{1}{4}$–$\frac{1}{3}$ in. long; bracts deltoid-cuspidate, nearly or quite as long as the pedicels; perianth white, $\frac{1}{4}$ in. long; segments keeled with green; stamens $\frac{2}{3}$ the length of the perianth; filaments subequal, lanceolate. *Thunb. Prodr.* 62; *Fl. Cap. edit. Schult.* 315; *Willd. Sp. Plant.*

ii. 117 ; *Roem. et Schultes, Syst. Veg.* vii. 529 ; *Kunth, Enum.* iv. 371 ; *Baker in Journ. Linn. Soc.* xiii. 278.

SOUTH AFRICA : without locality, *Thunberg !*

72. O. Galpini (Baker); bulb globose, $\frac{1}{2}$–$\frac{3}{4}$ in. diam.; produced leaves 1–3, erect, oblong or oblong-lanceolate, membranous, glabrous, 2–3 in. long, $\frac{1}{4}$–$\frac{1}{2}$ in. broad, narrowed to a short sheathing petiole; peduncle slender, 6–8 in. long; raceme lax, many-flowered, 2–3 in. long; pedicels ascending, lower $\frac{1}{4}$–$\frac{1}{2}$ in. long; bracts small, ovate-cuspidate; perianth $\frac{1}{4}$ in. long; segments oblong, white, with a distinct green keel; stamens distinctly shorter than the perianth; filaments lanceolate; style as long as the globose ovary.

COAST REGION : Queenstown Div. ; mountain sides, Queenstown, alt. 4000 ft., *Galpin,* 1552b !

73. O. trichophyllum (Baker in Engl. Jahrb. xv. Heft 3, 7); bulb globose, 1 in. diam. ; produced leaves ovate-lanceolate, thin, hairy, 1–1$\frac{1}{2}$ in. long, $\frac{1}{4}$ in. broad low down, narrowed to an acute point; peduncle slender, 4–5 in. long; raceme lax, 2–3 in. long; pedicels slender, ascending, the lower $\frac{1}{4}$ in. long ; bracts small, ovate, with a large cusp ; perianth campanulate, $\frac{1}{4}$ in. long ; segments oblong, white, with a distinct brown keel ; stamens shorter than the perianth ; filaments linear, uniform ; anthers small, oblong ; style as long as the globose ovary.

SOUTH AFRICA : without locality, *Ecklon and Zeyher,* 67 ! No specimen at Kew.

XLI. **ANDROCYMBIUM,** Willd.

Perianth polyphyllous, marcescent; segments subequal, with a distinct canaliculate claw and an acute lamina with incurved edges. *Stamens* 6, inserted at the apex of the claw of the perianth-segments; filaments filiform, thickened towards the base; anthers oblong, versatile, dehiscing laterally. *Ovary* sessile, 3-celled; ovules many, superposed ; styles 3, distinct; stigmas minute. *Capsule* ovoid, chartaceous, dehiscing septicidally. *Seeds* globose; testa brown, membranous; albumen firm in texture.

Rootstock an ovoid corm, with membranous tunics; stem short, simple or none; leaves crowded in the acaulescent species; flowers crowded into a globose capitulum, overtopped by the exterior bracts, white, green or purplish.

DISTRIB. Two species Mediterranean and two Tropical African.

Stem produced ; leaves not crowded :
　　Leaves linear ...　　...　　...　　...　　...　　... (1) **Dregei.**
　　Leaves lanceolate　　...　　...　　...　　.. 　　... (2) **melanthioides.**
Stem none ; leaves crowded :
　　Leaves circinate at the tip :
　　　　Leaves linear :
　　　　　　Leaves neither crisped nor ciliated　　... (3) **circinatum.**
　　　　　　Leaves crisped and ciliated　　...　　... (4) **crispum.**
　　　　Leaves ovate-lanceolate　　　...　　...　　... (5) **volutare.**

Leaves not circinate at the tip :
Perianth $\frac{1}{3}$–$\frac{3}{4}$ in. long :
Leaves short, lanceolate ... ... (6) cuspidatum.
Leaves long, lanceolate :
Styles $\frac{1}{3}$ in. long ... ... (7) leucanthum.
Styles $\frac{1}{4}$ in. long ... ... (8) natalense.
Leaves oblong-lanceolate ... ... (9) albomarginatum.
Leaves ovate lanceolate ... ... (10) Burchellii.
Leaves ovate ... ... (11) latifolium.
Perianth about an inch long :
Claw of perianth-segments as long as blade :
Capsule subglobose ... ... (12) eucomoides.
Capsule oblong ... ... (13) Burkei.
Claw of perianth-segments longer than blade ... ... (14) longipes.

1. **A. Dregei** (Presl, Bot. Bemerk. 116); corm ovoid, $\frac{1}{4}$ in. diam.; tunics brown-black, crustaceous; neck slender, 1–1$\frac{1}{2}$ in. long, with a brown membranous sheath; stem sometimes none, sometimes 1–1$\frac{1}{2}$ in. long; leaves 3–4, linear or subulate, the lower 3–4 in. long; capitulum few-flowered, $\frac{1}{4}$–$\frac{1}{3}$ in. diam.; bracts 6–8, greenish, ovate, –$\frac{1}{3}$ in. long; perianth $\frac{1}{3}$ in. long; claw much shorter than the ovate-cuspidate blade; stamens not exserted; capsule as long as the perianth; carpels narrowed gradually into a very short style. *Walp. Ann.* i. 875; *Baker in Journ. Linn. Soc.* xvii. 443. *Melanthium tenue, E. Meyer in Herb. Drège.*

WESTERN REGION: Little Namaqualand; Silver Fontein, near Ookiep, *Drège*, 2705! Twee Fontein, near Concordia, *Bolus*, 6566!

2. **A. melanthioides** (Willd. in Ges. Naturf. Fr. Berl. Mag. ii. 21); corm ovoid, $\frac{1}{2}$–1 in. diam.; tunics nearly black, firm in texture; stem 1–6 in. long above ground; leaves 2–3, linear or lanceolate, the lower 6–8 in. long; capitulum globose, many-flowered, $\frac{1}{2}$–1 in. diam.; bracts 4–8, ovate, acute, the inner whitish, scariose, 1–3 in. long, distinctly striped with green or brown, the outer more like the leaves in texture and furnished with a leafy point; pedicels $\frac{1}{4}$–$\frac{1}{2}$ in. long; perianth whitish, $\frac{3}{8}$–$\frac{1}{2}$ in. long; claw shorter than the ovate-cucullate blade; stamens as long as or a little longer than the blade; anthers yellowish, $\frac{1}{12}$–$\frac{1}{6}$ in. long; capsule as long as the perianth; styles $\frac{1}{6}$ in. long. *Schlecht. in Linnæa*, i. 89; *Kunth, Enum.* iv. 153; *Baker in Journ. Bot.* 1874, 244; *Journ. Linn. Soc.* xvii. 442.

VAR. *β*, **acaule** (Baker in Journ. Linn. Soc. xvii. 442); stem not produced; leaves and bracts crowded.

VAR. *γ*, **A. subulatum** (Baker in Journ. Bot. 1874, 245); leaves subulate, channelled down the face; claw of perianth-segments nearly or quite as long as the blade.

CENTRAL REGION: Somerset Div.; Fish River Heights, *Hutton!* near Little Fish River, *Schinz!* Graaff Reinet Div.; "Africas Hoogde," *Burke*, 285! *Zeyher*, 1712! summit of Cave Mountain, near Graaff Reinet, 4200 ft., *Bolus*, 464!

KALAHARI REGION: Transvaal; Crocodile River, *Oates!* Saddleback Range, near Barberton, 5000 ft., *Wood*, 4290! *Thorncroft*, 274! Moodies, near Barberton, 4500 ft., *Galpin*, 789! Houtbosch, *Rehmann*, 5768! Johannesberg, *E.S.C.A.*

Herb., 370! near Lydenburg, *Atherstone!* Var. β, Transvaal; near Pretoria, *Roe in Herb. Bolus*, 3042! Var. γ, Orange Free State; Bloemfontein, *Rehmann*, 3727!

EASTERN REGION: Natal; under the Drakensberg, *Keit*, 6!

3. **A. circinatum** (Baker in Journ. Linn. Soc. xvii. 443); corm small, ovoid; stems several to a corm, with underground necks 3–4 in. long; proper leaves 2, linear, complicate, 2–3 in. long, spirally recurved at the tip; capitulum sessile, few-flowered; bracts ovate, inconspicuously striped with brown, obtuse, or with a leafy point; perianth ½–⅝ in. long; claw as long as the ovate blade; stamens as long as the blade; anthers oblong, ⅛ in. long; capsule not seen; styles ¼ in. long. *Melanthium maculatum, E. Meyer in Herb. Drège.*

WESTERN REGION: Little Namaqualand; between Uitkomst and Geelbeks Kraal, 2000–3000 ft., *Drège*, 2706!

4. **A. crispum** (Schinz in Bull. Herb. Boiss. iv. 415); corm oblong; outer tunics firm, brown-black; stem shortly produced above ground; proper leaves linear, circinate at the apex, about 3 in. long, ½ in. broad at the base, crisped and densely ciliated; bracts ovate, acute, not ciliated, white or rose-red, with red lines and purple dots; perianth white, ⅔ in. long; stamens an inch long.

WESTERN REGION: Great Namaqualand; Han, near Bethany, *Purcell*. Not in Kew Herbarium.

5. **A. volutare** (Burch. Trav. i. 213); corm not seen; underground neck 1½–2 in. long; leaves 2, lanceolate or ovate-lanceolate, 3–4 in. long, recurved and more or less curled up spirally at the acute tip; stem none; capitulum few-flowered, ½ in. diam.; pedicels very short; bracts ovate, acute or obtuse, 1–1½ in. long, similar to the leaves in texture, not striped; perianth ½ in. long; claw filiform, as long as the deltoid-cuspidate blade; stamens longer than the blade; anthers oblong, ⅛ in. long; capsule subglobose, ⅓ in. long; styles slender, recurved, as long as the capsule. *Baker in Journ. Linn. Soc. xvii. 443.*

CENTRAL REGION: Tulbagh Div.; between Great and Little Doorn Rivers, *Burchell*, 1215! Frazerburg Div.; between Stink Fontein and Seldery Fontein, *Burchell*, 1400!

6. **A. cuspidatum** (Baker in Journ. Bot. 1874, 245); corm not seen; underground neck 1½–2 in. long; leaves 2–3, sessile, lanceolate, acute, 2–3 in. long, ⅓–½ in. broad; stem none; capitulum few-flowered, ½–¾ in. diam.; pedicels very short; outer bracts ovate, similar to the leaves in texture, not distinctly striped, 1–1½ in. long; perianth greenish, ¾ in. long; claw broad, channelled, as long as the ovate-lanceolate blade; stamens as long as the blade; anthers oblong, yellow, ⅙ in. long; capsule subglobose, ¼ in. long; styles as long as the carpels. *Baker in Journ. Linn. Soc. xvii. 444.*

CENTRAL REGION: Fraserburg Div.; by the Great Riet River, *Burchell*, 1376!

7. A. leucanthum (Willd. in Ges. Naturf. Fr. Berl. Mag. ii. 21); corm ovoid, ½–¾ in. diam.; tunics brown-black, firm in texture; underground neck 1–3 in. long; stem none; proper leaves 2–3, outer lanceolate, 6–9 in. long; capitulum many-flowered, 1–1½ in. diam.; outer bracts several, ovate, 2–3 in. long, not distinctly striped; pedicels short; perianth whitish, ⅓–⅝ in. long; claw nearly as long as the ovate-cucullate blade; stamens as long as the blade; anthers oblong, yellowish, ₁₂ in. long; capsule ⅓ in. long; carpels narrowed gradually into persistent styles ⅛ in. long. *Schlecht. in Linnæa*, i. 87; *Kunth, Enum.* iv. 153; *Baker in Journ. Bot.* 1874, 245; *Journ. Linn. Soc.* xvii. 443. *A. eucomoides, Sweet, Brit. Flow. Gard. t.* 165, *non Willd. A. punctatum, Baker in Gard. Chron.* 1874, i. 786. *Melanthium punctatum, Linn. Pl. Afr. Rar.* 10; *Amen. Acad.* vi. 87. *M. capense, Linn. Sp. Pl. ed.* 2, 483. *Thunb. Prodr.* 67; *Fl. Cap. edit. Schult.* 338. *Anguillaria capensis, Spreng. Syst.* ii. 146.

SOUTH AFRICA: without locality, *Zeyher*, 1720!
COAST REGION: Cape Div.; Lion Mountain and Devils Mountain, *Thunberg!* Simons Bay, *Wright!* Devils Mountain, 1000 ft., *Bolus*, 4725! Knysna Div.; near Groene Vallei, *Burchell*, 5628! Uitenhage Div., *Ecklon and Zeyher*, 122!
WESTERN REGION: Little Namaqualand, between Uitkomst and Geelbeks Kraal, 2000–3000 ft., *Drège*, 2709!
KALAHARI REGION: Basutoland, *Cooper!*

There is a drawing of this by Masson in the British Museum.

8. A natalense (Baker); corm small, globose; underground neck 2–3 in. long; proper leaves 2–3, lanceolate, erect, ½–1 ft. long; stem none or very short; capitulum few-flowered; pedicels ½–¾ in. long; outer bracts lanceolate, 1–3 in. long, similar to the leaves in texture; perianth greenish, ⅓ in. long; claw subulate, as long as the ovate blade; stamens as long as the blade; anthers small, oblong; capsule globose, ⅓ in. long; styles half as long as the capsule.

EASTERN REGION: Natal; Inanda, *Wood*, 200! *Mrs. K. Saunders!*

9. A. albomarginatum (Schinz in Bull. Herb. Boiss. iv. 415), corm not seen; stem not produced; proper leaves oblong-lanceolate, acute, about 5 in. long, margined with white; bracts shorter, ovate-lanceolate; perianth white, ⅔ in. long; claw of the segments as long as the blade; stamens as long as the perianth-segments.

SOUTH AFRICA: without locality, *Fleck*, 302a. Not in Kew Herbarium.

10. A. Burchellii (Baker in Journ. Bot. 1874, 246); corm not seen; underground neck 1½–2 in. long; proper leaves 2, spreading, ovate-lanceolate, obtuse, 2½–3 in. long; stem none; capitulum few-flowered, ½ in. diam.; pedicels very short; outer bracts broad-ovate, about an inch long, subscariose, not striped; perianth ⅝ in. long; claw broad, channelled, longer than the ovate blade; stamens twice as long as the blade; anther oblong, ⅙ in. long; capsule not seen. *Journ. Linn. Soc.* xvii. 444.

CENTRAL REGION: Fraserburg Div.; between Stink Fontein and Seldery Fontein, *Burchell*, 1401!

11. A. latifolium (Schinz in Bull. Herb. Boiss. iv. 415); corm subglobose ; outer tunics hard, black ; neck produced ; proper leaves ovate, 3 in. long by half as broad, abruptly narrowed at the base, subcoriaceous, obtuse or subacute, dotted with brown, ciliated ; bracts deep rose-red ; perianth pinkish-white ; segments $\frac{2}{3}$ in. long, with a lanceolate blade, and claw longer than the blade ; stamens an inch long.

WESTERN REGION : Great Namaqualand ; Han, near Bethany, *Purcell*. Not in Kew Herbarium.

12. A. eucomoides (Willd. in Ges. Naturf. Fr. Berl. Mag. ii. 21); corm ovoid, $\frac{3}{4}$–1 in. diam. ; tunics firm, brown-black ; underground stem 2–3 in. long ; stem none ; proper leaves 2–3, outer lanceolate, sometimes a foot long, inner shorter, ovate-lanceolate ; capitulum many flowered, $1\frac{1}{2}$–2 in. diam. ; pedicels stout, $\frac{3}{4}$–1 in. long ; outer bracts ovate, 2–3 in. long, firm in texture, not distinctly striped ; perianth greenish, $\frac{3}{4}$–1 in. long; claw broad, about as long as the ovate-cucullate blade ; stamens exserted ; anthers linear-oblong, $\frac{1}{6}$ in. long ; capsule subglobose, $\frac{1}{2}$ in. long; carpels narrowed suddenly into subulate styles, $\frac{1}{3}$–$\frac{1}{2}$ in. long. *Schlecht. in Linnæa*, i. 89 ; *Roem. et Schultes, Syst. Veg.* vii. 1526 ; *Kunth, Enum.* iv. 153 ; *Baker in Journ. Bot.* 1874, 245 ; *Journ. Linn. Soc.* xvii. 444. *Melanthium eucomoides, Jacq. Ic.* ii. 22, *t.* 452 ; *Bot. Mag. t.* 641. *Cymbanthes fœtida, Salisb. in Trans. Hort. Soc.* i. 329.

CENTRAL REGION : Fraserburg Div. ; near Sutherland, *Burchell*, 1339 ! at Stink Fontein, *Burchell*, 1395 !
WESTERN REGION : Little Namaqualand ; by the Hazenkraals River, *Drège*, 2710 !

13. A. Burkei (Baker in Journ. Bot. 1874, 246) ; corm not seen ; proper leaves 4–5, lanceolate, erect, firm in texture, $\frac{1}{2}$–$\frac{3}{4}$ in. broad low down, the outer a foot long ; capitulum few-flowered, globose, under an inch in diameter ; pedicels very short ; outer bracts ovate, acute, $1\frac{1}{2}$–2 in. long, similar to the leaves in texture ; perianth about an inch long ; claw as long as the ovate-lanceolate blade ; capsule oblong, above $\frac{1}{2}$ in. long ; style straight, erect, nearly as long as the capsule. *Journ. Linn. Soc.* xvii. 444.

KALAHARI REGION: Orange Free State ; near the Vaal River, *Burke !*

14. A. longipes (Baker in Journ. Bot. 1874, 246) ; corm globose, $\frac{1}{2}$ in. diam. ; underground neck 2 in. long ; proper leaves 4, lanceolate-acuminate, 5–10 in. long, under an inch broad low down ; stem none ; capitulum few-flowered, under an inch in diameter ; pedicels very short ; outer bracts few, lanceolate, similar to the leaves in texture ; perianth 1–$1\frac{1}{4}$ in. long ; claw slender, longer than the ovate-acuminate blade ; stamens shorter than the blade ; anthers oblong, $\frac{1}{12}$ in. long; capsule not seen. *Journ. Linn. Soc.* xvii. 445.

CENTRAL REGION: Somerset Div., *Bowker !*

I know nothing about *A. littorale* and *A. cucullatum*, mentioned by name only by Ecklon in *Top. Verz.* 6–7.

XLII. WURMBEA, Thunb.

Perianth gamophyllous, persistent; tube campanulate or cylindrical; segments linear or lanceolate, equal, patent, with two glandular foveoles above the base. *Stamens* 6, inserted at the base of the perianth-segments; filaments short, filiform; anthers oblong, versatile, introrsely affixed, dehiscing extrorsely near the margin. *Ovary* sessile, trilocular, 3-lobed at the apex; ovules many, superposed; styles short, subulate, divaricate; stigma small, capitate. *Capsule* globose, septicidally 3-valved. *Seeds* subglobose; testa thin, brown, depressed; albumen firm.

Rootstock a tunicated corm; stem simple, with few superposed narrow le of moderately firm texture; inflorescence spicate, without bracts.

DISTRIB. Also Australian and Tropical African.

Tube distinct, campanulate or cylindrical   ...   .   (1) capensis.
Tube very short ...   .   (2) Kraussii.

1. **W. capensis** (Thunb. Nov. Gen. Pl. i. 18, with pl.); corm small, ovoid; underground neck 2–3 in. long; stem ½–1 ft. long including the spike; produced leaves 3–4, lanceolate or linear, acuminate, 3–9 in. long, channelled down the face and clasping the stem at the base, the upper shorter with a dilated base; spike few- or many-flowered, lax or dense, 1–4 in. long; perianth ⅓–½ in. long; tube campanulate; segments lanceolate, acute, whitish, longer than the tube, with two distinct black glands on the face a short distance above the base; filaments ⅛–⅙ in. long; carpels narrowed gradually into falcate styles ⅛ in. long. *W. campanulata, Willd. Sp. Pl.* ii. 265; *Lam. Ill. t.* 270; *Ait. Hort. Kew. edit.* 2, ii. 325; *Kunth, Enum.* iv. 159; *Baker in Journ. Linn. Soc.* xvii. 435. *Melanthium monopetalum, Linn. fil. Suppl.* 213; *Bot. Mag. t.* 1291. *M. spicatum, N. L. Burm. Fl. Cap. Prod.* 11. *M. wurmbeum, Thunb. Prodr.* 67; *Fl. Cap. edit. Schult.* 338. *M. remotum, Ker in Bot. Mag. sub t.* 694; *Kunth, Enum.* iv. 162 (a lax, few-flowered form). *Wurmbea remota, Herb. Banks. ex Ker loc. cit.*

VAR. *a*, **purpurea** (Baker); flowers dark purplish-black; segments much longer than the campanulate tube. *W. campanulata, Willd. var. purpurea, Kunth, Enum.* iv. 160. *W. purpurea, Dryand. in Ait. Hort. Kew. edit.* 2, ii. 326. *W. capensis, Andr. Bot. Rep. t.* 221. *Melanthium spicatum, Ker in Bot. Mag. t.* 694. *M. revolutum, Ker in Bot. Mag. sub t.* 694 (form with recurved linear segments).

VAR. *β*, **marginata** (Baker); segments pale, with a distinct black edge, longer than the campanulate tube. *W. campanulata, Willd. var. marginata, Kunth,* iv. 160. *Melanthium marginatum, Desv. in Lam. Encyc.* iv. 29.

VAR. *γ*, **truncata** (Baker); perianth-tube broadly cyathiform; segments linear, pale with a blackish edge, about as long as the tube. *W. truncata, Schlecht. in Linnæa,* i. 93; *Kunth, Enum.* iv. 161.

VAR. *δ*, **longiflora** (Baker); more robust than the type; perianth-tube cylindrical, as long as or longer than the falcate lanceolate segments. *W. longiflora, Willd. Sp. Plant.* ii. 266; *Schlecht. in Linnæa,* i. 94; *Kunth, Enum.* iv. 161. *Melanthium tubiflorum, Soland. MSS.*

VAR. *ε*, **inusta** (Baker); a dwarf form, with brown perianth-segments not more than half as long as the cylindrical tube. *Melanthium inustum, Soland. ex Kunth, Enum.* iv. 161.

VAR. ζ, latifolia (Baker); leaves ovate-acuminate, 1–1¼ in. broad; spike lax; perianth-segments lanceolate, purplish-black, about as long as the infundibuliform tube.

SOUTH AFRICA: without locality, var. β, specimens at Kew without indication of collector! Var. γ, *Zeyher*, 1722! Var. ε, *Harvey*, 100!

COAST REGION: Malmesbury Div.; Groene Kloof and elsewhere, *Thunberg.* Albany Div.; near King William's Town, *Mrs. Barber!* Queenstown Div.; damp valleys amongst the mountains, *Mrs. Barber!* Var. α, Malmesbury Div.; Groene Kloof, *MacOwan*, 2563! *MacOwan and Bolus, Herb. Norm. Aust. Afr.*, 802! *Pappe!* Paarl Div.; between Paarl and Lady Grey Railway Bridge, below 1000 ft., *Drège!* Stellenbosch, *Sanderson*, 979! Var. γ, Cape Div.; near Riet Vallei, *Mund and Maire.* Cape Flats, *Bergius.* Simons Bay, *Wright!* Var. δ, Clanwilliam Div.; between Groene River and Watervals River, 2500–3000 ft., *Drège*, 2660b! Clanwilliam, *Mader*, 183! Tulbagh, *Thom!* Caledon Div.; Villiersdorp. *Grey!* Var. ζ, Riversdale Div.; near Zoetmelks River, *Burchell*, 6628! Port Elizabeth, *E S.C.A. Herb.*, 13!

CENTRAL REGION: Somerset Div.; summit of Bosch Berg, 4500 ft., *MacOwan*, 979! Aliwal North Div.; Witte Bergen, 7000–8000 ft., *Drège*, 3512!

EASTERN REGION: Griqualand East; Ingeli, near Kokstad, 6500 ft., *Tyson*, 1281!

2. **W. Kraussii** (Baker in Journ. Linn. Soc. xvii. 437); corm small, oblong; tunics produced as a sheath to its underground neck for 1½–2 in.; produced leaves 2, firm in texture, one basal linear or subulate, 3–4 in. long, the other a little below the base of the spike, much shorter, lanceolate, erect, clasping the stem; stem slender, 3–4 in. long including the spike; spike 3–6-flowered; perianth ¼ in. long; tube campanulate, very short; segments lanceolate, whitish; foveoles indistinct; stamens more than half as long as the perianth; carpels narrowed gradually into falcate styles 1/12 in. long.

EASTERN REGION: Griqualand East; Zuurberg Range, 5000 ft., *Tyson*, 1864! Natal; at the foot of Table Mountain, *Krauss*, 450! and without precise locality, *Gerrard*, 549!

XLIII. BÆOMETRA, Salisb.

Perianth polyphyllous, marcescent; segments subequal, oblanceolate, with a long convolute claw. *Stamens* 6, inserted at the apex of the claw of the perianth-segments; filaments short, subulate, tapering upwards; anthers linear-oblong, versatile, dehiscing along the margin. *Ovary* sessile, cylindrico-triquetrous; ovules many, superposed; styles 3, short, spreading, stigmatose laterally. *Capsule* subcoriaceous, cylindrico-triquetrous, septicidally 3-valved. Seeds many, globose or angled by pressure; testa brown, opaque; albumen cartilaginous.

DISTRIB. Endemic.

1. **B. columellaris** (Salisb. in Trans. Hort. Soc. i. 330); corm small, ovoid; tunics membranous, brown; stems simple, ½–1 ft. long; leaves several, dry, persistent, moderately firm in texture, lower lanceolate, 6–9 in. long, clasping the stem, upper growing gradually much shorter; flowers one or few in a simple raceme; pedicels short, ascending, not articulated; bracts subulate; perianth ½–¾ in. long, red outside, yellow within, with a black spot at the

base of the blade; stamens about half as long as the blade; capsule 1–2 in. long. *Kunth, Enum.* iv. 162; *Baker in Journ. Linn. Soc.* xvii. 446. *Tulipa breyniana, Linn. Sp. Plant.* ii. 438; *Thunb. Prodr.* 65. *Melanthium uniflorum, Jacq. Ic.* ii. 21, *t.* 450; *Desv. in Lam. Encyc.* iv. 30; *Gawl. in Bot. Mag. t.* 767. *M. flavum, Smith in R*ees' *Cyclop. No.* 7. *M. æthiopicum, Desv. in Lam. Encyc.* iv. 29. *Kolbea breyniana, Schlecht. in Linnæa,* i. 82. *Jania breyniana, Roem. et Schultes, Syst. Veg.* vii. 1528. (*Breyn. Cent. t.* 36; *Rudb., Elys.* 2, *f.* 11).

COAST REGION: Clanwilliam Div.; Clanwilliam, *Mader*, 184! Cape Div.; Wynberg, 50 ft., *Bolus*, 2836! Simons Bay, *Wright!* Table Mountain, *Ecklon*, 813! *MacOwan and Bolus, Herb. Norm. Aust. Afr.*, 501! Paarl Div.; by the Berg River, near Paarl, *Drège*, 307a! Tulbagh, *Thom!* Riversdale Div.; hills near Zoetmelks River, *Burchell*, 6769! and Bulb 236!

XLIV. DIPIDAX, Salisb.

Perianth polyphyllous, deciduous; segments subequal, oblanceo-late-unguiculate, with a pair of nectariferous spots at the base of the blade. *Stamens* 6, inserted on the claw of the perianth-segments; filaments short, subulate; anthers oblong, versatile, introrsely affixed, dehiscing along the margin extrorsely. *Ovary* sessile, 3-celled, 3-lobed; ovules many, superposed; styles 3, subulate, falcate, stigmatose internally. *Capsule* turbinate, septicidally 3-valved. *Seeds* subglobose; testa thin, brown; albumen firm in texture.

Rootstock a tunicated corm; leaves generally 3, superposed, firm in texture, persistent; flowers few or many, spicate, whitish.

DISTRIB. Endemic.

Lower leaves ovate or lanceolate   ...   .   (1) **ciliata.**
Lower leaves subulate-triquetrous   ...   .   (2) **triquetra.**

1. **D. ciliata** (Baker in Journ. Linn. Soc. xvii. 447); corm small, subglobose; tunics brown, membranous; stem simple, $\frac{1}{2}$–1 ft. long; leaves usually 3, ciliated with short spreading hairs, lower lanceolate, acuminate, 4–6 in. long, $\frac{1}{2}$–1 in. broad, upper much shorter, clasping the stem at the middle; spike 2–6 in. long, generally dense and many-flowered; perianth $\frac{1}{3}$–$\frac{1}{2}$ in. long; segments whitish or tinged with red, about $\frac{1}{12}$ in. broad; nectariferous spots and veins indistinct; anthers minute, oblong; filaments $\frac{1}{6}$ in. long; ovary turbinate, acutely triquetrous; styles slender, $\frac{1}{12}$ in. long; capsule $\frac{1}{2}$ in. long. *Melanthium ciliatum, Linn. fil. Suppl.* 213; *Thunb. Prodr.* 67; *Fl. Cap. edit. Schult.* 339; *Schlecht. in Linnæa.* i. 83; *Roem. et Schultes, Syst. Veg.* vii. 1544; *Kunth, Enum.* iv. 156. *M. capense, Willd. Sp.* ii. 267; *Lam. Ill. t.* 269. *M. punctatum, Mill. Dict.* viii. *No.* 3.

VAR. a, **secunda** (Baker); leaves narrow; flowers secund, white spotted with red, the segments denticulate on each side at the base of the claw. *Melanthium secundum, Desv. in Lam. Encyc.* iv. 28; *Ill. t.* 269, *fig.* 2; *Kunth, Enum.* iv. 156.

VAR. β, **Bergii** (Baker); leaves linear; flowers few, pale, not distinctly blotched. *Melanthium Bergii, Schlecht. in Linnæa,* i. 83; *Kunth, Enum.* iv. 157.

VAR. γ, **gracilis** (Baker) ; leaves narrow ; flower solitary.   *Melanthium gracile, Desv. in Lam. Encyc.* iv. 29.

VAR. δ, **garnotiana** (Baker, loc. cit.); leaves lanceolate ; perianth-segments white, distinctly foveolate at the base of the blade.   *Melanthium garnotianum, Kunth, Enum.* iv. 157.

VAR. ε, **rubicunda** (Baker) ; differs from the last by its reddish perianth-segments.   *Melanthium rubicundum, Willd. in Ges. Naturf. Fr. Berl. Mag.* ii. 22 ; *Schlecht. in Linnæa,* i. 85 ; *Kunth, Enum.* iv. 157.

VAR. ζ, **marginata** (Baker); lower leaf short, ovate or ovate-oblong ; perianth-segments distinctly foveolate on each side at the base of the blade. *Melanthium marginatum, Schlecht. in Linnæa,* i. 84.   *M. schlechtendalianum, Roem. et Schultes, Syst. Veg.* vii. 1546.

SOUTH AFRICA : without locality ; Var. *a, Sonnerat.*   Var. *β, Bergius.*   Var. γ, *Herb. Thouin,* Var. δ, *Garnot.*   Var. ζ, *Lichtenstein.*

COAST REGION : Cape Div. ; hills near Cape Town, *Thunberg.*   Table Mountain, *Ecklon,* 508 !   Camp Ground, *Bolus,* 3700 !   Camps Bay, *MacOwan,* 2557 !   *MacOwan and Bolus, Herb. Norm. Aust. Afr.,* 300 !   Paarl Div. ; between Paarl and Lady Grey Railway Bridge, *Dräge !*   Tulbagh Div., *Thom !* near Tulbagh Waterfall, 1500 ft.; *MacOwan and Bolus, Herb. Norm. Aust. Afr.,* 1392 ! near Ceres Road, *MacOwan, Herb. Aust. Afr.,* 1563 !   Stellenbosch, *Sanderson,* 963 !   Swellendam, *Kennedy !*   Riversdale Div. ; near Zoetmelks River, *Burchell,* 6689 !   Var. *β,* Cape Div. ; sea-shore and near Riet Vallei, *Zeyher,* 4300 ! Simons Bay, *Wright !*   sand-dunes near Cape Town, *Bolus,* 3758 !   Worcester Div. ?   *Cooper,* 3204 !   Var. ε, near Tulbagh, *Lichtenstein.*   Caledon, *Zeyher !*

2. **D. triquetra** (Baker in Journ. Linn. Soc. xvii. 447) ; corm small, globose ; tunics brown ; stem simple, 1–1½ ft. long ; leaves 3, not ciliated, lower placed near the base of the stem, subulate-trique-trous, 1–1½ ft. long, two upper contiguous, placed near the base of the spike ; uppermost very short, lanceolate-amplexicaul ; spike 1–6 in. long ; perianth-segments ⅓ in. long, ⅙ in. broad, with numerous brown veins and two purplish nectariferous blotches at the base of the blade ; anthers minute, purplish ; filaments ⅛ in. long; ovary turbinate, the cells not keeled acutely on the back ; styles slender, 1⁄16 in. long.   *Melanthium triquetrum, Linn. fil. Suppl.* 213 ; *Thunb. Prodr.* 67 ; *Fl. Cap. edit. Schult.* 340 ; *Schlecht. in Linnæa,* i. 86 ; *Roem. et Schultes, Syst. Veg.* vii. 1547 ; *Kunth, Enum.* iv. 157. *M. junceum, Jacq. Ic.* ii. 21, *t.* 451 ; *Sims in Bot. Mag. t.* 558 ; *Lodd. Bot. Cab. t.* 978.   *Dipidax rosea, Salisb. in Trans. Hort. Soc.* i. 330.

COAST REGION : Clanwilliam, *Mader,* 109 !   Cape Div. ; Cape Flats, *MacOwan and Bolus, Herb. Norm. Aust. Afr.,* 299 !   *Rogers !*   Stellenbosch, *Sanderson,* 962 !   Caledon Div., *Zeyher !*

XLV. ORNITHOGLOSSUM, Salisb.

Perianth polyphyllous, persistent ; segments equal, lanceolate, spreading or reflexing, unguiculate, with a nectariferous foveole at the top of the convolute claw.   *Stamens* 6, attached to the base of the perianth-segments ; filaments filiform ; anthers oblong, versatile, dorsifixed, dehiscing extrorsely near the margin.   *Ovary* oblong, sessile, 3-celled ; ovules many, superposed ; styles 3, long, subulate, stigmatose at the apex.   *Capsule* large, oblong, finally loculicidally

3-valved. *Seeds* subglobose; testa thick, fleshy; albumen cartilaginous; embryo minute.

DISTRIB. The same species in Tropical Africa.

1. O. glaucum (Salisb. Parad. t. 54); corm small, ovoid, with an
underground neck 2–3 in. long; stem about ¼ ft. long including the
inflorescence; leaves about 6, moderately firm in texture, amplexicaul,
lanceolate, the lowest 4–6 in. long, ¼–½ in. broad, the upper gradually
shorter; inflorescence corymbose or racemose; pedicels 1–2 in. long,
at first ascending, finally reflexed; bracts lanceolate, foliaceous;
perianth greenish-purple, ⅓–¾ in. long; stamens shorter than the
perianth-segments; anthers yellow, ⅛–¼ in. long; capsule about as
long as the perianth; styles ⅙–¼ in. long. *Schlecht. in Linnæa,* i. 90;
Roem. et Schultes, Syst. Veg. vii. 1536; *Kunth, Enum.* iv. 163;
Baker in Journ. Linn. Soc. xvii. 449. *O. viride, Ait. Hort. Kew.*
edit. 2, ii. 327. *Melanthium viride, Linn. fil. Suppl.* 213; *Thunb.*
Prodr. 67; *Andr. Bot. Rep. t.* 233; *Bot. Mag. t.* 994. *Lichten-*
steinia lævigata, Willd. in Ges. Naturf. Fr. Berl. Mag. ii. 20.
Cymation lævigatum, Spreng. Syst. ii. 142.

VAR. *a*, grandiflorum (Baker, loc. cit.); more robust; flowers larger; leaves
not crisped.

VAR. *β*, undulatum (Baker, loc. cit.); leaves much crisped; stems short;
flowers few, corymbose, about an inch long. *O. undulatum, Spreng. Cur. Post.*
143; *Sweet, Brit. Flow. Gard. t.* 131. *O. Lichtensteinii, Schlecht. in Linnæa,*
i. 91; *Roem. et Schultes, Syst. Veg.* vii. 1537; *Kunth, Enum.* iv. 163. *Lichten-*
steinia undulata, Willd. in Ges. Naturf. Fr. Berl. Mag. ii. 20. *Cymation*
undulatum, Spreng. Syst. ii. 142.

VAR. *γ*, Zeyheri (Baker, loc. cit.); acaulescent, with numerous, small, much-
crisped leaves, and a few, small, corymbose, brown-purple flowers.

SOUTH AFRICA: without locality; Var. *a, Zeyher,* 1724! Var. *γ, Zeyher,*
1721!

COAST REGION : Cape Div.; hills below Table Mountain, *Thunberg.* Camps
Bay, *Harvey!* Var. *β,* Clanwilliam Div.; Berg Vallei, below 1000 ft., *Drège!*
Zeyher, 1723!

CENTRAL REGION : Somerset Div., *Bowker,* 25! Graaff Reinet Div.; near
Graaff Reinet, 2500 ft., *Bolus,* 375! *Bowker,* 21! Fraserburg Div.; between
Zak River and Kopjes Fontein, *Burchell,* 1502! Colesberg Div.; hills south of
the Orange River, *Hutton!* Colesberg, 4500 ft., *Drège!* Var. *a,* Tulbagh Div.;
on the Wind-heuvel, Koedoes Mountains, *Burchell,* 1283! Prince Albert Div.;
by the Gamka River, *Burke! Drège!* Colesberg Div., *Shaw!* Var. *β,* Prince
Albert Div.; by the Gamka River, *Burke!*

KALAHARI REGION : Griqualand West; Dutoits Pan and Colesberg Kop,
Mrs. Barber, 18! on the plains, *Bowker,* 15! Var. *a,* at the Kloof Village,
Asbestos Mountains, *Burchell,* 2023!

XLVI. GLORIOSA, Linn.

Perianth polyphyllous, persistent; segments equal, unguiculate,
spreading or reflexed. *Stamens* 6, hypogynous; filaments filiform;
anthers linear-oblong, versatile, introrsely attached, dehiscing extrorsely near the margin. *Ovary* sessile, oblong, trilocular; ovules
many, superposed; style long, filiform, trifurcate at the apex;
branches stigmatose internally. *Capsule* large, coriaceous, septicidally

3-valved. *Seeds* globose; testa bright red, spongy; albumen firm in texture.

Rootstock tuberous; stem slender, usually elongated and scandent; leaves sessile, cirrhiferous at the apex; flowers few, large, showy, corymbose.

DISTRIB. Through Tropical Africa and Tropical Asia; five species.

Perianth-segments crisped ... ... (1) **superba.**
Perianth-segments plane ... ... (2) **virescens.**

1. **G. superba** (Linn. Sp. Plant. 305); stems slender, widely scandent; leaves membranous, sessile, ovate or ovate-lanceolate, with a cirrhiferous tip, central opposite or ternate, upper ovate, alternate; flowers few in a lax terminal corymb; peduncles cernuous at the apex, 3–6 in. long, ebracteate; perianth about 2 in. long; segments oblanceolate, acute, unguiculate, bright red or yellow, reflexing, much crisped, $\frac{1}{4}$–$\frac{1}{3}$ in. broad; stamens more than half as long as the perianth-segments; anthers $\frac{1}{4}$ in. long; style horizontal, $1\frac{1}{2}$–2 in. long. *Bot. Reg. t.* 77; *Andr. Bot. Rep. t.* 129; *Reich. Fl. Exot. t.* 51; *Wight, Ic. t.* 2047; *Baker in Journ. Linn. Soc.* xvii. 457. *Methonica superba, Lam. Encyc.* iv. 133; *Ill. t.* 247; *Red. Lil. t.* 26; *Kunth, Enum.* iv. 276.

KALAHARI REGION: Transvaal; near Barberton, 2500–3000 ft., *Galpin,* 760! Houtbosch, *Rehmann,* 5812! near Rhenoster Poort, *Nelson,* 537!

Also Tropical Africa and Tropical Asia.

2. **G. virescens** (Lindl. in Bot. Mag. t. 2539); stems slender, widely scandent when well developed; leaves generally ovate with a cirrhiferous apex, central 2–4-nate, upper scattered; flowers few in a corymb at the end of the branches; peduncles long, slender, ebracteate, cernuous at the tip; perianth reflexing, generally bright red or yellow, $1\frac{1}{2}$–2 in. long; segments oblanceolate, acute, unguiculate, not crisped, $\frac{1}{2}$ in. broad at the middle; stamens much shorter than the perianth-segments; style slender, about an inch long. *Baker in Journ. Linn. Soc.* xvii. 458. *Methonica superba, β, Lam. Encyc.* iv. 133. *M. virescens, Kunth, Enum.* iv. 277; *Hook. in Bot. Mag. t.* 4938. *M. virescens var. Plantii, Flore des Serres, t.* 865. *M. petersiana and M. platyphylla, Klotzsch in Peters' Reise Mossamb. Bot. t.* 54–55.

COAST REGION: sand-hills of Lower Albany, *Bowker!* King Williamstown Div.; Keiskamma, *Mrs. Hutton!*
EASTERN REGION: Pondoland, *Bachmann,* 254! Natal; near Durban, *Wood,* 1589! and without precise locality, *Plant,* 19! *Gerrard,* 736!

Also Tropical Africa.

XLVII. SANDERSONIA, Hook.

Perianth gamophyllous, persistent, globose, urceolate, with 6 short, broadly ovate segments, and at the base 6 saccate foveoles. *Stamens* 6, hypogynous, much shorter than the perianth-tube; filaments filiform; anthers oblong, versatile, introrsely attached, dehiscing extrorsely.

Ovary oblong, deeply laterally 3-lobed; ovules many, superposed; style short, cylindrical, trifid, its branches internally stigmatose. *Fruit* and *seeds* unknown.

DISTRIB. Endemic.

1. **S. aurantiaca** (Hook. in Bot. Mag. t. 4716); rootstock tuberous; stem erect, simple, 1–2 ft. long, leafy to the apex, except a short distance near the base; leaves sessile, alternate, ascending, linear or lanceolate, 2–4 in. long, $\frac{1}{4}$–$\frac{3}{4}$ in. broad; flowers solitary from the axils of a few of the upper leaves on slender, cernuous, ebracteate pedicels $\frac{1}{2}$–1 in. long; perianth bright yellow, $\frac{3}{4}$–1 in. long; stamens and pistil $\frac{1}{4}$–$\frac{1}{3}$ as long as the perianth. *Harvey, Cape Genera, edit.* 2, 403; *Baker in Journ. Linn. Soc.* xvii. 453.

EASTERN REGION: Pondoland, *Bachmann*, 255! Griqualand East; woods around Fort Donald, 5000 ft., *MacOwan and Bolus, Herb. Norm. Aust. Afr.*, 528! Natal, *Wood*, 459! *Sutherland! Sanderson! Gerrard*, 527!

XLVIII. LITTONIA, Hook.

Perianth marcescent, cut down nearly to the base into 6 equal, oblong-lanceolate, ascending segments. *Stamens* 6, hypogynous, much shorter than the perianth; filaments filiform; anthers linear-oblong, versatile, introrsely attached near the base, extrorsely dehiscent. *Ovary* sessile, oblong, trilocular, deeply trisulcate; ovules many, superposed; style short, erect, cylindrical, with 3 falcate branches stigmatose internally. *Capsule* large, coriaceous, septicidally 3-valved. *Seeds* not seen.

Rootstock tuberous; stem simple or branched, erect or sarmentose, leafy up to the top; flowers solitary from the axils of upper leaves.

DISTRIB. Four species in Tropical Africa and two in Arabia.

1. **L. modesta** (Hook. in Bot. Mag. t. 4723); stem slender, simple, short and suberect, or longer and sarmentose; leaves ovate or lanceolate or linear, cirrhiferous at the tip, central verticillate, upper opposite or alternate; flowers solitary from the axils of the upper leaves on slender, ebracteate, cernuous pedicels 1–2 in. long; perianth $\frac{3}{4}$–1$\frac{1}{4}$ in. long; segments oblong-lanceolate, acute; stamens and pistil about $\frac{1}{2}$ in. long; capsule 1$\frac{1}{2}$ in. long. *Harvey, Cape Gen. edit.* 2, 403; *Baker in Journ. Linn. Soc.* xvii. 458.

VAR. β, **Keiti** (*Leichtlin in The Garden*, 1885, xxviii. 116); stems longer, branched, sarmentose; flowers larger, as many as 50 to one stem. *L. Keiti, Leichtlin in The Garden*, 1883, xxiv. 87.

KALAHARI REGION: Orange Free State; Nelsons Kop, *Cooper*, 880! Transvaal, Houtbosch, 5813! Umvoti Creek, Barberton, *Galpin*, 761! near Lydenburg, *Atherstone!*
EASTERN REGION: Natal; Nottingham, *Buchanan!* near Tongaat, *M'Ken!* and without precise locality, *Gerrard*, 758! *Gueinzius!*

XLIX. WALLERIA, Kirk.

Perianth gamophyllous, finally deciduous; tube campanulate; segments equal, lanceolate, many-nerved, patent. *Stamens* 6, inserted at the throat of the perianth-tube; filaments very short, filiform; anthers connivent in a cone, lanceolate, basifixed, dehiscing by terminal pores. *Ovary* free, but included in the perianth-tube, globose, trilocular; ovules many, superposed; style subulate; stigma capitate. Mature *fruit* and *seeds* unknown.

Rootstock tuberous; stem erect, simple; leaves sessile, linear or oblong-lanceolate; pedicels axillary, 1-2-flowered.

DISTRIB. Three species, or probably varieties, known also in Angola and the Zambesi highlands.

1. **W. nutans** (Kirk in Trans. Linn. Soc. **xxiv.** 497, t. 52, fig. 1); stem slender, ½-1½ ft. long, decumbent at the base; leaves of the lower half of the stem small, rudimentary, of the upper half linear, sessile, alternate, ascending, 3–4 in. long; pedicels ascending, 1–2 in. long, cernuous at the apex, furnished with 1–2 linear foliaceous bracts; perianth with a greenish campanulate tube ⅛ in. long; segments lanceolate, acute, patent, ¼ in. long, laxly 5–7-nerved; anthers more than half as long as the perianth-segments. *W. Mackenzii var. nutans, Baker in Journ. Linn. Soc.* xvii. 499.

KALAHARI REGION: Transvaal; Houtbosch, *Rehmann,* 5811 !

ADDENDA.

31a. Moræa xerospatha (MacOwan Herb.); corm globose,
1 in. diam.; tunics of wiry strands; produced leaves 3–4,
narrowly linear, firm, glabrous, spreading, the lower above a foot
long; stem short; clusters of flowers 1–4, the lateral ones sessile;
spathes cylindrical, an inch long; valves lanceolate, scariose at the
flowering time; perianth fugitive, yellow; tube very slender, longer
than the limb; limb ⅓ in. long; segments oblanceolate-unguiculate,
subequal, spreading in the upper half; style-branches nearly as long
as the limb; appendages narrow, very acute.

Coast Region : Cape Div.; grassy places near Cape Town, *MacOwan*, 3118 !

3a. Homeria simulans (Baker); stem a foot or more long including the inflorescence, with a single, long, rigid, subterete leaf from
the base of the panicle; panicle copiously branched; peduncles
viscous below the spathes; spathes 2–3-flowered, cylindrical, an inch
long, strongly ribbed, with a brown scariose border, the inner valve
rather longer than the outer; perianth-limb yellow, ½ in. long;
segments oblong, obtuse, the 3 inner rather smaller than the others;
filaments united to the top; style-branches subpetaloid, emarginate
at the apex, the rather spreading tips stigmatose just within the
edge.

Coast Region : Cape Div.; open sandy places near Kenilworth, 100 ft., *Bolus*,
7931 !

Habit exactly resembling that of *Moræa viscaria, Ker.*

13. Aristea dichotoma var. macrocephala (Schlechter); heads
¾ in. diam.; bracts much larger than in the type, ¾ in. long, inner
white, with a narrow herbaceous centre.

Coast Region : Stellenbosch Div.; Lowrys Pass, 1500 ft., *Schlechter*, 7250 !

7a. Gladiolus oreocharis (Schlechter in Journ. Bot. 1896, 504);
corm unknown; stem slender, erect, glabrous, a foot long; leaves 3,
linear-setaceous, the lowest as long as the stem, the others shorter;
raceme secund, few-flowered; outer spathe-valve oblong, greenish-
brown, membranous at the tip; perianth-tube nearly straight, dilated
at the apex; segments oblong, subobtuse, the upper rather larger

than the others ; stamens rather shorter than the limb ; style reaching
to the tip of the segments.

SOUTH AFRICA : Matroos Berg, 6000–7000 ft., *Marloth*, 2265.   No specimen
in Kew Herbarium.

Near *G. gracilis, Jacq.*

9a. Gladiolus aureus (Baker) ; corm small, globose ; outer
tunics of strong, pale brown, parallel strands ; stem very slender,
½–1 ft. long, clothed with fine spreading hairs, bearing 2 hairy
leaves near the middle, the lower with a long sheath and a narrowly
linear, strongly-ribbed, erect, free tip, the other entirely sheathing ;
flowers 1–2, secund ; outer spathe-valves oblong or lanceolate, green,
¼–½ in. long ; perianth bright yellow, 1¼–1½ in. long ; tube funnel-
shaped above the slender base, slightly curved ; segments oblong,
acute, subequal, rather shorter than the tube ; filaments short ;
anthers large, linear ; style much overtopping the anthers.

COAST REGION : Cape Div. ; Foot of the Kommetjes Mountains, Cape
Peninsula, alt. 300 ft., *Bolus*, 7951 !

Near *G. trichonemifolius, Ker.*

58a. Gladiolus Flanagani (Baker) ; corm not seen ; produced
leaves 3, sheathing the base of the stem, ensiform, glabrous, ¼ ft.
long, ¼ in. broad at the middle, with thickened stramineous veins
and margins ; stem as long as the leaves, with a small leaf from the
middle ; spike lax, secund, 3-flowered ; outer spathe-valve oblong-
lanceolate, 1½ in. long, much tinged with crimson ; perianth bright
crimson ; tube curved, funnel-shaped, 1½ in. long ; segments oblong,
obtuse, as long as the tube, the lower as long as the upper, not
reflexing ; stamens and style nearly as long as the perianth-limb.

KALAHARI REGION : Basutoland ; Mont aux Sources, near the summit,
8500 ft., *Flanagan*, 1832 !   No specimen in Kew Herbarium.

Near *G. cardinalis, Curt.*, and *G. cruentus, T. Moore.*

62a. Gladiolus fusco-viridis (Baker) ; leaves 5–6 in a basal
rosette, ensiform, bright green, strongly ribbed, the lowest 12–15 in.
long, 1–1¼ in. broad at the middle ; stem, including inflorescence,
1½–2 ft. long, bearing 2 reduced leaves ; flowers 10–12 in a dense spike
5–6 in. long ; outer spathe-valves green, oblong-lanceolate, the lower
1½–2 in. long ; perianth-tube narrowly funnel-shaped, curved, 2 in.
long, ⅓ in. diam. at the throat ; 3 upper segments oblong, acute, as
long as the tube, with copious minute stripes of claret-brown on a
greenish ground ; 3 lower much smaller, recurved ; stamens and style
nearly as long as the concave upper segment.

SOUTH AFRICA : cultivated specimen !

Described from a plant that flowered at Kew in July, 1896, the exact locality
not known.

Nearly allied to *G. dracocephalus, Hook. fil.*

9a. Antholyza Steingroveri (Pax in Engl. Jahrb. xv. 156) ; stem
terete, glabrous, hollow, simply leafy, a foot long ; leaves narrowly

linear, glaucous, glabrous, 8 in. long, $\frac{1}{12}$ in. broad; spike dense, simple, secund; spathe-valves ovate-lanceolate, acute, glaucous, tinged with violet, nearly an inch long; perianth-tube slender, spurred, $\frac{1}{3}$ in. long, funnel-shaped and curved at the throat; upper segment oblong, acute, bright red, $1\frac{1}{4}$ in. long; lateral segments small, oblong-deltoid, acute, yellow, like the tube; segments not uniseriate. *Schinz in Bull. Herb. Boiss.* iv. *App.* iii. 50.

WESTERN REGION: Great Namaqualand; Aus, *Steingröver,* 47, *Schenk,* 156. No specimen in Kew Herbarium.

13a. Antholyza pulchrum (Baker); corm not seen; leaves erect, linear, firm, glabrous, much shorter than the stem, not strongly ribbed; stem terete, $1\frac{1}{4}$ ft., bearing 3 much-reduced leaves; spike lax, equilateral, 4 in. long; spathe-valves linear-oblong, $\frac{1}{3}$ in. long, the inner membranous, the outer firmer; perianth bright purple; tube subcylindrical, an inch long; segments obovate, subequal, spreading, $\frac{1}{2}$ in. long; anthers just exserted from the throat of the perianth-tube; style overtopping the anthers.

COAST REGION: Bredasdorp Div.; Elim, 500 ft., *Schlechter,* 7611!

Near *A. nervosa, Thunb.,* and *A. lucidor, Linn. fil.*

1a. Hypoxis rubella (Baker); corm globose, $\frac{1}{4}$–$\frac{1}{3}$ in. diam., with copious, very slender root-fibres; leaves about 4 to a corm, with a long, clasping, membranous base, and an erect subterete blade an inch long, ciliated on the margin with soft spreading hairs; flowers 1–2 to a corm; peduncles 1-flowered, much shorter than the leaves, glabrous, tinged with red; ovary obconic, glabrous, $\frac{1}{8}$ in. long; segments linear, glabrous, reddish, twice as long as the ovary; stamens very short.

KALAHARI REGION: Basutoland; near the summit of the Mont aux Sources, 9500 ft., submerged in water, *Flanagan,* 2024! No specimen in Kew Herbarium.

3a. Hypoxis monophylla (Schlechter, MS. in Herb. Kew); corm globose, $\frac{1}{4}$ in. diam., crowned with a dense ring of fibres; leaf solitary, spreading, narrowly linear, 1–$1\frac{1}{2}$ in. long, narrowed to the base, obscurely hairy; peduncle 1-flowered, 1–$1\frac{1}{2}$ in. long, with a clasping reduced leaf above the base; ovary glabrous, obconic, $\frac{1}{12}$ in. long; perianth-segments lanceolate, $\frac{1}{8}$ in. long, yellow, glabrous, the outer tinged with brown outside; stamens $\frac{1}{4}$ the length of the perianth.

COAST REGION: Bredasdorp Div.; Elim, 700 ft., *Schlechter,* 7615!

Near *H. curculigoides, Bolus.*

1a. Gethyllis pusilla (Baker); bulb the size of a pea; leaves about 4, produced before the flowers, narrowly linear, glabrous, spirally twisted, $1\frac{1}{2}$ in. long; perianth-tube very slender, straight, whitish, above an inch long; segments narrowly linear, reddish,

nearly an inch long; stamens 6; filaments filiform, $\frac{1}{6}$–$\frac{1}{4}$ in. long; anthers linear, $\frac{1}{6}$ in. long.

COAST REGION: Cape Div.; Kenilworth Racecourse, *Bolus*, 7981! No specimen in Kew Herbarium.

7a. Crinum acaule (Baker); bulb not seen; leaves linear, complicate, moderately firm, glabrous, 1½ ft. long, ¼ in. broad low down; stem 1-flowered, not produced above ground; spathe-valves linear, green; perianth-tube erect, cylindrical, 2 in. long; segments lanceolate, erect, recurved at the tip, twice as long as the tube, $\frac{1}{2}$–$\frac{3}{4}$ in. broad at the middle, keeled with pale red; stamens half as long as the perianth-segments; anthers linear-oblong, $\frac{1}{4}$–$\frac{1}{3}$ in. long. Style reaching to the tip of the perianth.

EASTERN REGION: Zululand; Sambaans territory, *Saunders*!

This was collected by Mr. Charles Saunders in 1896, and cultivated in his garden, a specimen and drawing being sent to Kew by Mrs. K. Saunders.

3a. Ammocharis taveliana (Schinz in Verh. Bot. Ver. Prov. Brandenb. xxxi. 214); bulb very large; tunics horizontally truncate at the apex; leaves oblong, acute, 15 in. or more long, above 2 in. broad; peduncle 8–10 in. long, nearly 1 in. diam.; umbels dense, globose, 4–5 in. diam.; pedicels 1½ in. long; spathe-valves oblong, an inch broad; perianth-tube above ¼ in. long; segments oblong-lanceolate, acute, 2 in. long; capsule the size of a pigeon's egg; pericarp thin, membranous.

WESTERN REGION: Great Namaqualand; near Bethany, *Schinz*, 16, *Schenk*, 358, *Pohle*.

Also Hereroland. No specimen in Kew Herbarium.

2a. Strumaria bidentata (Schinz in Bull. Herb. Boiss. iv. App. iii. 46); bulb subglobose, white; leaves unknown; peduncle 3–4 in. long; flowers 10–12 in an umbel; spathe-valves 2, purple, lanceolate-cuspidate; pedicels slender, erect; perianth ¼ in. long, white, tinged with red outside at the base; segments lanceolate; filaments shortly connate, bidentate and dilated at the base.

WESTERN REGION: Great Namaqualand; Orange river, *Schenk*, 232. No specimen in Kew Herbarium.

3a. Cyrtanthus Flanagani (Baker); leaves lorate, subobtuse, moderately firm, glabrous, nearly an inch broad; peduncle moderately stout, a foot long; flowers 4–7 in an umbel; spathe-valves lanceolate, 2 in. or more long; pedicels $\frac{3}{4}$–1¼ in. long; perianth white; tube 2 in. long, ½ lin. diam. low down, dilated gradually to $\frac{1}{6}$ in. at the throat; segments oblong, $\frac{3}{4}$ in. long; stamens not protruded from the perianth-tube, inserted a short distance below its throat.

KALAHARI REGION: Basutoland; slopes of Mont aux Sources, 8000-9000 ft., *Flanagan*, 1824! No specimen in Kew Herbarium.

6a. Cyrtanthus stenanthus (Baker); leaves 4 to a stem, linear, glabrous, erect, a foot long, $\frac{1}{6}$ in. broad; peduncle slender, rather longer than the leaves; flowers 5–6 to an umbel; spathe-valves linear, 1–1½ in. long; pedicels short; perianth red, 1¼ in. long; tube ½ lin. diam. low down, dilated gradually to 1 lin. at the throat; segments ovate, ⅛ in. long; anthers sessile a short distance below the throat of the perianth-tube.

KALAHARI REGION: Basutoland; slope of Mont aux Sources and Bester's Vlei Mountain, 6000–8000 ft., *Flanagan*, 2047! No specimen in Kew Herbarium.

17a. **Asparagus Fleckii** (Schinz in Bull. Herb. Boiss. iv. App. iii. 43); a squarrose undershrub with divaricate angled branches; upper internodes ⅕–⅓ in. long; leaves produced at the base into straight pungent spines; cladodia 6–10 in a cluster, subulate, hard, pungent, ⅓–½ in. long; pedicels 1–6-nate; flowers hermaphrodite.

KALAHARI REGION: Reit Fontein, *Fleck*, 238. No specimen in Kew Herbarium.

6a. **Kniphofia fibrosa** (Baker); old leaves splitting into copious fibres; produced leaves 6–8 to a stem, narrowly linear, 1½ ft. long, ⅛–⅙ in. broad low down, with recurved edges and a flat back with a slender midrib; peduncle slender, 1–1½ ft. long; raceme dense, oblong, 1–2 in. long, 1¼ in. diam.; pedicels very short; bracts ovate-lanceolate, the lower ¼ in. long; flowers all pale yellow, deflexed; perianth cylindrical, slender, ⅝ in. long; anthers small, oblong, blackish, finally exserted; style overtopping the anthers.

EASTERN REGION: Natal; Mahwaqua Mountain, 6000–7000 ft., *Evans*, 649! Near *K. gracilis*, *Harv.*, and *K. Evansii*, *Baker*.

17a. **Kniphofia Thodei** (Baker); old leaves splitting into fibres, produced few, stiffly erect, linear, under a foot long, ⅛ in. broad, with a thick midrib and scabrous margin; peduncle 1½ ft. long; flowers few in a short, dense raceme, all drooping, orange-scarlet; pedicels very short; bracts oblong-lanceolate, acute, scariose, white, ¼ in. long; perianth cylindrical, 1¼ in. long, ⅙ in. diam. at the throat; lobes oblong, obtuse; stamens not exserted from the perianth-tube.

KALAHARI REGION: Basutoland; grassy slopes of Caledon Range, 7000–8000 ft., *Thode*, 62! No specimen in Kew Herbarium.

K. natalensis var. condensata, *Baker*, was at first supposed to be a distinct species, and was named provisionally *K. Woodii*, Baker. This name was printed in the *Garden*, 1895, xlviii. 292.

20a. **Kniphofia primulina** (Baker); leaves many, ensiform, green, 3–4 ft. long, ½–1 in. broad low down, very acutely keeled, smooth on the margin; peduncle stout, stiffly erect, as long as the leaves; raceme dense, oblong, 3–4 in. long; pedicels very short, deflexed; bracts small, ovate; flowers all pale yellow; perianth subcylindrical, an inch long; lobes small, ovate; stamens and style much exserted.

EASTERN REGION: Natal, *Hort. Leichtlin!* Flowered in the Temperate House at Kew, January, 1897.

Near *K. natalensis, Baker.*

15a. Aloe glauca (Miller, Dict. edit. viii. No. 16); stem simple, finally ½–1 ft. long; leaves 30–40, densely rosulate, not recurved, lanceolate, 6–8 in. long, 1½–2 in. broad low down, narrowed gradually to the point, intensely glaucous, not spotted, obscurely lineate, ¼–⅓ in. thick, sparingly tubercled on the back towards the tip; marginal teeth unequal, deltoid, red-brown, ¹⁄₁₂–⅛ in. long; peduncle simple, 1–1½ ft. long; raceme simple, rather lax, ½–1 ft. long, 3½–4 in. diam.; pedicels 1–1½ in. long, cernuous at the apex; bracts deltoid, 6–9 lines long; perianth pale red, 15–16 lines long; tube very short; stamens not exserted. *Haw. in Trans. Linn. Soc.* vii. 18; *Syn.* 79; *Roem. et Schult. Syst. Veg.* vii. 690; *Salm-Dyck, Aloe,* sect. xvii. *fig.* 2; *Kunth, Enum.* iv. 520; *Baker in Journ. Linn. Soc.* xviii. 160. *A. rhodacantha, DC. Plantes Grasses, t.* 44; *Gawl. in Bot. Mag. t.* 1278.

VAR. **muricata** (Baker); leaves less glaucous, more spreading, tubercles on the back of the leaf more numerous, marginal teeth larger, bright red. *A. glauca var. spinosior, Haw. Revis.* 40. *Aloe muricata, Schultes, Obs.* 70.

SOUTH AFRICA : without locality.

Introduced into cultivation in 1831. Near *A. Serra, DC.*

3a. Bulbine concinna (Baker); tuber small, globose; leaves contemporary with flowers, very slender, short, subulate, glabrous; peduncle very slender, 3 in. long; raceme lax, few-flowered, an inch long; pedicels ascending, lower ¼–⅓ in. long; bracts small, ovate; perianth-segments linear, ¼ in. long, yellow, with a dark green keel; stamens less than half as long as the perianth; filaments densely bearded.

COAST REGION : Caledon Div.; Houw Hoek, 2500 ft., *Schlechter,* 7555!

Very near *B. minima, Baker.*

4a. Eriospermum Schlechteri (Baker); tuber with a long oblique neck clothed with brown membranous sheaths; leaf solitary, contemporary with the flowers, nearly sessile, oblong, 1–1½ in. long, acute, narrowed to the base, subcoriaceous glabrous; peduncle slender, 3–5 in. long; raceme few-flowered, lax, 1½–2 in. long; bracts minute, ovate, brown, membranous; pedicels ascending, lower ⅓–½ in. long; perianth ⅓ in. long; segments linear-oblong, whitish, with a red-brown keel; stamens less than half as long as the perianth; filaments as long as the linear anther.

COAST REGION : Caledon Div.; Bot River, 1000 ft., *Schlechter,* 7590!

Near *E. Haygarthi, Baker.*

8a. Eriospermum Fleckii (Schinz in Bull. Herb. Boiss. iv. App. iii. 37) ; leaf contemporary with the flowers, lanceolate, acute, green, glabrous, 1½ in. long, ⅓–½ in. broad ; raceme corymbose ; lower pedicels 3–4 in. long ; perianth ¼ in. long ; segments oblong, mucronate ; filaments thick, but not flattened.

KALAHARI REGION : Great Namaqualand ; Rehoboth, *Fleck*, 887. No specimen in Kew Herbarium.

16a. Eriospermum dissitiflorum (Schlechter) ; tuber oblong ; leaf single, contemporary with the flowers, long-petioled, ovate, 3–5 in. long, 1½–3½ in. broad, rounded at the base, membranous, glabrous ; peduncle 1–1½ ft. long ; raceme cylindrical ; pedicels short, ascending, the lower ⅓–½ in. long ; bracts small, membranous, persistent ; perianth campanulate, ⅛ in. long ; segments oblong, pure white, with a distinct green keel ; filaments short, ovate ; capsule globose, ¼ in. diam.

COAST REGION : Queenstown Div. ; summit of Queenstown Mountain Range, alt. 4300–4500 ft., *Galpin*, 1944 !

17a. Eriospermum roseum (Schinz in Bull. Herb. Boiss. iv. App. iii. 38) ; tuber globose, with a rose-red skin ; leaf single, contemporary with the flowers, cordate-suborbicular, cuspidate, glabrous, 2½–3 in. broad ; peduncle 2–3 in. long ; raceme lax, many-flowered, 1½–4 in. long ; pedicels patent, ¼–½ in. long ; bracts deltoid ; perianth campanulate, ¼–⅓ in. long ; segments oblong, obtuse, with a broad rose-red keel ; stamens half as long as the perianth ; filaments lanceolate.

WESTERN REGION : Great Namaqualand, *Fleck*, 888, *Schenk*, 368. No specimen in Kew Herbarium.

Also Hereroland.

20a. Eriospermum sprengerianum (Schinz in Bull. Herb. Boiss. iv. 416) ; tuber oblong, above an inch long ; leaf single, contemporary with the flowers, ovate, obtuse, glabrous, bright red at the margin, above an inch long ; peduncle about a foot long ; raceme ½ ft. long, dense upwards ; bracts ovate-lanceolate, red, deciduous ; perianth campanulate, $\frac{1}{10}$ in. long ; outer segments narrowed to the apex, spotted with red, inner not spotted, rounded or truncate at the apex ; filaments flattened, rather shorter than the perianth ; anthers oblong, $\frac{1}{15}$ in. long ; style short.

EASTERN REGION : Natal, *Hort. Dammann*. No specimen in Kew Herbarium.

7a. Ornithogalum Flanagani (Baker) ; bulb small, oblong ; sheath-leaf 1, basal, large ; produced leaf 1, narrowly linear, 4–5 in. long, erect, glabrous, strongly ribbed ; peduncle slender, 6–8 in. long ; raceme dense, few-flowered ; pedicels very short, ascending ; bracts ovate, scariose, the lower ¼ in. long ; perianth funnel-shaped, ¼ in. long ; segments oblanceolate, obtuse, white, not distinctly keeled; stamens half as long as the perianth; filaments slightly flattened; anthers small, oblong; style very short.

KALAHARI REGION: Basutoland; summit of the Mont aux Sources, alt. 9500 ft., *Flanagan*, 2028! Sent by Mr. Bolus, January, 1897.

Near *O. pilosum, Linn.*

55a. Ornithogalum Galpini (Baker); bulb globose, $\frac{1}{2}$ in. diam.; outer tunics membranous; produced leaves 2, subterete, not very slender, firm, erect, glabrous, 8–9 in. long; peduncle slender, 9–12 in. long; raceme dense, few-flowered; pedicels very short, ascending, at most $\frac{1}{8}$ in. long; bracts lanceolate, rather shorter than the flowers; perianth $\frac{1}{4}$ in. long; segments oblong, white, with a distinct reddish-brown keel; stamens nearly as long as the perianth; filaments all linear; style as long as the ovary.

COAST REGION: Queenstown Div.; summit of Andries Bergen, 6700 ft., in marshy ground, *Galpin*, 2272!

INDEX.

LONDON:
PRINTED BY GILBERT AND RIVINGTON, LD.,
ST. JOHN'S HOUSE, CLERKENWELL ROAD, E.C.